H. Monteith

CW01558940

*Hydrology and*
*Water Resources Engineering*

# Hydrology and Water Resources Engineering

K.C. Patra

Alpha Science International Ltd.

**Kanhu Charan Patra**
Associate Professor
Department of Civil Engineering
College of Engineering and Technology
O.U.A.T., Bhubaneswar-751 003, India

Copyright © 2001

Alpha Science International Ltd.
P.O. Box 4067, Pangbourne RG8 8UT, UK


Exclusive distribution in North America only by CRC Press LLC

All rights reserved. No part of this publication may be reproduced, stored in a retrieval
system or transmitted in any form or by any means, electronic, mechanical, photocopying,
recording or otherwise, without the prior written permission of the publishers.

ISBN 1-84265-023-8


Printed in India.

To the memories of those engineers and scientists,
who have enhanced our knowledge on water
resources and hydraulic engineering

# PREFACE

Water is precious for life on earth. There is an ever increasing demand for the supply of fresh water to the various sectors of the human needs. This has given rise to the problem of optimal management of water resources potential on all parts of the world, more so, in a developing country like India, where the distribution of the resource is highly uneven both in space and time. Hydrology plays the central role in the development and management of water resources and therefore, the protection of environment. As a result, hydrology forms a part of curricula at the undergraduate and postgraduate levels in agriculture, civil engineering, environmental engineering, geology and earth sciences, forestry and meteorology.

Based on my experience as an engineer in the water resources design, planning, development and management for more than a decade and a long teaching experience thereafter, I felt the need for a textbook written in simple and lucid style that can be easily followed by students who have not had an opportunity of the exposure to the subject before. A student can understand most parts of this book himself. At the same time, the book can serve as a source of information to the engineers and other professionals dealing with water resources planning, development and management. Therefore, the chapters have been designed and developed in a systematic way. The materials for the book are mostly taken from the class notes prepared for the graduate students and at the same time utilising my decade long experience from the field problems. Therefore, emphasis have been given on the applicability of the text materials to the field situations. Using data from the field, a large number of hydrologic design problems have been worked out for each concept of the chapters to illustrate the analysis and design procedure.

The text is arranged in ten chapters. Chapter 1, besides a brief resume to the history of hydrology, introduction to meteorology, discusses hydrologic cycle, cloud and availability of water on earth. Since statistics and probabilities play an important role in the estimation of hydrologic parameters, the same have been discussed in Chapter 2. Chapter 3 is essentially devoted to the precipitation, its measurement and analysis. Measurement and estimation of hydrologic losses like evaporation, evapotranspiration, interception, depression storage and infiltration are covered in detail in Chapter 4. Groundwater forms a major component in hydrologic cycle and therefore, the hydrologic aspects of ground water are discussed in Chapter 5. In Chapter 6, the aspects of stream flow measurement and the rainfall runoff process are discussed at length. Hydrographs are discussed in Chapter 7. Estimation of design flood for various types of catchments is given in Chapter 8, while Chapter 9 covers various methods of channel and reservoir routing. In Chapter 10, reservoir sedimentation and the aspects of cost benefit have been discussed. Sufficient problems are given at the end of each chapter for practice. Important references have been cited at the end of the book.

The author is grateful to Er. B.B. Singhsamant (Chief Engineer) and Er. N. C. Mohanty (Executive Engineer), Water Resources Department, Government of Orissa, for their active help and continuous encouragement while preparing the manuscript of this book. I am also to record my sincere thanks to Dr. S.K. Kar, Professor, Department of Civil Engineering, IIT Kharagpur and Dr. S.C. Mishra, Dean, College of Engineering, for their active help and support. I am also grateful to all the authors, whose works I have consulted.

Finally I thank my wife Gita and our children Sonam and Suman for their tolerance and understanding during the preparation of the manuscript.

I am thankful to N.K. Das and R. Mohan for their help in typing the manuscript and drawing figures in Autocad.

Any suggestion in improving the book will be thankfully received and will be incorporated in the next edition.

K.C. PATRA

# CONTENTS

# Chapter One

# Introduction

## 1.1 GENERAL

Hydrology deals with the origin, distribution and circulation of water in different forms in land phases and atmosphere. It is an interdisciplinary subject starting with meteorology through agronomy, forestry, geology, hydraulics and finally oceanography. Statistics, physics and chemistry help to formulate and understand the subject more conceptually. Solutions to hydrologic problems are usually carried out by borrowing techniques from several disciplines. Depending on the treatment of data, hydrology can be divided into various sub-branches as shown in Fig. 1.1.

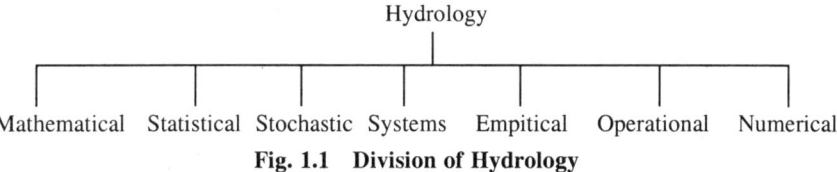

**Fig. 1.1  Division of Hydrology**

Broadly, hydrology is classified into two groups. The 'scientific hydrology' is concerned mainly with academic aspect, whereas, the 'engineering' or 'applied' hydrology includes: (i) estimation of water resources; (ii) study of transmission process like precipitation, evaporation, runoff and their interdependence; (iii) understanding the properties of water in nature and (iv) dealing with natural problems like droughts and floods. The distribution of water potential in India is highly uneven both in space and time. With the growth of population, the demand of water has also increased to many fold. The science of hydrology can be effectively employed to overcome these difficulties. Therefore, a lot of research and developmental opportunities do exist in this comparatively new branch of science. We would describe some of the aspects of hydrology, which should help a student to understand the subject. The book should also help field engineers or hydrologists to solve their problems connected with water resources engineering.

## 1.2  HISTORY OF HYDROLOGY

There is no evidence of the date of construction of the first water resource project on the earth. Water being the chief ingredient in life, all ancient civilisations flourished only near the sources of water and then probably collapsed either when the water supply failed or when they were devastated by floods. Man's quest for knowledge, his capacity for innovation and adventure have contributed greatly to his efforts at understanding and exploiting this resource right from Indus-valley civilisation. Archaeological evidence of the existence of wells in Mohenjodaro (3000 BC), the diversion weirs on the river Nile by king Means (3000 BC) and the kanats for irrigation in Persia were the few structures men believed to have constructed in the ancient times. Reference can also be found in Vedic literature (3000 BC), Arthashastra by Chanakya (300 BC), Brihatsamhita by Varahamihira (505–587 AD) and in lots of other ancient writings about the availability, measurement and use of the surface and ground water. In India, during Chanakya's period, kings probably maintained a network of gauges in their kingdoms for tax collection. Writings of ancient thinkers like Aristotle (350 BC), Plato (350 BC), Vitruvius (100 BC) and others suggest their scientific thinking of various hydrologic processes. Following Chow, the history of hydrology can be classified into the following eight periods.

**(a) Period of Speculation (Prior to 1400 AD):** During this period, many hydraulic structures were known to have been constructed. The important being the Arabian wells, Persian Kanats, Egyptian and Mesopotamian irrigation projects, Roman aqueducts, water supply and drainage projects in Indus valley and the Chinese irrigation systems. Vitruvius was probably the first great philosopher to think rationally about the hydrologic processes and storage during this period.

**(b) Period of Observation (1400–1600 AD):** During this period, Leonardo da Vinci correctly understood parts of the hydrologic cycle and put forth his ideas in writing.

**(c) Period of Measurement (1600–1700 AD):** In this century measurement of rainfall, runoff, evaporation and study of artesian wells were taken up by different thinkers. Correct predictions of some hydrologic phenomena were also made.

**(d) Period of Experimentation (1700–1800 AD):** New discoveries like Bernoulli's piezometer, Pittot tube, Waltman's current meter, Borda's tube, Bernoulli's theorem, Chezy's formula and D'Alembert's principle were made during this period. All these helped to quantify the total water resource potential of the world.

**(e) Period of Modernisation (1800–1900 AD):** Substantial contributions were made in the fields of both ground and surface water hydrology during this century. Important contributions being Darcy's law, Poiseuillie's equation of capillary flow, Dupuit-Thiem's well formula, Weir-discharge formula, Kutter's determination of Chezy's equation, Manning's equation, Price current meter and Dalton's law of vapour pressure.

**(f) Period of Empiricism (1900–1930 AD):** Thirty years following the early part of 20th century saw the science of hydrology largely empirical. A number of formulae were proposed and various governments undertook massive efforts in setting up organisations for measurement of hydrological parameters quantitatively in most of the important river basins of the world.

**(g) Period of Rationalisation (1930–1950 AD):** Most of the present concepts in hydrology date back to 1932, the year in which Le Ray K. Sherman rationalised the concept of hydrology by demonstrating the use of Unit Hydrograph for translating the rainfall excess into runoff hydrograph. The other great contributions during this period were Horton's (1933) determination of rainfall excess from infiltration characteristics, Theis (1935) well hydraulics, Gumbel's (1941) extreme value frequency distribution, Bernard's introduction of meteorology to hydrology and Einstein's theory of sediment load. A lot of laboratories of international repute were also established.

**(h) Period of Theorisation (Since 1950):** The advent of modern high speed computing machines, sophisticated instrumentation and emergence of modern fluid mechanics helped greatly in the formulation of new theories on hydrology. Many hydrologic activities of international level were also established. Some notables are Water Resources Development Centres (WRDC), World Meteorological Organisation (WMO), International Association of Scientific Hydrology (IASH) etc. In 1961 an International Hydrologic Decade (IHD) was proposed by IASH to co-ordinate international research and training programs in hydrology.

In India, National Institute of Hydrology (NIH) and Department of Hydrology at the University of Roorkee (besides other institutions) are promoting research and training programs and co-ordinating international activities.

## 1.3 METEOROLOGY (*METEOROS*-LOFTY; *LOGOS*-DISCOURSE)

It is that branch of science which deals with the entire gaseous envelope around the earth. Hydrology starts with precipitation of water vapour on the earth's surface. Therefore, it is desirable to know something about the factors, which bring about these changes. Atmosphere (*atoms*-vapour; *spheria*-sphere) is the gaseous envelope of the earth. It is a mixture of gases and the water vapour is of special interest to the hydrologists. The atmosphere exerts a weight of about 1.033 kg/cm$^2$, which is equivalent to 10.3 m depth of a water column. The atmospheric pressure is one atom (~ 1013.25 mb). Though water vapour in the atmosphere is limited to a few kilometres from earth's surface, it may be required to know the division of atmosphere into concentric layers. The layers with the variation of temperature and pressure across it are shown in Fig. 1.2.

Atmosphere enveloping the earth is divided into five concentric layers. *Troposphere* (a layer just above the earth's surface) contains all moisture, dust and almost three-fourth of total air mass. *Tropopause* is the top of troposphere, which separates it from the *stratosphere* layer above it. The height of troposphere varies from 8 km at poles to 16 km at the equator. In the stratosphere, temperature is almost constant. *Mesosphere* is just above stratosphere. Temperature in this zone is always higher than ground surface due to absorption of heat from sun's radiation. The fourth layer is known as *ionosphere*, which is 70–80 km above the earth's surface and is responsible for reflecting ordinary radio waves. This layer has a high concentration of free electrons. The outermost shell is called *exosphere*, where atmosphere loses all its properties. A sound knowledge of troposphere is important for a hydrologist as the weather is confined to this layer only.

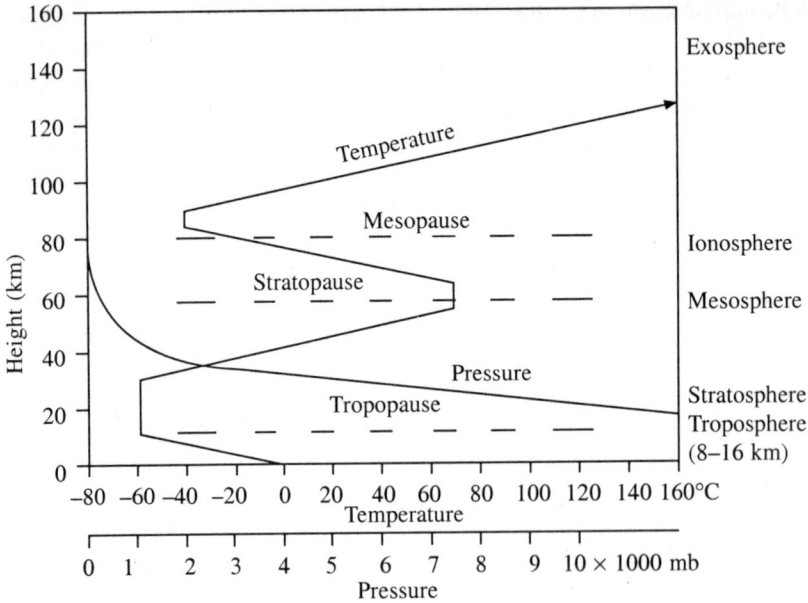

**Fig. 1.2   Variation of Temperature and Pressure with Height from Earth's Surface**

### 1.3.1   Lapse Rate

The rate at which temperature decreases with the increase of altitude at any particular time is called lapse rate. When a mass of air at the surface of earth is heated up, it expands and becomes lighter than the surroundings and then rises. Similarly, heavier air at higher altitudes sinks and contracts due to increase in pressure. Though no heat is added or subtracted from the system, the change in temperature takes place due to expansion and contraction only and this change in temperature for the dry air, known as dry adiabatic lapse rate is 10°C/km. The dry adiabatic lapse rate is higher than the normal air lapse rate which varies at the rate of 6.5°C/km up to the first 10 km from earth's surface and then almost at zero rate for the next 10 km. *Inversion* is the reverse of lapse rate. As air mass rises, it loses heat with altitude and at certain stage there is no further decrease in temperature even if it rises further unless there is condensation of some water vapour. This temperature at which the air mass just becomes saturated, if cooled at constant pressure without any moisture added or removed from it, is called *dew point*. After condensation level, the air mass continues to ascend due to the latent heat of condensation added to it. This change in temperature is called adiabatic saturation lapse rate, which is between 3 and 10°C/km of the altitude change.

Sometimes the term *normal temperature* meaning the mean of 30 records of the corresponding period picked up from each year is used. Temperature data published by India Meteorological Department (IMD) contains daily, weekly, monthly or annual normal records. For example, the normal temperature of 1st January is obtained by taking the average of 30 years of 1st January data from say 1970 to 1999. This may be either the minimum or maximum temperature for the day. On the other hand, the term 'mean' refers to the daily averages of maximum and minimum temperatures. For example, for 1st January 1999, the average of maximum and minimum temperatures for the day is the mean temperature.

### 1.3.2 Pressure

Atmospheric pressure is the pressure exerted by the gases of the atmosphere on the earth's surface due to their own weight. It decreases uniformly from the surface of earth and the rate of decrease depends on temperature and lapse rate. Again due to excess heat, air at the equator belt rises and sinks at around 30° latitude belts. Therefore the equator belt is of low pressure zone and the 30° north and south belts are the higher pressure zones (sub-tropical belt). At 60° north and south, the higher pressure air coming from poles meet the air coming from sub-tropical high pressure zones and causes it to rise to form a low pressure belt at this zone. Due to earth's rotation around its own axis the movement of a parcel of air is affected (coriolis force). This force, along with the fictional resistance on the earth's surface, forces the air at an angle across the isobars causing a counter clockwise movement around low pressure centres and clockwise movement around high pressure centres in the north hemisphere. A reversal of the situation can be seen in the southern hemisphere. At equator, the speed of the earth is 1670 km/h from west to east and at 60° latitude it is about 830 km/h. From the knowledge of physics, we can calculate a theoretical velocity of 2505 km/h of a parcel of air at rest, relative to earth's surface at the equator and moving east–northward direction to 60° latitudes. However, this does not happen because of frictional resistance the air has to overcome at the surface due to features of the earth and air densities. An oversimplified pattern of air circulation on earth's surface is shown in Fig. 1.3.

*Isobar* (line joining points of equal pressure) is drawn on a weather map in such a way that the lines pass through points of equal sea level pressure. A *front* is the border region of adjacent air masses, normally, the colder air forms a wedge above which warmer air ascends. The ascending warmer air cools with the lapse rate leading to the formation of cloud and precipitation. Front is of four types. A cold front is formed at the boundary between the advancing cold air and retreating warm air (Fig. 1.4a), whereas in a warm front, the advancing warm air meets the retreating cold air (Fig. 1.4b). In a stationary front, both the warm and cold masses cease to move. When the two cold fronts meet, the region between two fronts become warmer. This situation is conducive for the formation of cyclones. Both fronts get elevated and the condition is called occluded front.

### 1.3.3 Water Vapour

It is that gaseous stage of water which can be condensed. The amount of water vapour in the atmosphere is a meagre 1 part in 100000 of all the water on earth but it plays a vital role for the sustenance of life on earth. A large amount of water passes through the atmosphere due to the continuous process of vaporisation and condensation taking place globally. The process by which liquid is converted to vapour is called *evaporation* or *vaporisation*. *Condensation* is the reverse

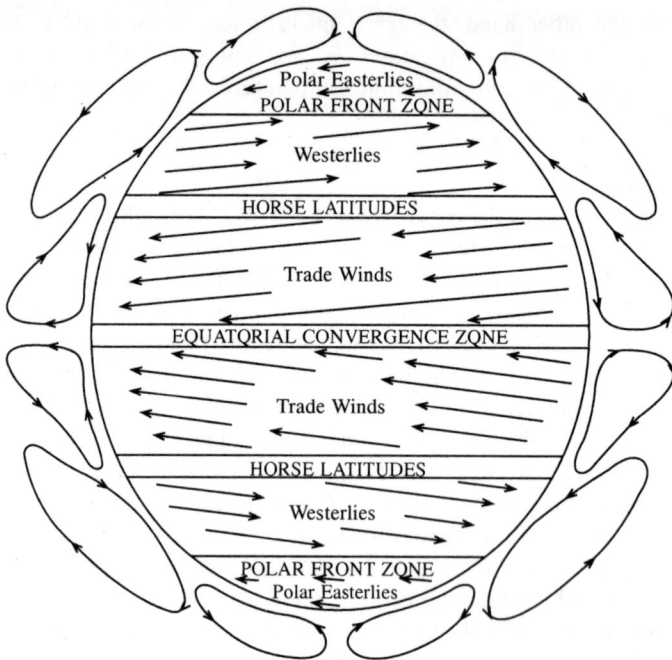

**Fig. 1.3   Schematic Representation of the General Circulation
(after Bergeron and Rossby, 1941)**

process by which water vapour is converted to liquid form. It follows from the Dalton's law of gaseous mixture, the partial pressure exerted by water vapour alone is called vapour pressure and is given by

$$e = p - p' \qquad (1.1)$$

where $e$ is the vapour pressure in pascals (1 $p_a$ = 1 N/m$^2$), $p$ is initial pressure of gaseous mix including water vapour in N/m$^2$ and $p'$ is the pressure of dry air.

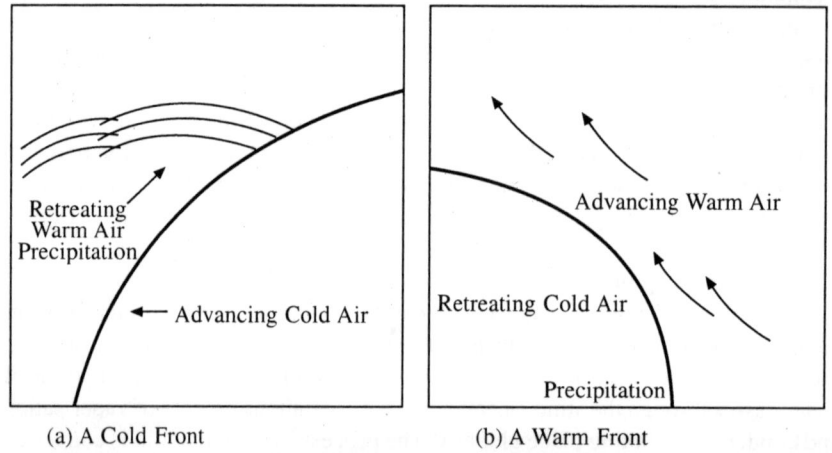

(a) A Cold Front  (b) A Warm Front

**Fig. 1.4   Cold and Warm Front**

At a given temperature, a given space can hold a certain maximum amount of water vapour independent of the presence of other gases. The space is said to be saturated at this stage. The vapour pressure is called saturation vapour pressure ($e_s$) which is the maximum vapour pressure possible at a given temperature. At this stage, the rate of evaporation and condensation are equal. Saturation vapour pressure over water surface can be related to air temperature $T$ as:

$$e_s = 611 \ e^{\frac{17.27\,T}{237.3+T}} \qquad (1.2)$$

where $e_s$ is in N/mm$^2$ and $T$ in °C. Variation of the saturation vapour pressure of water for various temperatures are given in Table 1.1.

**Table 1.1   Variation of Saturation Vapour Pressure with Temperature**

| Saturation vapour pressure (Pa) | 286 | 611 | 872 | 1227 | 1704 | 2337 | 3167 | 4243 | 5624 | 7378 |
|---|---|---|---|---|---|---|---|---|---|---|
| Temperature (°C) | −10 | 0 | 5 | 10 | 15 | 20 | 25 | 30 | 35 | 40 |

Molecular weight of air is 28.95 and that of water vapour is 18.0. This means that at the same temperature and pressure, the specific gravity of water vapour is 0.622 times than that of dry air. Since water vapour is lighter than air, it rises till it condenses. Vapour pressure is a partial pressure exerted by water vapour in air and is expressed in millibar (mb). One millibar is 1000 dynes/cm$^2$. Saturation vapour pressure can also be computed approximately from the equation

$$e_s = 0.0446 - 0.00064 \mid 1.8T + 48 \mid + 33.864 \ (0.00738T + 0.8072)^8 \quad (1.3)$$

where $e_s$ is in millibars and $T$ in °C. Equation (1.3) is applicable for temperatures ranging between −55 and 55°C only.

### 1.3.4   Precipitable Water

The amount of moisture available in a column of atmosphere extending from the surface of the earth is called precipitable water. Mass of precipitable water $P_w$ in kg between levels $h_1$ and $h_2$ can be calculated as

$$P_w = (h_2 - h_1) \ H_{av}\rho_{av} \qquad (1.4)$$

where $H_{av}$ is the average specific humidity of the air mass and $\rho_{av}$ the average air density over the interval $h_1$ and $h_2$.

*Specific humidity*, defined as the mass of water vapour per unit mass of moist air, can be calculated by

$$H_{av} = (0.622e)/p \qquad (1.5)$$

Total pressure $p$, i.e., the total pressure exerted by the moist air, is calculated as:

$$p = \rho_a R_a T \qquad (1.6)$$

where $\rho_a$ is the sum of densities of dry air and water vapour, $R_a$ the universal gas constant for moist air and $T$ the absolute temperature in degree Kelvin. For obtaining the average values of the parameters in equation (1.4), pressure and temperatures at other points are required to be calculated. Pressure at any other point, say at $h_2$, can be calculated from the known value at $h_1$ using the following form of the power law:

$$p_2 = p_1 \left( \frac{T_2}{T_1} \right)^{\frac{g}{\alpha R_a}} \qquad (1.7)$$

where $\alpha$ is the lapse rate and $g$ the gravitational acceleration. Temperature between the two heights $h_1$ and $h_2$ can be interpolated using relation

$$T_2 = T_1 - \alpha (h_2 - h_1) \qquad (1.8)$$

The subscripts 1 and 2 corresponds to the parameters at lower level (say ground level) and to a level higher than level one, respectively. Using equations (1.4) through (1.8), the precipitable water between the heights can be calculated. For calculating precipitable water for the entire atmosphere, the height (say up to 12 km) can be broken down to the increments of heights between 1 and 2 km and the precipitable water calculated for each segment of the column can be added to get the precipitable water for the entire column of the atmosphere. This practice is usually followed for better accuracy of the results.

---

**Example 1.1:** What will be the weight of 1 cubic meter of dry air at 20°C and at 1000 mb pressure?

**Solution**

Here　　　　　　　　　$T = 20°C = 273 + 20 = 293$ K

Gas constant $R_d = 287$ J/kg.k and pressure $p = 1000$ mb $= 100$ k pa $= 100 \times 1000$ pa

From equation (1.6)

$$p = \rho_d R_d T$$

or　　　　　　$\rho_d = p/R_d T = 100 \times 1000/(287 \times 293) = 1.189$ kg/m³.

**Example 1.2:** Calculate the mass of precipitable water between 0 and 1 km of column extending from earth's surface assuming the atmosphere to be saturated. Surface air pressure can be taken as 100.7 k pa. Surface air temperature is 32 °C. Take lapse rate as 6.5°C per 1000 m.

**Solution**

Here　　$h_1 = 0$ m, $T_1 = 273 + 32 = 305$ K, $P_1 = 100.7$kPa $= 100.7 \times 1000$kPa.

$h_2 = 1000$ m, $\alpha = 0.0065°C/m$, $R_a = 287$ J/kg.k

$T_2 = T_1 - \alpha (h_2 - h_1) = 32 - 0.0065(1000 - 0) = 25.5°C = 298.5$ K

$$p_2 = p_1 \left( \frac{T_2}{T_1} \right)^{\frac{g}{\alpha R_a}} = 100.7 \left( \frac{298.5}{305} \right)^{\frac{9.81}{0.0065 \times 287}}$$

$$= 100.7(298.5/305)^{5.2587} = 89.91 \text{ kPa}$$

Air density at zero level can be calculated from equation (1.6).

*At level* 1: The saturated vapour pressure is calculated as

$$p = \rho_a R_a T \quad \text{or} \quad \rho_a = p/R_a T = (100.7 \times 1000)/ (287 \times 305) = 1.15 \text{ kg/m}^3$$

*At level* 2: 1000 m above the level 1

$$p = \rho_a R_a T \quad \text{or} \quad \rho_a = p/R_a T = (89.91 \times 1000)/ (287 \times 298.5) = 1.05 \text{ kg/m}^3$$

Average density is $(1.15 + 1.05)/2 = 1.10 \text{ kg/m}^3$
Saturated vapour pressure at ground level is calculated from equation(1.2)

$$e_s = 611 \exp \{(17.27T)/(237.3 + T)\}$$

$$= 611 \exp \{(17.27 \times 32)/(237.3 + 32)\}$$

$$= 4756 \ P_a = 4.756 \text{ kPa.}$$

*At level* 2, the saturated vapour pressure is calculated as folows
The temperature at level 2 is 25.5°C. The value of saturated vapour pressure is calculated as

$$e_s = 611 \exp \{17.27T/(237.3 + T)\} = 611 \exp \{(17.27 \times 25.5)/(237.3 + 25.5)\}$$

$$= 3264 \text{ Pa} = 3.264 \text{ kPa}$$

Specific humidity at ground level is calculated from equation(1.5)

$$H_0 = (0.622 \ e)/p = (0.622 \times 4.756)/100.7 = 0.029 \text{ kg/kg of moist air}$$

and at 1000 meters above the zero elevation.

$$H_{1000} = (0.622e)/p = (0.622 \times 3.264)/89.91 = 0.0226 \text{ kg/kg of moist air}$$

Average specific humidity is $(0.029 + 0.0226)/2 = 0.0258$ kg/kg of moist air
Mass of precipitable water per square meter between 0 and 1000 m elevation is calculated by using equation (1.4)

$$P_w = (h_2 - h_1) \ H_{av} \ \rho_{av} = (1000 - 0) \times 0.0258 \times 1.1 = 28.36 \text{ kg.}$$

Thus, between 0 and 1000 m the quantity of precipitable water is 28.36 kg. For calculating precipitable water for the entire column extending from earth's surface, the above steps for 1–2 km, 2–3 km and up to a height of say 12 km is sufficient. Sum of all the results give the total precipitable water in the column of the atmosphere.

## 1.3.5  Latent Heat

Latent (or hidden) heat of vaporisation is the amount of heat required to convert a unit mass (1 g) of water to vapour without any change in its temperature. Since vaporisation and condensation are reverse processes, the same quantity of heat is released during condensation of water vapour to the water state. The latent heat ($L_h$) and temperature are related as

$$L_h = 597.3 - 0.564T \tag{1.9}$$

where $L_h$ is expressed in cal/g and $T$ is the temperature in °C . Equation (1.9) is valid up to 40°C. At zero degree, the latent heat of vaporisation is 597.3 cal/g. Similarly, the latent heat of fusion can be defined as the amount of heat required to convert unit mass (1 g) of water to ice at the same temperature. The same amount of heat is released while converting 1 g of ice to water at the same temperature. Approximately the latent heat of fusion is taken as 79.7 cal/g. Latent heat of sublimation is the heat required to convert 1 gm of ice to vapour without change in temperature. Obviously at 0°C its value is 597.3 + 79.7 = 677 cal/g.

### 1.3.6   Humidity

In general it indicates the amount of moisture present in the atmosphere. *Absolute humidity* is the mass of water vapour present per unit of volume of air at any instant and is expressed in gram per cubic meter. Since most of the atmosphere is unsaturated, it is useful to estimate the degree of saturation by *relative humidity* which is the ratio between the actual vapour pressure to saturation vapour pressure at the same temperature, i.e.

$$RH = 100e/e_s \qquad (1.10)$$

where $e$ and $e_s$ are the vapour pressure and absolute vapour pressure in $N/m^2$, respectively. Humidity is measured by *Psychometer* having a dry and a wet bulb thermometers. It can also be measured by *hair hygrometer* as the length of a hair varies with relative humidity.

Bolsen (1958) proposed the following equation for calculating relative humidity in percentage.

$$RH = 100 \times \left\{ \frac{112 - 0.1T + T_d}{112 + 0.9T} \right\}^8 \qquad (1.11)$$

where $T$ and $T_d$ are the air and dew point temperatures respectively. Equation (1.11) is valid for temperature ranging between –25 and 45°C. Relative humidity is a function of temperature. It does not give the amount of water vapour present in the atmosphere. For example, the moisture content of the atmosphere at 0°C and, say, 20°C with 100% relative humidity are entirely different. The term absolute humidity is defined as the mass of water vapour contained in a unit volume of space and is expressed in $g/m^3$. It is expressed in the same unit as the density of water vapour. Absolute humidity can be calculated as

$$e_{ah} = (m/R_a) \, e/T \qquad (1.12)$$

where $e$ is the vapour pressure (millibar), $T$ the absolute temperature (°C ) and $m$ the molecular weight of water.

---

**Exercise 1.3:** Compute the relative humidity, if the air temperature and the dew point temperatures are 33 and 18°C, respectively.

**Solution**
From equation (1.11)

$$RH = \{(112 - 0.1 \times 33 + 18)/(112 + 0.9 \times 33)\}^8$$

$$= \{126.7/141.7\}^8 = 0.408 = 40.8\%$$

---

### 1.4   CLOUD AND RAINDROP FORMATION

Meteorologists have agreed on an uniform system of cloud classification which is mainly based on the forms and positions of the clouds with respect to the earth in the atmosphere. The types arranged in order of their weather producing potential are as follows:

(1) Cirrus (Ci)

(2) Cirrostratus (Cs)

(3) Cirrocumulus (Cc)

(4) Altostratus (As)

(5) Altocumulus (Ac)

(6) Stratus (St)

(7) Stratocumulus (Sc)

(8) Nimbostratus (Ns)

(9) Cumulus (Cu)

(10) Cumulonimbus (Cb)

These clouds can be further subdivided into their species and varieties. Cirrus cloud gives the minimum precipitation which are found at great heights (up to 12 km) from the earth's surface, whereas cumulonimbus are low lying clouds associated with intense rainfall.

Diffusion of water vapour on super-cooled (upto –40°C) condensation nuclei helps to form visible size ice-crystals. This leads to the formation of initial cloud elements of size 10 to 50 $\mu$m in diameter. They are held in suspension even with small air currents of 0.5 cm/sec. A raindrop of 200 $\mu$m is the boundary zone below which it is considered as cloud element (droplet). Therefore, the droplet should grow in size for precipitation to occur.

In a cloud mass, the presence of a number of liquid droplets is much more than the number of ice-crystals. The difference of vapour pressure at both surfaces helps to grow (transfer of moisture to) the ice crystals. Due to the difference in the size of condensation nuclei varying between $10^{-2}$ and 10 $\mu$m, different sizes of raindrops are formed. Electric charges of the cloud droplets also help to attract neighbouring droplets to grow in size. Falling raindrops attain different terminal velocities depending on their size. Meteorological factors like temperature, relative humidity and wind speed govern the minimum size of the raindrop, which is formed at the cloud altitude so as to reach the earth's surface. Raindrops above say 440 $\mu$m diameter are formed to overcome the air resistance and evaporation. The cycle of condensation, falling, evaporation and rising back occurs on an average of ten times before a drop reaches a critical size which can be large enough to fall through the bottom of the cloud. It is estimated that 6–7 collisions between raindrops occur per kilometre of fall in the air, thus multiplying their size through the process called coalescence.

Thus collision between clouds and coalescence are considered as significant factors attributing to precipitation. For raindrops of 0.5 mm diameter the terminal velocity under normal condition is 200 cm/sec and for size 6 mm it can exceed 900cm/sec. Raindrops larger than 7 mm attains a fall velocity of 1000 cm/sec and break into smaller drops due to their own deformation while passing through air resistance.

## 1.5  HYDROLOGIC CYCLE

This cycle is a continuous process without any beginning or end. It is the focus of all activities, extending its scope from 12 km up into the atmosphere from the earth's surface to about 1 km below the earth's crust to the lithosphere through a maze of paths. It can be defined as "the sequence of cyclic events which correlates the movement of water from the atmosphere to the earth's surface and then to the large water bodies through surface and subsurface routes and finally going back to the atmosphere".

Thus a hydrologic cycle undergoes the complicated process of precipitation, interception, evaporation, transpiration, infiltration, percolation, runoff and various storages. The cycle accounting for the processes and storages can be illustrated differently. Illustratively, the sun, cloud, plants, hills and other related features can be shown which describe various relationships. In quantitative presentation, hydrologic cycle is pictured by different authors in different forms. The simplest is the fragmental presentation of a circle in *Pi* chart. Fig. 1.5 is considered as the most impressive form of presentation of the hydrologic cycle, correlating water in the atmosphere, hydrosphere and the lithosphere.

In atmosphere, water is available as water vapour. Various forms of water available in water bodies on the surface of the earth is the problem in hydrosphere and the availability of water below the ground is the interest of lithosphere. To understand the hydrologic cycle one has to explore from 1 km of lithosphere to about 12 km of the atmosphere. The system is very vast and complex. Different processes and storages (in Fig. 1.5) can be listed as:

|           | Surface and above | Sub-surfacerface |
|-----------|-------------------|------------------|
| Processes | Precipitation, Surface runoff, Stream flow, Evaporation | Infiltration, Percolation, Saline intrusion Evapotranspiration, Interflow, Base flow. |
| Storage   | Atmospheric moisture Interception (temporary), Surface detention, Channel storage and Storages in Large water bodies | Soil moisture (unsaturated zone), Ground water (saturated zone), Fixed water (deep percolation). |

There is no clear distinction between various processes and storages. Water, which is available in a stream at some place can percolate to ground water (influent) and many a times ground water and parts of the subsurface flow are available to the stream as delayed flow (effluent stream). The snowfall (precipitation) is held at the surface and at a later date (on melting) changes to surface water potential or adds to ground water potential.

Quantitative data confirming to the availability of the amount of water on various components of hydrologic cycle is still unexplored. Therefore, on a global scenario a precise estimation of the amount of water available for each process and storage of the cycle is unknown.

Sun is the prime source of energy. Solar radiation increases the temperature of air and landmasses resulting in evapo-transpiration from the land area and evaporation from water bodies. Average solar incoming radiation (insolation) is 116.4 cal/cm$^2$/h measured at the earth's atmosphere. Due to the absorption of radiation by the atmosphere, the distance of the surface of the earth from the sun (146 to 151 × 10$^6$ km) and the inclination of the earth with respect to its own axis, the actual incoming radiation at the earth's surface is 18 cal/cm$^2$/h. This magnitude also varies with seasons and the latitudes. This incoming radiation on an average increases the daily temperature of atmosphere by 1.4°C. But in reality it does not happen. This incoming radiation (short wave radiation due to ultraviolet rays) is converted to heat energy due to the presence of moisture and

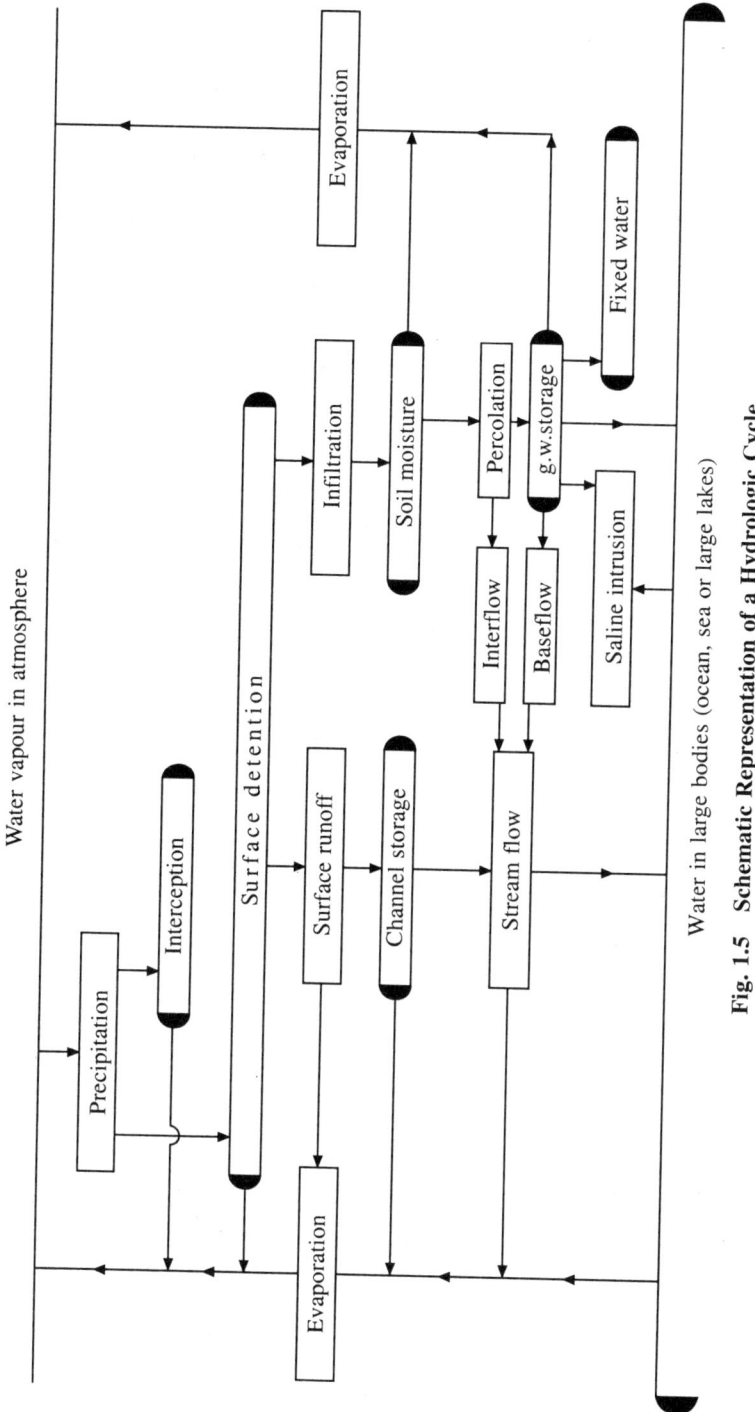

**Fig. 1.5 Schematic Representation of a Hydrologic Cycle**

carbon-dioxide gas. During nights this heat is radiated back by the long wave (infrared) radiation. Some more heat is transmitted by conduction and convection.

The evaporated water in atmosphere forms cloud. On condensation it falls as precipitation. Some precipitated water is intercepted by vegetation and evaporates back. The rest infiltrates into soil or flows down as surface runoff. The infiltrated water may join the stream later as subsurface flow or percolates further down to charge the ground water potential. The surface flow joins small streams and then through a network of channels discharge water to the large water bodies like oceans or seas. Part of the infiltrated water is available to the roots of trees and returns to the atmosphere through a process called transpiration. From most of the processes and storage evaporation takes place continuously.

A schematic diagram of the hydrologic cycle may look simple but in reality the problem is enormously complex and interrelated. It is more difficult to model the cycle on a regional basis than in a global scenario because of the encroachment of human activities on the natural environment.

It is this cycle of water, which maintains life on earth. Parameters governing the system are many and a minor change in the system causes drought and flood. Knowledge of the hydrologic cycle is a must for all the engineers in-charge of water resources in any way.

## 1.6   AVAILABILITY OF WATER ON EARTH

A world estimate of the total quantity of water is about $1400 \times 10^{15}/m^3$ of which about 96.56% is contained in large water bodies as saline water. This means that if the earth is assumed to be a flat surface, 2.745 km of water will stand over its entire area. The estimated break up of water potentials of the components of hydrologic cycle can approximately be taken as given in Table 1.2.

**Table 1.2   World Estimate of Water Quantities**

| Sl. No. | Source | Percentage of total water | Quantity of water | Percentage of fresh water |
|---------|--------|--------------------------|-------------------|--------------------------|
| (1) | (2) | (3) | (4) | (5) |
| 1. | Oceans (large water bodies) | 96.564 | $1351.9 \times 10^{15} \ m^3$ | 0.00 |
| 2. | Polar ice/glaciers | 1.730 | $24.22 \times 10^{15} \ m^3$ | 69.61 |
| 3. | Lakes | 0.0130 | $0.180 \times 10^{15} \ m^3$ | 0.261 |
| 4. | Rivers | 0.0002 | $0.0028 \times 10^{15} \ m^3$ | 0.006 |
| 5. | Atmosphere | 0.0010 | $0.014 \times 10^{15} \ m^3$ | 0.040 |
| 6. | Ground water | 1.6899 | $23.65 \times 10^{15} \ m^3$ | 30.00 |
| 7. | Soil moisture | 0.0010 | $0.014 \times 10^{15} \ m^3$ | 0.050 |
| 8. | Water in marshes | 0.0008 | $0.110 \times 10^{15} \ m^3$ | 0.030 |
| 9. | Biological | 0.0001 | $0.0014 \times 10^{15} \ m^3$ | 0.003 |
| | Total | 100.000 | $1400 \times 10^{15} \ m^3$ | 100.000% |

Water available in lakes, rivers and in the atmosphere (perceptible water) is a meagre 0.0142% of the total water potential of the world. Fresh surface water which can be utilised from lakes, rivers and other sources for beneficial purposes is very small. Out of this about 96% of the annual river water goes as waste into the oceans.

At any instant the atmospheric water content is only a meagre quantity. Over a year, enormous amount of water passes through this system. Of the fresh water resources about 70% is contained in polar ice and glaciers. Ground water available below 600 m can be taken as saline. Biological water held up by plants and animals is only 0.003% of the fresh water. This distribution of water, however, maintains a perfect balance to sustain life on earth. On a global scenario, water balance between precipitation and its components on an annual basis is given in Table 1.3.

**Table 1.3 Annual Water Balance over Land and Oceans**

|  | Ocean | Land |
|---|---|---|
| Area ($10^6$ km$^2$) | 361.3 | 148.8 |
| Annual precipitation(1000 km$^3$/mm) | 458/1278 | 119/800 |
| Annual Evaporation (1000 km$^3$/mm) | 505/1400 | 72/484 |
| Annual Runoff (1000 km$^3$/mm) | – | 47/316 |

The continental annual rainfall and runoff figures are given in Table 1.4.

**Table 1.4 Annual Rainfall and Runoff over Continents and in India**

| Continent | Total area ($\times 10^6$ km$^2$) | Annual average rainfall (mm/km$^3$) | Annual average run off (mm/km$^3$) | Runoff as % of rainfall | Evaporation as % of rainfall |
|---|---|---|---|---|---|
| (1) | (2) | (3) | (4) | (5) | (6) |
| Asia | 45.00 | 726/32670 | 293/13185 | 40.3% | 59.7 |
| Europe | 09.18 | 734/7193 | 320/3126 | 43.5% | 56.7 |
| North America | 20.70 | 670/13869 | 286/5941 | 42.8% | 57.2 |
| South America | 17.80 | 1648/29334 | 583/10377 | 35.4% | 64.6 |
| Africa | 30.30 | 686/20780 | 139/4212 | 20.3% | 79.7 |
| Australia | 8.70 | 736/6403 | 226/1966 | 30.7% | 69.3 |
| India | 3.28 | 1190/3903 | 510/1673 | 42.9% | 57.1 |

Amongst various continents, the depth of maximum runoff occurs in South America whereas the percentage of maximum runoff is observed in Europe. The runoff potential of India is 510 mm per year. One can call India as rich in surface water potential but in terms of its utilisation it is one of the poorest, utilising only 20% of the available resources.

---

**Exercise 1.4:** Calculate average time of residence of moisture in the atmosphere in a global contest and comment. Take data from Tables 1.2 and 1.3.

lution

In the atmosphere, volume of water present at any instant taken from Table 1.2 is 0.014 $\times 10^{15}$ m$^3$ or 14000 km$^3$.

Precipitation per year = Precipitation in ocean + Precipitation on the surface

$$= (458 + \overset{\circ}{1}19) \times 1000 \text{ km}^3/\text{year} = 577000/365 \text{ m}^3/\text{day}.$$

Therefore average time of residence of moisture in atmosphere

$$= 14000/(577000/365) = 8.86 \text{ days}.$$

Residence time of 8.86 days of the water vapour in the atmosphere means that the weather prediction is not possible beyond this day.

**Exercise 1.5:** The annual evaporation from a lake with surface area of 1500 hectare is 240 cm. Calculate daily average evaporation rate in hectare meter per day during the year.

**Solution**

Total volume of evaporation from the lake is $1500 \times 2.4 = 3600$ Ha. m.

Mean daily evaporation $= 3600/365 = 9.863$ Ha. m.

## 1.7 IMPORTANCE OF HYDROLOGY AND ITS APPLICATIONS IN ENGINEERING

Knowledge on hydrology is essential for engineers dealing with: (i) irrigation and drainage engineering, (ii) highway engineering, (iii) water supply engineering, (iv) water power engineering, (v) inland navigation and (vi) flood control.

Water is the most complex natural resource correlating its availability from the atmosphere to lithosphere through hydrosphere. The availability of water is highly uneven in space and time. Some of the basic things to be considered while planning and designing the engineering structures are:

(a) maximum flows which are expected to occur at a place.

(b) minimum reservoir capacity to be fixed to meet all water demands from a multipurpose reservoir.

(c) minimum flows which can occur during any dry period.

(d) possible regulation of floods at the downstream reaches once a hydraulic structure is erected.

(e) possible supply of water from a river to meet water supply demands for agriculture, hydropower generation, industrial supplies, domestic supplies, navigational requirements, recreational uses and aquaculture.

(f) environmental impacts of a hydraulic structure.

(g) study of ground water potential and its uses.

Improper assessment of water resources potential is disastrous. Many a times underestimation of the flood leads to overtopping of the dam and consequent failure of the structure. For the projects where water potential is overestimated, the system may not be in a position to fill up to the full reservoir level. Before designing any water resources related structure, evaluation of the hydrologic potential at the project site is a prerequisite. For this, collection and analysis of long term hydrological and meteorological data like rainfall, runoff, infiltration characteristics, temperature, humidity, cloud cover, wind speed and others for the area are essential. In a developing country like India, long term data are seldom available. Measurement of discharge data requires skill and many a time data at the required site are not available. For most of the river basins, however, India Meteorological Department (IMD) observes and maintains a long term hydrometeorological data. Based on these data, correct assessment of the hydrology can be made. This enables engineers to take decisions on the design of structures. Various agencies recording and maintaining different types of data are grouped as follows.

**Table 1.5   Availability of Data with the Departments/Agencies**

| Data | Agencies |
| --- | --- |
| River Discharge | Central Water Commission, State Government Irrigation/ Water Resources Departments |
| Rainfall | India Meteorological Department, State Government Agencies, Irrigation |
| | Departments, State Revenue Departments, Central Water Commission |
| Meteorological | India Meteorological Department, State Government Agencies, Irrigation |
| | Departments, Project Authorities, Agriculture Departments |
| Ground water | Central Ground Water Board, State Government Lift Irrigation Departments, Minor Irrigation Departments |
| Geological | Geological Survey of India, State Geology Directorates, State Soil Conservation Departments |
| Maps | Survey of India |

A hydrologist (or engineer) faces a difficult situation without a broad data base. Therefore, his work begins with the measurement and publication of data, analysis of the data to evaluate a definite principle correlating various data and then applying the principle to solve the problem of the region as a whole.

In the chapters to follow, various hydrologic principles and their applications in solving real engineering problems normally encountered by water resources planners and managers have been discussed.

**PROBLEMS**

1.1   List at least ten engineering activities where hydrological studies are essential.

1.2   Select five potential water using departments of your state/country and explain their hydrologic responsibilities.

1.3   Define hydrologic cycle. Sketch the cycle and tabulate the various processes and storages involved in the system.

1.4   Describe the importance of various hydrologic data in water resources engineering.

1.5   Calculate the relative humidity if air temperature is 27°C and dew point temperature is 13°C.

1.6   In the month of July 1999, a reservoir with water spread area of 1640 hectare has dropped its water level by 1.10 m. Rainfall during the month is 28 cm and evaporation is 22 cm. If the average inflow is 0.64 $m^3$/sec, calculate the average rate of drawl of water from the reservoir assuming no other losses.

1.7    A reservoir has the inflow and outflow rate of 18 and 7 $m^3$/sec respectively. Storage at 8.00 AM on a day is 230 Ha. m. Calculate the storage at 2.00 PM on the next day.

1.8    Calculate the due point temperature if the air temperature and relative humidity are 28°C and 68% respectively.

1.9    Over an area of 2.5 $km^2$ rainfall at the rate of 12 mm/h was recorded for a period of 2 days. Determine the volume of rainfall due to the storm in hectare meter, million cum, and $m^3$/sec.

1.10    Inflow at the rate of 3.5 $m^3$/sec enters a reservoir at the water spread area of 8 $km^2$. Determine the time required for the reservoir to fill up for a depth of 10 cm.

1.11    At 20°C how many calories of heat will be required to evaporate 5 litres of water? With the same amount of heat, what quantity of ice can be melt ? Refer the specific heat of ice from the tables.

1.12    Compute the heat required for vapourisation of water at 10°C, 20°C, 30°C and 40°C.

1.13    Calculate the weight of 10 $m^3$ of dry air at temperature of 20°C and 1000 mb pressure.

1.14    When the air temperature is 30°C and RH is 70%, obtain the dew point temperature.

# Chapter Two

# Statistics and Probabilities in Hydrology

## 2.1  INTRODUCTION

Performance of any water resource project depends on the correct prediction of future hydrologic events. Information on the past observed records helps to derive statistical parameters based on which the future occurrences are predicated. Laws of probability help to arrive at the desired objective. Observed records called historical data are the outcome of complex natural hydrologic phenomena. A hydrologic process is a change of this natural hydrologic phenomenon with time. A precipitations-runoff process is a complex natural phenomena and can be presented by a simple mathematical equation of the form $R = a + bP$ involving the parameters $a$ and $b$. Historical data help to estimate the values of $a$ and $b$ such that for any value of say precipitation $P$, the corresponding value of runoff ($R$) can be calculated. However, it is difficult to model a hydrologic system accurately.

In hydrology, the use of physical models to predict future sequences is less attempted than abstract models which describe the system in mathematical terms. A set of equations linking the input and output variables are described. Events like rainfall for a particular day are called variables, which may be the functions of space, time or may be pure random. Depending on the approach, an abstract model may be either deterministic, stochastic or probabilistic with the consideration of features given in Table 2.1.

### Table 2.1   Modelling Approaches

| | | |
|---|---|---|
| Deterministic | (i) | Sequencing of occurrence of the variable is considered. |
| | (ii) | Model is considered to follow a definite law of certainty but not the law of probability. |
| | (iii) | Chance of occurrence of variables is not considered. |
| Probabilistic | (i) | Sequence of occurrence of the variables is neglected, i.e., they are time independent. |
| | (ii) | Chance of occurrence of variables follows a definite probability distribution. |
| | (iii) | Variables are considered as pure random. |
| Stochastic | (i) | Sequence of occurrence of the variables is considered. |
| | (ii) | Probability distribution of the variables may or may not be time dependent. |
| | (iii) | Variables may or may not be pure random i.e., variables or events may or may not be dependent on each other. |

Most of the hydrologic models are either deterministic or stochastic. A deterministic model can make a forecast but a stochastic model makes predictions. For example, a deterministic model can forecast with reasonable accuracy the daily evaporations from a lake but the predictions of rainfall for a particular day at a particular place is purely stochastic as the random component is very large.

Sequencing of the occurrences of any particular hydrological data is called a *time series*. In a complete duration series, all data are considered. In a *partial duration* series, the data are so selected that their magnitudes always exceed a certain base value. For example, if all floods exceeding a certain base value (say *x*) are considered then it may so happen that from one year, several values are picked up while some other years may go unrepresentated. When all yearly or monthly largest or smallest values are considered, then the sequence is called an *extreme value* series.

A time series is said to be *stationary* if distribution characteristics like mean and standard deviation remain constant for the series. As an example, annual precipitation data of 50 years can be divided into 5 groups of say 10 years each, if the series is stationary then statistical parameters like mean and standard deviation should be the same for all the 5 groups. When yearly runoff values are considered, the series becomes stationary while monthly values are never stationary because of seasonal effect. Due to this, precipitations for the months of July and April in India are never comparable.

Hydrological phenomena like river discharge are *continuous*. Sometimes it is necessary to break down the series into intervals and group them to a *discrete series*. On the other hand, discrete variables can be made continuous by fitting continuous functions to the series. Such type of conversions are necessary depending on the method of analysis. Hydrology mainly deals with forecasting or predictions which can only be achieved by carrying out the statistical and probabilistic analysis of the historical data. Probabilistic analysis deals with the prediction of chance from the collected sample while statistics deals with the computation of parameters with the objective to extract essential informations from the set of data. The important characteristics of the sample are represented by a small set of statistics. In this chapter, a brief discussion on statistics and probabilities with their role in hydrology is discussed.

## 2.2   STATISTICAL PARAMETERS

The *method of moments* is commonly used for computation of statistical parameters from the observed data. We compute the statistical parameters like mean, variance, skewness and kurtosis from a short sample record out of the total length of the time series. The sample estimates are converted to an unbiased estimate and is used subsequently for further synthesis. To define the distributions of a series, more than two such parameters are essential, the important ones are described in the following paragraphs.

### 2.2.1   Central Tendency Parameters

Mean, median and mode are the central tendency parameters representing the concentration of the distribution about the central values. A sample distribution of data is shown in Fig. 2.1.

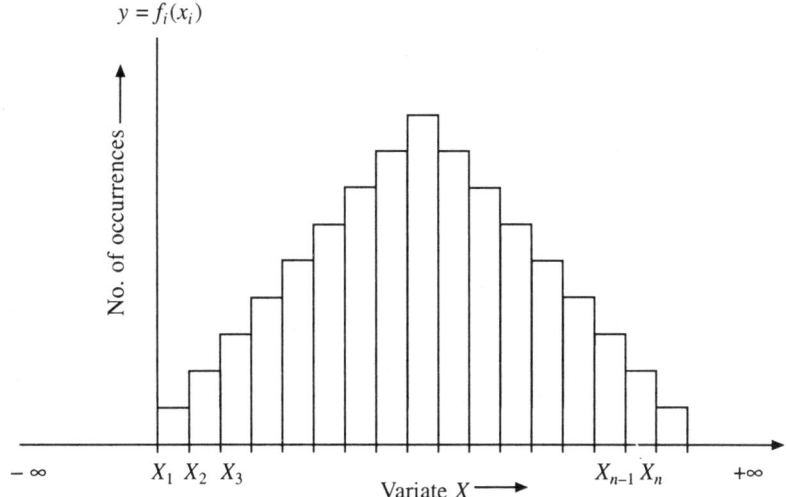

**Fig. 2.1 Distribution of Sample Data**

### 2.2.1.1 *Mean*

As we understand, mean is the average of all the records about which the distribution is equally weighed. The distance of mean from the origin is obtained by taking the first moment from the origin. It represents a location from origin the first moment about which all positive and negative points balance. Sum of all the departures from a mean should be equal to zero. We come across terms like arithmetic mean, geometric mean, harmonic mean and weighted mean. Methods to calculate these means are discussed below.

**Arithmetic Mean:** For a grouped series with $N$ data points, the mean is calculated by taking the first moment of the distribution of the sample from origin as shown in Fig. 2.1.

$$M_0 = \mu = \sum_{i=-\infty}^{+\infty} X_i f(X_i) \tag{2.1}$$

in which the summation ranges between $-\infty$ and $+\infty$, $f(X_i)$ is the number of occurrences of each event, $M_0$ the moment of the sample from origin, $\mu$ the population mean and $X_i$ the variate. For discrete variables, the above equation can be reduced to the following form.

$$X_{av} = \left(\frac{1}{N}\right) \sum_{i=1}^{N} X_i \tag{2.2}$$

where $X_{av}$ is the arithmetic mean, $X_i$ the variate, i.e., records used in the computation and $N$ the total number of records.

**Geometric Mean:** The sample estimate of the geometric mean is given by

$$X_{gm} = (X_1 \cdot X_2 \cdot X_3 \ldots X_n)^{1/N} \tag{2.3}$$

**Harmonic Mean:** The sample estimate of this mean can be calculated from the following relation

$$X_{hm} = \frac{N}{\sum\limits_{i=1}^{N}\left(\dfrac{1}{X_i}\right)} = \frac{N}{\left(\dfrac{1}{X_1} + \dfrac{1}{X_2} + \dfrac{1}{X_3} + \dots + \dfrac{1}{X_n}\right)} \tag{2.4}$$

**Weighted Mean:** For a sample, if different weights are attached to different variables, then the computed mean is called weighted mean. This mean is calculated from the following relation

$$X_w = \frac{w_1 X_1 + w_2 X_2 + \dots + w_n X_n}{w_1 + w_2 + \dots + w_n} \tag{2.5}$$

in which $w_i$ are the weights attached to each data $X_i$ of the sample. This mean is used while computing average catchment rainfall by assigning weights to the area of coverage of the respective raingauge stations. The arithmetic mean is used for most of hydrologic computations. For any given sample, the arithmetic mean gives the highest value followed by geometric mean and harmonic mean.

---

**Example 2.1:** A basin has the following stations. Calculate the mean rainfall of the basin from the following data for July 1997.

| Station | A | B | C | D | E | F | G | H | I |
|---|---|---|---|---|---|---|---|---|---|
| Rainfall (cm) | 51.8 | 32.0 | 28.7 | 43.4 | 38.6 | 50.5 | 59.6 | 31.5 | 31.7 |
| Area influenced by the station (sq. km) | 31.0 | 58.0 | 31.5 | 31.0 | 86.0 | 71.0 | 27.0 | 43.5 | 66.0 |

Prove that arithmetic mean is greater than the other two.

**Solution**

$$\text{Mean} = X_{av} = \frac{\Sigma X_i}{N} = \frac{51.8 + 32.0 + 28.7 + \dots + 31.7}{9} = \frac{367.8}{9} = 40.87 \text{ cm}$$

Geometric mean $X_{gm} = (X_1 \times X_2 \times X_3 \dots X_n)^{1/N} = (51.8 \times 32.0 \times \dots \times 31.7)^{1/9} = 39.6 \text{ cm}$

$$\text{Harmonic mean} = X_{hm} = \frac{N}{\sum\limits_{i=1}^{N}\left(\dfrac{1}{X_i}\right)} = \frac{N}{\left(\dfrac{1}{X_1} + \dfrac{1}{X_2} + \dfrac{1}{X_3} + \dots + \dfrac{1}{X_n}\right)}$$

$$= \frac{9}{\left(\dfrac{1}{51.8} + \dfrac{1}{32} + \dots \dfrac{1}{31.7}\right)} = \frac{9}{0.234225} = 38.42 \text{ cm}$$

$$\text{Weighted Mean} = \frac{w_1 X_1 + w_2 X_2 + \dots + w_9 X_9}{w_1 + w_2 + \dots + w_9}$$

$$= \frac{51.8 \times 31 + 32 \times 58 + \dots + 31.7 \times 66}{31 + 58 \dots + 66}$$

$$= \frac{17688}{445} = 39.75 \text{ cm}$$

It can be seen that $\qquad\qquad X_{av} > X_{gm} > X_{hm}.$

---

### 2.2.1.2 Median

It is defined as the middle value of the data for an odd number of samples or average of two central values for an even number of samples which divide the length of the sample into two equal parts. Median represents a better estimate of the central tendency parameters as fifty percent of the data are distributed equally on either side. For grouped data the median can be computed from the relation.

$$\text{Median} = M_d = L_1 + h\,\frac{N/2 - S_{bm}}{f_{\text{Median}}} \qquad (2.6)$$

in which $L_1$ is the lower boundary of the class containing median, $N$ the total number of data in the sample, i.e., cumulative frequency, $S_{bm}$ the sum of all frequencies below the median class, $f_{\text{Median}}$ is frequency of median class and $h$ the size of the median class interval.

---

**Example 2.2:** Calculate the median for the rainfall data for the nine rainy days in a month. The quantities of rainfall in millimeters for the 9 stations are 90, 72, 100, 12, 54, 63, 46, 46, and 34. Also find the median by taking the same series without the value 34 in the above.

**Solution**

To find the median, the data are to be arranged in increasing order. The series becomes 12, 34, 46, 46, 54, 63, 72, 90, 100. The middle value can be seen as 54.

Similarly for the even number of data without the value of 34, the series can be arranged in increasing order and the median can be found as

$$\text{Median} = \frac{(54 + 63)}{2} = \frac{117}{2} = 58.5$$

---

### 2.2.1.3 Mode

Mode is the variate which occurs most frequently. For continuous variables, it is the maximum value in the frequency distribution. Calculation of mode is usually carried out for the grouped data and is given as

$$\text{Mode} = L + \left[\frac{(f_1 - f_0)}{(f_1 - f_0) + (f_1 - f_2)}\right] h \qquad (2.7)$$

in which $L$ is the lower limit of modal class, $f_1$ frequency of the modal class, $f_0$ the frequency before the modal class, $f_2$ the frequency after the modal class and $h$ the length of interval of modal class. Modal class means the class with maximum frequency. For a symmetrical distribution of data points, the mean, median and mode are the same. For all practical calculations, a hydrologist generally prefers to use the arithmetic mean than median or mode because it is considered that mean represents the central tendency parameter better.

---

**Example 2.3:** At a station 55 days of rainfall were observed in a year. The rainfall data are grouped in the interval of 10 as given in Table 2.2. Find the mean, median and mode of the sample.

**Table 2.2   Rainfall Data for Example 2.3**

| Rainfall (cm) | Number of occurrences (frequency) of the rainfall |
|---------------|---------------------------------------------------|
| 41–50 | 5 |
| 51–60 | 8 |
| 61–70 | 10 |
| 71–80 | 11 |
| 81–90 | 12 |
| 91–100 | 4 |
| 101–110 | 3 |
| 111–120 | 2 |

**Solution**

Computation are carried out in Table 2.3.

**Table 2.3   Computation of Mean, Median and Mode of Example 2.3**

| Sl. No. | Rainfall Class | Class mean $(x_i)$ | Frequency $(f_i)$ | Cumulative frequency | $f_i x_i$ | Remarks |
|---------|----------------|--------------------|--------------------|----------------------|-----------|---------|
| (1) | (2) | (3) | (4) | (5) | (6) | (7) |
| 1 | 41–50 | 45 | 5 | 5 | 225 | *Median |
| 2 | 51–60 | 55 | 8 | 13 | 440 | class is |
| 3 | 61–70 | 65 | 10 | 23 | 650 | 55/2 = 27.5 |
| 4 | 71–80 | 75 | 11 | 34* | 825 | lying in the |
| 5 | 81–90 | 85 | 12** | 46 | 1020 | fourth group. |
| 6 | 91–100 | 95 | 4 | 50 | 380 | **Model class |
| 7 | 101–110 | 105 | 3 | 53 | 315 | has the freq- |
| 8 | 111–120 | 115 | 2 | 55 | 230 | uency of 12 |
| | Total | = | 55 | | 4085 | |

For grouped data mean $= \dfrac{\Sigma f_i X_i}{N} = \dfrac{4058}{55} = 74.27$ cm

Median $M_d = L_1 + h\,\dfrac{N/2 - S_{bm}}{f_{\text{Median}}} = 71 + \dfrac{55/2 - 23}{11}\,10 = 75.09$ cm

Mode $= 81 + \left[\dfrac{(12 - 11)}{\{(12 - 11) + (12 - 4)\}}\right] \times 10 = 81 + \dfrac{10}{9} = 82.11$ cm

Location of the central tendency parameters in the distribution graph of the rainfall is shown in Fig 2.2.

A relation between mean, median and mode can be expressed as

$$\text{Mean} - \text{Mode} = 3\,(\text{Mean} - \text{Median}) \tag{2.8a}$$

or

$$\text{Mode} = \text{Mean} - 3\,(\text{Mean} - \text{Median}) \tag{2.8b}$$

The relation is empirical and holds good for a large sample size. For the application of this empirical relationship, the class interval should be small. In the above example it does not hold good because of the limitation.

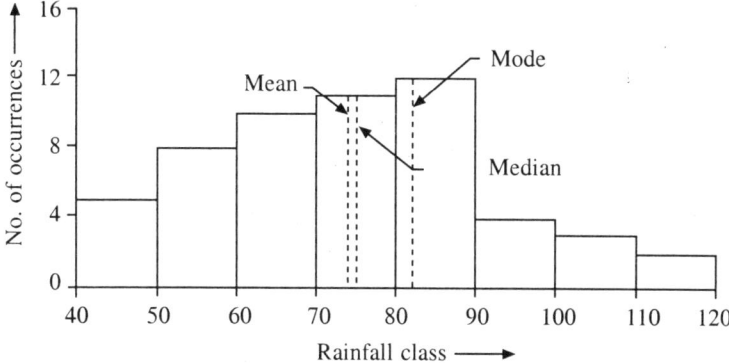

**Fig. 2.2 Rainfall distribution graph of Example 2.3**

### 2.2.2 Dispersion Characteristics

Variability of data is measured by the parameter *variance* which is the second moment of all the points of the sample about the mean. It is expressed in the following form

$$\sigma_n^2 = \frac{1}{N} \Sigma (X_i - X_{av})^2 = \left\{ \Sigma X_i^2 - \frac{(\Sigma X)^2}{N} \right\} \qquad (2.9)$$

in which $\sigma_n^2$ is the variance of the data and $N$ the number of data in the sample. When the sample length of 30 or less is available, the variance is not considered as the true representative of the population. For such a case estimation of variance is carried out by multiplying equation (2.9) by $N/(N-1)$ to arrive at the unbiased estimate of the parameter. Equation (2.9) is modified to the form

$$\sigma_{n-1}^2 = \frac{1}{(N-1)} \Sigma (X_i - X_{av})^2 \qquad (2.10)$$

and for a grouped data the relation is expressed in the following form.

$$\sigma_{n-1}^2 = \frac{1}{(N-1)} \Sigma f_i (X_i - X_{av})^2 \qquad (2.11)$$

#### 2.2.2.1 Standard Deviation

Variability is better represented by the parameter called standard deviation. It is the positive square root of the variance and is represented by

$$\sigma_n = \sqrt{\frac{1}{N} \Sigma (X_i - X_{av})^2} \qquad (2.12)$$

For an unbiased estimate, it is expressed in the form

$$\sigma_{n-1} = \sqrt{\frac{1}{(N-1)} \Sigma (X_i - X_{av})^2} \qquad (2.13)$$

depending on where the sample length used for the parameter estimation is more or less than 30. It has the same dimension as the variate. The higher is the value of the standard deviation, the larger is the spread of data from the mean.

### 2.2.2.2   Coefficient of Variation
It is a dimensionless measure of the variability and is represented as the ratio of the standard deviation to the mean of the sample.

$$C_v = \frac{\sigma}{X_{av}}$$
(2.14)

in which $C_v$ is the coefficient of variation

### 2.2.2.3   Standard Error of Parameter
Standard error $(S_e)$ of the standard deviation is given as

$$S_e = \frac{\sigma}{\sqrt{2N}}$$
(2.15)

Standard error of the mean $S_{em}$ is calculated from the relation

$$S_{em} = \frac{\sigma}{\sqrt{N}} = \frac{1}{N}\sqrt{\Sigma(X_i - X_{av})^2}$$
(2.16)

It represents the standard deviation of the sampling distribution of the statistical parameter.

### 2.2.2.4   Range of Data
Range is the difference between the largest and the smallest value of data in the given sample. For example, for an annual maximum series of say 30 years data, range is the difference between the largest and the smallest of the values among the 30 data. For a normally distributed time series Hurst and Klemes (1951, 1956, 1974) proposed range as

$$R = \sigma\left(\frac{n}{2}\right)^k \quad 0.5 < k < 1$$
(2.17)

in which $k$ is an exponent.

### 2.2.3   Skewness
It is defined as the third moment about the mean. Skewness represents symmetry of distribution of data about the mean. If the peak of the data are distributed to the right with a long tail to the left it is called the right or negatively skewed (Fig. 2.3a) and when the data are distributed to the left (Fig. 2.3b), it is called the left or positively skewed. Figure 2.3c shows symmetrical data distribution. Skewness is given as

$$\alpha\,(\text{or } \mu_3) = \frac{1}{N}\,\Sigma\,(X_i - X_{av})^3$$
(2.18)

when the data length is less than thirty then the unbiased estimate is obtained by

multiplying equation (2.18) by $\dfrac{N^2}{(N-1)(N-2)}$. The equation (2.18) is rewritten as

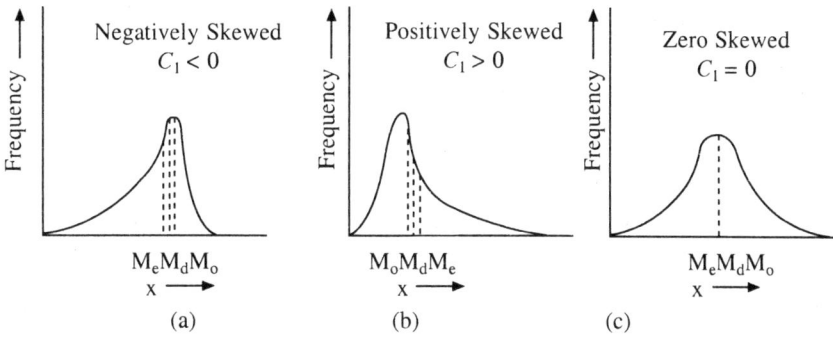

**Fig. 2.3   Skewness of Data**

$$\mu_3 = \frac{N}{(N-1)(N-2)} \Sigma (X_i - X_{av})^3 \qquad (2.19)$$

Parameter of statistical interest is skewness coefficient or coefficient of skewness. It is defined as the ratio of the third moment about mean to the cube of standard deviation.

$$C_s = \frac{\mu_3}{\sigma^3} \qquad (2.20)$$

For sample length exceeding 30 (>30) skewness can also be calculated using the relation

$$S_{ps} = \frac{\text{Mean} - \text{Mode}}{\text{Standard deviation}} = \frac{3(\text{Mean} - \text{Median})}{\text{Standard deviation}} \qquad (2.21)$$

Skewness calculated from relation (2.21) is called *Pearson's skewness*. For right skewed distribution, $C_s > 0.0$, for left skewed distribution $C_s < 0.0$ and for symmetrical distribution $C_s$ equals to zero. The position of mean, median and mode for positively, negatively and symmetrical distribution is shown in Fig. 2.3. If the data has large skewness due to the presence of a few extreme values, then the central tendency parameter like mean is affected substantially.

### 2.2.4   Kurtosis
This parameter is calculated by first taking the fourth moment of the data about the mean. The fourth moment is calculated as

$$\mu_4 = \left(\frac{1}{N}\right) \Sigma (X_i - X_{av})^4 \qquad (2.22)$$

For unbiased estimate, equation (2.22) is modified to the following form

$$\mu_4 = \left[\frac{N^2}{\{(N-1)(N-2)(N-3)\}}\right] (X_i - X_{av})^4 \qquad (2.23)$$

It represents the congestions or grouping of data at a central place. It is a measure of the peakedness because this value tends to become zero faster as it represents the fourth power deviation from mean. It approaches infinity faster when the deviation from mean increases. Kurtosis coefficient is defined as the ratio of fourth power about the mean to the square of variance, and is given as

$$\beta_2 = \frac{\mu_4}{\sigma^4} \tag{2.24}$$

The value of $\beta_2$ along with $\gamma_2 (= \beta_2 - 3)$ represents the grouping of data at the central place. For flat curve $\beta_2 < 3$ and $\gamma_2$ is negative and for peaked curve $\beta_2 > 3$ and $\gamma_2$ is positive. For symmetrical distribution $\beta_2 = 3$ and $\gamma_2 = 0$. The value of $\beta_2$ is always positive. Kurtosis of three types of sample data is shown in Fig 2.4.

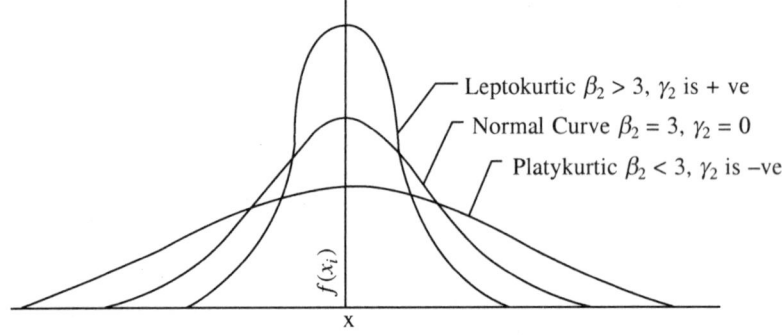

**Fig. 2.4   Kurtosis of Data**

---

**Example 2.4**   Yearly maximum rainfall records for 100 years are available for a station. Find mean, mode, median, variance, coefficient of variation, skewness and kurtosis of the data. Plot a curve between cumulative frequency and the mid class (Ogive curve).

**Table 2.4   Yearly Maximum Rainfall Records for Example 2.4**

| | | | | | | | | | |
|---|---|---|---|---|---|---|---|---|---|
| 77 | 78 | 80 | 93 | 96 | 97 | 98 | 98 | 98 | 97 |
| 84 | 95 | 83 | 93 | 82 | 91 | 80 | 88 | 84 | 86 |
| 100 | 100 | 101 | 102 | 106 | 86 | 82 | 87 | 109 | 104 |
| 75 | 89 | 99 | 96 | 94 | 93 | 92 | 90 | 86 | 78 |
| 79 | 84 | 83 | 87 | 88 | 89 | 75 | 76 | 76 | 79 |
| 80 | 81 | 89 | 99 | 104 | 100 | 103 | 104 | 107 | 110 |
| 110 | 106 | 102 | 107 | 103 | 101 | 101 | 101 | 86 | 94 |
| 93 | 96 | 97 | 99 | 100 | 102 | 103 | 107 | 107 | 108 |
| 109 | 94 | 93 | 97 | 98 | 99 | 100 | 97 | 87 | 86 |
| 94 | 96 | 97 | 98 | 100 | 105 | 106 | 103 | 85 | 84 |

**Solution**

The data can be grouped into the following classes.

### Table 2.5  Computation of Statistics for Example 2.4

| Class | Frequency $(f_i)$ | Relative frequency col. (2)/100 | Cumulative relative frequency | Mean class $(x_i)$ | $(f_i x_i)$ col. (2)× col (5) | $x_i^2$ col. (5)× col (5) | $f_i(x_i - x_{av})^2$ col. (2)×{(5) −(93.7)}$^2$ | $f_i x_i^2$ col. (2)× col (5)$^2$ | $f_i(x_i - x_{av})^3$ col. (2){(5)− −(93.7)} | $f_i(x_i - x_{av})^4$ col. (2){(5)− (93.7)}$^4$ |
|---|---|---|---|---|---|---|---|---|---|---|
| (1) | (2) | (3) | (4) | (5) | (6) | (7) | (8) | (9) | (10) | (11) |
| 75–79 | 09 | 0.09 | 0.09 | 77 | 693 | 5929 | 2510 | 53361 | −41917 | 7000017 |
| 80–84 | 12 | 0.12 | 0.21 | 82 | 984 | 6724 | 1643 | 80688 | −19219 | 224866 |
| 85–89 | 15 | 0.15 | 0.36 | 87 | 1305 | 7569 | 673 | 113535 | −4511 | 30227 |
| 90–94 | 12 | 0.12 | 0.48 | 92 | 1104 | 8464 | 35 | 101568 | −59 | 100 |
| 95–99 | 19 | 0.19 | 0.67 | 97 | 1843 | 9409 | 207 | 178771 | 683 | 2253 |
| 100–104 | 20 | 0.20 | 0.87 | 102 | 2040 | 10404 | 1378 | 208080 | 11436 | 94917 |
| 105–109 | 11 | 0.11 | 0.98 | 107 | 1177 | 11449 | 1946 | 125939 | 25579 | 344191 |
| 110–114 | 02 | 0.02 | 1.00 | 112 | 224 | 12544 | 670 | 25088 | 12257 | 224303 |
| Sum | 100 | 1.00 | | | 9370 | | 9062 | 887030 | −15451 | 1620874 |

$$\text{Mean} = \frac{9370}{100} = 93.70$$

$$\text{Median} = 95 + 5 \frac{100/2 - 48}{19} = 95 + \frac{10}{19} = 95.52$$

$$\text{Mode} = 100 + \left\{ \frac{5(20 - 19)}{(20 - 19) + (20 - 11)} \right\} = 100 + \frac{5}{10} = 100.50$$

$$\text{Variance} = \frac{\Sigma f_i X_i^2}{N} - \left( \frac{\Sigma f_i x_i}{N} \right)^2 = \frac{(887030)}{100} - \left( \frac{9370}{100} \right)^2$$

$$= 8870.30 - (93.7)^2 = 8870.3 - 8779.69 = 90.6$$

or $\quad$ $$\text{Variance} = \frac{1}{N} \Sigma f_i (X_i - X_{av})^2 = \frac{9062}{100} = 90.6$$

Standard deviation $$\sigma = (90.6)^{1/2} = 9.52$$

$$\text{Coefficient of variation } C_v = \frac{\sigma}{x_{av}} = \left( \frac{9.52}{93.70} \right) = 0.1017 \text{ or } 10.17 \%$$

$$\text{Skewness } \alpha = \frac{1}{N} (X_i - X_{av})^3 = - \frac{15451}{100} = - 154.51$$

$$\text{Coefficient of skewness } \frac{\alpha}{\sigma^3} = - \frac{154.51}{(9.52)^3} = - 0.179$$

For calculating kurtosis, the fourth moment about mean is given by

$$\mu_4 = \frac{1}{N} \Sigma f_i (X_i - X_{av})^4 = 16208.74$$

$$\beta_2 = \frac{\mu_4}{\sigma^4} = \frac{16208.74}{(9.52)^4} = 1.97$$

and $$\gamma_2 = \beta_2 - 3 = 1.97 - 3 = (-) 1.03$$

Now $\gamma_2$ is negative, therefore data is negatively skewed. Since $\beta_2 < 3$ and $\gamma_2$ is negative, the distribution is platykurtic or a flat curve.

Plot of the histogram of the rainfall data are shown in Fig. 2.5.

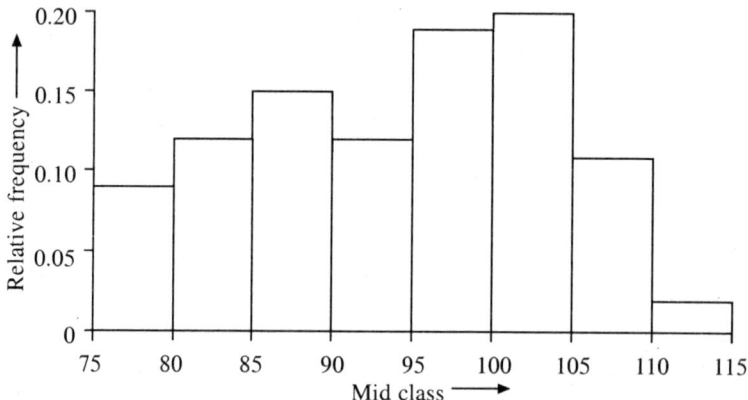

**Fig. 2.5   Relative Frequency Function of Example 2.4**

A frequency polygon is obtained by joining the top centres of the blocks forming rectangles of the histogram by straight lines. When a smooth curve is drawn passing through the centre points of the histogram block tops, we get a frequency curve.

**Ogive curve:** It is a curve drawn between cumulative values of the frequency against the mid classes. The plotted ogive curve for example 2.4 is shown in Fig 2.6.

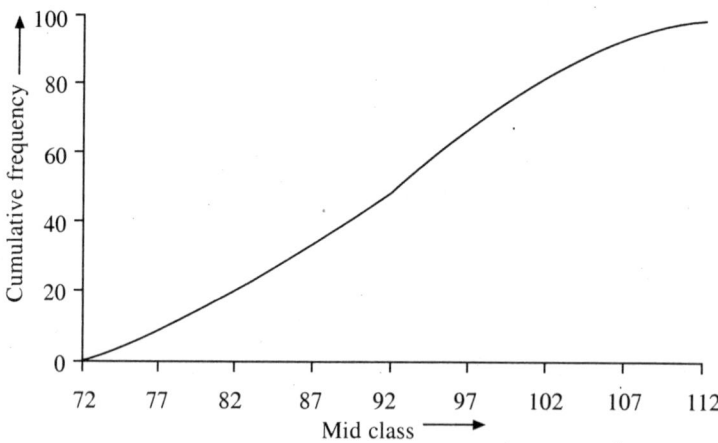

**Fig. 2.6   Ogive or Cummulative Frequency Curve of Example 2.4**

## 2.3   THEORETICAL PROBABILITY DISTRIBUTION

Statistical parameters describe the statistical distribution characteristics of a sample. A hydrologist must be in a position to predict hydrological events with their frequency of occurrence. This helps to assess a flood of a particular magnitude that can be expected in the life of the project. By fitting a frequency distribution to the set of hydrological data, the probability of occurrences of a random parameter can be calculated. Fitting of the distribution can be carried out either by

(1) method of moments, or
(2) method of maximum likelihood.

We will concentrate on the first method in which the moments of Probability Density Function (PDF) about its origin is equated with the moments of the sample data. To obtain PDF let us go back to example 2.4 in which the frequency histogram is plotted.

First the range of the random variables is divided into classes (discrete intervals $\Delta X$) The number of observations falling in each interval (frequency) is counted. A plot between the number of observations in each interval against class magnitude of the variate in abscissa gives the so called *frequency distribution*.

*Relative frequency function* is obtained by dividing the frequencies to the total number of observations. Sum of the values of the relative frequency at any point gives the cumulative frequency function. The relative frequency is also called as the probability $P$ of a function and the total probability for all variates should be ($\Sigma P = 1$) unity. In a limiting case, as the sample size becomes very large, i.e., $n \to \infty$, and $\Delta X \to 0$, the relative frequency function divided by the interval $\Delta X$ becomes the PDF. Plotting of relative frequency function and cumulative frequency function of example 2.4 is shown in Figs. 2.5 and 2.6. Depending on the probability distribution, the cumulative frequency curve can be linearised on a probability paper. This helps to predict the information required by a hydrologist by extending the straight line plot on either side.

There are many types of probability papers available. The selection of a particular type to suit the given data depends on the experience of the hydrologist and the nature of data handled. Probability papers are made on the basis of theoretical distributions. In the following, some of theoretical distributions normally used for hydrologic analysis are discussed.

### 2.3.1 Discrete Distribution

#### 2.3.1.1 Binomial Distribution
The Binomial expression of $(A + B)^n$ is written as

$$(A + B)n = {}^nC_0 A^n B^0 + {}^nC_1 A^{n-1}B^1 + {}^nC_2 A^{n-2}B^2 + \ldots + {}^nC_n A^0 B^n \quad (2.25)$$

The expression holds good for any value of $n$. Bernoulli experimented the tossing of coin by taking $P(= A)$ as probability of success and $q(1 - P)$ as probability of failure and $n$ the total number of trials of tossing. The trial of successive tossing of coin is mutually exclusive and therefore has the equal chances of its outcomes. The distribution is discrete and is represented by $(q + p)^n$ as

$$(q + p)^n = {}^nC_0 q^n p^0 + {}^nc_1 q^{n-1} p^1 + \ldots + {}^nc_n q^0 p^n \quad (2.26)$$

The probability density function is

$$p(x) = {}^nc_x p^x q^{n-x} = \left\{ \frac{n!}{[x!(n-x)!]} \right\} p^x q^{n-x} \quad (2.27)$$

Mean of probability is $np$ and variance of distribution is $n \cdot p \cdot q$. This type of distribution is well suited to rainy and non-rainy days, the events being mutually exclusive. In the above trial $x$ can be the number of rainy days out of $n$ days under consideration. The distribution is symmetrical if $p = q$; skewed to the right if $q > p$ and skewed to left if $q < p$.

---

**Example 2.5:**   For any day in July, there is 20% chance of rainfall. A building slab is to be cast which requires no rainfall during its casting period extending over 5 days. What is the probability of having no rainfall during these days and one rainy day during the period?

**Solution**
Use Binomial distribution
   Here $n = 5$, $p = 0.2$, $q = 0.8$, $x = 0$
Substituting the values in the equation, we get probability $p(x = 0)$ as

$$p(x = 0) = {}^n c_x p^x q^{n-x}$$

$$= \left[ \frac{n!}{\{x!(n-x)!\}} \right] p^x q^{n-x} = \left[ \frac{5!}{\{0!(5-0)!\}} \right] = (0.2)^0 (0.8)^5 = 0.328 \quad \text{(as } 0! = 1\text{)}$$

For probability of having one rainy day out of 5 days, we have $x = 1$.

   Therefore

$$p(x = 1) = \left[ \frac{5!}{\{1!(5-1)!\}} \right] = (0.2)^1 (0.8)^4 = \left\{ \frac{(5 \times 4 \times 3 \times 2 \times 1)}{(1 \times 4 \times 3 \times 2 \times 1)} \right\} 0.2(0.8)^4 = 0.41$$

**Example 2.6:**   During the construction period of 10 years of a reservoir, a coffer dam is required to be constructed with a capacity to take care of 5 year flood. (The concept of 5 year flood will be clear later). What is the probability that (a) the flood will not occur at all and (b) it will occur twice during the construction period?

**Solution**

Here $\qquad p = \dfrac{1}{5} = 0.2$, $n = 10$, $x = 0$, $q = (1 - p) = 0.8$

Probability that the flood will not occur in 10 years at all is given by

$$P(x = 0) = \left\{ \frac{10!}{0!(10-0)!} \right\} \times (0.2)^0 (0.8)^{10} = 0.107$$

Probability that it will occur twice in 10 years is given by

$$P(x = 2) = \left\{ \frac{10!}{2!(10-2)!} \right\} (0.2)^2 (0.8)^8$$

$$= \left\{ \frac{(10 \times 9)}{2} \right\} (0.2)^2 (0.8)^8 = 45 \times 0.04 \times 0.1678 = 0.302$$

---

### 2.3.1.2   *Poisson Distribution*
Poisson distribution is a limiting case of Binomial distribution in which (i) $n$ is large, (ii) $p$ is small and (iii) the product $p \cdot n$ is a finite, say $\lambda$, i.e., $p = \lambda/n$. The probability distribution function is given by

$$p(x) = \frac{\lambda^x e^{-\lambda}}{x!} \qquad \lambda > 0; x = 0, 1, 2, \ldots \qquad (2.28)$$

Both the mean and variance of this distribution is $\lambda$. The skewness is given by $1/(\lambda)^{0.5}$. In water resources engineering, Poisson distribution is finding increasing application in drought and flood studies when 100 years of flood or drought samples are available.

---

**Example 2.7:** Rainfall records of 80 years were scanned and it was found that the probability of precipitation of 300 mm in a day is 0.02. Find the probability of three precipitations of one day magnitude exceeding 300 mm in the next 10 years.

**Solution**

Here $\qquad\qquad\qquad\qquad p = 0.02$ and $n = 10$

Therefore $\qquad\qquad\qquad np = \lambda = 0.02 \times 10 = 0.20$

$$P(x = 3) = \frac{\{(0.2)^3 \cdot e^{-0.2}\}}{3!} = 0.0011$$

The probability of three precipitations exceeding 300 mm in the next 10 years is 0.0011 or 0.11 percent.

---

### 2.3.2 Continuous Distribution

In hydrology many events are considered as part of continuous processes. For such events, continuous distributions like the Normal, Log-normal, Gamma, Pearson type-III, Log-Pearson type-III, Extreme value type I (Gumbel) and type-III may be applied to the observed hydrologic variable. Brief theoretical concepts of these continuous distributions are discussed here and their applications to water resource problems are discussed subsequently.

#### 2.3.2.1 Normal Distribution

A random variable $x$ is said to have normal distribution with parameter $\mu$ (mean) and $\sigma^2$ (variance), when PDF is given by

$$f(x) = \frac{1}{\sigma\sqrt{2\pi}} \exp\left[ -\frac{1}{2} \left\{ \frac{(x - \mu)}{\sigma} \right\}^2 \right] \qquad -\infty < x < \infty \qquad (2.29)$$

where $\mu$ and $\sigma$ are the population mean and standard deviations and $f(x)$ is the PDF of variable $x$ varying between $-\infty$ and $+\infty$. This represents the density or the intensity of probability at that point. Frequency distribution curve for normal distribution has the following properties

(i)   It is a symmetrical and bell shaped curve.
(ii)  Top of the bell is above the mean of the sample.
(iii) Mean, median and mode for such a distribution coincide.

In equation (2.29) if $\dfrac{(x - \mu)}{\sigma}$ is replaced by a standard normal variate $t$ and integrated between ±∞, the equation represents a standard normal distribution. Since the variate, theoretically varies between −∞ to +∞ the area under the curve between any two of its limits gives the probability of occurrences of the event. The variable $t$ is normally distributed with zero mean and unit standard deviation.

To elaborate, let us take the histogram of example 2.4. A frequency curve can be obtained by passing a smooth curve through the top centres of all the histogram rectangle blocks. When the sample size is increased from 100 to say 1000 and the class interval chosen for the histogram rectangles can be reduced to smaller values from the present five, the data will be uniformly distributed on either side of the maximum histogram rectangle. The resulting frequency curve becomes continuous. Frequency ordinates can be substituted by relative frequency which is the ratio between the number of occurrences in the class interval ($n$) and the total occurrences ($N$). Relative frequency is taken as synonymous with probability and area of each rectangle of histogram represents probability of the interval. Total area under the curve becomes unit.

The area of the standard normal distribution curve for the ranges of $t$ can be obtained from any standard test book on statistics. The area of the curve for $t = \pm 1, 2$ and 3 is given in Fig. 2.7. Area under the standard normal curve for all standardised variables can be read from Table 2.6. For any cumulative probability, the value of $t$ can be found from the tables of area from which $x$, the variate is calculated from the inverse of the transfer function as

$$t = \frac{(x - \mu)}{\sigma} \qquad (2.30a)$$

or
$$x = \mu + \sigma t \qquad (2.30b)$$

Equation (2.30) is true for the population. For the sample data the equation can be modified to the following form

$$x = x_{av} + \sigma t \qquad (2.31)$$

where $x_{av}$ represents the average or mean of the sample. Normal distribution is usually applied to continuous random variables of time cumulative values of hydrologic variables of periods larger than one year. This means the annual

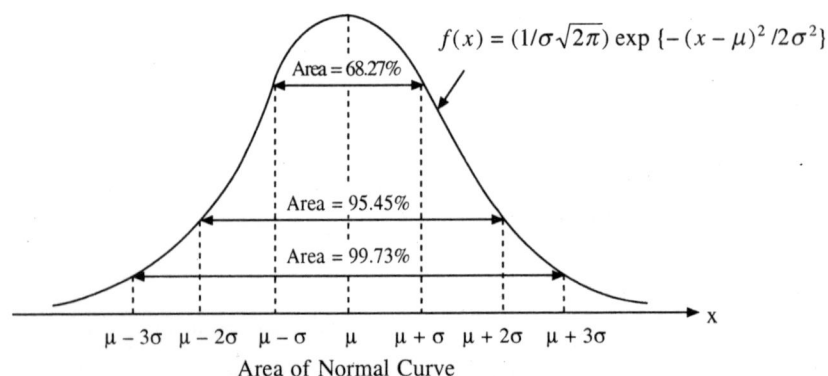

$$f(x) = (1/\sigma\sqrt{2\pi}) \exp \{-(x - \mu)^2 / 2\sigma^2\}$$

Area = 68.27%

Area = 95.45%

Area = 99.73%

$\mu - 3\sigma$   $\mu - 2\sigma$   $\mu - \sigma$   $\mu$   $\mu + \sigma$   $\mu + 2\sigma$   $\mu + 3\sigma$

Area of Normal Curve

**Fig. 2.7   Area Under the Normal Probability Distribution Curve**

Table 2.6  Area Under the Standard Normal Curve $t = (x - \mu)/\sigma$

| $t$ | 0.00 | 0.01 | 0.02 | 0.03 | 0.04 | 0.05 | 0.06 | 0.07 | 0.08 | 0.09 |
|------|--------|--------|--------|--------|--------|--------|--------|--------|--------|--------|
| 0.0 | 0.0000 | 0.0040 | 0.0080 | 0.0120 | 0.0159 | 0.0199 | 0.0239 | 0.0279 | 0.0319 | 0.0359 |
| 0.1 | 0.0398 | 0.0438 | 0.0478 | 0.0517 | 0.0557 | 0.0596 | 0.0636 | 0.0675 | 0.0714 | 0.0735 |
| 0.2 | 0.0793 | 0.0832 | 0.0871 | 0.0910 | 0.0948 | 0.0987 | 0.1026 | 0.1064 | 0.1103 | 0.1141 |
| 0.3 | 0.1179 | 0.1271 | 0.1255 | 0.1293 | 0.1331 | 0.1368 | 0.1406 | 0.1443 | 0.1480 | 0.1517 |
| 0.4 | 0.1554 | 0.1591 | 0.1628 | 0.1664 | 0.1700 | 0.1736 | 0.1772 | 0.1808 | 0.1884 | 0.1879 |
| 0.5 | 0.1915 | 0.1950 | 0.1985 | 0.2019 | 0.2054 | 0.2088 | 0.2123 | 0.2157 | 0.2190 | 0.2224 |
| 0.6 | 0.2257 | 0.2291 | 0.2324 | 0.2357 | 0.2389 | 0.2422 | 0.2454 | 0.2486 | 0.2518 | 0.2549 |
| 0.7 | 0.2580 | 0.2611 | 0.2642 | 0.2673 | 0.2704 | 0.2734 | 0.2464 | 0.2794 | 0.2823 | 0.2852 |
| 0.8 | 0.2881 | 0.2910 | 0.2939 | 0.2967 | 0.2995 | 0.3023 | 0.3051 | 0.3078 | 0.3106 | 0.3133 |
| 0.9 | 0.3159 | 0.3186 | 0.3212 | 0.3238 | 0.3264 | 0.3289 | 0.3315 | 0.3340 | 0.3365 | 0.3389 |
| 1.0 | 0.3413 | 0.3438 | 0.3461 | 0.3485 | 0.3508 | 0.3531 | 0.3554 | 0.3577 | 0.3599 | 0.3621 |
| 1.1 | 0.3643 | 0.3665 | 0.3686 | 0.3708 | 0.3729 | 0.3749 | 0.3770 | 0.3790 | 0.3810 | 0.3830 |
| 1.2 | 0.3849 | 0.3869 | 0.3888 | 0.3907 | 0.3925 | 0.3944 | 0.3962 | 0.3980 | 0.3997 | 0.4015 |
| 1.3 | 0.4032 | 0.4049 | 0.4066 | 0.4082 | 0.4099 | 0.4115 | 0.4131 | 0.4147 | 0.4162 | 0.4177 |
| 1.4 | 0.4192 | 0.4207 | 0.4222 | 0.4236 | 0.4251 | 0.4265 | 0.4279 | 0.4292 | 0.4306 | 0.4319 |
| 1.5 | 0.4332 | 0.4345 | 0.4657 | 0.4370 | 0.4382 | 0.4394 | 0.4406 | 0.4418 | 0.4430 | 0.4441 |
| 1.6 | 0.4452 | 0.4463 | 0.4474 | 0.4485 | 0.4495 | 0.4505 | 0.4515 | 0.4525 | 0.4535 | 0.4545 |
| 1.7 | 0.4554 | 0.4564 | 0.4573 | 0.4582 | 0.4591 | 0.4599 | 0.4608 | 0.4616 | 0.4625 | 0.4633 |
| 1.8 | 0.4641 | 0.4649 | 0.4656 | 0.4664 | 0.4671 | 0.4678 | 0.4686 | 0.4693 | 0.4699 | 0.4706 |
| 1.9 | 0.4713 | 0.4719 | 0.4726 | 0.4732 | 0.4738 | 0.4744 | 0.4750 | 0.4756 | 0.4762 | 0.4767 |
| 2.0 | 0.4772 | 0.4778 | 0.4783 | 0.4788 | 0.4793 | 0.4798 | 0.4803 | 0.4808 | 0.4812 | 0.4817 |
| 2.1 | 0.4821 | 0.4826 | 0.4830 | 0.4835 | 0.4838 | 0.4842 | 0.4846 | 0.4850 | 0.4854 | 0.4857 |
| 2.2 | 0.4861 | 0.4865 | 0.4868 | 0.4871 | 0.4875 | 0.4878 | 0.4881 | 0.4884 | 0.4887 | 0.4890 |
| 2.3 | 0.4893 | 0.4896 | 0.4898 | 0.4901 | 0.4904 | 0.4906 | 0.4909 | 0.4911 | 0.4913 | 0.4916 |
| 2.4 | 0.4918 | 0.4990 | 0.4922 | 0.4925 | 0.4927 | 0.4929 | 0.4931 | 0.4932 | 0.4934 | 0.4936 |
| 2.5 | 0.4938 | 0.4940 | 0.4941 | 0.4943 | 0.4945 | 0.4946 | 0.4948 | 0.4949 | 0.4951 | 0.4952 |
| 2.6 | 0.4953 | 0.4955 | 0.4956 | 0.4957 | 0.4959 | 0.4960 | 0.4961 | 0.4962 | 0.4963 | 0.4964 |
| 2.7 | 0.4965 | 0.4966 | 0.4967 | 0.4968 | 0.4969 | 0.4970 | 0.4971 | 0.4972 | 0.4973 | 0.4974 |
| 2.8 | 0.4974 | 0.4975 | 0.4976 | 0.4977 | 0.4977 | 0.4978 | 0.4979 | 0.4980 | 0.4980 | 0.4981 |
| 2.9 | 0.4981 | 0.4982 | 0.4983 | 0.4983 | 0.4984 | 0.4984 | 0.4985 | 0.4985 | 0.4986 | 0.4986 |
| 3.0 | 0.4986 | 0.4987 | 0.4987 | 0.4988 | 0.4988 | 0.4989 | 0.4989 | 0.4989 | 0.4980 | 0.4990 |

hydrological data can be approximated to normal distribution, but the daily, weekly or monthly data should not be attempted to fit this distribution. Distribution of the observed data of hydrologic variables is termed as *empirical distribution.* Fitting of the empirical distribution to normal distribution will be discussed subsequently.

### 2.3.2.2  *Log-Normal Distribution*

Log-normal distribution is a special case of normal distribution in which the variates are replaced by their logarithmic transformed values with base $e$. After logarithmic transformation, observed data is assumed to follow normal distribution and is analysed exactly in the same way as under normal distribution. If $z_i = \ln x_i$, where $x_i$ are the observed variates, than $\mu_z$ can be the mean of $z_i$ and $\sigma_z$ is the standard deviation of $z_i$. Following Chow, the statistical parameters for $x$-series can be obtained as

Mean
$$\mu_x = e^{\mu_z + \frac{\sigma_z^2}{2}}$$
(2.32)

Variance $\qquad\qquad\qquad\sigma_x^2 = \mu_x^2(e^{\sigma_z^2} - 1)$ $\qquad\qquad$ (2.33)

Probability density function $f(x)$ is given by

$$f(x) = \left(\frac{1}{x\sigma\sqrt{2\pi}}\right) e^{\left[-\frac{1}{2}\left\{\frac{(\ln x - \mu_z)}{\sigma_z}\right\}^2\right]} \quad x > 0, \mu_z = z_{av} \qquad (2.34)$$

Using the tables of standard normal distribution, the log-normal distribution may be fitted to the series.

### 2.3.2.3   Gamma Distribution
A continuous random variable of the series $x$ having probability density function (PDF) is said to have a gamma distribution with parameter $\lambda$ and $\gamma$ when

$$f(x) = \left\{\frac{(\lambda^\gamma x^{\gamma-1})}{\Gamma(\gamma)!}\right\} e^{-\lambda x} \quad x \rangle 0, \quad \lambda \rangle 0, \lambda = \frac{x_{av}}{\sigma^2}, \gamma = \frac{1}{c_v^2} \qquad (2.35)$$

with mean as $\gamma/\lambda$ and variance as $\gamma/\lambda^2$. The symbol $\Gamma$ is called gamma function. This distribution is useful to find the time taken for a particular event to occur using Poisson distribution. The special properties of this distribution are that the distributions are additive. This helps in formulating flood forecasting techniques when tributaries join the main river.

### 2.3.2.4   Pearson (Type III) Distribution
The basic equation defining probability density of Pearson type-III distribution is given as

$$f(x) = \left\{\frac{1}{\Gamma(b)}\right\} \{\lambda^b(x - c)^{b-1} e^{-\lambda(x-c)}\} \; x \geq c, x = \frac{\sigma}{\sqrt{b}}, b = \left(\frac{2}{c_s}\right)^2 \qquad (2.36)$$

This distribution reduces to gamma distribution by substituting $c = 0$. This skewed distribution has a long tail to the right which can be reduced to a normal distribution as a special case. It was Foster who applied the Pearson Type-III distribution to describe the probability distribution of annual maximum flood peaks.

### 2.3.2.5   Extreme Value Distribution
Extreme value type-I distribution is also known as Gumbel distribution since Gumbel first introduced it in 1941 for flood frequency analysis. Type-III of extreme value distribution is known as Weibull distribution which is used for studies of drought. Type II of extreme value distribution is known as Frechet distribution. This distribution is not widely used in hydrology.

*Type-I Distribution (Gumbel Distribution)*
Following the theory of the initial exponential distributions like normal distribution, the log-normal distribution can be made double exponential which converges to exponential function as $x$ increases. The cumulative probability of occurrence of an event is

$$F(x)e^{-e^{-\frac{(a+x)}{c}}} = e^{-e^{-y}} \tag{2.37}$$

where $y = (a + x)/c$, $a$ and $c$ are constants. Therefore probability of occurrences of an event equal to or larger than a value $x_0'$ is

$$P(x \ge x_0) = 1 - e^{-e^{-\gamma}}$$

The value of $c$ and $a$ in equation (2.37) is usually taken as $c = (\sqrt{6}/\pi)\sigma$ and $a = \gamma c - x_{av}$. The value of $\gamma$ is taken as 0.57721. Therefore $\gamma c = 0.450055\ \sigma$ and $a = 0.450055\sigma - x_{av}$. Rearranging equation (2.37) we get

$$y = \frac{(a + x)}{c} = -\ln\left[-\ln F(x)\right] \tag{2.38a}$$

or
$$y = -\ln\left\{\ln\left(\frac{T}{T-1}\right)\right\} \tag{2.38b}$$

where $T = 1/P$ is the reciprocal of probability. From equation (2.38), it can be seen that a straight line plot can be obtained by taking double exponential of the probability function $F(x)$ in abscissa and the value of $x$ in ordinate. Fitting of this theoretical frequency distribution to the empirical distribution is discussed in detail in the later part of this chapter.

### 2.3.2.6 Exponential Distribution
The exponential distribution is given as

$$f(x) = \lambda\, e^{-\lambda x} \tag{2.39a}$$

and the cumulative distribution function (CDF) is given by

$$F(x) = 1 - e^{-\lambda x} \tag{2.39b}$$

The distribution has a mean of $1/\lambda$ and variance of $1/\lambda^2$. The parameters $\lambda$ is the mean rate of occurrence of the event. The distribution is used to obtain the time interval between the occurrence of events like say floods, cyclone etc.

### 2.3.2.7 Log Pearson Type-III Distribution
This distribution is widely used in United States, India and other countries as the standard distribution for flood frequency analysis of annual maximum floods. The distribution has the added advantage of providing skew adjustment. The distribution can be reduced to log-normal distribution if skew adjusted is zero. The variates of the hydrologic series if represented by $x$ and if log $(x)$ follows the Pearson Type III distribution, then $x$ has the log-pearson Type III distribution. The PDF is given as

$$f(x) = \left(\frac{1}{ax\Gamma(b)}\right)\left\{\frac{y-c}{a}\right\}^{b-1} e^{\frac{-(y-c)}{a}} \tag{2.40}$$

where $\quad y = \log x$, mean $\mu_y = c + ab$, variance $= \sigma_y^2 = a^2 b$

It follows that if the log transferred series of the sample is fitted with Pearson type-III distribution then the $x_i$ (untransfered) sample should follow log-pearson type-III distribution. For skewed data, log-pearson type-III gives a better fit and is widely used. Zero skew reduces log person-III distribution to log normal and pearson type III to normal distribution. Method of fitting the above theoretical distribution to the observed data are discussed below.

## 2.4   FREQUENCY ANALYSIS

Determination of the frequency of occurrence of extreme hydrologic events like floods, droughts and severe storms are important in water resources planning and management. There is a definite relation between the frequency of occurrences and magnitude; the ordinary events occur almost regularly than the severe storms. Frequency or probability distribution helps to relate the magnitude of these extreme events with their number of occurrences such that their chance of occurrence with time can be predicted successfully.

To fit any of the theoretical distribution discussed in the previous section to the sampled historical records, the data are first analysed for obtaining the statistical information. Observed frequency distribution, which is also called as *empirical frequency distribution* is assumed to follow any of the theoretical distribution. Methods of moments as discussed under the section 2.2 are computed and statistical parameters are obtained from the sample data. Assuming these characteristics of statistical distributions to hold good for theoretical distributions, the general solution of the equation is obtained. This gives the necessary probability of occurrence of the event. To achieve this, the hydrologic data employed should be carefully selected such that the homogeneity, assumption of their independence and the minimum length of sample record are satisfied to suit the fitted distribution. To achieve this sometimes the annual maximum or minimum series are selected. The reasonable length of record for frequency analysis should be more than 20.

In hydrology the theoretical distribution equations discused before are less used because the probability distribution function for majority of the distributions are not readily invertible. There are two simplified methods available to fit theoretical distribution to empirical distribution. The methods are

(i) Chow's method using frequency factors
(ii) Graphical method using probability papers.

Before we discuss frequency factor methods, a general concept on *recurrence interval* or *return period* or *frequency* (*T*) must be known. Recurrence interval is defined as the average interval of time $T$ within which a flood (or any other extreme event) of given magnitude will be equalled or exceeded at least once. For an example let us take the probability of an event $P$ as 0.25. This means the possibility of occurrence of the event is 25% every year. Thus out of 100 years of life of the project, flood of this magnitude will exceed 25 years. Thus the average return period of the flood is 100/25 = 4 years. If $P$ is the probability in percentage then

Return period $$T = \frac{1}{P} \tag{2.41}$$

Probability of not occurrence of the event is $(1 - P) = 1 - \dfrac{1}{T}$ \hfill (2.42)

Probability of not occurrence in the life of $n$ years $\left(1 - \dfrac{1}{T}\right)^n$ \hfill (2.43)

Probability of occurrence is $\left\{1 - \left(1 - \dfrac{1}{T}\right)^n\right\}$ \hfill (2.44)

This probability of occurrence is called *risk*. For design of a project, if an engineer is allowed to take 5% risk in the life span of the project of, say, 40 years, then the return period of the event should be calculated as

$$0.05 = 1 - (1 - 1/T)^{40}$$

or $\qquad (1 - 1/T)^{40} = 1 - 0.05 = 0.95$

or $\qquad 1 - 1/T = 0.9987184$

or $\qquad T = 780 \text{ years}$

Return period is the average time interval of an event of a given magnitude being equalled or exceeded but not the actual time interval. When the hydrologic event say maximum or minimum values are arranged in some order (ascending or descending) than it can be seen that any variate can be represented as

$$X_T = x_{av} \pm \Delta x \tag{2.45}$$

where $x_{av}$ is the mean and $\Delta x$ is the departure from mean. Chow (1951) proposed the general equation for hydrologic frequency analysis as

$$X_T = x_{av} \pm \sigma k \tag{2.46}$$

where $X_T$ is the event of return period of $T$ years, $x_{av}$ the mean of data, $\sigma$ the standard deviation and $k$ the *frequency factor*. Value of the variate $X_T$ is the sum of the mean plus a departure represented by standard deviation times the frequency factor. Frequency factor is dependent on the recurrence interval $T$ and the probability distribution assumed for the series. From the theoretical distributions discussed in the previous section, a $k$-$T$ relationship is usually established. Standard books on statistics give this relation in the form of tables. Equation (2.46) holds good for all types of probability distributions depending on which suitable value of $k$ is selected. Since $k$ and $T$ are interdependent, they can be related mathematically or it can be simplified and presented in the form of tables. For skewed distribution, the coefficient of skewness and the length of sample is found to affect the $k$-$T$ relation. Procedural steps to compute probability of an event for any recurrence interval $T$ are outlined below.

(i) Compute mean $x_{av}$ and standard deviation $\sigma$ from the observed data (sample).

(ii) Select a type of distribution. This depends on the nature of distribution of the data and the experience of the hydrologist.

(iii)  From $k$-$T$ mathematical relation or from $k$-$T$ table, determine the value of $k$ for the required return period $T$.

(iv)  Value of $X_T$ for the return period $T$ can be computed using the relation given in equation (2.46).

Frequency analysis for some important theoretical distributions are given below.

### 2.4.1   Gumbel's Distribution

Gumbel probability distribution is widely used for extreme value analysis of hydrologic and meteorological data like floods, maximum rainfalls and other events. Equation (2.38) is reproduced here.

$$\frac{a + x}{c} = - \ln [- \ln F(x)] = - \ln \{\ln (1 - P)\}$$

We have substituted $F(x)$ as $= 1 - P$ because the probability of occurrence of the event equal to or greater than a value say $x_0$ is our concern.

$$\frac{a + x}{c} = - \ln [- \ln F(x)] = - \ln \{\ln (1 - P)\}$$

or  $X_T = - a - c \ln [\ln T - \ln (T - 1)]$

$$= - (\gamma c - x_{av}) - c \ln \{\ln T - \ln (T - 1)\} \qquad \text{where } a = (\gamma c - x_{av})$$

$$= x_{av} - \left(\frac{\sqrt{6}}{\pi}\right) \sigma \gamma - \left(\frac{\sqrt{6}}{\pi}\right) \sigma \ln \{\ln T - \ln (T - 1)\}$$

$$= x_{av} - \left(\frac{\sqrt{6}}{\pi}\right) \sigma \left[\gamma + \ln \left\{\ln \frac{T}{T - 1}\right\}\right]$$

$$= x_{av} - \left(\frac{\sqrt{6}}{\pi}\right) \sigma \left[0.57721 + \ln \left\{\ln \frac{T}{T - 1}\right\}\right] \qquad (2.47)$$

Comparing equation (2.47) with the general Chow's equation, we get

$$k = - \left(\frac{\sqrt{6}}{\pi}\right) \left[0.57721 + \ln \left\{\ln \frac{T}{T - 1}\right\}\right]$$

$$= - [0.45005 + 0.7797 \ln \{\ln T/(T - 1)\}] \qquad (2.48)$$

For the mean value under Gumbel's distribution, $k \times \sigma$ factor should be zero. Since $\sigma$ is not zero, $k = 0$. Substituting $k = 0$ in equation (2.48), the return period $T$ of the flood can be obtained as 2.33 years. Therefore, the average flood has a return period of 2.33 years if Gumbel's distribution is assumed.

The empirical relation (2.47) holds good when the record length is 100 years or more. However, to obtain the frequency factors for small sample length of the record, the value of frequency factors for Gumbel's distribution for various sample sizes are given in Table 2.7. From the sample size of length $n$, the frequency factor $k$ can be read from the table. The mean $x_{av}$, standard deviation $\sigma$ along with the frequency factor $k$ for the desired return period help to find out $X_T$ using equation (2.46).

**Table 2.7   Frequency Factor for the Gumbel Distribution**

| Sample size n | Return period or Recurrence Interval | | | | | | | | |
|---|---|---|---|---|---|---|---|---|---|
| | 5 | 10 | 15 | 20 | 25 | 50 | 75 | 100 | 1000 |
| 15 | 0.967 | 1.703 | 2.117 | 2.410 | 2.632 | 3.321 | 3.721 | 4.004 | 6.265 |
| 20 | 0.919 | 1.625 | 2.023 | 2.302 | 2.517 | 3.179 | 3.563 | 3.836 | 6.006 |
| 25 | 0.888 | 1.575 | 1.963 | 2.235 | 2.444 | 3.088 | 3.463 | 3.729 | 5.842 |
| 30 | 0.866 | 1.541 | 1.922 | 2.188 | 2.393 | 3.026 | 3.393 | 3.653 | 5.727 |
| 35 | 0.851 | 1.516 | 1.891 | 2.152 | 2.354 | 2.979 | 3.341 | 3.598 | |
| 40 | 0.838 | 1.495 | 1.866 | 2.126 | 2.326 | 2.943 | 3.301 | 3.554 | 5.576 |
| 45 | 0.829 | 1.478 | 1.847 | 2.104 | 2.303 | 2.913 | 3.268 | 3.520 | |
| 50 | 0.820 | 1.466 | 1.831 | 2.086 | 2.283 | 2.889 | 3.241 | 3.491 | 5.478 |
| 55 | 0.813 | 1.455 | 1.818 | 2.071 | 2.267 | 2.869 | 3.219 | 3.467 | |
| 60 | 0.807 | 1.446 | 1.806 | 2.059 | 2.253 | 2.852 | 3.200 | 3.4465 | |
| 65 | 0.801 | 1.437 | 1.796 | 2.048 | 2.241 | 2.837 | 3.183 | 3.429 | |
| 70 | 0.797 | 1.430 | 1.788 | 2.038 | 2.230 | 2.824 | 3.169 | 3.413 | 5.359 |
| 75 | 0.792 | 1.423 | 1.780 | 2.029 | 2.220 | 2.812 | 3.155 | 3.400 | |
| 80 | 0.788 | 1.417 | 1.773 | 2.020 | 2.212 | 2.802 | 3.145 | 3.387 | |
| 85 | 0.785 | 1.413 | 1.767 | 2.013 | 2.205 | 2.793 | 3.135 | 3.376 | |
| 90 | 0.782 | 1.409 | 1.762 | 2.007 | 2.198 | 2.785 | 3.125 | 3.367 | |
| 95 | 0.780 | 1.405 | 1.757 | 2.002 | 2.193 | 2.777 | 3.116 | 3.357 | |
| 100 | 0.779 | 1.401 | 1.752 | 1.998 | 2.187 | 2.770 | 3.109 | 3.349 | 5.261 |
| ∞ | 0.719 | 1.305 | 1.635 | 1.866 | 2.044 | 2.592 | 2.911 | 3.137 | 4.936 |

### 2.4.2   Pearson Type-III Distribution

This distribution is normally used for skewed data. Therefore, the frequency factor is related to the return period and skewness coefficient. Procedure to fit the observed empirical distribution to the theoretical Pearson distribution is discussed below.

(i)   Compute mean, standard deviation and coefficient of skewness of the sample.

(ii)   Multiply a factor equal to $(1 + 8.5/N)$ to the coefficient of skewness $C_s$ as suggested by Foster to overcome the difficulties of data length. If data for 100 years or more are available, the factor need not be multiplied.

(iii)   To compute flood for any desired return period, read the frequency factor $k$ from the standard $k$-$T$ table for Pearson type-III distribution (Tables 2.8A and 2.8B) corresponding to the skewness coefficient $C_s$ of the sample.

(iv)   Compute the flood of desired return period from Chow's general equation (2.46).

### 2.4.3   Log-Pearson Type-III Distribution

In this case the log transferred series with base 10 are assumed to follow Log-Pearson type-III distribution. To fit this distribution to the observed data, the first step is to take the logarithm of the series. Stepwise procedure to fit Log-Pearson type-III distribution to the observed series are outlined below.

(i) Transfer the observed data series to the logarithmic values. Generate $y_i$ series as $y_i = \log_{10} x_i$

(ii) Find mean $y_{av}$, standard deviation $\sigma_y$ and coefficient of skewness $C_s$ for the log transferred series.

(iii) Multiply the coefficient of skewness with a factor $(1 + 8.5/N)$ as suggested by Foster to overcome the short length of data. For records exceeding 100, the factor need not be multiplied. This is done because it is assumed that by multiplying the factor, the sample statistics of the skewness coefficient is converted to be the representative of the population.

(iv) To compute flood of required return period $T$, find frequency factor $k$ from $k$-$T$ relation Table (2.8a or 2.8b) corresponding to the skewness coefficient of the log transferred series.

(v) Knowing $\sigma_y$, $y_{av}$, and $k$ and using Chow's general equation (2.46), compute the event $(y_T)$ of desired return period $T$ in logarithmic scale.

(vi) By taking antilog of $y_T$ find $X_T$ which is the desired value of the event for the return period $T$.

U.S. water resources council however, does not incorporate the adjustment for skew as proposed by Foster. The procedure remains the same as described above except the multiplication of the factor $(1+8.5/N)$ to the skew coefficient. For zero skew, Pearson type-III distribution reduces to normal distribution and log-pearson type-III to lognormal distribution.

### 2.4.4  Normal Distribution

Normal distribution is a special case of Pearson type-III in which skewness coefficient is zero. Therefore, Tables 2.8(a) and 2.8(b) can be used to calculate the frequency factor $k$ by taking $C_s$ as zero. From these tables the row in which $C_s = 0$ is used to read $k$ factors. The procedure is the same as outlined for other frequency distributions. When the skewness coefficient is not zero but the distribution is assumed to follow normal then the procedure to solve such kind of problem using the area under standard normal curve is illustrated in Example 2.8. The area under standard normal curve is given in Table 2.6. Using Table 2.6 the procedure to find frequency factor $k$ for normal distribution and the event magnitude of the desired return period $T$ can be seen from the example. Abramowitz and Stegum (1965) used the following approximation to calculate the standard normal variate $t$

$$t = u - \frac{2.515517 + 0.8028534\,u + 0.0103284\,u^2}{1 + 1.4327884\,u + 0.18926\,u^2 + 0.001308\,u^3} \qquad (2.49)$$

where $\qquad u = \left[ \ln\left(\frac{1}{p^2}\right) \right]^{1/2}, \qquad p = 1/T, \qquad 0 < p \le 0.5 \qquad (2.50)$

For $p > 0.50$, $(1 - p)$ can be substituted in equation (2.50). The value of $t$ computed in equation (2.49) is assigned with negative sign. The value of $t$ calculated is equal to the frequency factor $k_T$. Error introduced by the above formulae in calculating $k_T$ is almost negligible. For example for $T = 50$, $p = 1/T = 0.02$ and

**Table 2.8(a) Frequency Factor for the Pearson Type-III Distribution with Positive Skew Coefficients (After Water Resources Council, 1967)**

| | Recurrence interval (Years) | | | | | | | | | |
|---|---|---|---|---|---|---|---|---|---|---|
| | 1.0101 | 1.0526 | 2 | 5 | 10 | 25 | 50 | 100 | 200 | 1000 |
| | Percent chance (z) | | | | | | | | | |
| Skew coefficient $C_s$ | 99 | 95 | 50 | 20 | 10 | 4 | 2 | 1 | 0.5 | 0.10 |
| 3.0 | −0.667 | −0.665 | −0.396 | 0.420 | 1.180 | 2.278 | 3.152 | 4.051 | 4.970 | 7.250 |
| 2.9 | −0.690 | −0.688 | −0.390 | 0.440 | 1.195 | 2.277 | 3.134 | 4.013 | 4.904 | – |
| 2.8 | −0.714 | −0.711 | −0.384 | 0.460 | 1.210 | 2.275 | 3.114 | 3.973 | 4.847 | – |
| 2.7 | −0.740 | −0.736 | −0.376 | 0.479 | 1.224 | 2.272 | 3.092 | 3.932 | 4.783 | – |
| 2.6 | −0.769 | −0.762 | −0.368 | 0.499 | 1.238 | 2.267 | 3.071 | 3.889 | 4.718 | – |
| 2.5 | −0.799 | −0.790 | −0.360 | 0.518 | 1.250 | 2.262 | 3.048 | 3.845 | 4.652 | 6.600 |
| 2.4 | −0.832 | −0.819 | −0.351 | 0.537 | 1.262 | 2.256 | 3.023 | 3.800 | 4.584 | – |
| 2.3 | −0.867 | −0.850 | −0.341 | 0.555 | 1.274 | 2.248 | 2.997 | 3.753 | 4.415 | – |
| 2.2 | −0.905 | −0.882 | −0.330 | 0.574 | 1.284 | 2.240 | 2.970 | 3.705 | 4.444 | 6.200 |
| 2.1 | −0.946 | −0.914 | −0.319 | 0.592 | 1.294 | 2.230 | 2.942 | 3.656 | 4.372 | – |
| 2.0 | −0.990 | −0.949 | −0.307 | 0.609 | 1.302 | 2.219 | 2.912 | 3.605 | 4.298 | 5.910 |
| 1.9 | −1.037 | −0.984 | −0.294 | 0.627 | 1.310 | 2.207 | 2.881 | 3.535 | 4.223 | – |
| 1.8 | −1.087 | −1.020 | −0.282 | 0.643 | 1.318 | 2.193 | 2.848 | 3.499 | 4.147 | 5.660 |
| 1.7 | −1.140 | −1.056 | −0.268 | 0.660 | 1.324 | 2.179 | 2.815 | 3.444 | 4.069 | – |
| 1.6 | −1.197 | −1.093 | −0.254 | 0.667 | 1.329 | 2.163 | 2.780 | 3.388 | 3.990 | 5.390 |
| 1.5 | −1.256 | −1.131 | −0.240 | 0.690 | 1.333 | 2.146 | 2.743 | 3.330 | 3.910 | – |
| 1.4 | −1.318 | −1.168 | −0.225 | 0.705 | 1.337 | 2.128 | 2.706 | 3.271 | 3.828 | 5.110 |
| 1.3 | −1.383 | −1.206 | −0.210 | 0.719 | 1.339 | 2.108 | 2.666 | 3.211 | 3.745 | – |
| 1.2 | −1.449 | −1.243 | −0.195 | 0.732 | 1.340 | 2.087 | 2.626 | 3.149 | 3.661 | 4.820 |
| 1.1 | −1.518 | −1.280 | −0.180 | 0.745 | 1.341 | 2.066 | 2.585 | 3.087 | 3.575 | – |
| 1.0 | −1.588 | −1.317 | −0.164 | 0.758 | 1.340 | 2.043 | 2.542 | 3.022 | 3.489 | 4.540 |
| 0.9 | −1.660 | −1.353 | 0.148 | 0.769 | 1.339 | 2.018 | 2.498 | 2.957 | 3.401 | 4.395 |
| 0.8 | −1.733 | −1.388 | −0.132 | 0.780 | 1.336 | 1.993 | 2.453 | 2.891 | 3.312 | 4.250 |
| 0.7 | −1.806 | −1.423 | −0.116 | 0.790 | 1.333 | 1.967 | 2.407 | 2.824 | 3.223 | 4.105 |
| 0.6 | −1.880 | −1.458 | −0.099 | 0.800 | 1.328 | 1.939 | 2.359 | 2.755 | 3.132 | 3.960 |
| 0.5 | −1.955 | −1.491 | −0.083 | 0.808 | 1.323 | 1.910 | 2.311 | 2.686 | 3.041 | 3.815 |
| 0.4 | −2.029 | −1.524 | −0.066 | 0.816 | 1.317 | 1.880 | 2.261 | 2.615 | 2.949 | 3.670 |
| 0.3 | −2.104 | −1.550 | −0.050 | 0.824 | 1.309 | 1.849 | 2.211 | 2.544 | 2.856 | 3.525 |
| 0.2 | −2.178 | −1.586 | −0.033 | 0.830 | 1.301 | 1.818 | 2.159 | 2.472 | 2.763 | 3.380 |
| 0.1 | −2.252 | −1.616 | −0.017 | 0.836 | 1.292 | 1.785 | 2.107 | 2.400 | 2.670 | 3.255 |
| 0.0 | −2.326 | −1.645 | −0.000 | 0.842 | 1.282 | 1.751 | 2.054 | 2.326 | 2.576 | 3.090 |

$u = 2.7971$. Substituting this value of $u$ in equation (2.49) we get $k_T = 2.054$. The same value of $k_T$ for $T = 50$ is obtained from person type III table for zero coefficient of variation ($C_s = 0$).

### 2.4.5 Log Normal Distribution

In this distribution, logarithmic values of sample data is assumed to follow normal distribution. The distribution is same as log Pearson type when $C_s = 0$.

**Table 2.8(b) Frequency Factor for the Pearson Type-III Distribution with Negative Skew Coefficients (After Water Resources Council, 1967)**

| Skew coefficient $C_s$ | 1.0101 | 1.0526 | 2 | 5 | 10 | 25 | 50 | 100 | 200 | 1000 |
|---|---|---|---|---|---|---|---|---|---|---|
| | 99 | 95 | 50 | 20 | 10 | 4 | 2 | 1 | 0.5 | 0.10 |
| 0.0 | −2.326 | −1.645 | 0.000 | 0.842 | 1.282 | 1.751 | 2.054 | 2.326 | 2.576 | 3.090 |
| −0.1 | −2.400 | −1.673 | 0.017 | 0.846 | 1.270 | 1.716 | 2.000 | 2.252 | 2.482 | 2.950 |
| −0.2 | −2.472 | −1.700 | 0.033 | 0.850 | 1.258 | 1.680 | 1.945 | 2.178 | 2.388 | 2.810 |
| −0.3 | −2.544 | −1.726 | 0.050 | 0.853 | 1.245 | 1.643 | 1.890 | 2.104 | 2.294 | 2.675 |
| −0.4 | −2.615 | −1.750 | 0.066 | 0.855 | 1.231 | 1.606 | 1.834 | 2.029 | 2.201 | 2.540 |
| −0.5 | −2.686 | −1.774 | 0.083 | 0.856 | 1.216 | 1.567 | 1.777 | 1.955 | 2.108 | 2.400 |
| −0.6 | −2.755 | −1.797 | 0.099 | 0.857 | 1.200 | 1.528 | 1.720 | 1.880 | 2.016 | 2.275 |
| −0.7 | −2.824 | −1.819 | 0.116 | 0.857 | 1.183 | 1.488 | 1.663 | 1.806 | 1.926 | 2.150 |
| −0.8 | −2.891 | −1.839 | 0.132 | 0.856 | 1.166 | 1.448 | 1.606 | 1.733 | 1.837 | 2.035 |
| −0.9 | −2.957 | −1.858 | 0.148 | 0.854 | 1.147 | 1.407 | 1.549 | 1.660 | 1.749 | 1.910 |
| −1.0 | −3.022 | −1.877 | 0.164 | 0.852 | 1.121 | 1.366 | 1.492 | 1.588 | 1.664 | 1.880 |
| −1.1 | −3.087 | −1.894 | 0.180 | 0.848 | 1.107 | 1.324 | 1.435 | 1.518 | 1.581 | − |
| −1.2 | −3.149 | −1.910 | 0.195 | 0.844 | 1.086 | 1.282 | 1.379 | 1.449 | 1.501 | − |
| −1.3 | −3.211 | −1.925 | 0.210 | 0.838 | 1.064 | 1.240 | 1.324 | 1.383 | 1.424 | − |
| −1.4 | −3.271 | −1.938 | 0.225 | 0.832 | 1.041 | 1.198 | 1.270 | 1.318 | 1.351 | 1.465 |
| −1.5 | −3.330 | −1.951 | 0.240 | 0.825 | 1.018 | 1.157 | 1.217 | 1.256 | 1.282 | − |
| −1.6 | −3.388 | −1.962 | 0.254 | 0.817 | 0.994 | 1.116 | 1.166 | 1.197 | 1.216 | − |
| −1.7 | −3.444 | −1.972 | 0.268 | 0.808 | 0.970 | 1.075 | 1.116 | 1.140 | 1.155 | − |
| −1.8 | −3.499 | −1.981 | 0.282 | 0.799 | 0.945 | 1.035 | 1.069 | 1.187 | 1.097 | 1.130 |
| −1.9 | −3.553 | −1.989 | 0.294 | 0.788 | 0.920 | 0.996 | 1.023 | 1.037 | 1.044 | − |
| −2.0 | −3.605 | −1.996 | 0.307 | 0.777 | 0.895 | 0.959 | 0.980 | 0.990 | 0.995 | − |
| −2.1 | −3.656 | −2.001 | 0.319 | 0.765 | 0.869 | 0.923 | 0.939 | 0.946 | 0.949 | − |
| −2.2 | −3.705 | −2.006 | 0.330 | 0.752 | 0.844 | 0.888 | 0.900 | 0.905 | 0.907 | 0.910 |
| −2.3 | −3.753 | −2.009 | 0.341 | 0.739 | 0.819 | 0.855 | 0.864 | 0.867 | 0.869 | − |
| −2.4 | −3.800 | −2.011 | 0.351 | 0.725 | 0.795 | 0.823 | 0.830 | 0.832 | 0.833 | − |
| −2.5 | −3.845 | −2.012 | 0.360 | 0.711 | 0.771 | 0.793 | 0.798 | 0.799 | 0.800 | − |
| −2.6 | −3.889 | −2.013 | 0.368 | 0.696 | 0.747 | 0.764 | 0.7968 | 0.769 | 0.769 | − |
| −2.7 | −3.932 | −2.012 | 0.376 | 0.681 | 0.724 | 0.738 | 0.740 | 0.740 | 0.741 | − |
| −2.8 | −3.973 | −2.010 | 0.384 | 0.666 | 0.702 | 0.712 | 0.714 | 0.714 | 0.714 | − |
| −2.9 | −4.013 | −2.007 | 0.390 | 0.651 | 0.681 | 0.683 | 0.689 | 0.690 | 0.690 | − |
| −3.0 | −4.051 | −2.003 | 0.396 | 0.636 | 0.660 | 0.666 | 0.666 | 0.666 | 0.667 | 0.668 |

*Recurrence interval (Years)* — column headers: 1.0101, 1.0526, 2, 5, 10, 25, 50, 100, 200, 1000. *Percent chance (z)* — 99, 95, 50, 20, 10, 4, 2, 1, 0.5, 0.10.

The procedure outlined for log Pearson type-III is followed and Table 2.8 is used to read $k$ values for different recurrence interval for $C_s = 0$.

Chow (1964) derived the frequency factors for log-normal distribution based on which the values of $k$ are given in Table 2.9. It is more appropriate to use Chow's frequency factors from Table 2.9 than using Pearson table for zero skew as it is found that the data may still have skewness in their distribution even after

**Table 2.9 Frequence Factor for Log-normal Distribution (After Chow 1964)**
**(Theoretical log normal frequencey factors)**

| | Probability in Percent equal to or greater than the given variate | | | | | | | | |
|---|---|---|---|---|---|---|---|---|---|
| | 99 | 95 | 80 | 50 | 20 | 5 | 1 | 0.1 | 0.01 |
| $C_S$ | − | − | − | − | + | + | + | + | $C_V$ |
| 0 | 2.33 | 1.65 | 0.84 | 0 | 0.84 | 1.64 | 2.33 | 3.09 | 3.72 | 0 |
| 0.1 | 2.25 | 1.62 | 0.85 | 0.02 | 0.84 | 1.67 | 2.40 | 3.22 | 3.95 | 0.033 |
| 0.2 | 2.18 | 1.59 | 0.85 | 0.04 | 0.83 | 1.70 | 2.47 | 3.39 | 4.18´ | 0.067 |
| 0.3 | 2.11 | 1.56 | 0.85 | 0.06 | 0.82 | 1.72 | 2.55 | 3.56 | 4.42 | 0.100 |
| 0.4 | 2.04 | 1.53 | 0.85 | 0.07 | 0.81 | 1.75 | 2.62 | 3.72 | 4.70 | 0.136 |
| 0.5 | 1.98 | 1.49 | 0.86 | 0.09 | 0.82 | 1.77 | 2.70 | 3.88 | 4.96 | 0.166 |
| 0.6 | 1.91 | 1.46 | 0.85 | 0.10 | 0.79 | 0.79 | 2.77 | 4.05 | 5.24 | 0.197 |
| 0.7 | 1.85 | 1.43 | 0.85 | 0.11 | 0.78 | 1.81 | 2.84 | 4.21 | 5.52 | 0.230 |
| 0.8 | 1.79 | 1.40 | 0.84 | 0.13 | 0.77 | 1.82 | 2.90 | 4.37 | 5.81 | 0.262 |
| 0.9 | 1.74 | 1.37 | 0.84 | 0.14 | 0.76 | 1.84 | 2.97 | 4.55 | 6.11 | 0.292 |
| 1.0 | 1.68 | 1.34 | 0.84 | 0.15 | 0.75 | 1.85 | 3.03 | 0.72 | 6.40 | 0.234 |
| 1.1 | 1.63 | 1.31 | 0.83 | 0.16 | 0.73 | 1.86 | 3.09 | 4.87 | 6.71 | 0.351 |
| 1.2 | 1.58 | 1.29 | 0.82 | 0.17 | 0.72 | 1.87 | 3.15 | 5.04 | 7.02 | 0.381 |
| 1.3 | 1.54 | 1.26 | 0.82 | 0.18 | 0.71 | 0.88 | 3.21 | 5.19 | 7.31 | 0.409 |
| 1.4 | 1.49 | 1.33 | 0.81 | 0.19 | 0.69 | 1.88 | 3.26 | 5.35 | 7.62 | 0.436 |
| 1.5 | 1.45 | 1.21 | 0.81 | 0.20 | 0.68 | 1.89 | 3.31 | 5.51 | 7.92 | 0.462 |
| 1.6 | 1.41 | 1.18 | 0.80 | 0.21 | 0.67 | 1.89 | 3.36 | 5.66 | 8.26 | 0.490 |
| 1.7 | 1.38 | 1.16 | 0.79 | 0.22 | 0.65 | 1.89 | 3.40 | 5.80 | 8.58 | 0.517 |
| 1.8 | 1.34 | 1.14 | 0.78 | 0.22 | 0.64 | 1.89 | 3.44 | 5.96 | 8.88 | 0.544 |
| 1.9 | 1.31 | 1.12 | 0.78 | 0.23 | 0.63 | 1.89 | 3.48 | 6.10 | 9.20 | 0.570 |
| 2.0 | 1.28 | 1.10 | 0.77 | 0.24 | 0.61 | 1.89 | 3.52 | 6.25 | 9.51 | 0.596 |
| 2.1 | 1.25 | 1.08 | 0.76 | 0.24 | 0.60 | 1.89 | 3.55 | 6.39 | 9.79 | 0.620 |
| 2.2 | 1.22 | 1.06 | 0.76 | 0.25 | 0.59 | 1.89 | 3.59 | 6.51 | 10.12 | 0.643 |
| 2.3 | 1.20 | 1.04 | 0.75 | 0.25 | 0.58 | 1.88 | 3.62 | 6.65 | 10.43 | 0.667 |
| 2.4 | 1.17 | 1.02 | 0.74 | 0.26 | 0.57 | 1.88 | 3.65 | 6.77 | 10.72 | 0.691 |
| 2.5 | 1.15 | 1.00 | 0.74 | 0.26 | 0.56 | 1.88 | 3.67 | 6.90 | 10.95 | 0.713 |
| 2.6 | 1.12 | 0.99 | 0.73 | 0.26 | 0.55 | 1.87 | 3.70 | 7.02 | 11.25 | 0.734 |
| 2.7 | 1.10 | 0.97 | 0.72 | 0.27 | 0.54 | 1.87 | 3.72 | 7.13 | 11.55 | 0.755 |
| 2.8 | 1.08 | 0.96 | 0.72 | 0.27 | 0.53 | 1.86 | 3.74 | 7.25 | 11.80 | 0.776 |
| 2.9 | 1.06 | 0.95 | 0.71 | 0.27 | 0.52 | 1.86 | 3.76 | 7.36 | 12.10 | 0.796 |
| 3.0 | 1.04 | 0.93 | 0.71 | 0.28 | 0.51 | 1.85 | 3.78 | 7.47 | 12.36 · | 0.818 |
| 3.2 | 1.01 | 0.90 | 0.69 | 0.28 | 0.49 | 1.84 | 3.81 | 7.65 | 12.85 | 0.857 |
| 3.4 | 0.98 | 0.88 | 0.68 | 0.29 | 0.47 | 1.83 | 3.84 | 7.84 | 13.36 | 0.895 |
| 3.6 | 0.95 | 0.86 | 0.67 | 0.29 | 0.46 | 1.81 | 3.87 | 8.00 | 13.83 | 0.930 |
| 3.8 | 0.92 | 0.84 | 0.66 | 0.29 | 0.44 | 1.80 | 3.89 | 8.16 | 14.23 | 0.966 |
| 4.0 | 0.90 | 0.82 | 0.65 | 0.29 | 0.42 | 1.78 | 3.91 | 8.30 | 14.70 | 1.000 |
| 4.5 | 0.84 | 0.78 | 0.63 | 0.30 | 0.39 | 1.75 | 3.93 | 8.60 | 15.62 | 1.081 |
| 5.0 | 0.80 | 0.74 | 0.62 | 0.30 | 0.37 | 1.71 | 3.91 | 8.86 | 16.45 | 1.155 |

transferring the data to their logarithmic values. Frequency factors $k$ in Chow's Table 2.9 is calculated from the relation

$$k = \frac{e^{\sigma_y k_y - \sigma_y^2/2} - 1}{[e^{\sigma^2} - 1]^{0.5}}$$

(2.51)

in which $y = \ln x$, $k_y = (y - y_{av})/\sigma_y$. In log normal distribution $C_v$ and $C_s$ are related as

$$C_S = 3C_v + C_v^3 \tag{2.52}$$

---

**Example 2.8:**  From the following annual maximum runoff records for a sub-basin in Orissa, compute the 50-year and 100 year runoff values based on the annual series using (i) Normal, (ii) Log-normal, (iii) Extreme value type-I (Gumbel), (iv) Pearson type-III and (v) Log-Pearson type-III distributions. The annual maximum runoff peaks are in terms of millimeters over the catchment.

**Table 2.10   Data for Example 2.8**

| Year | 1950 | 1951 | 1952 | 1953 | 1954 | 1955 | 1956 | 1957 | 1958 | 1959 |
|---|---|---|---|---|---|---|---|---|---|---|
| Run off (mm) | 113 | 94.5 | 76 | 87.5 | 92.7 | 71.3 | 77.3 | 85.1 | 122.8 | 69.4 |
| Year | 1960 | 1961 | 1962 | 1963 | 1964 | 1965 | 1966 | 1967 | 1968 | 1969 |
| Run off (mm) | 81 | 94.5 | 86.3 | 68.6 | 82.5 | 90.7 | 99.8 | 74.4 | 66.6 | 65 |
| Year | 1970 | 1971 | 1972 | 1973 | 1974 | | | | | |
| Run off (mm) | 91 | 106.8 | 102.2 | 87 | 84 | | | | | |

**Solution**

Mean, standard deviation, coefficient of variation and coefficient of skewness are computed in Table 2.11.

*For the series*

Mean
$$X_{av} = \frac{x_i}{n} = \frac{2170}{25} = 86.8 \text{ mm}$$

Standard deviation  $\sigma = \left\{ \dfrac{\Sigma (x_i - x_{av})^2}{n - 1} \right\}^{1/2} = \left\{ \dfrac{5156.7}{25 - 1} \right\}^{1/2} = 14.66 \text{ mm}$

Coefficient of variation
$$C_v = \frac{14.66}{86.8} = 0.1689$$

Coefficient of skewness  $C_s = \dfrac{\alpha}{\sigma^3} = \left( \dfrac{1}{\sigma^3} \right) \left\{ \dfrac{N}{(N-1)(N-2)} \right\} \Sigma (X_i - X_{av})^3$

or  $\qquad C_s = \left( \dfrac{1}{14.66^3} \right) \left\{ \dfrac{25}{(25-1)(25-2)} \right\} 41793.7 = 0.553$

*For the log transferred series*

Mean
$$y_{av} = \frac{48.381}{25} = 1.9327$$

Standard deviation
$$\sigma = \left\{ \frac{0.124358}{25 - 1} \right\}^{1/2} = 0.072$$

**Table 2.11  Calculation of Parameters for Example 2.8**

| Year | Annual runoff $x$ (mm) | $(x - x_{av})^2$ | $(x - x_{av})^3$ | Log of col. (2) | $\{Col. (5) - (1.9327)\}^2$ | $\{Col. (5) - (1.9327)\}^3$ | Rainfall arranged in descending order | Rank $m$ | Return period | Relative Frequency $m/(N+1)$ |
|---|---|---|---|---|---|---|---|---|---|---|
| (1) | (2) | (3) | (4) | (5) | (6) | (7) | (8) | (9) | (10) | (11) |
| 1950 | 113.0 | 686 | 17985 | 2.053078 | 0.0144 | 0.00174 | 122.8 | 1 | 26.00 | 0.0385 |
| 1551 | 94.5 | 59 | 457 | 1.975431 | 0.0018 | 0.00008 | 113.0 | 2 | 13.00 | 0.0769 |
| 1552 | 76.0 | 117 | -1260 | 1.880813 | 0.0027 | -0.00014 | 106.8 | 3 | 8.67 | 0.1154 |
| 1953 | 87.5 | 0 | 0 | 1.942008 | 0.00009 | 0.0000 | 102.2 | 4 | 6.50 | 0.1538 |
| 1954 | 92.7 | 35 | 205 | 1.967079 | 0.00118 | 0.0004 | 99.8 | 5 | 5.20 | 0.1923 |
| 1955 | 71.3 | 240 | -3724 | 1.85089 | 0.0634 | -0.0005 | 94.5 | 6 | 4.33 | 0.2308 |
| 1956 | 77.3 | 90 | -857 | 1.888179 | 0.00199 | -0.00009 | 94.5 | 7 | 3.71 | 0.2692 |
| 1957 | 85.1 | 3 | -5 | 1.929929 | 0.00001 | 0.0000 | 92.7 | 8 | 3.25 | 0.3077 |
| 1958 | 122.8 | 1296 | 46656 | 2.089198 | 0.02448 | 0.00383 | 91.0 | 9 | 2.89 | 0.3462 |
| 1959 | 69.4 | 303 | -5268 | 1.841359 | 0.00835 | -0.00076 | 90.7 | 10 | 2.60 | 0.3846 |
| 1960 | 81.0 | 34 | -195 | 1.908485 | 0.00059 | -0.00001 | 87.5 | 11 | 2.36 | 0.4231 |
| 1961 | 94.5 | 59 | 457 | 1.975431 | 0.00182 | 0.00008 | 87.0 | 12 | 2.17 | 0.4615 |
| 1962 | 86.3 | 0 | 0 | 1.936010 | 0.00001 | 0.00000 | 86.3 | 13 | 2.00 | 0.5000 |
| 1963 | 68.6 | 331 | -6029 | 1.836324 | 0.00930 | -0.00090 | 85.1 | 14 | 1.86 | 0.5385 |
| 1964 | 82.5 | 18 | -80.0 | 1.916453 | 0.00027 | -0.00000 | 84.0 | 15 | 1.73 | 0.5769 |
| 1965 | 90.7 | 15 | 9.0 | 1.957607 | 0.00062 | 0.00002 | 82.5 | 16 | 1.63 | 0.6154 |
| 1966 | 99.8 | 169 | 2197 | 1.99913 | 0.00441 | 0.00029 | 81.0 | 17 | 1.53 | 0.6538 |
| 1967 | 74.4 | 154 | -1907 | 1.871572 | 0.00374 | -0.00023 | 77.3 | 18 | 1.44 | 0.6923 |
| 1968 | 66.6 | 408 | -8242 | 1.823474 | 0.01194 | -0.00130 | 76.0 | 19 | 1.37 | 0.7308 |
| 1969 | 65.0 | 475 | -10360 | 1.812913 | 0.01436 | -0.00172 | 74.4 | 20 | 1.30 | 0.7692 |
| 1970 | 91.0 | 18 | 74 | 1.95904 | 0.00069 | 0.00002 | 71.3 | 21 | 1.24 | 0.8077 |
| 1971 | 106.8 | 400 | 8000 | 2.028571 | 0.00918 | 0.00088 | 69.4 | 22 | 1.18 | 0.8462 |
| 1972 | 102.2 | 237 | 3652 | 2.009450 | 0.00589 | 0.00045 | 68.6 | 23 | 1.13 | 0.8846 |
| 1973 | 87.0 | 0 | 0 | 1.939519 | 0.00005 | 0.00000 | 66.6 | 24 | 1.08 | 0.9231 |
| 1974 | 84.0 | 8 | -22 | 1.924279 | 0.00007 | -0.0000 | 65.0 | 25 | 1.04 | 0.9615 |
| Sum | 2170 | 5156.7 | 41793.7 | 48.381 | 0.124358 | 0.001762 | | | | |

Coefficient of variation $\quad C_v = \dfrac{0.072}{1.9327} = 0.0372$

Coefficient of skewness $C_s = \dfrac{1}{0.072^3} \left\{ \dfrac{(25 \times 0.001762)}{(24 \times 23)} \right\} = 0.197$

**Log Normal Distribution**

Following Chow's approach for log normal distribution

$$C_s = 3C_v + C_v^3 = 3 \times 0.1689 + 0.1689^3 = 0.572$$

From Table 2.9, the frequency factors for return periods of 50 and 100 years are obtained by taking the coefficient of skewness as 0.512. The values are read as $k_{50} = 2.470$ and $k_{100} = 2.710$.

$$X_{50} = 86.8 + 2.47 \times 14.66 = 123.01 \text{ mm}$$

$$X_{100} = 86.8 + 2.71 \times 14.66 = 126.53 \text{ mm}$$

Using the Pearson table of frequency distribution, the values of $k_{50}$ and $k_{100}$ are read from Table 2.8. The values are $k_{50} = 2.055$ and $k_{100} = 2.710$ assuming $C_s = 0$.

$y_{50} = 1.932737 + 2.055 \times 0.071983 = 2.0806621 \Rightarrow X_{50} = 10^{2.0806621} = 120.4 \text{ mm}$

$y_{100} = 1.932737 + 2.328 \times 0.071893 = 2.1003134 \Rightarrow X_{100}$

$= 10^{2.1003134} = 126.0 \text{ mm}$

Both approaches give nearly comparable results. However, the second approach should be used with caution as the value of $C_s$ is appreciable for the given data.

**Extreme Value Type-I**

From Gumbel's $k\text{-}T$ relation (Table 2.7) for sample length of 25, the values of $k_T$ are read as

$$k_{50} = 3.088 \text{ and } k_{100} = 3.729$$

Therefore, $\qquad X_{50} = 86.8 + 3.088 \times 14.66 = 132.0 \text{ mm}$

and $\qquad X_{100} = 86.8 + 3.729 \times 14.66 = 141.5 \text{ mm}$

**Pearson Type-III**

Coefficient of skewness $C_s = 0.553$.

From Table 2.8 for $C_s = 0.553$, the frequency factors for 50 and 100 year of return periods are $k_{50} = 2.315$ and $k_{100} = 2.70$

Flood for return period of 50 years is $\qquad X_{50} = 86.8 + 2.315 \times 14.66 = 120.7 \text{ mm}$

Flood for return period of 100 years is $\qquad X_{100} = 86.8 + 2.70 \times 14.66 = 126.4 \text{ mm}$

**Log Pearson Type-III Distribution**

For the log transferred data, the coefficient of skewness $C_s = 0.197$.

From Table 2.8, the frequency factors for return periods of 50 and 100 years are read by taking the coefficient of skewness as $C_s = 0.196$.

The frequency factors are $k_{50} = 2.15$ and $k_{100} = 2.47$

$y_{50} = 1.932737 + 2.15 \times 0.071983 = 2.0875005 \Rightarrow X_{50} = 10^{2.0875005} = 122.3 \text{ mm}$

$y_{100} = 1.932737 + 2.47 \times 0.071893 = 2.110535 \Rightarrow X_{100} = 10^{2.110535} = 129.00 \text{ mm}$

**Normal Distribution**

For 50 year return period, $T = 50$, $P(X > x)$ is the area of the normal curve bounded between 0 and $(1 - 1/T = 1 - 1/50 = 49/50 = 98\%)$ on either side. From the standard normal curve for area up to $+48\%$ $(98\%-50\%$ i.e, for $t = 0.48)$ $z = 2.054$

$$\ddot{X}_{50} = 86.8 + 2.054 \times 14.66 = 116.9 \text{ mm}$$

For 100 year return period $T = 100$ years, $p = 1 - 1/T = 1-1/100 = 99\%$
From the standard normal table for area up to $+49\%$ or 0.49 is $Z = 2.328$

$$X_{100} = 86.8 + 2.328 \times 14.66 = 120.43 \text{ mm}$$

The same results of $k_{50}$ and $k_{100}$ can be obtained from Table 2.8 and also from equations (2.49) and (2.50).

For comparison the results of Normal vs. Pearson type-III and Log normal vs. Log Pearson type-III along with the results of EVT-1 model are given as follows.

| Distribution type | Return period flood of | | Coefficient of skewness |
|---|---|---|---|
| | 50 years | 100 years | |
| Normal | 116.9 | 120.4 | 0 |
| Pearson Type III | 120.7 | 126.4 | 0.553 |
| Log Normal | 120.4 | 126.0 | 0 |
| Log Pearson | 122.3 | 129.0 | 0.196 |
| Extreme Value | 132.5 | 141.5 | – |

The effect of coefficient of skewness to the magnitude of the event with return period of 50 and 100 years can be seen from the above table.

**Example 2.9:** The mean annual flood peak at a site is 280 $m^3$/sec and the standard deviation 40 $m^3$/sec. What is the probability that a flood of 400 $m^3$/sec occurring in the river in the next 10 years?

**Solution**

Given that $\qquad x_{av} = 280 \text{ m}^3/\text{sec}, \sigma = 40 \text{ m}^3/\text{sec}$

$$X_T = 400 \text{ m}^3/\text{sec}, T = ?$$

Assume Gumbel distribution for the data. From Chow's relation

$$X_T = x_{av} \pm \sigma k \text{ or } 400 = 280 + 40k$$

This gives $k = 3.00$. Since the sample length is not given, it is assumed that the statistics are representative of the population. Therefore equation (2.48) can be used to obtain the return period of the variate $X_T = 400 \text{ m}^3/\text{sec}$.

$$k = -\left[0.45005 + 0.7797 \ln\left\{\ln\frac{T}{T-1}\right\}\right] \quad \text{or} \quad 3 = -0.45 - 0.7797 \ln\left\{\ln\frac{T}{T-1}\right\}$$

or $\qquad \ln\left\{\ln\frac{T}{T-1}\right\} = -4.4248$ or $\ln\{T/(T-1)\} = e^{-4.4248} = 0.01197$.

or $$\frac{T}{T-1} = 1.012049$$

from which we get $T = 84$ years. The probability of the event is $P = 1/T = 1/84 = 0.0119$. Probability of the event occurring in the next 10 years is

$$1 - (1 - 1/T)^{10} = 1 - (1 - 1/84)^{10} = 0.1128 = 11.3\%.$$

**Example 2.10:** For a project, India Meteorological Department (IMD) supplied 2-day storm precipitation of 50 and 100 year return periods as 42 cm and 48 cm respectively on the basis of analysis of 30 years data. If the authorities decided to design the project for 200 year return period storm, then estimate its magnitude.

**Solution**
Given that the sample size is 30 cm, $X_{50} = 42$ and $X_{100} = 48$.
From the table of Gumbel's distribution for $T = 50$ and 100 years $k_{50} = 3.026$ and $k_{100} = 3.653$. Using Chow's relation

$$X_{50} = 42 = x_{av} + \sigma 3.026$$

and      $X_{100} = 48 = x_{av} + \sigma 3.653$

Solving the two equations we get $\sigma = 9.57$ and $x_{av} = 13.04$

Now for $T = 200$ years $k_{200} = 3.883$,

Therefore,      $X_{200} = 13.04 + 3.883 \times 9.57 = 50.2$ cm

---

## 2.5   GRAPHICAL METHOD USING PROBABILITY PAPER
The basic idea of graphical method is to develop a linear relationship between the recurrence interval $T$ (or probability) and the event magnitudes. For such plottings, the recurrence interval is taken on abscissa and the event magnitudes as ordinate. The ordinate may be either ordinary scale or logarithmic scale. The scales are so selected that the observed data should plot close to a straight line. A linear relationship so developed helps to extrapolate or interpolate the relation between the event magnitudes and recurrence intervals. General procedure of plotting the observed data on a probability paper are outlined here.

 (i) Collect the required hydrological data (say annual maximum floods).
 (ii) Prepare a table in which the first column contains the year and the second column the corresponding hydrological data.
 (iii) In the next column arrange the data of col.(2) in decreasing order of magnitude.
 (iv) The rank of data $m$ is the position of the data in decreasing order, thus for $m = 1$ is the highest event, $m = 2$ for second highest event and so on till the last event when $m = N$. Here $N$ is the sample length.
 (v) Probability $P$ of the event being equalled or exceeded is calculated by any of the plotting-position formula given in Table 2.12.

There are a number of other formulae used for plotting data. However, Weibull's plotting-position is the most acceptable and widely used one. The disadvantage of California plotting position is that it produces 100% probability for $m = N$, which is difficult to plot in a probability paper. Weibull plotting formula plots the largest values of the sample at smaller return period. For normally distributed data Blom's equation can be taken as unbiased, where as for EVI distribution, Gringorten's method is the best suited.

**Table 2.12 Various Plotting Position Formulae**

| Sl. No. | Formula name | Probability $P$ of the event |
|---------|--------------|------------------------------|
| 1 | California (1923) | $m/N$ |
| 2 | Hazen (1914) | $(m - 0.5)/N$ |
| 3 | Weibull (1939) | $m/(N + 1)$ |
| 4 | Beard (1943) | $(m - 0.31)/(N + 0.38)$ |
| 5 | Chegodayev (1955) | $(m - 0.3)/(N + 0.4)$ |
| 6 | Blom (1958) | $(m - 3/8)/(N + 1/4)$ |
| 7 | Tukey (1962) | $(3m - 1)/(3N + 1)$ |
| 8 | Gringorten (1963) | $(m - 0.44)/(N + 0.12)$ |
| 9 | Cunnane (1978) | $(m - 0.4)/(N + 0.2)$ |
| 10 | Adamowski (1981) | $(m - 1/4)/(N + 1/2)$ |

(vi) Recurrence interval or the return period $T$ is entered in the next column. $T$ is computed from the relation $T = 1/P = (N +1)/m$.

(vii) A suitable probability paper is chosen. This depends on the type of data (say annual maximum or annual minimum) and the experience of the analyst. In the market all types of probability papers are available.

(viii) Plot the points on the probability paper by taking $T$ in abscissa and the magnitude of the event as ordinate. Ordinates of some probability papers have logarithmic scale. The abscissa represents probability. The aim of plotting the data on a suitable paper is that the distribution plots a straight line.

(ix) A straight line is fitted, which helps to extrapolate the event for any given recurrence interval.

Graphical methods are not suitable for larger extrapolations, the reason being, the errors in sampling may be magnified giving wrong results. For large extrapolations, graphical probability distributions should be used with caution. Graphical plotting rather should be used as a check to know the suitability of the probability distribution than for large extrapolations.

### 2.5.1 Construction of Probability Paper

A probability paper can also be constructed for any type of distribution knowing their $k$-$T$ relation. For various values of $k$, the corresponding values of $T$ are computed. In a rectangular plot, the abscissa consisting of $k$ values varying from 0 to +7 are marked and the corresponding $T$ values are noted. The ordinate scale plots either arithmetic or logarithmic values of the variate. This gives the probability paper for the selected probability distribution.

### 2.5.2 Selection of Type of Distribution

Extreme value distribution (Gumbel distribution) is used for the analysis of flood frequencies. Maximum one from 365 days of record is selected to represent annual maximum series. Preferably the data for consecutive 30 years or more are required for analysis. Extreme value type-III distribution is used for drought studies. When the coefficient of skewness for the sample $C_s$ is 1.139 and the coefficient of variation $C_v$ is 0.364, a log normal distribution can be preferred. Exponential distribution is more suitable for the extrapolation of partial duration series. For the collected sample, obtain the mean and standard deviation analytically. The speciality of normal probability paper is that, for relative frequency of 50%, 84.13% and 15.87%, the event magnitudes correspond to the mean $X_{av}$ ($X_{av} + \sigma$) and ($X_{av} - \sigma$) respectively. Therefore, while using a normal probability paper the mean is obtained corresponding to the relative frequency of 50%. Compute the standard deviation from the graph by reading data corresponding to 84.13% and 15.87% of the relative frequency (RF) and then using the relation

$$\sigma = \frac{\text{Discharge at RF of 84.13\%} - \text{Discharge at RF of 15.87\%}}{2}$$

This is because 50% probability corresponds the mean (as $k = 0$); 84.13% and 15.87% probability corresponds to $x_{av} + \sigma$ and $x_{av} - \sigma$ respectively as $k$ is $\pm 1$ at these stages.

## 2.6   CONFIDENCE LIMITS

The sample from which we derive the statistics like mean, standard deviation, skewness and other parameters may not be the true representative of their population values. Extrapolation of data to obtain the probability of occurrence of an event of larger return period may lead to errors due to the limitation of the sample record. Confidence limits can be laid on both upper and lower sides of the predicted distribution of the events. When curves known as control curves are drawn joining such confidence limit points at equidistant on both sides of the data of various return periods, we get confidence bands. All data lying within the confidence bands are reliable to the extent of probabilities on which the confidence limits are based. Following Gumbel, the confidence interval of an event can be estimated by the relation

$$X_{TU} = X_T + S\Delta x \tag{2.53}$$

and
$$X_{TL} = X_T - S\Delta x \tag{2.54}$$

$$X_{TU} > X_T > X_{TL}$$

If the sample size is more than 30, the distribution of statistics is approximated to normal distribution. From the area of normal distribution curve, we get $\mu \pm \sigma$ bounded for 68.72% of time, $\mu \pm 2\sigma$ bounded for 95.45% of time and $\mu \pm 3\sigma$ bounded for 99.73% of time. In other words, we can say that the values ($\mu + \sigma$), ($\mu + 2\sigma$), ($\mu + 3\sigma$) and ($\mu - \sigma$), ($\mu - 2\sigma$), ($\mu - 3\sigma$) are the upper and lower confidence limits of the estimate of the event at 68.72%, 95.45% and 99.73% confidence limits.

For confidence limits of 50, 68, 80, 90, 95 and 99%, we can derive the values of the parameter $S$ in equations (2.53) and (2.54) from the curve of the normal distribution. Table 2.13 gives the values of $S$ for various confidence levels.

Now, $\Delta x$ can be computed as

$$\Delta x = a \frac{\sigma_{n-1}}{\sqrt{n}} \tag{2.55}$$

**Table 2.13   Departure Parameter S for Various Confidence Limits**

| Confidence limit (Probability) in % | 50 | 68 | 80 | 90 | 95 | 99 |
|---|---|---|---|---|---|---|
| Values of S | 0.674 | 1.00 | 1.282 | 1.645 | 1.96 | 2.58 |

where $\dfrac{\sigma_{n-1}}{\sqrt{n}}$ represents the standard error of the mean and $n$ is the sample size. Coefficient $a$ (2.55) is given as

$$a = (1 + 1.3k + 1.1k^2)^{1/2} \qquad (2.56)$$

in which $k$ is the frequency factor. For sample size of less than 30, normally $\sigma_{n-1}$ should be computed. Procedure to draw confidence bands for the extreme value distribution is outlined here.

(1) Compute the mean $x_{av}$ and standard deviation $\sigma_n$ or $\sigma_{n-1}$ of the sample data.
(2) Compute the frequency factor $k$ or read the value of $k$ from $k$-$T$ table for the known probability distribution of the sample.
(3) Compute values for $X_T$ for different return periods $T$.
(4) Compute $a$ and $\Delta x$.
(5) Take a probability paper and plot the fitted line for $X_T$.
(6) Calculate the values of $X_{TU}$ and $X_{TL}$.
(7) In the same paper plot $X_{TU}$ and $X_{TL}$ against the values of $T$.

Confidence bands are drawn on both sides of the fitted line for confidence limits of 95%, 80% and other desired limits. Thus 99% confidence bends will be the outermost bend and 50% will be close to the fitted line. Further, as the return period $T$ increases, the limit of confidence increases, thereby more spreading of confidence bands forming part of the conical shape. 95% confidence level means out of 100 sample, 95 values will be inside the interval. It is the numerical boundary value beyond which, we discard the computed parameter as insignificant or unreliable. In hydrology, 80% and 95% significance level are normally chosen.

---

**Example 2.11:** Using data from example 2.8, calculate the 50 year and 100 year rainfall values graphically following:

(i) Gumbel Distribution
(ii) Log Pearson Type-III Distribution
(iii) Normal Distribution

**Solution**
As outlined in section 2.5, graphical solution to the above problem are carried out using probability papers. Arrangement of data and the computation of probabilities are shown in Table 2.11. In Gumbel probability paper, the discharge are plotted as ordinates against the return periods in abscissa and a best fit line is drawn through the plotted points. The plot is shown in Fig. 2.8. From the plot, rainfall for 50 and 100 year frequencies are read out.

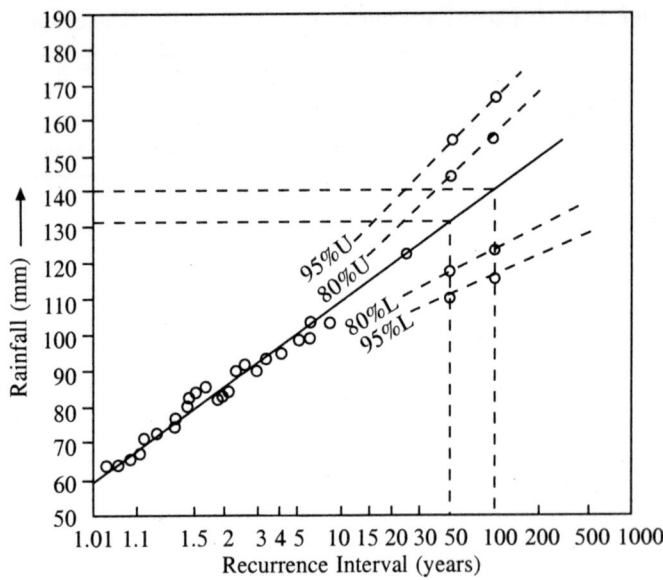

**Fig. 2.8  Gumbel's Distribution with Confidence Bands**

Gumbel Distribution :  $X_{50} = 132.0$ mm.

$X_{100} = 141.5$mm.

For Pearson type-III and Normal distribution, the relative frequency given in Table 2.11, col. (11) is plotted against the event magnitude ordinate to give the best fit line for the assumed probability distribution. From the plot, rainfall of 50 and 100 year frequencies are read as

Log-Pearson Distribution :  $X_{50} = 120.0$ mm

$X_{100} = 130.0$ mm

Normal Distribution :  $X_{50} = 117.0$ mm

$X_{100} = 120.0$ mm

The plots are shown in Figs. 2.9 and 2.10 respectively.

**Example 2.12:**  For the example 2.8, draw confidence bands at 80% and 95% level of significance for the return periods of 50 year and 100 year events. Use Gumbel distribution.

**Solution**

Frequency factors for 50 and 100 year extreme rainfalls are 3.088 and 3.729, respectively. Standard deviation $\sigma$ of the sample is 14.66 and length of sample $n = 25$; $X_{50} = 132.0$ mm and $X_{100} = 141.5$ mm.

Now, from equation (2.56)

$$a_{50} = (1 + 1.3k + 1.1k^2)^{1/2} = (1 + 1.3 \times 3.088 + 1.1 \times 3.088^2)^{1/2} = 3.938$$

$$a_{100} = (1 + 1.3 \times 3.729 + 1.1 \times 3.729^2)^{1/2} = 4.60$$

$$\Delta x_{50} = a_{50}\left(\frac{\sigma}{\sqrt{n}}\right) = \frac{3.938 \times 14.66}{\sqrt{25}} = 11.55$$

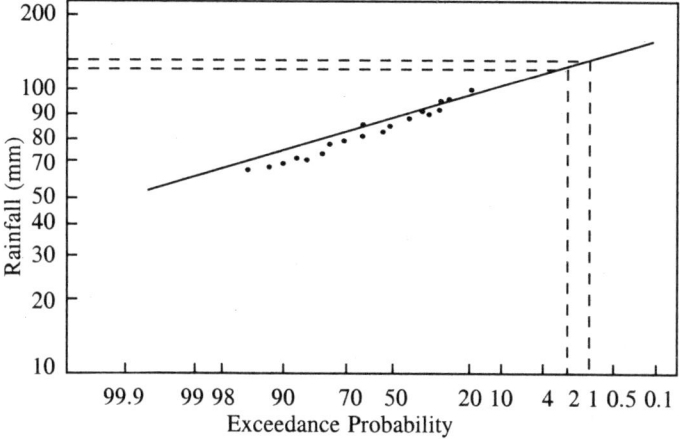

**Fig. 2.9   Log-Pearson Type-III Distribution**

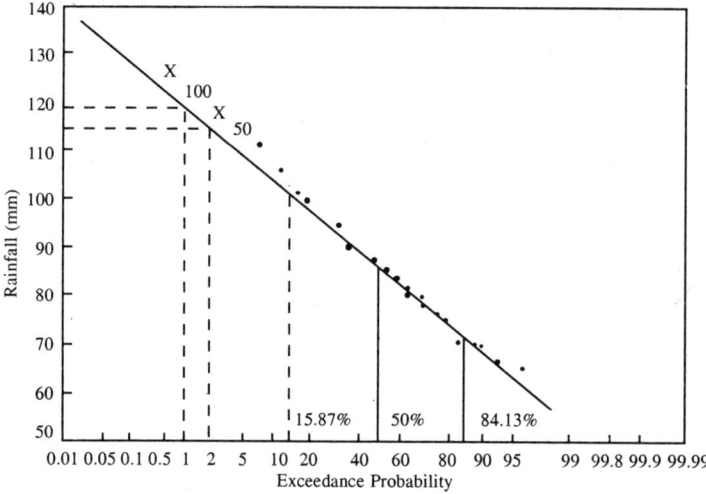

**Fig. 2.10   Normal Distribution Plot for Example 2.11 (This curve can show increasing trend to the right side if rainfall is arranged in ascending order of magnitude and relative frequency is calculated thereof)**

$$\Delta x_{100} = a_{100}\left(\frac{\sigma}{\sqrt{n}}\right) = \frac{4.60 \times 14.66}{\sqrt{25}} = 13.49$$

For 50 year return period flood at 80% level of significance

$$X_{TU} = X_T + S\,\Delta x = 132 + 1.282 \times 11.55$$

$$= 146.8\ (S \text{ is taken as } 1.282 \text{ from Table 2.13})$$

$$X_{TL} = 132 - S\,\Delta x = 132 - 1.282 \times 11.55 = 117.2$$

At 95% level of significance for the 50 year return period flood

$$X_{TU} = 132 + 1.96 \times 11.55 = 154.60$$

$$X_{TL} = 132 - 1.96 \times 11.55 = 109.4$$

For 100 year return period event at 80% confidence level.

$$X_{TU} = 141.5 + 1.282 \times 13.49 = 158.8$$

$$X_{TL} = 141.5 - 1.282 \times 13.49 = 124.4$$

At 95% level of significance for the 100 year return period flood.

$$X_{TU} = 141.5 + 1.96 \times 13.49 = 167.9$$

$$X_{TL} = 141.5 - 1.96 \times 13.49 = 115.1$$

The plotting points given below.

80% line upper are (50, 146.8), (100, 158.8)
80% line lower are (50, 117.2), (100, 124.4)
95% line upper are (50, 154.6), (100, 167.9)
95% line lower are (50, 109.4), (100, 115.1)

Plots of the confidence bands for 80% and 95% level of significance are shown in Fig 2.8.

## 2.7   REGRESSION AND CORRELATION

So far, the analysis and prediction of the hydrological data are limited to a single variable. However, in a hydrologic design, it is necessary to evaluate the relation between two or more variables simultaneously. For example, a mathematical relation between rainfall and corresponding runoff or river discharge at two or more tributaries along with that of the main river needs to be developed so that one event can be predicted from the knowledge of others. Thus a future sequence can be synthesized preserving the characteristics of the historical data. This can be achieved by regressing one set of variables on the other set. If runoff is represented by $y_i$ and the corresponding rainfall as $x_i$, then $y_i$ is regressed upon $x_i$ to get the necessary relation. We call $x_i$ as independent variable and $y_i$ as dependent. The two methods available to fit a curve between the given sets of data are graphical method and the analytical method.

### 2.7.1   Graphical Method

In graphical method, $x$ and $y$ coordinates are drawn to scale on a rectangular system and all the points are marked on it. The resulting plot is called *scatter diagram* (Fig. 2.11). A curve is traced by eye approximation such that it passes nearly through the mean of the spread of all points. This method is thus quick and practical. Knowing any value of $x$, the corresponding value of $y$ can be read from the graph. The limitation of such a procedure is that it may be subjected to large errors, if the best fit curve passing through the mean is not drawn. The term best fit will be discussed later.

### 2.7.2   Analytical Method

In analytical method we try to fit a curve passing through the scatter points of the

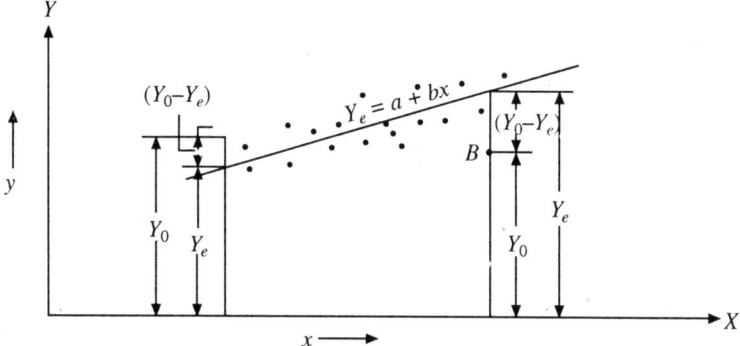

**Fig. 2.11    A Scatter Diagram between Two Variables $X$ and $Y$ (Scatter Points are Fitted with a Straight Line Curve)**

scatter diagram in such a way that the sum of squares of departures of observed points from the fitted function is the minimum. In Fig. 2.11 the error between the observed and estimated point, obtained after fitting the line is shown as $(y_0 - y_e)$. For $N$ sets of $(x_i, y_i)$ observations, the line $y = a + bx$, should be so selected that $\Sigma (y_0 - y_e)^2$ shall be the minimum, i.e., the curve should pass close to all the points in the diagram. The procedure to obtain the best values of $a$ and $b$ in the equation $y = a + bx$ should be such that the error function $S_e$ in the following equation is the minimum.

$$S_e = \sum_{i=1}^{N} (y_{0i} - y_{ei})^2 \tag{2.57}$$

This is called the *method of least squares*. Here $N$ is the number of sets of data, $y_{0i} (= y_i)$ is the observed data, and $y_{ei}$ the computed or fitted curve points corresponding to $x_i$. In the simple linear relation $y = a + bx$ the parameter $a$ is called intercept and $b$ the regression slope of the curve. Regression of $y$ on $x$ is not the same as $x$ on $y$. We have to evaluate the regression constants $a$ and $b$ from the observed $N$ values of $y_i$ and $x_i$. To achieve this, substitute $y_{ei}$ in equation (2.57) the estimated value of $a + bx_i$, the resulting equation becomes

$$S_e = \sum_{i=1}^{N} (y_i - a - bx_i)^2 \tag{2.58}$$

Since $S_e$ should be the minimum, differentiating the equation with respect to $a$ and equating to zero, we get

$$\Sigma y_i - Na - b \Sigma x_i = 0$$

or
$$\Sigma y_i = Na + b \Sigma x_i \tag{2.59}$$

again differentiating the equation with respect to $b$ and equating to zero, we get

$$\Sigma xy_i - a \Sigma x_i - b \Sigma x_i^2 = 0$$

or
$$\Sigma xy_i = a \Sigma x_i + b \Sigma x_i^2 \tag{2.60}$$

Since $x_i$ and $y_i$ are known, we can solve equations (2.59) and (2.60) for $a$ and $b$ as

$$a = \frac{\Sigma y_i \, \Sigma x_i^2 - \Sigma x_i \, \Sigma x_i y_i}{N \Sigma x_i^2 - (\Sigma x_i)^2} = \frac{\Sigma y_i - b \, \Sigma x_i}{N} \tag{2.61}$$

and

$$b = \frac{N \Sigma x_i y_i - \Sigma x_i \, \Sigma y_i}{N \, \Sigma x_i^2 - (\Sigma x_i)^2} \tag{2.62}$$

Greater is the value of $b$, faster is the rate of increase or decrease of $y_i$ with $x_i$.

### 2.7.3　Correlation

The degree of association between variables is known as correlation and the statistical parameter determining their relation is known as *correlation coefficient*. When two variables say rainfall $x_i$ and runoff $y_i$ are involved, we call it a simple correlation. Similarly, when more than two parameters are involved, we call it multiple correlation. Correlation coefficient should be attempted only when there is an association between the variables. At a place, the number of rainy days may perfectly correlate with the number of accidents over a year, but there is no inter dependence between the two. Such types of correlation is termed as *spurious correlation*. On the other hand, zero correlation coefficient between rainfall and the corresponding runoff does not mean that there is no association between the two, there may be a high non-linear correlation between them.

The correlation is said to be positive when $x_i$ increases with $y_i$ simultaneously. It is called negative when $y_i$ decreases with increase of $x_i$. For a bestfit linear regression equation, the scatter diagram should show that all points plot close to such a line. When a correlation coefficient of +1 is obtained, we call it perfectly associated. Such situation is expected when for example the diameter of various circles are taken as $x_i$ and area of the circles as $y_i$. On the other hand, when diameter of the circles ($x_i$) is correlated to the reciprocal of the area ($z_i = 1/y_i$), than we get correlation coefficient as −1, we call it negatively correlated. In a scatter diagram when all points are well distributed along some curve other than a straight line, the correlation is called non-linear and a non-linear equation should be tried. A positive negative and a weak correlation patterns of scatter diagram are shown in Fig. 2.12.

The sum of squares of deviations from the fitted equation given in equation (2.57) can be related to two other squares of deviations as

$$S_{od} = S_e + S_{ed} \tag{2.63}$$

where $S_{od}$ is the sum of squares of deviations of the observed values from their mean given by

$$S_{od} = \Sigma \, (y_{oi} - y_{mo})^2 \tag{2.64}$$

in which $y_{mo}$ is the mean of the observed series $y_{oi}$. This is the variance of the observed series. Similarly square of the deviation of the estimated values of $y_{ei}$ from the mean of the observed data $y_{0i}$ is given as

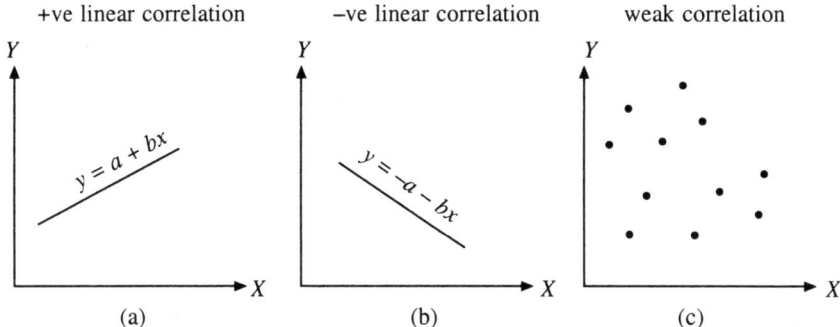

**Fig. 2.12   A Positive, Negative and Weak Correlation**

$$S_{ed} = \Sigma \, (y_{ei} - y_{mo})^2 \qquad (2.65)$$

sum of the error $(S_e)$ of the dispersion of observed $y_0$ values from their corresponding estimated values $y_e$ plus the dispersion of the estimated $y_{ei}$ values from the mean $y_{m0}$ is equal to the dispersion of the observed values of $y_0$ about the mean $y_{m0}$. The parameters that decide the goodness of fit of the curve are the standard error of estimate and the correlation coefficient. As has been discussed earlier, the standard error of estimate of $y_i$ from $N$ pairs of $(x_i \, y_i)$ is given as

$$S_{ee} = \sqrt{\frac{1}{N} \Sigma \, (y_{0i} - y_{ei})^2} = \sqrt{\frac{S_e}{N}} \qquad (2.66)$$

The parameter standard error of estimate measures the spread of points around the fitted curve. Greater is the value of $S_{ee}$, less accurate are the values determined from regression line. When $S_{ee} = 0$ all observed points must be on the fitted equation.

The correlation coefficient between $N$ pairs of $(x_i \, y_i)$ is given as

$$r = \frac{\text{Cov} \, (x_i, y_i)}{\sigma_x, \sigma_y} \qquad (2.67)$$

where Cov $(x_i y_i)$ is called the covariance of $x_i$ and $y_i$. This is the second moment of the cross product of $x_i$ and $y_i$ represented by

$$\text{Cov} \, (x_i, y_i) = \left\{ \frac{1}{N} \sum_{i=1}^{N} (x_i - x_{av})(y_i - y_{av}) \right\} \qquad (2.68)$$

The correlation coefficient can also be obtained from the following relation

$$r = \frac{\Sigma \, x_i y_i - N x_{av} y_{av}}{\sqrt{\Sigma \, x_i^2 - N x_{av}^2} \, \sqrt{\Sigma \, y_i^2 - N y_{av}^2}} = \frac{\Sigma \, x_i y_i - (\Sigma \, x_i \, \Sigma \, y_i)/N}{\sqrt{\Sigma \, x_i^2 - (\Sigma \, x_i)^2/N} \, \sqrt{\Sigma \, y_i^2 - (\Sigma \, y_i)^2/N}}$$

$$(2.69)$$

in which $x_{av}$ is the mean of $x$ variates and $y_{av}$ is the mean of $y$ variates. Correlation coefficient is an important parameter which signifies the association between $x_i$ and $y_i$. Square of the correlation coefficient

$$r^2 = \frac{S_{ed}}{S_{od}} = \frac{\text{Explained variance}}{\text{Total variance}} \qquad (2.70)$$

is called the *coefficient of determination*. Equation (2.70) explains the percentage of relationship between $x_i$ and $y_i$ by fitting the curve. If the fitted line pass through all the observed points, then $r = \pm 1$ and $S_e = 0$ which gives $S_{od} = S_{ed}$. On the other hand for a situation where no relationship exists between $x_i$ and $y_i$, $r = 0$. This gives $S_{ed} = 0$. For example if $r = 0.80$, then only $r^2 = 0.8 \times 0.8 = 0.64$ or 64% is explained due to their association and the rest are due to unexplained factors. In hydrology, the value of $r$ should be greater than 0.60.

The regression line $y_i$ on $x_i$ can also be represented as

$$y_i - y_{av} = r \frac{\sigma_y}{\sigma_x} (x_i - x_{av}) \qquad (2.71)$$

### 2.7.4   Significance of Parameters

Since the length of sample is usually small in comparison with the population (events like rainfall occurring since time immemorial), the value of $r$ should be tested for confidence limits. This can be achieved by computing standard deviation of the correlation coefficient as

$$S_r = \frac{(1 - r^2)}{\sqrt{N}} \qquad (2.72)$$

Now compute $\{(r + 3S_r)$ and $(r - 3S_r)\}$ or $\{(r + 4S_r)$ and $(r - 4S_r)\}$. If both limits have the same sign as $r$, we consider $r$ as significantly different from zero with respect to either of these empirial confidence limits. If these limits have different signs then $r$ is not considered as significantly different from zero.

The intercept $a$ and the coefficient $b$ should also be tested for their level of significance. For testing $b$, the $t$ distribution is given as

$$t = \left\{ r \frac{(b - \beta)}{b} \right\} \left\{ \frac{(N - 2)}{(1 - r^2)} \right\}^{1/2} \qquad (2.73)$$

where $\beta$ is the true value of $b$ and $t$ following $t$-distribution with $(N - 2)$ degrees of freedom. The value of $t$ can be obtained from standard $t$-distribution table.

To compute the test statistics take $\beta = 0$ and compute $t$ for the value of $r$ and degree of freedom as $(N - 2)$ from equation (2.73). If the $t$ value computed is greater than the corresponding values of $t$ at the tested level of significance, i.e., at 95% level or at 90% then $b$ is considered as significantly different from zero.

The intercept $a$ can be tested in the following way. The standardised variate of $a$ is given as

$$t = \left\{ r \frac{(a - \alpha)}{b} \right\} \left[ \frac{(N - 2)}{(1 - r^2)(\sigma_n^2 + x_{av}^2)} \right]^{1/2} \qquad (2.74)$$

When the line is assumed to pass through the origin, the value of $\alpha = 0$. At 95% level of significance, the value of $t$ is obtained from the table of students $t$-

distribution. At this level, both the values of $a$ are calculated from the relation (2.74). If the value of $a$ calculated from sample data and the value of $a$ from relation (2.74) are much different from zero then the hypothesis of the intercept being significantly different from zero is accepted. In other words, the hypothesis is tested by comparing the value of $t$ obtained from equation (2.74) and from students $t$ distribution table for the probability level that is $t$ from equation (2.74) > students table $t$ at the testing significance level.

### 2.7.5 Standard Forms of Bivariate Equations

Some standard form of equations used in the bivariate regression analysis in hydrology are given below.

(i)   Linear $y = a + bx$ (2.75)
(ii)  Exponential $y = be^{ax}$ (2.76)
(iii) Parabola $y = ax^b$ (2.77)
(iv)  Higher order equation $y = a_1 + a_2 + a_3x^2 + \ldots + a_{n+1}x^n$ (2.78)
(v)   Other forms of equations like

$$y = a + \left( \frac{b}{x} \right) \tag{2.79}$$

$$y = \frac{a}{(b + x)} \tag{2.80}$$

$$y = \frac{x}{(a + bx)} \tag{2.81}$$

$$y = c + be^{ax} \tag{2.82}$$

$$y = c + ax^b \tag{2.83}$$

$$y = \frac{c + b}{(x - a)} \tag{2.84}$$

$$y = \frac{c + x}{(a + bx)} \tag{2.85}$$

$$y = d + cx + be^{ax} \tag{2.86}$$

$$y = d^{cx}b^{ax} \tag{2.87}$$

$$y = d\,e^{cx} + b\,e^{ax} \tag{2.88}$$

or $\qquad\qquad\qquad y = e^{ax}\,(d \cos bx + c \sin bx)$ (2.89)

can be evaluated for the problem but the use of these equations are restricted because of the fact that by fitting a higher degree polynomial to the sample data, the curve may be made to pass through all the observed points but their extrapolation may not behave the same way as the fitted curve, giving errors in their extrapolation or interpolation.

Exponential and parabolic equations can be solved easily by taking logarithmic to both sides. By this the equation (2.77) can be reduced to a linear form of the type

$$\log_{10} y = \log_{10} a + b \log_{10} x \qquad (2.90)$$

and equation (2.76), after taking logarithmic with base $e$ reduces to

$$\ln (y) = \ln (b) + ax \qquad (2.91)$$

Thus a linear relation can be fitted to the equations (2.76) and (2.77) after taking logarithm of both $y$ and $x$ series. For quadratic parabola (polynomial of power 2) of the form $y = a + bx + cx^2$, it is required to evaluate for $a$, $b$ and $c$ as follows.

$$\sum y_i = aN + b \sum x_i + c \sum x_i^2 \qquad (2.92)$$

$$\sum xy_i = a \sum x_i + b \sum x_i^2 + c \sum x_i^3 \qquad (2.93)$$

$$\sum x_i^2 y_i = a \sum x_i^2 + b \sum x_i^3 + c \sum x_i^4 \qquad (2.94)$$

The summation is between $i = 1$ and $N$, where $N$ is the number of sample data. There are three unknowns and three equations. These can be solved easily. For higher degree polynomials solution to the equations can be obtained using matrix inversion. Standard computer programs are available for such type of solutions.

For hydrological studies the value of $n$ in equation (2.78) should preferably be restricted up to 4. The polynomial which gives minimum standard error of estimate should be considered suitable for extrapolation or interpolation of the data.

---

**Example 2.13:** The weighted catchment rainfall for the months of June and July alongwith the runoff for July at Lawkera site of IB river in Orissa are given below. Develop (i) a linear relationship and (ii) a non-linear relationship between rainfall and runoff for the month of July. Also calculate the correlation coefficient $r$. Test that it is significantly different from zero and therefore a good correlation exists.

**Table 2.14    Rainfall Runoff Data at Lawkera Station for Example 2.13**

| Sl. No. | Year | Rainfall of June (mm) | Rainfall of July (mm) | Runoff of July (mm) |
|---------|------|-----------------------|-----------------------|---------------------|
| (1) | (2) | (3) | (4) | (5) |
| 1. | 1978 | 229.2 | 344.4 | 187.1 |
| 2. | 1979 | 560.5 | 661.2 | 427.8 |
| 3. | 1980 | 153.2 | 369.6 | 122.9 |
| 4. | 1981 | 111.8 | 297.7 | 80.7 |
| 5. | 1982 | 85.2 | 372.2 | 194.9 |
| 6. | 1983 | 127.7 | 462.6 | 211.3 |
| 7. | 1984 | 109.8 | 302.4 | 125.12 |
| 8. | 1985 | 414.3 | 656.7 | 309.4 |
| 9. | 1986 | 281.4 | 446.7 | 221.4 |
| 10. | 1987 | 144.1 | 325.7 | 46.8 |
| 11. | 1988 | 346.9 | 336.5 | 247.9 |

**Solution**

Assume a linear relation of the form $y = a + bx$, where $y$ is the runoff and $x$ the rainfall of July, respectively (in mm). To fit an equation to the given data the calculations are carried out in the following table.

**Table 2.15   Calculation for Example 2.13 for Developing Linear Relation**

| Sl. No. | Year | Rainfall $(x)$ of July (mm) | Runoff $(y)$ of July (mm) | $(x)(y)$ | $x^2$ | $y^2$ |
|---|---|---|---|---|---|---|
| (1) | (2) | (3) | (4) | (5) | (6) | (7) |
| 1. | 1978 | 344.4 | 187.1 | 64437.24 | 118611.3 | 35006.41 |
| 2. | 1979 | 661.2 | 427.8 | 282861.3 | 437185.4 | 183012.80 |
| 3. | 1980 | 369.6 | 122.9 | 45423.84 | 136604.1 | 15104.41 |
| 4. | 1981 | 297.7 | 80.7 | 24024.39 | 88625.3 | 6512.49 |
| 5. | 1982 | 372.2 | 194.9 | 72541.78 | 138532.8 | 37786.01 |
| 6. | 1983 | 462.6 | 211.3 | 99747.38 | 213998.7 | 44647.69 |
| 7. | 1984 | 302.4 | 125.2 | 37860.48 | 91445.7 | 15675.04 |
| 8. | 1985 | 656.7 | 309.4 | 203182.9 | 431254.8 | 95728.36 |
| 9. | 1986 | 446.7 | 221.4 | 98899.38 | 199540.8 | 49017.96 |
| 10. | 1987 | 325.7 | 46.8 | 15242.76 | 106080.4 | 2190.24 |
| 11. | 1988 | 336.6 | 247.9 | 83418.35 | 113232.2 | 61454.41 |
| Sum | | 4575.7 | 2175.4 | 1025639 | 2075112 | 546336 |

The equations are

$$\Sigma y = aN + b \Sigma x$$

$$\Sigma yx = a \Sigma x + b \Sigma x^2$$

Substituting the data and solving we get,

$$a = \frac{\Sigma y_i \Sigma x_i^2 - \Sigma x_i \Sigma x_i y_i}{N \Sigma x_i^2 - (\Sigma x_i)^2} = \frac{(2175.4 \times 2075112 - 4575.7 \times 1025639)}{(11 \times 20755112 - 4575.7^2)}$$

$$= -\frac{1.7912542 \times 10^8}{1886456} = -94.95$$

$$b = \frac{N \Sigma x_i y_i - \Sigma x_i \Sigma y_i}{N \Sigma x_i^2 - (\Sigma x_i)^2} = \frac{(11 \times 1025639 - 2175.7 \times 4575.7)}{1886476} = \frac{1329229}{1886456} = 0.704$$

The regression equation is $y = -94.95 + 0.704$,

$$\text{Now } r = \frac{\Sigma x_i y_i - (\Sigma x_i \Sigma y_i)/N}{\sqrt{\Sigma x_i^2 - (\Sigma x_i)^2 /N} \sqrt{\Sigma y_i^2 - (\Sigma y_i)^2 /N}}$$

$$r = \left\{ \frac{1025639 - (2175.4 \times 4575.7)/11}{[\{2075112 - (4575.7^2)/11\}^{1/2} \{546336 - 2175.4^2 /11\}^{1/2}]} \right\} = 0.855$$

**Test for $r$**

Standard deviation of correlation coefficient $= S_r = \dfrac{(1 - r^2)}{N^{0.5}} = \dfrac{1 - 0.855^2}{11^{0.5}} = 0.081$

$$r + 3S_r = 0.855 + 3 \times 0.081 = + 1.098$$

$$r - 3S_r = 0.855 - 3 \times 0.081 = + 0.612$$

$$r + 4S_r = 0.855 + 4 \times 0.081 = + 1.179$$

$$r - 4S_r = 0.855 - 4 \times 0.081 = + 0.531$$

Since all are positive, the value of $r$ is significantly different from zero.

For a relation of type $y = ax^b$ take logarithim of both sides as

$$\log (y) = \log (a) + b \log (x)$$

This is of the form $Y = A + BX$. Calculation are carried out in Table 2.16.

**Table 2.16    Development of a Non-linear Relation for Example 2.13**

| Sl. No. | Year | Rainfall of July mm ($x$) | Runoff of July mm ($y$) | Log of ($x$) | Log of ($y$) | ($x$)($y$) | $x^2$ | $y^2$ |
|---|---|---|---|---|---|---|---|---|
| (1) | (2) | (3) | (4) | (5) | (6) | (7) | (8) | (9) |
| 1 | 1978 | 344.4 | 187.1 | 2.537063 | 2.272073 | 5.764394 | 6.436689 | 5.162319 |
| 2 | 1979 | 661.2 | 427.8 | 2.820332 | 2.631240 | 7.420974 | 7.954277 | 6.923428 |
| 3 | 1980 | 369.6 | 122.9 | 2.567731 | 2.089551 | 5.365409 | 6.593247 | 4.366227 |
| 4 | 1981 | 297.7 | 80.7 | 2.473778 | 1.906873 | 4.717183 | 6.119581 | 3.636166 |
| 5 | 1982 | 372.2 | 194.9 | 2.570776 | 2.289811 | 5.886594 | 6.608891 | 5.243238 |
| 6 | 1983 | 462.6 | 211.3 | 2.665205 | 2.324899 | 6.196335 | 7.013321 | 5.405157 |
| 7 | 1984 | 302.4 | 125.2 | 2.480581 | 2.097604 | 5.203279 | 6.153286 | 4.399943 |
| 8 | 1985 | 656.7 | 309.4 | 2.650015 | 2.345177 | 7.016709 | 7.016709 | 6.202691 |
| 9 | 1986 | 446.7 | 221.4 | 2.650015 | 2.345177 | 6.214758 | 7.022584 | 5.499850 |
| 10 | 1987 | 325.7 | 46.8 | 2.512817 | 1.670245 | 4.197023 | 6.314253 | 2.789721 |
| 11 | 1988 | 336.5 | 247.9 | 2.526985 | 2.394276 | 6.050301 | 6.385653 | 5.732560 |
| Sum | | 4575.7 | 2175.4 | 28.62265 | 24.51227 | 64.03296 | 74.62934 | 55.36131 |

$$\log a = \frac{(\Sigma y \, \Sigma x^2 - \Sigma x \, \Sigma xy)}{\{N \Sigma x^2 - (\Sigma x)^2\}} = \frac{(24.51 \times 74.63 - 28.62 \times 64.03)}{(11 \times 74.63 - 28.62^2)} = \frac{-3.3573}{1.8286} = -1.839$$

$$a = \log^{-1} (-1.839) = 10^{-1.839} = 0.0145$$

$$b = \frac{\{N \Sigma xy - \Sigma x \, \Sigma x\}}{\{N \Sigma x^2 - (\Sigma x)^2\}} = \frac{(11 \times 60.03 - 28.62 \times 24.51)}{1.8256} = \frac{2.8538}{1.8215} = 1.563$$

The developed equation is $y = 0.0145x^{1.563}$

## 2.8    MULTIVARIATE LINEAR REGRESSION AND CORRELATION

In hydrology, problems with multiple variables are frequently encountered. In flood forecasting we come across a situation in which discharge in a main river is to be computed from the known discharges of all tributaries joining the river upstream to it. If two tributaries say $x_1$ and $x_2$ contribute to the main river x, then the equation for discharge can be formulated as

$$x = b_0 + b_1 x_1 + b_2 x_2 \tag{2.95}$$

in which $b_0$ is called the intercept, $b_1$ the multiple regression coefficient between $x$ and $x_1$ when $x_2$ is constant and $b_2$ is the multiple regression coefficient between $x$ and $x_2$ keeping $x_1$ as constant. By least square, we can evaluate the coefficients $b_0$, $b_1$ and $b_2$ as

$$\Sigma x = N b_0 + b_1 \Sigma x_1 + b_2 \Sigma x_2 \tag{2.96}$$

$$\Sigma x x_1 = b_0 \Sigma x_1 + b_1 \Sigma x_1^2 + b_2 \Sigma x_1 x_2 \tag{2.97}$$

$$\Sigma x x_2 = b_0 \Sigma x_2 + b_1 \Sigma x_1 x_2 + b_2 \Sigma x_2^2 \tag{2.98}$$

where $N$ is the number of data in the sample. Solving equations (2.96, 2.97 and 2.98) simultaneously for the three unknowns $b_0$, $b_1$ and $b_2$, respectively, the curve for the multivariate regression equation is formulated.

Equation of the type

$$x = a x_1^{b_1} \cdot x_2^{b_2} \cdot x_3^{b_3} \dots x_n^{b_n} \tag{2.99}$$

can also be evaluted. Such equations can be converted to straight line form by logarithmic transformation. The equation then reduces to a multiple-linear relation, the solution of which can be obtained by the method of least square.

### 2.8.1 Multiple Correlation Coefficient

The association between the independent variables ($x_1$, $x_2$ etc.) taken together with the variable $x$ is expressed in multiple correlation coefficient as

$$r = \frac{\sigma_{est}}{\sigma_x} \tag{2.100}$$

where $\sigma_{est}$ is the standard deviation of estimated values. It is obtained by substituting the observed $x_1$ and $x_2$ values in equation (2.95) and computing the estimated series of $x_i$ as $x_{iest}$. From the series $x_{iest}$, the standard deviation $\sigma_{est}$ is computed. The symbol $\sigma_x$ in equation (2.100) represents the standard deviation of the $x_i$ series.

In multiple correlation, sometimes, a hydrologist is interested to know the correlation between the dependent variable $x_i$ and a particular independent variable out of say ($x_1$, $x_2$, $x_3$ ... $x_n$). This can be achieved by the following relation

$$r_{1-i}^2 = \frac{(\rho_1^2 - \rho_{1-i}^2)}{(1 - \rho_{1-i}^2)} \tag{2.101}$$

in which $\rho_1$ is the multiple correlation coefficient between dependent variable $x$ and all independent variables $x_1$, $x_2$ and $x_3$ and $\rho_{1-i}$ is the multiple correlation coefficient between $x$ and all independent variables except $x_1$, with which the relation is required. This is called the *partial correlation* and the coefficient is called *partial correlation coefficient*. Calculation of multiple correlation between the variables $x$ and ($x_1$, $x_2$, ... $x_n$) is given in equation (2.100). In the same way, the multiple correlation coefficient between $x$ and ($x_2$, $x_3$, ... $x_n$ except $x_1$) can also be calculated. It can be easily seen that $r_1$ is always greater than $r_{1-i}$.

**Example 2.14:**  Solve Example 2.13, if the rainfall of June and July are considered to be contributing to the runoff for the month of July. Develop an equation for this.

**Solution**

The problem is a multiple regression analysis. It can be solved by taking a relation of the type

$$y = a + bx_1 + cx_2$$

where $x_1$ is the rainfall for June, $x_2$ is the rainfall for July and $y$ is the runoff for July. By the method of least square

$$\Sigma\, y = aN + b\, \Sigma\, x_1 + c\, \Sigma\, x_2$$

$$\Sigma\, yx_1 = a\, \Sigma\, x_1 + b\, \Sigma\, x_1^2 + c\, \Sigma x_1 x_2$$

$$\Sigma\, yx_2 = a\, \Sigma x_2 + b\, \Sigma\, x_1 x_2 + c\, \Sigma\, x_2^2$$

Substituting data from Table 2.17 we get

$$11a + 2564b + 4576c = 2175 \quad \text{or} \quad a + 233.1b + 416c = 197.73 \qquad \text{(A)}$$

$$2564a + 830220b + 1224871c = 651078 \text{ or } a + 323.8b + 477.7c = 253.93 \quad \text{(B)}$$

$$4576a + 1224871b + 2075112c = 1025639 \text{ or } a + 267.7b + 453.5c = 224.1 \quad \text{(C)}$$

$$\text{Subtracting (A) from (B) we get} \quad 90.7\, b + 61.7c = 56.2 \qquad \text{(D)}$$

$$\text{Subtracting (C) from (B) we get} \quad 56.1b + 24.2c = 29.8 \qquad \text{(E)}$$

From (D) we get $b + 0.68c = 0.62$

and from (C) $b + 0.43c = 0.53$

Solving     $0.25c = 0.09$   or    $c = 0.09/0.25 = 0.36$

and              $b = 0.62 - 0.68 \times 0.36 = 0.375$

$$a = 197.73 - 416 \times 0.36 - 233.1 \times 0.375 = -39.50$$

The equation becomes      $y = -39.5 + 0.375x_1 + 0.36x_2$

**Example 2.15:**  From the fitted relation of Example 2.14, calculate the multiple correlation coefficient between the rainfall of June-July with runoff of July.

**Solution**

For calculation details refer Table 2.18.

$$\text{Variance of the series} = \sigma_{n-1}^2 = \left\{ \frac{511443.1}{(11-1)} \right\} = 226.15^2$$

$$\text{Variance of the estimated series} = \sigma_{est}^2 = \left\{ \frac{499161.3}{(11-1)} \right\} = 223.42^2$$

$$\text{Correlation coefficient } r = \left\{ \frac{223.42^2}{226.15^2} \right\}^{1/2} = 0.988$$

**Table 2.17  Computation of the Development of Multiple Regression Equation for Example 2.14**

| Sl. No. | Year | Rainfall of June mm $(x_1)$ | Rainfall of July mm $(x_2)$ | Runoff of July mm $(y)$ | $(x_2)(y)$ | $x_2^2$ | $y^2$ | $(x_1)(y)$ | $(x_1)(x_1)$ | $(x_1)(x_2)$ |
|---|---|---|---|---|---|---|---|---|---|---|
| (1) | (2) | (3) | (4) | (5) | (6) | (7) | (8) | (9) | (10) | (11) |
| 1 | 1978 | 229.2 | 344.4 | 187.1 | 64437.24 | 118611.3 | 35006.41 | 42883.32 | 52532.64 | 78936.48 |
| 2 | 1979 | 560.5 | 661.2 | 427.8 | 282861.3 | 437185.4 | 183012.8 | 239781.9 | 314160.2. | 370602.6 |
| 3 | 1980 | 153.2 | 369.6 | 122.9 | 45423.84 | 136604.1 | 15104.41 | 18828.28 | 23470.24 | 56622.86 |
| 4 | 1981 | 111.8 | 297.7 | 80.7 | 24024.39 | 88625.29 | 6512.49 | 9022.26 | 12499.24 | 33282.86 |
| 5 | 1982 | 85.2 | 372.2 | 194.9 | 72541.78 | 138532.8 | 37986.01 | 16605.48 | 7259.04 | 31711.44 |
| 6 | 1983 | 127.7 | 462.6 | 211.3 | 97747.38 | 213998.7 | 44647.69 | 26983.01 | 16307.29 | 59074.02 |
| 7 | 1984 | 109.8 | 302.4 | 125.2 | 37860.48 | 91445.8 | 15675.04 | 13746.96 | 12056.04 | 33203.52 |
| 8 | 1985 | 414.3 | 656.7 | 309.4 | 203182.9 | 431254.8 | 95728.36 | 128184.4 | 171644.4 | 272070.8 |
| 9 | 1986 | 281.4 | 446.7 | 221.4 | 98899.38 | 199540.8 | 49017.96 | 62301.96 | 79185.96 | 125701.3 |
| 10 | 1987 | 144.1 | 325.7 | 46.8 | 15242.76 | 106080.4 | 2190.24 | 6743.88 | 20764.81 | 46933.37 |
| 11 | 1988 | 346.9 | 336.5 | 247.9 | 83418.35 | 113232.2 | 61454.41 | 85996.51 | 120339.6 | 116731.8 |
| Sum | | 2564.1 | 4575.7 | 2175.4 | 1025639 | 2075112 | 546335.8 | 651077.9 | 830219.6 | 1224871 |

**Table 2.18    Computation of Multiple Correlation Coefficient for Example 2.15**

| Sl. No. | Year | Rainfall of June mm $(x_1)$ | Rainfall of July mm $(x_2)$ | Run off of July mm $(y)$ | Computed run off* | $(y - y_{av})^2$ | $(y_{com} - y_{av\ com})^2$ |
|---|---|---|---|---|---|---|---|
| (1) | (2) | (3) | (4) | (5) | (6) | (7) | (8) |
| 1 | 1978 | 229.2 | 334.4 | 187.1 | 170.4 | 113.7 | 741.4 |
| 2 | 1979 | 560.5 | 661.2 | 427.8 | 408.7 | 183012.8 | 167051.6 |
| 3 | 1980 | 153.2 | 369.6 | 122.9 | 151.0 | 15104.4 | 22802.8 |
| 4 | 1981 | 111.8 | 297.7 | 80.7 | 109.6 | 6512.5 | 12011.5 |
| 5 | 1982 | 85.2 | 372.2 | 194.9 | 126.4 | 37986.0 | 15987.6 |
| 6 | 1983 | 127.7 | 462.6 | 211.3 | 174.9 | 44647.7 | 30598.2 |
| 7 | 1984 | 109.8 | 302.4 | 125.2 | 110.5 | 15675.0 | 12218.9 |
| 8 | 1985 | 414.3 | 656.7 | 309.4 | 352.3 | 95728.4 | 124097.3 |
| 9 | 1986 | 281.4 | 446.7 | 221.4 | 226.8 | 49018.0 | 51455.0 |
| 10 | 1987 | 144.1 | 325.7 | 46.8 | 131.8 | 2190.2 | 17368.5 |
| 11 | 1988 | 346.5 | 336.5 | 247.9 | 211.7 | 61454.4 | 44828.5 |
| Sum | | 2564.1 | 4575.7 | 2175.4 | 2174.3 | 511443.1 | 499161.3 |
| Mean | | | | 197.76 | 197.66 | | |

## 2.9  ANALYSIS OF TIME SERIES

When a series of observations are arranged with respect to time of their occurrences in a systematic order, the resulting series is called *time series*. This series in hydrology is considered as *time-homogeneous*, when identical events in the series are likely to occur at the same time. Due to a large number of caustic factors affecting the phenomena, a hydrologic time series is never time homogeneous. By *stationary* we mean, the time series segments drawn from the same population should have the same expected value of statistical parameters for each section. Annual series may be stationary but daily or monthly series are never stationary. Properties like trend, periodicity and persistence make the departure of a time series from its true homogeneity. They should be identified, quantified and removed as they are deterministic in nature. The residuals are the stochastic components due to the property of randomness. A random stochastic component is said to be present when serial correlation coefficients of different lags are zero. A time series $x_t$ is modeled as

$$x_t = x_{pt} + x_{Tt} + x_{ot} + x_{jt} + E_t \tag{2.102}$$

in which $x_{Tt}$ is the component due to *trend*, $x_{pt}$ the component due to *periodicity*, $x_{ot}$ the component due to *oscillations*, $x_{jt}$ is the component due to *jump* and $E_t$ is the stochastic component. Presence of trend, long term oscillation and seasonality for a time series are shown in Fig. 2.13.

Knowledge on analysis of time series is a must for a hydrologist. In a developing country like India, generation of data is an essential requirement for any water resources project planning and management. An effective generating technique breaks up the data into its components, analyse the mechanism underlying the

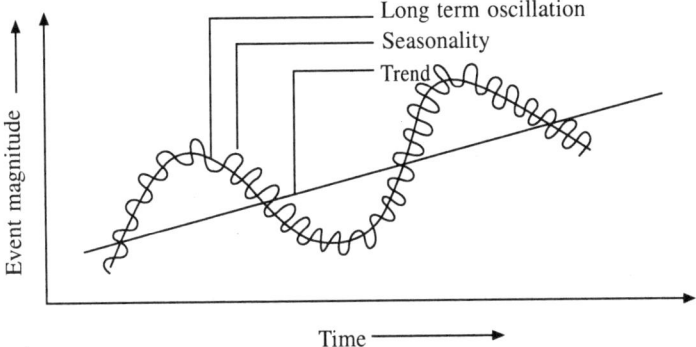

**Fig. 2.13 Trend, Long Term Oscillations and Seasonality in Time Series**

formation of each constituent and preserve them, while generating the future sequence of the series. Thus, properties of each constituents must be preserved while combining the constituents to form the future time series. We are assuming here that the pattern or the system that has been identified will continue. When the constituents of the time series are not properly identified or the pattern identified does not continue in future, a wrong generating technique is incorporated. This may lead to erroneous results and must be checked and correcteded.

In a hydrologic time series the constituents like trend, periodicity, oscillation and jump are deterministic in nature which can be quantified and removed. The residual stochastic component is studied and modeled suitably and will be discussed later.

A hydrologic time series can be classified into the following groups (Fig. 2.14).

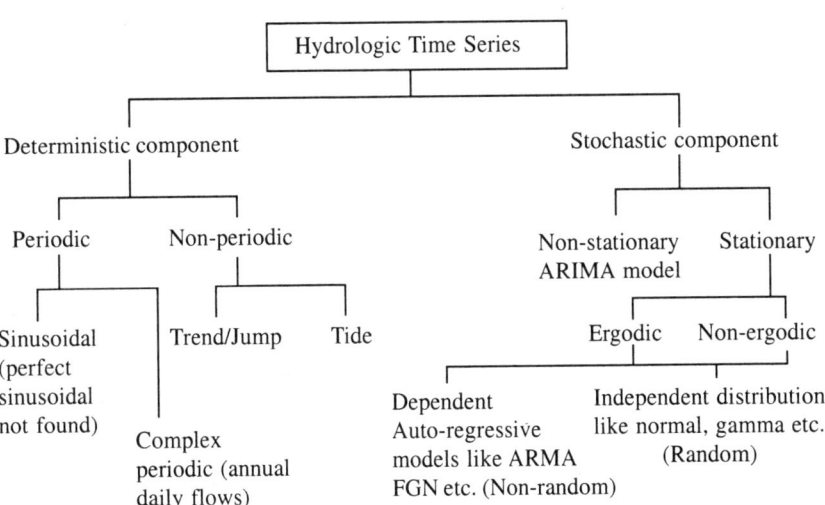

**Fig. 2.14 Classification of Hydrologic Time Series**

## 2.9.1   Trend Analysis

A steady increase or decrease of the time series characteristics is known as trend. Natural or man-made changes like deforestation, urbanisation, large scale landslide, large changes in watershed conditions are responsible for the introduction of trend in the time series. Tests for detecting the presence of trend are

> Turning Point Test
> Kendal's Rank-Correlation Test
> Regression Test for Linear Trend

### 2.9.1.1   Turning Point Test

In this test, we try to find out how many turning points are there in a sample data. We consider a turning point to exist when $x_i$ is either greater than both preceding and succeeding values or less than both. Thus, any of the conditions for a variate $x_{i-1} < x_i > x_{i+1}$ or $x_{i-1} > x_i < x_{i+1}$ gives a turning point. The procedure is outlined as follows.

(i) Arrange the data in order of their occurrence.
(ii) Apply either of the conditions $x_{i-1} < x_i > x_{i+1}$ or $x_{i-1} > x_i < x_{i+1}$ and ascertain how many turning points are there in the series.
(iii) Let the total number of turning points be $p$.
(iv) Expected number of turning points in the series is

$$E(P) = \frac{2(N-2)}{3} \qquad (2.103)$$

where $N$ is the total number of data.

(v) Variance of $P$ is $\mathrm{Var}\,(P) = \dfrac{(16N - 29)}{90}$ \qquad (2.104)

(vi) Expressing $P$ in standard normal form $Z = \dfrac{\{P - E(P)\}}{\{\mathrm{Var}\,(P)\}^{1/2}}$ \qquad (2.105)

(vii) Test it at 5% level of significance, that is, take the value of Z as ± 1.96 at 5% level of significance.
(viii) If Z is less than ± 1.96, we say that at 5% level of significance there is no reason to suspect that the sample events are other than a random sequence.

### 2.9.1.2   Kendal's Rank-Correlation Test

The procedure of this method is as follows:

(i) Pick up the first value of the series $x_i$ and compare it with the rest of series $x_2, x_3 \ldots x_n$. Find out how many times it is greater than the rest, i.e., compare $x_1$ and $x_2$, if $x_1 > x_2$ then mark it as 1 or say take $P_{1.1}$ as 1. Compare it with $x_3$ and if it is again greater than $x_3$, take $P_{1.2}$ as 2. Complete the process for $x_1$ with rest of series and let $P_{1\mathrm{ex}}$ is all the expected value for $x_1$. Repeat it for $x_2, x_3 \ldots x_n$ and let $P_{2\mathrm{ex}}, P_{3\mathrm{ex}}$ are the corresponding values.

(ii) Find $P = P_{1\mathrm{ex}} + P_{2\mathrm{ex}} + \ldots + P_{n\mathrm{ex}}$ \qquad (2.106)

(iii)  Maximum value of $P$ can be $P_{max} = \dfrac{n(n-1)}{2}$ $\qquad$ (2.107)

(iv)  $E(P) = \dfrac{n(n-1)}{4}$ $\qquad$ (2.108)

(v)  Kendal's $\tau$ is computed as

$$\tau = \left[\left\{\frac{4P}{n(n-1)}\right\} - 1\right], \qquad E(\tau) \text{ should be zero} \qquad (2.109)$$

(vi)  Variance of $\tau = \text{Var}\,(\tau) = \left[\dfrac{\{2(2n+5)\}}{9n(n-1)}\right]$ $\qquad$ (2.110)

(vii)  Standard test for statistics of $Z = \dfrac{\tau}{\{\text{Var}\,(\tau)\}^{1/2}}$ $\qquad$ (2.111)

(viii)  Test for the hypothesis at 5% level of significance of Z, i.e., $Z = \pm\,1.96$.

### 2.9.1.3  *Regression Test for Linear Trend*
A regression line of the order $y = a + bx$ is fitted to the sample data. If $b$ is found significantly different from zero then we assume trend to be present in the data. A trend line is fitted and the residuals are computed. The residuals are tested for their dependence. Trend analysis should be carried out for annual series only.

### 2.9.1.4  *Estimation and Removal of Trend*
The following methods are normally used to quantify and remove any trend component present in a hydrologic series.

(i)  Moving average method
(ii)  Least square method

*Moving Average Method*
By this method we first try to smoothen the irregularities of the time series and then subtract the smoothened ordinates from the corresponding sample events. The result of the subtracted values are the trend free data. The procedure is as follows.

(i)  Write the years and the corresponding events against the years in a row.
(ii)  A moving average of 3 or 5 is normally used to smoothen the series. For moving average of 3 the equation is

$$x_2' = \frac{(a_1 x_1 + a_2 x_2 + a_3 x_3)}{3}$$

$$x_3' = \frac{(a_1 x_2 + a_2 x_3 + a_3 x_4)}{3}$$

$$x_{n-1}' = \frac{(a_1 x_{n-2} + a_2 x_{n-1} + a_3 x_n)}{3} \qquad (2.112)$$

In this process the first and last data of the series are lost. The coefficients $a_1$, $a_2$ and $a_3$ are determined by fitting a polynomial to the series. For this, data can be arranged in three columns as $(x_1\ x_2\ x_3)$, $(x_2,\ x_3,\ x_4)$, $(x_3,\ x_4,\ x_5)$ and so on. By least square method, the coefficients $a_1$, $a_2$ and $a_3$ can be computed. The simplest is when $a_1 = a_2 = a_3 = 1$. But it should not be applied as it is not weighing properly to the just preceding and succeeding values.

(iii) By using the fitted equation, the smoothened values are written. For $m = 3$, $(n - 2)$ values of the $x_i'$ are obtained. The moving average values range from $x_2$ to $x_{n-1}$.

(iv) In the next row, the events with trend removed is obtained by subtracting events of step (i) from (iii).

While applying moving average method to a time series, an oscillatory movement into the random component is introduced. The oscillations already present with period less than or equal to $m$ is also dampened out. This problem was first noticed by Slutzky and Yule and is known as *Slutzky — Yule effect*. Care must be taken to discuss the oscillatory property of the time series while ascertaining trend by moving average method.

*Least Square Method*
As discussed before, the sample data is fitted with a polynomial. The difference between the polynomial and the observed values can be taken as the trend free data. The polynomial should be fitted by least square method. When a straight line equation of the form $Y = a + bX$ is fitted to the data, it should be seen that the coefficient $b$ in the equation is not significantly different from zero. High value of $b$ indicates the presence of trend in the data.

---

**Example 2.16**: The observed runoff series at Sundargarh site from 1978 to 1987 are given. Check by turning point test, whether there is any trend in the data. Remove the trend if (any) present in the series.

Runoff (mm)   652   153   732   500   357   499   676   526   482   and   514.

**Solution**
Expected number of turning points $P = 5$ (Table 2.19)

$$E(P) = \frac{2(N-2)}{3} = \frac{2(10-2)}{3} = \frac{16}{3} = 5.33$$

Variance of   $$P = \frac{(16\ N - 29)}{90} = \frac{(16 \times 10 - 29)}{90} = 1.455$$

Standard normal of $P$ represented by   $$Z = \frac{\{P - E(P)\}}{\{Var\,(p)\}^{1/2}} = \frac{5 - 5.33}{(1.455)^{0.5}} = -0.276$$

Since $Z = -0.276$ is within the range of $X = \pm 1.96$ at 5% level of significance, we suspect that there is no trend in the series.

**Table 2.19   Testing the Presence of Trend of Example 2.16**

| Year | Data | Test for turning point | | | Presence of turning point | Cumulative turning point |
|------|------|------|------|------|------|------|
| (1) | (2) | (3) | | | (4) | (5) |
| 1978 | 652 | – | – | – | – | |
| 1979 | 153 | 652 | 153 | 732 | 1 | 1 |
| 1980 | 732 | 153 | 732 | 500 | 1 | 2 |
| 1981 | 500 | 732 | 500 | 357 | – | – |
| 1982 | 357 | 500 | 357 | 499 | 1 | 3 |
| 1983 | 499 | 499 | 499 | 676 | – | – |
| 1984 | 676 | 499 | 676 | 526 | 1 | 4 |
| 1985 | 526 | 676 | 526 | 482 | – | – |
| 1986 | 482 | 526 | 482 | 574 | 1 | 5 |
| 1987 | 574 | – | – | – | – | – |

*Kendal's rank correlation test*

By comparing                        652 with rest of series we get $P_{1ex} = 7$
Similarly comparing                 153 with rest of series we get $P_{2ex} = 0$
                                    732 with rest of series, we get $P_{3ex} = 7$
                                    500 with rest of series we get $P_{4ex} = 3$
                                    357 with rest of series we get $P_{5ex} = 0$
                                    499 with rest of series we get $P_{6ex} = 1$
                                    676 with rest of series we get $P_{7ex} = 3$
                                    526 with rest of series we get $P_{8ex} = 2$
                                    482 with rest of series we get $P_{9ex} = 0$

Total number of $P = 23$

Maximum value of $P = \dfrac{n(n-1)}{2} = \dfrac{(10 \times 9)}{2} = 45$

$$E(P) = \frac{n(n-1)}{4} = \frac{10 \times 9}{4} = 22.5$$

Kendal's $\tau = \left[\left\{\dfrac{4P}{n(n-1)}\right\} - 1\right] = \left\{\dfrac{4 \times 23}{(10 \times 9)} - 1\right\} = 0.022$

Variance of $\tau = \left[\dfrac{\{2(2n+5)\}}{9n(n-1)}\right] = \dfrac{2(2 \times 10 + 5)}{\{9 \times 10 \times (10-1)\}} = \dfrac{50}{(90 \times 9)} = \dfrac{5}{81} = 0.0617$

Standard test for statistics $Z = \dfrac{\tau}{\{Var\,(\tau)\}^{1/2}} = \dfrac{0.022}{0.0617^{0.5}} = 0.0894$

At 5% level of significance $Z = \pm 1.96$. Since $Z$ is within the range of $\pm1.96$, we do not expect trend to be present in the data.

## 2.9.2   Oscillation

Small oscillations may be found around the trend line. Sometimes the oscillations are cyclic. In the cyclic type, the maximum and minimum values occur at constant amplitude and at fixed time intervals. However, all oscillations are well distributed about their mean value. Oscillatory movement can be long term or short term. Sometimes wave length of oscillations are very long and some times they are very short such that they form part of the seasonality. Their detection, quantification and removal are not proposed in this book as this component is assumed to be absent in most of the hydrologic data. A long term oscillatory movement with the seasonality or periodic component is shown in Fig. 2.13.

### 2.9.3  Jump

A sudden change in the catchment characteristics may lead to a jump in the statistical parameters of a time series. A jump in the mean values is deterministic which can be quantified and removed from a time series. From inspection and trial, a jump in the mean can be located and quantified. For this a long term data covering either side of the time of occurrence of jump is essential. A jump in a time series is shown in Fig. 2.15.

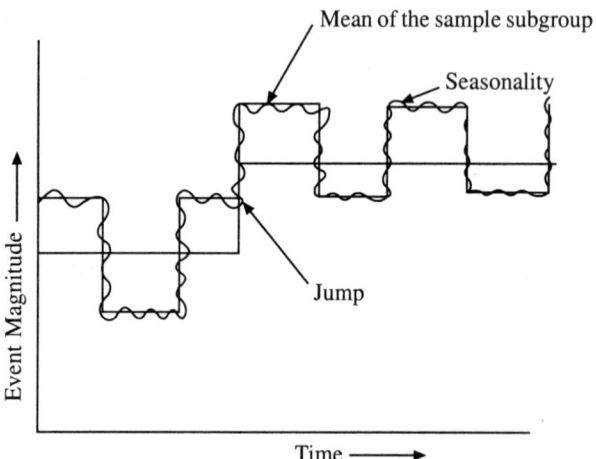

**Fig. 2.15   A Jump in the Mean of the Data**

### 2.9.4  Periodicity or Seasonality

Monthly and seasonal hydrologic events exhibit a regular oscillatory form of variations due to the rotation of earth around the sun. The dry and wet seasons maintain a nearly constant length of time in their occurrences. A sinusoidal curve can be superimposed on the time versus magnitude curve. This helps to determine the periodic component or the component due to seasonality by harmonic analysis. A time series of monthly magnitudes can be modelled as

$$X_t = \mu_t + \sigma_t \varepsilon_t \tag{2.113}$$

where $\mu_t$ is the population mean, $\sigma_t$ is the population standard deviation with their sample estimates of $m_t$ and $s_t$ respectively and $\varepsilon_t$ the stochastic component. The series to be modeled should be free from trend and oscillations. The periodic component $\mu_t$ of the mean and $\sigma_t$ of standard deviation is computed from the following equations

$$\mu_i = m_x + \sum_{i=1}^{k} \alpha_t \sin\left(\frac{2\pi it}{w}\right) + \sum_{i=1}^{k} \beta_t \cos\left(\frac{2\pi it}{w}\right) \qquad (2.114)$$

$$\sigma_i = s_x + \sum_{i=1}^{k} \alpha_t \sin\left(\frac{2\pi it}{w}\right) + \sum_{i=1}^{k} \beta_t \cos\left(\frac{2\pi it}{w}\right) \qquad (2.115)$$

$$\alpha_i = \frac{2}{w} \sum_{t=1}^{w} m_t \sin\left(\frac{2\pi it}{w}\right), \qquad w = 1, 2 \dots 6 \qquad (2.116)$$

$$\beta_i = \frac{2}{w} \sum_{t=1}^{w} m_t \cos\left(\frac{2\pi it}{w}\right), \qquad w = 1, 2 \dots 6 \qquad (2.117)$$

$$m_i = \frac{1}{m} \sum_{t=1}^{n} x_i \qquad (2.118)$$

where $n$ is the number of years of data, $\alpha_t$ and $\beta_t$ are the harmonic coefficients and $w$ is the number of months in a year. For fitting the model as per the above equations, it is required to determine the value of $k$, i.e., the number of significant harmonic required to explain the major part of the variance $\sigma_t$ of the periodic parameter. For monthly data, it is sufficient to fit a maximum of six harmonics of periodic parameter for a time series and should be tested for the significance. Variance $h_t$ explained by each harmonic is given by

$$h_t = (\alpha_t^2 + \beta_t^2)/2$$

$$\Delta p_t = (\text{variance of } h_t)/s^2 \qquad (2.119)$$

which represents part of the variation of periodic component explained by *i*th harmonic, $s$ is variance of the standard deviation of the monthly data. Having decided the number of harmonics to be fitted to the 12 estimated values of periodic mean and periodic standard deviation, the next step is to standardise them so as to remove the periodicity of the time series by the equation

$$y_{pt} = \frac{(x_{pt} - \mu_t)}{\sigma_t} \qquad (2.120)$$

Since equation (2.120) is approximately a standard variable, its mean $y_{av}$ and standard deviation $s_y$ are found and the standardised stochastic component is obtained from the equation as

$$\varepsilon_{pt} = \frac{(y_{pt} - y_{av})}{s_y} \qquad (2.121)$$

After removing the components due to the seasonality, $n_p$ values of $\varepsilon_{p,t}$ are leftout from the residuals of equation (2.121). The main object of this exercise is to study the residual series $\varepsilon_{p,t}$. The serial correlation coefficient of different lags should be zero if the residuals are independent.

### 2.9.5   Serial Correlation

A serial correlation is the association between the successive terms in the same series $x_t$ by lagging them suitably as per the requirement. The method of calculation of correlation coefficient between two variables $(x_i, y_i)$ can be extended to the individual series of either $x_i$ or $y_i$ lagged by $k$ distance apart. As an example for a runoff series of $N$ periods, it may be desired to find the relation or dependence between the same series with lag or spacing of say $k$ time apart. The correlation referred as auto-correlation and is defined as

$$\rho_k = \frac{\text{Cov}\,(x_i, x_{i+k})}{\sigma_{xi}\sigma_{x+k}} \tag{2.122}$$

Equation (2.122) gives an idea of the dependence or association between the values of the same series. When $k$ is taken as one and the event represent the magnitude of monthly runoff, then the serial correlation defines the association or dependence between the consecutive monthly values of the time series. This shows the effect of the flow of June on July, July on August and so on for the lag of $k = 1$ unit of time. The following simplified equation can be used to obtain the serial correlation coefficient of lag or spacing of $k$ time apart.

$$r_k = \frac{\dfrac{1}{(n-k)}\Sigma X_i X_{i+k} - \dfrac{1}{(n-k)^2}\Sigma X_i \Sigma X_{i+1}}{\left\{\dfrac{1}{(n-k)}\Sigma X_i^2 - \dfrac{1}{(n-k)^2}(\Sigma X_i)^2\right\}^{1/2}\left\{\dfrac{1}{(n-k)}\Sigma X_{i+k}^2 - \dfrac{1}{(n-k)^2}(\Sigma X_{i+k})^2\right\}^{1/2}} \tag{2.123}$$

in which $n$ is the length of the data, $k$ the lag distance in the serial correlation, $x_i$ is the variate. The summation should be carried out from $i = 1$ to $n - k$. For $k = 0$, the correlation coefficient $r_0 = 1$ and for other values of $k$, the value of $r$ lies between $\pm 1$. If the series is random $r_k = 0$ for all lags of $k$.

A plot between the lags $k$ in abscissa against the serial correlation $r_k$ is defined as *correlagram*. At 95% significance level, a tolerance band on either side of the plotted points (of $r_k$ vs. $k$) can be drawn using the following relation.

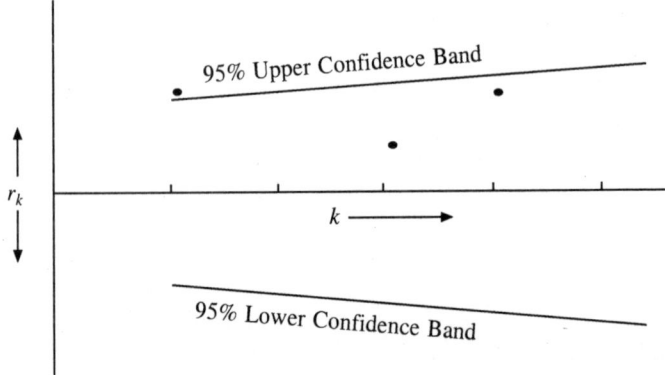

**Fig. 2.16   A Correlagram**

$$r_k = \frac{-1}{(n-k)} \pm 1.645 \left\{ \frac{\sqrt{n-k-1}}{n-k} \right\} \qquad (2.124)$$

A plot of such a curve is shown in the Fig. 2.16. In the following example the dependence of a series is tested for various lags.

---

**Example 2.17:** For the peak daily river discharge shown in col. 2 of the table below, find out the serial correlation coefficient of lag 1 and 3.

**Solution**
Refer Table 2.20 for detail calculations.

**Table 2.20   Serial Correlation Coefficient of Lag 1 and 3 ($k = 1$ and $3$)**

| Sl. No. | $x_i$ (given) | $x_{i+1}$ | $x_i x_{i+1}$ | $x_i^2$ | $x_{i+1}^2$ | $x_{i+3}$ | $x_i x_{i+3}$ | $x_{i+3}^2$ |
|---|---|---|---|---|---|---|---|---|
| (1) | (2) | (3) | (4) | (5) | (6) | (7) | (8) | (9) |
| 1 | 10.12 | 19.61 | 198.45 | 102.41 | 384.55 | 46.36 | 469.16 | 2149.25 |
| 2 | 19.61 | 29.03 | 569.28 | 384.55 | 842.74 | 48.06 | 942.46 | 2309.76 |
| 3 | 29.03 | 46.36 | 1345.83 | 842.74 | 2149.25 | 49.64 | 1441.05 | 2464.13 |
| 4 | 46.36 | 48.06 | 2228.06 | 2149.25 | 2309.76 | 58.49 | 2711.60 | 3421.08 |
| 5 | 48.06 | 49.64 | 2385.7 | 2309.76 | 2464.13 | 59.40 | 2854.76 | 3528.36 |
| 6 | 49.64 | 58.49 | 2903.44 | 2464.13 | 3421.08 | 58.13 | 2885.57 | 2379.10 |
| 7 | 58.49 | 59.40 | 3474.31 | 3421.08 | 3528.36 | 52.74 | 3084.76 | 2781.51 |
| 8 | 59.40 | 58.13 | 3452.92 | 3528.36 | 3379.10 | 48.84 | 2901.10 | 2385.35 |
| 9 | 58.13 | 52.74 | 3065.78 | 3379.10 | 2781.51 | 45.16 | 2625.15 | 2039.43 |
| 10 | 52.74 | 48.84 | 2575.82 | 2781.51 | 2385.35 | 43.98 | 2319.51 | 1934.24 |
| 11 | 48.84 | 45.16 | 2205.61 | 2385.35 | 2039.43 | 42.97 | 2098.65 | 1846.42 |
| 12 | 45.16 | 43.98 | 1986.14 | 2039.43 | 1934.24 | 39.24 | 1772.08 | 1539.78 |
| 13 | 43.98 | 42.97 | 1889.82 | 1934.24 | 1846.42 | 37.27 | 1639.13 | 1389.05 |
| 14 | 42.97 | 39.24 | 1684.14 | 1846.42 | 1539.78 | 36.63 | 1573.99 | 1341.76 |
| 15 | 39.24 | 37.27 | 1462.47 | 1539.78 | 1389.05 | 35.80 | 1404.79 | 1281.64 |
| 16 | 37.27 | 36.63 | 1365.2 | 1389.05 | 1341.76 | 34.22 | 1275.38 | 1171.01 |
| 17 | 36.63 | 35.80 | 1311.35 | 1341.76 | 1281.64 | 31.38 | 1149.45 | 984.70 |
| 18 | 35.80 | 34.22 | 1225.08 | 1281.64 | 1171.01 | 29.40 | 1052.52 | 864.36 |
| 19 | 34.22 | 31.38 | 1073.82 | 1171.01 | 984.70 | 27.40 | 937.63 | 750.76 |
| 20 | 31.38 | 29.40 | 922.57 | 984.70 | 864.36 | 25.25 | 792.35 | 637.56 |
| 21 | 29.40 | 27.40 | 805.56 | 864.36 | 750.76 | 23.27 | 684.14 | 541.49 |
| 22 | 27.40 | 25.25 | 691.85 | 750.76 | 637.56 | 21.57 | 591.02 | 465.26 |
| 23 | 25.25 | 23.27 | 587.57 | 637.56 | 541.49 | 21.30 | 537.83 | 453.69 |
| 24 | 23.27 | 21.57 | 501.93 | 541.49 | 465.26 | 20.61 | 479.59 | 424.77 |
| 25 | 21.57 | 21.30 | 459.44 | 465.26 | 453.69 | 19.67 | 424.28 | 386.91 |
| 26 | 21.30 | 20.61 | 438.99 | 453.69 | 424.77 | 18.37 | 391.28 | 337.46 |
| 27 | 20.61 | 19.67 | 405.40 | 424.77 | 386.90 | 17.08 | 352.02 | 291.73 |
| 28 | 19.67 | 18.37 | 361.34 | 386.90 | 337.46 | 16.00 | 314.72 | 256.00 |
| 29 | 18.37 | 17.08 | 313.76 | 337.46 | 291.73 | 15.17 | 278.67 | 230.13 |
| 30 | 17.08 | 16.00 | 273.28 | 292.73 | 256.00 | 14.44 | 246.64 | 208.51 |

*(Contd.)*

| (1) | (2) | (3) | (4) | (5) | (6) | (7) | (8) | (9) |
|-----|-----|-----|-----|-----|-----|-----|-----|-----|
| 31 | 16.00 | 15.17 | 242.72 | 256.00 | 230.13 | 13.89 | 222.24 | 192.93 |
| 32 | 15.17 | 14.44 | 219.05 | 230.13 | 208.51 | | | |
| 33 | 14.44 | 13.89 | 200.57 | 208.52 | 192.93 | | | |
| 34 | 13.89 | | | | | | | |

| | | | | | | | | |
|-----|-----|-----|-----|-----|-----|-----|-----|-----|
| Sum | 1110.49 | 1100.37 | 42829.3 | 43124.9 | 43215.4 | 1051.7 | 40453.5 | 41988.1 |

| | | | | | | | | |
|-----|-----|-----|-----|-----|-----|-----|-----|-----|
| Sum/ $(N - 1)$= | 33.66 | 32.34 | 1297.8 | 1306.8 | 1309.6 | – | – | – |

| | | | | | | | | |
|-----|-----|-----|-----|-----|-----|-----|-----|-----|
| Sum/ $(N - 3)$= | 5.82 | – | – | 1391.1 | – | 33.9 | 1305.0 | 1354.5 |

Serial correlation coefficient of lag 1 is found from equation (2.123) as

$$r_1 = \frac{(1297.8 - 33.66 \times 32.34)}{\left(1306.8 - \frac{1110.5^2}{33^2}\right)^{1/2}\left(1309.6 - \frac{1051.7^2}{33^2}\right)^{1/2}} = \frac{175.6}{(13.2 \times 17.1)} = 0.78$$

$$r_3 = \frac{(1305.0 - 35.82 \times 33.9)}{\left\{\left(\frac{1391.1 - 1110.5^2}{31^2}\right)^{1/2}\left(\frac{1354.5 - 1051.7^2}{31^2}\right)^{1/2}\right\}} = 90.7/(10.4 \times 14.3) = 0.61$$

It can be noted that the observed data shows a good serial correlation at lag 1 equal to 0.78 and when the time shifts to the third day, the correlation becomes weaker. For any gauging site it is quite acceptable that the dependence of say today's flow is very much dependent on yesterday's flow but is less dependent on the previous third day's flow. The dependence of the above series can be tested at 5% level of significance. The array of auto-correlation coefficients $r_1$, $r_2$, $r_3$, ..., $r_n$ can be plotted as ordinates against $k$ as abscissa, giving rise to autocorrelagram.

### 2.9.6   Stochastic Component

In equation (2.102) $E_t$ is the component due to randomness. We need to study the properties of this component. When the stochastic component is serially dependent, the correlation coefficient of different logs will not be equal to zero. When it happens a random series may be generated, preserving the statistical parameters of the stochastic component. Most of the stochastic components of hydrologic series exhibit dependence with the previous values.

## 2.10   DEPENDENCE MODELS

Dependence of the stationary stochastic component is tested by serial correlation coefficient of different lags given in equation (1.123). Once dependence is established, any of the following models can be used for regenerating the future sequence preserving the inherited properties of the observed data:

(i)  Markov Process
(ii)  Moving Average Model
(iii)  Reversible Process of Harmonic Analysis

Markov process and the moving average model are outlined here. The following dependence models are mostly used in hydrology.

ARIMA Model (Auto-Regressive Integrated Moving Average Model)
ARMA Mode (Auto-Regressive Moving Average Model)

The ARMA model consists of auto regressive (AR) model + a moving average (MA) model. The moving average model in its general form is given as

$$Z_t = f(Z_t, Z_{t-1}, Z_{t-2}, Z_{t-3} \ldots) + a_t \tag{2.125}$$

in which $Z_t, Z_{t-1}, Z_{t-2} \ldots$ are the stochastic component after removal of trend and periodicity. The auto-regressive model can be expressed as

$$Z_i = \Sigma \ Q_i Z_{i-1} + a_t \tag{2.126}$$

The two models can be combined together to give rise to the so called ARMA model of the form

$$Z_t = \overset{p}{\underset{i=1}{\Sigma}} Q_i Z_{t-1} + a_t + \overset{q}{\underset{j=1}{\Sigma}} Q_j a_{t-j} \tag{2.127}$$

$$\underbrace{\qquad\qquad}_{\text{AR}(p)\text{th model}} \quad \underbrace{\qquad\qquad}_{\text{MA}(q)\text{th model}}$$

Out of the two expressions forming the equation (2.127), the first two terms of the right hand side are the auto regressive (AR) $p$th model and the second and third terms together are the moving average (MA) $q$th model. An AR model of $p$th order suggests that the value of $Z_t$ at any time $t$ can be estimated from the weighted sum of $p$ values at times $(t - 1)$, $(t - 2)$ ... $(t - p)$ and a random component $a_t$. If $p = 1$, it depends on the just preceding value, and so on.

With the interest to generate sequences with periods less than annual, values of the hydrologic variables become seriously dependent on a deterministic as well as random part, which is essentially stochastic in nature. The generation of sequences of monthly values of discharge involves non-stationary, both in mean and standard deviation as the annual cycle is associated with the sun. Thomas-Fiering model allows for this non-stationarity as well as stationarity in the month-to-month correlation structure.

### 2.10.1 Thomas-Fiering (T-F) Model

Russian mathematician A.A. Markov (1856–1922) introduced the concept of a chain process called Markov process for forecasting the flow. The Markov model is a linear auto regressive scheme (stochastic process) in which probability of being a particular state in a given time period is dependent on the actual state in the preceding time period. The first order Markov model is

$$x_t - \mu = \alpha_1(x_{t-1} - \mu) + v_t \tag{2.128}$$

in which $x_t$ is the dependent stationary stochastic series, $\mu$ the mean, $\alpha_1$ the auto regressive coefficient, $v_t$ is the independent stationary stochastic component. The concept of Markov process was utilised by Julian (1961), who related the generated annual flow series in the following way.

$$x_t = r \cdot x_{t-1} + E(y)e^t \tag{2.129}$$

in which $r$ is the first order serial correlation coefficient for the event, say, the runoff and $E(y)$ the random uncorrelated component due to annual rainfall. Brittan (1961) slightly modified the previous model and generated annual flow sequence using the following Markov chain model

$$x_t = rx_{t-1} + (1 - r) x_{av} + s_x(1 - r^2)^{1/2}\varepsilon_p \tag{2.130}$$

in which $x_{av}$ is the mean annual flow computed from historical record, $s_x$ the standard deviation of historical runoff data, $\varepsilon_p$ is the random variate which is assumed to be normally distributed with mean = 0 and standard deviation = 1.

Thomas and Fiering (1962) used the above Markov chain model for generating monthly flows by taking into consideration the serial correlation of monthly flows. The model allows a month-to-month correlation structure. The model is widely used to generate flow series for monthly or seasonal periods, where persistence is present. At a river gauging site we can always assume that the July flows are always dependent on June values. There is a definite dependence between the sequence of flows between the consecutive monthly or say 10-daily periods. Since annual values are never dependent with the previous values, the model is not suitable for generating yearly flow series. The model uses the following recursion equation.

$$\frac{(Q_{j+i} - Q_{av\ j+1})}{S_{j+1}} = r_j \frac{(Q_j - Q_{avj})}{S_j} + t_p \sqrt{1 - r^2} \tag{2.131}$$

or $\qquad Q_{p+j+1} = Q_{av\ j+1} + b_j (Q_{py} - Q_{av} + {}_j) + t_p S_{j+1} \sqrt{1 - r^2} \tag{2.132}$

where $Q_{j+1}$ and $Q_j$ are the discharge volumes during $(i + 1)$th and $j$th months respectively, $Q_{av\ j+1}$ and $Q_{avj}$ the mean monthly discharge volumes during $(j + 1)$th and $j$th months, respectively, $S_{j+1}$ and $S_j$ standard deviations for $(j + 1)$th and $j$th months respectively, $r_j$ the correlation coefficient between the $j$th and $(j + 1)$th months, $t_p$ the random independent variate with zero mean and unit variance, $p$ the year, $j$ the month, i.e., $j = 1$ stands for January and so on. Now

$$b_j = r_j \left( \frac{S_{j+1}}{S_j} \right) \tag{2.133}$$

$$Q_{av,j} = \frac{\Sigma\ Q_{p,j}}{N} \tag{2.134}$$

$$S_j = \left[ \frac{\Sigma\ (Q_{pj} - Q_{avj})^2}{(N - 1)} \right]^{1/2} \tag{2.135}$$

$$r_j = \frac{\Sigma\ Q_{p,j}Q_{p,j+1} - (\Sigma\ Q_{p,j}\ \Sigma\ Q_{p,j+1})/N}{\sqrt{\Sigma\ Q_{p,j}^2 - (\Sigma\ Q_{p,j})^2/N}\ \sqrt{\Sigma\ Q_{p,j+1}^2 - (\Sigma\ Q_{p,j+1})^2/N}} \tag{2.136}$$

where $N$ is the number of years for which the record is available. If the unit of time considered as month then equation (2.132) indicates that the runoff of a given month is equal to the mean monthly flow plus a constant times the runoff of the previous month and a random component. In tropical countries like India, there can be practically no flow during the dry season in a stream. To generate synthetic discharge sequences for this case, the Thomas-Fiering model is used with the following modifications.

(i)   Calculate mean and standard deviation of each monthly flows.
(ii)  Find out the correlation coefficients between all the successive months. Thus, there will be twelve values of mean, standard deviation and correlation coefficients.
(iii) If $N$ is the number of years of data available, then let $n_j$ be the number of years out of $N$ for which the flow is available for the months. To illustrate, if 15 years of data are available then for June, say let there be 12 years for which flow is available. The value of $n_j = n_6 = 12$. Here $j$ is taken as 6 for June. Calculate $p_j = n_j/N$ for all the months.
(iv)  Fit the Thomas-Fiering model to these successive pairs of months. The synthetic sequences of monthly flows are as follows.
(v)   For month $j$, choose a pseudo-random number rectangularly distributed over (0, 1). The pseudo-random numbers generated sequentially are available in statistical hand books. These numbers are continuously matched with the $P_j$ values for all the months. For the month of say July, the value of $p_j$ is always 1, therefore flow is definitely to be generated for this month. For any month if the number is less than $p_j$ but greater than zero, then flow is to occur in the month $j$, otherwise no flow is to occur.
(vi)  If no flow is to occur in the month $j$, then generation of the flow for the month may not be carried out.
(vii) If flow is to occur in the month $j$ and flow also has already occurred in the month $j - 1$, use the regression equation of *T-F* model to obtain the flow for the month $j$ .
(viii) Sometimes negative values are generated while using *T-F* model. Where negative values are encountered, it is recommended that these values should be retained and used to derive the subsequent values in the sequence. Once the generated sequence is completed, all the negative values in the generated sequence should be replaced by zero. Observed flow data is available up to say December 1997 and it is required to generate flow sequence for a period of 10 years beginning with January 2000. A sample calculation of the T-F model for the month of January 2000 can be represented as

$$Q_{Jan} = Q_{av\ Jan} + b_{D\text{-}J}\ (Q_{Dec} - Q_{av\ D}) + t_i S_j\ (l - r_{DJ}^2)^{1/2} \qquad (2.137)$$

When *T-F* model is fitted to monthly stream flows it is observed that some values in the generated sequence becomes negative. Possible reasons for this are (i) wide variation in the flow values in the same month for different years and (ii) short length of the observed sample data which may not be the true representative of the population sample. The negative values in the generated sequence can be overcome if *T-F* model is fitted to the logarithmic transferred flows. The zero values are replaced by 1, wherever it is encountered in the observed flow sequence, so that $\log_{10}(1)$ is zero. The procedure is exactly the same as that of modified *T-F* model except that after generating the flows, antilogarithm of the generated flow is taken.

Sometimes it is observed that a few generated values become too high. Though negative values are not encountered, very high values obtained in the month of say June gives some reservation in the acceptance of the results. To overcome the above, square root transferred observed values is subjected to *T-F* model. Square of the results obtained is the generated flow series.

The model should be applied to generate flow series with caution when data of less than 12 years are available.

---

**Example 2.18:** River flow at a site from January 1989 to December 1994 are given in Table 2.21. Generate the probable flow sequence for the next 10 years using Thomas-Fiering model.

**Solution**
See Tables 2.21-2.24.
For the given data, equation for Thomas-Fiering model for months of December and January are as follows

$$Q_{Dec} = 12.2 + (0.14)(Q_{Nov} - 81.2) + t_i \, 15.31 \, (1 - 0.958^2)^{1/2}$$

$$Q_{Jan} = 0.80 + (-0.05)(Q_{Dec} - 12.2) + t_i \, 1.96 \, (1 - 0.039^2)^{1/2}$$

---

## 2.11   CHI-SQUARE TEST OF GOODNESS OF FIT

Goodness of fit between the observed events and the fitted distribution can be tested by chi-square test. A relation between observed number of occurrences $O_i$ and expected number of occurrences $P_{ei}$ can be developed in the following form

$$\xi_v^2 = \sum_{i=1}^{v} \frac{(Q_i - P_{ei})^2}{P_{ei}} = \sum_{i=1}^{v} \frac{\Delta Z_i^2}{P_{ei}} \qquad (2.138)$$

and is known as chi-square test for goodness of fit. Here $\xi_v^2$ distribution has $v$ degrees of freedom. The value $v$ is equal to $(N - h - 1)$, where $N$ is the total number of sample data, $h$ is the number of parameters used in filling the proposed distribution. We subtract 1, because here the chi-square parameter is also computed. The value of $\xi^2$ for various degrees of freedom and cumulative distribution (percentage level of significance) are given in Table 2.25. In hydrology, 95% level of confidence is considered as the typical value. A null hypothesis is proposed in which we compare the value of $\xi^2$ obtained from equation (2.138) with the limiting value given in Table 2.25. The fitted probability distribution is accepted if the value of $\xi^2$, obtained from the equation is less than its corresponding value read from Table 2.25 for the degree of freedom of $(N - h - 1)$.

**Table 2.21 Computation of Statistics from River Flow Data (unit = $\times 10^6$ m$^3$)**

| Year | J$^+$ | F | M | A | M | J | J | A | S | O | N | D |
|---|---|---|---|---|---|---|---|---|---|---|---|---|
| (1) | (2) | (3) | (4) | (5) | (6) | (7) | (8) | (9) | (10) | (11) | (12) | (13) |
| 1989 | 0 | 0 | 0 | 0 | 0 | 1.6 | 56.9 | 15.8 | 22.1 | 83.6 | 71.3 | 10.8 |
| 1990 | 0 | 0 | 0 | 0 | 75.1 | 9.0 | 117.7 | 283.8 | 282.0 | 137.7 | 1.6 | 0.00 |
| 1991 | 0 | 0 | 0 | 11.5 | 21.8 | 21.7 | 56.6 | 98.1 | 138.4 | 161.7 | 258.1 | 32.2 |
| 1992 | 0 | 0 | 0 | 0 | 267.3 | 138.9 | 13.9 | 55.2 | 278.7 | 97.8 | 6.4 | 0.00 |
| 1993 | 0 | 0 | 0 | 0 | 18.2 | 120.8 | 100.4 | 272.4 | 403.1 | 532.4 | 149.3 | 30.2 |
| 1994 | 0 | 0 | 0 | 1.7 | 0 | 90.7 | 29.6 | 26.5 | 34.5 | 86.8 | 0.6 | 0.00 |
| Mean | 0.8 | 0 | 0 | 2.2 | 63.7 | 63.7 | 62.5 | 125.3 | 210.0 | 183.3 | 81.2 | 12.2 |
| SD | 1.96 | 0 | 0 | 4.62 | 103.5 | 60.5 | 40.0 | 121.8 | 169.2 | 173.8 | 104.4 | 15.3 |
| $r^*$ | −0.39 | 0 | 0 | 0 | −0.239 | 0.514 | −0.394 | 0.89 | −0.89 | 0.62 | 0.43 | 0.96 |
| $b_j$ | −0.05 | 0 | 0 | 0 | −5.359 | 0.3005 | −0.261 | 2.72 | 1.26 | 0.63 | 0.26 | 0.14 |

*Correlation Coefficient, $^+$January, SD = Standard Deviation

Table 2.22    Generation of Yield Series by Thomas-Fiering Model (unit = × 10⁶ m³)
(January 1995 to December 2004)

| Year | J | F | M | A | M | J | J | A | S | O | N | D |
|------|------|------|------|------|------|------|------|------|------|------|------|------|
| (1) | (2) | (3) | (4) | (5) | (6) | (7) | (8) | (9) | (10) | (11) | (12) | (13) |
| 1st | 0.459 | 0 | 0 | 0.303 | 5.272 | 5.03 | 1.734 | 0 | 11.767 | 13.669 | 11.91 | 2.095 |
| 2nd | 0 | 0 | 0 | 0.684 | 0 | 3.318 | 4.067 | 6.373 | 14.321 | 15.032 | 40 | 0 |
| 3rd | 0.053 | 0 | 0 | 0 | 16.595 | 15.118 | 0.286 | 0 | 5.701 | 17.033 | 5.682 | 0.739 |
| 4th | 0 | 0 | 0 | 0 | 10.572 | 17.892 | 5.494 | 20.015 | 29.555 | 5.148 | 7.616 | 0 |
| 5th | 0 | 0 | 0 | 0 | 0 | 0 | 8.955 | 15.123 | 20.631 | 15.949 | 10 | 0 |
| 6th | 0 | 0 | 0 | 0.442 | 0 | 1.18 | 4.646 | 11.321 | 11.603 | 6.386 | 0 | 1.135 |
| 7th | 0 | 0 | 0 | 0 | 0 | 9.139 | 7.593 | 19.982 | 23.724 | 9.926 | 20.108 | 0 |
| 8th | 0 | 0 | 0 | 0.098 | 0 | 7.835 | 6.287 | 14.357 | 24.524 | 8.243 | 15.905 | 0 |
| 9th | 0 | 0 | 0 | 0 | 0 | 3.298 | 6.238 | 8.948 | 9.931 | 0 | 5.33 | 0 |
| 10th | 0 | 0 | 0 | 0 | 0 | 7.18 | 4.632 | 5.486 | 16.065 | 0 | 23.108 | 0 |
| Mean | 0.051 | 0 | 0 | 0.153 | 3.244 | 7.00 | 4.99 | 10.16 | 16.78 | 9.14 | 13.97 | 0.397 |
| S.D. | 0.144 | 0 | 0 | 0.243 | 5.866 | 5.81 | 2.578 | 7.313 | 7.56 | 6.288 | 11.553 | 0.719 |

**Table 2.23  Generation of Yield Series by Thomas-Fiering Model Applied to Logarithmic Values (unit = $\times 10^6\,\mathrm{m}^3$)**

| Year | J | F | M | A | M | J | J | A | S | O | N | D |
|---|---|---|---|---|---|---|---|---|---|---|---|---|
| (1) | (2) | (3) | (4) | (5) | (6) | (7) | (8) | (9) | (10) | (11) | (12) | (13) |
| 1st | 0.565 | 0 | 0 | 0.19 | 1.316 | 2.127 | 1.894 | 1.44 | 6.023 | 10.083 | 7.37 | 1.391 |
| 2nd | 0 | 0 | 0 | 0.474 | 0 | 0.913 | 3.383 | 4.366 | 8.7 | 38.14 | 18.55 | 0 |
| 3rd | 0.178 | 0 | 0 | 0 | 17.8 | 22.488 | 1.354 | 3.404 | 6.943 | 9.428 | 24.643 | 0 |
| 4th | 0 | 0 | 0 | 0 | 3.691 | 176.428 | 4.064 | 35.326 | 56.358 | 10.726 | 1.154 | 0.33 |
| 5th | 0 | 0 | 0 | 0 | 0 | 0.17 | 10.575 | 5.283 | 7.325 | 7.59 | 9.119 | 0 |
| 6th | 0 | 0 | 0 | 0.268 | 0 | 0.419 | 3.959 | 9.759 | 9.418 | 4.317 | 0.533 | 0.274 |
| 7th | 0 | 0 | 0 | 0 | 0 | 7.672 | 7.018 | 19.1 | 19.958 | 15.623 | 7.02 | 0 |
| 8th | 0 | 0 | 0 | 0.115 | 0.161 | 7.002 | 4.94 | 9.137 | 18.144 | 12.187 | 1.89 | 0 |
| 9th | 0 | 0 | 0 | 0 | 0 | 0.908 | 5.498 | 3.606 | 3.99 | 6.772 | 0.115 | 0 |
| 10th | 0 | 0 | 0 | 0 | 0 | 3.754 | 3.692 | 2.953 | 7.481 | 15.88 | 0.05 | 0 |
| Mean | 0.074 | 0 | 0 | 0.105 | 0.297 | 25.181 | 4.638 | 9.441 | 14.434 | 13.074 | 7.045 | 0.199 |
| SD | 0.181 | 0 | 0 | 0.162 | 5.571 | 55.44 | 2.652 | 10.452 | 15.616 | 9.532 | 8.481 | 0.437 |

Table 2.24   Yield Series by Thomas-Fiering Model Applied to Square root Transferred Values (unit= $\times 10^6$ m$^3$)

| Year | J | F | M | A | M | J | J | A | S | O | N | D |
|------|------|------|------|-------|--------|--------|--------|--------|--------|--------|--------|-------|
| (1) | (2) | (3) | (4) | (5) | (6) | (7) | (8) | (9) | (10) | (11) | (12) | (13) |
| 1st | 0.517 | 0 | 0 | 0.123 | 2.014 | 3.389 | 2.003 | 0.71 | 10.492 | 11.997 | 11.289 | 1.755 |
| 2nd | 0 | 0 | 0 | 0.523 | 0 | 1.973 | 3.823 | 5.462 | 12.141 | 36.679 | 15.946 | 0 |
| 3rd | 0.081 | 0 | 0 | 0 | 15.835 | 19.058 | 0.992 | 3.054 | 8.377 | 9.105 | 16.592 | 0 |
| 4th | 0 | 0 | 0 | 0 | 8.739 | 28.187 | 4.234 | 21.851 | 32.349 | 8.988 | 2.598 | 0.266 |
| 5th | 0 | 0 | 0 | 0 | 0 | 0.027 | 9.653 | 10.046 | 13.57 | 8.198 | 12.476 | 0 |
| 6th | 0 | 0 | 0 | 0.241 | 0 | 0.688 | 4.490 | 11.232 | 10.422 | 1.344 | 1.354 | 0.168 |
| 7th | 0 | 0 | 0 | 0 | 0 | 8.834 | 7.012 | 18.844 | 20.018 | 16.751 | 9.755 | 0 |
| 8th | 0 | 0 | 0 | 0.02 | 0.322 | 8.59 | 5.263 | 10.88 | 20.615 | 13.198 | 4.902 | 0 |
| 9th | 0 | 0 | 0 | 0 | 0 | 1.959 | 5.925 | 5.6 | 5.688 | 6.038 | 0.05 | 0 |
| 10th | 0 | 0 | 0 | 0 | 0 | 5.988 | 4.081 | 3.635 | 11.953 | 19.51 | 0.122 | 0 |
| Mean | 0.06 | 0 | 0 | 0.092 | 2.691 | 7.868 | 4.748 | 9.131 | 14.563 | 13.479 | 7.506 | 0.219 |
| SD | 0.163 | 0 | 0 | 0.174 | 5.363 | 9.103 | 2.452 | 6.899 | 7.797 | 10.571 | 6.467 | 0.548 |

**Table 2.25  Chi-Square Distribution Percentages**

| Degree of freedom (y) | 0.005 | 0.010 | 0.025 | 0.05 | 0.10 | 0.20 | 0.30 | 0.40 | 0.50 | 0.60 | 0.70 | 0.80 | 0.90 | 0.95 | 0.975 | 0.990 | 0.995 | Y |
|---|---|---|---|---|---|---|---|---|---|---|---|---|---|---|---|---|---|---|
| 1 | 0.04393 | 0.3175 | 0.02982 | 0.02393 | 0.0158 | 0.642 | 0.148 | 0.275 | 0.455 | 0.708 | 1.07 | 1.64 | 2.71 | 3.84 | 5.02 | 6.63 | 7.88 | 1 |
| 2 | 0.0100 | 0.0201 | 0.0506 | 0.103 | 0.211 | 0.446 | 0.713 | 1.02 | 1.39 | 1.83 | 2.41 | 3.22 | 4.61 | 5.99 | 7.38 | 9.21 | 10.6 | 2 |
| 3 | 0.0717 | 0.115 | 0.216 | 0.352 | 0.584 | 1.00 | 1.42 | 1.87 | 2.37 | 2.95 | 3.67 | 4.64 | 6.25 | 7.81 | 9.35 | 11.3 | 12.8 | 3 |
| 4 | 0.207 | 0.297 | 0.484 | 0.711 | 1.06 | 1.65 | 2.19 | 2.75 | 3.36 | 4.04 | 4.88 | 5.99 | 7.78 | 9.49 | 11.1 | 13.3 | 14.9 | 4 |
| 5 | 0.412 | 0.554 | 0.831 | 1.15 | 1.61 | 2.34 | 3.00 | 3.66 | 4.35 | 5.13 | 6.06 | 7.29 | 9.24 | 11.1 | 12.8 | 15.1 | 16.7 | 5 |
| 6 | 0.676 | 0.872 | 1.24 | 1.64 | 2.20 | 3.07 | 3.83 | 4.57 | 5.35 | 6.21 | 7.23 | 8.56 | 10.6 | 12.6 | 14.4 | 16.8 | 18.8 | 6 |
| 7 | 0.989 | 1.24 | 1.69 | 2.17 | 2.83 | 3.82 | 4.67 | 5.49 | 6.35 | 7.28 | 8.38 | 9.80 | 12.0 | 14.1 | 16.0 | 18.5 | 20.3 | 7 |
| 8 | 1.34 | 1.65 | 2.18 | 2.73 | 3.49 | 4.59 | 5.53 | 6.42 | 7.34 | 8.35 | 9.52 | 11.0 | 13.4 | 15.5 | 17.5 | 20.1 | 22.0 | 8 |
| 9 | 1.73 | 2.09 | 2.70 | 3.33 | 4.17 | 5.38 | 6.39 | 7.36 | 8.34 | 9.41 | 10.7 | 12.0 | 14.7 | 16.9 | 19.0 | 21.7 | 23.6 | 9 |
| 10 | 2.16 | 2.56 | 3.25 | 3.94 | 4.87 | 6.18 | 7.27 | 8.30 | 9.34 | 10.5 | 11.8 | 13.4 | 16.0 | 18.3 | 20.5 | 23.2 | 25.2 | 10 |
| 11 | 2.60 | 3.05 | 3.82 | 4.57 | 5.58 | 6.99 | 8.15 | 9.24 | 10.3 | 11.5 | 12.9 | 14.6 | 17.3 | 19.7 | 21.9 | 24.7 | 26.8 | 11 |
| 12 | 3.07 | 3.57 | 4.40 | 5.23 | 6.30 | 7.18 | 9.03 | 10.2 | 11.3 | 12.6 | 14.0 | 15.8 | 18.5 | 21.0 | 23.3 | 26.2 | 28.3 | 12 |
| 13 | 3.57 | 4.11 | 5.01 | 5.89 | 7.04 | 8.63 | 9.93 | 11.1 | 12.3 | 13.6 | 15.1 | 17.0 | 19.8 | 22.4 | 24.7 | 27.7 | 29.8 | 13 |
| 14 | 4.07 | 4.66 | 5.36 | 6.57 | 7.79 | 9.47 | 10.8 | 12.1 | 13.3 | 14.7 | 16.2 | 18.2 | 21.1 | 23.7 | 26.1 | 29.1 | 31.3 | 14 |
| 15 | 4.60 | 5.23 | 6.26 | 7.26 | 8.55 | 10.3 | 11.7 | 13.0 | 14.3 | 15.7 | 17.3 | 19.3 | 22.3 | 25.0 | 27.5 | 30.6 | 32.8 | 15 |
| 16 | 5.14 | 5.81 | 6.91 | 7.96 | 9.31. | 11.2 | 12.6 | 14.0 | 15.3 | 16.8 | 18.4 | 20.5 | 23.5 | 26.3 | 28.8 | 32.0 | 34.3 | 16 |
| 17 | 5.70 | 6.41 | 7.56 | 8.67 | 10.1 | 12.0 | 13.5 | 14.9 | 16.3 | 17.8 | 19.5 | 21.6 | 24.8 | 27.6 | 30.2 | 33.4 | 35.7 | 17 |
| 18 | 6.26 | 7.01 | 8.23 | 9.39 | 10.9 | 12.9 | 14.4 | 15.9 | 17.3 | 18.9 | 20.6 | 22.8 | 26.0 | 28.9 | 31.5 | 34.8 | 37.2 | 18 |
| 19 | 6.84 | 7.63 | 8.91 | 10.1 | 11.7 | 13.7 | 15.4 | 16.9 | 18.3 | 19.9 | 21.7 | 23.9 | 27.2 | 30.1 | 32.9 | 36.2 | 38.6 | 19 |
| 20 | 7.43 | 8.26 | 9.59 | 10.9 | 12.4 | 14.6 | 16.3 | 17.8 | 19.3 | 21.0 | 22.8 | 25.0 | 28.4 | 31.4 | 32.2 | 37.6 | 40.0 | 20 |
| 21 | 8.03 | 8.90 | 10.3 | 11.6 | 13.2 | 15.4 | 17.2 | 18.8 | 20.3 | 22.0 | 23.9 | 26.2 | 29.6 | 32.7 | 35.5 | 38.9 | 41.1 | 21 |
| 22 | 8.64 | 9.54 | 11.0 | 12.3 | 14.0 | 16.3 | 18.1 | 19.7 | 21.3 | 23.0 | 24.9 | 27.3 | 30.8 | 33.9 | 36.8 | 40.3 | 42.8 | 22 |
| 23 | 9.26 | 10.2 | 11.7 | 13.1 | 14.8 | 17.2 | 19.0 | 20.7 | 22.3 | 24.1 | 26.0 | 28.4 | 32.0 | 35.2 | 38.1 | 41.6 | 44.2 | 23 |
| 24 | 9.89 | 10.9 | 12.4 | 13.8 | 15.7 | 18.1 | 19.9 | 21.7 | 23.3 | 25.1 | 27.1 | 29.6 | 33.2 | 36.4 | 39.4 | 43.0 | 45.6 | 24 |
| 25 | 10.5 | 11.5 | 13.4 | 14.6 | 16.5 | 18.9 | 20.9 | 22.6 | 24.3 | 26.1 | 28.2 | 30.7 | 34.4 | 37.7 | 40.6 | 44.3 | 46.9 | 25 |
| 26 | 11.2 | 12.2 | 13.8 | 15.4 | 17.3 | 19.8 | 21.8 | 23.6 | 25.3 | 27.2 | 29.2 | 31.8 | 35.6 | 38.9 | 41.9 | 45.6 | 48.3 | 26 |
| 27 | 11.8 | 12.9 | 14.6 | 16.2 | 18.1 | 20.7 | 22.7 | 24.5 | 26.3 | 28.2 | 30.3 | 32.9 | 36.7 | 40.1 | 43.2 | 47.0 | 49.6 | 27 |

*(Contd)*

| Degree of free-dom (y) | 0.005 | 0.010 | 0.025 | 0.05 | 0.10 | 0.20 | 0.30 | 0.40 | 0.50 | 0.60 | 0.70 | 0.80 | 0.90 | 0.95 | 0.975 | 0.990 | 0.995 | Y |
|---|---|---|---|---|---|---|---|---|---|---|---|---|---|---|---|---|---|---|
| 28 | 12.5 | 13.6 | 15.3 | 16.9 | 18.9 | 21.6 | 23.6 | 25.5 | 27.3 | 29.2 | 31.4 | 34.0 | 37.9 | 41.3 | 44.5 | 48.3 | 51.0 | 28 |
| 29 | 13.1 | 14.3 | 10.6 | 17.7 | 19.8 | 22.5 | 24.6 | 26.5 | 28.3 | 30.3 | 32.5 | 35.0 | 39.1 | 42.6 | 45.7 | 49.6 | 52.3 | 29 |
| 30 | 13.8 | 15.0 | 16.8 | 18.5 | 20.6 | 23.4 | 25.5 | 27.4 | 29.3 | 31.3 | 33.5 | 36.3 | 40.3 | 43.8 | 47.0 | 50.9 | 53.7 | 30 |
| 35 | 17.2 | 18.5 | 20.6 | 22.5 | 24.8 | 27.8 | 30.2 | 32.3 | 34.3 | 36.5 | 38.9 | 41.8 | 46.1 | 49.8 | 53.2 | 57.3 | 60.3 | 35 |
| 40 | 20.7 | 22.2 | 24.4 | 26.5 | 29.1 | 32.3 | 34.9 | 37.1 | 39.3 | 41.6 | 44.2 | 47.3 | 51.8 | 55.8 | 59.3 | 63.7 | 66.8 | 40 |
| 45 | 24.3 | 25.9 | 28.4 | 30.6 | 33.4 | 36.9 | 39.6 | 42.0 | 44.3 | 46.8 | 49.5 | 52.7 | 57.5 | 61.7 | 65.4 | 70.0 | 73.2 | 45 |
| 50 | 28.0 | 29.7 | 32.4 | 34.8 | 37.7 | 41.4 | 44.3 | 46.9 | 49.3 | 51.9 | 54.7 | 58.2 | 63.2 | 67.5 | 71.4 | 76.2 | 79.5 | 50 |
| 100 | 67.3 | 70.1 | 74.9 | 77.9 | 82.4 | 87.9 | 92.1 | 95.8 | 99.3 | 102.9 | 106.9 | 11.7 | 118.5 | 124.3 | 129.6 | 135.6 | 140.2 | 100 |

**Example 2.19:** For the data of example 2.3, use chi-square test to determine whether the Normal distribution fits the sample adequately.

**Solution**

All the pertinent calculation for the example are carried out in Table 2.26.

**Table 2.26   Chi-Square Test of Goodness of Fit**

| Sl. No. | Rainfall class interval | Frequency $(f_i)$ | Relative frequency $(f_i/n) =$ col.(3)/55 | Cumulative relative frequency $\Sigma$ col. (4) | $Z_i$ of normal distribution | Cumulative probability of variate $z_i$ $F(x_i)$ | Incremental probability $p(x_i)$ | chi-square $\xi^2$ |
|---|---|---|---|---|---|---|---|---|
| (1) | (2) | (3) | (4) | (5) | (6) | (7) | (8) | (9) |
| 1 | 41–50 | 5 | 0.091 | 0.091 | −1.363 | 0.087 | 0.087 | 0.010 |
| 2 | 51–60 | 8 | 0.145 | 0.236 | −0.802 | 0.212 | 0.125 | 0.176 |
| 3 | 61–70 | 10 | 0.182 | 0.418 | −0.239 | 0.405 | 0.193 | 0.034 |
| 4 | 71–80 | 11 | 0.200 | 0.618 | 0.329 | 0.629 | 0.224 | 0.141 |
| 5 | 81–90 | 12 | 0.218 | 0.836 | 0.883 | 0.811 | 0.182 | 0.392 |
| 6 | 90–100 | 4 | 0.073 | 0.909 | 1.445 | 0.926 | 0.115 | 0.843 |
| 7 | 101–110 | 3 | 0.055 | 0.964 | 2.007 | 0.948 | 0.022 | 2.722 |
| 8 | 111–120 | 2 | 0.036 | 1.000 | 2.570 | 0.995 | 0.047 | 0.141 |
| Total | | 55 | 1.000 | | | | | 4.459 |

Relative frequency of col. (4) is calculated by dividing the number of occurrences of the event of col. (3) in the class interval by the total number of events occurred in the sample, i.e., 55.

Mean of the sample is 74.27.

Variance = $\{5(45 - 74.27)^2 + 8(55 - 74.27)^2 + ... + 2(115 - 74.27)^2\}/55$

= $\{4284 + 2971 + 859 + 6 + 1382 + 1719 + 2833 + 2318\}/55 = 315.85$

Standard Deviation = $(315.85)^{0.5} = 17.78$

$z_i$ of col. (6) is calculated as (for row 4) $= \dfrac{(x - \mu)}{\sigma} = \dfrac{(80 - 74.27)}{17.78} = 0.329$

For col. (7) table for the area under standard normal curve is used.

For $z_i = 0.329$, $F(x_i) = 0.50 + 0.129$ (from table 2.6) = 0.629

For $z_i = -1.363$, $F(x_i) = 0.50 - 0.413$ (from table 2.6) = 0.087 and so on

For checking the goodness of fit, the chi-square test statistics is calculated in col. (9). For example, 0.141 in row (4) is

$$\xi^2 = \frac{\{55(0.20 - 0.224)^2\}}{0.224} = 0.141$$

From the table of $\xi^2$ distribution for degrees of freedom of $(8 - 2 - 1) = 5$, $\xi^2_{0.95} = 11.1$. Since 4.459 < 11.1, the hypothesis cannot be rejected at 95% confidence level, that is, fitting of the normal distribution to the data is accepted.

## CONCLUSIONS

All the generating techniques discussed above help a hydrologist to create records longer than the historical. An acceptable method should be the one in which the historical data and generated data are significantly not different under any statistical test.

We use the previous models to (i) generate synthetically the flow series, (ii) for flood/drought forecasting, (iii) filling the missing data, (iv) evaluation of a system.

Knowledge on statistics and probabilities of hydrologic data is a must for all hydrologists and engineers dealing with water resources potential. This is the only tool to overcome the difficulties like (i) errors in data, (ii) gaps in data records, (iii) paucity of information on extreme cases of flood and drought. By sequential generation it is possible to generate as many combinations of hydrologic sequences as desired for evaluation of a system by changing the parameters.

Discussion in this chapter is limited to those statistical and probability analysis which are exclusively used for hydrological analysis. However, for more detailed information on the subjects, standard test books on statistics and probabilities may be referred.

### PROBLEMS

2.1   Explain how the method of moments is used to describe the characteristics of hydrologic data? How should we calculate the characteristics of the sample?

2.2   Calculate the mean, standard deviation, coefficient of variation, skewness coefficient and kurtosis from the following river discharge data recorded in $m^3$/sec.

56, 58, 84, 92, 112, 126, 131, 160, 170, 186, 192, 195, 197, 231, 250, 257, 263, 273, 276, 281, 290, 296, 300, 320, 350, 264, 380, 420, 425, 435, 455, 458, 490, 560, 570, 576, 665, 680, 700.

2.3   Plot the relative frequency curve of the data of problem 2.2.

2.4.   (a)   Sediment yield ($S$) in ton and catchment runoff depth (D) in cm are related in the form of an equation $S = aD^b$. Obtain the best fit curve from the data given below

| Runoff (D) | 1.6 | 5.2 | 0.1 | 2.5 | 1.4 | 3.0 | 2.5 | 1.0 |
|---|---|---|---|---|---|---|---|---|
| | 0.4 | 2.6 | 4.6 | 2.3 | 1.5 | | | |
| Sediment (S) | 460 | 950 | 6 | 720 | 210 | 505 | 325 | 35 |
| | 70 | 430 | 475 | 760 | 360 | | | |

Also find out the degree of association between the variables.

2.4   (b)   What will be the sediment yield for runoff of 5, 4, 3 and 2 cms? How much variance is explained by fitting the curve?

2.5   Collect the annual peak discharge for the last 20 years of the nearest river site of your locality. Calculate all the statistical parameters. Find the flood of 200 and 500 year return periods.

2.6   Discuss the types of correlations used in hydrology. Give examples for each.

2.7   From the following data of annual runoff depths in cm over a catchment, find if there is any trend in the data. Remove the trend by moving average method.

36, 43, 44, 40, 35, 39, 41, 47, 45, 39, 52, 48.

2.8    Fourteen years of weighted rainfall and corresponding runoff ($y$) for the month of July are given below. Develop a best fit linear or a nonlinear relation between them. In linear form, is the relation $y = a + bx$ and $x = a + by$ the same? Calculate the parameters $a$, $b$, $r$ and standard error using the two forms of equation.

| Weighted rainfall (cm) | 12.5 | 11.1 | 5.9 | 18.2 | 25.3 | 17.0 | 7.7 | 4.9 | 8.8 |
|---|---|---|---|---|---|---|---|---|---|
| | 9.6 | 10.3 | 16.2 | 5.8 | 2.8 | | | | |
| Run off (cm) | 10.8 | 4.1 | 1.3 | 14.4 | 10.1 | 2.7 | 3.6 | 1.2 | 2.4 |
| | 3.6 | 3.3 | 12.4 | 3.8 | 1.8 | | | | |

2.9    What is the probability that a 50 year flood will occur in the next 15 years?

2.10   In normal distribution, calculate area of the curve for standard variate between – 2.0 to +2.0.

2.11   What is the probability that a 5 year flood will not occur at all in 20 years and it will occurrence in 25 years?

2.12   From the relationship between risk ($R$) and return period ($T$), plot a curve for $n = 10$, 50 and 100. Can you use this curve?

2.13   Discuss the scope and limitations of frequency studies.

2.14   Outline the procedure to carry out frequency analysis of floods or storms of return periods greater than the sample length.

2.15   Find out the return period of a 400 m$^3$/sec design flood for a structure if 20 years data at the site gives the mean and standard deviation as 114 and 48 m$^3$/sec respectively.

2.16   Estimate the peak flood for return periods of 100, 200 and 500 years from the following observed data.

| Year(19-) | 76 | 77 | 78 | 79 | 80 | 81 | 82 | 83 | 84 | 85 |
|---|---|---|---|---|---|---|---|---|---|---|
| | 86 | 87 | 88 | 89 | 90 | | | | | |
| Q(m$^3$/sec) | 180 | 341 | 252 | 233 | 312 | 415 | 188 | 177 | 252 | 333 |
| | 126 | 402 | 322 | 191 | 286 | | | | | |

Use (i) Gumbel and (ii) Log-Normal distribution.

2.17   Using the data of problem 2.16, estimate the 200 year return period flood using Log-Pearson and Normal distribution. Compare the results with the Gumbel and Log-Normal distribution and comment.

2.18   For design of a structure, a 100 year flood is considered. If the life of the structure is taken as five years, what risk the designer is accepting for the work.

2.19   Rainfall, runoff and evaporation observed during nine storms over a basin are given below. Find out the regression relation of the form $y = a + bx_1 + cx_2$ and $y = ax_1^b x_2^c$. Find the multiple correlation coefficient

| Storm date | 24/4 | 2–3/7 | 14/7 | 28–29/7 | 4/8 | 10/8 | 21–23/8 | 29/8 | 1–2/9 |
|---|---|---|---|---|---|---|---|---|---|
| Rainfall (cm) | 7.8 | 11.2 | 9.1 | 13.3 | 3.6 | 4.2 | 9.6 | 5.1 | 2.6 |
| Runoff (cm) | 2.1 | 5.3 | 6.4 | 9.6 | 0.6 | 0.9 | 2.6 | 2.4 | 1.5 |
| Evaporation (cm) | 0.5 | 1.2 | 0.2 | 0.8 | 0.6 | 0.5 | 1.1 | 0.5 | 0.4 |

2.20   The following informations are recorded from a storm.

| Time (min) | 0–10 | 10–20 | 20–30 | 30–40 | 40–50 | 50–60 |
|---|---|---|---|---|---|---|
| Rainfall intensity (cm/h) | 3 | 6 | 4 | 7 | 3 | 1 |

Calculate the first three moments of the data about the origin.

2.21   Compute the first two moments about the centroid of the rainfall histogram of problem 2.20.

2.22   A linear relation of the form $y = a + bx$ is to be fitted to the sample of $n$ values. Determine theoretically the parameters $a$ and $b$ by the method of least square.

2.23   Let an equation of the form $y = a + bx + cx^2$ be fitted to the data of $n$ values of variable $x$ and $y$. How can you calculate the parameters $a$, $b$ and $c$ theoretically by the method of least square?

2.24   Annual rainfall for two stations Berhampur and Gopalpur in Orissa from 1965 to 1992 are given below. Calculate the explained and unexplained variance and the standard error of estimate for the data by fitting a linear relation between them.

Annual rainfall        149  157  164  154  143  148  142  161  138  139  155  152  156
at Berhampur (cm) 140  138  142  149  152  133  154  164  125  145  144  167  139
                               133  155
Annual rainfall        161  170  164  169  149  160  152  173  154  165  163  158  151
at Gopalpur  (cm)  146  162  172  165  144  173  180  139  160  155  173  144  134
                               160  169

2.25   Find out the area under standard normal curve for (a) $z$ between 0 and 1.5, (b) $z$ between $-2.0$ and $+1.0$.

2.26   What is the probability that a 4 year flood will occur 4 times in a period of 12 years? What is its probability that it will not occur at all in 12 years? During 12 years how many floods of this magnitude may occur?

2.27   Calculate mean, mode and median of the following data of September 1998.

| Temp °C | 38 | 29 | 31 | 36 | 33 | 26 | 25 |
|---------|----|----|----|----|----|----|----|
| No. of days | 3 | 5 | 9 | 2 | 6 | 2 | 3 |

2.28   The following discharge data are observed at a gauging site

| Discharge (cumecs) | No. of occurrences |
|--------------------|--------------------|
| 500-2000 | 18 |
| 2000-3500 | 12 |
| 3500-5000 | 10 |
| 5000-6500 | 5 |
| 6500-8000 | 3 |
| 8000-9500 | 0 |
| 9500-11000 | 2 |
| 11000-12500 | 2 |

Using the method of moments, fit the normal distribution to the above data. Use Chi square test to determine whether the normal distribution adequately fits the data.

2.29   Following data of annual precipitation in cms from 1936 to 1995 are recorded at a station.

48.7   44.1   42.8   48.4   34.2   32.4   46.4   38.9   37.3   50.6   44.8   34.0
45.6   37.3   43.7   41.8   41.1   31.2   35.2   35.1   49.3   44.2   41.7   30.8
53.6   34.5   50.3   43.8   21.6   47.1   31.2   27.0   37.0   46.8   26.9   25.4
23.0   56.5   43.4   41.3   46.0   44.3   37.8   29.6   35.1   49.7   36.6   32.5
61.7   47.4   33.9   31.7   31.5   59.6   50.5   38.6   43.4   28.7   32.0   51.8

Calculate the mean, standard deviation and coefficient of skewness of the data by grouping them at the interval of 5.0. Find the same statistical parameters without grouping them and compare.

2.30   Annual values of ground water table fluctuations, rainfall and depth of water withdrawn from a watershed draining an area of 100 sq. km are given below.

Obtain a regression relation of the form $y = a + bx_1 + cx_2$ and $y = ax_1^b x_2^c$. Find the multiple correlation coefficient

| Storm date | 1988 | 1989 | 1990 | 1991 | 1992 | 1993 | 1994 | 1995 | 1996 |
|---|---|---|---|---|---|---|---|---|---|
| Rainfall (cm) | 140 | 85 | 103 | 119 | 98 | 150 | 124 | 99 | 68 |
| Depth of water pumped out (cm) | 11 | 13 | 9 | 12 | 8 | 9 | 13 | 12 | 14 |
| Water table Fluctuations (cm) | 50 | 32 | 47 | 50 | 60 | 90 | 57 | 52 | 24 |

2.31   A linear relation of the form $y = a - bx$ are to be developed for a set of data. Determine the parameters $a$ and $b$ by the method of least square.

2.32   Annual runoff at a gauging site is assumed to follow normal disdtribution. From the records of 40 years it is found that themean discharge is 2220 m³/sec and standard deviation as 540 m³/sec. What is the probability that in an year the annual runoff is greater than 2555 m³/sec? If in a year, flood between 1500 and 2500 m³/sec are expected, find the probability of its occurrence.

# Chapter Three

# Precipitation

## 3.1 INTRODUCTION

Precipitation is that part of atmospheric moisture, which reaches the earth's surface in different forms. Hydrologists start working when the precipitation reaches the ground. This connects hydrology with meteorology. There is a great variation of precipitation in space and time. In India, atmospheric moisture (or weather system) causes good precipitation during June-October and nearly dry weather during the remaining periods. Some of the precipitation that might get intercepted while reaching the ground by trees and buildings and evaporates back is called the *initial loss*. The other part meets requirements like *depression storage* and *infiltrates* into the ground. The excess rainfall flows in streams to large water bodies. Factors like soil-type, vegetation, geology and topography of the area largely determine the quantity of rainfall excess available as stream flow from the precipitable water.

Nearly one-fourth of the total precipitation that falls on land reaches large water bodies as direct runoff. The balance three-fourths of water returns back to the atmosphere at different times as evaporation.

The essential requirements for precipitation to occur are: (i) moisture in the atmosphere, (ii) presence of nuclei around which condensation of vapour takes place, (iii) dynamic cooling responsible for condensation of water vapour and (iv) precipitation product must reach the ground in some form. In the atmosphere, condensation nuclei are available in plenty. Dust particles and smoke wastes form excellent nuclei. When such a nucleus collects freezing moisture from the cloud and grow to a size greater than 1 mm, it forms a rain drop and falls. Years back, *cloud seeding* was attempted for making artificial rain in India. Cloud seeding should be attempted only when there is sufficient moisture in the atmosphere and dynamic cooling exists. In the process, artificial nuclei like portland cement, salt powder, carbon-dioxide powder, ice-powder, clay or silver iodide powder is introduced into the cloud by aircraft, balloon, rocket, projectile and other devices. Due to rapid industrialisation, cloud seeding is no more attempted. Owing to the great variability of meteorological processes, cloud seeding have so far provided inconsistent results.

Location of a region with respect to general circulation, latitude, topography, orographic features and distance from moisture source are the responsible factors for rainfall to occur. Orographic barriers force a precipitating system to rain on its windward side and the leeward side receives less precipitation. Such precipitation distributions are noticed in western ghat mountain ranges and at the south of

Himalayan ranges of India. West of western ghats and the north-eastern states of India usually receive heavy precipitation due to the orographic features. Precipitation decreases with altitude and, therefore, we can say that higher is the altitude lesser is the precipitation. When the prevailing wind moves over an ocean containing warm water, it picks up high moisture. This gives very high precipitation. During low pressure time, India and the adjoining areas experience such type of heavy precipitations. While other conditions remain same, a place far away from any coast receives less precipitation than one closer to sea. Location of a place with respect to the prevailing wind is very effective in weather systems. Formation of a desert in western part of India and the floods occurring frequently in north-eastern part are part of this system. Presence of orographic features and oceans adds to the advantage.

Here precipitation measurement, network design, analysis and presentation of precipitation data are discussed to the extent a partitioning civil engineer dealing with water resource projects ought to know.

## 3.2   FORMS OF PRECIPITATION

Any product of atmospheric water must reach the surface of earth after condensation. However, fog and frost are not part of precipitation as they are not falling moisture. Some common forms of precipitation explained below are: (i) rain, (ii) snow, (iii) drizzle, (iv) glaze, (v) sleet, (vi) hail and (vii) dew.

**(i) Rain:** When precipitation reaches the surface of earth in the form of droplets of water, we call it rain. The size of drops vary from 0.5 mm to 6 mm as drops larger than this size are found to breakup during their fall in the air. Rain is considered as light if intensity of rainfall is up to 2.5 mm/h, moderate from 2.5 to 7.5 mm/h and heavy over 7.5 mm/h.

**(ii) Snow:** It is precipitation in the form of ice-crystals, normally hexagonal in shape. Snow reaches the earth's surface either separately or combines together to form flakes. The density of snow is usually 0.10 g/cm$^3$, which means that 10 cm of snowfall is equivalent to 1.0 cm of rainfall.

**(iii) Drizzle:** Drizzle is defined as water droplets of size less than 0.5 mm. It reaches the ground with intensity less than 1.00 mm/h. These water droplets are so light that they appear to be floating in air.

**(iv) Glaze:** It is the drizzle, which freezes immediately in contact with cold objects of the earth's surface.

**(v) Sleet:** Where rain falls through air of subfreezing temperature, the drops freeze to form grains of ice, called sleet. Sometimes snow and rain precipitates simultaneously. The rain drops under this circumstance are half frozen.

**(vi) Hail:** It is the precipitating rain in the form of any irregular form of ice with size ranging from 5.0 mm to 50 mm or above. Cummulonimbus convective clouds with strong vertical currents are responsible for the formation of hail. The density of hails are normally 0.8 gm/cm$^3$. While falling they combine together to form bigger sizes.

**(vii) Dew:** During nights when surface of the objects on earth cools by radiation, the moisture present in atmosphere condenses on the surface of these objects forming water droplets called dew.

### 3.3   TYPES OF PRECIPITATION
Moisture is always present in the atmosphere and there is no shortage of condensation nuclei in the present days due to rapid industrialisation. Adiabatic cooling of the moist air through lifting is the main cause of condensation. Precipitation, normally classified according to the factors responsible for lifting the air mass, is of following types: (i) convective, (ii) orographic, (iii) cyclonic and (iv) thunder storms.

*(i) Convective*
Unequal heating at the surface of earth is the main reason for this type of precipitation. In summer days air in contact with surface of the earth gets heated up, expands and rises due to lesser density. Surrounding air rushes to replace it and in turn gets heated up and rises. In the process, increasing quantities of water vapour are taken up by the air. When the air mass reaches to a great height, temperature falls below the dew point. At this stage condensation releases latent heat of 539 cal/g of water. This heat is added to the air at the height of condensation causing further heating to its upper air. This forces the air mass to move up. As more moist air from the surrounding joins the system, more energy is added. The vertical air currents develop tremendous velocities and dynamic cooling takes place. Depending on the moisture content, cooling and other factors, the precipitation intensity varies from light showers to cloud bursts amounting to 300 mm or more rainfall per hour. Such type of precipitation covers limited aerial extent and is normally found near equatorial zone. Sometimes strong upward wind currents exceeding 150 kmph freezes the rain drops to form hail. Multiple rise and fall of hail due to very strong upward currents may add to its size. Hail storms in India during summer are very common.

*(ii) Orographic*
Orographic or mountain-range barriers cause lifting of the air masses. Dynamic cooling takes place causing precipitation on the side of the blowing wind as shown in Fig. 3.1. Precipitation is normally heavier on the windward side and lighter on leeward side. In India, heavy precipitation in Himalayan region and at the western coast are mainly due to orographic features associated with the south west wind carrying sufficient quantity of moisture, while passing over Arabian sea. Orographic precipitation gives medium to high intensity rainfall and continues for longer duration.

*(iii) Cyclonic*
A cyclone is a low pressure area surrounded by a larger high pressure area.

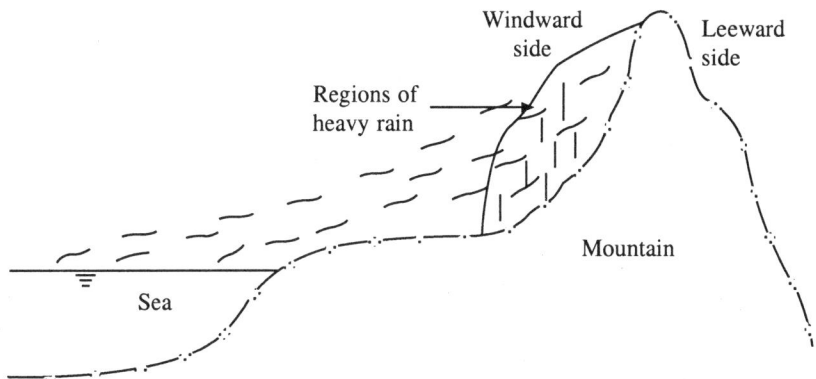

**Fig. 3.1   Precipitation due to Orographic Features**

When low pressure occurs in an area, especially over large water bodies, air from the surrounding rushes, causing the air at the low pressure zone to lift. Such a type of cyclone is called *Tropical cyclone* or simply cyclone in India, *Typhoon* in south-east Asia and *Hurricane* in America. The cyclone centre is called eye and is a calm area. This zone is surrounded by a very strong wind zone with wind speed sometimes exceeding 300 km/h. Aerial extent of the cyclone extends to a few hundred kilometers. The pressure distribution diagram for this zones in the form of *Isobars* (lines joining equal pressure points with respect to mean sea level) are closely spaced. There is steep decrease in pressure towards the eye.

Tropical cyclones originate near equator at 5 to 10 degree latitude and move towards higher latitudes in a path guided by number of weather factors. Sea water temperature between 25 and 27°C is the most favorable for formation of cyclones. The system derives its energy from sea vapour and grows in size. Once the cyclone crosses over to land, the energy source is cutoff, it becomes week and disappears gradually. The rainfall is normally heavy in the entire zone travelled by a cyclone. In northern hemisphere, cyclones move in anti-clockwise direction and in southern hemisphere they move clockwise. Cyclonic storms move at the rate of about 30–50 km/h and give medium to high intensity rainfall over a larger area.

A cyclone formed outside the tropical zone near the boundary between warm and cold air is called *extratropical cyclone*. Aerial extent of such cyclones are large, while wind and precipitation characteristics are lesser than the tropical cyclones.

An *anticyclone* is an area of high pressure in which winds tend to blow spirally outward in clockwise direction in the northern hemisphere and anti-clockwise in southern hemisphere. Weather is normally calm and such anticyclones are not associated with rain. General pattern of wind movement in the typical seasons is shown in Fig. 3.2.

*(iv) Thunder Storms s*
An air mass which moves from sea to land gets increased friction over land. These air masses rise gradually as they move inland, giving rise to condensation

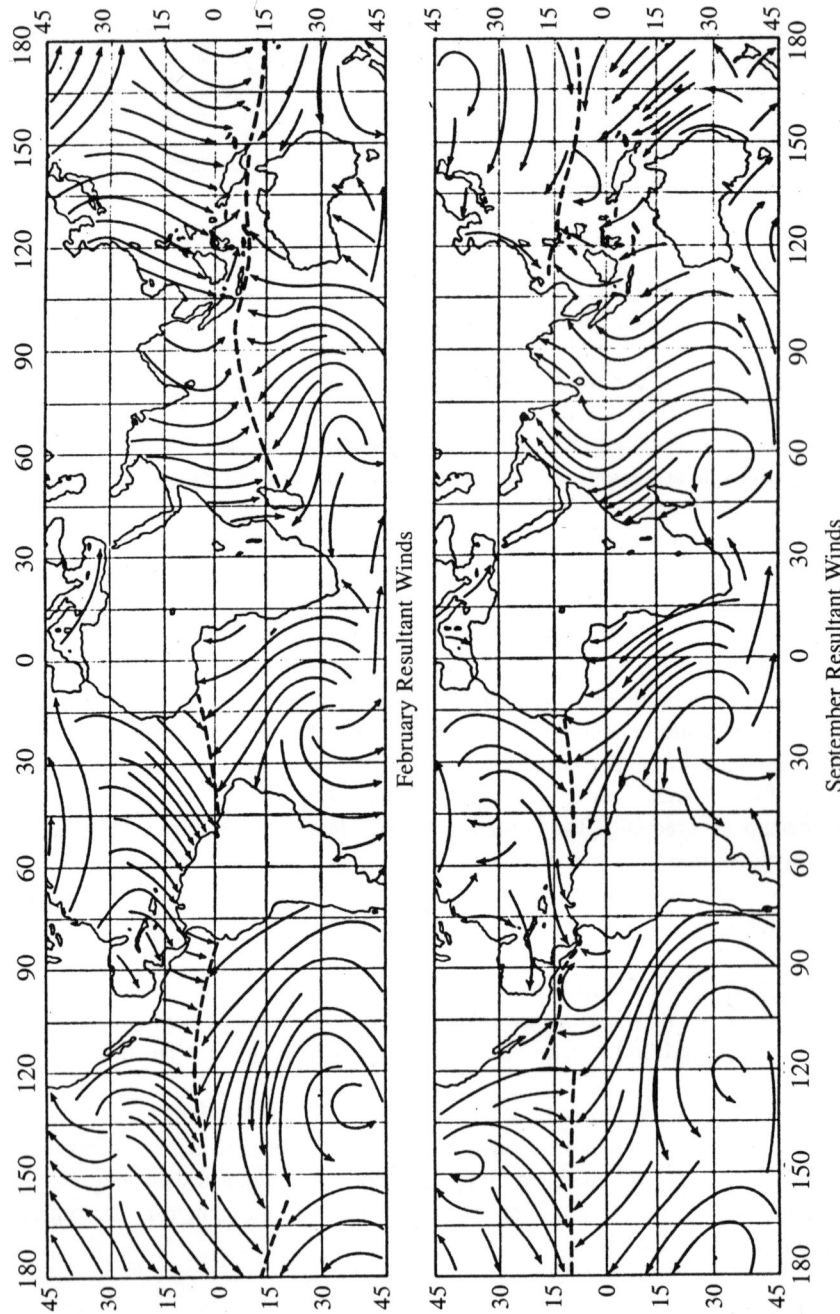

February Resultant Winds

September Resultant Winds

**Fig. 3.2 General Pattern of Wind Movement in the Typical Monsoon and Non-monsoon Months.**

and precipitation over a limited area. Winter rainfall in southern part of India and Indonesia are mainly due to this process. Sometimes thunder storms result in very intense rainfall.

## 3.4 RAINFALL IN INDIA

India gets most of its rainfall from June to October. Due to its tropical belt (north of the equator) and its size, the country experiences four distinct weather periods. Both in space and time, there is considerable variation in the amount of precipitation, its frequency, intensity and duration. Annual normal rainfall map of India is shown in Fig. 3.3.

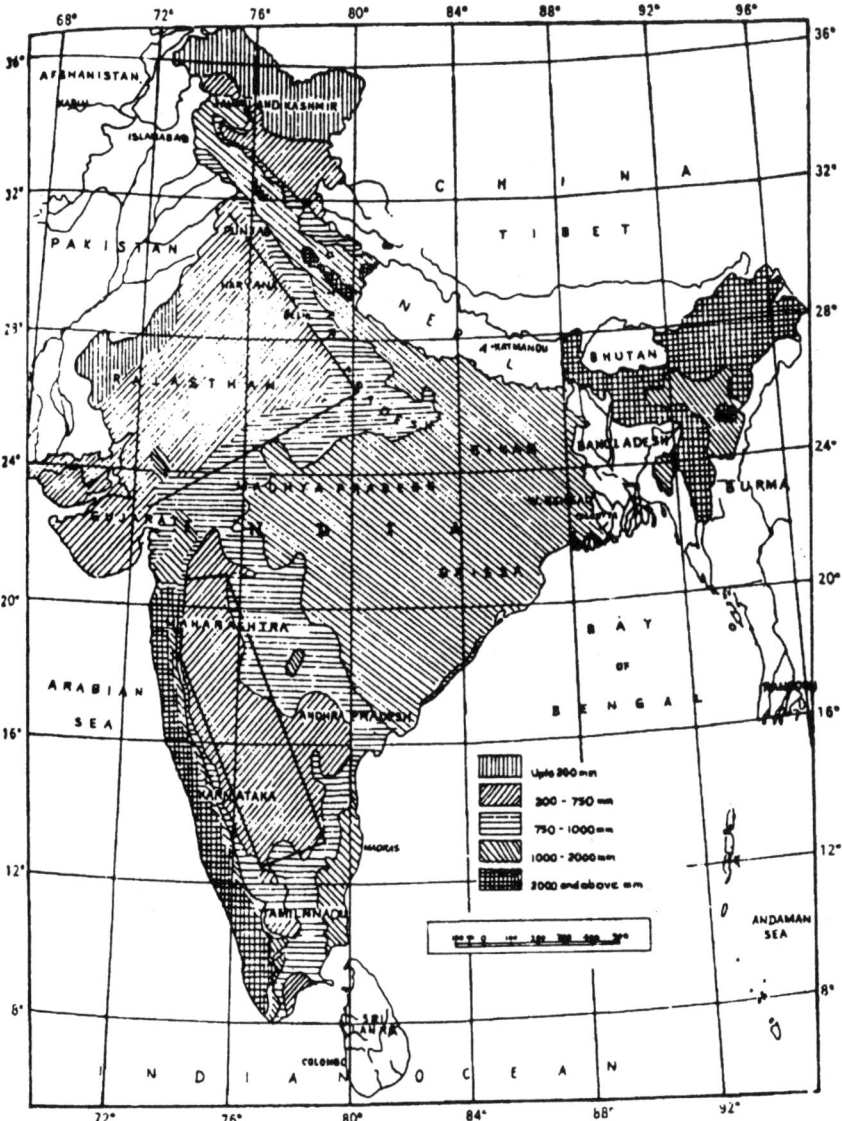

**Fig. 3.3   Annual Normal Rainfall of India.**

*Monsoon*

Except Jammu and Kashmir and south-eastern part of the Indian peninsula, the entire country gets more than 75% of annual rainfall due to monsoon winds which extends from June to October. Monsoon winds originate in the Indian ocean and proceed to the sub-continent. The dates of onset and withdrawal of monsoon in different parts of India are shown in Fig. 3.4. It touches the Kerala state by the end of May and covers the entire country by 7th of July. This wind advances from south-western direction and breaks up into two distinct branches; viz. (i) Bay of Bengal branch and (ii) Arabian sea branch.

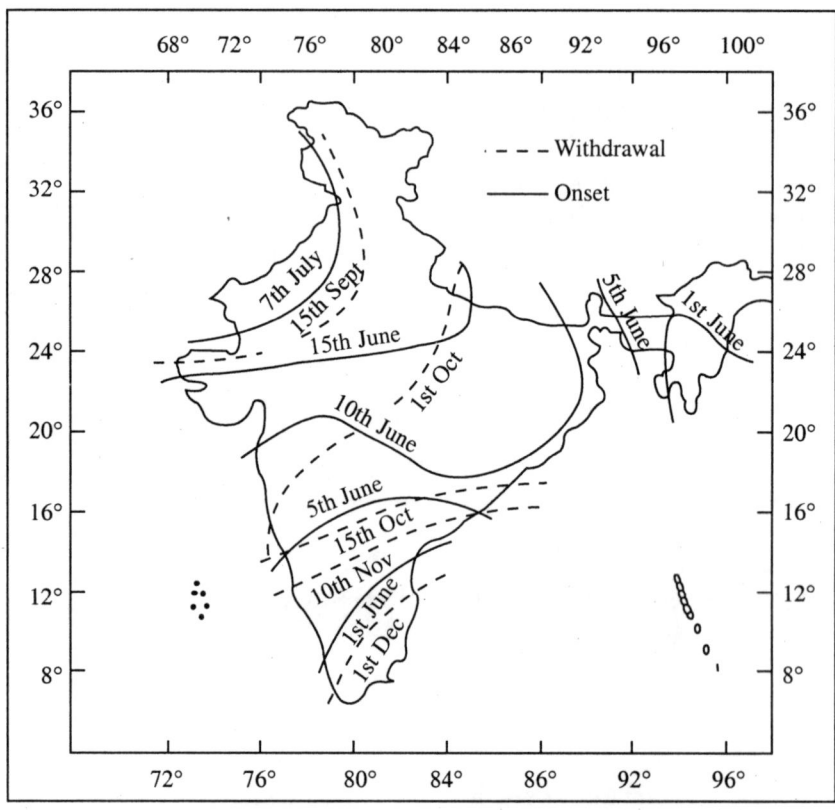

**Fig. 3.4   Dates of Onset and Withdrawal of Monsoon Rainfall in India**

The Bay of Bengal branch reaches the north-eastern states of the country and causes heavy precipitation at the same time when the Arabian sea branch touches Kerala. The former branch moves westward and causes rain in West Bengal, Orissa, Bihar, Uttar Pradesh and Delhi. The Arabian sea branch moves north-ward causing heavy precipitation in western ghats. Some residual moisture of this wind moves to the south-eastern part of the country and causes rain in parts of Tamil Nadu. The Arabian sea branch advances further causing precipitation over central India and parts of north-western states like Punjab, Haryana and Himachal Pradesh.

The two branches form a low pressure region called *monsoon trough* extending from Bay of Bengal to Rajasthan. The position of this trough generally determines the rainfall pattern of the entire central and northern parts of the country. During monsoon, wind speeds of 30–50 kmph are normally experienced. Variability of rainfall of the country is so wide that the north-eastern states get 200-400 cm of rainfall, west of western ghat regions 200-300 cm, Orissa and West Bengal 120-160 cm, Punjab, Uttar Pradesh, Haryana region gets 100-120 cm of rainfall per year during these months. Depressions usually form in the Arabian sea and Bay of Bengal in these months at a frequency of 2–3 numbers per month and move inland along the trough causing additional rainfall in the states of Orissa, West Bengal, Andhra Pradesh, Maharshtra and Gujarat.

*Post-Monsoon (October-November)*
In October and November, the south-western monsoon retreats and a north-eastern wind becomes active. Several tropical cyclones are also formed in the Bay of Bengal and in the Arabian sea. These cyclones are associated with intense rainfall causing heavy loss to life and property in the south-eastern states of the Indian peninsula. Winds blowing over Bay of Bengal touch the south-eastern part of the southern peninsula causing light rain.

*Winter Rainfall (December-February)*
During this period a few low pressure pockets are still created in the Bay of Bengal causing some rainfall in the eastern coasts of southern India. At this time, disturbances of extra-tropical origin called Western disturbances travel eastward and cause precipitation in the form of snow and rain in Jammu and Kashmir and parts of northern region of north India.

*Summer Rainfall (March-May)*
Some precipitation occurs in parts of the eastern coast and north-eastern states due to localised convective currents. At times, low pressure pockets develop over Bay of Bengal and cause cyclonic storms in the states of West Bengal, Orissa and Andhra Pradesh. Precipitation in these months are called pre-monsoon showers.

## 3.5 MEASUREMENT OF RAINFALL
Precipitation is measured as depth of water equivalent from all forms that would accumulate on a horizontal surface if there are no losses. The vertical depth of water is expressed in millimeters and tenths in metric system and in inches and hundredths in FPS system. Precipitation data is a basic input for the study of any water resources system and should be measured extensively. Due to its great variability in space and time this natural parameter is recorded continuously. Rainfall is collected and measured in instruments called *rain gauges*. If snow is the form of precipitation, then it is collected in snow-gauges, melted and its water equivalent is recorded. A gauge in its simplest form is a horizontal circular opening aperture of known cross sectional area in the form of a cylindrical vessel. The circular opening leads its catch to a collecting and measuring jar.

e types of instruments generally used for measurement of rainfall are: (i) non-recording gauge, (ii) recording gauge and (iii) weather radars.

### 3.5.1 Non-Recording Rain Gauge

Various types of non-recording rain gauges are available. A *Symon's* type or a *Standard gauge* is recommended by India Meteorological Department (IMD) and is extensively used under Indian Conditions. IS: 5225–1969 is followed for specification of a standard type of gauge. IS: 4986–1968 gives details of installation and measurement of non-recording type rain gauge. The two types of gauges are alike but differ in their size proportion and material. A standard gauge is made up of reinforced fiberglass polyester material of different combinations of collectors and bottles. Details of installation of a standard type of rain gauge is shown in Fig. 3.5.

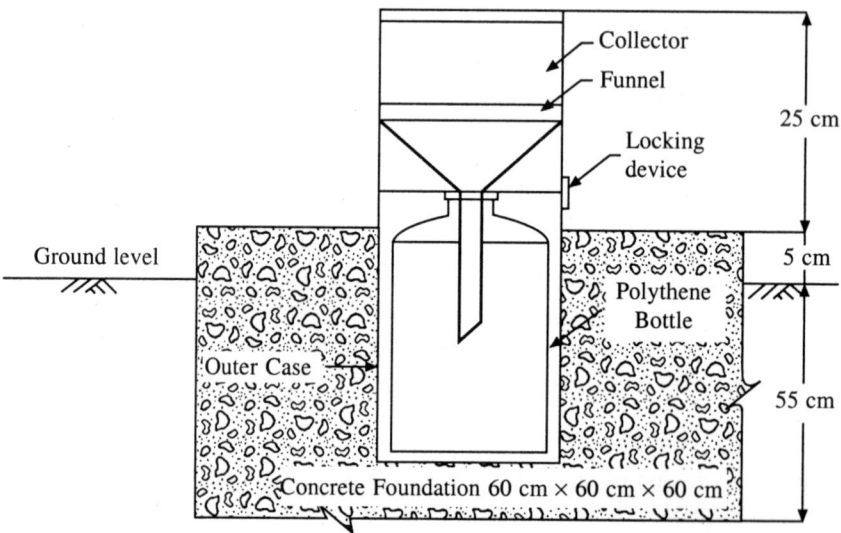

**Fig. 3.5   A Standard Gauge (Non-recording) Adopted by IMD**

The circular collector opening has an area of either 100 or 200 cm² from which rain enters into the receiving vessel through a funnel. The collectors are interchangeable. Top of the circular opening is placed at a standard height of 30 cm above ground level. The metal container should be fixed to a concrete block of 60 cm × 60 cm × 60 cm as shown in Fig. 3.5.

The rain catch collected in the bottle is taken out at 8.30 AM (India Standard Time) and poured into a graduated measuring glass jar (chosen accordingly for 100 cm or 200 cm collector), which gives directly the depth of rainfall for the day. Degree of accuracy of a graduated jar is 0.1mm. If precipitation on a day is very heavy then more readings should be taken and summed up to give the rainfall depth for the day along with its final observation taken at 8.30 AM. Such a system can also be used for snow measurement but the snow should be melted to arrive at the equivalent depth of water recorded at 8.30 AM. This system needs the service of an attending observer who monitors the gauge at regular intervals, usually daily and hourly during continuous and heavy precipitations.

Since the above system does not record rain but simply collects, it is called a non-recording gauge. The ratio of volume of water collected in $cm^3$ divided by the area of opening of the gauge mouth in $cm^2$ gives the depth of rainfall for the day. If any non-standard measuring jar is used, then it should be calibrated to give directly the depth of rainfall in centimeters and millimeters.

### 3.5.2 Recording Type Rain-Gauges

These rain gauges give a continuous record of rainfall at a place over time. Such gauges give all the required informations of a storm like the onset and cessation of rain, i.e., duration of the storm, intensity and the cumulative rainfall. The recording gauges are commonly installed along with a non-recording type gauge for the purpose of checking and calibration. Many types of recording gauges are available in the market. One may use a *tipping bucket type*, *weighing bucket type* or *syphon* type gauge on the consideration of their merits and demerits to suit the conditions prevailing over the site. When such a gauge is fitted with an electronic device to transmit rainfall data to a base station then it is known as *Telemetering gauge*.

**(a) Tipping Bucket Type:** A Stevens tipping bucket type of rain gauge consists of a 200 mm collector that directs the rain water through a funnel into a two-compartmental bucket. The size of each bucket is 0.25 mm of rain. Once rain water fills up a bucket, it over-balances and the water tips down to the casing of the container bringing thereby the second bucket to its measuring position beneath the funnel. A tipping bucket rain gauge is shown in Fig. 3.6. Tipping of the buckets actuates an electric circuit which records the number of tips during rain. This type of gauge can be installed at an inaccessible area from where electric pulses generated due to the tipping of the buckets is recorded at the control room far away from the gauge location. Disadvantages of such a type of gauge are (i) when tipping of buckets takes place, rainfall at that instant is not recorded, (ii) very high intensity rainfall gives close signals, which can make it difficult to record the number of tips and (iii) calibration of tips may change due to rusting and dirt accumulation.

**(b) Weighing Bucket Type:** This type of gauge can be used for recording rainfall as well as snow. Rain is collected in a receiver bucket supported on a spring balance. A mechanical lever arm of the balance is connected with a pen which touches a clock mounted drum with a graph paper. As it rains, the weight of the bucket gradually increases. This changes the position of the pan of the balance. With time the pen marks a line on the continuously moving graph paper. The record shows the accumulation of precipitation over time. The recording can be taken after 24 hours or 7 days depending on the clock and drum size. Such gauges are normally used in USA and are becoming increasingly popular. A sketch of the weighing bucket type rain gauge is shown in Fig. 3.7.

Disadvantages of such a gauge are (i) when very heavy precipitation occurs,

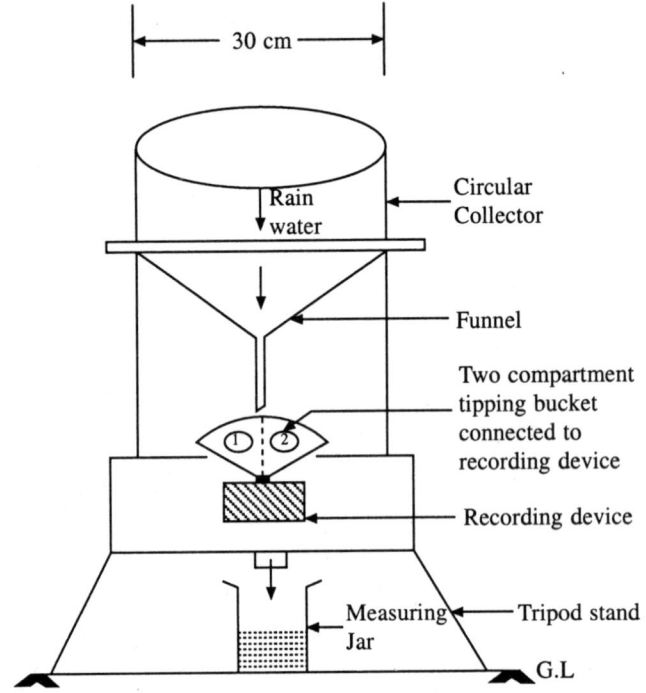

**Fig. 3.6.   A Tiping Bucket Type Recording Rain Gauge**

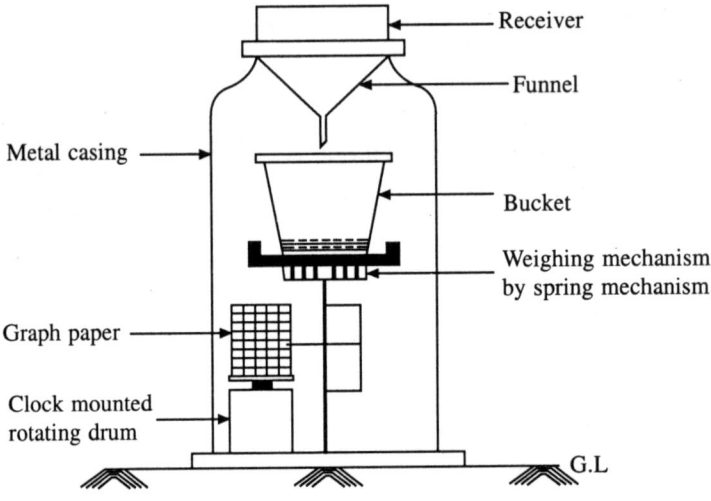

**Fig. 3.7   A Weighing Bucket Type Rain Gauge**

there is good chance that the bucket will overflow and (ii) such instruments are costly. When the pen reaches the upper end of the graph paper it reverses its move, thus continuously recording the precipitation over time.

**(c) Syphon (Float) Type:** Specifications regarding syphon type rain gauge is given in IS: 5235–1969. This type of rain gauge is very popular in India. Such a rain gauge is shown in Fig. 3.8. Rain entering the gauge is led to a float

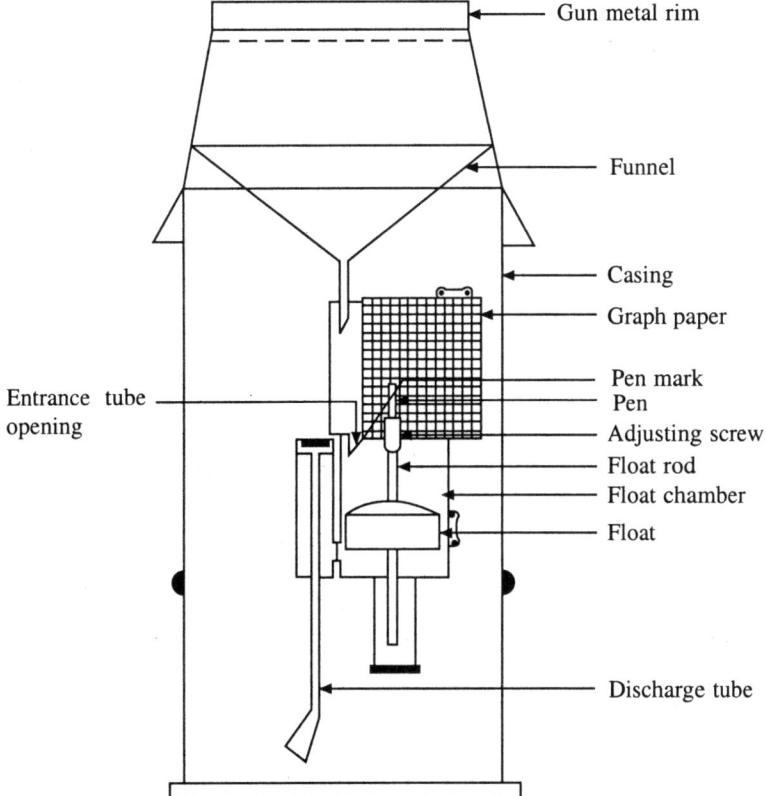

**Fig. 3.8   A Syphon (Float) Type Automatic Rain Gauge**

chamber through a funnel. With increase in rainwater in the chamber, the float rises. A pen mounted on the float through a lever system touches a graph chart warped around the circumference of the drum. The drum is mounted on a mechanical clock. Clock of the rotating drum is wound either for 24 hours or 7 days after which the graph chart is to be replaced. By the time the pen reaches the top of the graph the float also reached the top of the chamber. At this point syphonic action takes place in the chamber and all the water in the chamber below the float empties. The pen comes back to its original zero position. If there is no rainfall, the pen moves horizontally over the graph paper at that level. One syphonic action means 10 mm of rainfall and the time taken to collect the depth of rain can be noted from the horizontal axis of the graph paper. A used graph paper specified by Indian Standard for syphon type gauge is shown in Fig. 3.9.

Disadvantages of such gauges are: (i) they are costlier than other non-recording type and (ii) mechanical defects sometimes give erroneous results. There are many advantages of this type of recording gauge. All types of information about a storm can be obtained from the cumulative plot of graph paper. The beginning

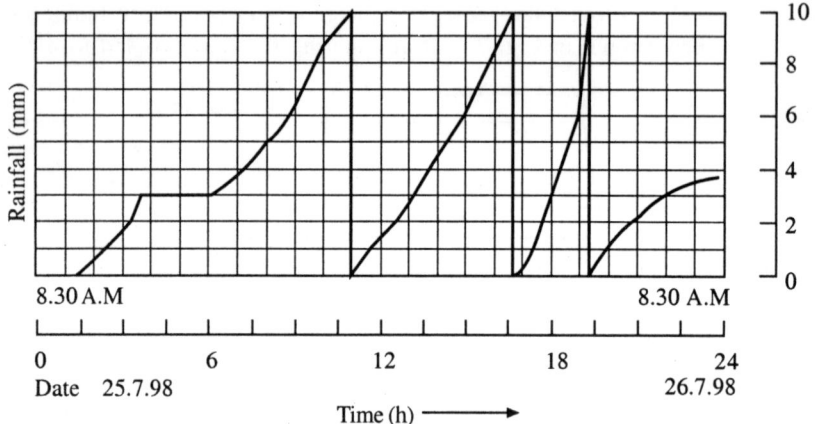

**Fig. 3.9   Rainfall Mass Curve From a Syphon Gauge**

and the end of the storm, its intensity, duration, distribution of rain and the depth of total storm precipitation can easily be obtained from the plot of the graph.

### 3.5.3   Weather Radar

Measurement of rainfall by a weather radar is based on the principle that the amount of power returned from hydrometeor (rain drops) target is related to the intensity of fall. The radar equation for scattering of electromagnetic pulses by airborne liquid or solid particles is given by

$$P_r = \frac{P_t h A_e}{8\pi r^2} \frac{\pi^4}{\lambda^4} \left( \frac{E-1}{E+2} \right)^2 \Sigma D^6 \tag{3.1}$$

where $P_t$ is the transmitted power, $P_r$ the average received power, $h$ the pulse width, $A_e$ the effective aperture of the antenna, $r$ the range of the target, $\lambda$ the wave length, $E$ the complex dielectric constant or the complex refractive index of the material of which the particle is composed and $\Sigma D^6$ is the effective radar reflectivity factor which is taken as the symbol $Z_e$. For a given radar, all the terms of equation (3.1) are constant which can be combined together to give a coefficient $C$. Symbol $D$ represents diameter of the rain drop. With the substitutions equation (3.1) reduces to the following form

$$P_r = \frac{C \cdot Z_e}{r^2} \tag{3.2}$$

To use this equation for rainfall measurement, it is necessary to find a relationship between $Z_e$ and rate of precipitation, i.e., between drop size distribution and rainfall. The drop size distribution is converted to volume distribution by taking into account the fall speed of drops of different sizes. The drop size and corresponding fall speeds (terminal velocities of rain drops) gives the intensity of rainfall at any instant. It is seen from the above radar equation that the average power received from a target is inversely proportional to the square of range of target being explored. This is called range attenuation. By measuring $P_r$ in watts,

$Z_e$ can be calculated as $E$ is known for the radar. Knowing $Z_e$ which is equal to $\Sigma D^6$, we can calculate rain drop size $D$. An empirical formula relating $\Sigma D^6$ and rainfall rate $R$ is given by

$$Z_e = \Sigma\, D^6 = aR^b = 200R^{1.6} \qquad (3.3)$$

where $D^6$ is measured in $mm^6/m^3$ and R is the rainfall intensity (mm/h.). Thus knowing the reflective value, precipitation rate can be calculated. Meteorological radars normally operate on 10 cm wavelength. The range (radius) of operation of a radar is 200 km. All weather radars use one degree beam width. Radars have the following distinct advantages over other systems of rainfall measurement.

(i) Total aerial rainfall can be computed in near real time by the use of on-line computer at the weather station.
(ii) The stream flow (flood) can be predicted immediately from the use of rainfall and suitable rainfall-runoff model by the same computer.
(iii) The movement of weather system can be tracked and shown.
(iv) It can forecast time of onset, cessation and intensity of rainfall.
(v) Solid precipitation can be detected and measured.
(vi) Dangerous meteorological phenomena can be detected.

### 3.5.4 Totalisers

There are types of other gauges available for measuring rainfall for hydrologic purposes. A *totaliser* has a large container with a small funnel like opening which can hold rain catch of appreciable quantity. This can be used to record rain depth at a remote place at the interval of one week, a fortnight or a month depending on the size of the container. A *standpipe* gauge or a *storage gauge* can be used for recording snow as precipitation at remote locations. The standpipe of 305 mm diameter should be high enough to store the anticipated depth of snow between the measuring intervals. An antifreeze chemical like calcium chloride solution can be kept in the storage of the gauge to melt the snow for recording its water equivalent. For both types of gauges, a thin layer of oil is spread to prevent evaporation.

### 3.6 NETWORK DESIGN

Ideally, a basin should have as many number of gauges possible to give a clear representative picture of the aerial distribution of the precipitation. Factors like economy, topography, accessibility and rainfall variability govern the number of stations for a basin. There is no definite rule as to how many gauges are needed for a complete ungauged basin. To begin with a few pilot gauges can be fixed and after a few years of data are available, statistical analysis can be carried out to check the adequacy of the system. World Meteorological Organisation (WMO) recommends certain densities of gauges to be followed for different types of catchments, but IS:4987–1968 recommends different densities of gauges, sufficient from practical considerations for India. Table 3.1 gives their recommendations.

The present density of rain gauge network in India is 630 sq. km per gauge. A High Level Committee on floods recommended a minimum of one station per 500 sq. km. Israel has the highest network density in the world with one gauge

**Table 3.1   Network of Precipitation Stations**

| Type of regions | Minimum area for one station under ideal condition in sq. km. | Area to be covered under difficult condition per station in sq. km. |
|---|---|---|
| **WMO Recommendations (1969)** | | |
| 1. Flat regions of temperate mediterranean and tropical zones. | 600–900 | 900–3000 |
| 2. Mountainous regions of temperate mediterranean and tropical zones. | 100–250 | 250–1000 |
| 3 Small mountainous regions with irregular precipitation. | 25 | – |
| 4. Arid and polar Zones. | 1500–10,000 | – |
| **Indian Standard (IS: 4987–1968)** | | |
| 1. Plain area. | 520 | – |
| 2. Regions of average elevation 1000 m. | 260–390 | – |
| 3. Predominantly hilly areas with heavy rainfall. | 130 | – |

covering 26 sq. km and Vietnam has the lowest with one gauge covering nearly 2600 sq. km.

### 3.6.1   Optimum Number of Rain Gauge Stations

Records from all the existing gauges of a basin help to fix the optimum number of stations. The following statistical analysis help to obtain number of gauges for a basin optimally on the basis of an assigned percentage of error in estimating the mean aerial rainfall.

$$N = \left\{ \frac{C_v}{E_p} \right\}^2 \qquad (3.4)$$

where $N$ is the optimal number of stations, $E_p$ the allowable percentage of error in the estimation of mean aerial rainfall, $C_v$ is the coefficient of variation of the rainfall from the existing stations in percentage. Coefficient of variation can be calculated in the following steps from the data of existing $n$ stations.

(i) Calculate the mean of rainfall from the equation $P_{av} = (1/n) \sum P_i$

(ii) Calculate the standard deviation as $\sigma_{n-1} = \left\{ \frac{1}{(n-1)} \sum (P_i - P_{av})^2 \right\}^{\frac{1}{2}}$

(iii) Compute the coefficient of variation as $C_v = \dfrac{\sigma_{n-1} \times 100}{P_{av}}$

If the allowable percent of error in estimating the mean rainfall is taken higher, then a basin will require fewer numbers of gauges and vice-versa. The allowable percentage of error $E_p$ is normally taken as 10%. While computing the value of $C_v$ if its value comes less than 10%, we can assume the existing stations to be sufficient for the basin. In case $N > n$, the additional stations required for the basin can be found as $(N - n)$. Annual rainfall values are normally used in the above analysis. Additional stations are to be established at the appropriate locations giving an even distribution over the basin.

### 3.6.2 Ideal Location for a Rain Gauge Station
While setting up any rain Gauge station the following points should be noted.

(i) The site should be on a level ground, i.e., slopping ground, hill tops or hill slopes are not suitable.
(ii) The site should be an open space.
(iii) Horizontal distance between the rain gauge and the nearest objects should be twice the height of the objects.
(iv) Site should be away from continuous wind forces.
(v) Other meteorological instruments and the fencing of the site should maintain the step (iii) above.
(vi) The site should be easily accessible.
(vii) The gauge should be truly vertical.
(viii) Ten percent of total number of rain gauge stations of any basin should be self-recording.
(ix) The observer must visit the site regularly to ensure its proper readiness for measurement.

Precipitation measurements are susceptible to the following errors, which can be (a) in measurement, (b) mistakes in recording, (c) instrumental errors, (d) initial loss in wetting the gauge (which is equal to 0.25 mm per precipitation or may be 25 mm/year), (e) error due to rain drop splash, (f) evaporation from the gauge and (g) loss of 1.5% of rainfall if gauge is inclined 10 degree from its true vertical position.

---

**Example 3.1:** A sub-basin has six numbers of rain gauges. Annual rainfall recorded by the gauges are given below. Considering 10% error in the estimation of mean annual rainfall, calculate optimum number of gauges required for the sub-basin and check if the present network is sufficient.

| Rain gauge name | A | B | C | D | E | F |
|---|---|---|---|---|---|---|
| Annual rainfall (cm) | 102 | 77 | 84 | 53 | 66 | 80 |

**Solution**
From the given data, the statistical parameters are calculated in the following table.
Optimum number of stations required for the sub-basin are found to be 5. Since there are already 6 gauges in the basin no more gauge is required to be installed.

**Table 3.2  Network Design for Example 3.1**

| Sl. No. | Station | Annual rainfall | $(P - P_{av})$ | $(P - P_{av})^2$ | Remarks |
|---|---|---|---|---|---|
| (1) | (2) | (3) | (4) | (5) | (6) |
| 1 | A | 102 | 25 | 625 | $\sigma_{n-1} = \left\{ \dfrac{1380}{6-1} \right\}^{0.5} = 16.61$ |
| 2 | B | 77 | 0 | 0 | $C_v = 100 \times \dfrac{16.6}{77} = 21.57$ |
| 3 | C | 84 | 7 | 49 | |
| 4 | D | 53 | 24 | 576 | $N = \left( \dfrac{21.57}{10} \right)^2 = 4.6$ |
| 5 | E | 66 | 11 | 121 | Say $N = 5$ |
| 6 | F | 80 | 3 | 9 | |
| Total | | 462 | | 1380 | |
| Mean | | $P_{av} = 77$ | | | |

**Example 3.2:** A sub-basin with area of 1038 sq. km has 7 stations. The normal annual rainfall depths for all the seven stations are given below. Determine the optimum number of raingauge stations to be established in the basin if it is desired to limit the error in the mean value of rainfall to 10%. Indicate how you are going to distribute the additional rain gauge stations (if required). Is it possible to have zero percent error in the estimate of the mean value.

| Stations | $A_1$ | $A_2$ | $A_3$ | $A_4$ | $A_5$ | $A_6$ | $A_7$ |
|---|---|---|---|---|---|---|---|
| Normal annual rainfall (cm) | 62 | 94 | 62 | 47 | 32 | 88 | 70 |

**Solution**

The solution is obtained from Table 3.3.

**Table 3.3  Network Design for Example 3.2**

| Sl. No. | Station Name | Normal annual rainfall (cm) | $(P - P_{av})^2$ | Remarks |
|---|---|---|---|---|
| (1) | (2) | (3) | (4) | (5) |
| 1 | $A_1$ | 62 | 09 | $\sigma_{n-1} = \left\{ \dfrac{2826}{7-1} \right\}^{\frac{1}{2}} = 21.70$ |
| 2 | $A_2$ | 94 | 841 | $C_v = \dfrac{21.7 \times 100}{65} = 33.39\%$ |
| 3 | $A_3$ | 62 | 09 | $E_p = 10\%$ |
| 4 | $A_4$ | 47 | 324 | $N = \left( \dfrac{C_v}{E_p} \right)^2 = \left( \dfrac{33.39}{10} \right)^2 = 11.15$ |
| 5 | $A_5$ | 32 | 1089 | Say $N = 11$ stations. |
| 6 | $A_6$ | 88 | 529 | Existing stations $(n) = 7$ |
| 7 | $A_7$ | 70 | 25 | Additional stations |
| | | | | required $= (N - n) = 11 - 7 = 4$ |
| $n = 7$ | | $\Sigma P = 455$ | 2826 | |

For location of these additional 4 stations, the principles laid down by IMD are to be followed. The best way is to draw a Thiessen polygon for the area and find out the area represented by each station. The density per station with 11 stations comes to 94 sq. km. The orographic features and the density is to be kept in mind while locating these additional 4-stations, as there is large variation in the normal annual rainfalls from 32 cm to 94 cm in the sub-basin. The term Thiessen-polygon will be discussed afterwards. It is not possible to have zero percent error in the estimation of mean value as this will give the number of gauges required for the basin as infinite. This is not practicable.

## 3.7 CONSISTENCY OF RAINFALL DATA

Rainfall data reported from a station may not be consistent always. Over the period of observation of rainfall record, there could be (i) unreported shifting of the rain gauge site by as much as 8 km aerially or 30 m in elevation, (ii) significant construction work in the area might have changed the surroundings, (iii) change in observational procedure incorporated from a certain period or (iv) a heavy forest fire, earthquake or landslide might have taken place in that area. Such changes at any station are likely to affect the consistency of data from a station. Use of double mass-curve checks the consistency of the record and helps to correct the rain gauge data for the station. In this method, the accumulated annual rainfall of a particular station is compared with the concurrent accumulated values of mean rainfall of groups of 5 to 8 surrounding base stations. The basis of such an exercise is that a group of sample data (for any period) drawn from its population will be the same. The procedure for double mass-curve analysis is discussed as follows:

(i) The doubtful station, say A, is marked and the group of stations surrounding it are identified.

(ii) A table is prepared in which the first column represents the year in decreasing order, i.e., it starts with the latest year of station A.

(iii) Yearly precipitation values of station A are written in second column.

(iv) In the third column the cumulative rainfall of second column are entered.

(v) Mean yearly precipitation of the group of stations surrounding station A are computed and entered in the fourth column against the year of col. 1.

(vi) In column five, cumulative precipitation of the group of stations of column four are computed.

(vii) A graph is plotted taking the cumulative rainfall of the group of station as abscissa and cumulative rainfall of the station A as ordinate. Consecutive points are joined by straight line.

(viii) If the consistency of the station A has undergone changes from any year, then it can be noticed from the slope of the plot. The line joining the initial points of the graph are extended by a dotted line and correction ($c/c_1$) as shown in the Fig. 3.10 is computed.

(ix) Rainfall of subsequent years from the year of deviation (marked $x$ in the figure) are corrected by multiplying the correction factor.

This exercise helps to bring the older rainfall data of station A to the new environment. Corrections are to be applied to the data of station A only when the change of slope of the double mass-curve is observed for more than five years.

As shown in Fig. 3.10 corrections to the data from 1983 to 1977 are applied for station A. Unless the change is significant, i.e., exceeds 10% of the original slope, it should be confirmed whether the deviation of the line is not part of the usual scatter. Correction should be applied for change in slope exceeding 10% of the original line.

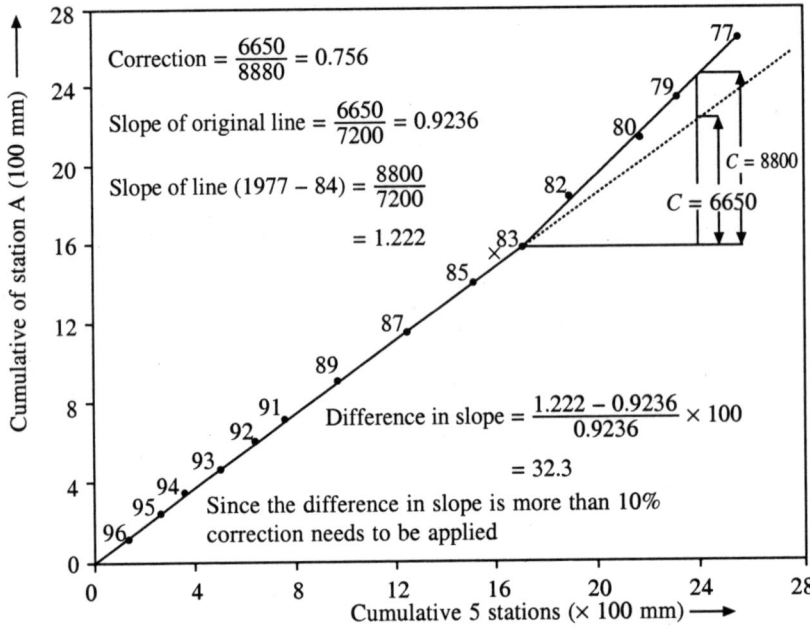

**Fig. 3.10   Double Mass Curve Analysis for Checking Consistency of Rainfall Data**

**Example 3.3:** Annual rainfall of station A and the average annual rainfall of five surrounding stations from 1996 to 1977 are given below. Check the consistency of data of station A. If data is found inconsistent, then correct the inconsistent data.

**Table 3.4   Data for Example 3.3**

| Year | Annual avg. precipitation of station A (mm) | Annual avg. ppt. of 5 stations surrounding station (A) (mm) |
|------|---------------------------------------------|------------------------------------------------------------|
| 1996 | 1430 | 1410 |
| 1995 | 1100 | 1260 |
| 1994 | 1170 | 1100 |
| 1993 | 1100 | 1230 |
| 1992 | 1200 | 1150 |
| 1991 | 1220 | 1430 |
| 1990 | 1280 | 1150 |
| 1989 | 750 | 950 |
| 1988 | 1120 | 1230 |
| 1987 | 1250 | 1350 |
| 1986 | 1380 | 1440 |

*(Contd)*

| Year | Annual avg. precipitation of station A (mm) | Annual avg. ppt. of 5 stations surrounding station A (in mm) |
|------|---------------------------------------------|-------------------------------------------------------------|
| 1985 | 1210 | 1360 |
| 1984 | 1760 | 1730 |
| 1983 | 1400 | 1080 |
| 1982 | 1240 | 970 |
| 1981 | 1760 | 1320 |
| 1980 | 1480 | 1350 |
| 1979 | 1740 | 1410 |
| 1978 | 1420 | 1270 |
| 1977 | 1580 | 1260 |

**Solution**

Cumulative annual precipitation of station A and the annual average precipitation of group of five surrounding stations are calculated in the following table.

**Table 3.5 Checking the Consistency of Data for Example 3.3**

| Year | Annual rainfall of (A) | Cumulative value of (A) | Mean yearly ppt. of 5 stations | Cumulative of col. (4) | Corrected value of station A from 1984 | Remarks |
|------|------------------------|-------------------------|--------------------------------|------------------------|----------------------------------------|---------|
| (1) | (2) | (3) | (4) | (5) | (6) | (7) |
| 1996 | 1430 | 1430 | 1410 | 1410 | | Slope of |
| 1995 | 1100 | 2530 | 1260 | 2670 | | correction |
| 1994 | 1170 | 3700 | 1100 | 3770 | | = 6650/8800 |
| 1993 | 1100 | 4800 | 1230 | 5000 | | = 0.756. |
| 1992 | 1200 | 6000 | 1150 | 6150 | | From 1983 |
| 1991 | 1220 | 7220 | 1430 | 7580 | | onwards |
| 1990 | 1280 | 8500 | 1150 | 8730 | | correction |
| 1989 | 750 | 9250 | 950 | 9680 | | is applied as |
| 1988 | 1120 | 10370 | 1230 | 10910 | | the slope |
| 1987 | 1250 | 11520 | 1350 | 12260 | | changes from |
| 1986 | 1380 | 13000 | 1440 | 13700 | | that year |
| 1985 | 1210 | 14210 | 1360 | 15060 | | |
| 1984 | 1760 | 15970 | 1730 | 16790 | 1760 | |
| 1983 | 1400 | 17370 | 1080 | 17870 | 1058 | |
| 1982 | 1240 | 18610 | 970 | 18840 | 937 | |
| 1981 | 1760 | 20370 | 1320 | 20160 | 1330 | |
| 1980 | 1480 | 21850 | 1350 | 21510 | 1119 | |
| 1979 | 1740 | 23590 | 1410 | 22920 | 1315 | |
| 1978 | 1420 | 25010 | 1270 | 24190 | 1074 | |
| 1977 | 1580 | 26590 | 1260 | 25450 | 1194 | |

A graph (Fig. 3.10) is plotted taking the cumulative annual rainfall of the five surrounding stations in abscissa and the cumulative annual rainfall of the station A in ordinate. The consecutive points are joined by a straight line. Deviation of the straight line plot is noticed from the year 1983. Slope of the line is found as 0.756. Therefore correction is applied to all the data from 1983 to 1977. The corrected data is entered in column (6) of Table 3.5.

## 3.8   ESTIMATING MISSING DATA

Failure of any rain gauge or absence of observer from a station causes short break in the record of rainfall at the station. These gaps are to be estimated first before we use the rainfall data for any analysis. The surrounding stations located within the basin help to fill the missing data on the assumption of hydro-meteorological similarity of the group of stations. The general equation of the weightage transmission of the rainfall of the nearby stations to the missing station ($x_i$) can be represented as

$$P_{xi} = \frac{\sum\limits_{i=1}^{n} a_i p_i}{\sum\limits_{i=1}^{n} a_i} \tag{3.5}$$

where $P_i$ is the normal rainfall of $i$th surrounding station, $i = 1, 2, 3, \dots, n$ are the surrounding gauge numbers which are used for filling the gaps, $a_i$ the weighing factor of the station $p_i$ and $p_{xi}$ is the data required to be filled up. The methods mostly used in hydrology for filling the missing data are discussed below.

### 3.8.1   Arithmetic Mean Method

This method is used when: (i) the normal annual rainfall of the missing station say $x$ is within 10% of the normal annual rainfall of the surrounding stations, (ii) data of at least three surrounding stations, called index stations, are available within the basin, (iii) the index stations should be evenly spaced around the missing station and should be as close as possible, (iv) The missing rainfall data of station $x$ is computed by simple arithmetic average of the index stations in the form

$$p_x = \frac{1}{n}(p_1 + p_2 \dots + P_n) \tag{3.6}$$

in which $P_1, P_2, \dots, P_n$ are the precipitations of index stations and $P_x$ is that of the missing station, $n$ is the number of index stations. This is the simplest form of equation where $a_1, a_2, \dots, a_n = 1$. The word normal means average of 30 years of data, i.e., 30 values of the latest records. For example, for a station when the last 30 years of June month rainfall is averaged, we call it as normal rainfall for the month of June for that station.

### 3.8.2   Normal Ratio Method

This method is used when the normal annual precipitation of the index stations differ by more than 10% of the missing station. The rainfall of the surrounding index stations are weighed by the ratio of normal annual rainfalls by using the following equation.

$$P_x = \frac{N_x}{n}\left(\frac{P_1}{N_1} + \frac{P_2}{N_2} + \dots + \frac{P_n}{N_n}\right) \tag{3.7}$$

where $P_1, P_2, \dots, P_n$ are the rainfall data of index stations, $N_1, N_2, \dots, N_n$ the normal annual rainfall of index stations, $P_x$ and $N_x$ the corresponding values for the missing station $x$ in question and $n$ is the number of stations surrounding the station $x$. Comparing equations (3.7) and (3.5), the weight coefficient $a_i$ for the $i$th station is given as

$$a_i = \Sigma \left( \frac{N_x}{nN_i} \right) \tag{3.8}$$

### 3.8.3 Regression Method

A multiple linear regression of the form

$$P_x = a_0 + a_1 P_1 + a_2 P_2 + a_3 P_3 + \ldots + a_n P_n \tag{3.9}$$

may be established. The coefficients $a_0, a_1, \ldots, a_n$ can be calculated by least square method. The equation can be used to compute rainfall $P_x$ of the missing station. Use of this method is advantageous over the previous two and can be effectively used when a digital computer is available at a site. A random error component $\varepsilon_1 \times S_{yx}$ may be added to equation (3.9) if large amount of missing data is to be estimated, where $\varepsilon_1$ is the normal random number with zero mean and unit standard deviation. It can be selected from random number tables given in any standard mathematical book. $S_{yx}$ is the standard error of estimate. The standard error of estimate can be estimated from the equation

$$S_{yx} = \sqrt{\sum_{i=1}^{n} (Y_i - Y_{ei})^2} \tag{3.10}$$

in which $Y_{ei}$ is the estimated value of $Y_i$ for given $X_i$ and $n$ is the number of data. Addition of the last term $\varepsilon_t \times S_{yx}$ maintains the standard deviation of the estimated value of $P_x$ close to the observed standard deviation. Use of the above method is not popular because of the nature of computations involved.

### 3.8.4 Inverse Distance (US Weather Service) Method

In this method, a rectangular coordinate system is superimposed over the map marked with rain gauge station in such a way that the origin (0, 0) represents the missing station as shown in Fig. 3.11. The surrounding index stations lie within the quadrants to the point for which rainfall is to be estimated. The distance of index stations from the missing station gives a weightage of the station by which missing rainfall is estimated. The following relation may be used.

$$P_x = \frac{\sum\limits_{i=1}^{n} P_i \times W_i}{\sum\limits_{i=1}^{n} W_i} \tag{3.11}$$

where $W_i = 1/D^2$, $D^2 = (\Delta X^2 + \Delta Y^2)$ is the distance of the station $i$ in $X$ and $Y$ coordinates taking missing rainfall station at (0, 0) position. This is the most acceptable method and is widely used for determining the missing rainfall for any scientific analysis. However, the limitation is that it estimates missing rainfall between the highest and lowest values of the index stations.

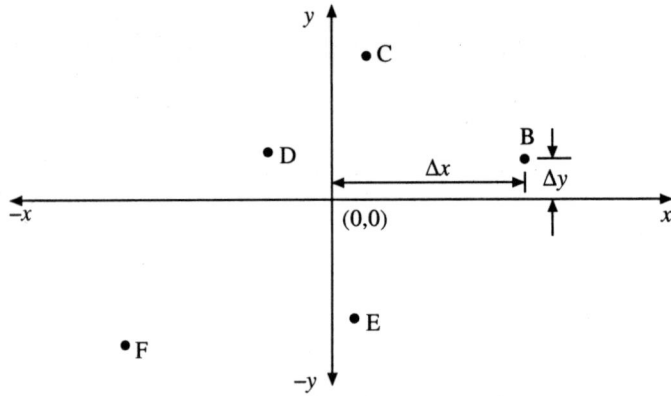

**Fig. 3.11   Estimation of Missing Rainfall by U.S Weather Service Method**

Other methods using various *numerical interpolations* of Isohyetal method, Thiessen polygon method, Station-year-method, Graphical method and Rational method have their own limitations for computation of the missing data. Estimation of missing daily rainfall data is less reliable than annual values and such estimates should be subjected to further statistical analysis. In plain terrain any of the above methods can be suitably used but in hilly regions, the normal ratio method or linear regression method can be more effectively used for better results.

**Example 3.4:** In a river basin, a station A was inoperative during a storm, while stations B, C and D surrounding A were in operation, registering 12.3, 14.8 and 11.9 cm of precipitation. Mean annual precipitation at the four stations A, B, C and D are 1290, 1510, 1680 and 1375 mm respectively. Estimate the missing storm precipitation of station A by all the methods you know. The coordinates of B, C and D are (6, 4), (8, −6) and (−4, 4), respectively, y where as the coordinate of A is (0, 0) as shown in Fig. 3.12.

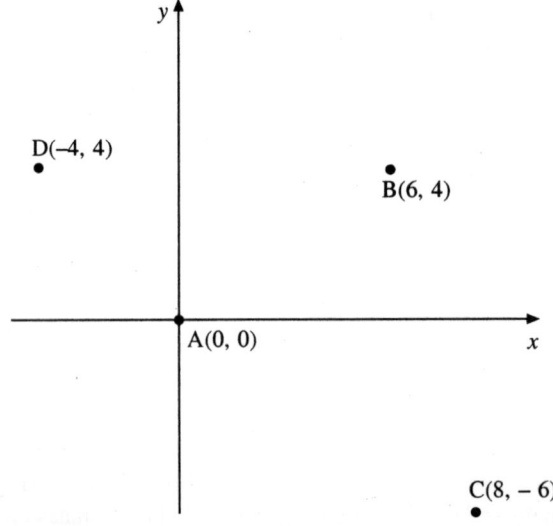

**Fig. 3.12   Location of the Index Stations of A, B, C and D of Example 3.4**

**Solution**

(a) Since the normal annual rainfall of station A is 1290 mm and that of C is 1680 mm which exceeds beyond 10% of station A, application of arithmetic mean method is not acceptable.

(b) By using normal ratio method, the average precipitation is calculated as

$$P_x = \frac{N_x}{n}\left(\frac{P_1}{N_1} + \frac{P_2}{N_2} + \ldots + \frac{P_n}{N_n}\right) = \frac{1290}{4}\left\{\frac{123}{1510} + \frac{148}{1680} + \frac{119}{1375}\right\}$$

$$= 82.6 \text{ mm} = 8.26 \text{ cm}$$

(c) By using regression techniques, let us fit a simple linear equation of the form $Y = a + bX$ (Table 3.6).

**Table 3.6  Use of Regression Technique for Filling Missing Data**

| Y | X | $Y^2$ | $X^2$ | XY |
|---|---|---|---|---|
| 12.3 | 151.0 | 151.29 | 22801 | 1857.3 |
| 14.8 | 168.0 | 219.04 | 28224 | 2486.4 |
| 11.9 | 137.5 | 141.61 | 18906 | 1636.3 |
| Sum = 39 | 456.5 | 511.94 | 69931 | 5980 |

which gives

$$a = \frac{\Sigma Y \Sigma X^2 - \Sigma X \Sigma XY}{\{N\Sigma X^2 - (\Sigma X)^2\}} = \frac{(39 \times 69931 - 456.5 \times 5980)}{(3 \times 69931 - 456.5^2)} = \frac{-2561}{1400.75} = -1.8283$$

$$b = \frac{N\Sigma XY - \Sigma X \Sigma Y}{\{N\Sigma X^2 - (\Sigma X)^2\}} = \frac{(3 \times 5980 - 39 \times 46.5)}{1400.75} = \frac{136.5}{1400.75} = 0.09744$$

The equation becomes $Y = -1.8283 + 0.09744 X$

When $X = 129.0$ cm, $Y = -1.8283 + 0.09744 \times 129.0 = 10.74$ cm

The missing data for the storm is 107.4 mm.

(d) For the stations with the given coordinates the values of $D$ are

$$D_b = 6^2 + 4^2 = 52, D_c = 8^2 + (-6)^2 = 100 \text{ and } D_d = (-4)^2 + 4^2 = 32$$

$$W_b = \frac{1}{52} = 0.01923, W_e = \frac{1}{100} = 0.01, W_d = \frac{1}{32} = 0.03125$$

$$P_x = \frac{\Sigma P_i W_i}{\Sigma W_i} = \frac{(0.01923 \times 12.3 + 0.01 \times 14.8 + 0.03125 \times 11.9)}{(0.01923 + 0.01 + 0.03125)} = 12.5 \text{ cm}$$

**Example 3.5:** Normal rainfall of 6 rain gauge stations $A_1, A_2, A_3, A_4, A_5$ and $A_6$ are 122, 98, 72, 116, 135 and 110 cms respectively. During a particular storm station $A_2$ was inoperative due to some mechanical defect. The location of the stations on a map are given in Fig. 3.13 and bracket terms give the rainfalls. Calculate the missing rainfall of station $A_2$ by all the methods and compare the results.

**Solution**

From Fig. 3.13, coordinates of the points are read as $A_1 (2, 12)$, $A_2(0, 0)$, $A_3(-4, -5)$, $A_4(-12, 2)$, $A_5(8, -8)$, $A_6(10, 1)$. The sign of the coordinate system has no bearing as the

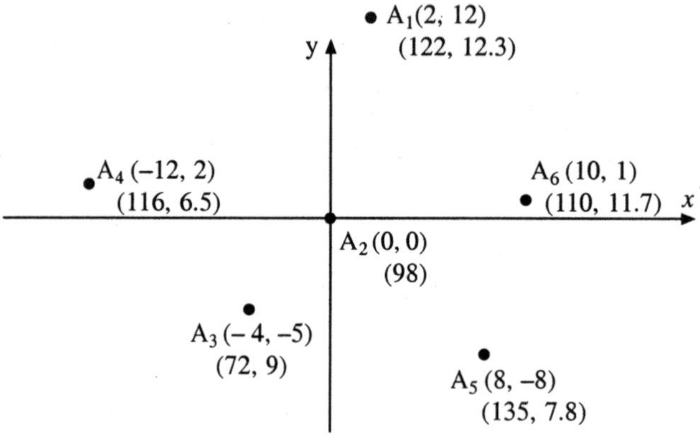

$\bullet$ A$_1$(2, 12)

(122, 12.3)

y

.A$_4$ (–12, 2)

(116, 6.5)

A$_6$ (10, 1)

$\bullet$ (110, 11.7)   x

A$_2$ (0, 0)

(98)

$\bullet$

A$_3$ (– 4, –5)

(72, 9)

$\bullet$

A$_5$ (8, –8)

(135, 7.8)

**Fig. 3.13   Location of Stations for Example 3.5**

square of $\Delta X$ and $\Delta Y$ are taken for calculation. The missing rainfall by various methods are discussed below.

### (i) Arithmetic Mean Method
The normal annual rainfall of the missing station is 98 cm. The surrounding stations do not fall within 10% range, i.e., within 107.8 cm to 88.2 cm. We cannot apply the arithmetic mean method to this problem.

### (ii) Normal Ratio Method
By normal ratio method the storm rainfall for the station is calculated as

$$P_2 = \frac{98}{5}\left\{\frac{12.3}{122} + \frac{9.0}{72} + \frac{6.5}{116} + \frac{7.8}{135} + \frac{11.7}{110}\right\} = 19.6 \times 0.446 = 8.74\ \text{cm}$$

### (iii) Inverse Distance Method
Solution is given in the following table.

**Table 3.7   Inverse Distance Method of Filling the Missing Data**

| Station name | Storm rainfall (cm) | $\Delta X$ (measured | $\Delta Y$ from map) | $D^2 = \Delta X^2 + \Delta Y^2$ | $W_i = 1/D^2$ | $(P_iW_i)$ |
|---|---|---|---|---|---|---|
| A1 | 12.3 | 2 | 12 | 148 | 0.00676 | 0.0831 |
| A2 | — | 0 | 0 | 0 | 0.00000 | 0.0000 |
| A3 | 9.00 | 4 | 5 | 41 | 0.02440 | 0.2196 |
| A4 | 6.5 | 12 | 2 | 148 | 0.00676 | 0.0439 |
| A5 | 7.8 | 8 | 8 | 128 | 0.00781 | 0.0609 |
| A6 | 11.7 | 10 | 1 | 101 | 0.00990 | 0.1158 |
| | | | | | 0.05563 | 0.5233 |

Missing rainfall is $\dfrac{\Sigma(P_iW_i)}{\Sigma W_i} = \dfrac{0.5233}{0.05563} = 9.41\ \text{cm}$

## 3.9   PRESENTATION OF PRECIPITATION DATA
Rainfall is usually presented in the form of the following graphs. Such graphs are useful for analysis and design purposes.

### 3.9.1 Moving Average Curve

In hydrology, rainfall data are plotted chronologically between time in x-axis and precipitation in y-axis. A rain event is associated with randomness. To overcome the random component, a simple moving average of order 3 or 5 is used. This helps to isolate the trend in the rainfall data. If there is any dry or wet cyclic trend associated with precipitation, then such a trend can be clearly visible from the plot. From the graph, the wet period mean, overall mean and dry period mean can be identified. Such a method is applicable to annual series. If $x_1$, $x_2$, $x_3$, $x_4$, $x_5$ are annual precipitation at a station and a 5 year moving average is applied to the series then the following moving mean are computed.

$$X_1 = \frac{(x_1 + x_2 + x_3 + x_4 + x_5)}{5}$$

Similarly
$$X_2 = \frac{(x_2 + x_3 + x_4 + x_5 + x_6)}{5}$$

and
$$X_3 = \frac{(x_3 + x_4 + x_5 + x_6 + x_7)}{5} \tag{3.12}$$

Such a curve can be presented from third year onwards only. For example, if data are available from 1961 to 1996, then a 5 year moving average can be represented from 1963 to 1994. The first two years data of 1961-1962, and the last two years data of 1995-1996 are lost in the moving average process. In hydrology a moving average of more than 5 is usually not applied as some of the cyclic trend associated with the data are smoothened out. A plot of moving average is shown in Fig. 3.14.

### 3.9.2 Mass Curve

When the cumulative rainfall taken as ordinate is plotted against time in abscissa, the resulting plot is a mass curve. Plot of a mass curve gives information about rainfall intensity, duration, magnitude, onset and cessation of precipitation of any storm. All self recording rain gauges automatically record the mass curve of precipitation at a place over time. Therefore, all information about the storm at the place is known from the graph record. A typical mass curve is shown in Fig. 3.15.

### 3.9.3 Rainfall Hyetograph

Rainfall intensity means the ratio of rainfall depth with time and is expressed in cm/hour (cm/h) or mm/hour (mm/h). From the mass curve, a plot between the intensity of rainfall with time can be obtained. During a storm, intensity always changes with time. On a mass curve any two points can be marked and the depth of rainfall ($\Delta y$) between these two points are noted from the y-axis. Time between these two points ($\Delta t$) are recorded from x-axis. The depth divided by time, i.e., $\Delta x/\Delta t$ is the intensity of rainfall for the period under consideration. Usually $\Delta t$ selected is either 1, 2 or 3 hours. For larger cyclonic storms spreading over a few days $\Delta t$ may be selected for 6 or 12 hours.

When the plot of rainfall intensity with time is presented in the form of a bar

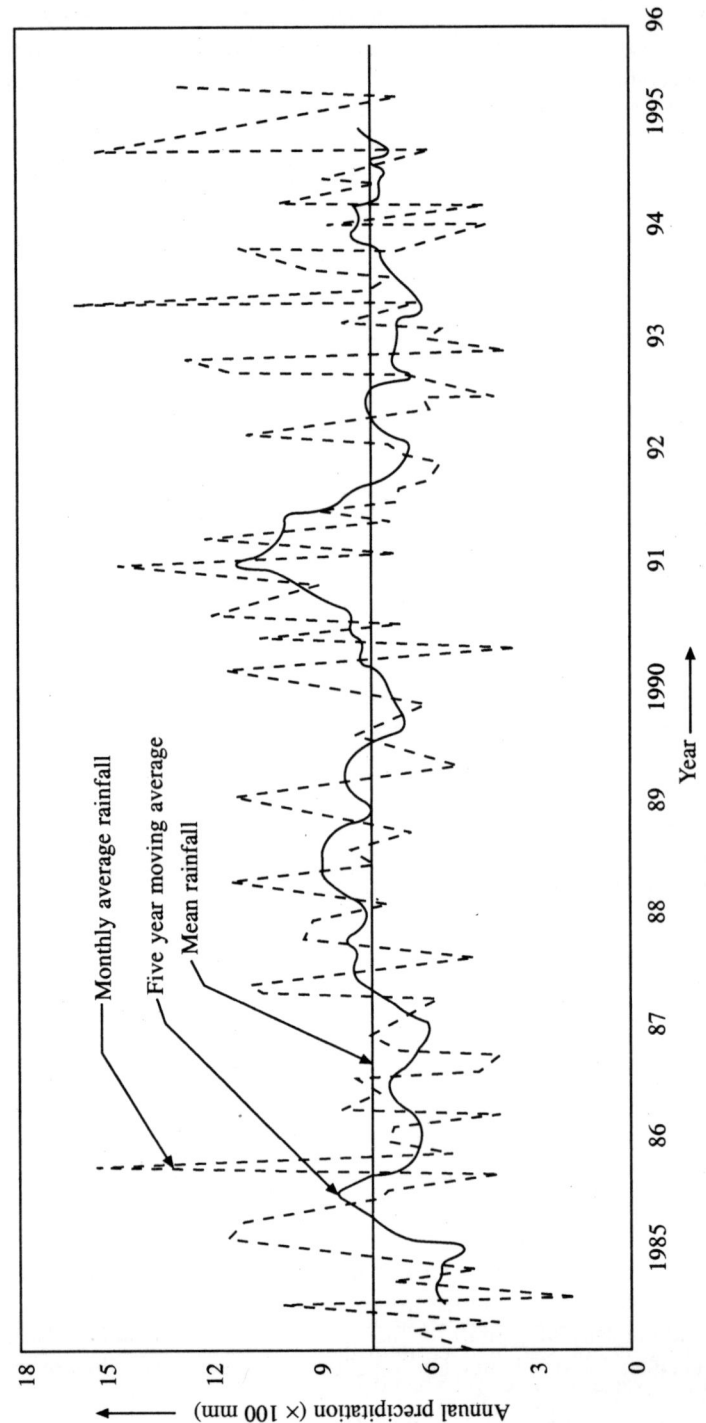

**Fig. 3.14   A Five Year Moving Average Model Applied to Monthly Data of a Station in Orissa**

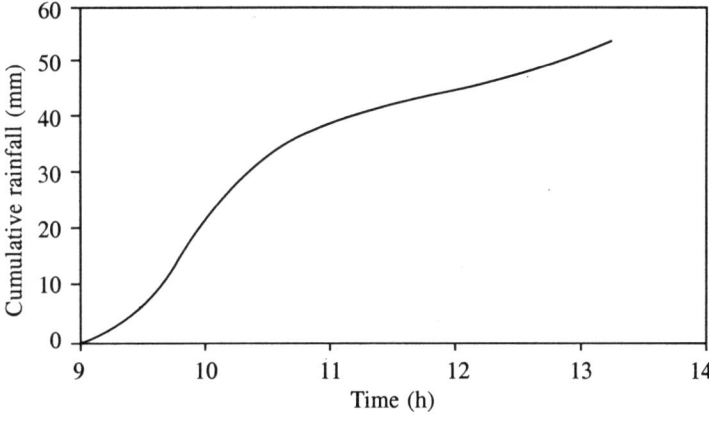

**Fig. 3.15   A Mass Curve**

graph then such a graph is known as hyetograph (Fig. 3.16). The plot is very useful for flood studies and calculation of rainfall loss indics.

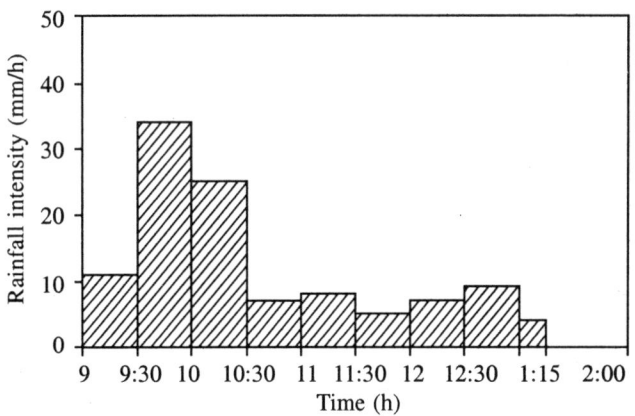

**Fig. 3.16   Rainfall Hyetograph**

### 3.9.4   Intensity-Duration-Frequency Curves

An intensity-duration and frequency curve is a three parameter curve in which duration is taken on x-axis, intensity on y-axis and the return period or frequency as the third parameter. By fixing the return period of say 10-, 50-, 100-years or any other period, a particular curve between intensity and duration can be obtained for the area. Through such a curve, an exponential equation of the following order can be fit.

$$I = CT^a (D + b)^{-d} \tag{3.13}$$

in which $T$ is the frequency or the return period of the storm of intensity $I$ cm/h and duration $D$-hours. Other terms in the equation are constants. For Indian conditions $a$ varies between 0.15 and 0.7, $b$ from 0.10 to 1.05, $c$ from 3 to 15 and $d$ from 0.75 to 1.25. Plot of such a curve is shown in Fig. 3.17.

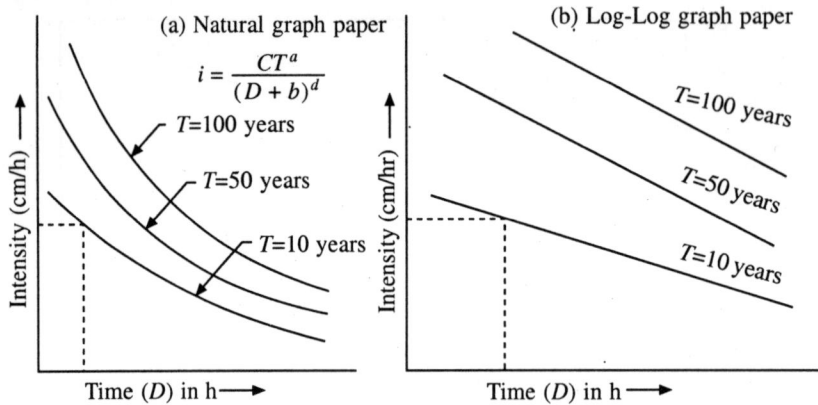

Fig. 3.17   Rainfall Intensity-Duration-Frequency Curve

**Example 3.6:** Rainfall recorded by a self recording rain gauge for the flood event on 28th August, 1996 at a station is given below. Construct a rainfall mass curve and plot the hyetograph for the event.

| Time(h) | 9.0 | 9.5 | 10.0 | 10.5 | 11.0 | 11.5 | 12.0 | 12.5 | 13.0 | 13.25 |
|---------|-----|-----|------|------|------|------|------|------|------|-------|
| Rainfall (cm) | 0.0 | 0.55 | 2.25 | 3.5 | 3.85 | 4.25 | 4.5 | 4.85 | 5.3 | 5.4 |

**Solution**
The cumulative rainfall is plotted on $y$-axis against time on $x$-axis (Fig. 3.15). Rainfall intensities are calculated in Table 3.8.

Table 3.8   Calculation of Rainfall Intensity for Example 3.6

| Time (h) | Cumulative rainfall (mm) | Rainfall depth (mm) | Rainfall intensity (mm/h) |
|----------|--------------------------|---------------------|---------------------------|
| (1) | (2) | (3) | (4) |
| 9.0 | 0.0 | – | – |
| 9.5 | 5.5 | 5.5 | 11.0 |
| 10.0 | 22.5 | 17.0 | 34.0 |
| 10.5 | 35.0 | 12.5 | 25.0 |
| 11.0 | 38.5 | 3.5 | 7.0 |
| 11.5 | 42.5 | 4.0 | 8.0 |
| 12.0 | 45.0 | 2.5 | 5.0 |
| 12.5 | 48.5 | 3.5 | 7.0 |
| 13.0 | 53.0 | 4.5 | 9.0 |
| 13.15 | 54.0 | 1.0 | 4.0 |

Rainfall intensity vs. time is plotted in Fig. 3.16.

## 3.10   MEAN AERIAL RAINFALL

A rain gauge records rainfall at a geographical point. In most of the hydrologic analysis, average depth of precipitation over the area under consideration is required to be computed on hourly, daily, storm period, ten-day, monthly or yearly basis. There are many methods available in literature for computation of average precipitation over the basin. However, depending on the accuracy and objective of the analysis any of the followings methods can be used:

(i) Arithmetic average, (ii) Thiessen polygon, (iii) Isohyetal, (iv) Grid point, (v) Orographic or (vi) Isopercental method.

Other methods available in the literature include the triangular mean weight, trend surface analysis, reciprocal distance square, modified polygon, analysis of variance and the double Fourier series. These methods are not extensively used in hydrology. The rigorous mathematical calculations involved in these methods do not support the accuracy achieved in arriving at the mean aerial rainfall over a basin.

### 3.10.1   Arithmetic Mean Method
This method is suitably applied for a basin where the gauges are uniformly distributed and the individual gauge catches do not vary much from the mean. The basin should be a reasonably flat area. The assumption made is that all gauges weigh equally. This method gives fairly good results if the topographic influences on precipitation and aerial representativeness are considered while selecting the gauge site. It is the simplest form in which the average depth of precipitation over the basin is obtained by taking simple arithmetic mean of all the gauged amounts within the basin

$$P_{av} = \frac{P_i + P_2 + ... + P_n}{n} = \frac{1}{n} \sum_{i=1}^{n} P_i \qquad (3.14)$$

where $P_1$, $P_2$, ... , $P_n$ are the precipitation recorded by $n$ number of gauges located within the basin. This method gives a rough estimate of the average precipitation. It does not account for the topographic and other influences. For use of this method, no gauge station located outside the boundary of the watershed should be considered.

### 3.10.2   Thiessen Polygon Method
In this method, weightage is given to all the measuring gauges on the basis of their aerial coverage on the map, thus eliminating the discrepancies in their spacing over the basin. All the stations in and around the basin are considered and a linear variation in the precipitation between two gauge stations is assumed. Procedure for delineating area for each station in the basin is described below.

(a) All the gauges in and around the basin are accurately marked on a map drawn to scale.
(b) Consecutive stations are joined by dotted straight lines, forming triangles.
(c) Perpendicular bisectors are drawn to these dotted lines such that the bisectors form a polygon around each station. For a basin having large number of stations, the approach should be to start from one end of the map and traverse to the other end gradually forming polygons. The boundary of the map and the perpendicular bisectors cutting them form polygons for the periphery stations represented by *abcc'* for station A and *cdec'* for

station B and so on as shown in Fig. 3.18, while for inner stations, the bisectors only form the polygons represented by stations G, F and K in the figure.

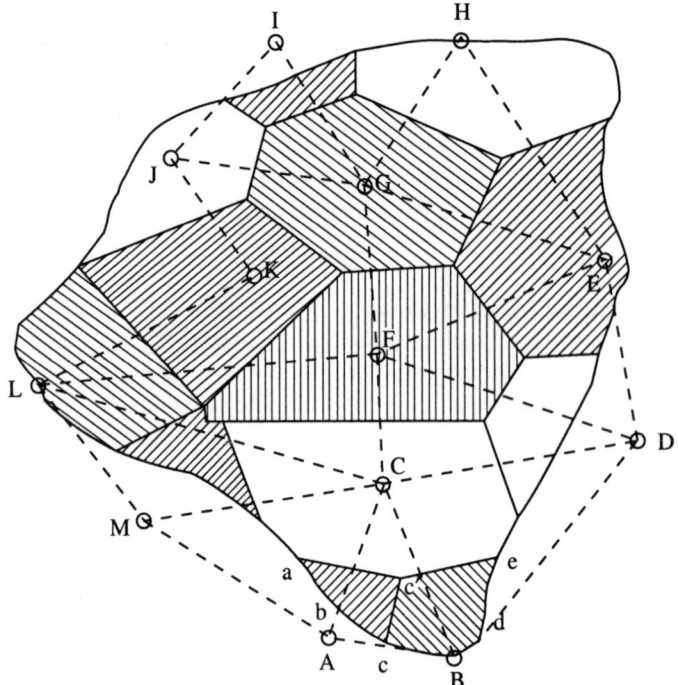

**Fig. 3.18** **A Thiessen Polygon Map (A, B, C are joined to form a triangle. Bisectors of AB, BC and AC meet at point c′. Bisector lines are thick lines joining basin boundary.)**

(d) Each station on the map is thus enclosed by a polygon. A polygon represents an area for which the station rainfall is the representative.

(e) Area of each polygon is measured by planimetering. Sum of the areas of all the polygons must be equal to the total area of the basin.

(f) Thiessen weights are computed by dividing the area of each polygon by the total area of the basin. Thus, sum of Thiessen weights for all stations should be equal to unity. If there are seven stations in and around the basin then seven thiessen polygons are drawn. Sum of all the seven theissen weights must be equal to unity.

(g) The average precipitation is computed from the relation

$$P_{av} = \frac{A_1 P_1 + A_2 P_2 + A_3 P_3 + \ldots + A_n P_n}{A_1 + A_2 + A_3 + \ldots + A_n}$$

or $\qquad P_{av} = P_1 W_1 + P_2 W_2 + P_3 W_3 \ldots + P_n W_n \qquad (3.16)$

in which $P_1, P_2, \ldots, P_n$ represents precipitation at stations 1, 2, 3, . . ., n, and $A_1, A_2, \ldots, A_n$, represents the area of polygons representing the corresponding stations, $A$ is the total area of basin which is the sum of all the polygons and $W_1, W_2, \ldots, W_n$ are Thiessen weights computed as $W_1 = A_1/A$, $W_2 = A_2/A$, $\ldots$, $W_n = A_n/A$ such that $W_1 + W_2 + \ldots + W_n = 1.00$.

This method suffers from the following limitations.

(i) If the network of a basin is changed, i.e., if there is an addition or removal of a station from the basin, then a new thiessen diagram is to be drawn.

(ii) Orographic features are not accounted for.

(iii) A linear variation of precipitation between two stations is assumed, whereas the precipitation is influenced by a large number of meteorological and catchment characteristics.

(iv) Topographic influences and other barriers are not considered.

However the advantage of this method is that, it is much more accurate than the previous methods and the procedure of computation becomes simple, once the areas of the polygons are measured. This method is popularly applied to most of the field problems.

### 3.10.3 Isohyetal Method

This method gives more accurate results of the average rainfall of a basin. An experienced analyst takes care of the orographic features and storm characteristics while drawing contours of equal rainfall depths of a basin. The resulting map can represent the actual rainfall pattern of the storm over the watershed. An *Isohyet* is a line joining points of equal rainfall magnitude on a map. Steps in determining the average precipitation of an area are given below.

(i) A map of the basin is drawn to scale.

(ii) All gauge stations in and around the basin are accurately located on the map.

(iii) Depth of precipitation recorded at each station are marked on the map.

(iv) Isohyets are drawn by eye approximation interpolating the distances between stations on consideration of orographic, storm characteristics and other factors that affect the rainfall variability at the place. They follow the principles of elevation contours drawn on a map. Isohyets do not cross each other. Over a large area, all isohyets form closed contours. Closely spaced isohyet contours indicate that the precipitation gradient is more. Area bounded by the highest closed contour represents the eye of a cyclonic storm.

(v) Area between successive isohyets within the basin is measured. (planimetered).

(vi) Average precipitation between two successive isohyets multiplied by the area bounded by them should be computed for all isohyets covering the area.

(vii) Sum of all such products over the entire basin divided by the total area of the basin gives average precipitation. This can be computed using the following equation.

$$P_{av} = \frac{A_1 \dfrac{P_1 + P_2}{2} + A_2 \dfrac{P_2 + P_3}{2} + \dots + A_{n-1} \dfrac{P_{n-1} + P_n}{2}}{A_1 + A_2 + \dots A_{n-1}} \qquad (3.17)$$

in which $P_1, P_2, P_3, \dots, P_n$ are the isohyetal values such that $P_1$ and $P_2$ bound the area $A_1$, $P_2$ and $P_3$ bound the area $A_2$ and so on. Sum of all such sub-areas bounded by all isohyets, i.e., $A_1 + A_2 + \dots + A_n$ is the total area of the basin as shown in Fig. 3.19. If a linear interpolation between stations is used while drawing isohyets, then the result of Thiessen-Polygon method and this method will essentially be the same. Isohyetal method gives more accurate results than the previous two. To achieve this an experienced hydrologist or the engineer should draw the isohyets with due consideration of the topographical and other influences. Thorough knowledge of the topography of the basin helps to draw isohyets more accurately. Demerit of such a system is that for each storm a separate isohyet is to be drawn. If monthly rainfall is to be computed by isohyetal method then for each month of each year one set of isohyet is to be drawn thus multiplying the work many fold. Since this method permits consideration of topographic features, it is invariably used to calculate the average precipitation during storms for flood studies of various projects.

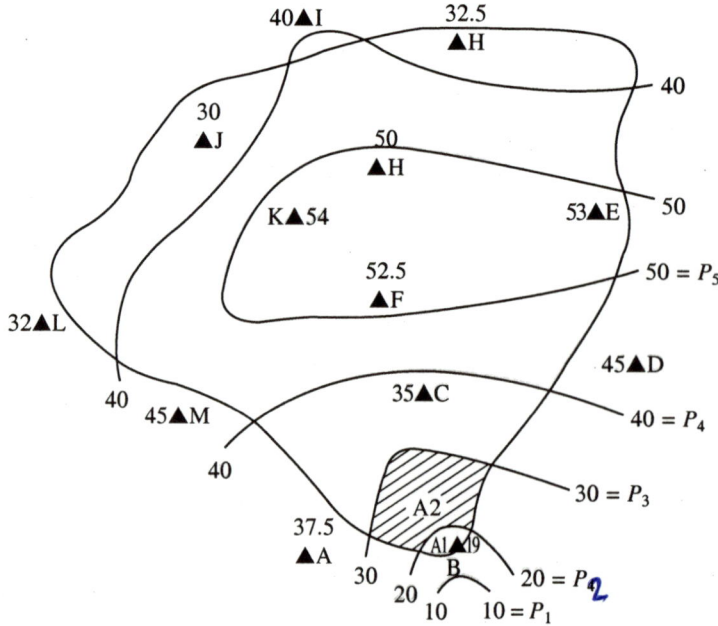

**Fig. 3.19   An Isohyetal Map**

### 3.10.4   Grid Point Method

The following procedure may be followed to obtain average catchment precipitation using grid point method.

(i)  A map of the basin is drawn to scale and all gauging stations are marked accurately on it.

(ii) Depth of precipitation recorded at each station are marked on the map.

(iii) A uniform grid of desired spacing is superimposed on the map.

(iv) Precipitation at the corners of grid points are estimated suitably by considering the weather and storm characteristics of the basin.

(v) Average precipitation of the four grid corners multiplied by area of the grid gives average precipitation volume for the grid.

(vi) Sum of all such products divided by the total area of the basin gives the average aerial precipitation.

The grid point method of calculating average precipitation has certain advantages over other methods. Use of a digital computer is very much needed as the computational procedure is repetitive and lengthy. While using a computer, the basin area needs to be digitised first and the procedure outlined above is adopted. Basin map drawn on autocad is very much useful in calculating the average precipitation of a basin in the shortest time.

### 3.10.5 Orographic Method

This method is identical to the isohyetal method except that the lines of equal elevation above sea level are considered as an additional parameter while drawing the isohyets. The rest of the procedure outlined in method 3.10.3 remains the same.

### 3.10.6 Isopercental Method

This method gives most accurate estimate of the average rainfall when the catchment is affected by orographic and other topographical features. The method is similar to the isohyetal method of estimation of catchment average rainfall. The procedure to calculate the average storm of say monthly rainfall $R_i$ over the catchment are outlined below.

(i) Obtain a map of the basin to scale showing the positions of all rain gauge stations.

(ii) Collect the storm or monthly rainfall data $R_i$ and the normal annual rainfall $N_i$ for the marked stations.

(iii) Convert the storm or monthly rainfall values $R_i$ as the percentage of the normal annual values $N_i$ using the relation $X_i = 100 R_i/N_i$.

(iv) Draw the isopercental lines, i.e., lines joining the percentage of same rainfall over the basin as thick lines by selecting a suitable interval of isopercentages. This can be drawn following the rules of drawing of the contours.

(v) On the same map draw isohyetal lines by considering the normal annual rainfalls over the basin. Let these lines be shown as dotted lines.

(vi) Locate the points where the isopercental lines cut the annual isohyets.

(vii) Calculate the rainfall at the points of intersection of step (6). This can be done by multiplying the values of $(X_{nli}/100)$ and $(N_i)$ in the units of depth of rainfall, where $X_{nli}$ is the value of $X_i$ at the paints of intersection.

(viii) Now draw the isohyetal patterns of the area on a separate map by considering these intersection points and their depth of rainfall.

(ix) Obtain the catchment rainfall over the basin by considering the isohyets as per the procedure outlined in the Section 3.10.3.

The method though gives the most promising alternative of the calculation of average catchment rainfall in a mountainous regions, the method lacks its popularity due to repetitive and difficult nature of computations involved. The data requirement for this method is also more.

---

**Example 3.7:** For a catchment of 590 sq. km rainfall for the month of July along with their areas of influence for the stations are given below. Calculate average precipitation over the basin by (a) Thiessen-polygon method, (b) Isohyetal method.

| Station | A | B | C | D | E | F | G | H | I | J | K | L | M |
|---|---|---|---|---|---|---|---|---|---|---|---|---|---|
| Rainfall for 7/1994 (cm) | 37.5 | 19 | 35 | 45 | 53 | 52.5 | 56 | 32.5 | 40 | 30 | 54 | 32 | 45 |
| Area of influence (km²) (Fig. 3.18) (as per Thiessen-Polygon) | | 3 | 8 | 91 | 15 | 69 | 79 | 88 | 70 | 6 | 30 | 66 | 55 | 10 |

| Isohyet (Fig.3.19) | (< 20 cm) | (20-30 cm) | (30-40 cm) | (40-50 cm) | (50-60 cm) |
|---|---|---|---|---|---|
| area (km²) | 3 | 44 | 82 | 242 | 219 |

**Solution**

Calculation of average precipitation by Thiessen-polygon method is carried out in the following table (see Fig. 3.18).

**Table 3.9   Average Precipitation by Thiessen-Polygon Method for Example 3.7**

| Station | Rainfall (cm) | Area of influence (km²) | Weightage of each station col. (3)/590 | $P_i W_i$ = col. (2) × col. (4) |
|---|---|---|---|---|
| (1) | (2) | (3) | (4) | (5) |
| A | 35.5 | 3 | 0.0051 | 0.063 |
| B | 19.0 | 8 | 0.0136 | 0.258 |
| C | 35.0 | 91 | 0.1542 | 5.398 |
| D | 45.0 | 15 | 0.0254 | 1.144 |
| E | 53.0 | 69 | 0.1169 | 6.198 |
| F | 52.5 | 79 | 0.1339 | 7.030 |
| G | 50.0 | 88 | 0.1491 | 7.458 |
| H | 32.5 | 70 | 0.1186 | 3.856 |
| I | 40.0 | 6 | 0.0102 | 0.017 |
| J | 30.0 | 30 | 0.0509 | 1.525 |
| K | 34.0 | 66 | 0.1119 | 6.041 |
| L | 32.0 | 55 | 0.0932 | 2.9833 |
| M | 45.0 | 10 | 0.0170 | 0.7627 |
| Sum | | 590 | 1.000 | 43.124 |

Average precipitation for July 1994 by Thiessen-Polygon method is found to be 43.1 mm.

(b) Using isohyetal method, average precipitation is calculated in Table 3.10. (refer Fig. 3.19).

**Table 3.10   Average Precipitation by Isohyetal Method for Example 3.7**

| Isohyet interval (cm) | Mean precipitation | Area bounded (km²) | Weightage of the area | Av. Precipitation col. (2) × col. (4) |
|---|---|---|---|---|
| (1) | (2) | (3) | (4) | (5) |
| <20 | 20 | 3 | 0.0051 | 0.102 |
| 20–30 | 25 | 44 | 0.0746 | 1.865 |
| 30–40 | 35 | 82 | 0.1390 | 4.865 |
| 40–50 | 45 | 242 | 0.4102 | 18.455 |
| 50–60 | 55 | 219 | 0.3712 | 20.416 |
| | | 590 | 1.0000 | 45.73 |

Average precipitation by isohyetal method is found to be 45.73 cm.

## 3.11   DEPTH-AREA-DURATION (DAD) CURVE

Depth of precipitation of a storm is related to the area of its coverage and duration of the storm. DAD analysis is carried out to obtain a curve relating the depth of precipitation $D$, area of its coverage $A$ and duration of occurrence of the storm $D$. A DAD curve is a graphical representation of the gradual decrease of depth of precipitation with the progressive increase of the area of the storm, away from the storm center, for a given duration taken as the third parameter. This gives a direct relationship between depth, area and duration of precipitation over the region for which the analysis is carried out. The purpose of DAD analysis is to determine the maximum precipitating amounts that have occurred over various sizes of drainage area during the passage of storm periods of say 6 h, 12 h, 24 h or other durations. There are two methods available for carrying out DAD analysis. The first method, known as *mass curve method* is not extensively used by hydrologists. The second method called the *incremental-isohyetal method* is more popular. Steps in carrying out DAD analysis of a region by the second method are as follows.

(i)   All the major storms of the area are identified.

(ii)   The duration for all the storms are noted (if duration chosen is say 1-day, then all storms occurring for one day are selected. When a storm has occurred for 3 days, then the maximum one day precipitation out of the 3 days is noted).

(iii)   Isohyetal patterns for all 1 day storms are prepared on maps.

(iv)   One 1 day storm is taken up and the area bounded within the highest isohyet is planimetered. This is called the eye-area of the storm. Next the area bounded between the largest and the second largest isohyet is planimetered. The depth of precipitation covering up to the second largest isohyet is obtained by the relation $d_2 = (P_{m1}A_1 + P_{m2}A_2)/(A_1 + A_2)$, where $P_{m1}$ is mean precipitation bounded by the highest isohyetal area $A_1$ and $P_{m2}$ is the mean precipitation between the first and second highest isohyets covering area $A_2$. Similarly for the area covering up to the third isohyet, the progressive depth of precipitation can be obtained by the relation

$d_3 = (P_{m1}A_1 + P_{m2} A_2 + P_{m3}A_3)/(A_1 + A_2 + A_3)$ where $P_{m3}$ is the mean precipitation between the second and third highest isohyet covering area $A_3$ between them. The procedure is repeated to cover the remaining isohyets of the area.

(v) All the area-depth precipitations are recorded in a table.

(vi) The steps from (iv) to (v) are repeated for all other 1 day storm experiences of the area.

(vii) A graph is plotted between area as abscissa and maximum average depths of precipitation as ordinate covering the depth-area data of all 1 day storms of step (vi).

(viii) Such an exercise may also be taken up for 6-h, 12-h, 2- and 3-day storms of the region. The curves are plotted on the same paper as in step (vii) or separate plots may be made for each duration.

(ix) If a semi-log graph paper is used with area plotted on log scale then the curve will plot close to a straight line.

The resulting plot is the DAD curve for the region. DAD analysis requires all the previous storm informations and involves tremendous computational efforts. Use of computers reduces the work load to a great extent. India Meteorological Department on request can carry out DAD analysis for any portion of the country. Depending on the availability of data DAD analysis for any region and for any other periods can be prepared. DAD curves for 12-h, 1 and 2-day storms are shown in Fig. 3.20.

Interesting points to note from the graph are (a) when the area of the storm increases, the depth of precipitation decreases (b) when the duration of the storm increases for the given area, the depth of precipitation increases. Use of DAD graph is further discussed in the flood chapter.

---

**Example 3.8:** From the isohyetal map shown in Fig. 3.19 and the sub-areas between the isohyets given in Example 3.7, prepare a DAD curve for the basin if the rain depths are recorded from a 2-day storm.

**Solution**

On the assumption of the isohyetal patterns of Fig 3.19 resulting from a 2-day storm, the following information are computed in the tabular form, which helps to draw the DAD curve.

**Table 3.11   Depth-Area-Duration Analysis**

| Isohyet interval (cm) | Area enclosed (km$^2$) | Average isohyetal depth | Incremental volume (cm-km$^2$) | Cumulative volume (cm-km$^2$) | Cumulative area (km$^2$) | Average depth (cm) |
|---|---|---|---|---|---|---|
| (1) | (2) | (3) | (4) | (5) | (6) | (7) |
| 50 > | 219 | 50 | 10950 | 10950 | 219 | 50.00 |
| 40–50 | 242 | 45 | 10890 | 21840 | 416 | 47.38 |
| 30–40 | 82 | 35 | 2870 | 24710 | 543 | 45.51 |
| 20–30 | 44 | 25 | 1100 | 25810 | 587 | 43.97 |
| < 20 | 3 | 20 | 60 | 25870 | 590 | 43.85 |

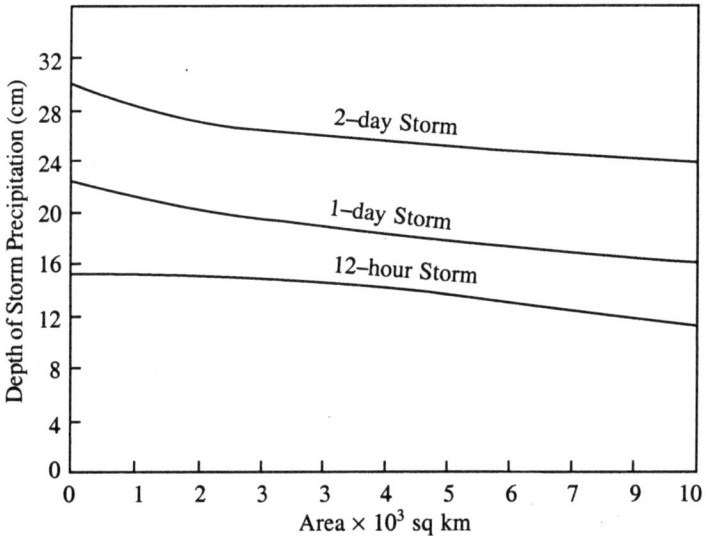

**Fig. 3.20   Depth-Area-Duration Curves of Storm Rainfall**

A plot between the area of column (6) as abscissa vs. the depth of precipitation of column (7) as ordinate, yields the depth-area curve, shown in Fig. 3.21. The third parameter is the duration of 2-day storm. Ideally, such a curve should be plotted from the data of good number of 2-day storms selected from the region and picking up the informations in cols. (6) and (7) in decreasing order from the combined results of all the recorded storms.

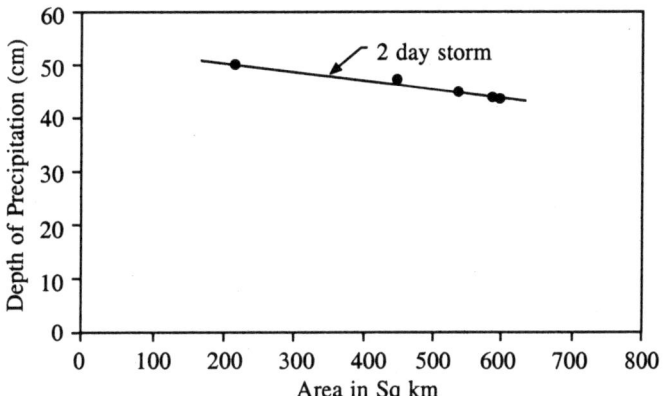

**Fig. 3.21   A 2-day DAD Curve for Example 3.8**

## 3.12   DESIGN STORM

A storm is a period of heavy rainfall associated with high wind speed, caused mainly due to the development of a low pressure zone over large water bodies and moving inland. All water resource projects must take due care for such type of unprecedented precipitations. Design storm is a storm producing a critical depth of rainfall, which is considered for design of a structure in terms of its potential of producing flood acceptable for the safety of the structure. Three terms used in defining a design storm are: (i) *Statistical or Frequency Based Storm* (FBS), (ii) *Probable Maximum Precipitation* (PMP) and (iii) *Standard Project Storm* (SPS). Precipitation record is more extensively available than discharge data at any place of interest of a basin. Therefore, for a developing country like India, more emphasis is given to estimate the design storm for a project, from which an appropriate rainfall-runoff model helps to estimate the design flood.

### 3.12.1  Statistical Storm

For design of structures like small culverts, bridges and minor dams, the probability of occurrence of a storm of desired return period is the interest of the engineer in-charge. Precipitation at a place is a random event and when arranged in time, it constitutes a so-called *time-series*. Annual maximum values of say 24 or 48 h rainfalls, when arranged in time, gives an annual maximum series for the duration of 24 or 48 h. Frequency analysis is carried out of this series to compute probability of occurrence of the event. Annual maximum series usually conform to Gumbel,

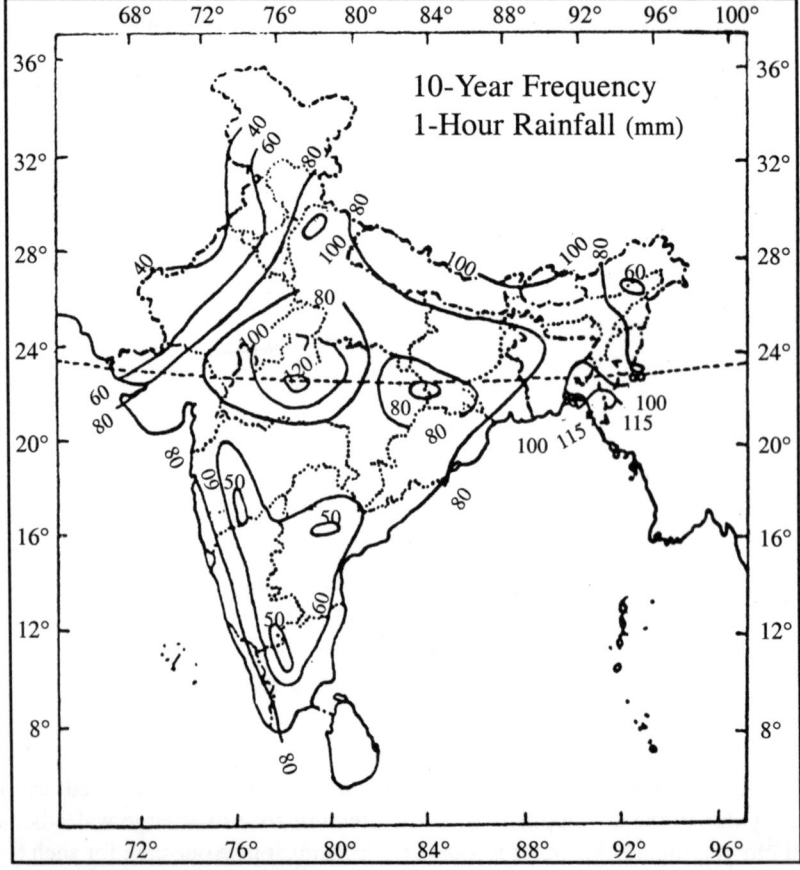

**Fig. 3.22   Map Showing Rainfal Characteristics for Runoff Calculation for 10-year 1-hour Rainfall (mm)**

Pearson type-III, Log-Pearson or Log-Normal distribution. Computation of magnitude of storm of various return periods are discussed in detail in Chapter 2. Rainfall contours for 10, 25 and 50-year return periods for 1 and 6-h duration storms for India are given in Figs. 3.22 through 3.27.

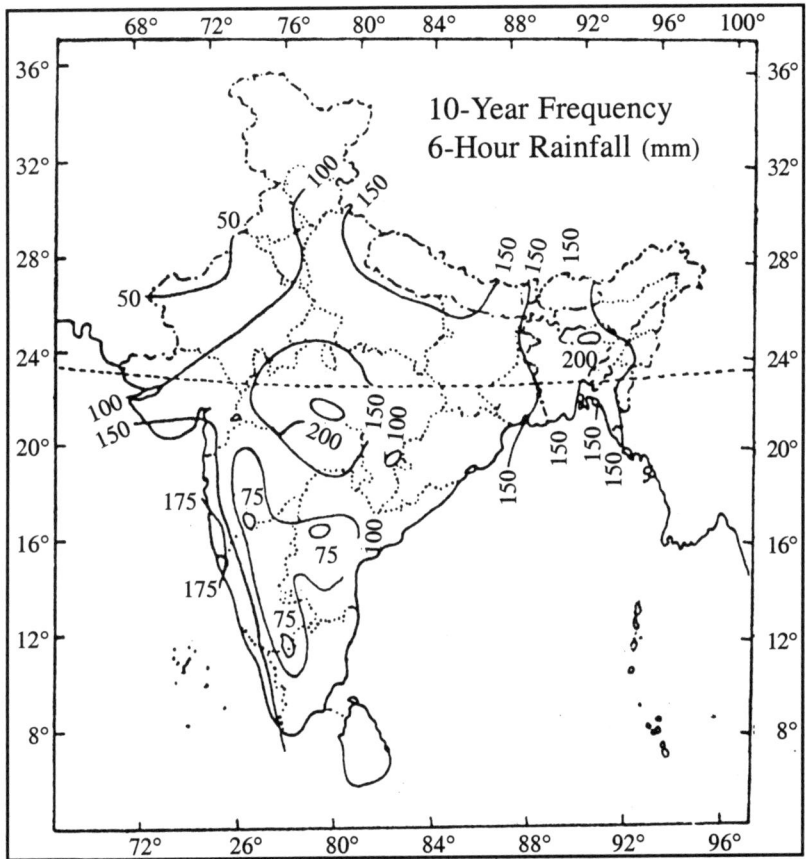

**Fig. 3.23** **Map showing rainfall characteristics for runoff calculation for 10-year, 6-hour rainfall (mm)**

---

**Example 3.9:** Compute 50 and 100 year extreme rainfalls at a place from the data of 72 h annual maximum storms in millimeters observed from 1968–93.

| Year | 1968 | 1969 | 1970 | 1971 | 1972 | 1973 | 1974 | 1975 | 1976 | 1977 | 1978 | 1979 |
|------|------|------|------|------|------|------|------|------|------|------|------|------|
|      | 1980 | 1981 | 1982 | 1983 | 1984 | 1985 | 1986 | 1987 | 1988 | 1989 | 1990 | 1991 |
|      | 1992 | 1993 |      |      |      |      |      |      |      |      |      |      |
| 72h storm | 676 | 756 | 1290 | 293 | 397 | 454 | 810 | 824 | 362 | 755 | 556 | 1098 |
|      | 602 | 1226 | 1100 | 431 | 553 | 429 | 368 | 664 | 1643 | 882 | 1964 | 565 |
|      | 1375 | 1735 |      |      |      |      |      |      |      |      |      |      |

**Solution**

Mean and standard deviation of the annual maximum series for 72-h storm are calculated as:

Mean $X_{av}$ = 838.77 mm
Standard Deviation = $\sigma_{n-1}$ = 459.71 mm

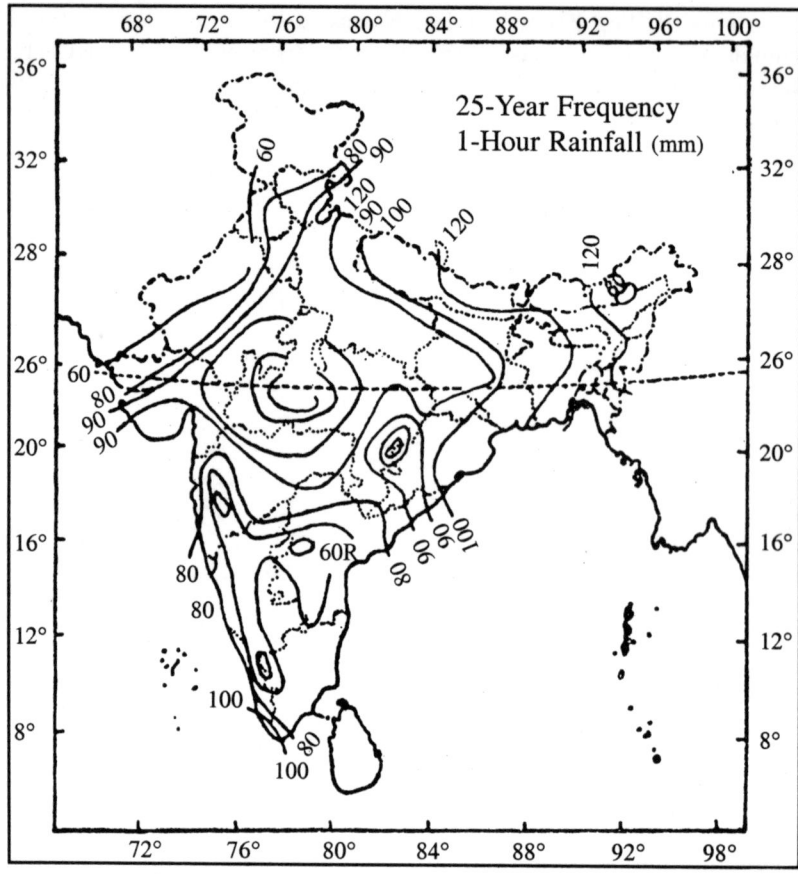

**Fig. 3.24   Map Showing Rainfall Characteristics for Runoff Calculation for 25-year, 1-hour Rainfall (mm)**

For annual maximum series, Gumbel's method of frequency analysis gives the most acceptable results. Using Gumbel's table, the frequency factors are obtained.

For $T = 50 \Rightarrow k_{50} = 3.076$ and for $T = 100 \Rightarrow k_{100} = 3.714$

The storm rainfalls for 50 and 100 year return periods are calculated as follows:

$$X_{50} = 838.77 + 3.076 \times 459.71 = 2252.8 \text{ mm}$$
$$X_{100} = 838.77 + 3.714 \times 459.71 = 2546.1 \text{ mm}$$

The storms for 50 and 100 year return periods are 2252.8 and 2546.1 mm respectively.

**Example 3.10:** Maximum 1-day storm rainfall for 25 years for a place is given. Calculate the 50, 100 and 1000 year return period rainfall by the suitable methods of extrapolation.

| Year | 1971 | 1972 | 1973 | 1974 | 1975 | 1976 | 1977 | 1978 | 1979 | 1980 | 1981 | 1982 |
|---|---|---|---|---|---|---|---|---|---|---|---|---|
| | 1983 | 1984 | 1985 | 1986 | 1987 | 1988 | 1989 | 1990 | 1991 | 1992 | 1993 | 1994 |
| | 1995 | | | | | | | | | | | |
| Rainfall (mm) | 84 | 86.3 | 87 | 94.5 | 102.2 | 81.0 | 106.8 | 69.4 | 91 | 122.8 | 65 | 85.1 |
| | 66.6 | 77.3 | 74.4 | 99.8 | 90.7 | 82.5 | 68.6 | 113.0 | 94.5 | 76.0 | 87.5 | 92.7 |
| | 71.3 | | | | | | | | | | | |

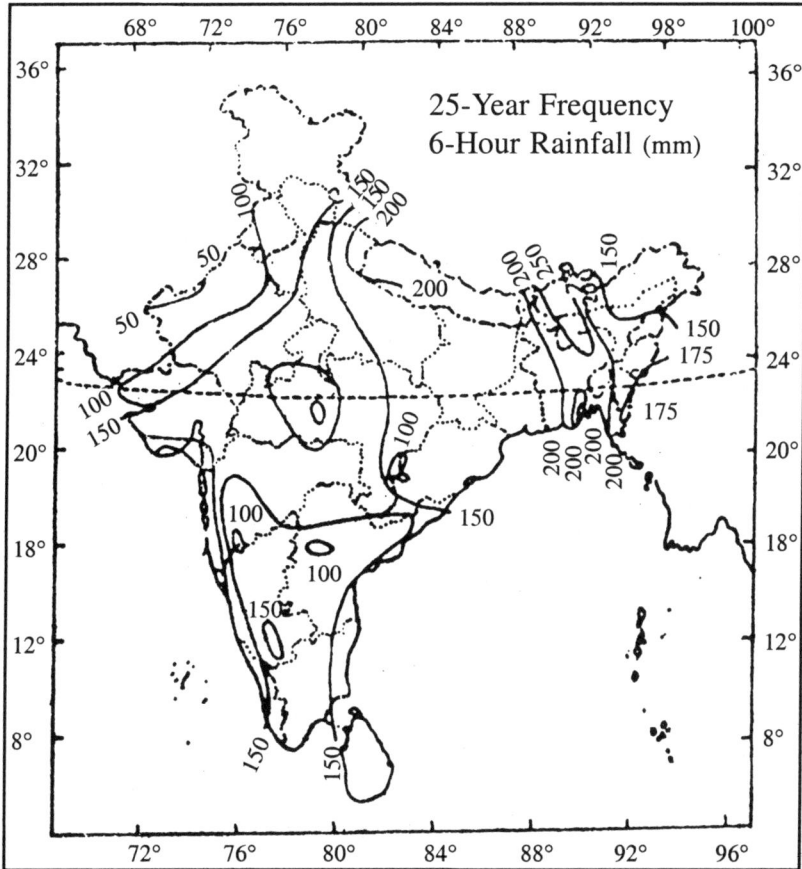

**Fig. 3.25  Map Showing Rainfall Characteristics for Runoff Calculation for 25-year, 6-hour Rainfall (mm)**

**Solution**

The mean, standard deviation and coefficient of skewness for the given data are obtained as (details of estimation is not shown here),

Mean $X_{av}$ = 86.8 mm

$$S.\,D(\sigma) = \sqrt{\frac{5156.66}{(25-1)}} = 14.66 \text{ mm},$$

$$C_v = \frac{14.44}{86.88} = 0.1689$$

$$\text{Coefficient of skewness} = \left[\frac{N}{\{(N-1)(N-2)\}}\right]\frac{\mu_3}{\sigma^3} = \frac{4179376}{(14.66)^3} \times \frac{25}{24 \times 23} = 0.600$$

*For log transferred series*

Mean is calculated as $z_{av}$ = 1.932737

Second moment $\mu_2 = \Sigma\,(x - z_{av})^2 = 0.124357$

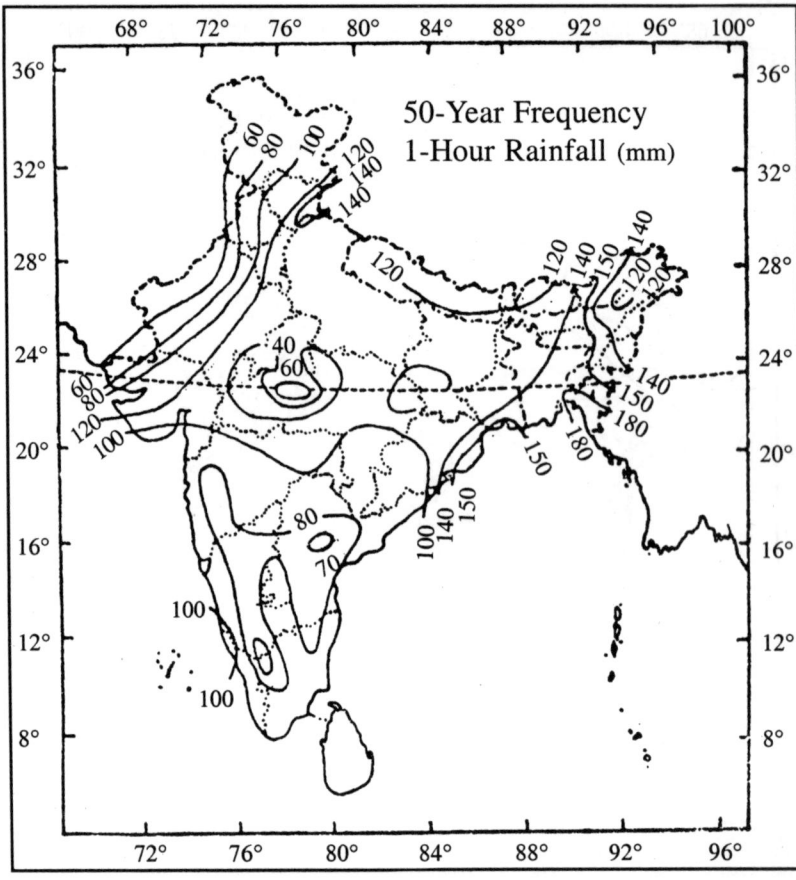

**Fig. 3.26   Map Showing Rainfall Characteristics for Runoff Calculation for 50-year, 1-hour Rainfall (mm)**

Standard Deviation $\sigma_{n-1} = \sqrt{\dfrac{0.124375}{25}} = 0.072$

Third moment $m_3 = \sum (x - z_{av})^3 = 0.001762$

Coefficient of skewness $C_s = \dfrac{0.001762 \times 25}{0.072^3 \times 24 \times 23} = 0.214$

**Gumbel's Method**

From using Gumbel's table (Table 2.7) we get

$k_{50} = 3.088$ ; $X_{50} = 86.8 + 3.088 \times 14.66 = 132.07$ mm

$k_{100} = 3.729$ ; $X_{100} = 86.8 + 3.729 \times 14.66 = 141.47$ mm

$k_{1000} = 5.842$ ; $X_{1000} = 86.8 + 5.842 \times 14.66 = 172.44$ mm

**Pearson Type-III Distribution**

Assuming the data to follow Pearson type-III distribution, the frequency factors for the distribution are taken from Pearson $k$-$T$ table. For coefficient of skewness of 0.60, the frequency factors are obtained from Table 2.8(a) as

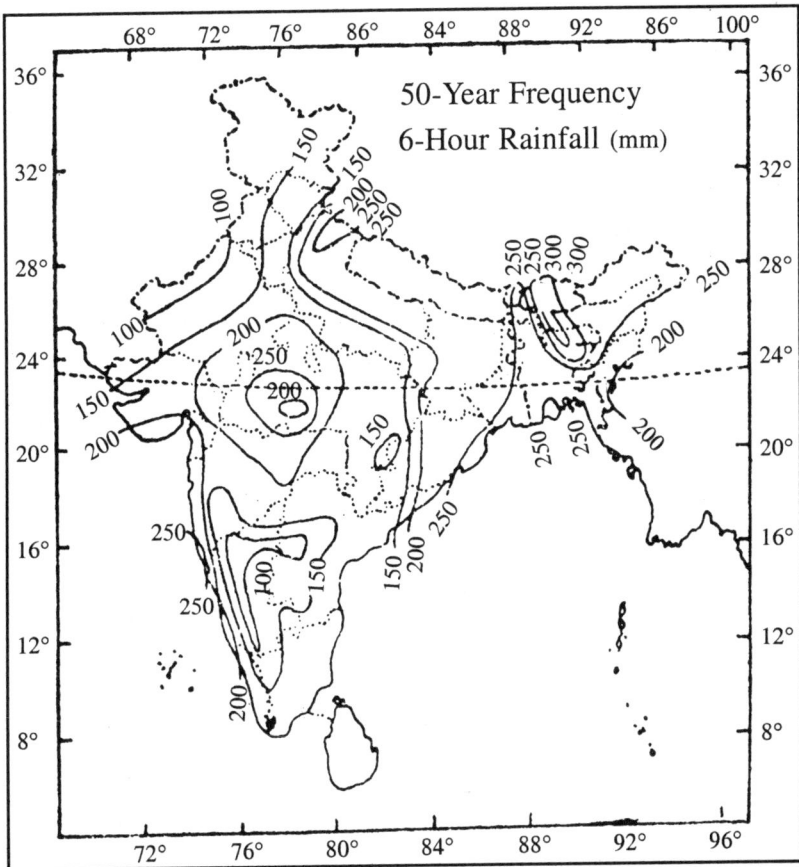

**Fig. 3.27    Map showing Rainfall Characteristics for Runoff Calculation for 50-year, 6-hour Rainfall (mm)**

$k_{50} = 2.359$ ; $X_{50} = 86.8 + 2.359 \times 14.66 = 121.4$ mm

$k_{100} = 2.755$ ; $X_{100} = 86.8 + 2.755 \times 14.66 = 127.2$ mm

$k_{1000} = 3.96$ ; $X_{1000} = 86.8 + 3.96 \times 14.66 = 144.9$ mm

**Log-Pearson-III Distribution**

The frequency factors are obtained from the Pearson distribution table (Table 2.8a). For $C_s = 0.214$

$k_{50} = 2.17$, $y_{50} = 1.932737 + 2.17 \times 0.072 = 2.088977 \Rightarrow X_{50} = 122.73$ mm

$k_{100} = 2.48$, $y_{100} = 1.932737 + 2.48 \times 0.072 = 2.111297 \Rightarrow X_{100} = 129.21$ mm

$k_{1000} = 3.40$, $y_{1000} = 1.932737 + 3.40 \times 0.072 = 2.177537 \Rightarrow X_{1000} = 150.50$ mm

**Log-normal Distribution**

Obtain the frequency factors from Pearson table for $C_s = 0.0$ as

$k_{50} = 2.054$ ; $y_{50} = 1.932737 + 2.054 \times 0.072 = 2.0625 \Rightarrow X_{50} = 120.4$ mm

$k_{100} = 2.326$ ; $y_{100} = 1.932737 + 2.326 \times 0.072 = 2.100209 \Rightarrow X_{100} = 125.95$ mm

$k_{1000} = 3.09$ ; $y_{1000} = 1.932737 + 3.09 \times 0.072 = 2.155217 \Rightarrow X_{1000} = 142.96$ mm

**Normal Distribution**

For $T = 50$ years, $1 - \dfrac{1}{T} = 0.98$, $z = 98\% - 50\% = 48\% = 0.48$

For $z = 0.48$, $k_{50} = 2.055$

$\therefore X_{50} = x_{av} + 2.055 \times \sigma_{n-1} = 86.8 + 2.055 \times 14.66 = 116.93$ mm

Similarly for $T = 100$ years, $1 - \dfrac{1}{100} = \dfrac{99}{100} = 99\%$.

Corresponding to $z = 0.49$, the value of $k_{100} = 2.326$

$\therefore X_{100} = 86.8 + 2.326 \times 14.66 = 120.90$ mm

For 1000 years $1 - \dfrac{1}{T} = \dfrac{999}{1000} = 99.9\%$

From the table for $z = 99.9 - 50 = 49.9$ or $0.499$

$k_{1000} = 3.090$

$\therefore X_{1000} = 86.8 + 3.090 \times 14.66 = 132.160$ mm

**Table 3.12    Graphical Estimation of Storm Magnitudes for Various Return Periods for Example 3.10**

| Year | Rainfall (mm) | Decreasing order of rainfall (mm) | Rank (m) | $\dfrac{m}{N+1} = p$ | $T = \dfrac{1}{p}$ |
|------|------|------|------|------|------|
| (1) | (2) | (3) | (4) | (5) | (6) |
| 1971 | 84.0 | 122.8 | 1 | 0.0385 | 26.0 |
| 1972 | 86.3 | 113.0 | 2 | 0.0769 | 13.0 |
| 1973 | 87.0 | 106.8 | 3 | 0.1154 | 8.8 |
| 1974 | 94.5 | 102.2 | 4 | 0.1538 | 6.5 |
| 1975 | 102.2 | 99.8 | 5 | 0.1923 | 5.20 |
| 1976 | 81.0 | 94.5 | 6 | 0.2307 | 4.33 |
| 1977 | 106.8 | 94.5 | 7 | 0.2692 | 3.71 |
| 1978 | 69.4 | 92.7 | 8 | 0.3076 | 3.28 |
| 1979 | 91.0 | 91.0 | 9 | 0.3461 | 2.89 |
| 1980 | 122.8 | 90.7 | 10 | 0.3846 | 2.60 |
| 1981 | 65.0 | 87.5 | 11 | 0.4231 | 2.36 |
| 1982 | 85.1 | 87.0 | 12 | 0.4615 | 2.16 |
| 1983 | 66.6 | 86.3 | 14 | 0.5385 | 1.85 |
| 1984 | 77.3 | 85.1 | 15 | 0.5769 | 1.73 |
| 1985 | 74.4 | 84.0 | 16 | 0.6154 | 1.63 |
| 1986 | 99.8 | 82.5 | 17 | 0.6538 | 1.53 |
| 1987 | 90.7 | 81.0 | 18 | 0.6923 | 1.44 |
| 1988 | 82.5 | 77.3 | 19 | 0.7308 | 1.37 |
| 1989 | 68.6 | 76.0 | 20 | 0.7692 | 1.30 |
| 1990 | 113.0 | 74.4 | 21 | 0.8077 | 1.24 |
| 1991 | 94.5 | 71.3 | 22 | 0.8461 | 1.18 |
| 1992 | 76.0 | 69.4 | 23 | 0.8846 | 1.13 |
| 1993 | 87.5 | 68.6 | 24 | 0.9230 | 1.08 |
| 1994 | 92.7 | 66.6 | 25 | 0.9615 | 1.04 |
| 1995 | 71.3 | 65.0 | 26 | 0.9630 | 1.04 |

## Using Gumbel Probability Paper

A graph is plotted between the recurrence interval (see Table 3.12 for data) in X-axis and the rainfall as ordinate on a Gumbel probability paper. From the plot (Fig. 3.28), values of $X_{50}$, $X_{100}$, $X_{1000}$ are read as $X_{50}$ = 125.5 mm, $X_{100}$ = 134.0 mm and $X_{1000}$ = 159.0 mm.

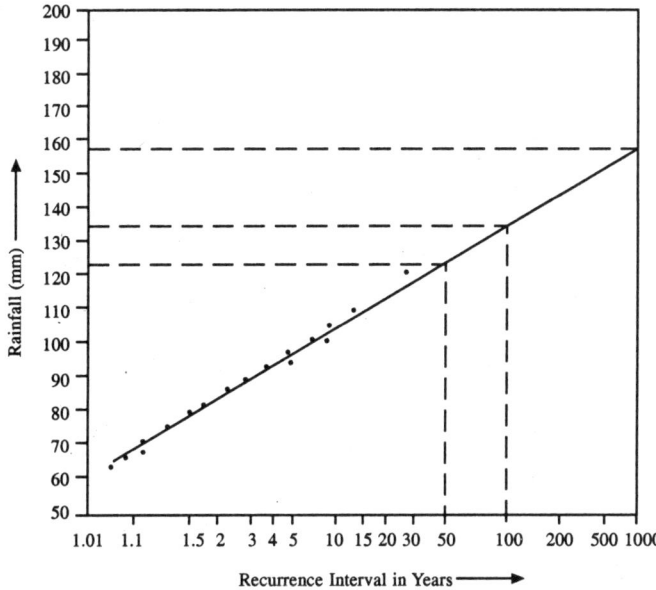

**Fig. 3.28  Fitting of Gumbel Distribution to Extreme Rainfall Data**

## Using Theoretical Equations (Gumble's Method)
Using Gumbel's equation

$$X_{50} = x_{av} - \sigma \frac{\sqrt{6}}{\pi}\left[0.57721 + \ln\left(\ln\frac{T}{T-1}\right)\right]$$

$$= 86.8 - \sigma\frac{\sqrt{6}}{\pi}\left[0.577217 + \ln\left(\ln\left(\frac{50}{49}\right)\right)\right]$$

$$= 86.8 - 0.77970 \times 14.66[0.57721 - 3.90794] = 124.80 \text{ mm}$$

similarly $X_{100} = 86.8 - 0.77970 \times 14.66\left(0.57721 + \ln.\ln\frac{100}{99}\right) = 132.78$ mm

and $\quad X_{1000} = 86.8 - 0.7797 \times 14.66\left(0.57721 + \ln\left(\ln\frac{1000}{999}\right)\right) = 159.15$ mm

**Example 3.11**   Probability of occurrence of rainfall on any day from June to October is 0.20. What is the probability that 5 out of 20 days in the month of July will remain dry?

**Solution**

Here $n = 20$, $x = 5$, $p = 0.2$, $q = 1 - 0.2 = 0.8$

The probability of 5 out of 20 days in July remaining dry can be obtained by considering binomial distribution. Thus

$$P(x) = {}^nc_x p^x q^{n-x} = \left\{ \frac{n!}{x!(n-x)!} \right\} p^x q^{n-x}$$

$$P(x) = \frac{20!}{5! \times (20-5)!} \times 0.25^5 \times 0.8^{(20-5)}$$

$$= \frac{20 \times 19 \times 18 \times 17 \times 16}{1 \times 2 \times 3 \times 4 \times 5} \, 0.25^5 \times 0.8^{15} = 0.1746.$$

Therefore the probability that 5 out of 20 days in July will be dry is 0.1746 or say 17.5%.

**Example 3.12:** A river is dry for 5% days in a year. A crop requires 115 days of water supply from this river for its base period of 120 days. What is the probability that the crop will fail in a year? What is the probability of failure in 4 years?

**Solution**
For the present problem $n = 120$ and $p = 0.05$

Therefore $\lambda = np = 120 \times 0.05 = 6$

Using Poisson's distribution for the problem for $x = 1.0$

$$P(x) = \frac{\lambda^x e^{-\lambda}}{x!} = \frac{6^1 e^{-6}}{1!} = 0.0025 \times 6 = 0.01487$$

The probability of failure of the crop in that year is 0.01487 or say 1.5%.

Probability of failure of the crop in 4 years is obtained by taking $x = 4$

$$P(x) = \frac{6^4 \times e^{-6}}{4!} = 0.135$$

Probability of failure of the crop in 4 years is 13.5%.

### 3.12.2 Probable Maximum Precipitation (PMP)

Failure of a major water resource project leads to immense loss of life and property. No risk can be taken in designing such a structure and therefore no probability can be realistically attached to the storm or flood events for such projects. To keep the probability of failure virtually zero, the maximum possible precipitation that can reasonably be accepted in the location is estimated. At any location there is a physical upper limit of the amount of precipitation possible at a time. Probable maximum precipitation (PMP) is defined as the estimate of the extreme maximum rainfall of a given duration that is physically possible over the basin under critical hydrological and meteorological conditions. Using a suitable rainfall runoff model this precipitation is used to compute flood (considered as design flood for the project). Virtually no risk is taken due to such estimate of maximum precipitation. Two available methods of PMP estimation are: (i) statistical procedure and (ii) meteorological approach. The statistical approach of PMP uses the following Chow's equation

$$\text{PMP} = \overline{P} + k\sigma \tag{3.18}$$

in which $\overline{P}$ is the mean of annual maximum values, $\sigma$ the standard deviation and $k$ is the frequency factor which varies with rainfall duration and is found to vary between 5 and 30. The approach should not be interpreted to imply that a specific probability is assigned to PMP. This method gives a rough estimate of the magnitude of the event.

In meteorological approach, the storm experience of the basin is maximised by taking all the storms of the basin and adjoining areas which are meteorologically homogeneous. Steps involved in obtaining PMP are (i) depth-area-duration analysis of major storms of the region which is considered transposable to the new basin of interest, (ii) maximisation of the storm and (iii) enveloping the maximised values of all the storms to obtain DAD curve of PMP.

Procedure of Depth-Area-Duration DAD analysis is covered in Section 3.11. Storm transposition and maximisation are discussed in this section.

### Storm Transposition

Storm transposition technique is applied to those project areas which do not have adequate rainfall data, i.e., no severe storm has occurred over the project area under investigation. Storm transposition implies the application of outstanding or major rain storms from one area to another. It's main purpose is to increase the storm experience of a basin by considering not only the storms which have occurred over or near the basin in the past but also those which have resulted in heavy rainfall on the adjoining areas that are hydrometeorologically homogeneous. A meteorologically homogeneous area defined as the one, which is affected by the same moisture source, experiences the same type of storms, having similar topographic features and same orientation to seasonal winds. The chief factors which affect homogeneity of the area are

(i) Distance from sea
(ii) Direction of prevailing wind
(iii) Mean annual temperature
(iv) Topography

Various steps in storm transposition are

1. Identify where and when the heaviest storms occurred in the meteorologically homogeneous regions of the project catchment.
2. Identify synoptic situations associated with the storms.
3. Prepare isohyetal patterns of selected rain storms.
4. Transfer the isohyetal maps of each transposable storm to transparent sheets.
5. Superimpose each of the transparent sheets on the problem basin one by one in the most critical manner. The pattern producing maximum depth of precipitation over the catchment is selected as transposable.

The storm yielding maximum depths of precipitation over the basin is considered for further adjustments to obtain the most representative rain depth of the design storm for the problem basin. The following precautions are to be taken while transposing the storm.

(a) Both areas should be meteorologically homogeneous for the typical storm transposed.
(b) If the terrain is hilly then such orographic rainstorms are never transposed to a place far away from the region.
(c) Large latitudinal shift may involve considerable change in air mass characteristics of the storm. Therefore, such large shifts should not be attempted.
(d) Axis of the storm should not be rotated more than 20°. The direction of rotation should be towards the axis of normal storm isohyetal patterns.

### Storm Maximization

Storm maximization is to ascertain by how much the rainfall from a particular storm would have increased by physically possible increase in the meteorological factors which produce the storm. The factors are

1. Efficiency of the mechanism which causes the moisture present in the atmosphere to precipitate.
2. Moisture content of the air mass responsible for the storm in question.

The following maximizations are generally carried out to find the probable maximum precipitation.

### (1) Wind Maximization

Storms in Indian subcontinent are caused due to cyclonic depressions in the Bay of Bengal and Arabian sea. Therefore wind maximisation is not carried out as it is assumed that rain storms are generally associated with high wind speed.

### (2) Maximization Due to Transposition, Barrier and Topography

On transposition, it is necessary to obtain storm depth for change in the moisture regime due to the relocation. If $W_s$ is the precipitable water associated with the storm in the original location and $W_m$ is the precipitable water corresponding to maximum due point temperature in the new location, then the moisture maximization factor together with adjustment for relocation is taken as $W_m/W_s$.

Barrier adjustment is applied to the storms if transposition is carried across mountainous regions of lesser elevation (below 600 m). This accounts for loss in moisture in the ascending air on windward slopes of the mountain. The adjustment factor is $W_{crest}/W_{low}$, where $W_{crest}$ and $W_{low}$ represent precipitable water at the crest and at the foothill.

When storms are transposed from an area close to sea to the project basin far away from sea, the moisture content in the air would be less due to the moisture source being far away. The correction factor for this adjustment is $W_{ml}/W_s$, where $W_{ml}$ is the precipitable water at the new location level .

For computation of PMP for the Indian region, corrections for wind speed, distance from coast, barrier and topography are combined together and can be taken as unit.

### (3) Moisture Maximization

Moisture adjustment of an observed storm is carried out to determine the rainfall which would result if the moisture available to the storm is maximum over the project basin, considering mechanical efficiency to be the same. The *Moisture adjustment factor* (MAF) is defined as the ratio of likely maximum total precipitable water (i.e., maximum total moisture in a column of unit cross section of atmosphere extending from the surface of earth) in the project region to the total precipitable water prevailing at the time of storm in its original location and is expressed as

$$MAF = \frac{W_{max}}{W_{storm}} \tag{3.19}$$

where $W_{max}$ is the likely maximum precipitable water at the project location and $W_{storm}$ is the precipitable water during the storm period. For all practical purposes the moisture in the atmosphere is considered to be limited from 1000 mb (at mean sea level) to a height of 300 mb level. Atmosphere is assumed to contain very less moisture beyond this level.

To find MAF, data of extreme persisting dew point in the project area and the maximum persisting 12-h dew point at the time of storm are essential. *Dew point* is the temperature at which air would become saturated, if cooled at constant atmospheric pressure. We assume that a column of air has pseudo-adiabatic lapse rate for its dew point temperatures. Without transposition, MAF is taken as the ratio of the precipitable water corresponding to maximum dew point, to the precipitable water corresponding the dew point prevailing during the storm. Dew to transposition, MAF is modified to the ratio between precipitable water corresponding to the maximum dew point temperature in the new location of the project region $W_{max}$ and the precipitable water corresponding to the storm dew point $W_{storm}$. Steps involved in computing MAF are:

*Computation of Storm Precipitable Water at Original Location*

1. Obtain the representative storm dew point temperature from meteorological data and average elevation of the original area from toposheet.
2. Reduce the dew point temperature from the elevation of original site to the mean sea level by using Fig. 3.29 or from pseudo-adiabatic lapse rate tables converted from such figures.
3. Compute the depth of storm precipitable water between 1000 and 300 mb using Fig. 3.30 or from tables converted from such figures.
4. Compute the depth of storm precipitable water between 1000 mb (zero elevation) and the average elevation of the storm location from the same figure.
5. Depth of storm precipitable water $W_{storm}$ from original storm area elevation to 300 mb elevation is obtained by subtracting step (4) from (3).

*Computation of Maximum Precipitable Water at Project Location*

6. Obtain the persisting extreme due point temperature of the project area from meteorological records. Find out the average elevation of the project area.
7. Repeat steps (2) to (5) for the location at project site with data of step (6) to compute $W_{max}$.

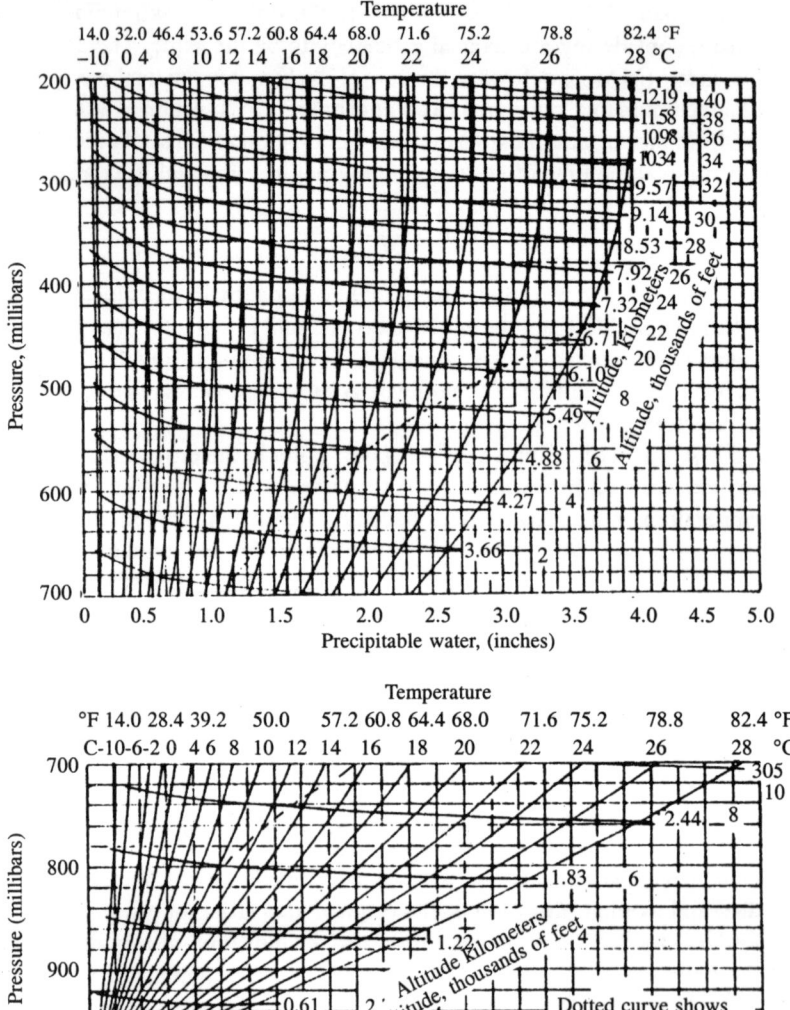

**Fig. 3.29** **Depths of recipitable water in a column of air of any height above the 1000-milibar level as a function of the 1000-milibar dewpoint, assuming saturation and pseudoadiabatic lapse rate (U.S. National Weather Service).**

8. Compute MAF.
9. PMP = Catchment averaged storm depth after transposition × MAF × other adjustment factors.

The following example illustrates the procedure.

---

**Example 3.13:** Compute the probable maximum precipitation at a project site from the following data.

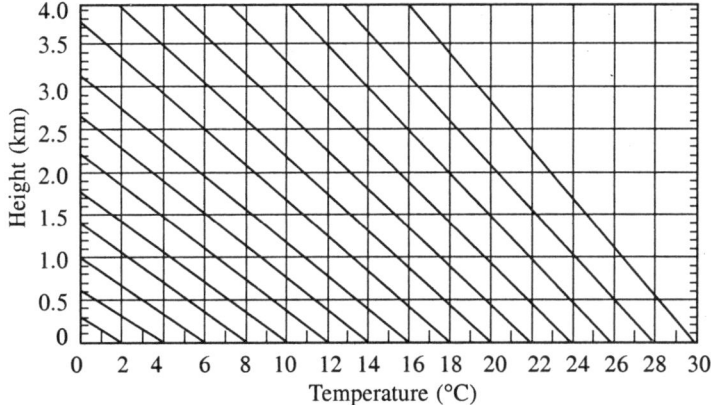

**Fig. 3.30 Pseudoadiabatic Diagram for dew-point Reduction to 1000 mb at the Mean Sea Level (height zero).**

1. Maximum dew point temperature persisting for a period of 12 h during the storm is 22°C.
2. Extreme persisting dew point temperature in the project area is 28°C.
3. Adjustment factors for distance from coast, barrier and topography is 1.00.
4. Maximum recorded rainfall averaged over the catchment after transposition is 230 mm.
5. Average elevation of the original storm site is 127.5 m and the average project area is also at the same elevation.

**Solution**

**At storm location**
1. Maximum persisting representative storm dew point temperature = 22°C
2. Representative storm dew point temperature reduced to 1000 mb level, i.e., from 127.5 m elevation to sea level = 22.5°C.
3. Depth of storm precipitable water between 1000 and 300 mb = 64.5 mm.
4. Depth of storm precipitable water between 1000 mb and original storm location = 4 mm.
5. Depth of storm precipitable water between 300 mb level and original storm level $W_{storm}$ = 64.5 – 4.00 = 60.5 mm.

**At project location**
6. Persisting extreme dew point temperature in project area = 28°C.
7. Extreme dew point temperature reduced to 1000 mb level = 28.3°C.
8. Depth of extreme precipitable water between 1000 and 300 mb level = 103 mm.
9. Depth of extreme precipitable water between 1000 mb and average project area level = 5 mm.
10. Depth of extreme precipitable water between 1000 mb level and average project area level ($W_{max}$) = 103 – 5 = 98 mm
11. Moisture adjustment factor (MAF) = $W_{max}/W_{storm}$ = 98/60.5 = 1.62
    Probable maximum precipitation (PMP) = average storm rainfall over catchment × MAF × other adjustment factors = 230 mm × 1.62 × 1.00 = 372.6 mm.

### 3.12.3 Standard Project Storm (SPS)

It is defined as the largest storm the region has experienced in the period of available rainfall records. The procedure to find PMP may be applied to SPS except that maximisation of the selected major transposable storms are not carried out.

Sometimes exceptionally high storms in the region are not accounted while obtaining SPS on the consideration that it may be a freak event. This event may not be part of the general characteristic of the region. SPS is often used to compute *Standard Project Flood* (SPF), where failure of the structure would have somewhat less disastrous effect. Standard project storm is used to compute design flood for structures like permanent barrages, minor dams (with capacity less than 60 m$^3$), important bridges and other structures. Peak discharge computed from SPS is generally 40 to 60% of probable maximum flood calculated from PMP for the same drainage basin.

## PROBLEMS

3.1   (a) What are the essential requirements for precipitation to occur?
      (b) Describe various types of precipitation.
      (c) What do you understand by the term 'forms of precipitation'?
          What forms are important for water resources engineering?
3.2   (a) What are the various methods available to record rainfall depths at a place?
      (b) What considerations you will make for selecting a precipitation site?
3.3   Explain how are you going to check the consistency of data obtained from a station?
3.4   Describe how you are going to supplement a missing rainfall data?
3.5   How is the density of rain gauge to affect the accuracy of precipitation over a basin?
3.6   What are the various methods available to calculate average precipitation over a basin?
3.7   What do you understand by *PMP*? How are you going to calculate *PMP* over your project catchment?
3.8   Write the time-space-frequency relationship of storms over a region.
3.9   Describe rainfall in India.
3.10  A pentagonal plot consists of a rectangle and a triangle. The triangle is attached to one side of the rectangle forming the pentagon. The sides of rectangle are 3, 5, 3 and 5 km and that of the triangle are 5, 4 and 4 km. If the rainfall at the corners of the rectangle is 12, 8.0, 13.5, 11 cm and that at the vortex of the triangle is 7 cm, resulting from the storm, calculate the average precipitation.
3.11  For the month of August, the isohyetal patterns of rainfall over a catchment gives the following information.

| Isohyet Interval (cm) | 0–3 | 3–6 | 6–9 | 9–12 | 12–15 | 15–18 | 18–21 | <24 |
|---|---|---|---|---|---|---|---|---|
| Area bounded between Isohyets ( km$^2$) | 18 | 42 | 88 | 110 | 72 | 51 | 33 | 12 |

Calculate the average precipitation for the month of August for the catchment.
3.12  There are four rain gauge stations neighbouring a gauge 'A', which was inoperative during a storm. The records show that the storm rainfall for the four stations are 13.7, 14.1, 14.5 and 12.6 cms and the respective normal precipitation of the stations are 140, 146, 157 and 122 cms. If the normal rainfall of station A is 131cm, calculate storm precipitation of station A.

3.13 The following data from a self recording gauge during a storm are obtained.

| Time (h) | 10.15 | 10.3 | 11.0 | 11.3 | 12.0 | 12.3 | 12.45 | 13.0 |
|---|---|---|---|---|---|---|---|---|
| Cumulative rainfall (cm) | 2.2 | 4.0 | 5.1 | 6.7 | 8.8 | 9.1 | 9.5 | 9.7 |

Plot the rainfall hyetograph. Also plot an intensity-duration curve from the data.

3.14 To check the consistency of a station '$X$' the following data are collected from the neighbouring stations.

| Year | 1994 | 1993 | 1992 | 1991 | 1990 | 1989 | 1988 | 1987 | 1986 |
|---|---|---|---|---|---|---|---|---|---|
| | 1985 | 1984 | 1983 | 1982 | 1981 | 1980 | 1979 | 1978 | 1977 |
| | 1976 | 1975 | | | | | | | |
| July ppt. | 17.1 | 14.3 | 17.5 | 16.0 | 19.1 | 16.7 | 19.4 | 14.5 | 16.1 |
| of station X | 20.0 | 13.9 | 15.7 | 14.1 | 13.0 | 10.1 | 15.1 | 14.2 | 13.9 |
| | 13.1 | 14.0 | | | | | | | |
| Seven | 14.2 | 13.1 | 14.7 | ·14.8 | 16.2 | 15.6 | 15.1 | 11.8 | 12.9 |
| station avg. | 19.4 | 15.7 | 16.5 | 15.6 | 14.4 | 11.2 | 13.6 | 16.4 | 13.4 |
| precipitation | 14.3 | 12.9 | | | | | | | |
| for July (cm) | | | | | | | | | |

Is the data of station '$X$' consistent? If not, correct the data and find out the mean precipitation over the sub-basin.

3.15 Allowing 7% error in estimating the mean rainfall data, calculate the minimum number of additional stations required over the existing 6 number of gauges to represent the basin adequately. The annual rainfall data for stations are 1280, 1440, 1200, 1090, 1660 and 1030 mm respectively.

3.16 Annual rainfall in cm for station 'A' from 1964 to 1993 are as follows.
130.0, 136.1, 132.3, 127.9, 140.1, 129.3, 128.5, 136.6, 125.6, 139.3, 118.5, 116.3, 137.7, 132.1, 133.6, 135.5, 130.3, 122.7, 120.2, 128.8, 138.3, 131.7, 121.1, 117.7, 112.3, 139.6, 141.1, 133.3, 129.2, 126.6.
Calculate the rainfall of return periods of 10-, 60-, 100-, 200- and 400 years. What is the probability of rainfall of magnitude 140 cm?

3.17 A station A was inoperative during September of 1999. Six neighbouring stations were operational during the period. The coordinates of the stations on a rectangular system superimposed on the basin map with station A coinciding (0, 0) position are given below.

| Station | Rainfall (cm) | Coordinates of the station | |
|---|---|---|---|
| | | $\Delta x$ | $\Delta y$ |
| B | 8.3 | 4.0 | 5.0 |
| C | 10.1 | 6.0 | 8.0 |
| D | 7.7 | 9.0 | 6.0 |
| E | 12.2 | 3.0 | 4.0 |
| F | 11.3 | 4.0 | 6.0 |
| G | 12.4 | 5.0 | 7.0 |
| A | – | 0.0 | 0.0 |

Calculate the missing rainfall of station 'A'.

3.18 Plot the annual rainfall data of Problem 3.16. Apply a simple 3-year moving average over it. Also plot the mean of the series and comment on the nature of the data.

3.19 What do you understand by network design? Is it possible to design a rain gauge network with zero percent error in its mean? What are the standards prescribed by IMD and WMO in network design?

3.20 Annual precipitations in mm for five stations are given as follows.

| Year | A | B | C | D | E |
|------|------|------|------|------|------|
| 1985 | 1522 | 1177 | 1230 | 1042 | 1197 |
| 1986 | 1810 | 1683 | 1609 | 1552 | 1612 |
| 1987 | 1550 | 1352 | 1505 | 1204 | 1306 |
| 1988 | 1290 | 1275 | 1186 | 1222 | 1173 |
| 1989 | 1460 | 1157 | 1296 | 1403 | 1398 |
| 1990 | 1199 | 1122 | 1354 | 1451 | 1315 |
| 1991 | 1104 | 1060 | 1225 | 1003 | 1080 |
| 1992 | 948 | 1076 | 1120 | 901 | 962 |
| 1993 | – | 1156 | 1302 | 1412 | 1388 |
| 1994 | 1273 | 1231 | 1248 | 1178 | 1112 |

Find the missing rainfall of station A for the year 1993. Check the consistency of data by double mass curve method for stations A and C. Suggest any corrections.

3.21 An experimental rectangular plot of 10 km × 12 km has five rain gauge stations. Fit a coordinate system to the plot such that the side 10 km represents the abscissa. The storm rainfall and coordinates of the stations are as follows.

| Station | Station coordinate | Normal annual rainfall (cm) | Storm rainfall (cm) |
|---------|--------------------|-----------------------------|----------------------|
| A | (1, 3) | 128 | 12 |
| B | (8, 11) | 114 | 11.4 |
| C | (3, 10) | 136 | 13.2 |
| D | (5, 8) | 144 | 14.6 |
| E | (7, 5) | 109 | ? |

Compute the missing rainfall of station E. Find the average rainfall of the plot by Thiessen Polygon and isohyetal method.

3.22 Compute the *PMP* at a project site from the following data.
Maximum dew point temperature persisting for 12 h during storm = 18 °C
Extreme Persisting due point temperature in the project area = 24 °C
Maximum average catchment rainfall due to storm after transposition = 240 mm
Average elevation of the original storm area = 120 m
Average elevation of the project area = 160 m
Take other adjustment factors = 1.0

3.23 Is there any relationship that exists between area and rainfall depth, intensity and duration of the storm? Describe them individually.

3.24 (a) What do you understand by standard project storm? How it is different from probable maximum precipitation?

(b) What do understand by a 100-year-2-day storm at a place?

# Chapter Four

# Losses From Precipitation

## 4.1 INTRODUCTION

Losses from precipitation in engineering is defined as the quantity which does not yield for hydropower generation, irrigation, domestic water supply, navigation and other uses. Thus, for a surface water resource engineer the difference between precipitation and runoff in a stream is, a loss, which can be taken as the sum of the losses of (i) evaporation, (ii) transpiration, (iii) interception, (iv) depression storage and (v) infiltration. These losses form a major portion of the hydrologic cycle. Evaporation is a major loss followed by transpiration and infiltration. Infiltration, which might be a loss to a surface water hydrologist is a major gain to those dealing with ground water potential and utilisation. Mathematically,

$$\text{Precipitation} - \text{Surface runoff} = \text{Total loss} \qquad (4.1a)$$

$$\text{Total loss} = \text{Evaporation} + \text{Transpiration} + \text{Interception}$$
$$+ \text{Depression Storage} + \text{Infiltration} \qquad (4.1b)$$

This chapter discusses above hydrologic losses to solve most of the day-to-day field problems.

## 4.2 EVAPORATION AND ITS ESTIMATION

It is a continuous natural process by which a substance changes from liquid to gaseous state. For arid regions, where the net annual evaporation is very significant, the loss can be as high as 90% of the annual precipitation (Table 4.1). Therefore, evaporation may be defined as loss of water to the atmosphere over the period under consideration. The main source of evaporation is solar radiation. One gram of water requires about 597 calories of heat at 0°C or one gram of ice at 0°C requires about 677 calories for evaporation. This heat is known as the latent heat of water and is supplied by the sun. When water surface gets heated, the kinetic energy of water molecules increases. A stage comes when the surface tension and cohesive force cannot hold the molecules together and they get projected up into the atmosphere. The equivalent molecular weight of air is 28.95 and that of water vapour is 18, i.e., water vapour is 62% lighter than air. This helps water vapour to rise into the atmosphere to a height where it condenses. There is always a denser layer of water vapour next to the water surface which decreases with elevation. As more molecules join the atmosphere its density near the water surface further increases. Some projected water molecules collide with the water vapour in the atmosphere close to the water surface and return back. However, there is net evaporation taking place during warm periods. By this process, temperature of the water surface is maintained at lower level. In a hydrologic cycle, evaporation takes place from all stages of its process and storages. Even the falling raindrops evaporate. A hydrologist normally limits his area of interest to that part of evaporation occurring in land phases of hydrologic cycle. Importance of evaporation and its potential can be gauged from water budget of continents and oceans (Table 4.1).

**Table 4.1   Evaporation from Continents**

| Continent | Precipitation (mm/year) | Evaporation (mm/year) | % Loss by evaporation (mm/year) | Runoff (mm/year) |
|---|---|---|---|---|
| (1) | (2) | (3) | (4) | (5) |
| Africa | 690 | 430 | 62.3 | 260 |
| Asia | 600 | 310 | 51.7 | 290 |
| Australia | 470 | 420 | 89.4 | 50 |
| Europe | 640 | 390 | 60.9 | 250 |
| North America | 660 | 320 | 48.5 | 340 |
| South America | 1630 | 700 | 42.9 | 930 |
| Land area | 730 | 420 | 57.5 | 310 |

Average annual rainfall in India is about 1120 mm which is equal to 370 million hectare-m of water. Total runoff by all the rivers of the country is 170 million hectare-m. If annual ground water recharge is 37 million hectare-m, then the balance of 163 million hectare-m is lost to atmosphere as evaporation and transpiration. Part of the 37 million hectare-m is also lost to atmosphere as transpiration. The factors responsible for evaporation are

(i) Meteorological: (a) vapour pressure, (b) solar radiation, (c) air temperature, (d) wind velocity and (e) atmospheric pressure.
(ii) Nature of evaporating surface and
(iii) Quality of water.

Following Dalton's law, the rate of evaporation $E$ (mm/day) can be related to vapour pressure as

$$E = c(e_s - e_a)$$

or
$$E = (e_s - e_a)(a + bv) \tag{4.2}$$

where $e_s$ is the saturation vapour pressure at water surface temperature in millibar, $e_a$ the vapour pressure of air in millibar, $a$, $b$, $c$ are constants and $v$ is the wind speed in km/h. For evaporation to continue, $e_a$ should be less than $e_s$. If $(e_s - e_a)$ is large, evaporation is high and when $e_s = e_a$, evaporation is zero. Increase in air temperature increases the evaporation rate though not always proportionaly. For the same temperature, colder months have less evaporation than summer months due to the combined effect of other environmental parameters. Wind removes the overlying vapour from an evaporating body thereby increasing the rate of evaporation. There is a relation between wind speed and size of water bodies or evaporating surface. High wind speed is not necessary to remove water vapour from a small water body. Increase in atmospheric pressure decreases the rate of evaporation as can be noticed in deep valleys. Soluble salts reduce the rate of evaporation to the extent of the percentage of salts present in the water. For example, the Bay of Bengal has salt concentration of 3.15%, its evaporation rate is nearly 3% less than the fresh water. Different evaporating surfaces like soil, barren land, forest area, houses and lakes affect evaporation to the extent they have the potential. Black cotton soils help to evaporate the soil water faster than red soil because such soils have the potential to absorb incoming radiation more effectively. Evaporation from wet soil is faster and it reduces gradually as soil becomes drier. Deep water bodies evaporate slower than shallow water bodies in summer and in winter season they evaporate faster.

It is rather impossible to measure evaporation directly in field. Evaporation from water surface is estimated by different methods and its values are correlated to field data. The methods available to estimate evaporation losses from surfaces of large water bodies are:

(i) Measurement using evaporation pans
(ii) Empirical equations
(iii) Water balance method
(iv) Energy budget method
(v) Mass transfer method.

The methods are discussed in the following sections.

### 4.2.1 Measurement Using Evaporation Pans

Evaporation measurement using pans is the most reliable and the best of all the available methods. An evaporimeter as specified by IS:5973–1970 for Indian conditions is a pan of 1.22 m in diameter and 0.255 m deep. It is a modified version of the US Weather Bureau class A pan since the latter is made of galvanised iron and is not painted white outside. The Indian Standard pan is made up of 0.90 mm thick copper sheet with hexagonal wire netting of galvanised iron mesh covering it to protect its water from birds. Details of an IS specified pan is given in Fig. 4.1. The pan is placed over a wooden platform of 10 cm height so that circulation of air is possible all around the pan. This also helps to thermally insulate the pan completely from ground. Water level in the pan is recorded by a point gauge arrangement placed inside a stilling basin. Normally, an evaporation pan is placed along with other weather measuring instruments for recording humidity, temperature, rainfall, wind speed and other parameters. Measurement is taken at least once a day by adding water to the pan by a calibrated cylindrical glass jar to bring the water level to the previous position. This gives directly the evaporation depth over the time lapse. If there is rainfall exceeding the depth of evaporation, then water is taken out of the pan in the same way by the measuring jar and knowing the depth of rainfall from the rain gauge the evaporation depth is found out by subtraction.

Other prominent pans available are: (i) Sunken pan (Colorado Bureau of

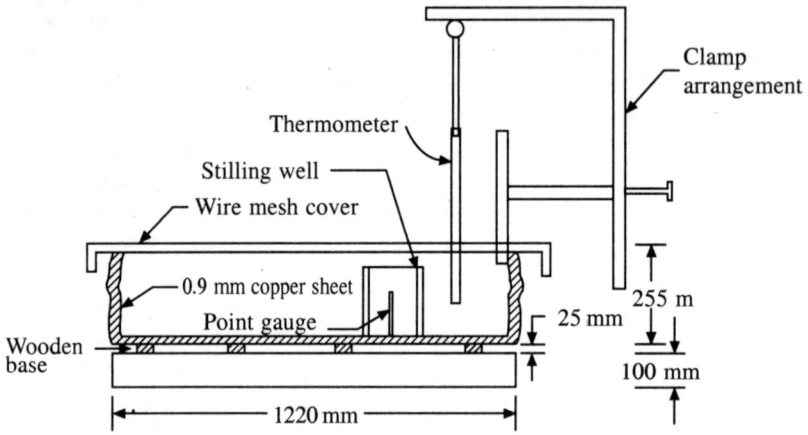

**Fig. 4.1   Indian Standard Evaporation Pan**

Plant Industry Pan), (ii) US Geological Survey Floating pan, (iii) US Weather Service Class A Land Pan and (iv) GGI-3000 pan (for details see World Meteorological Organisation reports for different type of pans used in various countries). Sometimes an *Atmometer* is used for measuring evaporation. This essentially consists of a porous bulb drawing water from a container. Evaporation takes place from the bulb. A coefficient when multiplied to the depth of evaporated water from the porous bulb gives the required evaporation data for the day.

Evaporation recorded by a pan differs from that of a lake or reservoir due to: (i) depth of exposure of pan above ground, (ii) colour of the pan, (iii) height of the rim, (iv) heat storage and heat transfer capacity with respect to reservoir, (v) pan diameter, (vi) variation in vapour pressure, wind speed and water temperature (unless it is a floating pan which records the rate of evaporation from a lake directly). We have to reduce the evaporation recorded by the pan to that of the lake or reservoir by multiplying a pan coefficient between 0.60 and 0.80.

$$\text{Lake evaporation} = \text{Pan coefficient} \times \text{Pan evaporation} \qquad (4.3)$$

A relation between pan diameter and ratio of pan evaporation to lake evaporation is given in Table 4.2.

**Table 4.2   Relation Between Pan Diameter and Pan Coefficient to Convert to Lake Evaporation**

| Pan diameter (m) | 4.0 | 3.0 | 2.0 | 1.5 | 1.0 | 0.5 | 0.1 |
|---|---|---|---|---|---|---|---|
| Ratio of pan evaporation to evaporation from lakes of 7.3 km² | 1.16 | 1.18 | 1.21 | 1.28 | 1.33 | 1.45 | 1.8 |
| Pan coefficient | 0.86 | 0.85 | 0.83 | 0.78 | 0.75 | 0.70 | 0.56 |

Lake evaporation is used as the potential value for computation of crop water requirement by climatic approach. Evaporation from lake is considered the same as that of a saturated soil covered with vegetation.

**Example 4.1:** Annual pan evaporation from an observatory in Orissa is 120 cm. The reservoir water spread area varies from a maximum of 10.8 sq. km in the beginning of January to a minimum of 3.6 sq. km in May and is back to a level of 10.8 sq. km at the end of December. Calculate loss of water from the reservoir during the year. Assume pan coefficient as 0.80.

**Solution**
Mean area of water spread is computed by cone formula as

$$A_m = \frac{1}{3} \times \frac{5}{12} \{A_1 + A_2 + (A_1 A_2)^{0.5}\} + \frac{1}{3} \times \frac{7}{12} \{A_1 + A_2 + (A_1 A_2)^{0.5})$$

$$= \frac{1}{3} \times \frac{5}{12} \{10.8 + 3.6 + (10.8 \times 3.6)^{0.5}\} + \frac{1}{3} \times \frac{7}{12} \{10.8 + 3.6 + (10.8 \times 3.6)^{0.5}\}$$

$$= 2.866 + 3.862 = 6.728 \text{ km}^2.$$

Therefore, annual volume of water lost from the reservoir assuming a pan coefficient of 0.80 is calculated as

$$V_1 = (6.728 \times 10^6 \text{ m}^2) \times \left(\frac{120}{100} \text{ m}\right) \times 0.80 = 6.459 \times 10^6 \text{ m}^3.$$

Total quantity of water lost by evaporation from the reservoir is 6.459 M · m³.

**Example 4.2:** Calculate the daily lake evaporation from the following data from a class-A pan. Assume pan coefficient as 0.75.

| Date | 7/6/97 | 8/6/97 | 9/6/97 | 10/6/97 | 11/6/97 |
|---|---|---|---|---|---|
| Rainfall (mm) | 6 | 0 | 16 | 3 | 5 |
| Water added (mm) | +8 | +12 | −5 | +10 | +9 |

**Solution**
Daily pan evaporation is the sum of the rainfall and the water added or taken out from the pan. Calculation is carried out daily in the following table.

**Table 4.3   Calculation of Daily Evaporation from a Lake**

| Day | Water added/taken out (mm) | Rainfall (mm) | Pan evaporation (mm) | Lake evaporation (mm) |
|---|---|---|---|---|
| (1) | (2) | (3) | (4) | (5) |
| 7.6.97 | +8 | 6 | 14 | 10.50 |
| 8.6.97 | +12 | 0 | 12 | 9.00 |
| 9.6.97 | −5 | 16 | 11 | 9.25 |
| 10.6.97 | +10 | 3 | 13 | 9.75 |
| 11.6.97 | +9 | 5 | 14 | 10.50 |

## 4.2.1.1   Network Design

As mentioned earlier, evaporation is measured along with other meteorological parameters at the same station. World Meteorological Organisation (WMO) specifies the following minimum densities of evaporimeters which are followed by most of the countries.

1. Arid Zone: One station covering maximum of 30,000 km².
2. Humid temperate climate: Minimum one station covering 50,000 km².
3. Cold regions: Minimum one station covering 1,00,000 km².

Present density of evaporation gauges in India is one for every 15,000 km² which is higher than the WMO recommended density. As per guidelines of India Meteorological Department, a hydrological observatory should have the following additional instruments:

(i) Ordinary rain-gauge
(ii) Recording rain-gauge
(iii) Thermometers for recording minimum and maximum temperatures (*Stevenson box*)
(iv) Dry and Wet bulb thermometers for recording humidity
(v) Pan evaporimeter
(vi) Wind anemometer
(vii) Sunshine recorder

### 4.2.2 Empirical Equations

Many empirical equations have been proposed for the computation of lake evaporation, correlating the saturation vapour pressure at water surface temperature $e_s$ mmHg and the actual vapour pressure of evaporating air at certain height above water surface $e_a$ mmHg. Evaporation $E$ (mm/day) from lake may be calculated by any suitable equation given in Table 4.4.

**Table 4.4 Lake Evaporation Calculation by Empirical Equations**

| Sl. No. | Name of equation | Evaporation rate (mm/day) | Terms used |
|---|---|---|---|
| (1) | (2) | (3) | (4) |
| 1. | General equation | $E = k\, f(u)\, (e_s - e_a)$ | $k$ is a coefficient, $f(u)$ is function of wind speed. $e_a$ and wind speed should be measured at the same height. (4.4) |
| 2. | Meyer's formula USA, small lake (1915) | $E = \left(1 + \dfrac{U}{16}\right) \cdot C \cdot (e_s - e_a)$ | $U$ is monthly mean wind speed in km/h at 9 m above ground. $C = 0.36$ for large deep lakes and 0.50 for shallow lakes. (4.5) |
| 3. | Rhower's formula (USA 1931) | $E = 0.771(1.465 - 0.000732P)$ $\times (0.44 + 0.0733U)\,(e_s - e_a)$ | $P$ is the mean barometric reading in mmHg and $U$ the mean wind velocity at 0.6 m above ground in km/h. (4.6) |

*(Contd.)*

| (1) | (2) | (3) | (4) |
|---|---|---|---|
| 4. | Penman's formula (England, applicable for small tanks) | $E = 8.9 \, (1 + 0.15U)(e_s - e_a)$ | $U$ is measured at 2 m above ground level. (4.7) |
| 5. | USBR Formula | $E = 0.833 \, (4.57 \, t + 43.3)$ | $E$ is mm/month, $t$ mean annual temperature in °C. (4.8) |
| 6. | Lake Mead formula (mm/day) | $E = 0.046 \times t \times w$ $\times [1 - 0.03(t_a - t_s)](e_s - e_a)$ | $t$ is number of days of evaporation, $t_a$ average air temp. in °C+1.9°C $t_s$ is average water surface temperature °C, $w$ is 1.85$U$, $e_a$ and $e_s$ in mb. (4.9) |
| 7. | Lake Hefner | $E = 0.028 \, U \, (e_s - e_a)$ | $e_a$ and $e_s$ in mb. (4.10) |
| 8. | Fitzgerald | $E = (10.2 + 3.14U)(e_s - e_a)$ | (4.11) |
| 9. | Shahtin Mamboub's equation | $E = (3.5 + 0.53U)(e_s - e_a)$ | $U$ is measured at 2 m above ground. (4.12) |
| 10. | Kuzmin formula (mm/month) | $E = (152.4 + 19.8U)(e_s - e_a)$ | $U$ is measured at 8 m above ground. (4.13) |
| 11. | Marciano and Harbeck's formula | $E = 0.918 \, U \, (e_s - e_a)$ | $U$ is measured at 8 m above ground. (4.14) |

*$U$ is the wind speed in km per hour.

Wind speed at any height $h_1$ up to 500 m above ground is calculated using the (1/7)th power law given as $U_{hi} = U_h \left( \dfrac{h_1}{h} \right)^{1/7}$, where $U_h$ is the wind speed measured at height $h$. Among all the above equations, Meyer's formula is widely used for the computation of lake evaporation. Saturation vapour pressure of water surface temperature can be taken from Table 4.5.

---

**Example 4.3:** A reservoir with average surface spread of 3.3 km² in December has the water surface temperature of 22.5°C and relative humidity of 35%. Wind velocity measured at 2.0 m above the ground at a nearby observatory is 15 km/h. Calculate average evaporation loss from the reservoir in mm/day and the total depth and volume of evaporation loss for December.

**Solution**

For water surface temperature of 22.5°C (Table 4.5), the saturation vapour pressure is
$$e_s = 20.44 \text{ mmHg.}$$

$$\text{Relative humidity } (R_H) = 35\% = 0.35$$

To use Meyer's equation, wind speed is to be converted to a height of 9 m above ground by (1/7)th power law

$$U \text{ (at 9 m above ground level)} = 15 \times (9/2)^{1/7} = 18.6 \text{ km/h.}$$

Saturation vapour pressure of air $e_a = e_s \times R_H = 20.44 \times 0.35 = 7.15$ mmHg.
Using Meyer's equation, evaporation loss is computed as

$$E = C\left(1 + \frac{U}{16}\right)(e_x - e_a) = 0.36\left(1 + \frac{18.6}{16}\right)(20.44 - 7.15) = 10.34 \text{ mm/day}.$$

Depth of evaporation in December $= 31 \times 10.34 = 320.5$ mm
Volume of evaporated water from the reservoir for December will be

$$(320.5 \times 3.3 \times 10^6) \times 10^{-3} = 1.0578 \text{ M m}^3 = 105.8 \text{ hectare m}.$$

**Table 4.5  Saturation Vapour Pressure of Water**

| Temperature (°C) | Saturation vapour pressure $e_s$ (mmHg) | (millibar) | Slope of plot between (1) and (2) |
|:---:|:---:|:---:|:---:|
| (1) | (2) | (3) | (4) |
| 0.0 | 4.58 | 6.11 | 0.30 |
| 5.0 | 6.54 | 8.72 | 0.45 |
| 7.5 | 7.78 | 10.37 | 0.54 |
| 10.0 | 9.21 | 12.28 | 0.60 |
| 12.5 | 10.87 | 14.49 | 0.71 |
| 15.0 | 12.79 | 17.05 | 0.80 |
| 17.5 | 15.00 | 20.00 | 0.95 |
| 20.0 | 17.54 | 23.38 | 1.05 |
| 22.5 | 20.44 | 27.95 | 1.24 |
| 25.0 | 23.76 | 31.67 | 1.40 |
| 27.5 | 27.54 | 36.71 | 1.61 |
| 30.0 | 31.81 | 42.42 | 1.85 |
| 32.5 | 36.68 | 48.89 | 2.07 |
| 35.0 | 42.81 | 57.07 | 2.35 |
| 37.5 | 48.36 | 64.46 | 2.62 |
| 40.0 | 55.32 | 73.14 | 2.95 |
| 42.5 | 62.18 | 84.23 | 3.25 |
| 45.0 | 71.2 | 94.91 | 3.66 |

### 4.2.3  Water Balance Method

Water balance or *water budget* method balances all the incoming, outgoing and stored water in a lake or reservoir over a period of time. The equation in its simplest form is

$$\Sigma \text{ Inflow} - \Sigma \text{ Outflow} = \text{Change in storage} + \text{Evaporation loss}$$

or
$$E = \Sigma I - \Sigma O \pm \Delta S \qquad (4.15)$$

It can be more generalised by taking all the factors of inflow and outflow. Equation (4.15) can be rewritten as

$$E = (P + I_{sf} + I_{gf}) - (O_{sf} + O_{gf} + T) \pm \Delta S \qquad (4.16)$$

where $P$ is the precipitation, $I_{sf}$ the surface inflow, $I_{gf}$ the ground water inflow, $O_{sf}$ the surface water outflow, $O_{gf}$ the ground water outflow, $T$ the transpiration loss (may be neglected), $\Delta S$ is the change in storage. Measurement of all quantities is possible except $I_{gf}$, $O_{gf}$ and $T$ (if present). Therefore, this equation fails to give accurate results since ground water inflow and outflow are very difficult to measure for a lake or reservoir. It may give fairly good result if considered annually but should not be used for daily estimation of evaporation. This equation is good for theoretical considerations or may be applied to water tight lakes located on impervious rocks for budgeting annual water.

### 4.2.4 Energy Budget or Energy Balance Method

Like water balance, energy balance or *energy budget* for lakes or reservoirs can be carried out to calculate lake evaporation. This method uses the conservation of energy by incorporating all the incoming, outgoing and stored energy of a lake in the following form.

$$H_{li} + H_{si} - H_{so} - H_{lo} = H_i + H_{if} + H_s + H_e + H_{lr} + H_{gf} \qquad (4.17)$$

$H_{li}$ = Long wave radiation incident on the surface of water
$H_{si}$ = Solar radiation (heat) or net energy received at water surface by short wave
$H_{so}$ = Reflected solar energy
$H_{lo}$ = Reflected long wave radiation
$H_i$ = Increase in stored heat energy of water
$H_{if}$ = Net energy convected/ conducted from and out of the system by flow of water
$H_s$ = Energy conducted from water mass to air as sensible head
$H_e$ = Energy used for evaporation = $\rho \cdot E \cdot L_H$
$H_{lr}$ = Long wave radiation emitted from water
$H_{gf}$ = Heat flux into ground water
$\rho$ = Density of water (cm$^3$/g)
$E$ = Evaporation rate (cm/day)
$L_H$ = Latent heat of vaporisation (cal/g)

The above energy terms are expressed in calories per unit surface area per day. A water body and the energy terms are shown in Fig. 4.2. Daily calculation of lake evaporation by this method is unreliable due to difficulties in measuring all the parameters involved in the equation, but good estimate can be obtained if applied to monthly or yearly values. We can measure or calculate all other terms except the sensible heat transfer between water surface and atmosphere $H_s$. It can be measured by

$$\beta = H_s/H_e \qquad (4.18)$$

Equation (4.18) is Known as *Bowen's ratio*, defined as the ratio between heat lost by conduction to heat lost by evaporation. Estimate of $\beta$ can be made from the relation

$$\beta = c \cdot p \cdot \frac{(T_s - T_a)}{[100\,(e_s - e_a)]} \qquad (4.19)$$

where $p$ is the atmospheric pressure in millibars, $T_s$ and $T_a$ are the water surface and air temperatures in °C, $c$ is a constant varying between 0.58 and 0.66 (average

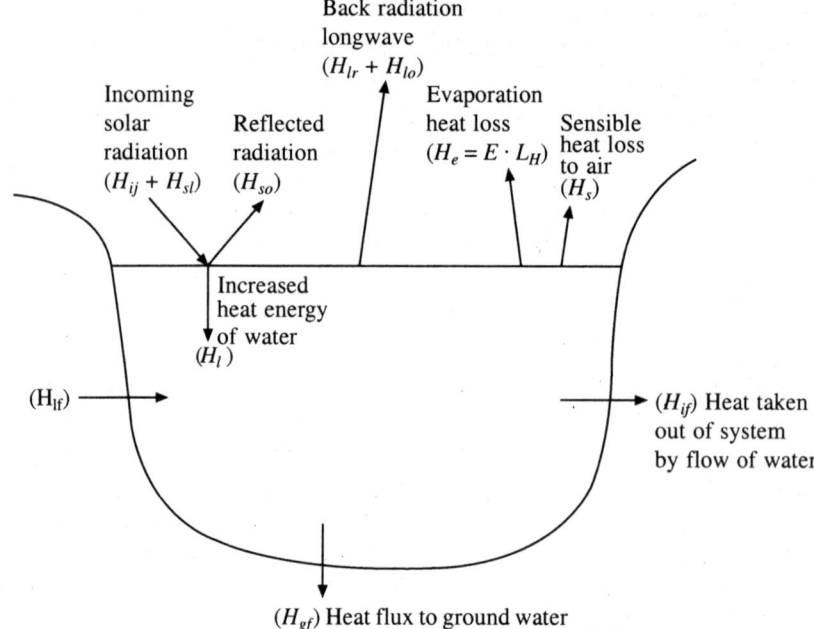

Fig. 4.2    **Energy Budget for a Lake**

value being 0.61), $e_s$ and $e_a$ are the saturation vapour pressure at water surface and air temperatures in millibar. From equation (4.18) we get

$$H_s = \beta \rho L_H E \qquad (4.20)$$

Therefore

$$H_e + H_s = \rho L_H (1 + \beta) E$$

or

$$E = \frac{(H_e + H_s)}{[\rho L_H (1 + \beta)]} \qquad (4.21)$$

From equation (4.17) we get

$$E = \frac{(H_{li} + H_{si} - H_{so} - H_{lo} - H_i - H_{if} - H_{lr} - H_{gf})}{[\rho L_H (1 + \beta)]} \qquad (4.22)$$

$$= \frac{(H_n - H_{if} - H_{gf} - H_i)}{[\rho L_H (1 + \beta)]}$$

where $H_n$ is the net radiation given by

$$H_n = H_{li} + H_{si} - H_{so} - H_{lo} - H_{lr} \qquad (4.23)$$

Values of radiation from sun and sky ($H_{li} + H_{si}$) are available for different latitudes. Reflected solar radiation is dependent on factors like spectral wave length and turbidity of water and air. For water it can be taken between 5 to 15% of incident radiation. Neglecting $H_{si}$, net effect of all the long wave radiation $H_{lr}$ is given in calories per square centimeter per day as the sum of incoming long wave radiation from atmosphere $H_{li}$ plus reflected long wave radiation $-H_{lo}$ and long wave radiation emitted by water $-H_{lr}$ as

or
$$H_{lr} = H_{li} - (H_{lo} + H_{lr}) = (1 - r)\, H_{li} \qquad (4.24)$$

The incident solar energy is calculated by the equations proposed by Penman (1948) as

$$H_{li} = (0.18 + 0.55n/N)\, I_o \qquad (4.25)$$

where $I_o$ is the solar radiation received at earth's outer surface in (cal/cm²-day), $n$ is the actual number of bright sunshine hours, $N$ is the possible maximum number of hours of sunshine at the place given in Table 4.6. $I_o$ is converted to mm of evaporable water/day and is given in Table 4.7. The net outgoing thermal radiation expressed by Penman is given as

$$H_{net} = \sigma T^4\, [0.56 - 0.092\, e_a^{0.5}](0.1 + 0.90n/N) \qquad (4.26)$$

in which $\sigma$ is the *Stefan-Boltzmann constant* $= 2 \times 10^{-9}$ mm of water/day, $T$ is the water surface temperature in ° Kelvin, $H_{if}$ can be calculated from the relation

$$H_{if} = \frac{W_{sh} H_e T_e}{L_H} \qquad (4.27)$$

in which $W_{sh}$ is the specific heat of water in cal/g °C. $H_n$ can be calculated from the relation

$$H_n = H_{li}\, (1 - r) - H_{net} \qquad (4.28)$$

Net energy added into lake water, $H_{if}$ is measured by knowing all the volume of water in-flowing and coming out of the lake and their corresponding temperatures during the period of water budget. This term should sum all channel inflows and outflows, evaporation, condensation, rainfall, seepage and other losses. The term increase in stored heat energy of water for the lake or reservoir is a difficult parameter to obtain, which can be computed by knowing precisely the average temperature of lake water and the volume of water at the beginning and end of the budget period.

Application of the energy-budget principle gives good results for water tight lakes and it may give highly erroneous and confusing results for other lakes. This is due to the fact that accurate determination of such large number of parameters from a vast and complex system may not be possible. The method was applied by Penman to calculate evapotranspiration.

---

**Example 4.4:** Calculate for the month of August the solar radiation incident on earth's surface and the net outgoing thermal (long wave) radiation for a place 20°N latitude. Take number of sunshine hours recorded at the near by IMD observatory as 11.5 and mean air temperature as 25°C. Take relative humidity for August as 98%.

**Solution**
Maximum number of sunshine hours ($N$) for 20°N can be read from Table 4.6 as 12.8.
Mean monthly solar radiation incident at the earth's outer space (extra terrestrial radiation)

**Table 4.6    Mean Monthly Values of Possible Sunshine Hours (N)**

| Month | North Latitude in Degrees | | | | | |
|-------|------|------|------|------|------|------|
|       | 0°   | 10°  | 20°  | 30°  | 40°  | 50°  |
| Jan.  | 12.1 | 11.6 | 11.1 | 10.4 | 9.6  | 8.6  |
| Feb.  | 12.1 | 11.8 | 11.5 | 11.1 | 10.7 | 10.1 |
| Mar.  | 12.1 | 12.1 | 12.0 | 12.0 | 11.9 | 11.8 |
| Apr.  | 12.1 | 12.4 | 12.6 | 12.9 | 13.2 | 13.8 |
| May   | 12.1 | 12.6 | 13.1 | 13.7 | 14.4 | 15.4 |
| Jun.  | 12.1 | 12.7 | 13.3 | 14.1 | 15.0 | 16.4 |
| Jul.  | 12.1 | 12.6 | 13.2 | 13.9 | 14.7 | 16.0 |
| Aug.  | 12.1 | 12.4 | 12.8 | 13.2 | 13.8 | 14.5 |
| Sept. | 12.1 | 12.9 | 12.3 | 12.4 | 12.5 | 12.7 |
| Oct.  | 12.1 | 11.9 | 11.7 | 11.5 | 11.2 | 10.8 |
| Nov.  | 12.1 | 11.7 | 11.2 | 10.6 | 10.0 | 9.4  |
| Dec.  | 12.1 | 11.5 | 10.9 | 10.2 | 9.1  | 8.1  |

**Table 4.7    Mean Monthly Solar Radiation Incident on Earth's Outer Space (Extra Terrestrial Radiation) in mm of Evaporable Water per day (After Criddle 1958)**

| Month | North Latitude in Degrees | | | | | | | | | |
|-------|------|------|------|------|------|------|------|------|------|------|
|       | 0°   | 10°  | 20°  | 30°  | 40°  | 50°  | 60°  | 70°  | 80°  | 90°  |
| Jan.  | 14.5 | 12.8 | 10.8 | 8.5  | 6.0  | 3.6  | 1.3  | –    | –    | –    |
| Feb.  | 15.0 | 13.9 | 12.3 | 10.5 | 8.3  | 5.9  | 3.5  | 1.1  | –    | –    |
| Mar.  | 15.2 | 14.8 | 13.9 | 12.7 | 11.0 | 9.1  | 6.8  | 4.3  | 1.8  | –    |
| Apr.  | 14.7 | 15.2 | 15.2 | 14.8 | 13.9 | 12.7 | 11.1 | 9.1  | 7.8  | 7.9  |
| May   | 13.9 | 15.0 | 15.7 | 16.0 | 15.9 | 15.4 | 14.6 | 13.6 | 14.6 | 14.9 |
| Jun.  | 13.4 | 14.8 | 15.8 | 16.5 | 16.7 | 16.7 | 16.5 | 17.0 | 17.8 | 18.1 |
| Jul.  | 13.5 | 14.8 | 15.7 | 16.2 | 16.3 | 16.1 | 15.7 | 15.8 | 16.5 | 16.8 |
| Aug.  | 14.2 | 15.0 | 15.3 | 15.3 | 14.8 | 13.9 | 12.7 | 11.4 | 10.6 | 11.2 |
| Sept. | 14.9 | 14.9 | 14.4 | 13.5 | 12.2 | 10.5 | 8.5  | 6.8  | 4.0  | 2.6  |
| Oct.  | 15.0 | 14.1 | 12.9 | 11.3 | 9.3  | 7.1  | 4.7  | 2.4  | 0.2  | –    |
| Nov.  | 14.6 | 13.1 | 11.2 | 9.1  | 6.7  | 4.3  | 1.9  | 0.1  | –    | –    |
| Dec.  | 14.3 | 12.4 | 10.3 | 7.9  | 5.5  | 3.0  | 0.9  | –    | –    | –    |

$I_o$ in mm of evaporable water per day is read from Table 4.7 as 15.3. Solar radiation incident on earth's surface is given as

$$H_{li} = (0.18 + 0.55 \ n/N) \ I_o = \left( 0.18 + 0.55 \times \frac{11.5}{12.8} \right) \times 15.3 = 10.3 \text{ mm of water/day}$$

For calculating the long wave radiation, we have $\sigma = 2 \times 10^{-9}$ mm of water/day

$T = 25 + 273 = 298$ °K, $e_s$ for $T_a = 25°C$ is read from Table 4.5 as 23.76

$e_a = 23.76 \times$ Relative Humidity $= 23.76 \times 0.98 = 23.28$

$H_{net} = \sigma T^4 \{0.56 - 0.092 \ e_a^{0.5}\}\{0.1 + 0.9n/N\}$

$\qquad = 2 \times 10^{-9} \times 298^4\{0.56 - 0.092 \times 23.28^{0.5}\}\{0.1 + 0.9 \times 11.5/12.8\}$

$\qquad = 15.77 \times 0.11515 \times 0.9086 = 1.65$ mm of water/day.

### 4.2.5 Mass Transfer Method

Accurate estimation of the amount of water vapour transferred to atmosphere from a lake surface is still investigated. The equation proposed by Thornthwaite and Holzman (1939) takes the following form

$$E = \frac{0.000119 \ (e_1 - e_2)(u_2 - u_1)}{p \times \left[\ln\left(\dfrac{h_2}{h_1}\right)\right]^2} \qquad (4.29)$$

where $E$ is m/sec, $u_2$ and $u_1$ are the velocities of wind in m/sec at heights $h_2$ and $h_1$ m respectively, $e_2$ and $e_1$ are vapour pressure of air in pascal (Pa) at height(s) $h_2$ and $h_1$, $p$ is mean atmospheric pressure in pa (1 N/m$^2$ = 1 Pa; 1 KPa = 10 mb) between lower height $h_1$ and upper height $h_2$. These terms should be precisely measured with very sophisticated instruments as minor variation in their measurement may lead to erroneous results. Height $h_1$ is taken close to water surface level.

---

**Example 4.5:** For air temperature of 25°C, relative humidity of 98%, air pressure of $101.3 \times 10^3$ Pa and wind speed of 3 m/sec measured at 2 m above the water surface, calculate the evaporation loss from the water surface. Take saturation height $h_1$ as 0.03 m.

**Solution**

Here $e_1$ (= $e_s$) for air temperature of 25°C at the level close to water surface is read from Table 4.5 as 31.67 mb = 3167 Pa.

$\qquad e_2$ = Relative humidity $\times e_s$ = 0.98 $\times$ 3167 = 3104 Pa.

$\qquad u_1$ = 0 at water surface level, $u_2$ = 3 m/sec

using the equation (4.29)

$$E = \frac{0.000119 \ (e_1 - e_2)(u_2 - u_1)}{p \times \left[\ln\left(\dfrac{h_2}{h_1}\right)\right]^2}$$

$$E = \frac{0.000119 \ (3167 - 3104)(3 - 0)}{101.3 \times 10^3 \left[\ln\left(\dfrac{2}{0.03}\right)\right]^2} = \frac{0.0225}{1786681} = 1.2588 \times 10^{-8} \text{ m/sec}$$

$$\qquad = 1.2588 \times 10^{-8} \times 1000 \times 24 \times 3600 \text{ mm/day} = 1.2 \text{ mm/day}$$

---

## 4.3 METHODS TO REDUCE RESERVOIR EVAPORATION

In arid and semi-arid regions, evaporation of water from reservoirs is a substantial loss, both in terms of economy and natural resource. Table 4.8 gives a rough idea about the substantial loss of the valuable water resources by evaporation from the Indian reservoirs.

**Table 4.8 Reservoir Evaporation Losses (cm)**

| Months | Jan. | Feb. | Mar. | Apr. | May | Jun. | Jul. | Aug. | Sept. | Oct. | Nov. | Dec. | Total |
|---|---|---|---|---|---|---|---|---|---|---|---|---|---|
| **Evaporation loss from** | | | | | | | | | | | | | |
| (i) South and Central India | 10 | 10 | 18 | 23 | 25 | 18 | 15 | 15 | 15 | 13 | 10 | 10 | 182 |
| (ii) North India | 7 | 9 | 13 | 25 | 26 | 24 | 18 | 14 | 14 | 13 | 9 | 8 | 170 |
| (iii) Hirakud dam | 10.5 | 9.5 | 18.4 | 22.9 | 26.3 | 17.8 | 15.8 | 15.8 | 15.2 | 13.1 | 10.2 | 10.5 | 186 |
| (iv) Bhakra dam | 5 | 6 | 11 | 19 | 26 | 23 | 13 | 10 | 11 | 10 | 7 | 5 | 146 |

Annual evaporation losses from reservoirs in India vary between 150 and 200 cm except the extremes in the states of Rajasthan and Jammu and Kashmir, where the figures are 300 and 50 cm, respectively. Annual evaporation loss of 170 cm can be taken as standard for Indian reservoirs. Taking the average annual surface area of Hirakud reservoir as 500 km$^2$ (at MWL the surface area is 725 km$^2$) the annual evaporation loss from the project is about 85 million m$^3$, which is a huge quantity. By any standard a country cannot afford to lose such amount of storage in terms of its economic output by way of utilizing this resource for the beneficial purposes like irrigation, hydropower generation, drinking water supply and other uses. Such losses should be reduced to a minimum. Various measures to reduce the reservoir or pond evaporation are discussed as follows.

*(i) Reduction in Surface Area*
A site where the depth of reservoir is more to store the same quantity of water, should be preferred. By this the surface area of reservoir is reduced which reduces evaporation loss from the reservoir. Thus a large reservoir is preferable than going for number of small reservoirs.

*(ii) Providing Mechanical Covers*
Covers like rafts or polyethylene may be provided for small agriculture ponds in arid zones, as has been adopted in small ponds in Gujarat and Rajasthan.

*(iii) Spreading Chemical Films*
Spreading chemicals like cetyl-alcohol ($C_{16}$-$H_{33}$-OH) or stearly alcohol (octa-decamol) over the reservoir surface, reduces lake evaporation considerably. With only 0.015 micron thickness, it can effectively reduce evaporation from a reservoir by 20-50 %. About 2.2 kg of cetyl alcohol is required per hectare of water area. The alcohol is attached to water to one side and repels water on other side, i.e., this film allows rain drops to go into the reservoir but does not allow water molecules to escape through it. However, wind, oxidation and birds may disturb the layer requiring regular replenishments.

*(iv) Proper Design of Outlets*
Outlets like canal and power house intakes should be designed such that only

warmer surface water is taken out from a reservoir. This will reduce evaporation to a great extent.

*(v) Creating Forest Around the Reservoir*
Forest creation helps to retard wind speed over the reservoir surface and gives cooling around the area, which effectively reduces evaporation.

*(vi) Removing Trees like Phreatophytes*
Such trees should be removed from the vicinity of reservoirs, as they act like water pipes from reservoirs resulting in excess evaporation of water.

*(vii) Creating Underground Reservoirs*
An underground reservoir, has the least exposure to the atmosphere and therefore evaporation loss from such storages is minimum.

## 4.4 EVAPOTRANSPIRATION (CONSUMPTIVE USE) AND ITS ESTIMATION

A hydrologist determines the total amount of evaporation from a watershed, is taken as the sum of evaporation losses from barren land and the plants (transpiration). Evaporation and transpiration together is called *evapotranspiration*, which is the total water lost to atmosphere over a period of time as water vapour from a watershed.

*Transpiration* is the process by which water is lost to atmosphere as water vapour from the body of plants. As much as 99% of the total water received by a plant through its roots is lost to the atmosphere by this process. Transpiration is associated with *photosynthesis* of plants and is therefore a process of day light hours. All factors responsible for evaporation do apply to transpiration along with the type and density of plants. Water from soil is taken up by plant roots through the membrane by a process called *osmosis*, that occurs due to greater concentration of salts in plant roots with respect to outside. *Chloroplasts* are green cells in plant-leaf. They prepare food in presence of sun light and $CO_2$ and leave water through small openings called *stomata*. The density of stomata may vary from 8000 to 12000 per sq. cm. Transpiration can be measured in a laboratory by a *phytometer* or *potometer* vessel. Measurement of transpiration alone is not the interest of a hydrologist rather it is the combined effect of evaporation and transpiration. Consumptive use and evapotranspiration can be considered same except that the former considers the additional water to make plant tissue.

*Transpiration ratio* (TR) is the ratio of total weight of water transpired by a plant during its growth period to the mass of dry matter produced. It is an important parameter defining the type of plant in terms of water requirement. For wheat TR is between 300 and 600 and for rice between 600 and 800, i.e., rice requires almost two times the water required by wheat to produce the same quantity of dry material, including roots.

As already defined, evapotranspiration is the total loss of water from land as evaporation and from plants as transpiration from a watershed. Obviously *potential evapotranspiration* (PET) means the rate of evapotranspiration from a fully vegetated watershed when sufficient moisture is always available to completely meet the requirements. The *actual evapotranspiration* (AET) from an area is always less than or equal to PET depending on the specific situation. As will be discussed in detail in Chapter 5, the water holding capacity of soil against gravity is termed as *field capacity* (FC) beyond which any drop of water will percolate down to join ground water table. On the other hand, when water available in the soil is so less that plants can no more extract water necessary for sustenance and wilt, the moisture content is termed *wilting point* (WP). The *available water*, which is the difference between FC and WP should always be present in the soil for plants to grow.

A relation between AET/PET and available moisture can be developed for different types of soils on the basis of experimental results. For the same AET/PET ratio, sandy soil has more available moisture than clayey soil. In other words, for the same percent of available moisture, the ratio of AET/PET will be less for sandy soil than for clayey soil (Fig. 4.3).

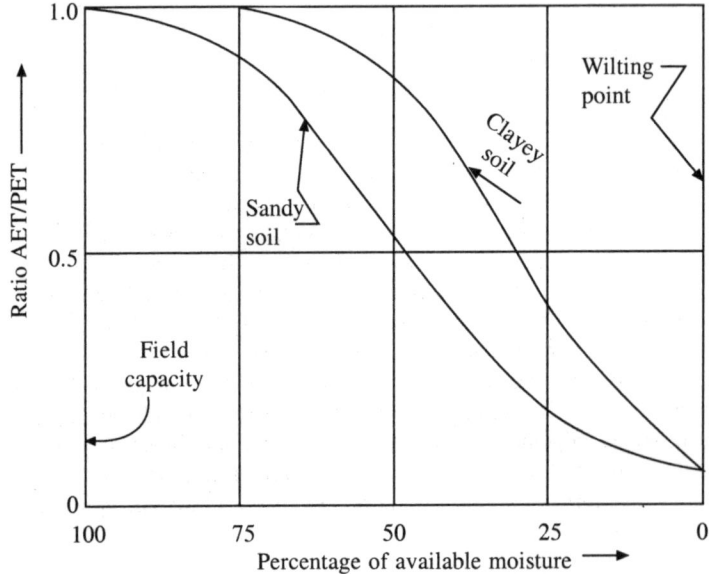

**Fig. 4.3   Available Moisture for Sandy and Clayey Soil**

Evapotranspiration can be estimated by: (i) Experimental field measurement by (a) instruments called Lysimeters, (b) using field plots, (c) studies concerning ground water table fluctuations and (ii) Climatic approach.

### 4.4.1   Experimental Measurement

A *lysimeter* is a special tank containing soil which is set in the same surrounding as that of the field. Plants are grown in it as in the neighborhood. A lysimeter is buried to the level of natural soil. The diameter of lysimeters may vary from 0.60 to 3.3 m and depth from 1.8 to 3.3 m. Details of a lysimeter are shown in

Fig. 4.4. Arrangements are made to weigh the lysimeter whenever reading is required to be taken. Outflow from the lysimeter is measured by a metering

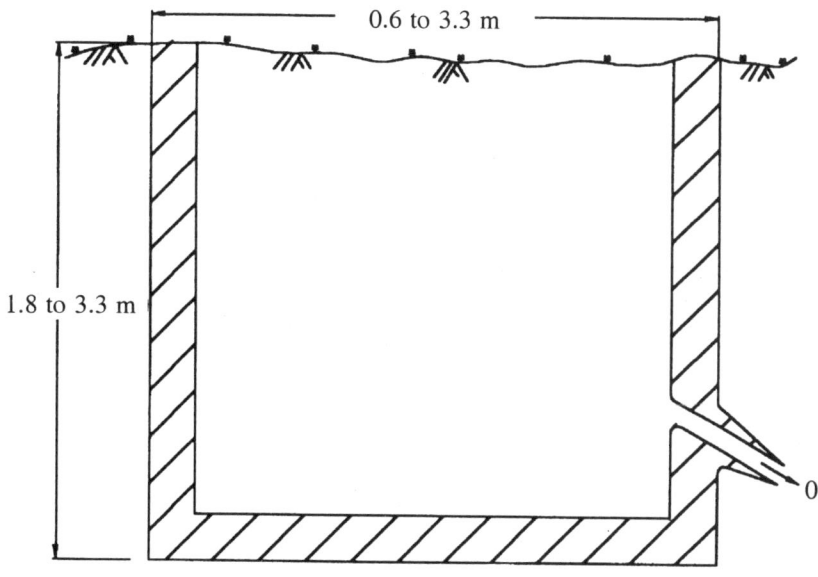

**Fig. 4.4  A Lysimeter**

device. If $P$ is precipitation to the system, $W$ the water supplied, and $O$ the quantity of water drained out of the system, then evapotranspiration is calculated as

$$P + W = O + ET + \Delta S \qquad (4.30)$$

where $ET$ is the evapotranspiration and $\Delta S$ the change in soil moisture storage. $P$, $W$ and $O$ are measured directly. Change in soil moisture $\Delta S$ is quantified by any of the suitable probes discussed in Chapter 5. When arrangement for weighing a lysimeter is made, then $\Delta S$ can be found out by the difference of the weight of the lysimeter before and after the experiments over the period of days. Knowing all other quantities of equation (4.30), $ET$ can be computed. To produce results very close to field, soil conditions, moisture content, plant types and the methods of water application should be properly chosen such that they represent the surrounding natural condition. Lysimeters are expensive to maintain and time consuming. Therefore, the measurement of evapotranspiration through lysimeter is not carried out except for experimental purposes.

### 4.4.2  Climatic Approaches
Some empirical and theoretical equations are derived on the basis of regional relationship between measured $ET$ and climatic factors. The following methods are the combination of some empirical, analytical and theoretical approaches.

### 4.4.2.1  Blaney-Criddle Method
Blaney and Criddle (1962) proposed an empirical relation which is used largely

by irrigation engineers to calculate crop water requirement of various crops. Estimation of potential evapotranspiration (consumptive use) is carried out by correlating it with sunshine temperature. Sunshine at a place is dependent on latitude of the place and varies with month of the year. Table 4.9 gives the values of percentages of monthly day time for use in Blaney-Criddle equation. PET for a crop during its growing season is given by

$$PET = \Sigma \, K.F \qquad (4.31)$$

where
$$F = (0.0457T_m + 0.8128) \, P \qquad (4.32)$$

Here $K$ is the monthly crop coefficient to be determined from experimental data, $F$ the monthly consumptive use factor, $PET$ the potential evapotranspiration in cm, $T_m$ the mean monthly temperature in °C, $P$ is the monthly percentage of hours of bright sunshine in the year.

**Table 4.9    Monthly Day time Percentage Hours ($P$) to be used by Blancy-Criddle Formula**

| North Latitude (deg.) | J | F | M | A | M | J | J | A | S | O | N | D |
|---|---|---|---|---|---|---|---|---|---|---|---|---|
| 0 | 8.5 | 7.66 | 8.49 | 8.21 | 8.50 | 8.22 | 8.5 | 8.49 | 8.21 | 8.5 | 8.22 | 8.5 |
| 5 | 8.32 | 7.56 | 8.47 | 8.29 | 8.66 | 8.40 | 8.68 | 8.60 | 8.23 | 8.42 | 8.06 | 8.30 |
| 10 | 8.13 | 7.47 | 8.45 | 8.37 | 8.81 | 8.60 | 8.86 | 8.71 | 8.25 | 8.34 | 7.91 | 8.10 |
| 15 | 7.94 | 7.36 | 8.43 | 8.44 | 8.89 | 8.80 | 9.05 | 8.83 | 8.28 | 8.26 | 7.75 | 7.88 |
| 20 | 7.74 | 7.25 | 8.41 | 8.52 | 9.15 | 9.00 | 9.23 | 8.96 | 8.30 | 8.18 | 7.58 | 7.66 |
| 25 | 7.53 | 7.14 | 8.39 | 8.61 | 9.33 | 9.23 | 9.45 | 9.09 | 8.32 | 8.09 | 7.40 | 7.42 |
| 30 | 7.30 | 7.03 | 8.38 | 8.72 | 9.53 | 9.49 | 9.67 | 9.22 | 8.33 | 7.99 | 7.19 | 7.15 |
| 35 | 7.05 | 6.88 | 8.35 | 8.83 | 9.76 | 9.77 | 9.93 | 9.37 | 8.36 | 7.87 | 6.97 | 6.86 |
| 40 | 6.76 | 6.72 | 8.33 | 8.95 | 10.02 | 10.08 | 10.22 | 9.54 | 8.39 | 7.75 | 6.72 | 6.52 |
| 45 | 6.42 | 6.54 | 8.30 | 9.09 | 10.33 | 10.47 | 10.57 | 9.75 | 8.41 | 7.60 | 6.42 | 6.13 |
| 50 | 5.98 | 6.30 | 8.24 | 9.24 | 10.68 | 10.91 | 10.99 | 10.00 | 8.46 | 7.45 | 6.10 | 5.65 |
| 55 | 5.42 | 6.01 | 8.16 | 9.41 | 11.13 | 11.53 | 11.56 | 10.34 | 8.52 | 7.25 | 5.65 | 5.03 |
| 60 | 4.67 | 5.65 | 8.08 | 9.65 | 11.74 | 12.34 | 12.31 | 10.70 | 8.57 | 6.68 | 4.31 | 4.22 |

Charts are available to read the values of $K$ for different crops at various agriculture research stations to be used in the above procedure. Table 4.10 gives the values of $K$.

**Table 4.10    Monthly Crop Coefficient Factor $K$ to be Used for Blaney-Cridle Method**

| Crop | Rice | Wheat | Maize | Sugar-cane | Cotton | Potato | Corn | Vegetable | |
|---|---|---|---|---|---|---|---|---|---|
| | | | | | | | | Light | Dense |
| (1) | (2) | (3) | (4) | (5) | (6) | (7) | (8) | (9) | (10) |
| Value of K | 1.1 | 0.65 | 0.65 | 0.90 | 0.65 | 0.70 | 0.75 | 0.80 | 1.20 |
| Range of K | 0.85–1.3 | 0.5–0.75 | 0.5–0.75 | 0.8–1.0 | 0.5–0.9 | 0.65–0.75 | 0.65–0.85 | 0.7–1.0 | 1.1–1.4 |

**Example 4.6:** Use Blaney-Criddle method to calculate consumptive use (PET) for rice crop grown from January to March (Dalua Rabi crop) in Orissa at a latitude 22° N from the following data taken from a nearby observatory. Find the net irrigation demand for rice using the given rainfall during crop period.

| Month | January | February | March |
|---|---|---|---|
| Mean temperature °C | 12 | 16 | 24 |
| Rainfall (mm) | 8 | 20 | 16 |

**Solution**

For rice crop, monthly crop coefficient $K$ of equation (4.31) may be taken as 1.10. Mean monthly sunshine hours for latitude of 22°N for the months of January, February and March are obtained from Table 4.9 and tabulated below.

**Table 4.11   Blaney-Criddle Method of Computation of Consumptive use of Rice Crop for Example 4.6**

| Month | Mean monthly temp $(T_m)$ | Monthly % $(P)$ of day time hours from Table 4.9 | Monthly consump-tive use factor $(F)$ | $K$ | PET $(4) \times (5)$ | Effective rainfall at 80% (cm) | Depth of irrigation demand $(6) - (7)$ (cm) |
|---|---|---|---|---|---|---|---|
| (1) | (2) | (3) | (4) | (5) | (6) | (7) | (8) |
| January | 12 | 7.62 | 10.37 | 1.1 | 11.40 | 0.64 | 10.76 |
| February | 16 | 7.20 | 11.12 | 1.1 | 12.23 | 1.60 | 10.63 |
| March | 24 | 8.40 | 16.04 | 1.1 | 17.64 | 1.28 | 16.36 |

$F$ for col. (4) for January = $(0.0457 \, T_m + 0.8128) \times P$

$= (0.0457 \times 12 + 0.8128) \times 7.62 = 10.37$ cm.

$F$ (February) = $(0.0457 \times 16 + 0.8128) \times 7.2 = 11.12$ cm

$F$ (March) = $(0.0457 \times 24 + 0.8128) \times 84 = 16.04$ cm

The net irrigation demand = 10.76 + 10.63 + 16.36 = 37.75 cm.

### 4.4.2.2   Penman Method

Penman developed a theoretical formula based on the principles of both energy-budget and mass-transfer approaches to calculate potential evapotranspiration. A simple energy budget neglecting all minor losses can be written as

$$\text{ET} = \frac{(HA + \alpha E_a)}{(A + \alpha)} \qquad (4.33)$$

where $H$ is the heat budget of an area with crops which is the net radiation in mm of evaporable water per day, ET the daily evaporation from free water surface in mm/day, $\alpha$ is a constant (called psychrometric constant whose value is 0.49 mmHg/°C or 0.66 mb/°C), $A$ the slope of the saturated vapour pressure vs. temperature curve at mean air temperature given in Table 4.5, $E_a$ is the drying power of air which includes wind velocity and saturation deficit and is estimated from the relation

$$E_a = 0.002187(160 + u_2) (e_s - e_a) \tag{4.34}$$

where $u_2$ is the mean wind speed in km/day measured 2 m above the ground, $e_s$ is saturation vapour pressure at mean air temperature in mm Hg (given in Table 4.5), $e_a$ is actual vapour pressure in the air in mm of mercury and $H$ is the daily net radiation in mm of evaporable water and is estimated from the energy budget theories using the relation

$$H = H_a(1 - r)(0.29 \cos \phi + 0.55n/N) - \sigma T_a^4(0.56 - 0.092 \sqrt{e_a})(0.10 + 0.9n/N)$$
$$\tag{4.35}$$

where $H_a$ is the extraterrestrial solar radiation received on a horizontal surface in mm of evaporable water per day (whose value for different latitudes are given in Table 4.7), $\phi$ the latitude of the place where PET is to be computed, $r$ is the reflection coefficient whose values for close crops may be taken as 0.15-0.25, for barren land 0.05-0.45 and for water surface 0.05, $n$ is the actual duration of bright sunshine which is a function of latitude and is an observed data at a place, $N$ is the maximum possible hours of bright sunshine available at different location (given in Table 4.6), $\sigma$ is the Stefan-Boltzman constant = $2.01 \times 10^{-9}$ mm/day, $T_a$ is the mean air temperature in °K = $(273 + °C)$ and $e_a$ is the actual vapour pressure in mm of Hg. The wind speed measured at any other height $z$ can be reduced to 2 m height by the relation

$$u_2 = u \left(\frac{2}{z}\right)^{0.143} \tag{4.36}$$

Equation (4.36) is known as $(1/7)^{th}$ power law. Knowing all other data from the table and measuring $n$, $e_a$ $u_2$, at the place, *PET* can easily be calculated from the relation given by Penman. This method is finding its increasing application for crop water estimation by various countries, including India.

---

**Example 4.7:** Using Penman's formula calculate consumptive use of rice for the month of February. Take the following data

Wind velocity measured at 2 m height = 30 km/day
Elevation of the area                              = 220 m
Relative humidity for February           = 50%
Latitude                                               = 22°N
Mean monthly temperature                 = 16°C

**Solution**
From Table 4.5, for temperature of 16°C, $e_s$ = 13.67 mmHg
Slope of the saturated vapour pressure vs. temp. curve A = 0.86 mm per °C
From Table 4.7, $H_a$ = 11.94 mm of water per day for 22°N latitude
From Table 4.6, $N$ = 11.42 h for the latitude of 22 °N.
Monthly percentage of day time hours = 7.20 h (from Table 4.9)

$$\therefore \qquad \frac{n}{N} = \frac{7.2}{11.42} = 0.63$$

Vapour pressure in air $e_a = e_s \times R_H = 13.67 \times 0.50 = 6.88$ mmHg

Drying power of air $E_a = 0.002187 (160 + U_2) (e_s - e_a)$

$$= 0.002187 (160 + 30) (13.67 - 6.88) = 2.84 \text{ mm/day}$$

The reflection coefficient for close crop like paddy is 0.20.

Take $\quad \alpha = 0.49, \sigma = 2.01 \times 10^{-9}$ mm/day, $T_a = 273 + 16 = 289°K$

$\therefore \quad H = H_a (1 - r) (0.29 \cos \phi + 0.55 n/N)$

$$- \sigma T_a^4 (0.56 - 0.092 \sqrt{e_a})(0.1 + 0.9 \, n/N)$$

or $\quad\quad H = 11.94 (1 - 0.2)(0.29 \cos 22° + 0.55 \times 7.2/11.42)$

$$- 2.01 \times 10^{-9} \times 289^4 (0.56 - 0.092\sqrt{6.88}) \times (0.1 + 0.9 \times 7.2/11.42)$$

$$= 11.94 \times 0.8 \times (0.269 + 0.347) - 14.02 (0.56 - 0.24) (0.1 + 0.568)$$

$$= 5.88 - 3.00 = 2.88 \text{ mm of water/day}$$

PET $= (AH + \alpha E_a)/(A + \alpha) = (0.86 \times 2.88 + 0.49 \times 2.84)/(0.86 + 0.49)$

$$= 3.864/1.35 = 2.865 \text{ mm/day}$$

$$= 2.865 \times 28 \times 1/10 \text{ cm/month} = 8.03 \text{ cm for February.}$$

Consumptive use of rice for February is 8.03 cm = 80.3 mm of water.

### 4.4.2.3 Christiansen Method

In this method, various climatological data are connected with PET in such a way that absence of any climatic data is taken care by the average value given as default option. General equation by this approach is

$$PET = 0.473 \, H_a C \quad\quad\quad (4.37)$$

where PET is in mm/day, $H_a$ is the extraterrestrial radiation converted to mm of water evaporated (Table 4.7) and $C$ is a coefficient, which can be determined from the relation.

$$C = C_T C_H C_U C_S C_E C_M \quad\quad\quad (4.38)$$

$$C_T = 0.393 + 0.02796 (T_m) + 0.000119 (T_m)^2 \quad\quad (4.39)$$

$T_m$ is the mean daily air temperature in °C with a default option 20°C

$$C_U = 0.708 + 0.0034 \, U - 0.0000038 \, U^2 \quad\quad (4.40)$$

$U$ is the daily wind movement in km/day with default option of 96.56 km/day.

$$C_H = 1.25 - 0.0087 \, H_m + 0.000075 \, H_m^2 - 8.5 \times 10^{-11} H_m^4 \quad (4.41)$$

$H_m$ is the mean relative humidity measured at noon with default option of 40%.

$$C_s = 0.542 + 0.0085 S - 7.8 \times 10^{-6} S^2 + 6.2 \times 10^{-8} S^3 \quad (4.42)$$

$S$ is the percentage of bright sunshine hours taken as $100n/N$ with a default option 80%.

$$C_E = 0.97 + 0.0000984E \qquad (4.43)$$

$E$ is the elevation of the place with default option of 304.88 m.

$C_m$ is the monthly vegetative coefficient determined empirically. This can be taken from Table 4.12 without much error even though the values are derived empirically for the latitude of 40° in USA by Christiansen.

**Table 4.12  Value of Vegetative Coefficient $C_m$**

| Month | J | F | M | A | M | J | J | A | S | O | N | D |
|---|---|---|---|---|---|---|---|---|---|---|---|---|
| $C_m$ | 1.08 | 1.06 | 0.93 | 0.89 | 0.88 | 0.87 | 0.88 | 0.91 | 0.99 | 1.07 | 1.13 | 1.16 |

**Example 4.8:** Using Christiansen method, calculate the consumptive use of paddy for the month of February using data of Example 4.7.

**Solution**
The coefficients are calculated as follows.
The temperature coefficient $C_T$ is calculated from equation (4.39) as

$$C_T = 0.393 + 0.02796 \times 16 + 0.000119 \times 16^2 = 0.87$$

Wind movement coefficient is calculated from equation (4.40) as

$$C_U = 0.708 + 0.0034 \times 30 - 0.0000038 \times 30^2 = 0.813$$

Coefficient for relative humidity is calculated from equation (4.41) as

$$C_H = 1.25 - 0.0087 \times 50 + 0.000075 \times 50^2 - 8.5 \times 10^{-10} \times 50^4 = 0.949$$

The sunshine has coefficient $C_s$ calculated from equation (4.42) taking $n = 7.2$ and $N = 11.42$ as per problem 4.6. It gives $S = (n/N) \times 100 = 63\%$

$$C_S = 0.542 + 0.0085 \times 63 - 7.8 \times 10^{-6} \times 63^2 + 6.2 \times 10^{-8} \times 63^3 = 0.892$$

Elevation coefficient $C_E$ is calculated from the equation (4.43) as

$$C_E = 0.97 + 0.0000984 \times 220 = 0.992$$

$C_m$ is assumed to be close to the value given in Table 4.12 and its value is read from the table as 1.00. The coefficient $C$ is calculated from equation (4.38) as

$$C = 0.87 \times 0.813 \times 0.949 \times 0.892 \times 0.992 \times 1.00 = 0.594$$

Value of $H_a$ read from table 4.7 as 11.94 mm of water/day. PET is calculated from equation (4.37) as

$$PET\ (C_u) = 0.473 \times H_a \times C$$

$$= 0.473 \times 11.94 \times 0.594 = 3.355 \text{ mm of water/day}$$

$$= 3.355 \times 28 \times 1/10 = 9.4 \text{ cm for February.}$$

### 4.4.2.4  Thornthwaite Equation
Thornthwaite (1948) developed an exponential relationship between mean monthly temperature and mean monthly consumptive, given as

$$PET = 1.62\ R_f \left( \frac{10 T_m}{T_e} \right)^a \qquad (4.44)$$

where $R_f$ is the reduction factor (See Table 4.13), $T_m$ the mean monthly temperature in °C, $a$ is a constant which can be computed from the relation

$$a = 0.4923 + 0.01792\ T_e - 0.0000771\ T_e^2 + 0.000000675\ T_e^3 \qquad (4.45)$$

where $T_e$ is the annual temperature efficiency index given by

$$T_e = \sum_{j=1}^{12} \left( \frac{T_m}{5} \right)^{1.514} \qquad (4.46a)$$

For one period, say for one month, $T_e$ is calculated as

$$T_e = \left( \frac{T_m}{5} \right)^{1.514} \qquad (4.46b)$$

The above calculations are made for a month of 30 days and for each day 12 h of evapotranspiration is considered. Since the two factors vary from 28 to 31 days and with latitude, the values of 12 h a day is not constant, it can be multiplied with factors from Table 4.13 depending on the month and latitude of the place.

**Table 4.13   Reduction Factor $R_f$ for PET to be Used in Equation (4.44)**

| Latitude | J | F | M | A | M | J | J | A | S | O | N | D |
|---|---|---|---|---|---|---|---|---|---|---|---|---|
| | | | | | | Month | | | | | | |
| 0°N | 1.04 | 0.94 | 1.04 | 1.01 | 1.04 | 1.01 | 1.04 | 1.04 | 1.01 | 1.04 | 1.01 | 1.04 |
| 10°N | 1.00 | 0.91 | 1.03 | 1.03 | 1.08 | 1.06 | 1.08 | 1.07 | 1.02 | 1.02 | 0.98 | 0.99 |
| 20°N | 0.95 | 0.90 | 1.03 | 1.05 | 1.13 | 1.11 | 1.14 | 1.11 | 1.02 | 1.00 | 0.93 | 0.94 |
| 30°N | 0.90 | 0.87 | 1.03 | 1.08 | 1.18 | 1.17 | 1.20 | 1.14 | 1.03 | 0.98 | 0.89 | 0.88 |
| 40°N | 0.84 | 0.83 | 1.03 | 1.11 | .1.24 | 1.25 | 1.27 | 1.18 | 1.04 | 0.96 | 0.83 | 0.81 |
| 50°N | 0.74 | 0.78 | 1.02 | 1.15 | 1.33 | 1.36 | 1.37 | 1.25 | 1.00 | 0.92 | 0.76 | 0.70 |

**Example 4.9:** Using Thornthwaite equation, calculate the consumptive use of paddy for the month of February. Take data from Example 4.7

**Solution**
Since PET is required to be calculated for February only, equation (4.46b) is used.

$$T_e = \left( \frac{T_m}{5} \right)^{1.514} = \left( \frac{16}{5} \right)^{1.514} = 5.818$$

$$a = 0.4923 + 0.01792 \times 5.818 - 0.0000771 \times 5.818^2 + \ldots$$

$$= 0.4923 + 0.1043 - 0.0026 = 0.594$$

Reduction factor for February at latitude of 22°N is 0.895 (from Table 4.13)

$$PET = 1.6 \times 0.895 \left( \frac{10 \times 16}{5.818} \right)^{0.594} = 10.38 \text{ cm.}$$

Proceeding in the same way, PET for other months can be calculated and added up. If a crop is grown from 15th February then the value of PET is to be reduced by (13/28) to arrive at the value for the month.

### 4.4.2.5   Hargreaves Method

Hargreaves proposed monthly consumptive use coefficient $k$ for various crop groups which when multiplied with monthly pan evaporation values gives PET.

$$PET\ (E_t) = \Sigma\, k \cdot E_p \tag{4.47}$$

where $k$ is the monthly consumptive use coefficient, which is dependent on the crop grown, $E_p$ the pan evaporation in mm and *PET* in mm.

Consumptive use coefficient $k$ is different for different crops and is different for the same crop at different places. For the same crop it is different during stages of its growth. Various research stations in India report values of $k$ for different crops. In absence of figure for specific crop, data from Table 4.14 may be used with caution.

**Table 4.14   Values of Hargreaves Monthly Consumptive Use Coefficient $k$**

| Crop Group | Important Crops Under the Group | $E_t/E_p$ |
|---|---|---|
| Group A | Potato, Cotton, Maize, Bean, Peas, Jower, Beat. | 0.20–1.00 |
| Group B | Tomato, Olive, Plumes and some delicious fruits. | 0.15–0.75 |
| Group C | Onions, Grapes, Melons, Carrots, Hops | 0.12–0.60 |
| Group D | Wheat, Barley, Celery and other grass type plants | 0.10–0.90 |
| Group E | Pesters, Plantin, Orchard crops, etc. | 0.70–1.10 |
| Group F : | Oranges, fruits, citrus crops | 0.60–0.60 |
| Group G : | Sugarcane, Alfalfa etc. | 0.50–1.00 |
| Paddy : | Maximum at 50% of growth is | 0.80–1.30 |

$E_t\ (C_u)$ is the evapotranspiration and $E_p$ is the pan-evaporation.

$\dfrac{E_t}{E_p} = k$ is the consumptive use coefficient. Hargreaves $k$ for crop groups are given in Table 4.15, which is primarily dependent on the crop and the percentage of growth period of the crop. For dark vegetation plants $k$ should be taken higher for the same group and for lighter crop, it is lesser. Tall plants has higher value of $k$ than small plants of same degree of greenness and density.

**Example 4.10:** Use Hargreaves method to calculate crop water requirement for paddy (Dalua-Rabi crop) to be grown from January 6 to March 26. Class-A pan evaporation values for the months are 10, 10 and 14 cm respectively. Rainfall during the three months can be taken as 11, 13 and 28 mm. Calculate the gross irrigation requirement at the head of field if irrigation efficiency is 85%.

**Table 4.15 Hargreaves Average Value of Consumptive use Coefficient ($k = E_t/E_p$)**

| % of crop growing season | Values of $k$ — Group | | | | | | | | | Wheat | | Cotton | Maize |
|---|---|---|---|---|---|---|---|---|---|---|---|---|---|
| | A | B | C | D | E | F | G | Rice | L | P | P | L |
| 1 | 2 | 3 | 4 | 5 | 6 | 7 | 8 | 9 | 10 | 11 | 12 | 13 |
| 0 | 0.2 | 0.15 | 0.12 | 0.08 | 0.90 | 0.60 | 0.50 | 0.80 | 0.14 | 0.30 | 0.22 | 0.40 |
| 5 | 0.2 | 0.15 | 0.12 | 0.08 | 0.90 | 0.60 | 0.55 | 0.90 | 0.17 | 0.40 | 0.22 | 0.42 |
| 10 | 0.36 | 0.27 | 0.22 | 0.15 | 0.9 | 0.6 | 0.60 | 0.95 | 0.23 | 0.51 | 0.23 | 0.47 |
| 15 | 0.50 | 0.38 | 0.30 | 0.19 | 0.9 | 0.6 | 0.65 | 1.00 | 0.33 | 0.62 | 0.24 | 0.54 |
| 20 | 0.64 | 0.48 | 0.38 | 0.27 | 0.9 | 0.6 | 0.70 | 1.05 | 0.45 | 0.73 | 0.26 | 0.63 |
| 25 | 0.75 | 0.56 | 0.45 | 0.33 | 0.9 | 0.6 | 0.75 | 1.10 | 0.60 | 0.84 | 0.35 | 0.75 |
| 30 | 0.84 | 0.63 | 0.50 | 0.40 | 0.9 | 0.6 | 0.80 | 1.14 | 0.72 | 0.92 | 0.58 | 0.85 |
| 35 | 0.92 | 0.69 | 0.55 | 0.46 | 0.9 | 0.6 | 0.85 | 1.17 | 0.81 | 0.96 | 0.80 | 0.96 |
| 40 | 0.97 | 0.73 | 0.58 | 0.52 | 0.9 | 0.6 | 0.90 | 1.21 | 0.88 | 1.10 | 0.95 | 1.04 |
| 45 | 0.99 | 0.74 | 0.60 | 0.58 | 0.9 | 0.6 | 0.95 | 1.25 | 0.90 | 1.10 | 1.03 | 1.07 |
| 50 | 1.00 | 0.75 | 0.60 | 0.65 | 0.9 | 0.6 | 1.00 | 1.30 | 0.91 | 1.00 | 1.08 | 1.09 |
| 55 | 1.00 | 0.74 | 0.60 | 0.71 | 0.9 | 0.6 | 1.00 | 1.30 | 0.90 | 0.91 | 1.08 | 1.10 |
| 60 | 0.99 | 0.74 | 0.60 | 0.77 | 0.9 | 0.6 | 0.95 | 1.30 | 0.89 | 0.80 | 1.07 | 1.11 |
| 65 | 0.96 | 0.72 | 0.58 | 0.82 | 0.9 | 0.6 | 0.90 | 1.25 | 0.86 | 0.65 | 1.05 | 1.10 |
| 70 | 0.91 | 0.68 | 0.55 | 0.88 | 0.9 | 0.6 | 0.85 | 1.20 | 0.83 | 0.51 | 1.00 | 1.07 |
| 75 | 0.85 | 0.64 | 0.51 | 0.90 | 0.9 | 0.6 | 0.80 | 1.15 | 0.80 | 0.40 | 0.93 | 1.04 |
| 80 | 0.75 | 0.56 | 0.45 | 0.90 | 0.9 | 0.6 | 0.75 | 1.10 | 0.76 | 0.30 | 0.85 | 1.00 |
| 85 | 0.60 | 0.45 | 0.36 | 0.80 | 0.9 | 0.6 | 0.70 | 1.00 | 0.71 | 0.20 | 0.73 | 0.97 |
| 90 | 0.46 | 0.35 | 0.28 | 0.70 | 0.9 | 0.6 | 0.55 | 0.9 | 0.65 | 0.12 | 0.62 | 0.89 |
| 95 | 0.28 | 0.21 | 0.17 | 0.60 | 0.9 | 0.6 | 0.50 | 0.8 | 0.58 | 0.10 | 0.50 | 0.81 |
| 100 | 0.20 | 0.20 | 0.17 | 0.60 | 0.9 | 0.6 | | 0.2 | 0.51 | 0.10 | 0.40 | 0.70 |
| Seasonal value | — | — | — | — | — | — | — | — | 0.61 | 0.61 | 0.68 | 0.86 |

P = Poona, L = Ludhiana.

Local values of $k$ should be used always whenever available.

For each crop group an average value of $k$ is given. These values should be increased if there are taller plants in the field or for darker green vegetation on the basis of experience and judgement.

**Solution**

Referring to Table 4.15, for rice, value of $k$ for the percent of growing season are found and entered in col. (5) of Table 4.16. $E_t$ values are computed in col. (6) (Table 4.16). Crop period for paddy is taken as 80 days from 6 January to 26 March.

**Table 4.16   Consumptive use of Paddy for Example 4.10**

| Date | No of days upto mid point | % of growing season | Pan evapora- tion $E_p$ in (cm) | $k$ from Table 4.15 | $E_t = k.E_p$ (cm) | Rainfall during the month (cm) | Net Irri. requirement = col. (6) – 80% ×col.(7) (cm) | Gross Irri. Require- ment = col. (8)/ efficiency |
|---|---|---|---|---|---|---|---|---|
| (1) | (2) | (3) | (4) | (5) | (6) | (7) | (8) | (9) |
| 6 --31 Jan. | 13 | 100(13/80) = 16.25 | 10 | 1.01 | 10.1 | 1.1 | 9.22 | 10.85 |
| 1 --28 Feb | 40 | 100(40/80) = 50.0 | 10 | 1.30 | 13.0 | 1.3 | 11.96 | 14.07 |
| 1 --26 Mar. | 67 | 100(67/80) = 83.75 | 14 | 1.02 | 14.28 | 2.8 | 12.04 | 14.16 |
| Total 80 days | | | | | 37.38 | | 33.22 | 39.08 |

Consumptive use for paddy = 37.38 cm
Net irrigation requirement = consumptive use – effective rainfall = 33.22 cm
Gross irrigation demand = Net irrigation demand/efficiency = 39.08 cm.

### 4.4.2.6   Lowry-Johnson Method

Lawry and Johnson proposed a linear relation between effective heat and PET as follows:

$$PET = (0.004755D_m + 24.4) \qquad (4.48)$$

in which $D_m$ is the accumulated degree days of maximum daily temperatures for the growing season in °F (above 32°F), PET is in cm. Conversion from °F to °C can be made using the simple conversion formula

$$°F = \frac{(°C \times 9)}{5} + 32 \qquad (4.49)$$

**Example 4.11:** Solve example 4.9 by Lowry-Johnson method if the maximum daily average temperatures for January, February and March are 20, 26 and 30°C, respectively.

**Solution**

$$20°C = \frac{(20 \times 9)}{5} + 32 = 68°F, \quad 26°C = \frac{(26 \times 9)}{5} + 32 = 78°F$$

and $$30°C = \frac{(30 \times 9)}{5} + 32 = 86°F$$

Accumulated degree-days of the maximum daily temperatures in the growing season

$$D_m = 68 \times 26 + 78.8 \times 28 + 86 \times 26 = 1768 + 2206.4 + 2236 = 6210.4 \text{ degree days}$$

∴   PET = 0.004755 × 6210.4 + 24.4 = 53.93 cm

Consumptive use for the given paddy crop is 53.90 cm.

### 4.4.2.7   Jensen-Haise Method

Jensen and Haise (1963) developed the following relation for the computation of PET after analysing more than 3000 observations spread over 35 years

$$PET = C_c \, (T - T_c) \, H_a \tag{4.50}$$

where $H_a$ is the solar radiation in mm/day. The coefficient $C_c$ is calculated as

$$C_c = \frac{1}{C_I + 7.6 \, C_H} \tag{4.51}$$

$C_H$ is calculated from the relation

$$C_H = 50 \text{ mb}/(e_2 - e_1) \tag{4.52}$$

and

$$C_I = 38 - (2°C \times \text{Elevation in m})/305 \tag{4.53}$$

where $e_2$ and $e_1$ are the saturation vapour pressures at the mean maximum and mean minimum temperatures for the hottest month of the year, $T$ the mean air temperature, $T_c$ is given as

$$T_c = -2.5 - 0.14 \, (e_2 - e_1) - \frac{\text{Catchment elevation (m)}}{550} \tag{4.54}$$

---

**Example 4.12:** Calculate PET using Jensen–Haise method from the following data.
Mean air temperature = 25°C
Latitude of the place = 30°N
Mean maximum temperature of the hottest month = 30°C
Mean minimum temperature of the hottest month = 20°C
Elevation of the catchment = 150 m
Crop period is April.

**Solution**
From Table 4.5, $e_2$ for 30°C = 42.42 mb and $e_1$ for 20°C = 23.4 mb

$$C_H = \frac{50}{(42.42 - 23.4)} = 2.63 \quad \text{and} \quad C_I = 38 - \frac{(2 \times 150)}{305} = 36.03$$

$$C_c = 1/(36.03 + 7.6 \times 2.63) = 0.01785$$

$$T_c = -2.5 - 0.14 \, (42.42 - 23.40) - 150/550 = -5.44$$

$H_a$ is read from Table 4.7 as 14.8 mm of water/day for April at 30 °N

$$PET = 0.01785 \times [25 - (-5.44)] \times 14.80 = 8.04 \text{ mm/day}$$

$$= (8.04/10) \times 30 = 24.1 \text{ cm for April}$$

---

### 4.4.2.8   Turc Method

On extensive survey and analysis of rainfall, temperature and evaporation data, Turc (1961) proposed the following equation for the computation of evapotraspiration (ET).

$$\text{ET (mm/year)} = \frac{P}{\sqrt{0.9 + (P/C_T)^2}} \qquad (4.55)$$

where $P$ is the rainfall in mm for a year, $C_T$ is evaporation constant of air given as

$$C_T = 300 + 25t + 0.05t^3 \qquad (4.56)$$

where $t$ is the mean temperature in °C. Incorporating humidity, the proposed equation can be modified to the following form.

$$\text{ET} = \frac{0.013\ T(58.2\ H_c + 50)}{T + 15} \qquad \text{for } R_H > 50\% \qquad (4.57)$$

or $\qquad \text{ET} = \dfrac{0.013\ T(58.2 H_c + 50)}{T + 15}\left(1 + \dfrac{50 - R_H}{70}\right) \qquad \text{for } R_H < 50\% \quad (4.58)$

where $R_H$ is the relative humidity in percentage and ET in mm/day.

---

**Example 4.13**: For the month of February for paddy crop, calculate consumptive use from the following data.

> Relative humidity = $R_H$ = 50%
> Mean temperature = 16°C
> Latitude = 22°N

**Solution**
For latitude of 22°N the value of $H_a$ is 11.94 mm of water/day (Table 4.7).
   Now using equation (4.58)

ET = {0.013 × 16/(16 + 15)} × (58.2 × 11.94 + 50){1+ (50 − 50)/70} = 5.0 mm/day

For August, PET = 31 × 5 = 155 mm = 15.5 cm

---

## 4.5   INTERCEPTION
Interception is that portion of precipitation which, while falling, is intercepted by aerial portion of vegetation, buildings and other objects above the surface of earth and evaporates back to the atmosphere. The adhesive force between the water drops and the vegetation holds back the drops of water against gravity until they grow in size to over weigh and slip down. Interception loss is measured as the volume of water and is an important factor to account for when a basin experiences large number of small storms. For flood studies, interception is not considered as significant loss. This is taken care by subtracting a lumpsum quantity of 10 mm from the initial block of the storm histogram. Factors on which interception depends are (a) intensity and duration of the storm, (b) density of trees, (c) type of trees (i.e. long or short, coniferous or deciduous, area of canopy) and other obstructions, (d) season of the year and (e) wind velocity at the time of precipitation. The part of precipitation retained by the aerial portion of vegetation and other objects and is either absorbed by them or evaporates back is called *interception loss*. The portion of precipitation which drips down to the ground is known as *through fall*. Rainfall in excess of the capacity of

bushes or trees to hold is available to ground surface either as through fall or may reach the ground by running down the stem of trees as *stem flow*. Therefore, the net rainfall available at ground surface is the gross rainfall minus the part of intercepted water which finally evaporates.

Interception is mostly satisfied from initial precipitation. Between 0.05 to 0.25 mm of rainfall is heldup by tree foliage before dripping takes place. Steam flow varies between 0.01 to 0.025$P$, where $P$ is the depth of precipitation in mm. Total interception during a storm is the sum of initial interception and the portion of the rainfall evaporating from tree leaf surface before it reaches the ground.

$$\text{Interception loss (mm/h) } I_l = I_s + R_v E D_s \qquad (4.59)$$

where $I_s$ is the initial loss or interception storage (0.2 to 1.2 mm), $R_v$ the ratio of total evaporating leaf surface to its projected surface in a horizontal plane, $E$ the rate of evaporation during rain in mm/h and $D_s$ the duration of storm in hours. The plot between rainfall in abscissa and interception loss in percentage is an exponentially decaying curve (Fig. 4.5), whose nature varies for different different types of trees. Total quantity of interception may be as high as 25% of total precipitation in a year for a dense forest canopy. Horton proposed a straight line relation between precipitation $P$ in mm and total interception loss $I_l$ for ash tree in the following form.

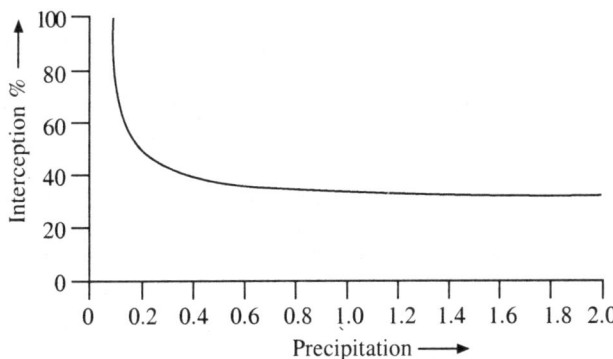

**Fig. 4.5   A Typical Interception Curve (After Horton 1919)**

$$I_l = 0.38 + 2.3P \qquad (4.60)$$

Depending on the vegetal cover and precipitation, a general form of equation for interception loss can be written as

$$I_l = aP + b\,(1 - e^{-p/b}) \qquad (4.61)$$

where $a$ and $b$ are constants which depend on the factors of infiltration loss, $a$ varies between 0.01 and 0.2 and $b$ between 2.5 and 38% of rainfall, $P$ is the precipitation depth in mm. When the intensity of rainfall is 25 mm/h interception rate varies between 15% for soyabean crop and 57% for tall grass.

---

**Example 4.14:** Calculate the interception loss for a tropical forest area in India ($a = 0.11$

and $b$ = 15% of $P$). The intensity of precipitation is low but uniform over the area. Total depth of storm rainfall is 4 cm.

**Solution**
Here $a$ = 0.11, $b$ = 0.15$P$, $P$ = 4.0 cm

$$\text{Interception loss } I_l = aP + b(1 - e^{-p/b})$$

$\therefore$   $I_l = 0.11P + 0.15\,P\,(1 - e^{-p/0.15P}) = 0.2598\,P = 1.039$ cm = 10.4 mm.

---

## 4.6   DEPRESSION STORAGE

Any natural ground surface generally has numerous shallow depressions of varying size, shape and depth. When precipitation fall, these depressions form miniature reservoirs detaining water temporarily. Water from these storages either evaporates or infiltrates into the ground charging the ground water reservoir. Storage from such depressions are not measurable due to practical difficulties. After filling all the small depressions, overland flow from the area takes place. Due to change in land use pattern, depression storages also change with time, which makes it almost a non-measurable quantity. It is measured indirectly by measuring all other terms of equation (4.1).

Hicks tried to measure the loss due to depression storage by analysing periods of high rainfall and suggested values of 2.5 mm for clayey, 3.8 mm for loamy and 5.0 mm for sandy soils.

A general form of depression storage curve can be given as

$$V_{SD} = K\,(1 - e^{-Pe/K}) \tag{4.62}$$

where $K$ is the capacity of the basin to store water in its depressions, $V_{SD}$ the depression storage volume, $P_e$ the volume of precipitation in excess of infiltration and interception. It is an exponentially decaying curve and attains the value of $K$ when the ratio $P_e/K$ increases. The important factors affecting depression storage are (a) land form, (b) soil characteristics, (c) topography, (d) antecedent precipitation index and (e) man made disturbances, like terrace farming. Collected water at these depressions never reaches the outlet of a basin. Depression storage helps to reduce soil erosion and increase soil moisture content. Therefore, farmers are encouraged to go for terrace farming to conserve soil and rain water for beneficial uses. But this water is a definite loss to an engineer-in-charge of water resource project. The sum of infiltration and depression storage can vary from 10 to 50 mm per storm depending on their intensity, duration and other characteristics.

## 4.7   INFILTRATION AND ITS ESTIMATION

Infiltration is as the entry or the passage of water into the soil through soil surface. It is a major loss of precipitation affecting runoff of a basin. This term should be properly understood and quantified. Losses like interception, depression storage and evaporation during precipitation are small, which cannot change the runoff of a basin significantly during major floods, but infiltration is a major process continuously affecting the magnitude, timing and distribution of surface runoff at any measured outlet of a basin. It is responsible for the growth and nourishment of life on earth. Infiltration process is initiated by creation of hydrogen bond between soil particles and the water. The adhesive force of attraction between soil and water, the surface tension, capillarity and gravitational forces help to force more water between the pores of soil particles as more water is added to the system due to rain.

*Infiltration capacity* is the maximum rate at which a given soil can absorb water under a given set of conditions at a given time. At any instant the actual infiltration $f_t$ can be equal to infiltration capacity $f_0$ only when the rainfall intensity is greater than $f_0$, otherwise actual infiltration will be equal to the rate of rainfall. This can be observed during low intensity rainfall when there is no surface runoff produced due to precipitation. Once water enters into the soil, the process of transmission of water within the soil known as *percolation* takes place, thus removing the water from near the surface to down below, charging the ground water reservoir. Infiltration and percolation are directly interrelated. When percolation stops, infiltration also stops. During any storm, infiltration is the maximum at the beginning of the storm, decays exponentially and attains a constant value $f_c$ as the storm progresses. The effect of infiltration is to

(i)  reduce flood magnitude
(ii)  delay the time of arrival of water to the channel
(iii)  reduce the soil erosion
(iv)  recharge to the ground water reservoir
(v)  fill the soil pores with water to its field-capacity, which subsequently supply water to the plants
(vi)  avail the ground water during the non-rain periods in the channels
(vii)  help to supply water to plants

### 4.7.1  Factors Affecting Infiltration
Factors affecting infiltration depend on both meteorological and soil medium characteristics. These are

*(a) Surface Entry*
If a soil surface is bare, the impact of raindrops causes inwashing of finer particles and clogs the surface. This retards infiltration. An area covered by grass and other bushy plants has better infiltration capacity than a barren land.

*(b) Percolation*
For infiltration to continue, water that has entered the soil must be transmitted down by the force of gravity and capillary actions. When percolation rate $(P_r)$ is slow, the infiltration rate is bounded by the rate of percolation. This depends on the factors like type of soil, its composition, permeability, porosity, stratification, presence of organic matter and presence of salts. A graph between infiltration capacity vs. grain size of the medium is given in Fig. 4.6.

*(c) Antecedent Moisture Condition*
Infiltration depends on the presence of moisture in the soil. For the second storm in succession, the soil will have lesser rate of infiltration than the first maiden

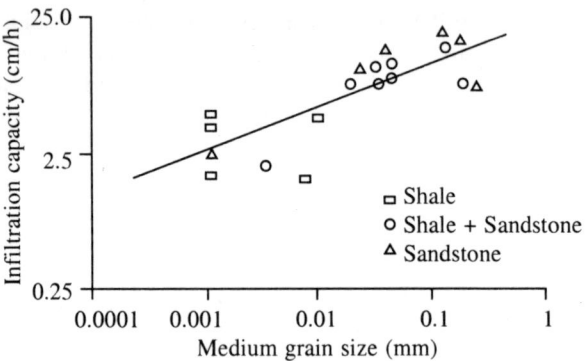

**Fig. 4.6  Infiltration Capacity vs Medium Grain Size**

storm of the season. Except sandy soil most other soils have swelling ingredients, which swells in presence of water and reduce infiltration rate to the extent of their presence.

*(d) Climate Conditions*

Temperature affects the viscosity of water. Variation of infiltration with soil temperature is shown in Fig. 4.7. Flow of water within the body of the soil is laminar; the flow being directly related to viscosity. In summer therefore, less viscous water causes more infiltration than in winter. In sub-zero temperature, water present in soil pores gets crystallised, thus blocking the passage. Other climatic factors may not influence infiltration rate to the extent, temperature does, and therefore, temperature can be considered as the only vibrant climatic factor affecting infiltration.

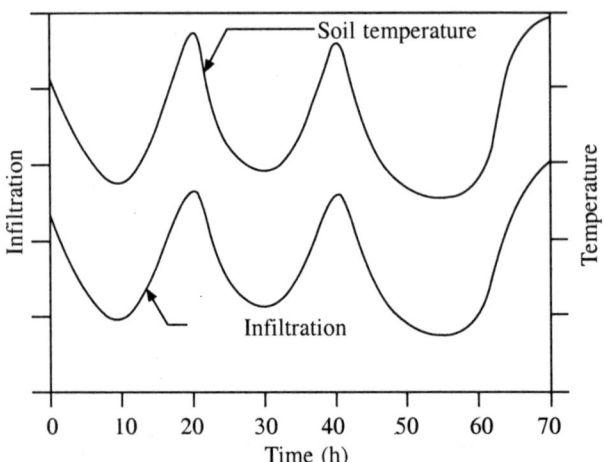

**Fig. 4.7  Variation of Soil Temperature and Infiltration with Time**

*(e) Rainfall Intensity and Duration*

During heavy rainfall, the top soil is affected by mechanical compaction and by the inwash of finer materials. This leads to faster decrease in the rate of infiltration than with low intensities of rainfall. Duration of rain affects to the extent that when the same quantity of rain falls in *n* number of isolated storms instead of a continuous one, the infiltration will be higher in the former case.

### (f) Human Activities

When crops are grown or grass coveres a barren land, the rate of infiltration is increased. On the other hand construction of roads, houses, overgrazing of pastures and playgrounds reduce infiltration capacity of an area considerably.

### (g) Depletion of Ground Water Table

Position of ground water table should not be very close to the surface for infiltration to continue. The quantity of infiltrated water entering into the soil should be drained out fully from the top soil zone so that there is some space available for the infiltrated water to store during the next rain.

### (h) Quality of Water

Water containing silt, salts and other impurities affect the infiltration to the extent they are present. Salts present affect the viscosity of water and may also react chemically with soil to form complexes which obstruct the porosity of soil, thereby affecting infiltration. Silts clog the pore spaces retarding infiltration rate considerably.

### (i) Vegetation

Soil covered with vegetation has greater infiltration than barren land. Because of growth and decay of roots and bacterial activities, dense natural forest provide good infiltration than sparsely planted crops.

### (j) Grain Size of Soil Particles

Other factors remaining the same, infiltration rate is directly proportional to the grain size diameter. When swelling minerals like illite and montmorillonite are present in soils, the infiltration rate reduces.

### (k) Catchment Parameters

A correlation between the *drainage density* (length of all streams of a basin divided by its catchment area) and infiltration can be established for various basins. Such curves exhibit negative correlation. When the drainage density increases, infiltration capacity decreases. A similar type of correlation can be obtained between the runoff rate ($m^3$/ sq. km) and infiltration capacity (cm/h). A relation between the sediment yield rate in ha.m/sq. km and infiltration capacity in cm/h also exhibit an inverse proportionality.

Three distinct approaches to determine infiltration capacity of a basin are available.

  (i) Experiment by a single or double cylindrical infiltrometer
  (ii) Rainfall simulator
  (iii) Rainfall-runoff analysis.

### 4.7.2   Field Measurement Using Infiltrometers

Two types of infiltrometers used are: single cylindrical and concentric double cylindrical.

*Single Tube Infiltrometer*

It consists of a hallow metal cylinder 30 cm in diameter and 60 cm long driven into the ground such that 10 cm of it projects above ground level. Water is poured at the top such that a head of 7 cm within the infiltrometer is maintained above ground level. A graduated jar or burette is used to add water, (Fig. 4.8), to give directly the volume of water added over time. A plot of time in abscissa against rate of water added in mm/h gives an infiltration capacity curve for the area. The setup resembles to the flooding type of irrigation situation in the field which can be a possible representation of real local conditions. Sufficient precautions should be taken to drive the cylinder into the ground with minimum disturbance to the soil structure. In a single infiltrometer, the major criticism is that water spreads out immediately beyond the bottom of the cylinder which does not represent a true infiltration condition of the field.

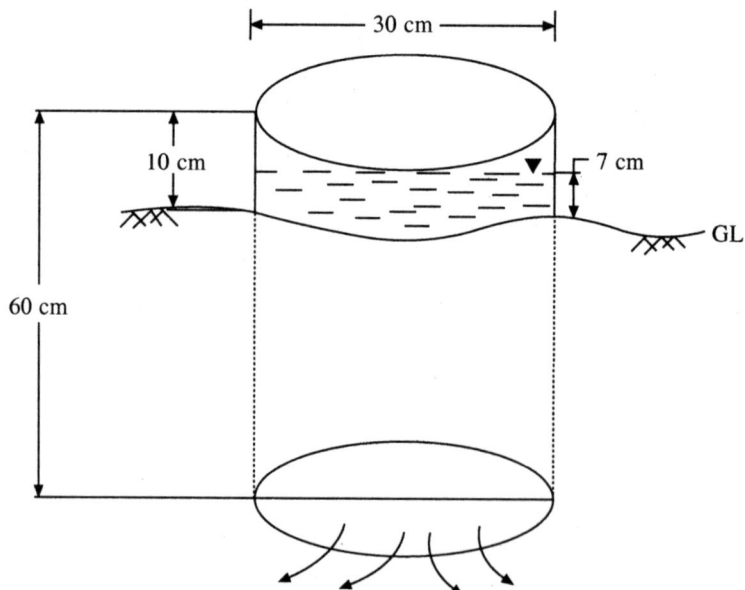

**Fig. 4.8   A Single Tube Infiltrometer**

*Double Tube Infiltrometer*

To overcome the objections of a single ring infiltrometer a set two concentric hollow cylinders of same length are used (Fig. 4.9). Water is added to both the rings to maintain the same height. Reading of the burette for the inner cylinder is taken as infiltration capacity of the soil. The outer cylinder is maintained to prevent spreading of water from the inner one. The important disadvantages still prevalent in these types of infiltrometers are

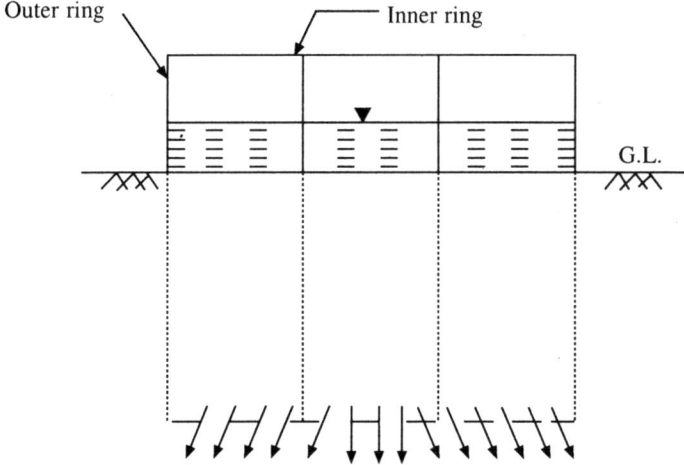

**Fig. 4.9  A Double Tube Infiltrometer**

(i) The size effect: Larger diameter infiltrometers give more accurate and always lesser value of infiltration than smaller diameter type.
(ii) Boundary effect.
(iii) Disturbance of original soil due to driving of the rings.

In the market a single ring portable infiltrometer is available which gives infiltration results quite acceptable for all practical purposes.

*Rainfall Simulator*
Rainfall simulators are used to overcome the difficulties of infiltrometers. A rainfall simulator consists of a sprinkler with nozzles capable of producing artificial rain of various intensities, drop sizes and durations. A field plot of 1.8 m in width and 3.65 m or more length is selected on which the nozzles spray water at height of 2 m or more to the field. The terminal velocity of rainfall is assumed to be close to raindrops though a height of 5–6 m may be the ideal situation. Rainfall intensities of 44.5 mm/h or multiple of it are generally created under such conditions. Arrangement is made to collect the runoff from the plot which can be measured ($S_{Rd}$). Prior to the experiment a run is made covering the plot with a polyethylene sheet to know the average rate of rainfall ($P_d$). Test run starts after removal of the polythene cover from the field and continues till a steady state of runoff from the plot is obtained. The following water budget equation is used to estimate infiltration rate from the experiment.

$$F_d = P_d - S_{Rd} - S_{ol} \tag{4.63}$$

where $F_d$ is the depth of infiltrated water in mm (to be computed), $P_d$ the simulated rainfall depth in mm (measured), $S_{Rd}$ the surface runoff depth in mm (measured), $S_{ol}$ the depression storages, surface detention, abstraction and other losses in mm. When steady state is reached, the analytical run is carried out, the volume $S_{ol}$ being no more effective, the constant rate of infiltration is calculated from the relation $F_d = (P_d - S_{Rd})$. When rainfall stops, runoff from the plot continues and the value of depression storage plus surface detention can be measured from the recession between stop of rainfall and the last drop of water flowing out of the plot as runoff. Abstraction losses (if any) can also be measured from the plot at the initial stage. Normally rainfall simulator gives less rate of infiltration than the flooding type infiltrometers.

### 4.7.3   Rainfall-Runoff Analysis

A plot between runoff from a watershed vs. time is known as Hydrograph. Analysis of a hydrograph to determine infiltration is a practical field solution, which considers the variations of rainfall, land forms, land slope, size of catchment, soil characteristics, vegetation types, depression storage, surface detention and all other factors affecting infiltration and as they occur in a field. A network of rain gauges gives the average storm precipitation over a basin. The runoff from the basin can be measured at the outlet or at the desired location by gauging the stream. The difference between precipitation and the corresponding runoff averaged over the basin over time should give the total loss. This should form the basis for the estimation of infiltration. During storms, evapotranspiration is negligible. Depression storage and interception can be estimated as discussed under sections 4.6 and 4.5, respectively, or they can be considered as lumpsum and subtracted from the rainfall hyetograph at the beginning of the storm. The balance gives a true representation of the basin infiltration losses. To simplify the procedure, some theoretical curves are proposed, which divide the rainfall histogram blocks between runoff and total losses.

An ideal infiltration capacity curve proposed by Horton (1933, 39) is

$$f_t = f_c + (f_o - f_c)\ e^{-kt} \qquad (4.64)$$

where $f_t$ is the infiltration capacity at any time $t$ from the beginning of the storm in mm/h, $f_c$ the infiltration rate in mm/h at the final steady stage when the soil profile becomes fully saturated, $f_0$ the maximum initial value when $t = 0$ in mm/h at the beginning of the storm, $k$ an empirical constant depending on soil cover complex, vegetation and other factors and $t$ the time lapse from the onset of the storm. Values of $f_0$, $f_c$ and $k$ are dependent on number of factors like soil characteristics and climatic conditions. Even for the same catchment the values vary from storm to storm and from season to season. A typical curve of $f_t$ separating the rainfall intensity histogram between infiltration and surface runoff is shown in Fig. 4.10.

Depending upon the situation, a number of methods are available to accurately determine the infiltration curve. For small watersheds (for urban hydrology) the procedure may be entirely different to the analysis carried out for large watersheds.

Since we are interested in determining the infiltration rate for storms of large basins, an average infiltration rate known as *infiltration index* is assumed. This type of assumption underestimates infiltration rate during the initial part of the storm and slightly overestimates towards the end of it. The assumption holds fairly good for analysis of high intensity and long duration storms or in a situation, when the area is fairly saturated before the onset of the storm. Flood producing storms follow both the situations.

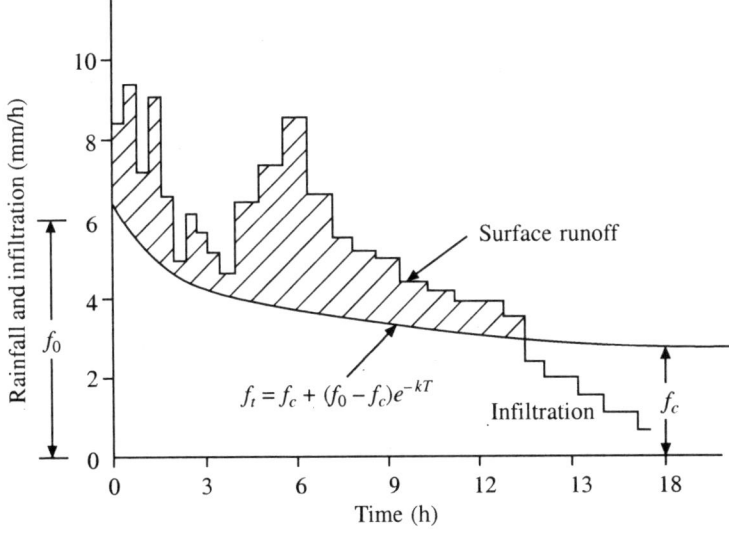

**Fig. 4.10 A plot of Infiltration Capacity Curve**

Difficulties with the theoretical approach to infiltration and practical difficulties to determine the values of $f_0$ and $k$, led to the use of *infiltration indices* based on an empirical approach. The indices commonly used to estimate direct runoff depths from rainfall intensity hyetographs are given after Example 4.15.

---

**Example 4.15:** From a double tube infiltrometer with inside ring diameter of 30 cm, the following observations were taken. Plot the infiltrations capacity curve and find the constant rate of infiltration that the experimental field have towards the end.

| Time (min.) | 0 | 2 | 5 | 10 | 20 | 30 | 50 | 80 | 120 | 150 |
|---|---|---|---|---|---|---|---|---|---|---|
| Cumulative volume of | 0 | 130 | 280 | 510 | 680 | 900 | 1040 | 1190 | 1280 | 1343 |

**Solution**

$$\text{Area of the infiltrometer} = \frac{\pi}{4} 30^2 = 706.5 \text{ sq. cm}$$

Solution is obtained in Table 4.17.

A graph between rate of infiltration in mm/min as ordinate vs. time as abscissa is plotted in Fig. 4.11.

---

### 4.7.4 Infiltration Indices

The following indices are mostly used for computation of infiltration rate from rainfall runoff data.

#### 4.7.4.1 Φ-Index

The average rate of rainfall above which the rainfall volume equals to runoff volume is called Φ-index. Procedure to derive Φ-index is described below. A schematic diagram showing the meaning of the Φ-index is given in Fig. 4.12.

**Table 4.17　Determination of Infiltration Rate Using Infiltrometers**

| Time (min.) | Time increment (min.) | Volume of water added (ml) | Depth of water added = col(3)/area (cumulative) | Increment depth (cm) | Rate of infiltration (mm/min) = $10 \times$ col. (5)/(2) |
|---|---|---|---|---|---|
| (1) | (2) | (3) | (4) | (5) | (6) |
| 0 | 0 | 0 | 0 | 0 | 0 |
| 2 | 2 | 130 | 0.184 | 0.184 | 0.92 |
| 5 | 3 | 280 | 0.396 | 0.212 | 0.71 |
| 10 | 5 | 510 | 0.722 | 0.326 | 0.65 |
| 16 | 6 | 680 | 0.962 | 0.240 | 0.40 |
| 30 | 10 | 900 | 1.274 | 0.312 | 0.31 |
| 50 | 20 | 1040 | 1.472 | 0.198 | 0.10 |
| 80 | 30 | 1190 | 1.684 | 0.212 | 0.07 |
| 120 | 40 | 1280 | 1.811 | 0.127 | 0.03 |
| 150 | 30 | 1343 | 1.901 | 0.090 | 0.03 |

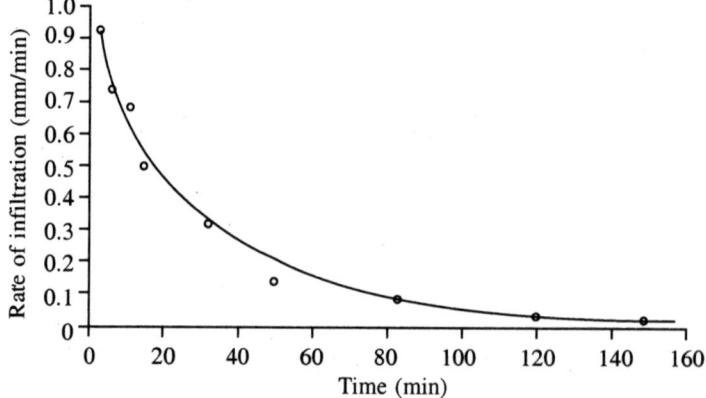

**Fig. 4.11　Plot of Infiltration Curve for Example 4.15**

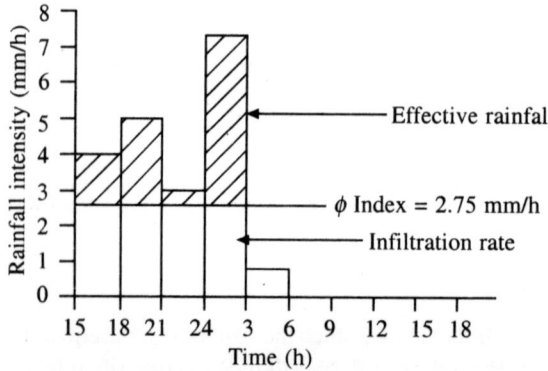

**Fig. 4.12　Effective Rainfall Using $\phi$-Index**

1. Storm producing runoff for the basin is identified.
2. All rainfall data in and around the basin are collected for the selected storm along with data of atleast one self recording gauge.
3. Corresponding storm runoff data for the basin are also collected.
4. Plot the runoff hydrograph and separate the base flow by any of the suitable method to estimate the direct runoff hydrograph. Compute the depth of direct runoff in mm by dividing the area of the basin by the total volume of the direct runoff. The concept of hydrograph and the separation of base flow is discussed in Section 7.3.
5. Find average aerial rainfall of the basin by any of the suitable method described in Section 3.10.
6. Compute time distribution percentages of storm rainfall on the basis of the self recording rain gauge data (or average time distribution if more than one SRRG data are available) for the basin under consideration. This is required because the non-recording rain gauges record rainfall for 24 h, whereas we require rainfall intensities over the basin for shorter periods of 2 to 6 h to prepare rainfall intensity histogram.
7. Distribute the average basin rainfall of step (5) to smaller convenient time unit of say 2 to 6 h according to the percentage distribution of rainfall of step (6).
8. Estimate initial losses due to interception and depression storage in mm and subtract it from the first initial rainfall intensity histogram blocks.
9. Assume a trial value (depth) of $\Phi$-index in mm/h and compute the total rainfall excess depth by summing up the excess depth of rainfall beyond this trial index line in the histogram. Compare it with the direct runoff depth obtained in step (4).
10. If rainfall excess is more, than increase the assumed value of $\Phi$-index of step (9) and repeat step (9) till rainfall excess in mm equals to direct runoff depth in mm of step(4).
11. Repeat the procedure for other storms, if under study, to obtain a general $\Phi$-index value for the basin.

Central Water Commission suggests a relation of the form

$$\Phi = I - R \tag{4.65}$$

in which $$R = a\, I^{1.2} \tag{4.66}$$

$I$ is the rainfall intensity in mm/h on a daily (24 h) basis, $R$ the runoff in mm. For computation of $R$ in equation (4.66), take $I$ in mm/day and for computation of $\Phi$ in equation (4.65), divide $R$ from equation (4.66) by 24 to convert it to mm/h. Values of coefficient $a$ vary from 0.17 to 0.50. For sandy soils it may be taken as 0.20, coastal alluvium 0.25 to 0.30, silt 0.35, red and clayey soil 0.40, black cotton soil 0.45 and hilly soil 0.50.

---

**Example 4.16:** Calculate $\Phi$-index of a storm from the following data. Plot the rainfall histogram and mark the $\Phi$-index on the plot.

Catchment area = 430 sq. km
Volume of direct-runoff after separation of base flow = 10.75 M.m$^3$
Runoff started at 3.00 pm on 17/08/98

| Time of rainfall (h) | 15 | 18 | 21 | 24 | 03 |
|---|---|---|---|---|---|
| Depth of rainfall (cm) | 1.2 | 1.5 | 0.9 | 2.2 | 0.2 |

**Solution**

$$\text{Depth of direct runoff } \frac{(10.75 \times 10^6 \times 100)}{(430 \times 10^6)} = 2.5 \text{ cm} = 25 \text{ mm}$$

Rainfall has occurred for 15 h between 15th hour of 17.8.98 to 6th hour of 18.8.98. Period of rainfall excess is from 15 h to next day 3 AM = 12 h.

Total loss $(12 \times 2.75 + 2) = 35$ mm of rainfall during the storm, which is the sum of all losses due to infiltration, depression storage, interception and evaporation during 12 h of storm period. Storm histogram and $\Phi$-index are plotted in Fig. 4.12.

**Table 4.18    Computation of $\Phi$-Index for Example 4.16**

| Time Period | Precipitation depth (mm) | Precipitation intensity (mm/h) | Rainfall excess for $\Phi$ = 2.8mm/h col. (2) – 2.8 $\times$ 3mm | Second trial $\Phi$ = 2.75 mm/h Rainfall excess =col. (2) – 2.75 $\times$ 3 |
|---|---|---|---|---|
| (1) | (2) | (3) | (4) | (5) |
| 15–18 | 12 | 4 | 3.6 | 3.75 |
| 18–21 | 15 | 5 | 6.6 | 6.75 |
| 21–24 | 9 | 3 | 0.6 | 0.75 |
| 0–3 | 22 | 7.33 | 13.6 | 13.75 |
| 3–6 | 2 | 0.67 | – | – |
| Total | 60 mm | | 24.4 mm | 25 mm |

$\therefore \Phi$ index = 2.75 mm/h

### 4.7.4.2    W-Index

This index is considered as an improvement over $\Phi$-index in the sense that surface storage and interception losses are considered in its computation. It is defined as the average rate of infiltration which equals to the rate of precipitation minus surface runoff and retention during time $t$ and is expressed as

$$W = (P - S_{Ro} - D_R)/t \qquad (4.67)$$

where $P$ is the total depth of storm rainfall in mm, $S_{Ro}$ the total depth of surface runoff in mm, $D_R$ the total depth of surface retention (depth of depression storage plus interception loss) and $t$ the time in hour during which rainfall rate exceeds infiltration rate. Obviously when $D_R = 0$, during heavy and longer storms, $W$-index and $\Phi$-index become the same.

Sometimes $W_{min}$-index is used instead of $W$-index when the soil condition is very wet so that the soil infiltration rate is almost constant and infiltration is at the minimum rate for the basin. Both $W_{min}$-index and $W$-index vary from storm to storm. For design flood, which is usually very large, the two values are nearly the same.

For practical use, refinement of infiltration index may not be justified due to the fact that rainfall producing storms are comparatively large and they occur in wet seasons. Initial losses for such storms are negligible in comparison to the depth of precipitation from them. The assumptions in their estimation and the limitations involved in the measurement of the parameters are subject to small errors which do not necessitate refining indices using a sophisticated model.

On the basis of experience, a more direct and simpler method is invariably adopted for field problems. While estimating design floods from Probable Maximum Precipitation (PMP) or Standard Project Storm (SPS), a constant index value of 1.00 mm/h is usually assumed for Indian catchments in absence of any data. This index is deducted from the rainfall intensities to give rainfall excess producing direct runoff volume for the design flood. This is because the catchment is assumed to be fully wet by the time the design storm producing design flood is assumed to be prevalent over the basin. Details of $\Phi$-index computation are given in example 4.16 and also in the subsequent chapters.

### 4.7.4.3 $F_{av}$-Index

Here, we assume an average infiltration loss throughout the storm. The time period is so selected that only period of excess rain or continuous supply of infiltration is considered, as shown in Fig. 4.13. The procedure is the same as discussed under $\Phi$-index method except that the period of $F_{av}$ is considered on experience and judgement, which may be beyond the time of rainfall excess as shown in the figure. An adjustment for period of no-rain may be assumed. The procedure may be more practical but requires adjustment of time-intensity histogram on judgement for period of no-rain, initial loss and extra period for $F_{av}$, beyond the time of rainfall excess. A good knowledge on soil of the basin, catchment and storm characteristics is required for such adjustments to be made. The figure is self explanatory for computation of $F_{av}$-index.

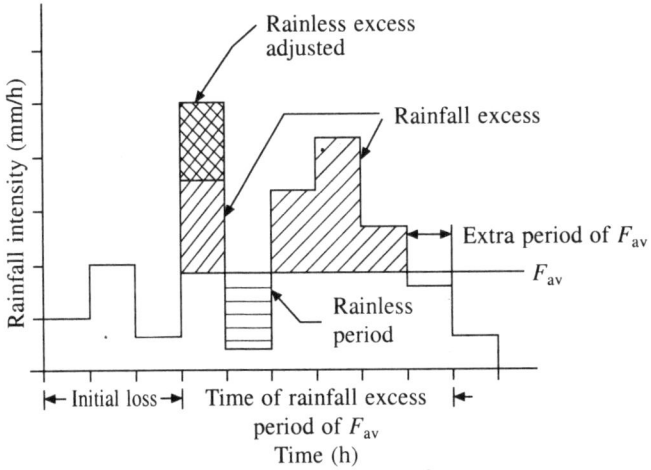

**Fig. 4.13 A Plot of $F_{av}$ Curve**

### 4.7.5   Mass Curve Method

For large water sheds, available rainfall and runoff data can be used effectively to obtain the average infiltration rate of a basin. An accurate estimate of infiltration over the storm period for large catchments may be of academic interest but for field engineers the method generally followed is based on the following :

1. Rainfall data covering the entire basin for the storm is collected. The storm must have covered the entire basin under consideration.
2. Average storm precipitation over the basin is calculated by Thiessen-polygon or any other suitable method.
3. Cumulative of the basin average rainfall is computed and plotted on a graph paper (as mass curve) against the corresponding time in abscissa.
4. Runoff of the basin measured at its outlet is used. Base flow is separated by a suitable method. Direct runoff hydrograph is divided into blocks of time units as considered for rainfall histogram and its equivalent depths found out by dividing the catchment area of the basin.
5. Cumulative depths of runoff of step (4) is computed and plotted in the same graph paper of cumulative average rainfall of step (3).
6. The difference in ordinates for different time blocks of the two cumulative plots of step (3) and (5) are read out from the graphs. This cumulative difference is plotted in the same graph paper.
7. The plot of the cumulative graph of step (6) will be irregular in shape. Smoothen the graph. This graph represents the cumulative value of infiltration and other losses.
8. Find the loss rate at different times from the slope of this curve. For storms producing floods, other losses can be neglected.
9. Plot infiltration rate at different times obtained in step (8) and plot a smooth curve between the infiltration loss rate and time, which gives an infiltration rate curve for the storm.

---

**Example 4.17:** Ordinates of a direct runoff hydrograph are given below.

| Time (h) | 0 | 6 | 12 | 18 | 2.4 | 30 | 36 | 42 | 48 | 54 | 60 |
|---|---|---|---|---|---|---|---|---|---|---|---|
| DRH (m³/sec) | 0 | 48 | 130 | 195 | 162 | 108 | 65 | 39 | 27 | 12 | 0 |
| Average cumulative rainfall (cm) | 0 | 3.7 | 10.4 | 18.3 | 18.3 | | | | | | |

Compute the rate of infiltration for the basin. Take the catchment area of the basin as 200 sq. km.

**Solution**

Cumulative rainfall and runoff volumes are computed in the table below.

The highest rate of infiltration and other losses is 2.56 cm/h, which tends to attain a constant rate of 1.65 cm/h towards the end of the storm This highest recorded loss rate usually is at the beginning of the storm. But due to precipitation patterns and the type of movement of storm in the basin, the basin characteristics has given rise to the loss rate as calculated in col. (7) of the above table.

**Table 4.19   Infiltration Rate of a Basin Due to a Single Storm**

| Time | cumulative avg. rainfall (cm) | Runoff (m³/sec) (cm) | Depth of runoff (cm) | cumulative depth of runoff (cm) | Difference (2) – (5) (cm/h) | Infiltration rate col. (6)/6 |
|------|------|------|------|------|------|------|
| (1) | (2) | (3) | (4) | (5) | (6) | (7) |
| 0 | 0 | 0 | | | | |
| | | | 0.26* | 0.26 | 3.44 | 0.57 |
| 6 | 3.7 | 48 | | | | |
| | | | 0.95 | 1.21 | 9.19 | 1.53 |
| 12 | 10.4 | 130 | | | | |
| | | | 1.74 | 2.95 | 15.35 | 2.56 |
| 18 | 18.3 | 195 | | | | |
| | | | 1.91 | 4.86 | 13.44 | 2.24 |
| 24 | 18.3 | 162 | | | | |
| | | | 1.44 | 6.30 | 12.0 | 2.00 |
| 30 | 18.3 | 108 | | | | |
| | | | 0.92 | 7.72 | 11.08 | 1.47 |
| 36 | 18.3 | 65 | | | | |
| | | | 0.56 | 7.78 | 10.52 | 1.75 |
| 42 | 18.3 | 39 | | | | |
| | | | 0.35 | 8.13 | 10.17 | 1.70 |
| 48 | 18.3 | 27 | | | | |
| | | | 0.21 | 8.34 | 9.96 | 1.66 |
| 54 | 18.3 | 12 | | | | |
| | | | 0.06 | 8.40 | 9.9 | 1.65 |
| 60 | 18.3 | 0 | | | | |

$$* \ \frac{6 \times 3600 \, (0 + 48)}{2} \ \frac{100}{200 \times 10^6} = 0.26 \text{ cm}$$

### 4.7.6   Analytical Models

Study of literature on infiltration shows that considerable efforts have been made to model the infiltration conceptually. The models consider the soil matrix as a conceptual system. A single or group of elements are considered to represent the system. Various methods are available to solve the problem of computing the output from these elements. Thus infiltration is simulated by the systems approach. Basing on this approach, models were developed by Green and Ampt (GA-1911), Philip (two Term Model (1957)), Holtan (1961), Overton (1964), Huggins and Monke (1966) and SCS model (1972). The US Soil Conservation Service (SCS) model is discussed in section 6.10 for estimating runoff for small catchments. Discussion of all models are beyond the scope of this book.

### 4.8   WATERSHED LEAKAGE

A basin is separated from the adjoining basins by a ridge line which represents the line of the highest contour of the area. Surface water flow during any storm is considered to be generated entirely from the precipitation falling over the basin area. However, there are instances when infiltrated water from storms may not recharge ground water or join the stream as interflow but finds its way to the adjacent basin or to the sea, if there is a hydraulic conduit through sub-surface. Water balance studies carried out basin-wise may give serious errors if there is any substantial flow of water to another basin by major faults, fissures and other geological formations, as shown in Fig. 4.14. For small basins, this type of loss should be properly accounted for, while balancing water for such basins.

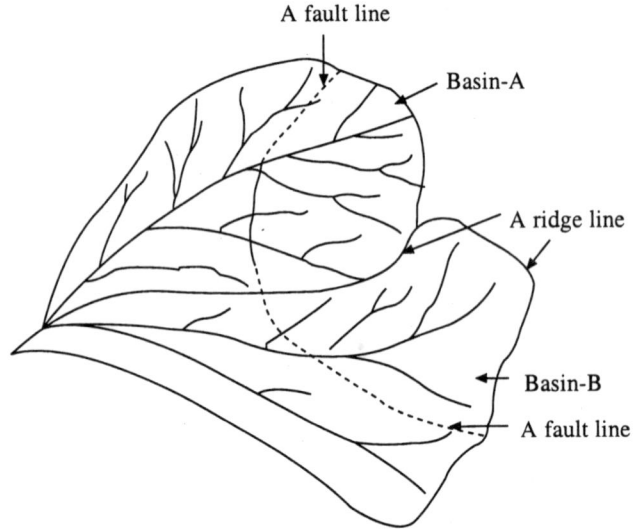

**Fig. 4.14    A Fault Line Showing Watershed Leakage**

## PROBLEMS

4.1  How does evaporation takes place? What are the factors that affect the evaporation process? How are you going to reduce evaporation from a near by tank?

4.2  What are the methods available to measure evaporation? Write five empirical approaches to determine lake evaporation in mm/day.

4.3  Explain energy budget method of computing lake evaporation. What are its limitations?

4.4  Define evapotranspiration. How it is different from evaporation?

4.5  What are the various climatic approaches to determine evaporation?

4.6  Write Blaney-Criddle method of calculating evapotranspiration. Is it the same as the consumptive use? Can evapotranspiration be estimated from pan evaporation data?

4.7  Describe with a neat sketch an IS specified evaporimeter.

4.8  Explain the balanced equation for precipitation and describe the terms (i) Interception and (ii) Depression storage.

4.9  What do you understand by the term infiltration? How can you measure it in the field?

4.10  Explain the terms $\Phi$-index, $W$-index, $F_{av}$-index and infiltration capacity.

4.11  Briefly explain the following:
   (a) watershed leakage
   (b) rainfall-runoff analysis for infiltration

(c) potential evapotranspiration and actual evapotranspiration
(d) actual infiltration rate and infiltration capacity
(e) depression storage
(f) difference between field capacity and wilting point
(g) rainfall simulators.

4.12 Rice was grown over an area of 420 hectares for the period covering from 20th June to 20th of November. Climatic data during the period is given below.

| Month | Mean temperature ($^\circ$C) | Sunshine hours (%) | Consumptive use coefficient | Rainfall (cm) |
|---|---|---|---|---|
| June | 32.5 | 9.8 | 1.17 | 14.1 |
| July | 31.2 | 9.2 | 1.23 | 18.4 |
| August | 29.7 | 8.9 | 1.22 | 19.2 |
| September | 28.4 | 8.6 | 1.12 | 10.4 |
| October | 27.2 | 8.3 | 1.01 | 6.3 |
| November | 26.5 | 8.0 | 0.86 | 1.3 |

If the canal efficiency is 75%, compute the volume of water to be drawn from the reservoir for the irrigation during the entire crop period.

4.13 A watershed of 48 sq. km produces a runoff of 2 M m$^3$ from the rainfall pattern of the storm given below. Calculate $\Phi$-index.

| Time (h) | 0 | 2 | 4 | 6 | 8 | 10 | 12 | 14 |
|---|---|---|---|---|---|---|---|---|
| Rainfall (mm) | 0 | 1.15 | 2.3 | 5.9 | 5.1 | 3.05 | 0.9 | 0 |

4.14 Rate of infiltration from the beginning of a storm are given below.

| Time (min) | 0 | 30 | 60 | 90 | 120 | 150 | 180 | 210 | 240 |
|---|---|---|---|---|---|---|---|---|---|
| Rate of infiltration (mm/h) | 821 | 54 | 20 | 22 | 16 | 14 | 12 | 8 | 8 |

Fit a infiltration capacity curve of exponential form.

4.15 A reservoir had the minimum water spread area of 6.3 km$^2$ on 1st June, 1995, increased to 16.8 km$^2$ on 31st October, 1995 and decreased to 6.7 km$^2$ on 31st May, 1996. If a class A pan evaporation data from a hydromet station located close to the reservoir from June to May are 210, 160, 160,160, 180, 155, 150, 140, 155, 220, 240, 240 and 250 mm, respectively, calculate water loss through evaporation from the reservoir for the year 1995–96.

4.16 Using Penman method, calculate the consumptive use of paddy for the month of July from the following data.
Wind velocity recorded at 4 m height = 80 km/day
Elevation of the place = 140 m
Relative humidity = 80%
Latitude = 20.5° North
Average temperature during July = 26°C

4.17 Calculate the consumptive use of paddy for July using data of problem 4.16 by three other approaches and compare the figures. Discuss the results. Which result you will accept for your project and why?

4.18 An isolated 3-h storm occurred over an area of 120 hectare has the following information:

| Partial area of catchment (hectare) | $\Phi$-index (cm/h) | Rainfall (cm) | | |
|---|---|---|---|---|
| | | 1st h | 2nd h | 3rd h |
| 36 | 0.9 | 0.6 | 2.4 | 1.3 |
| 18 | 1.1 | 0.9 | 2.1 | 1.5 |
| 66 | 0.5 | 1.0 | 2.1 | 0.9 |

What is the total rainfall on the catchment due to the storm? Estimate the runoff from the catchment taking the above $\Phi$-index values. What runoff will be produced by a rainfall 3.3 cm in 3-h uniformly distributed all over the catchment?

4.19 A reservoir has the following informations for the month of August, 1999.
Average surface area = 21.7 sq. km
Average inflow = 28.3 m$^3$/sec
Average outflow = 23.6 m$^3$/sec
Rainfall during the month = 18.4 cm
Change in storage = 17.2 M.m$^3$
The reservoir is assumed to be water tight having no seepage loss. Calculate reservoir evaporation for the month.

4.20 Cumulative rainfall during a storm are:

| Time (h) | 0 | 1 | 2 | 3 | 4 | 5 | 6 | 7 | 8 |
|---|---|---|---|---|---|---|---|---|---|
| Rainfall (mm) | 0 | 7 | 16 | 22 | 32 | 40 | 52 | 68 | 70 |

Assume an initial abstraction loss of 10 mm and a constant infiltration loss rate of 5.0 mm/h. Calculate the storm runoff volume from the catchment of 122 sq. km.

4.21 Compute daily evaporation using Meyers formula from the following data.
Relative Humidity = 66%
Wind Speed = 70 km/h
Water Temperature = 18°C
Air Temperature = 32°C

4.22 At an observatory there was rainfall of 18 mm during a day. Water of 4 mm depth has to be removed from the pan at the end of the day from the evaporimeter to bring the water level to the initial position. Assuming a pan coefficient of 0.75, calculate the volume of water lost from a nearby reservoir, having the water spread area of 7.2 sq. km during the day.

# Chapter Five

# Ground Water

## 5.1  INTRODUCTION

Ground water is an important segment of the hydrologic cycle and constitutes about one-third of world's fresh water reserves. Being relatively free from the pathological organisms, turbidity, radio-chemical and biological contaminations, it has distinct advantages over surface water resources. It is the most dependable resource. Its chemical composition is nearly constant unless there is sudden contamination from industries or mines and is available almost everywhere on land phases. Therefore, this major replenishable resource of water interests all the engineers connected with water supply and irrigation. Utilisation of this resource can be traced back to the Mohenjodaro civilisation (about 3000 BC), where brick-lined wells were constructed for storage and supply of drinking water.

Ground water is available in the saturated zone below earth's surface. All water available in the sub-surface zone includes water from saturated and unsaturated zone in the permeable formations of earth's crust. This chapter discusses the elementary idea about ground water hydrology which would help a hydrologist or practising engineer dealing with water resources and irrigation engineering to understand the part of hydrology in lithosphere.

Water present in the lithosphere is called sub-surface water. The source is either precipitation or seepage from large water bodies like reservoirs, lakes, sea or ocean.

## 5.2  ZONING OF SUBSURFACE

The subsurface extending from the earth's surface to the hard rock is classified into the following zones:

(i)  Aeration or unsaturated zone includes: (a) Soil water zone, (b) Intermediate zone and (c) Capillary zone.
(ii)  Saturated zone.

### 5.2.1  Aeration Zone

This zone extends from the top of soil surface to the water table (top of saturated zone). Soil of this zone is only partially saturated. The three sub-zones of this zone are shown in Fig. 5.1.

(a)  **Soil Water Zone:** This is the zone from where plants extract water. It

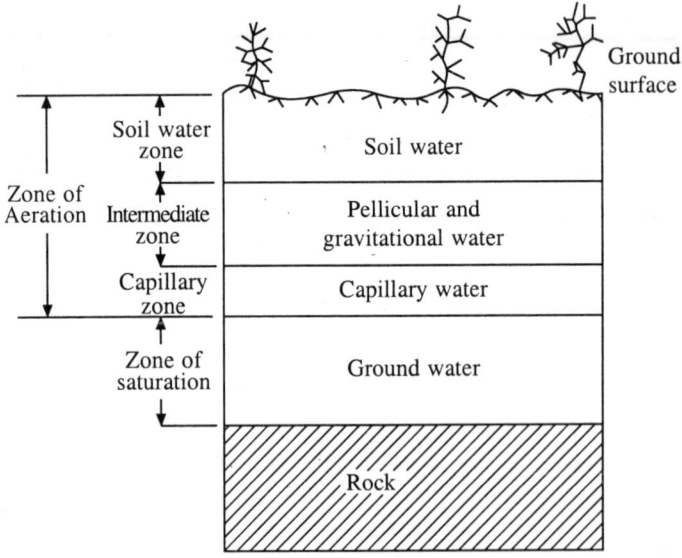

**Fig. 5.1  Division of Subsurface Water Profile**

extends from the top of soil to a few meters below the ground upto which roots of plants extend. The depth of roots can vary from a few cm (grass) to almost 30 m (Phreatophytes). A reasonable depth of 80–100 cm can be considered as root zone. This water is lost to the atmosphere through evaporation. There are three types of water present in this zone. *Hygroscopic water* (Fig. 5.2a) remains absorbed on the surface of soil particles. It condenses on soil particles and the volume of soil increases at the rate of 1% for sand, 7% for silt and 17% for clay soil. *Capillary water* (Fig. 5.2b) is held as liquid rings around a group of soil particles with an air bubble at the centre, made possible due to the surface tension of water. In the process of drainage of water through soil during rainfall, the excess water present is called *gravitational water* (Fig. 5.2c).

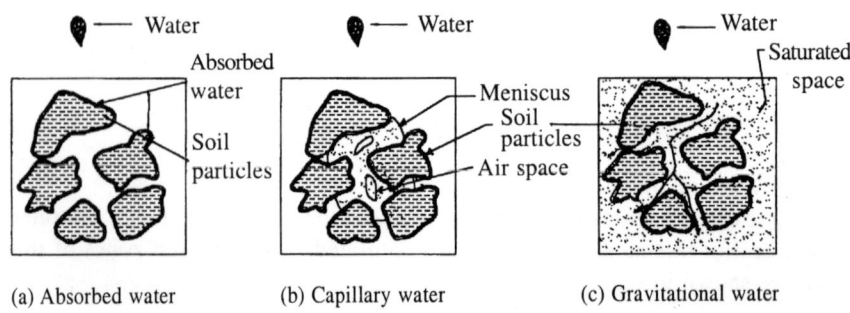

**Fig. 5.2  Types of Water Present in Soil**

**(b) Intermediate Zone:** This zone extends from below the soil water zone to the top of capillary fringe and forms the conduit through which all infiltrated water passes to recharge the ground water. The depth of this zone varies from zero (at certain places in rainy season) to a few meters in dry season.

**(c) Capillary Zone:** This zone extends from the top of saturation zone to a height determined by the nature of soil. Depending on the soil material, capillary zone height varies from 2 cm to about 20 m as shown in Table 5.1.

**Table 5.1   Capillary Zone Height for Various Soil Medium**

| Type of soil | Diameter of soil particle (mm) | No. of particles per cm$^3$ | Total surface area of soil particles (cm$^2$) | Capillary zone height (cm) |
|---|---|---|---|---|
| Clay | < 0.002 | $1.25 \times 10^{11}$ | 15700 | $\geq 1000$ |
| Silt | 0.002–0.06 | $1.25 \times 10^8$ | 1570 | 600–1000 |
| Sand | 0.06–2.0 | $10^6 - 10^3$ | 314–31.4 | 100–600 |
| Gravel (Medium) | 10 | 1 | 3.14 | 2–12 |

| Soil | Gravel | | Sand | | | Silt | | Clay |
|---|---|---|---|---|---|---|---|---|
| Composition | | Coarse | Medium | Fine | Coarse | Medium | Fine | |
| Size limit (mm) | >2 | 0.6 | | 0.2 | 0.06 | 0.02 | | 0.006 0.002< |

### 5.2.2   Saturated Zone

In saturated zone, water completely fills all the pore spaces between the soil grains, forming a huge underground reservoir of water. The top level of this water is known as *Ground Water Table*. Water table is thus defined as the locus of all points in the saturated zone where the water has atmospheric pressure. The top level of this water reservoir is measured from the soil surface or from mean sea levels (MSL). This water level can be seen in open wells.

## 5.3   WATER BEARING MATERIALS

The earth materials form potential water bearing reservoirs. On the basis of their capacities to hold water between their intergranular spaces, these materials can be classified into following four categories

*(a) Aquifuge*

These are the geological materials which neither allow water to be stored nor to flow through them. A solid layer of granite rock without any fracture or fault is a good example of aquifuge. This layer has negligible porosity and permeability.

*(b) Aquiclude*

Soil materials which have the property to store water due to good number of pores in them but passage of water through them is not possible are called aquiclude. Porosity helps to storage water in the soil media whereas, permeability helps water to flow through the soil medium. A thick layer of clay has good porosity but negligible permeability. These terms will be discussed in detail in later sections.

### (c) Aquitard

The formation layer which can store water and allow a small quantity to flow through it over a long period is called aquitard. Such type of layers transmit significant quantity of water for regional migration of ground water but not enough to supply to individual wells. A thin layer of clay or sandy clay forms such a formation.

### (d) Aquifer

Aquifers are natural formations below soil surface which allows water to be stored in them and also allow good quantity of water to pass through. Such layers have good porosity and permeability and yield significant quantity of water to be important economically. Layers of sand and gravel form the best type of aquifer material.

Since economy of extraction and availability are the two criterion on which we classify a saturated formation, in water scanty zones, an aquitard may be classified as aquifer. Krusemann and Deridder (1972) classified aquifers as: (i) confined, (ii) semi-confined, (iii) semi-unconfined and (iv) unconfined depending on their nature of occurrence.

**(i) Confined Aquifer:** Aquifer materials lying between layers of aquiclude or aquifuge, where the ground water is more than atmospheric pressure form confined aquifer. The water level in a well tapping confined aquifer will rise above the bottom of the upper confined layer. The condition of a confined aquifer may be *Hyperpiestic* if the water in a well rises above the ground level and is also called *flowing artesian well*. In a *Mesopiestic condition*, water level on tapping will rise to a level below the ground level but above the regional water table. A *Hypopiestic condition* prevails when water level on tapping lies below the regional water table (confined with negative head) but above the bottom of upper confined layer. Recharge of a confined aquifer takes place from the area where the confined zone of the aquifer touches the natural ground surface. Fig. 5.3 shows a confined aquifer showing all the above conditions.

**(ii) Semi-confined Aquifer:** A confined aquifer in which there is vertical leakage of water from upper or lower confining layers due to their finite permeability is known as semi-confined aquifer. Horizontal flow for such aquifers are negligible. Such aquifers are also called *leaky aquifer*, e.g., a layer of sand covered by sandy-clay-silt on one or both sides.

**(iii) Semi-unconfined Aquifer:** It is a state between unconfined and semi-confined aquifer. In this type appreciable quantity of water can pass through the upper or lower confined layers. This becomes part of the aquifer and even horizontal flow through this aquitard should be considered. For example, a sand bed overlaid by a sandy-clay or clayey-sand.

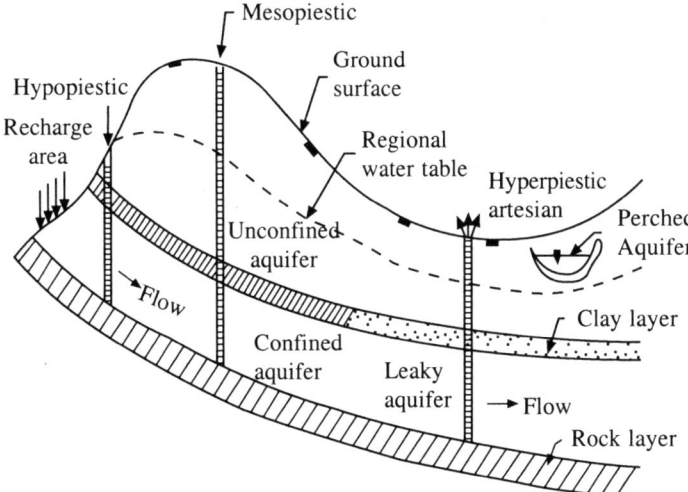

**Fig. 5.3   Types of Aquifers**

**(iv) Unconfined Aquifer:** The aquifer partly filled with water in which the water table forms the upper surface of saturation layer is called unconfined aquifer. Since the material of this layer extends upto the ground surface, the water table fluctuates with season. In rainy season, precipitation recharges it and hence the water level rises.

**(v) Perched Aquifer:** At special locations (Fig. 5.3), lenses of rock formation or clay layer of finite aerial extent appear in the unconfined zones. The saturated formation held by this source is like an impervious strata holding limited quantity of water. The water table of this aquifer is called perched aquifer. Many a times this water table is confused as the part of general water table but the wells at this location dry out early. Table 5.2 summarises the above types of aquifers.

**Table 5.2   Classification of Aquifers**

| Sl. No. | Covering Layer | Aquifer Type |
|---------|----------------|--------------|
| 1. | Impervious | Confined |
| 2. | Semi-pervious so that horizontal flow can be neglected but there is vertical flow. | Semi-confined |
| 3. | Less pervious than the main part of the aquifer but horizontal flow is not negligible. | Semi-unconfined |
| 4. | Same as main part of aquifer. | Unconfined. |

## 5.4   AQUIFER PROPERTIES AFFECTING GROUND WATER

Some of the important aquifer properties affecting ground water formations are:

### 5.4.1   Aquifer Material Properties

**(1) Porosity:** It a measure of the volume of void spaces in geological formations. It is defined as the ratio of pore volume to the bulk volume of the sample, given as

$$\alpha = \frac{V_P}{V_b} \tag{5.1}$$

where $\alpha$ is the porosity expressed in decimal fraction or in per cent, $V_b$ the bulk volume or total volume of the sample and $V_P$ the volume of pores in the sample. Porosity can also be expressed as

$$\alpha = \frac{V_P}{V_b} = \frac{V_b - V_g}{V_b} = 1 - \frac{V_g}{V_b} \tag{5.2}$$

or

$$\alpha = \frac{\rho_g - \rho_b}{\rho_g} = 1 - \frac{\rho_b}{\rho_g} \tag{5.3}$$

where $V_g$ is the grain volume, $\rho_b$ the bulk density and $\rho_g$ the grain density. Grain size, shape, arrangement, degree of compaction and consolidation are the main factors affecting porosity. Pores may be formed at the time of origin of the material (*primary porosity*) or after the formation of the material (*secondary porosity*), due to weathering actions of sun, rain, wind and other agents. Porosity is classified as low, medium or high depending on the ranges of pore volumes. Table 5.3 gives the representative values of various earth materials.

**Table 5.3   Characteristics of Some Common Earth Formation Materials**

| Material | Specific yield (%) | Coefficient of Permeability (cm/sec) | Porosity (%) | Porosity range classification |
|---|---|---|---|---|
| (1) | (2) | (3) | (4) | (5) |
| Clay | 1–10 | $10^{-5} - 10^{-9}$ | 45–55 | |
| Silt | 5–10 | $0.001 - 10^{-7}$ | 35–50 | High 5–20% |
| Sand fine | 10–20 | 0.001–0.05 | 30–35 | |
| Sand mixed | 15–30 | 0.005–0.01 | 20–35 | Medium 5–20% |
| Gravel | 15–30 | 1–100 | 30–40 | |
| Sand stone | 5–15 | 0.001–0.0001 | 10–20 | |
| Shale | 0.5–5 | $10^{-10} - 10^{-14}$ | 1–10 | Low 0–5 % |
| Lime stone | 0.5–5 | 0.001–0 00001 | 1–10 | |

**(2) Specific Yield ($S_y$):** It is defined as the volume of water that can be drained out by the force of gravity from a fully saturated aquifer per unit volume of aquifer material. It is expressed in percentage as

$$S_y = \frac{V_y}{V_b} \times 100 \tag{5.4}$$

where $V_y$ is the volume of water drained out, $V_b$ the total volume of saturated sample. Since all the water from saturated sample cannot be removed by the action of gravity due to surface tension and molecular attraction between aquifer materials and water, specific yield is always less than porosity. For sand and gravel, specific yield is very high and for silt and clay it is low (Table 5.3); because smaller is the grain size of the aquifer material more is the number of grains per unit volume. Thus the molecular attraction between water and aquifer material increases giving rise to less specific yield. Therefore grain size, shape, distribution, compaction, chemical and thermal conditions of the aquifer, time of drainage and other factors govern the specific yield from an aquifer. Specific yield for alluvial aquifer is 10–20%, i.e., from a fully saturated alluvial aquifer only 10–20% of total volume of soil equivalent of water can be drained out from the sample.

**(3) Specific Retention ($S_r$):** It is the volume of water which cannot be drained out from a saturated aquifer material by the force of gravity per unit volume of aquifer material. It is expressed in percentage as

$$S_r = \frac{V_r}{V_b} \times 100 \tag{5.5}$$

Since $V_r + V_y = V_p$, i.e., volume of water retained plus volume of water drained out is the total volume of pores in the material. We can redefine porosity as

$$a = S_r + S_y \tag{5.6}$$

**(4) Storage Coefficient or Storativity ($S$):** It is defined as the volume of water released from or taken into storage of an aquifer per unit surface area per unit decline or rise of the water table (Fig. 5.4). For unconfined aquifer, storage coefficient is taken as specific yield because, most of the water from storage is released by the action of gravity with a negligible part from the compression of the aquifer or expansion of the water. The value of storage coefficient varies from 0.1 to 0.25. Storage coefficient for confined aquifer is given as

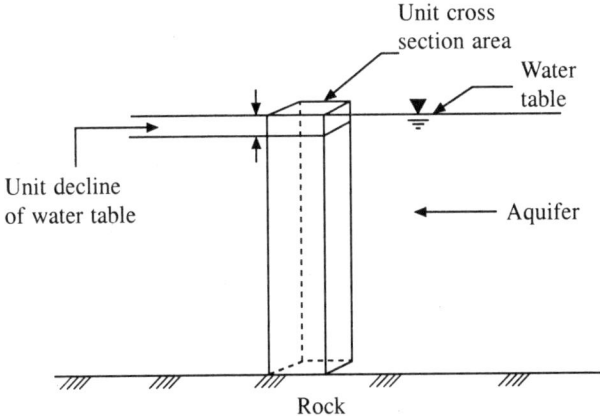

**Fig. 5.4 Sketch showing the Definition of Storage Coefficient**

$$S = \alpha \cdot w \cdot h \left( \frac{1}{E_w} + \frac{1}{E_s} \right) \tag{5.7}$$

where $E_w$ is the bulk modulus of elasticity of water which can be taken as $2.1 \times 10^9$ N/m$^2$, $E_s$ the bulk modulus of elasticity of aquifer materials (150 N/m$^2$ for loose sand, 100 N/m$^2$ for medium clay, 1500 N/m$^2$ for dense sandy-gravel, more than 30000 N/m$^2$ for rock), $w$ ($=\rho g$) the specific weight of water which can be taken as 1000 kg/m$^3$ and $h$ is the aquifer thickness in m. For confined aquifer storage coefficient varies from 0.001 to 0.00001.

---

**Example 5.1:** For an aquifer located in Orissa, a zone of 930 sq. km is bounded by confined aquifer of 22 m thick. The average maximum and minimum piezometric level variations range between 5–12 m. Taking storage coefficient as 0.001, calculate the annual rechargeable ground water storage from the area. In a year about 250 days of pumping from 40 wells irrigate this area. Calculate the average well yield.

**Solution**

Annual rechargeable ground water storage

= Area × variation in piezometric surface × Storage coefficient

= $930 \times 10^6$ m$^2$ × (12–5) m × 0.001

= $6.51 \times 10^6$ m$^3$ = 6.51 Mm$^3$

Average well yield/day × Pumping days × Number of wells

= Annual fluctuations in piezometric head or Average well yield/day

= Annual fluctuation volume of water/(Pumping days × No. of wells)

= 6.51 Mm$^3$/(250 × 40) = $6.51 \times 10^6$ m$^3$/10000

= 651 m$^3$/day or 27.125 m$^3$/h or 7.5 liters/sec

**Example 5.2:** In an unconfined aquifer covering 2 km$^2$, the original water table was 12.3 m below ground level. Pumping of 1 Mm$^3$ of water from the area dropped the water table to 15.1 m below ground level. Calculate (i) specific yield of the aquifer, (ii) specific retentions if porosity of aquifer material is 23%.

**Solution**

Volume of water pumped out = Aquifer area × Change in water table × Specific yield.

or      specific yield $(S_y) = \dfrac{10^6 \text{ m}^3}{\{2 \times 10^6 \text{ m}^2 \times (15.1 - 12.3)\text{m}\}} = 0.1786$ or 17.86%

Specific retention = Porosity – Specific yield = 23 – 17.87 = 5.13%.

---

### 5.4.2   Ground Water Flow Parameters

*(1) Darcy's Experiment and the Coefficient of Permeability*

Ground water is always under motion. Henry Darcy (1856) conducted a laboratory experiment in which sand was confined in an apparatus as shown in the Fig. 5.5. Applying Bernoulli's equation between points (1) and (2) it was found that

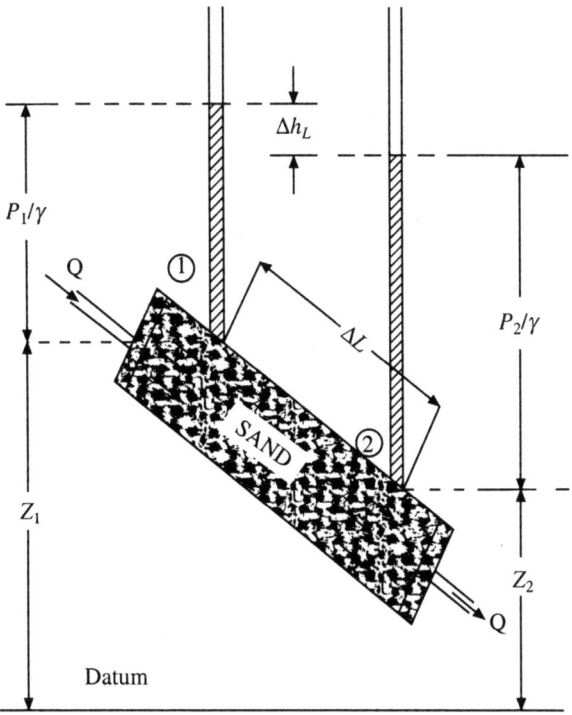

**Fig. 5.5   Flow through Sand Column Apparatus (After Henry Darcy, 1856)**

$$\text{Head loss} = \Delta h = \left( Z_1 + \frac{P_1}{\gamma} \right) - \left( Z_2 + \frac{P_2}{\gamma} \right) \tag{5.8}$$

where $Z_1$ and $Z_2$ are the elevations of the section centers from the reference level, $P_1$ and $P_2$ the pressures, $\gamma$ the specific weight of water. For flow through soil or earth medium the kinetic energy component $V^2/2g$ between sections (1) and (2) are very small and therefore neglected in the equation. This is because the total flow path between (1) and (2) is many times larger than the length $\Delta L$ (Fig. 5.5).

This is due to the circular path water has travelled to move around the sand particles. Darcy observed that

(a)  Discharge and the velocity of flow of water in sand is proportional to the head loss $V \propto \Delta h$.

(b)  Velocity of flow is inversely proportional to the length of sample under measurement $V \propto \dfrac{1}{\Delta L}$.

Therefore $\qquad\qquad V \propto \dfrac{\Delta h}{\Delta L} \quad \text{or} \quad V = \dfrac{K \times \Delta h}{\Delta L}$

Since discharge $Q = \text{Area} \times \text{Velocity} = A \times V$

$$Q = K \frac{\Delta h}{\Delta L} A = K \cdot i \cdot A \qquad (5.9)$$

or
$$K = \frac{Q}{iA} = \frac{Q}{\left\{ \left( \frac{\Delta h}{\Delta L} \right) \cdot A \right\}} \qquad (5.10)$$

where $V$ is the apparant velocity of water ($=Ki$). This velocity is apparent because it does not represent the actual variation from point to point in the medium but represents the overall value of $Q/A$. Actual velocity $V_a$ of flow is roughly taken as $V_a/\alpha$ since water has to flow round the soil particles, $i$ is the hydraulic gradient $= (-) \Delta h/\Delta L$, $\Delta h$ represents drop in the piezometric head in a distance $\Delta L$ measured along the flow direction, $A$ is the area of cross section perpendicular to the flow direction, $Q$ the discharge, $K$ is the constant of proportionately known as *coefficient of permeability* or *hydraulic conductivity* (has the unit of velocity). Since ground water flow is laminar in nature, this equation is applied to most of the ground water flow problems. It is generally given a negative sign because the head drops over length $\Delta L$.

From the above relation, the coefficient of permeability can be defined. A medium is said to have a unit hydraulic conductivity (coefficient of permeability) if a unit volume of fluid flow occurs at field temperature and at prevailing kinematic viscosity normal to unit area of cross section with unit hydraulic gradient. This definition is applicable to field flow conditions. For laboratory results, U.S. Geological Survey has defined coefficient of permeability $K_s$ as the flow of water at 60°F in gallons/day through a medium having 1 sq. ft cross sectional area under hydraulic gradient of one feet per feet. $K$ depends on fluid properties like viscosity and temperature and medium properties like grain size, shape, distribution, packing, specific surface, stratification etc. We can therefore analyse permeability as a product of medium and fluid properties given by

$$K = (C \cdot d_{rep}^2) \left( \frac{w}{\mu} \right) \qquad (5.11)$$

where $C$ is the shape factor constant which depends on the grain properties constituting the medium ($= f_s \times f_\alpha$), $f_s$ being the grain shape factor, $f_\alpha$ the grain porosity factor, $d_{rep}$ the mean partical size of the soil medium, $w$ the unit weight of water ($= \rho g$), $\rho$ the density of water and $\mu$ the dynamic viscosity of water. The first bracket term of equation (5.11) represents medium properties and the second bracket term the properties of water. We call the medium properties component ($Cd_{rep}^2 = K_i$) as the specific or *intrinsic* or *specific permeability* having dimension of length². Rearranging equation (5.10) we get

$$Q = K \cdot i \cdot A = K \cdot \left( \frac{\Delta h}{\Delta L} \right) \cdot A = C \cdot d_{rep}^2 \left( \frac{w}{\mu} \right) \left( \frac{\Delta h}{\Delta L} \right) A$$

$$= K_i \left( \frac{w}{\mu} \right) \left( \frac{\Delta h}{\Delta L} \right) A \qquad (5.12a)$$

or
$$K_i = \left(\frac{Q}{A}\right)\left\{\frac{\mu}{w}\bigg/\frac{\Delta h}{\Delta L}\right\} = \frac{\mu(Q/A)}{w(\Delta h/\Delta L)} \tag{5.12b}$$

$K_i$ is in cm$^2$. Sometimes $K_i$ is expressed in Darcy unit (= $0.987 \times 10^{-8}$ cm$^2$).

Laboratory or standard value of coefficient of permeability $K_s(k)$ and field coefficient of permeability $K_f$ are related only through temperature affecting viscosity. Therefore the equation

$$\frac{K_s}{K_f} = \frac{\mu_f}{\mu_{60f}} \tag{5.13}$$

is used to convert laboratory value of coefficient of permeability to the corresponding field value. In equation (5.13) $\mu_f$ is the viscosity at field temperature and $\mu_{60f}$ the viscosity at 60°F or 17.5°C. Some typical values of $K$ are given in Table 5.3. Since the apparent velocity of seepage is represented by

$$V = K \cdot i = C \cdot d_{\text{rep}}\left(\frac{\mu}{w}\frac{\Delta h}{\Delta L}\right) \tag{5.14}$$

the value can be related dimensionally to Reynold's number (Re) in the following form

$$\text{Re} = \frac{(\rho V \cdot d_m)}{\mu} \tag{5.15}$$

where $d_m$ is the effective size of particles representing 10% of the finer material, $V$ the total velocity (it can be taken as sum of $u$, $v$ and $w$ in $x$, $y$ and $z$ directions respectively by taking $u = -K (\Delta h/\Delta x)$, $v = -K (\Delta h/\Delta y)$ and $w = -K (\Delta h/\Delta z)$). Reynold's number for ground water flow varies from 1 to 10. It should be emphasised that the velocity of flow within the sample space from point to point vary due to the flow of water round the earth particles. However, the overall velocity $V$ between two sections (1) and (2) is calculated using Darcy's law. The actual velocity of flow can be observed in the field by injecting dye tracer at the higher water table end and noting the time taken $t$ by the tracer to travel the length $l$ ($V_a = l/t$).

*(2) Determination of Coefficient of Permeability*
Among all the aquifer properties, determination of coefficient of permeability (hydraulic conductivity) is the most important information explaining the aquifer type. Estimation of this factor can be done by any of the following:

(i) Empirical relations
(ii) Laboratory methods
(iii) Tracer technique
(iv) Geophysical methods
(v) Field pumping test.

Fair and Hatch (1957) developed empirical relation for computing the intrinsic or specific permeability using the following form of equation

$$\text{Specific permeability } K_i = \frac{1}{m\left\{ \dfrac{(1-\alpha)^2}{\alpha^3} \left[ \dfrac{\psi}{100} \; \Sigma \; \dfrac{P}{d_m} \right]^2 \right\}} \tag{5.16}$$

where $m$ is the packing factor (may be taken as 5), $\psi$ the sand shape factor (varies between 6 and 7.7), $\alpha$ the porosity, $P$ the percentage of sample material held between adjacent sieves, $d_m$ the geometric mean of the adjacent sieves.

In laboratory method, a *constant head permeameter* is generally used for coarse-grained materials like gravel and coarse sand. A *variable head permeameter* is used for low permeable materials like clay, fine sand and silt. Coefficient of permeability can be directly determined from the experiments. However, the results may not be the true representative of the field data due to disturbance of sample from the *situ* condition.

In tracer technique, tracer materials like salt, dye or radioactive isotope is allowed to mix at a point in the ground water flow and the time ($\Delta t$) taken for the tracer to reach at the other point of interest is noted. Knowing the distance ($\Delta L$), the average velocity of flow $V_{av}$ is determined as $V_{av} = \Delta L / \Delta t$ from which $K$ is obtained from darcy's law.

Geophysical methods involve indirect approximation of $K$ value from electric resistivity data. Electrical resistivity of a granular aquifer is directly related to the hydraulic conductivity as $K = A \cdot R$, where $A$ is a constant for an aquifer and $R$ the resistivity value.

In field pumping test, pumping is carried out from the existing wells under steady or unsteady conditions. From the well discharge and drawdowns accurate value of $K$ is determined. This aspect is discussed in detail in Section 5.8.

*(3) Coefficient of Transmissibility (T)*

The capacity of an aquifer to transmit water through its entire thickness is represented by the coefficient of transmissibility and is equal to the product of coefficient of permeability $K$ of the aquifer material and saturated thickness $B$ of the aquifer

$$T = K \cdot B \tag{5.17}$$

where $T$ is expressed in m²/sec or litre/day/m width or m²/day. Equation (5.9) gives

$$Q = K \cdot i \cdot A = K \cdot i \cdot l \cdot B = K \cdot B \cdot i \cdot l = T \cdot i \cdot l \tag{5.18}$$

where $l$ is the width of the aquifer. The dimension of $T$ is given as $L^2 T^{-1}$.

*(4) Specific Storage ($S_s$)*

Specific storage of a saturated aquifer is the amount of water released from unit volume of aquifer per unit decline in hydraulic head, i.e.

$$S_s = S/h$$

Therefore $\qquad\qquad\qquad\qquad S = S_s h \tag{5.19}$

Equation (5.19) is analogous to the equation $T = Kh$. If an aquifer area has variation in their aquifer properties in all three directions then the best term to use is the hydraulic conductivity $K$ and specific storage $(S_s)$. If this variation is not considered, then $T$ and $S$ may be used in the computations.

*(5) Hydraulic Diffussivity (D)*
It is a parameter which includes consideration of both transmissibility ($T$) and storage coefficient ($S$) in the following way

$$D = \frac{T}{S} = \frac{K \cdot h}{S_s h} = \frac{K}{S_s} \tag{5.20}$$

*(6) Hydraulic Resistance (C)*
This factor describes the resistance to leakage from aquiclude or aquitard. In Fig. 5.6, $h'$ is the saturated thickness of semi-pervious layer of an aquiclude or aquitard and $K'$ is the hydraulic conductivity of this layer. The ratio $K'/h'$ is called leakage. $h'/K'$ is called the hydraulic resistance $C$ of the confining layer with the dimension of time.

$$\text{Leakage factor} = \lambda = \sqrt{T \times C} = \sqrt{K_i \times h_i \times C} \tag{5.21}$$

If $C$ is finite and small, then it is a case of leaky aquifers. If $\lambda$ is high, the resistance to leakage is high.

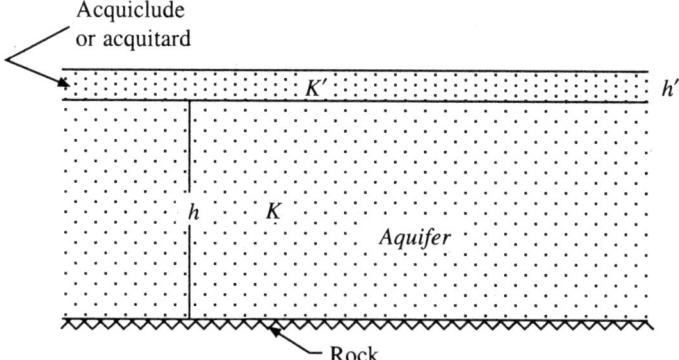

**Fig. 5.6   Leakage from the Overlaying Layer of Aquiclude or Aquitard**

### 5.4.3   Stratification of Layers

*(1) Horizontal Stratification Parallel to Flow Directions*
When a medium is stratified into horizontal layers having different hydraulic properties for each layer, (Fig. 5.7a), then the equivalent coefficient of permeability $K_{TH}$ is computed as

$$K_{TH} = \frac{\Sigma K_{xi} B_i}{\Sigma B_i} \tag{5.22a}$$

and in vertical direction it is given as

$$K_{TV} = \frac{\Sigma\, B_i}{\Sigma\, (B_i/K_{yi})} \qquad (5.22b)$$

where $K_{TH}$ and $K_{TV}$ are the average permeability of the aquifer of thickness $B$ in horizontal and vertical directions, $K_i$ and $B_i$ are the coefficients of permeability and the thicknesses of the layers forming the total thickness $(B = \Sigma\, B_i)$ through which water flows. Due to stratification each strata has different permeability $K_i$. The summation is carried out for the number of layers forming the aquifer depth $B$. In this case the stratification is horizontal forming layers parallel to each other as shown as in Fig. 5.7a. Water flow lines are along the aquifer stratification. Equivalent transmissivity of the aquifer is given by

$$T_H = \Sigma\, K_i B_i \qquad (5.23)$$

*(2) Vertical Stratification Normal to Flow Line*
If the stratification is vertical normal to the direction of flow line, (Fig. 5.7b),

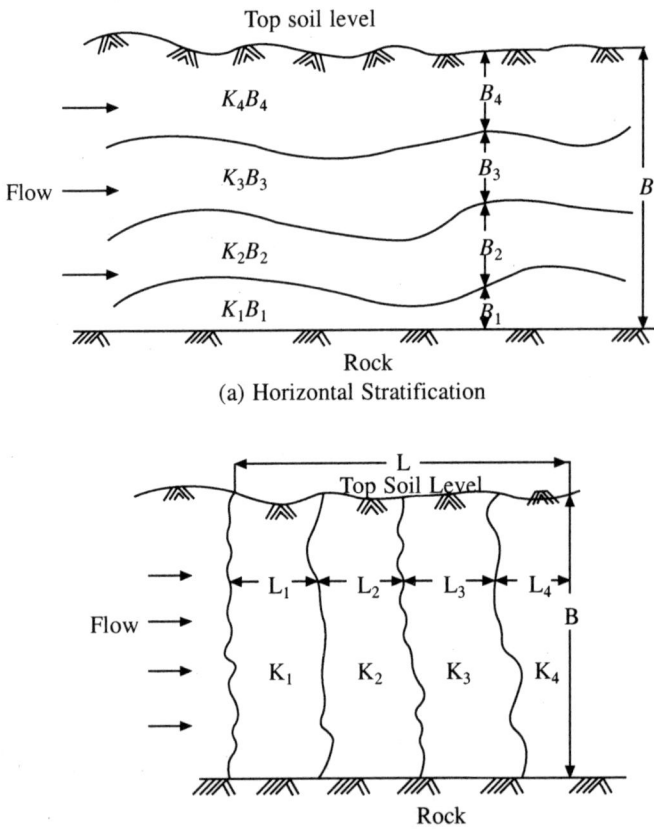

(a) Horizontal Stratification

(b) Vertical Stratification

**Fig. 5.7   Horizontal and Vertical Stratification of Aquifers.**

then the equivalent coefficient of permeability and transmissivity $T_{TV}$ (=$K_{TV}B$) can be calculated using equation (5.22b) but the length of seepage is sum of all the layer thicknesses taken together as shown in Fig. 5.7b.

---

**Example 5.3:** During soil exploration the following layers are encountered

| Layer name | A | B | C | D | E |
|---|---|---|---|---|---|
| Layer Thickness (m) | 0.5 | 2.0 | 2.8 | 3.6 | 3.0 |
| Coefficient of Permeability (m/day) ($K_x = K_y$) | 1.3 | 2.8 | 1.8 | 0.5 | 2.0 |

Find the equivalent coefficient of permeability both in vertical and horizontal directions.

**Solution**
Equivalent coefficient of permeability in vertical direction is given by

$$K_{TV} = \frac{\Sigma B_i}{\Sigma (B_i / K_{yi})}$$

$$= \frac{0.5 + 2.0 + 2.8 + 3.6 + 3.0}{0.5/1.3 + 2.0/2.8 + 2.8/1.8 + 3.6/0.5 + 3.0/2.0} = \frac{11.90}{11.35} = 1.048$$

Similarly the equivalent coefficient of permeability in horizontal direction is given by

$$K_h = \frac{(\Sigma B_i K_{xi})}{\Sigma B_i}$$

$$= \frac{0.5 \times 1.3 + 2.0 \times 2.8 + 2.8 \times 1.8 + 3.6 \times 0.5 + 3.0 \times 2.0}{0.5 + 2.0 + 2.8 + 3.6 + 3.0} = \frac{19.09}{11.90} = 1.606$$

---

### 5.4.4 Calculation of Flow from Flownet

If the flow lines and equipotential lines below a dam or a weir structure can be drawn as in Fig. 5.8, then the flow per unit width of the structure seeping below can be computed from

$$q = \frac{n_f}{n_e} K \cdot h \qquad (5.24)$$

where $n_f$ is the number of flow channels formed due to flow lines, $n_e$ the number of equipotential drops formed due to construction of equipotential lines, $h$ the water head difference between the upstream and downstream of the reservoir (Maximum Water Level–Tail Water Level), and $K$ is the coefficient of permeability. If at the foundation of the reservoir, $K$ in the horizontal and vertical directions are different then the equivalent $K = \sqrt{K_h \times K_v}$ may be used. $K_h$ and $K_v$ are the coefficient of permeability in horizontal and vertical directions. Flow lines and equipotential lines must cut each other orthogonally.

---

**Example 5.4:** A barrage was noticed to discharge considerable flow below its foundation. Flow lines and equipotential lines were drawn below the barrage cutting each other orthogonally. There are 30 flow channels and 45 equipotential drops constructed due to

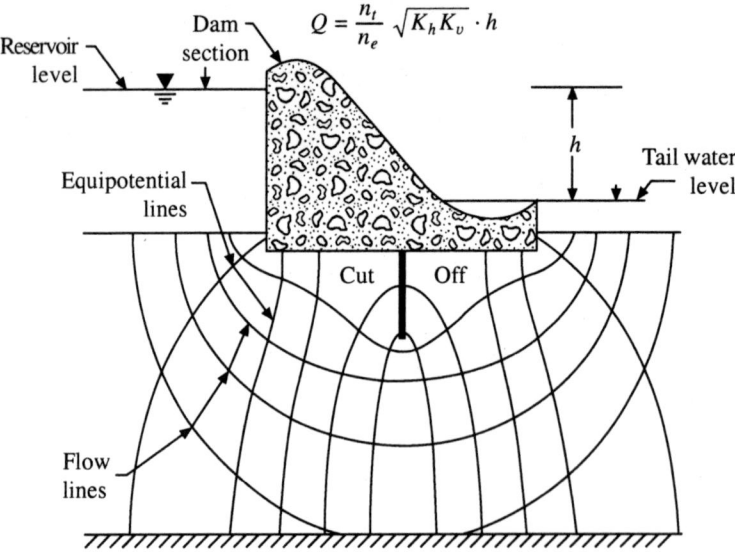

$$Q = \frac{n_f}{n_e} \sqrt{K_h K_v} \cdot h$$

**Fig. 5.8 Flownet under a dam**

the flow net. If the coefficient of permeability is 0.8 cm/sec and the head difference between upstream water level and the downstream tail water level is 18 m, calculate discharge per unit width of the barrage. Also calculate total discharge if total width of barrage is 1.0 km.

**Solution**

Total number of flow channel $n_f = 30$
Total number of equipotential lines $n_e = 45$
Coefficient of permeability $K = 0.8$ cm/sec = 0.8/100 m/sec

Discharge per unit width $q = \left(\dfrac{n_f}{n_e}\right) \cdot K \cdot h = \dfrac{30}{45} \times \dfrac{0.08}{100} \times 18 = 0.0096 \text{ m}^2/\text{sec}$

Total width of the barrage is 1 km = 1000 m

∴      Total discharge $Q = 0.0096 \times 1000 = 9.60 \text{ m}^3/\text{sec}$

**Example 5.5:** A confined aquifer 1.0 km wide discharges 0.03 m³/day/km to a dry river in the month of May. Calculate transmissivity, if the slope of the piezometric surface is 0.75 m/km.

**Solution**

Here hydraulic gradient $\dfrac{\Delta h}{\Delta l} = 0.75$ m/km
Discharge $Q = 0.03$ m³/day/km
Length $l = 1$ km
Transmissivity $T = K \cdot B$
Discharge $Q = K \cdot i \cdot A = T \cdot i \cdot l$

Therefore               $0.03 \text{ m}^3/\text{day} = T \times 0.75 \times 1.0$

or $\qquad T = 0.03/0.75 = 0.04$ (m³/day) × (km × m/km) = 0.04 m²/day.

Therefore transmissivity $T$ is 0.04 m²/day.

**Example 5.6:** A tracer took 18 h to travel from a well A to B 150 m away from it. Map of the water table contour shows a difference of 0.8 m in their water table elevations. The aquifer is made up of mixed sand with porosity of 35%. Calculate (i) coefficient of permeability, (ii) intrinsic permeability and (iii) Reynold's numbers if mean particle size of the medium is 1 mm.

**Solution**
(i) Actual velocity of flow of water in the medium is observed by the tracer. Therefore,

$$V = \frac{\text{Distance}}{\text{Time}} = \frac{(150 \times 100)}{(18 \times 60 \times 60)} = 0.23 \text{ cm/sec.}$$

Hydraulic gradient $i = \dfrac{\Delta h}{\Delta l} = \dfrac{0.8}{150} = 0.0053$

Apparent velocity of flow (Darcy's velocity)

= Porosity × Actual velocity of tracer in the medium

= 0.23 × 0.35 = 0.0805 cm/sec

From Darcy's law $V = 0.0805$ cm/sec = $K \cdot i$

Or $\qquad K = \dfrac{V}{i} = \dfrac{0.0805}{0.0053} = 15.19$ cm/sec.

(ii) Intrinsic permeability (from equation (5.12a)) = $K_i = \dfrac{K \cdot \mu}{w} = \dfrac{K \cdot \mu}{\rho \cdot g}$

$$= (15.19 \times 0.0114)/(1 \times 981) = 1.765 \times 10^{-4} \text{ cm}^2$$

$$= \frac{1.765 \times 10^{-4}}{9.87 \times 10^{-9}} = 17880 \text{ Darcy}$$

($\mu$ = kinematic viscosity = 0.0114 cm²/sec = 1.14 centistoke)

Reynold's number is given as $R_e = \dfrac{\rho \cdot V \cdot d_m}{\mu} = \dfrac{0.999 \times 0.0805 \times 0.10}{0.0114} = 0.71$

Or $\qquad\qquad\qquad$ say $R_e = 1$

---

## 5.5 STEADY FLOW EQUATION

During continued pumping when there is no further lowering of water level from a well the condition of flow is termed as steady state. Initially the water level may drop but with the passage of time the level tries to attain a steady state. For derivation of equation for such condition, we assume that (i) the fluid is not compressible, (ii) medium is homogeneous and isotropic and (ii) the flow is irrotational. A medium is said to be *isotropic* when its properties at any point are the same in all directions emanating from that point.

$$K_{x1} = K_{y1} = K_{z1}$$

$$K_{x2} = K_{y2} = K_{z2} \qquad\qquad (5.25)$$

$$...$$

A medium, whose properties do not vary from one point to another in the spacing is said to be *homogeneous*.

$$K_{(x_1 \cdot y_1 \cdot z_1)} = K_{(x_2 \cdot y_2 \cdot z_2)} \quad \text{and} \quad K_{x_1} = K_{x_2}, K_{y_1} = K_{y_2}, K_{z_1} = K_{z_2} \qquad (5.\ 26)$$

...

A homogeneous and isotropic medium is shown in Fig. 5.9. Movement of ground water is said to be *steady* or *time independent* when its characteristics are not affected by lapse of time, i.e., $\partial f/\partial t = 0$ or $\partial^2 f/\partial t^2 < 0$, where $f$ is the functional and $t$ the time.

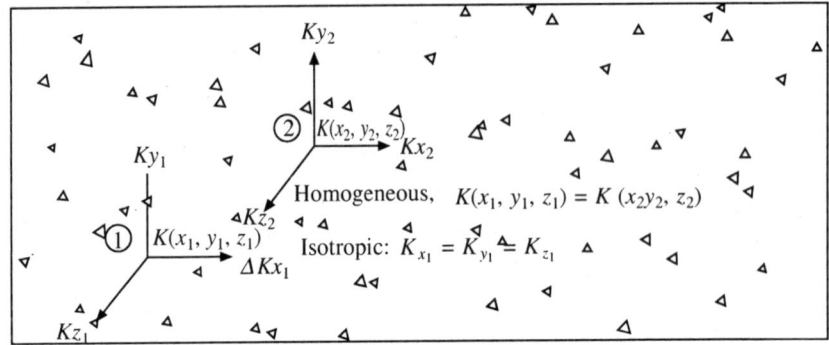

**Fig. 5.9   A Homogeneous and Isotropic Medium**

For derivation of steady state of flow, the principle of conservation of mass is used. Consider a representative parallel piped elemental volume in porous media of dimensions $\Delta x$, $\Delta y$ and $\Delta z$ in $x$, $y$ and $z$ directions respectively, as shown in Fig. 5.10. Let $u$, $v$ and $w$ be the components of velocity of flow of water in the medium in $x$, $y$ and $z$ directions, respectively.

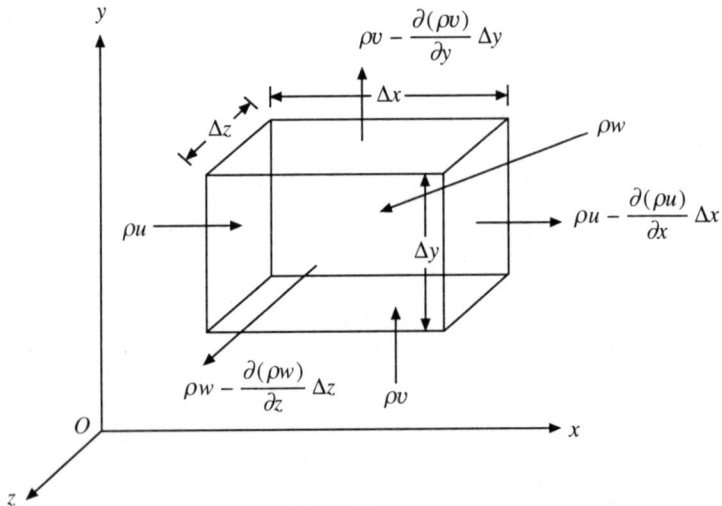

**Fig. 5.10   Flow of Ground Water through an Element**

Consider the flow parameters in $x$ direction.

Mass of fluid entering = Density × Velocity × Area of cross section across $x$ direction

$$= (\rho \cdot u) \, \Delta y \cdot \Delta z \tag{5.27}$$

Mass of fluid coming out of the element in $x$ direction

$$= \left\{ \rho u - \frac{\partial(\rho u) \, \Lambda x}{\partial x} \right\} \Delta y \, \Delta z \tag{5.28}$$

Change of fluid quantity in $x$ direction is obtained by subtracting equation (5.28) from (5.27)

$$= \left\{ \frac{\partial(\rho u)}{\partial x} \right\} \Delta x \, \Delta y \, \Delta z \tag{5.29a}$$

This is equal to the mass of fluid stored in the element $\Delta x \, \Delta y \, \Delta z$.

Similarly, mass of fluid stored in $y$ direction $= \left\{ \dfrac{\partial(\rho v)}{\partial y} \right\} \Delta x \, \Delta y \, \Delta z$    (5.29b)

and mass of fluid stored in $z$ direction $= \left\{ \dfrac{\partial(\rho w)}{\partial z} \right\} \Delta x \, \Delta y \, \Delta z$    (5.29c)

Total change in fluid quantity between its inflow and outflow in the elemental volume $\Delta x \, \Delta y \, \Delta z$

$$= \left\{ \frac{\partial(\rho u)}{\partial x} \right\} \Delta x \, \Delta y \, \Delta z + \left\{ \frac{\partial(\rho v)}{\partial y} \right\} \Delta x \, \Delta y \, \Delta z + \left\{ \frac{\partial(\rho w)}{\partial z} \right\} \Delta x \, \Delta y \, \Delta z$$

$$= \Delta x \, \Delta y \, \Delta z \left[ \left\{ \frac{\partial(\rho u)}{\partial x} \right\} + \left\{ \frac{\partial(\rho v)}{\partial y} \right\} + \left\{ \frac{\partial(\rho w)}{\partial z} \right\} \right] \tag{5.30}$$

Change in the mass of the element in time $t = \left\{ \dfrac{\partial(\rho \alpha)}{\partial t} \right\} \Delta x \, \Delta y \, \Delta z$    (5.31)

Equating (5.30) and (5.31) we get

$$\Delta x \, \Delta y \, \Delta z \left[ \left\{ \frac{\partial(\rho u)}{\partial x} \right\} + \left\{ \frac{\partial(\rho v)}{\partial y} \right\} + \left\{ \frac{\partial(\rho w)}{\partial z} \right\} \right] = \left\{ \frac{\partial(\rho \alpha)}{\partial t} \right\} \Delta x \, \Delta y \, \Delta z$$

$$\tag{5.32}$$

or $$\left[ \left\{ \frac{\partial(\rho u)}{\partial x} \right\} + \left\{ \frac{\partial(\rho v)}{\partial y} \right\} + \left\{ \frac{\partial(\rho w)}{\partial z} \right\} \right] = \left\{ \frac{\partial(\rho \alpha)}{\partial t} \right\} \tag{5.33}$$

If the flow is steady then there is no change in the density with respect to time

$$\left\{ \frac{\partial(\rho\alpha)}{\partial t} \right\} = 0$$

Porosity remains unchanged as the medium is assumed incompressible. Fluid can also be considered as incompressible. Therefore $\rho$ can be taken as constant. Equation (5.33) reduces to the form

$$\left\{ \frac{\partial u}{\partial x} + \frac{\partial v}{\partial y} + \frac{\partial w}{\partial z} \right\} = 0 \qquad (5.34)$$

From Darcy's law

$$u = -K(\partial h/\partial x), \ v = -K(\partial h/\partial y) \text{ and } w = -K(\partial h/\partial z)$$

where $h$ is the piezometric head. Substituting the differential forms of $u$, $v$ and $w$ in equation (5.34), we get

$$-K\left( \frac{\partial^2 h}{\partial x^2} + \frac{\partial^2 h}{\partial y^2} + \frac{\partial^2 h}{\partial z^2} \right) = 0 \qquad (5.35)$$

Since $K \neq 0$ 
$$\frac{\partial^2 h}{\partial x^2} + \frac{\partial^2 h}{\partial y^2} + \frac{\partial^2 h}{\partial z^2} = 0 \qquad (5.36a)$$

or 
$$\Delta^2 h = 0 \qquad (5.36b)$$

Equation (5.36) is called *Laplace equation* and is the fundamental equation of all potential flow problems. This partial differential equation representing steady flow of water in a homogeneous and isotropic medium and can be solved by numerical models from the known boundary values.

## 5.6 UNSTEADY FLOW EQUATION

For unsteady flow, the element representing the volume $\Delta x \, \Delta y \, \Delta z$ undergoes compaction of the aquifer material and water. Incorporating the specific storage $S_s$ to right hand side of equation (5.33), it can be rewritten as

$$\frac{\partial(\rho u)}{\partial x} + \frac{\partial(\rho v)}{\partial y} + \frac{\partial(\rho w)}{\partial z} = S_s \frac{\partial h}{\partial t} \qquad (5.37)$$

From Darcy's law, $u = -K_x \dfrac{\partial h}{\partial x}, v = -K_y \dfrac{\partial h}{\partial y}$ and $w = -K_z \dfrac{\partial h}{\partial z}$.

Substituting the values of $u$, $v$ and $w$

$$\frac{\partial}{\partial x}\left( -K_x \frac{\partial h}{\partial x} \right) + \frac{\partial}{\partial y}\left( -K_y \frac{\partial h}{\partial y} \right) + \frac{\partial}{\partial z}\left( -K_z \frac{\partial h}{\partial z} \right) = S_s \frac{\partial h}{\partial t} \qquad (5.38)$$

Equation (5.38) represents the general form equation of the piezometric head in space $(x, y, z)$ and time $t$ for non-homogeneous, anisotropic and confined aquifer. For homogeneous, anisotropic and confined medium the equation is written as

$$K_x \frac{\partial^2 h}{\partial x^2} + K_y \frac{\partial^2 h}{\partial y^2} + K_z \frac{\partial^2 h}{\partial z^2} = S_s \frac{\partial h}{\partial t} \qquad (5.39)$$

For homogeneous, isotropic and confined aquifer the equation is

$$\frac{\partial^2 h}{\partial x^2} + \frac{\partial^2 h}{\partial y^2} + \frac{\partial^2 h}{\partial z^2} = \left(\frac{S_s}{K}\right)\frac{\partial h}{\partial t} \tag{5.40}$$

Assuming the aquifer to be of constant thickness $B$, the equation (5.40) becomes

$$\frac{\partial^2 h}{\partial x^2} + \frac{\partial^2 h}{\partial y^2} + \frac{\partial^2 h}{\partial z^2} = \left(\frac{S_s}{K}\right)\frac{\partial h}{\partial t} = \left(\frac{S}{T}\right)\frac{\partial h}{\partial t} \tag{5.41}$$

where $S = S_s \times B$, $T = K \times B$, $T$ is the transmissibility, $S$ the storage coefficient, $S_s$ the specific storage, $K$ the coefficient of permeability, $h$ the piezometric head and $x$, $y$, $z$ are the coordinates. Equation (5.41) is essentially a nonlinear form.

For an unconfined aquifer, the storage coefficient $S$ is virtually the same as the specific yield $S_y$ as most of water is released by gravity drainage and negligible part comes from compression of aquifer and expansion of water.

Therefore, be it a steady or unsteady flow situation for unconfined aquifer, equation (5.41) can be approximated as

$$\frac{\partial^2 h}{\partial x^2} + \frac{\partial^2 h}{\partial y^2} + \frac{\partial^2 h}{\partial z^2} = 0 \tag{5.42}$$

which is written in the form of the famous *Laplace three dimensional* equation. Converting equation (5.41) into polar coordinate system, we get

$$\frac{\partial^2 h}{\partial x^2} + \frac{\partial h}{\partial r} = \frac{S}{T}\frac{\partial h}{\partial t} \tag{5.43}$$

General solution to the unsteady flow equation can be obtained by assuming aquifer having infinite thickness. Solution to the flow equation is discussed in detail in section 5.8.

## 5.7 GROUND WATER FLOW PROBLEMS

Laplace equation can be applied to a number of steady state flow problems to obtain the discharge and other related information for various ground water flow situations. Some important ground water flow situations are discussed in the following sections.

### 5.7.1 Steady Flow in Unconfined Aquifer

Such a situation arises when two streams with water depths $h_1$ and $h_2$ from a common datum are separated by the aquifer materials (Fig. 5.11). The following assumptions are made while deriving the flow equation between them.

(i) Velocity of flow is proportional to $\dfrac{dh}{dx}$ (tangent) instead of $\dfrac{dh}{ds}$ (sine), where $s$ is the distance along the flow path.

(ii) Flow is horizontal and uniform at any vertical section.

(iii) Distortion of flow field and velocity distribution near the phreatic surface are to be neglected.

For unidirectional flow, discharge per unit width for the condition given in Fig. 5.11 is

$$q = -Kh \, \frac{dh}{dx}$$

where $h$ is the height above the impervious datum and $x$ the direction of flow. Integrating the above equation, we get

$$\int q \, dx = \int - Kh \, dh \quad \text{or} \quad qx = -K\frac{h^2}{2} + C_1$$

Boundary condition for $x = 0$, $h = h_1$ may be used to evaluate the integration constant $C_1$. Applying the condition we get $C_1 = \dfrac{Kh_1^2}{2}$.

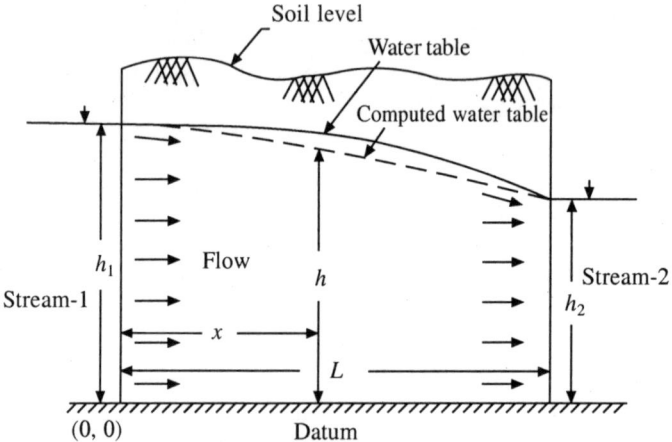

**Fig. 5.11  Steady one Dimensional Flow Between two Water bodies in Unconfined Aquifer**

Therefore
$$q = \frac{K}{2x} (h_1^2 - h^2) \qquad (5.44)$$

This equation, proposed by Dupuit, indicates that for a given $q$, the phreatic line representing water table is parabolic.

Now, when $h = h_2$, $x = L$.

Therefore
$$q = \frac{K}{2L} (h_1^2 - h_2^2) \qquad (5.45)$$

From equations (5.44) and (5.45) we get

$$h = \left\{ h_1^2 - (h_1^2 - h_2^2) \frac{x}{L} \right\}^{1/2} \qquad (5.46)$$

It is found that the computed water table following Dupuit's assumptions are always lower than the actual water table. Therefore, for accurate results $\dfrac{dh}{ds}$

should not be assumed as $\frac{dh}{dx}$. Again the velocity of flow for such condition is not horizontal but slightly downward. However, equations (5.44) and (5.45) give quite satisfactory results of discharge $q$ and coefficient of permeability $K$.

---

**Example 5.7:** Two rivers A and B (Fig. 5.12), are separated by an aquifer formation of 5.3 km. Compute the seepage flow per unit length of the river if $K$ is 18 m/day.

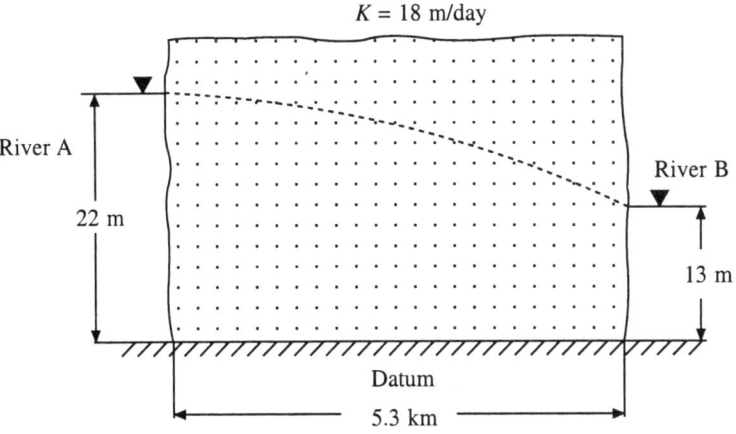

**Fig. 5.12   Flow Condition for Example 5.7**

**Solution**
Here $h_1 = 22$ m, $h_2 = 13$ m, $K = 18.0$ m/day and length of aquifer $L = 5.3$ km $= 5300$ m.
From equation (5.45)

$$q = \left(\frac{K}{2L}\right)(h_1^2 - h_2^2) = \left\{\frac{18}{(2 \times 5300)}\right\}(22^2 - 13^2) = 0.535 \text{ m}^3/\text{day}$$

---

### 5.7.2   Steady Flow in Confined Aquifer of Constant Thickness

A confined aquifer of depth $B$ is connected to two water systems, as shown in Fig. 5.13. Depth of water from the impervious datum are $h_1$ and $h_2$, respectively. These depths are the piezometric heads at distance $x = 0$ and $x = L$. Flow between water body (1) and (2) is considered unidirectional. Equation (5.42) can be written as

$$\frac{\partial^2 h}{\partial x^2} = 0$$

Integrating we get $\dfrac{\partial h}{\partial x} = C_1$

Again integrating $\qquad\qquad h = C_1 x + C_2$

Substituting the boundary conditions $h = h_1$ at $x = 0$, we get $h_1 = C_2$

$\therefore$ $\qquad\qquad\qquad\qquad h = C_1 x + h_1$

Again using the second boundary condition for $x = L$, $h = h_2$

$$h_2 = C_1 \cdot L + h_1 \quad \text{or} \quad C_1 = \frac{(h_2 - h_1)}{L} = (-)\frac{(h_1 - h_2)}{L}$$

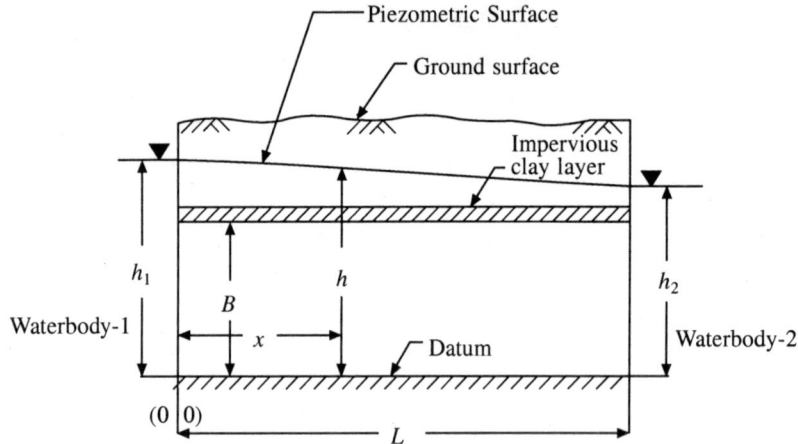

**Fig. 5.13   Steady one Dimensional Flow in Confined Aquifer of Constant Thickness**

$\therefore$ 
$$h = -(h_1 - h_2)\frac{x}{L} + h_1 \quad \text{or} \quad h = h_1 - \left\{\frac{(h_1 - h_2)}{L}\right\}x \qquad (5.47)$$

Differentiating (5.47)

$$\frac{\partial h}{\partial x} = \frac{-(h_1 - h_2)}{L}$$

From Darcy's law we get

$$q = -K \cdot B\frac{dh}{dx} = K \cdot B\left(\frac{h_1 - h_2}{L}\right) = T\left\{\frac{(h_1 - h_2)}{L}\right\} \qquad (5.48)$$

Thus knowing the piezometric heights at $h_1$ and $h_2$, $L$ distance apart, discharge per unit width of the aquifer can be computed from (5.48).

---

**Example 5.8:** Solve Example 5.7 when a horizontal clay layer of 1 m thick is located at a height 7 m from the bottom impervious layer as shown in Fig. 5.13. Take $K$ in the upper unconfined region and lower confined region as the same ($K_{uc} = K_{lc} = 18$ m/day).

**Solution**
Assume no flow to occur through the layer of clay of 1 m deep. The aquifer below it is considered as confined aquifer of 7 m uniform thickness. The layer above the clay layer is the unconfined aquifer with water head at river (1) side as $(22 - 7 - 1) = 14$ m and river at side (2) is $(13 - 7 - 1) = 5$ m.

Flow through upper unconfined aquifer is given by equation (5.45)

$$q_u = K\frac{(h_1^2 - h_2^2)}{2L} = \frac{18\,(14^2 - 5^2)}{(2 \times 5300)} = 0.2904 \text{ m}^3 \text{/day/m of river}$$

Flow through the lower confined aquifer (equation (5.48))

$$q_c = \frac{KB\,(h_1 - h_2)}{L} = \frac{18 \times 7 \times (22 - 13)}{(5300)} = 0.214 \text{ m}^3\text{/day/m of river}$$

Total discharge $= 0.294 + 0.2140 = 0.\,5044$ m³/day/m of river

$\therefore$ 
$$q_t = 0.5044 \text{ m}^3\text{/day/m of river length}$$

---

### 5.7.3 Steady Flow in a Confined Aquifer of Variable Thickness

As shown in Fig. 5.14, thickness of the confining zone $B$ changes with distance $x$. Discharge per unit width is given as

$$q = -KB\frac{dh}{dx} \quad \text{or} \quad dh = -q\frac{dx}{KB}$$

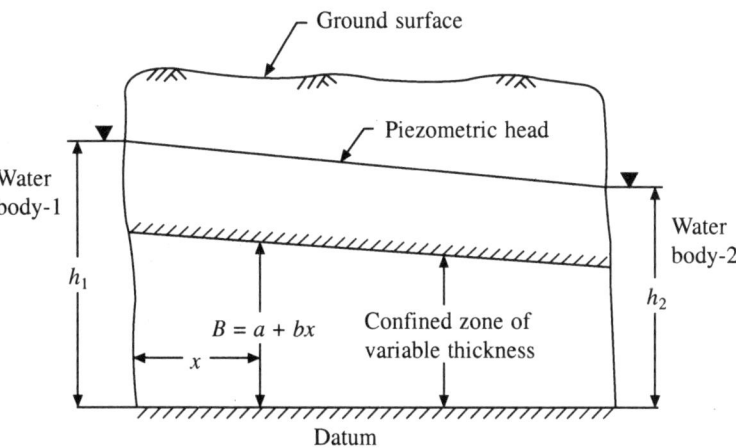

**Fig. 5.14   Steady one Dimensional Flow through a Confined Aquifer of Variable Thickness**

Let the depth of confined layer $B$ vary linearly and be represented by $B = a + bx$, where $x$ is the coordinate distance shown in the figure.

$$dh = -q\frac{dx}{K(a + bx)}$$

Integrating, we get

$$h = -\frac{q}{Kb}\ln(a + bx) + C \tag{5.49}$$

The integration constant $C$ can be determined by knowing the boundary conditions. For solving such problems the thickness $h_1$ and $h_2$ at the two ends of the aquifer should be known from which the coefficients $a$ and $b$ can be computed. Knowing $a$ and $b$, $C$ and discharge $q$ can easily be evaluated.

---

**Example 5.9:** Solve example 5.8, if a clay layer 1 m thick is sloping from 10 m level at $h_1$ end of the aquifer to 6 m level at the other end $h_2$.

#### Solution

Consider the upper unconfined aquifer portion

$h_1 = 22 - 10 - 1 = 11$ m

$h_2 = 13 - 6 - 1 = 6$ m

$$q_u = \left(\frac{K}{2L}\right)(h_1^2 - h_2^2) = \left(\frac{18}{2 \times 1300}\right)(11^2 - 6^2) = 0.1443 \text{ m}^3/\text{day/m length}$$

and for the lower confined aquifer portion, the depth of confinement at any distance $x$ from the higher elevation end is given as

$$B = a + bx = 10 - \{(10 - 6)/5300\} x = 10 - 0.000755x$$

from which we get $a = 10$ and $b = -0.000755$.

From equation (5.49)       $h = -\left(\dfrac{q}{Kb}\right) \ln(a + bx) + C$

for $h = 22$, $x = 0$.

$\therefore$        $22 = -\dfrac{q}{(18 \times -0.000755)} \ln(10) + C$    or    $22 = 169.4q + C$

for $x = 5300$, $h_2 = 13$

$\therefore$        $13 = -\dfrac{q}{(18 \times -0.000755)} \ln(6) + C$    or    $13 = 131.8q + C$

Solving the two equations

$$22 = 169.4\ q + C$$

and                      $13 = 131.8\ q + C$

We get $9 = 37.6\ q$ or $q = 0.2394$ m³/day/m length.

Total discharge = 0.1443 + 0.2394 = 0.3837 m³/day/m length

Hence, discharge between section (1) and section (2) is 0.3837 m³/day/m length.

### 5.7.4   Unconfined Flow with Recharge from Rainfall

Part of the rain which infiltrates into the soil, percolates further down to recharge the ground water. As shown in Fig. 5.15, flow problem for such a condition can

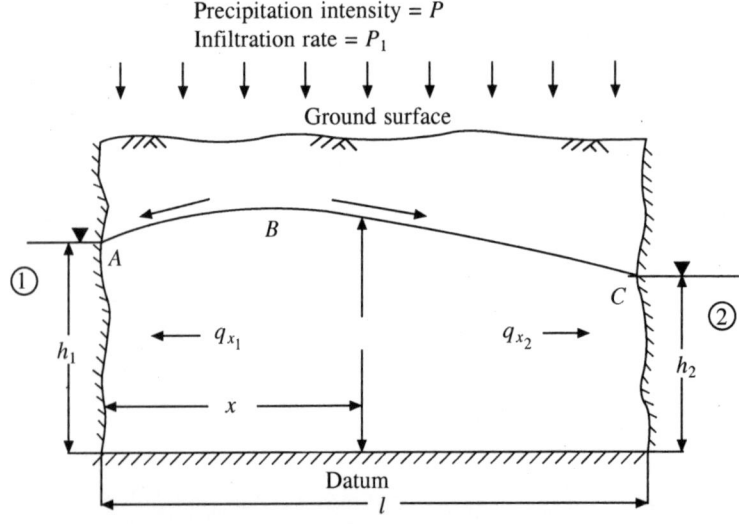

**Fig. 5.15   Steady One Dimensional Flow between Two Water Bodies with Rainfall as Recharge**

be analysed by taking $P_1$ as the part of rain water joining the flow. Since only one-dimensional steady flow is considered, $P_1 dx$ can be taken as the recharge to ground water flow due to rainfall per unit width, which is the component of flow due to precipitation $P$ per unit width. Flow is established due to $P_1$. Therefore

$$dq = P_1\, dx \qquad \text{(i)}$$

For unidirectional flow

$$q = -K \cdot h \cdot \frac{dh}{dx} = -\frac{1}{2} K \frac{d(h^2)}{dx} \quad \text{or} \quad \frac{dq}{dx} = -\frac{K}{2} \frac{d^2(h^2)}{dx^2} \qquad \text{(ii)}$$

and from equation (i)

$$\frac{dq}{dx} = P_1$$

$\therefore$

$$P_1 = -\left(\frac{K}{2}\right)\left(\frac{d^2 h^2}{dx^2}\right)$$

or

$$-\left(\frac{K}{2}\right) \cdot \frac{d}{dx}\left(h \cdot \frac{dh}{dx}\right) = P_1 \qquad \text{(iii)}$$

Similarly, it can also be shown that $\dfrac{d^2 h^2}{dx^2} + \dfrac{d^2 h^2}{dy^2} = \dfrac{-2 P_1}{K}$

Integrating equation (iii) twice we get $Kh^2 + P_1 x^2 = C_1 x + C_2$ (5.50 a)

Now we can use boundary conditions. For $x = 0$, $h = h_1$ which gives $C_2 = Kh_1^2$ Again when $x = l$, $\to h = h_2$ substituting we get

$$Kh_2^2 + P_1 l^2 = C_1 l + Kh_1^2$$

or

$$C_1 = (K h_2^2 + P_1 l^2 - K h_1^2)/l = (K/l)(h_2^2 - h_1^2) + P_1 l$$

$\therefore$

$$Kh^2 + P_1 x^2 = (Kx/l)(h_2^2 - h_1^2) + P_1 lx + Kh_1^2$$

or

$$h^2 = h_1^2 - \frac{x}{l}(h_1^2 - h_2^2) + \frac{P_1}{K}(l - x)x \qquad \text{(5.50b)}$$

or

$$h = \sqrt{h_1^2 - \frac{x}{l}(h_1^2 - h_2^2) + \frac{P_1}{K}(l - x)x} \qquad \text{(5.51)}$$

Equation (5.51) is presented as ABC in Fig. 5.15. The curve reaches a peak at B for a certain value of $x$ which can be located by differentiating equation (5.50) w.r.t. $x$ and equating to zero. This is the position of the ground water divide at $\dfrac{dh}{dx} = 0$, i.e., at location B. Discharge per unit width is given as $q_x = -Kh.\dfrac{dh}{dx}$. Differentiating equation (5.50) and substituting the value of $\dfrac{dh}{dx}$, at a location $x$ (Fig. 5.15), $q_x$ is obtained as

$$q_x = \frac{K}{2l}(h_1^2 - h_2^2) + P_1(x - l/2) \tag{5.52}$$

Discharge to the left side water body-1 is obtained by substituting $x = 0$ and $h = h_1$ in equation (5.52). Similarly for $x = l$, $h = h_2$ water discharge to right side water body is obtained.

---

**Example 5.10:** Solve Example 5.7 when rainfall rate of 10 mm/h occurs in the area between the rivers. Consider rainfall rate as less than infiltration rate and that soil is saturated.

**Solution**
Refer to Fig. 5.15 where the rate of rainfall is taken as $P = 10$ mm/h $= 0.24$ m/day, which is taken equal to the percolation rate. From equation (5.50a)

$$h^2 = \left(\frac{P_1}{K}\right)x^2 + C_1 x + C_2$$

Taking reference from the left bottom corner we have $x = 0$, $h = 22$, from equation (5.50)

$$22^2 = -0 + 0 + C_2 \quad \text{or} \quad C_2 = 22^2 = 484$$

Again when $x = 5300$ and $h = 13$, $13^2 = -\left(\frac{0.24}{18}\right)(5300)^2 + C_1\, 5300 + 484$ which gives

$$\Rightarrow \qquad C_1 = \frac{\left\{13^2 + 0.24\, \dfrac{(5300)^2}{18} - 484\right\}}{5300} = 70.61$$

Discharge per unit width $q_x$ is given by

$$q_x = \frac{K(h_1^2 - h_2^2)}{2L} + \frac{P_1}{2}(2x - l)$$

at $\qquad\qquad x = 5300$ m

$$q_{5300} = \frac{18\,(22^2 - 13^2)}{(2 \times 5300)} + \left(\frac{0.24}{2}\right)(-5300 + 2 \times 5300)$$

$$= \frac{18 \times 315}{10600} + \frac{5300 \times 0.24}{2} = 0.535 + 636$$

or $\qquad\qquad q_{5300} = 636.5$ m$^3$/day/m width.

Again for $x = 0$, $q_0 = \dfrac{18\,(22^2 - 13^2)}{(2 \times 5300)} + \left(\dfrac{0.24}{2}\right)(-5300) = 0.535 - 636.0 = -635.5$

It is negative showing the flow to the other side of the divide.

---

### 5.7.5   Drainage Using Tiles

In water logged areas tile drains are extensively used to drain out water. Tiles are laid at spacing of say 1 m apart. As shown in Fig. 5.16, if $P$ is the rainfall or irrigation rate per unit area then $P_1(\leq P)$ is the rate of recharge to the ground water.

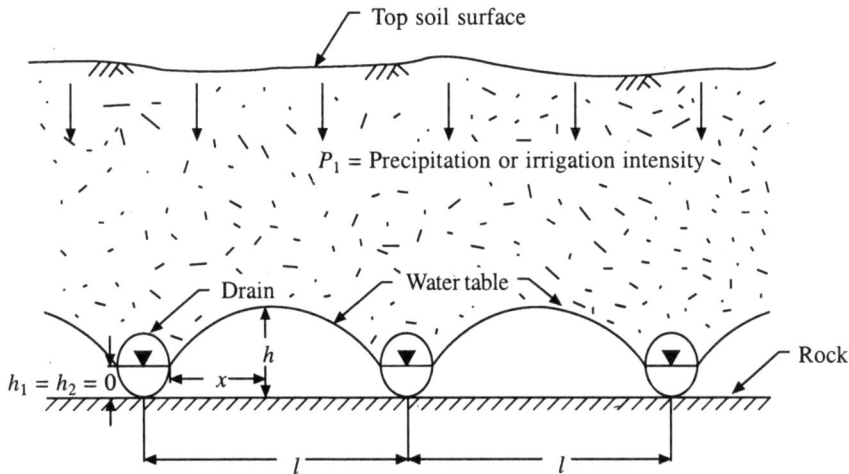

**Fig. 5.16   Drainage of Excess Irrigation Water through Rows of Tiles in Water Logged Area**

Assuming $P$ to be constant, use the relation derived for unconfined aquifer with uniform recharge $P_1$ to solve the problem. Taking $h_1 = h_2 = 0$, i.e., the tile diameters are small enough, the depth of flow in them can be taken as zero. Water table profile depth $h$ can be obtained by substituting $h_1 = h_2 = 0$ in equation (5.51)

$$h = \sqrt{\frac{P_1}{K}(l-x)x} \quad \text{or} \quad h^2 = \frac{P_1}{K}(l-x)x \qquad (5.53)$$

where $x$ is measured from the centre of the tiles. Water table height is maximum at half the length between tiles, i.e., for $x = \dfrac{l}{2}$

$$h_{\max} = \sqrt{\frac{P_1}{K}\frac{1}{2}(l-l/2)} = \frac{l}{2}\{P_1/K\}^{1/2} \qquad (5.54)$$

and from equation (5.52) the discharge per unit length of drain is given by

$$2q_x = \left(\frac{K}{2l}\right)(h_1^2 - h_2^2) - P_1(l/2 - x) = 2\{0 - p_1(l/2 - l)\} = P_1 l \quad (5.55)$$

---

**Example 5.11:** To drain out water from an orchard field, tile drains were provided at 50 m centre to centre. Equivalent recharge from irrigation is found to be 15 mm/h. If coefficient of permeability is 20 m/day, calculate discharge entering a drain per unit length. Also calculate the maximum height of water table profile in between the drains.

**Solution**
Discharge entering a drain per unit length is obtained from equation (5.55)

$$q = P_1 \cdot l = \left(\frac{15}{1000}\right) \times 24 \ (\text{m/day}) \times 50 = 18 \ \text{m}^2/\text{day}$$

Maximum height of water table occurs when $x = l/2$, from equation (5.53) or    $h_{max}^2 = \left(\dfrac{P_1}{K}\right)\left(l - \dfrac{l}{2}\right)\left(\dfrac{l}{2}\right) = \left(\dfrac{P_1}{K}\right)\left(\dfrac{l^2}{4}\right)$

or    $h_{max} = \dfrac{l}{2}\sqrt{\left(\dfrac{P_1}{K}\right)} = \left(\dfrac{50}{2}\right) \times \sqrt{\left\{\left(\dfrac{15 \times 24}{1000}\right)\Big/20\right\}} = 25 \times 0.1342 = 3.354 \text{ m}$

Water table will rise to a height of 3.35 m from the centre line of the tile at a place mid point between the tile lines.

### 5.7.6    Flow Through Leaky Aquifer

For a semi-confined aquifer, the leakage factor defined in equation (5.21) is

$$\lambda = \sqrt{TC} = \sqrt{T/(K'/h)} \quad \text{or} \quad \lambda^2 = T/(K'/h)$$

where $T$ is the transmissivity of the aquifer, $h$ and $K'$ are the thickness and the hydraulic conductivity of the semi-permeable layer, respectively. Leakage factor is an index of leakage, i.e., it represents the vertical percolation through a semipermeable layer from above or below it. A large leakage factor means the leakage is small. It has the dimension of length, expressed in meters.

As shown in Fig. 5.17 steady unidirectional flow in a leaky aquifer is analysed by taking leakage into consideration. The Laplace equation is modified by combining expressions of ground water flow in the aquifer with the quantity of leakage received from the aquitard, the expression for steady flow is

$$\frac{\partial^2 h}{\partial x^2} + \frac{(h_1 - h)}{l^2} = 0 \qquad (5.56)$$

and for unsteady flow

$$\frac{\partial^2 h}{\partial x^2} + \frac{(h_1 - h)}{l^2} = \left(\frac{S}{T}\right)\frac{\partial h}{\partial t} \qquad (5.57)$$

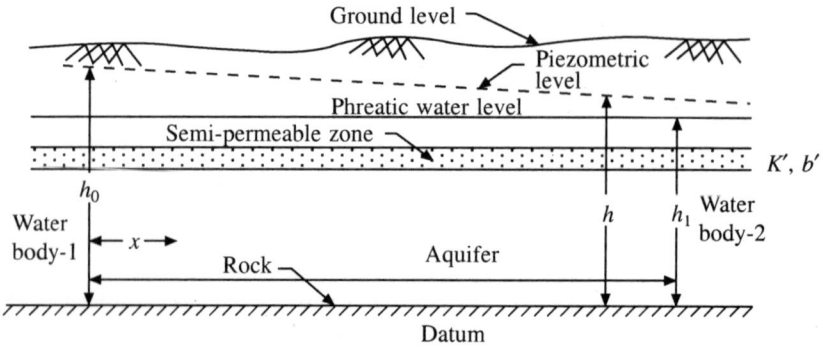

**Fig. 5.17    One dimensional flow in a leaky aquifer**

On integration, equation (5.56) yields

$$h_1 - h = C_1\, e^{x/l} + C_2\, e^{-x/l}$$

where $h$ is the height of piezometric surface due to confined aquifer from impermeable bottom, $h_1$ the height of phreatic or water table surface due to upper unconfined aquifer above the leakage layer, $C_1$ and $C_2$ are integration constants, $l$ (=$\lambda$) is the *leakage length* or *leakage factor*. Assuming the aquifer to be infinite and the phreatic surface (due to the upper aquifer) horizontal, the following boundary conditions may be taken.

$x = 0$, $h = h_0$, i.e, discharge is $Q_0$ for $x = 0$ from which we get

$x = h = h_1$ ($h_1$ is constant)

$$h_1 - h = (h_1 - h_0)\, e^{-x/l} \tag{5.58}$$

If $q'$ is the discharge per unit width of the leaky aquifer

$$q' = -\left(\frac{Kb}{l}\right)(h_1 - h_0)e^{-x/l} \tag{5.59}$$

Combining the above two expressions we get

$$h_1 - h = -\left(\frac{l}{Kb}\right)q_0\, e^{-x/l} \tag{5.60}$$

where $b$ is the thickness of the aquifer. Leakage factor distribution over this is given by

$$q_x = q_0\, e^{-x/l} \tag{5.61}$$

---

**Example 5.12:** For an area underlain by a two aquifer systems (leaky aquifer) the piezometric heads in the confined aquifer recorded at two places 1 km apart are 50 and 47.66 m, respectively. The phreatic surface in the area is 45.meter. Estimate the leakage factor for the area.

**Solution**

Here $\qquad h_1 = 45$ m, $h_0 = 50$ m, $x = 1$ km $= 1000$ m

From equation (5. 58)

$\qquad h_1 - h = (h_1 - h_0)\, e^{-x/l}$ or $(45 - 47.66) = (45 - 50)\, e^{-x/l}$

or $\qquad -2.66 = -5\, e^{-x/l}$

or $\qquad e^{-1000/l} = 0.532$ or $-1000/l = -0.6311$

or $\qquad l = 1585$ m

The leakage factor $l$ is found to be 1585 m.

**Example 5.13:** An aquitard of 1 m thick covers an aquifer of 20 m. Pumping test confirmed a leakage of 40 mm /year. Piezometer surface in the semiconfined aquifer is 15 m below the water table of the unconfined aquifer located above the aquitard layer. Determine the leakage, hydraulic resistance and leakage factor. Permeability of the aquifer is 15 m/day. A well located in the system pumps 4000 m³/day. How much area will be affecting the recharge?

**Solution**

Through aquitard, recharge occurs to the semiconfined layer and it can be found using Darcy's law. Recharge per day through the leaky aquifer is 40 mm/year = 0.0001096 m/day.

$$\text{hydraulic gradient} = \frac{\Delta h}{\Delta d} = \frac{15}{1} = 15$$

Flow from top layer to bottom, through leakage zone = 0.0001096 m/day = $K' \cdot i \cdot A$

∴ Flow per unit area is $K'i$ , 0.0001096 = $K' \times 15$

or $$K' = 7.306 \times 10^{-6} \text{ m/day}$$

$$\text{Leakage} = \frac{K'}{h'} = \frac{7.306 \times 10^{-6}}{1} = 7.306 \times 10^{-6} \text{ m/day (as thickness aquitard = 1 m)}$$

$$\text{Hydraulic resistance , } C = \frac{b'}{K'} = \frac{1}{(7.306 \times 10^{-6})} = 136875 \text{ days}$$

Transmissivity of the confined aquifer, $T = K \times b = 15 \times 20 = 300 \text{ m}^2/\text{day}$

$$\text{Leakage factor, } \lambda = \sqrt{T \times C} = \sqrt{300 \times 136875} = 6408 \text{ m}$$

Area affected will be in ratio of the rate of recharge to the rate of pumping.

∴ 0.0001096 × area affecting recharge = 4000 m³/day

Area affecting recharge = 36.5 km²

---

### 5.7.7 Flow into Infiltration Galleries

Infiltration galleries are the tunnel like structures constructed at the level of hard rock. If $h_0$ and $h_1$ are the heights of free water levels above the rock level (Fig. 5.18), then the piezometer height drops between $h_0$ to $h_1$ due to pumping of water from the gallaries. Applying the Laplace 1D equation for steady flow $q$ per unit length we get

$$q = Ki \, (h \times 1) = Kh \frac{dh}{dx} \quad \left( \text{as } i = \frac{dh}{dx} \right)$$

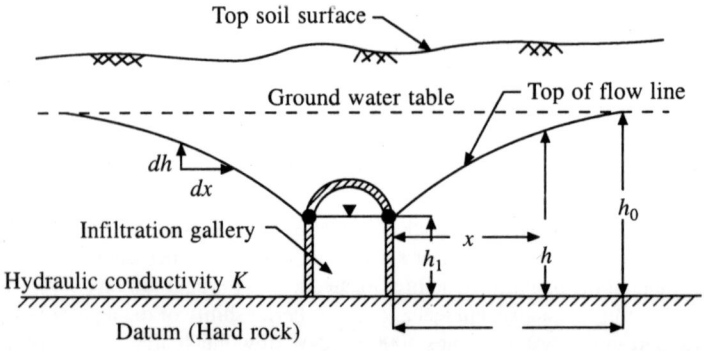

**Fig. 5.18 A Horizontal Infiltration Gallery**

Integrating the equation, we get $qx = \dfrac{Kh^2}{2} + C$

Substituting the boundary values $x = 0$ for $h = h_1$ which gives

$$C = -Kh_1^2/2 \quad \text{or} \quad q = \dfrac{K}{2x}(h^2 - h_1^2) \qquad (5.62a)$$

Again at $h = h_0$, $x = l$, we get

$$q = \dfrac{K}{2l}(h_0^2 - h_1^2) \qquad (5.62b)$$

The total flow into the gallery is contributed from both sides, which is equal to $2qL$, where L is the length of infiltration gallery. Equation to phreatic line can be obtained combining equation (5.62a) and (5.62b) as

$$h = \sqrt{h_1^2 + \dfrac{x}{l}(h_0^2 - h_1^2)} \qquad (5.62c)$$

---

**Example 5.14:** At an infiltration gallery of 200 m long, observed $K = 40$ m/day, influence of drawdown of the gallery is 300 m, height of zero influence from a common datum is 10 m, drawdown at the gallery is 7 m. Obtain (i) total flow into the gallery and (ii) equation of phreatic line and the location of ground water table at 50 m from the gallery.

**Solution**

Here $h_0 = 10$ m, $h_1 = 10 - 7 = 3$ m, $K = 40$ m/day, $L = 200$ m, $l = 300$ m

$\therefore$ $\quad q = \dfrac{K}{2l}(h_0^2 - h_1^2) = \dfrac{40}{2 \times 300}(10^2 - 3^2) = 6.067$ m$^2$/day/m

Total flow into the gallery $Q = 2qL = 2 \times 6.067 \times 200 = 2426.7$ m$^3$/day

Equation to phreatic line $h = \sqrt{3^2 + \dfrac{x}{300}(10^2 - 3^2)} = \sqrt{9 + 0.333x}$

At 50 m away from the gallery $h_{50} = \sqrt{9 + 0.333 \times 50} = 5.05$ m

---

## 5.8 WELL HYDRAULICS

The objective of ground water studies is to determine the quantity of water that can be safely withdrawn from a given aquifer over a hydrological year. Wells or infiltration galleries constitute major works through which most of the ground water supplies for irrigation, domestic and other needs are accomplished. A well should be properly designed and located in an aquifer to yield optimally at minimum cost. Knowledge of well hydraulics is a must for all hydrologists and engineers dealing with ground water resources. Certain terms commonly used in well hydraulics are discussed below.

*(1) Drawdown*
For an unconfined aquifer, the position of water table is represented by piezometric head which can be seen at any open well. When pumping removes water from

a well, the level of water near the well is lowered to a new position. This lowering of the water table at any location near the well from the original static stage before pumping is called *drawdown* (Fig. 5.19). This drawdown is essential because, it creates necessary hydraulic gradient in the system forcing water to flow into the well.

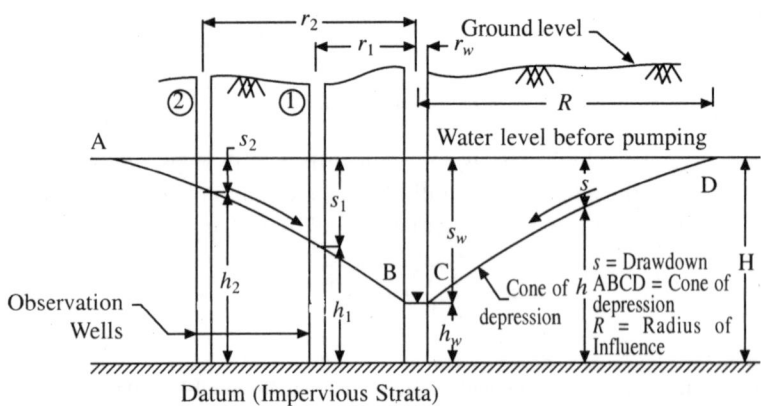

**Fig. 5.19   Radial Flow into a Well for an Unconfined Aquifer**

*(2) Cone of Depression*
The drawdown is maximum at the well and reduces gradually till it becomes almost zero at some distance away from the well. The factors that cause variation of drawdown with distance from well are discharge rate, aquifer characteristics and time for which the well is pumped. Thus in a homogeneous and isotropic medium, the resulting conical shape of the water table around the well due to radial flow of water into it, resulting from a constant rate of pumping is known as cone of depression (Fig. 5.19).

*(3) Radius of Influence (R)*
The radial distance, shown as $R$ in Fig. 5.19, from the centre of well to the limit where the drawdown is zero is called radius of influence. Water to the extent of $R$ from the well is under motion to fill up the pumping rate. The aerial extent of the area with radius $R$ is called *area of influence*. Usually $R$ ranges upto 300 m for unconfined aquifer.

Pumping out of water from such wells gives all the necessary information on aquifers. Information on drawdown, aquifer properties, recharge, radius of influence and other data can be determined by considering either (i) steady flow or (ii) the unsteady flow into the well for both confined or unconfined cases.

**5.8.1   Steady Radial Flow Into a Well**
A steady stage is said to be reached when the drawdown or cone of depression expands no more in a well during pumping. Once this stage is reached after lapse of certain time from the start of pumping, water level does not decline further. When pumping stops recuperation or recovery takes place gradually till the cone of depression is completely filled up. Gradual change in drawdown during initial

pumping and also during the recovery stage after pumping is an unsteady phenomena.

### 5.8.1.1 Steady Radial Flow into a Well in Unconfined Aquifer

The problem is solved under the following Dupuit's assumptions.

(i) Aquifer is infinite aerial extent.
(ii) Aquifer is homogeneous and isotropic so that the coefficient of permeability and transmissibility are taken constant for the whole aquifer.
(iii) Water table/piezometric head at the beginning of pumping is horizontal.
(iv) Aquifer is of uniform thickness.
(v) Well is fully penetrated, drawing water from the entire saturated thickness.
(vi) Pumping continues till a steady stage (equilibrium condition) is reached. The drawdown remains constant.
(vii) Flow is laminar. This means the flow lines are radial and horizontal.
(viii) Hydraulic gradient is taken as the slope of the free surface.

For such a flow condition, we may have to either compute the discharge from the well knowing aquifer characteristics or the aquifer characteristics may have to be evaluated from a pumping test. The above assumptions are never truly satisfied but the results are reliable and hence accepted universaly.

A well with a radius $r_w$ penetrates completely into an unconfined aquifer (Fig. 5.19). Two more observation wells are driven at $r_1$ and $r_2$ distance from the centre of the main well. Pumping in the main well is so adjusted that an equilibrium stage is reached. Absence of further drawdown indicates that the rate of pumping is equal to rate of yield to the well. At the observation wells $s_1$ and $s_2$ are the drawdowns located at $r_1$ and $r_2$ distances away from the main well. From Darcy's law the rate of radial flow $Q$ at a place $r$ distance away from the pumping well is given as

$$Q = K i A = K \frac{dh}{ds} A = K \frac{dh}{dr} A$$

Area contributing the flow is the cylindrical area of radius $r$ from which water is discharged into the well. Hence

$$Q = K \frac{dh}{dr} 2 \pi r h \quad \text{or} \quad \frac{dr}{r} = \left( \frac{2 \pi K}{Q} \right) h \, dh$$

Integrating between the limits $r_1$ to $r_2$ and $h_1$ to $h_2$

$$\ln \left( \frac{r_2}{r_1} \right) = \frac{\pi K}{Q} (h_2^2 - h_1^2)$$

$$\Rightarrow \qquad Q = \frac{\pi K (h_2^2 - h_1^2)}{\ln (r_2/r_1)} \quad \text{or} \quad Q = \frac{\pi K (H^2 - h_w^2)}{\ln (R/r_w)} \qquad (5.63)$$

$$\Rightarrow \qquad K = \frac{Q \ln (r_2/r_1)}{\pi (h_2^2 - h_1^2)} \qquad (5.64)$$

$$\Rightarrow \qquad K = \frac{2.3\, Q \log_{10}\, (r_2/r_1)}{\pi\,(h_2^2 - h_1^2)} \qquad\qquad (5.65)$$

Since near the test well Dupuit's assumption fails to hold good, the observation wells, $r_1$ and $r_2$ should be reasonably away from the test well. Since $(h_1 + s_1) = (h_2 + s_2) = H$, the above equations can also be written as

$$K = \frac{2.3\, Q \log\left(\dfrac{r_2}{r_1}\right)}{\{\pi\, 2\, H(S_1 - S_2)\}} \qquad\qquad (5.66)$$

$$[\text{as } h_1 + h_2 = 2H,\; h_2^2 - h_1^2 = 2H(s_1 - s_2)]$$

$$\Rightarrow \qquad Q = \frac{2\pi\, KH(s_1 - s_2)}{\left\{ 2.3\, Q \log\left(\dfrac{r_2}{r_1}\right)\right\}} \qquad\qquad (5.67)$$

$$\Rightarrow \qquad Q = \frac{2\pi\, T(s_1 - s_2)}{\left\{ 2.3\, Q \log\left(\dfrac{r_2}{r_1}\right)\right\}} \qquad\qquad (5.68)$$

where $T$ is the coefficient of transmissibility $(=KH)$, $H$ the total depth of piezometric surface above the impervious strata before the beginning of pumping. Use of the equations involving drawdown $s$ should be avoided as far as possible because it tries to overestimate the discharge.

---

**Example 5.15:** Pumping at a rate of 1500 lpm from a 30 cm diameter test well penetrating into 60 m of an unconfined aquifer gives drawdown of 2.0 and 1.10 m in observation wells located respectively at 120 and 160 m away from it. Calculate (i) hydraulic conductivity of the aquifer, (ii) drawdown of the pumping well.

**Solution**
Here discharge $Q$ = 1500 lit/min = 1.5 m³/min = 1.5 × 24 × 60 m³/day = 1.5/60 m³/sec

$$h_2 = H - s_2 = 60 - 1.1 = 58.9 \text{ m}$$
$$h_1 = H - s_1 = 60 - 2.0 = 58.\,0 \text{ m}$$

$$Q = 1.5/60 = \frac{\pi\, K(58.9^2 - 58.0^2)}{\left\{ 2.303\, \log\left(\dfrac{160}{120}\right)\right\}} = 1148.15\, K$$

$$\Rightarrow \quad K = \frac{1.5}{(60 \times 1148.15)} = 2.1774 \times 10^{-5} \text{ m/sec} = 0.002177 \text{ cm/sec} = 1.88 \text{ m/day}$$

Hydraulic conductivity = 1.88 m/day

Again, $\qquad Q = \dfrac{\pi\, K\,(H^2 - h_w^2)}{\ln\,(R/r_w)}$

Assuming $R$ as 300 m

$$Q = \frac{\pi \times 1.88\,(60^2 - h_w^2)}{2.203\,\ln\,(300/0.15)} \Rightarrow (24 \times 60 \times 1.5) = 0.776\,(60^2 - h_w^2)$$

$$\Rightarrow \qquad (60^2 - h_w^2) = 2783.5 \Rightarrow h_w = \{60^2 - 2783.5\}^{1/2} = 28.5 \text{ m}$$

Drawdown at the well is $(60 - 28.5) = 31.4$ m

---

### 5.8.1.2 Steady Radial Flow into a Well in Confined Aquifer

For a confined aquifer of thickness $B$ under Dupuit's assumptions (Fig. 5.20), Darcy's law for steady flow equation can be written as

$$Q = KiA = K\left(\frac{dh}{dr}\right)(2\pi rB)$$

or
$$\frac{dr}{r} = \left\{\frac{(2\pi KB)}{Q}\right\} dh$$

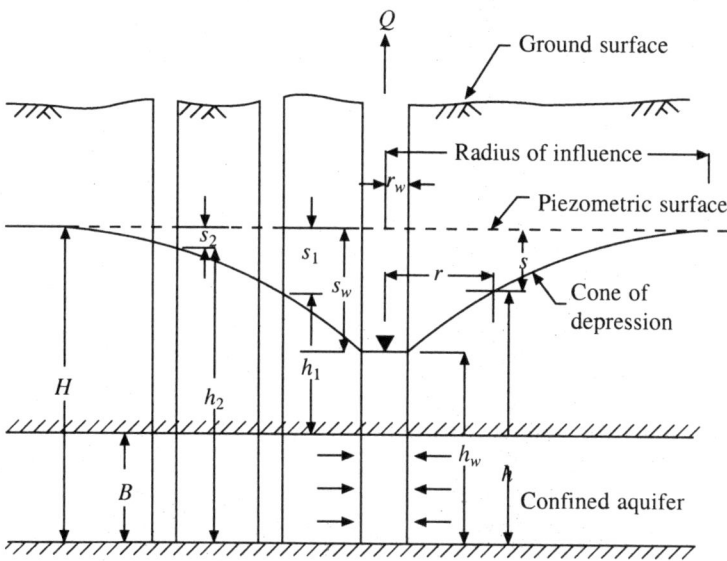

**Fig. 5.20   Radial Flow into a Confined Well**

where area of the cylindrical surface is the area from which water is discharged to the well. The well receives water from an area of $2\pi rB$, where $B$ is the depth of confinement. Integrating between the limits $(r_1, r_2)$ and $(h_1, h_2)$

$$\ln\left(\frac{r_2}{r_1}\right) = \left\{\frac{(2\pi KB)}{Q}\right\}(h_2 - h_1)$$

$$\Rightarrow \qquad Q = \frac{2\pi\,KB\,(h_2 - h_1)}{\ln\,(r_2/r_1)} = \frac{2\pi T\,(s_1 - s_2)}{\ln\,(r_2/r_1)} \qquad (5.69)$$

$$\Rightarrow \qquad K = \frac{\left\{Q \ln\left(\frac{r_2}{r_1}\right)\right\}}{\{2\pi B (h_2 - h_1)\}} = \frac{\left\{Q \ln\left(\frac{r_2}{r_1}\right)\right\}}{\{2\pi B (s_2 - s_1)\}} \qquad (5.70)$$

Equation (5.69) is popularly known as *Thiem's equation* and is valid under Dupuit's assumptions. When observation wells are not available, then $r_2$ can be taken as $R$ and $r_1$ as $r_w$. Following equation can be taken as an approximation for obtaining the values of $Q$, $T$ or $K$.

$$Q = \frac{2\pi T (H - h_w)}{\ln\left(\dfrac{R}{r_w}\right)} = \frac{2.72 T (H - h_w)}{\log_{10}\left(\dfrac{R}{r_w}\right)} \qquad (5.71)$$

Important conditions for above equation (5.71) to be valid are (i) the well must penetrate to the full depth of confined aquifer, (ii) the aquifer is of uniform thickness, (iii) flow should be steady and (iv) the aquifer should be homogeneous and isotropic.

---

**Example 5.16:** Total thickness of a confined aquifer is 20 m and it is asumed that the well penetrates into the full depth of the aquifer. Calculate the coefficient of transmissibility for discharge of 1.5 m³/min, $h_1 = 58.0$ m at $r_1 = 120$ m and $h_2 = 58.9$ m at $r_2 = 160$ m. Take well diameter = 30 cm.

**Solution**

Here
$$h_1 = 58.0 \text{ m and } h_2 = 58.9 \text{ m}$$

$$Q = 1500 \text{ lpm} = 1.5 \text{ m}^3/\text{min}$$

From equation (5.69)

$$Q = \frac{2\pi T (h_2 - h_1)}{\ln\left(\dfrac{r_2}{r_1}\right)} = \frac{2\pi T (58.9 - 58.0)}{\ln\left(\dfrac{160}{120}\right)} = 19.66 T$$

Since
$$\frac{1.5}{60} = 19.66 T \text{ or } T = \frac{1.5}{(60 \times 19.66)} = 0.00172 \text{ m}^2/\text{sec}$$

$$= 110 \text{ m}^2/\text{day} = 0.0763 \text{ m}^2/\text{min}$$

$$K = \frac{T}{B} = \frac{110}{20} = 5.5 \text{ m/day}$$

Applying the relation between $r_2$ and $r_w$ we get

$$Q = \frac{2\pi(0.0763)(58.9 - h_w)}{\ln(160/0.15)} = 0.068724 \,(58.9 - h_w)$$

$$58.9 - h_w = \frac{1.5}{0.068724} = 21.85$$

$$h_w = 58.9 - 21.85 = 37.1 \text{ m}$$

Drawdown at the well is $(60 - 37.1) = 22.90$ m

---

### 5.8.1.3 *Flow into a Well with Sloping Water Table*

For sloping water table case shown in Fig. 5.21, the discharge equations can still be used with reasonable accuracy subject to the following.

(i) There must be four observation wells located in a straight line with the discharging or the testing well.

(ii) First two observation wells are located at distance $r_1$ upstream and down stream of the testing well. Therefore, mean of the two drawdown heights is $h_1 = (h_1' + h_1'')/2$. Similarly the second set of two observation wells are located exactly $r_2$ distance upstream and downstream of the testing well and mean of the observed heights is $h_2 = (h_2' + h_2'')/2$. Here $h_1'$ and $h_1''$ represents the depths of water from a commom datum in the two observation wells on the two sides of the pumping well at distance $r_1$. Similarly, $h_2'$ and $h_1''$ represent depths of water in two observation wells on the two sides of pumping well at distances $r_2$.

(iii) All the five wells should again be so located that they are in a straight line and along the direction of initial slope of water table or piezometric surface.

(iv) The position of first two observation wells at $r_1$ distance should be reasonably away from the test well so that stream lines are not greatly distorted.

(v) The equations for steady radial flow into a well can be used to compute the well parameters on the discharge with reasonable accuracy.

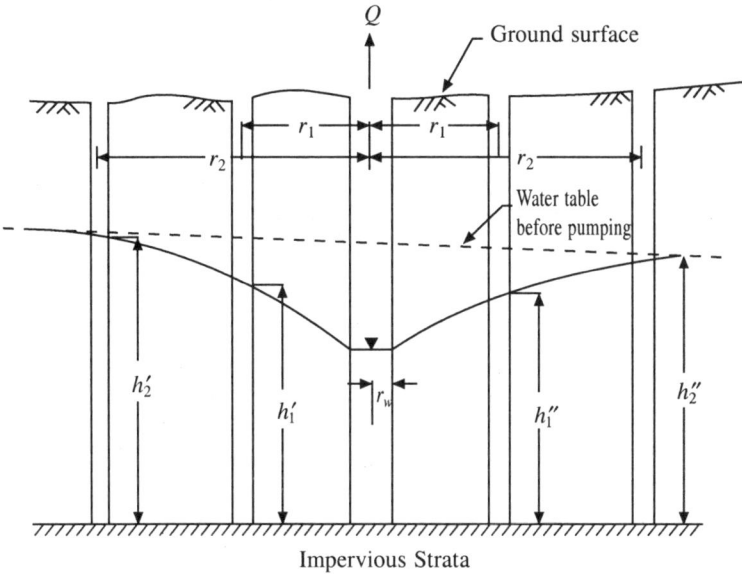

**Fig. 5.21 Well with Sloping Water Table**

### 5.8.2 Partially Penetrated Well

It is a well which does not penetrate fully into the thickness of the aquifer. For such cases the flow pattern becomes three dimensional as the vertical component of the flow cannot be neglected due to the fact that from the bottom of well water is forced to come out by pumping. This gives rise to increased length of the flow line for a partially penetrated well.

*For Confined Aquifer*

Following Kozeney (1933), discharge from a partially penetrating well of depth $B'$, Fig. 5.22a of confinement height $B$ in the aquifer is given as

$$Q_p = Q\varepsilon \left[ 1 + 7 \left\{ \sqrt{\frac{r_w}{2\varepsilon B}} \cos \left( \frac{\pi\varepsilon}{2} \right) \right\} \right] \tag{5.72}$$

where $Q_p$ is the discharge in a partially penetrated well, $Q$ the discharge of a fully penetrated well in confined aquifer, $\varepsilon$ the ratio $B'/B$, $B'$ the length of partial penetration of the well, $B$ the thickness of the aquifer and $r_w$ the radius of the well. Discharge in a partially penetrating well for a given drawdown is less than that of a fully penetrating well due to the increased loss of head.

*For Unconfined Aquifer*

Equation (5.72) can be applied to an unconfined aquifer, where $\varepsilon$ is the ratio

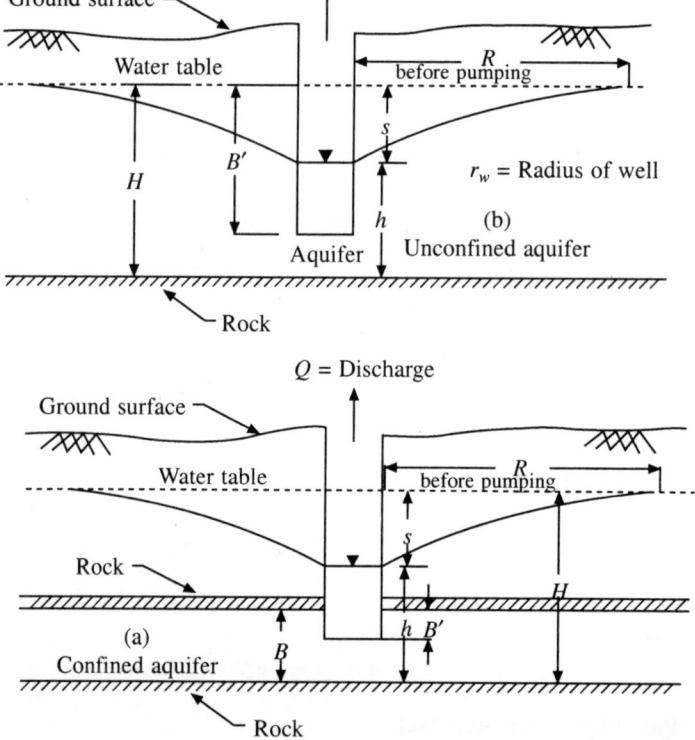

**Fig. 5.22   Partial Penetration of Wells**

$B'/H$, $Q$ the discharge for unconfined aquifer, $H$ the height of water table from the hard strata as shown in Fig 5.22b.

---

**Example 5.17:** A 40 cm well penetrates 40 m into an aquifer. Screen is provided for the first 15 m. The yield from the well is 2500 lpm with the drawdown at the well as 4 m. If the length of screen is increased to 25 m find the new discharge.

**Solution**
Here the length of screen $B' = 15$ m
Aquifer depth $B = 40$ m

The length ratio $$\varepsilon = \frac{B'}{B} = \frac{15}{40} = 0.375$$

$$Q_p = 2500 \text{ lpm} = 2.5 \text{ m}^3/\text{min}$$

Radius of the well $r_w = 20$ cm $= 0.20$ m.
If $Q$ is the discharge of the well with complete penetration and complete screening then, using equation (5.72)

$$Q_p = Q\varepsilon\left[1 + 7\left\{\sqrt{\frac{r_w}{2\varepsilon B}}\cos\left(\frac{\pi\varepsilon}{2}\right)\right\}\right]$$

$\therefore$ $$2.5 = Q \times 0.375\left[1 + 7\left\{\sqrt{\frac{r_w}{2 \times 0.375 \times 40\varepsilon B}}\cos\left(\frac{\pi \times 0.375}{2}\right)\right\}\right]$$

$$= Q \times 0.375 \,(1 + 0.5715 \times 0.831) = 0.553\, Q$$

$$Q = 2.5/0.553 = 4.519 \text{ m}^3/\text{min}$$

For 25 meter screening, $\varepsilon = \dfrac{25}{40} = 0.625$

$$Q_p = 4.519 \times 0.625\left[1 + 7\left\{\sqrt{\frac{0.2}{2 \times 0.625 \times 40}}\cos\left(\frac{\pi \times 0.625}{2}\right)\right\}\right]$$

$$= 4.519 \times 0.625 \,(1 + 0.4427 \times 0.555) \quad \text{or} \quad Q_p = 3.518 \text{ m}^3/\text{min}$$

$\therefore$ Discharge for 25 m screen is 3.518 m$^3$/min or 3518 lpm.

---

## 5.9 UNSTEADY FLOW INTO A WELL

This condition is encountered under two circumstances (a) from the instant of pumping the well to the time when a steady state is reached, i.e., when the cone of depression is expanding or (b) when the water level or piezometric head (for confined aquifer) of an aquifer is gradually recovering (period between the stoppage of pumping to its recovery).

### 5.9.1 Confined Aquifer

When a fully penetrating well is pumped, the cone of depression increases continuously till the equilibrium between the rate of discharge and rate of inflow into the well is reached. During this period of unsteady flow, the general differential equation for confined aquifer of constant thickness $B$ is given by equation (5.41) and is reproduced as

$$\frac{\partial^2 h}{\partial x^2} + \frac{\partial^2 h}{\partial y^2} + \frac{\partial^2 h}{\partial z^2} = \left(\frac{S}{KB}\right)\frac{\partial h}{\partial t} = \left(\frac{S}{T}\right)\frac{\partial h}{\partial t}$$

which represents the drawdown at any distance from the pumping well. In polar coordinate system, it can be expressed as

$$\frac{\partial^2 h}{\partial x^2} + \frac{\partial h}{\partial r} = \frac{S}{T}\frac{\partial h}{\partial r} \tag{5.73}$$

Under the same assumptions as Dupuit proposed for the steady flow condition, Theis (1935) proposed a solution to equation (5.73) under the assumption that water is released from the storage in the aquifer in immediate response to decline in water table. The drawdown is given by

$$H - h = s = \left(\frac{Q}{4\pi T}\right)\int_u^\infty \left(\frac{e^{-u}}{u}\right) du = \left(\frac{Q}{4\pi T}\right) W(u) \tag{5.74}$$

where the parameter $u$ is defined as

$$u = \left(\frac{r^2 \cdot S}{4T \cdot t}\right) \tag{5.75}$$

where $S$ is the storage coefficient, $t$ the time from start of pumping, $T$ the transmissibility of the aquifer, $Q$ the discharge under steady stage, $H$ the initial piezometric head and $h$ the height of drawdown curve at distance r from the aquifer. The well function $W(u)$ is defined as

$$W(u) = \int_u^\infty e^{-u} \cdot du = -0.5772 - \ln(u) + u - \frac{u^2}{2.2!} + \frac{u^3}{3.3!} - \dots \tag{5.76}$$

The well function can be expanded by Euler's series. Solution to the unsteady radial flow equation (5.74) into the well can be obtained by any of the following methods: Theis method, Jacob's method and Chow's method.

### 5.9.1.1    Theis Method
Theis (1935) proposed a graphical method to solve equation (5.74) as follows:

1. A plot between $u$ vs. $W(u)$ called *type curve* is available, which can also be plotted from the data table between $u$ and $W(u)$. Values between $u$ and $W(u)$ are given in Table 5.4 and a plot between them is shown in Fig. 5.23. Such curves plot $u$ in abscissa and $W(u)$ in the ordinate scale on a log-log paper.
2. From the field data a plot between drawdown $s$ vs. $r^2/t$ is obtained on a log-log paper by taking $(r^2/t)$ on abscissa and $s$ as ordinate.
3. The two graphs of steps (1) and (2) shall use the same scale.
4. Superimpose the two curves. They will fit (overlap) for some length of the curve.

**Table 5.4    Values of $F(u)$, $W(u)$ and $u$**

| $F(u)$ | $W(u)$ | $u$ | $F(u)$ | $W(u)$ | $u$ | $F(u)$ | $W(u)$ | $u$ |
|--------|--------|-----|--------|--------|-----|--------|--------|-----|
| (1) | (2) | (3) | (4) | (5) | (6) | (7) | (8) | (9) |
| – | 0.0000125 | 9.0 | 0.913 | 1.92 | 0.09 | 2.76 | 6.44 | 0.0009 |
| – | 0.0000377 | 8.0 | 0.956 | 2.03 | 0.08 | 2.81 | 6.55 | 0.0008 |
| – | 0.0001155 | 7.0 | 1.000 | 2.15 | 0.07 | 2.87 | 6.69 | 0.0007 |
| – | 0.0003601 | 6.0 | 1.060 | 2.30 | 0.06 | 2.94 | 6.84 | 0.0006 |
| 0.0734 | 0.00114 | 5.0 | 1.130 | 2.47 | 0.05 | 3.01 | 7.02 | 0.0005 |
| 0.0898 | 0.00378 | 4.0 | 1.210 | 2.68 | 0.04 | 3.11 | 7.25 | 0.0004 |
| 0.1170 | 0.0130 | 3.0 | 1.330 | 2.96 | 0.03 | 3.23 | 7.53 | 0.0003 |
| 0.1570 | 0.0489 | 2.0 | 1.490 | 3.55 | 0.02 | 3.41 | 7.94 | 0.0002 |
| 0.2590 | 0.2190 | 1.0 | 1.770 | 4.04 | 0.01 | 3.70 | 8.63 | 0.0001 |
| 0.2760 | 0.260 | 0.9 | 1.820 | 4.14 | 0.009 | 3.75 | 8.74 | 0.00009 |
| 0.3010 | 0.311 | 0.8 | 1.870 | 4.26 | 0.008 | – | 10.94 | $1 \times 10^{-5}$ |
| 0.3270 | 0.374 | 0.7 | 1.920 | 4.39 | 0.007 | – | 11.63 | $5 \times 10^{-6}$ |
| 0.3600 | 0.454 | 0.6 | 1.990 | 4.54 | 0.006 | – | 13.23 | $1 \times 10^{-6}$ |
| 0.4010 | 0.560 | 0.5 | 2.070 | 4.73 | 0.005 | – | 15.54 | $1 \times 10^{-7}$ |
| 0.4550 | 0.702 | 0.4 | 2.16 | 4.95 | 0.004 | – | 17.84 | $1 \times 10^{-8}$ |
| 0.5320 | 0.906 | 0.3 | 2.28 | 5.23 | 0.003 | – | 20.15 | $1 \times 10^{-9}$ |
| 0.6470 | 1.22 | 0.2 | 2.46 | 5.64 | 0.002 | – | 22.45 | $1 \times 10^{-10}$ |
| 0.8740 | 1.80 | 0.1 | 2.75 | 6.33 | 0.001 | | | |

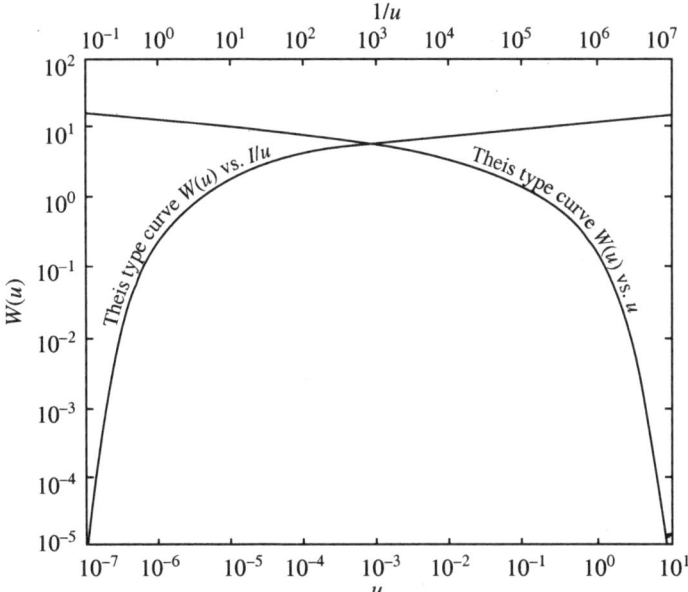

**Fig. 5.23    Theis Type Curves for $W(u)$ Versus $u$ and $W(u)$ vs. $1/u$**

5. Mark a point 'A' where perfect superimposition of the two curves is observed. Read coordinates on both the curves to get $\log(s)$, $\log(r^2/t)$, $u$ and $W(u)$.

6. Substitute the values in equation

$$S = \left(\frac{4T \cdot t}{r^2}\right) u \quad \text{and} \quad T = \left(\frac{Q}{4\pi s}\right) W(u)$$

to get the aquifer parameters. While matching the two curves, less importance should be given to the lower part of curve to overcome initial errors.

### 5.9.1.2   Cooper Jacob Method

Since function $u$ and time $t$ are inversely proportional, the value of $u$ decreases with increase in pumping time. Jacob (1947) used this approach to drop the higher values of equation (5.76) to arrive at an acceptable value of the well function $W(u)$. For values of $u \leq 0.01$ or large values of $t$, i.e., $t \geq (r^2 S / 0.01\, u\, t)$ higher terms of the series given in equation (5.76) are neglected. Drawdown $s$ can be approximated to the following equation by taking only the first two terms.

$$T = \left(\frac{Q}{4\pi s}\right)\{-0.5772 - \ln(u)\} = \left(\frac{Q}{4\pi s}\right)\ln\left(\frac{1}{1.784u}\right)$$

$$= \left(\frac{Q}{4\pi s}\right)\ln\left(\frac{4Tt}{1.784 r^2 S}\right)$$

or
$$s = \left(\frac{Q}{4\pi T}\right)\ln\left(\frac{4Tt}{1.784 r^2 S}\right) = \left(\frac{2.3Q}{4\pi T}\right)\log_{10}\left(\frac{2.25\,Tt}{r^2 S}\right) \qquad (5.77)$$

For a given pumping test from a well, if $s_1$ and $s_2$ are the drawdowns at times $t_1$ and $t_2$ respectively, then

$$s_2 - s_1 = \Delta s_1 = \left(\frac{2.3Q}{4\pi T}\right)\log\left(\frac{t_2}{t_1}\right) \qquad (5.78)$$

from which $T$ can be computed. Again from equation (5.77) for zero drawdown, i.e., for $s = 0.0$

$$0 = \left(\frac{2.3Q}{4\pi T}\right)\log_{10}\left(\frac{2.25T \cdot t}{r^2 S}\right)$$

Since     $2.3Q / 4\pi T \neq 0$, $\log_{10}\left(\frac{2.25T \cdot t}{r^2 S}\right) = 0.0$

or taking anti-logarithm     $\left(\frac{2.25T \cdot t}{r^2 S}\right) = 1$ \qquad (5.79)

From which we can compute storage coefficient $S = \left(\frac{2.25T \cdot t}{r^2}\right)$.

This method is simpler than Theis method and is extensively applied to calculate aquifer properties.

### 5.9.1.3   Chow's Method

Chow (1952) solved equation (5.74) by introducing a function $F(u)$, which entails the solution of the equation of the following form

$$F(u) = W(u) \frac{e^u}{2.3} \qquad (5.80)$$

in which $W(u)$ and $u$ are already defined. The relation between $F(u)$, $W(u)$ and $u$ is usually available in the form of a graph or table (see Table 5.4). The procedure to solve the unsteady flow problems using Chow's method is described as follows:

1. Observe the drawdowns $s$ of the observation well during the passage of time $t$ from the start of pumping.
2. Plot a graph between the drawdown and time by taking time $t$ in logarthmic scale in abscissa and drawdown $s$ as ordinate in ordinary scale. Draw a smooth curve passing through the points.
3. Select any arbitrary point (say A) on the curve of step (2) and read the coordinates of $s$.
4. At the selected point A, draw a tangent to the curve. Consider one log cycle of time and record the drawdown difference $\Delta s$.
5. Compute the value of $F(u)$ as $= s/\Delta s$. Obtain the value of $u$ and $W(u)$ from Table 5.4, by knowing $F(u)$.
6. Obtain the values of formation constants $S$ and $T$ from the following relations.

$$S = \left(\frac{4Tt}{r^2}\right)u \quad \text{and} \quad T = \left(\frac{Q}{4\pi s}\right)W(u)$$

---

**Example 5.18:** A well penetrates into a confined aquifer 30 m deep. Pumping is done at the rate of 2000 lpm. Hydraulic conductivity of the aquifer is 25 m/day. Calculate the drawdown of the well at 75 m away from it after a lapse of 10 h of pumping. Take storage coefficient $S$ as 0.003.

**Solution**

Transmissibility $T = KB = \dfrac{25}{24 \times 60 \times 60} \times 30 = 0.00868$ m²/sec.

From equation (5.75)

$$u = \frac{r^2 S}{4Tt} = (75^2 \times 0.003)/(4 \times 0.00868 \times 10 \times 60 \times 60) = 0.0135$$

The well function $W(u)$ given in equation (5.76) is

$$W(u) = -0.5772 - \ln(u) + u - \frac{u^2}{2 \times 2!} + \dots$$

$$= -0.5772 - \ln(0.0135) + 0.0135 - \frac{0.0131^2}{2. \times 2!} + \dots$$

$$= -0.5772 + 4.305 + 0.0135 - 0.000045$$

or $\qquad\qquad W(u) = 3.741$

From equation (5.74), drawdown is given by $s = \dfrac{Q}{4\pi T} W(u)$

where $Q = 2000$ lpm $= 2.00$ m³/min $= 2/60$ m³/sec.

Drawdown at 75 m from the testing well is given as

$$s_{75} = \frac{2/60}{4 \times \pi \times 0.00868} \times 3.741 = 1.143 \text{ m}$$

**Example 5.19:** The drawdown-observation time recorded at an observation well, situated 60 m from the pumping well are as follows:

| Time (min) | 2 | 4 | 6 | 8 | 10 | 20 | 50 | 100 | 240 |
|---|---|---|---|---|---|---|---|---|---|
| Draw down(m) | 0.2 | 0.8 | 1.1 | 1.8 | 2.4 | 3.7 | 5.1 | 6.0 | 7.0 |
| $r^2/t$ | 1800 | 900 | 600 | 450 | 360 | 180 | 72 | 36 | 15 |

If the well discharge is 2000 lpm, calculate transmissibility and storage coefficient of the aquifer by (i) Theis method, (ii) Cooper Jacob method and compare the results.

**Solution**

(i) *Theis Method*
Drawdown vs. $r^2/t$ is ploted on a log-log paper (Fig. 5.24). The curve is superimposed on Theis type curve between $u$ and $W(u)$. A matching point $A$ is marked on the curve. The coordinates of the point are read from the plot as

$$s = 6.8 \text{ m}, \quad \frac{r^2}{t} = 28 \text{ m}^2/\text{min}, \quad u = 0.0105, \quad W(u) = 3.8$$

$$Q = 2000 \text{ lpm} = 2 \text{ m}^3/\text{m} = 2 \times 60 \times 24 = 2880 \text{ m}^3/\text{day}$$

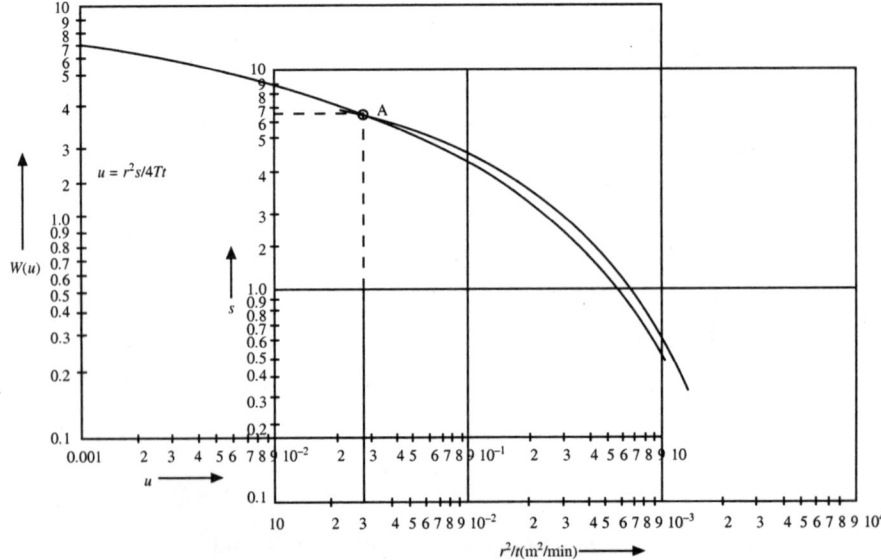

**Fig. 5.24    Theis *u* vs. *W(u)* Curve Superimposed on Drawdown vs. *r²/t* Curve Obtained from Field Pumping Test.**

Substituting the values in the equation for transmissibility we get

$$s = \left(\frac{Q}{4\pi T}\right) W(u) \quad \text{or} \quad T = \frac{Q}{4\pi s} W(u)$$

or $\qquad T = \frac{2}{4 \times \pi \times 6.8} \, 3.8 = 0.0889 = \text{m}^2/\text{min}$

Storage coefficient $\quad S = \frac{4T}{r^2/t} u = \frac{4 \times 0.0889}{28} \, 0.027 = 0.00033$

### (ii) *Cooper Jacob Method*

A graph is plotted between the logarithmic of time in abscissa against the drawdown as ordinate. A straight line is fitted passing through the scatter points (Fig. 5.25). Two points A and B are marked on the curve and the following information is read from the plot.

At point A $s_1 = 2.0$, $\log (t_1) = 0.90$ or $t_1 = 7.94$ min

At point B $s_2 = 4.2$, $\log (t_2) = 1.40$ or $t_2 = 25.12$ min

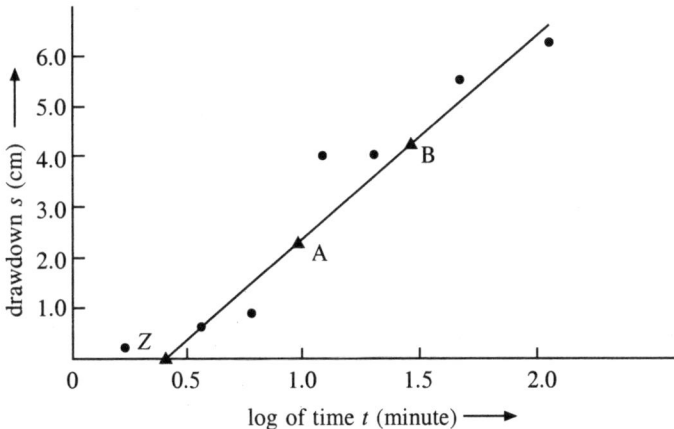

**Fig. 5.25   Plot of time vs. Drawdown by Jacobs Method for Example 5.18**

Therefore $\qquad s_2 - s_1 = 4.2 - 2.0 = \dfrac{2.3 \times Q}{4\pi T} \log \dfrac{t_2}{t_1}$

or $\qquad 2.2 = \dfrac{2.3 \times 2}{4\pi T} \log \dfrac{25.12}{7.94} = \dfrac{0.1831}{T}$

$\Rightarrow \qquad T = \dfrac{0.1831}{2.30} = 0.0832 \text{ m}^2/\text{min} = 120 \text{ m}^2/\text{day}$

Now, time corresponding to zero drawdown is obtained by extending the straight line curve to meet the X-axis at Z. Therefore,

$$\log (t_0) = 0.43 \quad \text{or} \quad t_0 = 2.70 \text{ min}$$

$$S = \frac{2.25 T t}{r^2} = 2.25 \times 0.0832 \times \frac{270}{60^2} = 1.4 \times 10^{-4}$$

The storage coefficient is $1.4 \times 10^{-4}$ or $S = 0.00014$.

### 5.9.2   Unsteady Radial Flow into a Well in Unconfined Aquifer

For unconfined aquifer, Theis assumption of immediate recovery to the well is not possible in fine grained formations. For small drawdowns, the equation for steady state known as Thiem or Theis formula can be used without much error. The well equation

$$Q = \frac{\pi K (h_2^2 - h_1^2)}{\ln(r^2/r_1)}$$

can be approximated to the form

$$Q = \frac{2\pi \cdot K \cdot H(s_1 - s_2)}{2.303 \log\left(\dfrac{r_2}{r_1}\right)} = \frac{2\pi T(h_2 - h_1)}{2.303 \log_{10}\left(\dfrac{r_2}{r_1}\right)} = \frac{2\pi T(H - h_w)}{2.303 \log_{10}\left(\dfrac{R}{r_w}\right)} \quad (5.81)$$

$$[h_2^2 - h_1^2 = (h_2 + h_1)(h_2 - h_1) = 2H(s_1 - s_2) \text{ as } (h_2 + h_1)$$

$$= 2H \text{ and } h_2 - h_1 = s_1 - s_2]$$

Transmissibility coefficient $T$ is assumed to remain constant here. For large drawdowns exceeding 10% of aquifer thickness, equation (5.81) proposed by Jacob (1944) may still be used but the aquifer should have the property of good drainage. For poor drainage aquifers (delay in the drawdown) Boulton's procedure may be adopted. From field test data, a graph between time $t$ in days and drawdown gives a curve on a log-log plot as shown in Fig. 5.26.

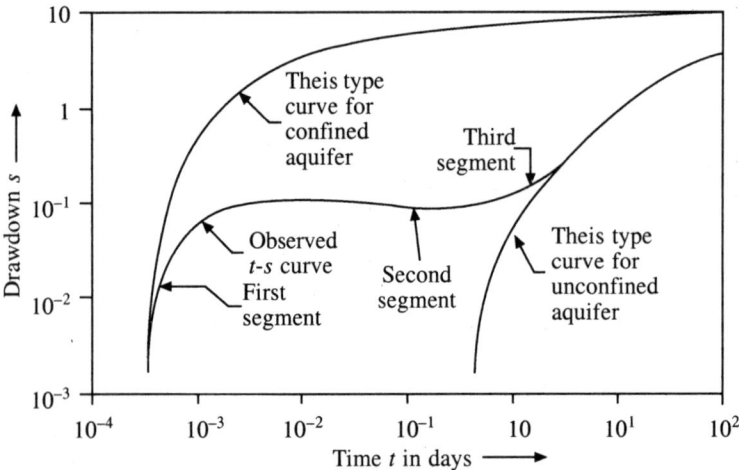

**Fig. 5.26   Delayed Yield in Unconfined Aquifer (after USBR 1977)**

*First Segment of the Curve*: This part of the curve represents part of drawdown within the first few minutes of pumping. It is similar to the Theis type curve for confined aquifer. Due to sudden release of pore pressure, aquifer material gets compacted and the entrapped air escapes. Water level also drops faster. Use of

Theis equation for confined aquifer gives the value of transmissibility coefficient $T$ for this stage.

*Second Segment*: This segment of the drawdown-time curve is relatively flat due to contribution of gravity drainage for the part of aquifer above cone of depression.

*Third Segment*: In this zone of drawdown-storage curve, an equilibrium is reached between gravity drainage and decline of water table so that departure from the Theis curve becomes progressively smaller.

For unconfined case, pumping time should be sufficiently long so that, the third segment of the curve is reached. As a guideline, the value of minimum time of pumping is given in Table 5.5. Once pumping is continued till the minimum time, i.e., to reach the third segment of the curve given in Fig. 5.26, the unsteady flow equation can be used to evaluate the aquifer parameters $S_y$.

**Table 5.5   Minimum Time of Pumping for Unsteady Flow (Unconfined Aquifer)**

| Aquifer Material | Minimum Pumping time (h) | Delay Index $(1/\alpha)$ min |
|---|---|---|
| (1) | (2) | (3) |
| Coarse sand | 4 | 10 |
| Medium sand | 4 | 60 |
| Fine sand | 30 | 200 |
| Very fine sand/silt | 170 | 1000 |
| Silt/clay | 170 | 2000 |

## 5.10   SPACING OF WELLS (WELL INTERFERENCE)

When the cones of influence of two or more wells overlap, the wells are said to interfere with one another. Many a times wells are dug at close spacings under the following two conditions.

(i)  For lowering of water table or piezometric surface. Such situations arise in (a) mining areas or
(b) dewatering of water logged area for foundation excavation.
(ii)  To meet excessive demand of water for irrigation or domestic supply.

For a safe margin, spacing between two tube wells should be more than two times the radius of influence. If such spacing cannot be provided then, by the principle of superposition, drawdown at any point in the system of the wells can be evaluated by summing up the drawdown components of each individual wells using equation .

$$s_t = s_1 + s_2 + \dots + s_n \tag{5.82}$$

where $s_1, s_2, \dots, s_n$ are the drawdown components at individual wells numbered as 1, 2, 3, ..., $n$, which are in the interference. When all the wells are pumped simultaneously, a single long and large curve of depression is formed (Fig. 5.27). Dotted lines shows cone of depression of individual wells. If $Q_1, Q_2 \dots, Q_n$ are respectively the discharge rates from wells $r_1, r_2, \dots, r_n$ distance away from the reference well, then for confined aquifer

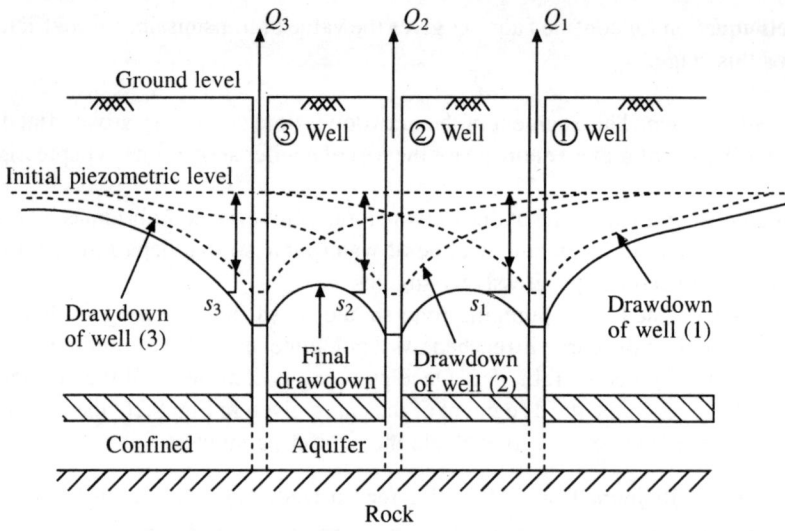

**Fig. 5.27   Well Interference and the Resulting Drawdown Due to 3 Wells.**

$$H - h = s_t = \sum_{i=1}^{n} \frac{Q_i}{2\pi KB} \ln (R_i/r_i) \qquad (5.83)$$

where $s_t$ is the sum of all drawdowns of individual wells, $H$ the original piezometric height, $h$ the height of piezometer due to combined effect of all $n$ wells, $R_i$ the distance of the *ith* well where its drawdown is considered as zero, $r_i$ the distance of reference well form the testing well and $Q_i$ is the pumping rate of the *i*th well. For unconfined aquifer

$$H^2 - h^2 = \left(\frac{\Sigma Q_i}{\pi K}\right) \ln \left(\frac{R_i}{r_i}\right)$$

or $$s_t = (H - h) = \left(\frac{\Sigma Q_i}{\pi K}\right) \ln \left(\frac{R_i}{r_i}\right) \Big/ (H - h) \qquad (5.84)$$

For all practical purposes take $R_i$ as 250-300 m. For unequilibrium (unsteady) condition

$$s_t = \frac{Q_1 W(u_1)}{4\pi T} + \frac{Q_2 W(u_2)}{4\pi T} + \dots + \frac{Q_n W(u_n)}{4\pi T} \qquad (5.85)$$

is used where $u_i = \left(\dfrac{r_i^2 S}{4T t_i}\right)$, $t_i$ is the time since start of pumping of *i*th well with discharge $Q_i$

---

**Example 5.20:** A confined aquifer with transmissibility of 1550 m²/day and storage coefficient of $4.75 \times 10^{-4}$ is pumped at the rate of 2880 m³/day. Determine the drawdown distribution around the pumping well. Find out the radius of influence after 1 day pumping. When four wells located 10 m away also operate simultaneously what will be the additional drawdown.

**Solution**

Here $t = 1$ day, $T = 1550$ m²/day, $S = 4.75 \times 10^{-4}$, $Q = 2880$ m³/day.

Take $r = 5$ m.

$$u = \left( \frac{r_i^2 S}{4Tt} \right) = \frac{(5^2 \times 4.75 \times 10^{-4})}{(4 \times 1550 \times 1)} = 1.912 \times 10^{-6}$$

from $u$ vs. $W(u)$ table we get $W(u) = 21.8$.

Substituting in Theis equation $s = \dfrac{QW(u)}{4\pi T} = \dfrac{2880 \times 21.8}{4 \times \pi \times 1550} = 3.22$ m

Taking various distances, the values of drawdowns are calculated and plotted in a semi log graph. It plots a straight line as shown in Fig. 5.28. The plotted drawdown curve is extended to obtain zero drawdown at a distance of 3000 m from the pumping well. This is the radius of influence of the well. From the graph drawdown at 10 m from the pumping well is read as 1.6 m.

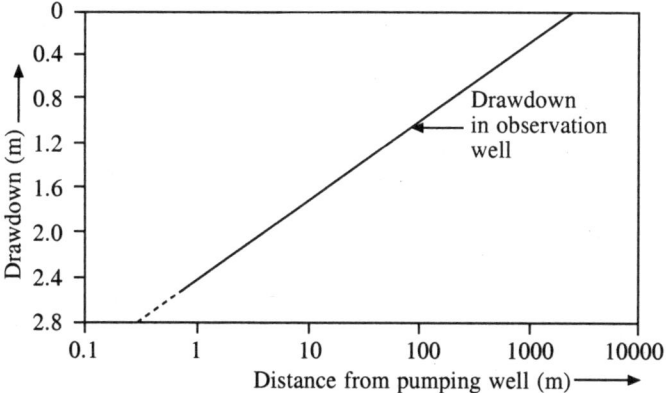

**Fig. 5.28   Distance–drawdown plot for Example 5.20**

Due to 4 wells pumping water simultaneously located at 10 m away from each other, the total drawdown will be $4 \times 1.60 = 6.4$ m.

## 5.11   WELL LOSS

Pumping from a well forces water to rush to the pump intake. Turbulent flow is established at the slots of the well strainer and in the well. As seen in Fig. 5.29 the total drawdown in the well is the sum of the drawdowns of the porous media of the aquifer ($s_w$) plus the loss through well screen and casing ($s_{wl}$). In other words, total drawdown $s_t$ is given as

$$s_t = \text{Aquifer loss} + \text{Well loss} = s_w + s_{wl} \tag{5.86}$$

For confined aquifer in steady state, drawdown of the porous media of the aquifer is

$$s_w = \frac{Q \ln (R/r_w)}{4\pi T}$$

$\Rightarrow$ $\qquad\qquad\qquad s_w \propto Q$  or  $s_w = AQ$ $\qquad\qquad\qquad$ (5.87)

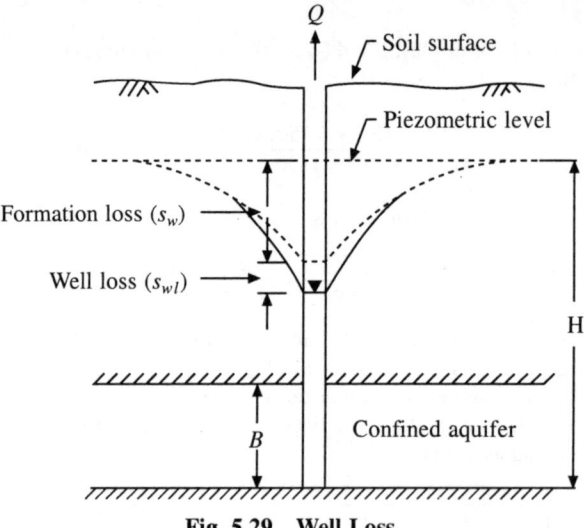

**Fig. 5.29    Well Loss**

Similarly loss through the screen and casing is found to be proportional to square of discharge or $s_{wl} \propto Q^2$ or

$$s_{wl} = BQ^2 \tag{5.88}$$

where A and B are the constants of proportionality. Combining equations (5.87) and (5.88), the total loss is obtained as

$$s_t = A \cdot Q + B \cdot Q^2 \tag{5.89}$$

$AQ$ is generally called the *formation loss* and $BQ^2$ the *well loss*.

A plot of $\frac{s_t}{Q}$ against $Q$ (in abscissa) is a straight line, slope of which gives the coefficient $B$ and the intercept on $\frac{s_t}{Q}$ (ordinate axis) gives the coefficient $A$.

Performance of a well is measured by *specific capacity*, which is defined as the discharge per unit draw down. Specific capacity is usually expressed in lpm/ m depends on the values of transmissibility, drawdown, discharge, time of pumping, well diameter and other parameters. A high specific capacity means the well is efficient. Specific capacity is also given by

$$\frac{Q}{s_t} = \cfrac{1}{\left(\cfrac{1}{4\pi T}\right) \ln \left\{\cfrac{2.25 Tt}{r_w^2 S}\right\} + BQ} \tag{5.90}$$

Abnormally high value of $BQ^2$ (well loss) indicates clogging of the well strainers.

### 5.12   WELL ADJACENT TO A STREAM
When a well is located very near to a stream, the fundamental assumption of the medium being of infinite aerial extent does not hold good. The stream forms a barrier or boundary to the cone of depression. Boundary effect can be assessed

by considering an *image well* located at a distance from the boundary equal to the distance of the real well but on the opposite side. Pumping or recharging into the image well is considered at the same rate as the original well.

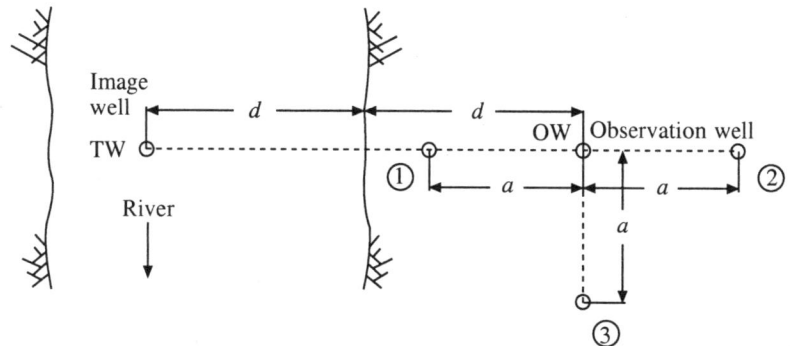

**Fig. 5.30 Wells Located Close to a River**

In Fig. 5.30, OW is the original well located at a distance $d$ from the stream. IW is the image well. Applying Jacob's flow equation for unsteady state for a confined non-leaky aquifer with no recharge from any other source, the drawdown at any instant $t$ is

$$s = \frac{2.3\,Q}{4\pi T}\log\left(\frac{2.25\,T \cdot t}{r^2 \cdot S}\right) = \frac{2.3\,Q}{4\pi T}\log\left(\frac{C}{r^2}\right) \tag{5.91}$$

where $C = \frac{2.25\,T \cdot t}{S}$. Drawdown at the well due to pumping and image well is

$$s_w = \left(\frac{2.3\,Q}{4\pi T}\right)\left\{\log\left(\frac{C}{r_w^2}\right) - \log\left(\frac{C}{4d^2}\right)\right\} = \left(\frac{2.3\,Q}{4\pi T}\right)\left\{\log\left(\frac{2d}{r_w}\right)^2\right\}$$

or
$$s_w = \left(\frac{2.3\,Q}{4\pi T}\right)\left\{\log\left(\frac{2d}{r_w}\right)\right\} \tag{5.92}$$

If flow is steady then the component due to time $t$ is omitted and the drawdown $s$ (instead of $s_w$) becomes

$$s = \frac{2.3\,Q}{4\pi T}\log\left(\frac{R}{r_w}\right) \tag{5.93}$$

where $R = 2d$ and $d$ the distance of the observation well from the river bank.

For observation wells located at (1), (2) and (3) positions (Fig. 5.30), the drawdowns are computed by extending equation (5.93) as

$$s_1 = \frac{2.3\,Q}{4\pi T}\left\{\log\left(\frac{2d-a}{a}\right)\right\}, \quad [R = 2d - a] \tag{5.94}$$

$$s_2 = \frac{2.3\,Q}{4\pi T}\left\{\log\left(\frac{2d+a}{a}\right)\right\}, \quad [R = 2d + a] \tag{5.95}$$

$$s_3 = \frac{2.3\,Q}{4\pi T}\left\{\log\left(\frac{\sqrt{4d^2 + a^2}}{a}\right)\right\}, \quad [R = \sqrt{4d^2 + a^2}] \qquad (5.96)$$

Drawdown of the observation well at any location around the pumping well can be calculated from the above approach.

---

**Example 5.21:** A 20 cm well located 100 m from a river is pumped at the rate of 1.2 m³/ min. The transmissibility of the aquifer is 0.01 m²/sec. Calculate the drawdown in the pumping well and also at the observation wells located 50 m away from the pumping well at the locations marked in the Fig. 5.30.

**Solution**

Here                         discharge $Q = 1.2$ m³/min = 0.02 m³/sec

Distance of the well from the river $d = 100$ m

Radius of the well $r = 0.1$ m

At the pumping well only the drawdown can be calculated from equation (5.93)

$$s_w = \left(\frac{2.3Q}{4\pi T}\right)\left\{\log\left(\frac{2d}{r_w}\right)\right\} = \left[\frac{2.3 \times 0.02}{2 \times \pi \times 0.01}\right]\left\{\log\left(\frac{2 \times 100}{0.1}\right)\right\} = 2.42 \text{ m}$$

At the observation well located 50 m away from the pumping well in the line towards the river at location (1), drawdown is

$$s_1 = \left(\frac{2.3Q}{4\pi T}\right)\left\{\log\left(\frac{2d - a}{a}\right)\right\} = \left[\frac{2.3 \times 0.02}{2 \times \pi \times 0.01}\right]\left\{\log\left(\frac{2 \times 100 - 50}{50}\right)\right\} = 0.35 \text{ m}$$

At the observation well 50 m away from pumping well opposite from the river at location (2) shown in Fig. 5.30, the drawdown is

$$s_2 = \left(\frac{2.3Q}{4\pi T}\right)\left\{\log\left(\frac{2d + a}{a}\right)\right\} = \left[\frac{2.3 \times 0.02}{2 \times \pi \times 0.01}\right]\left\{\log\left(\frac{2 \times 100 + 50}{50}\right)\right\} = 0.51 \text{ m}$$

At the observation well 50 m away from pumping well in the line parallel to the river at location 3 (Fig. 5.30), the drawdown is

$$s_3 = \left(\frac{2.3Q}{4\pi T}\right)\left\{\log\left(\frac{\sqrt{4d^2 + a^2}}{a}\right)\right\}$$

$$= \left[\frac{2.3 \times 0.02}{2\pi \times 0.01}\right]\left\{\log\left(\frac{\sqrt{4 \times 100^2 + 50^2}}{50}\right)\right\} = 0.45 \text{ m}$$

---

## 5.13   SEA WATER INTRUSION

Pumping of fresh water from a place close to sea causes serious water quality problems as sea water tries to occupy the space vacated by fresh water. Since fresh water is lighter than sea-water the problem becomes more aggravated. In Fig. 5.31, ground water table meets the sea water level at the surface level of sea. A plane of interface between fresh water and sea water is maintained. Based on the principle of equal pressure at the interface surface, Ghyben and

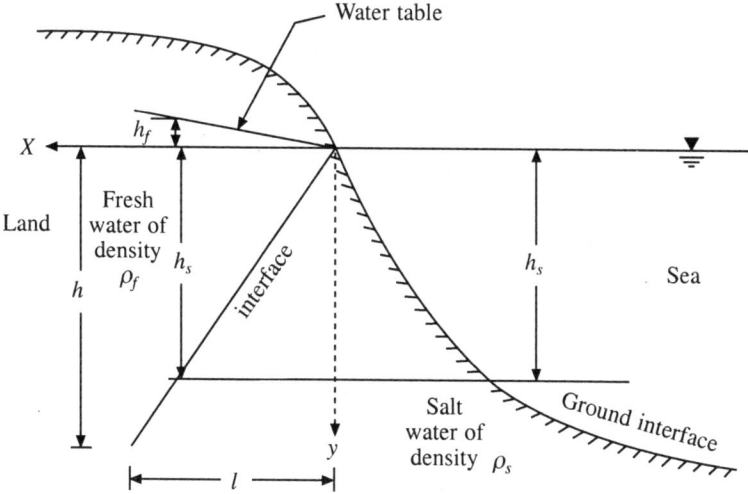

**Fig. 5.31 Ghyben-Herzberg Relation of Fresh-Sea Water Division**

Herzberg found that the interface will occur at a depth $h_s$ below the mean sea level as

$$\rho_s g h_s = \rho_f g (h_s + h_f)$$

or
$$h_s = \frac{\rho_f \cdot h_f}{(\rho_s - \rho_f)} \qquad (5.97)$$

where $h_s$ is the depth of interface below mean sea level, $h_f$ the elevation of ground water surface above mean sea level, $\rho_f$ the density of fresh water, $\rho_s$ density of sea water. Taking specific gravity of fresh water as 1.000 and that of sea water as 1.025 we get

$$h_s = 40\, h_f \qquad (5.98)$$

This means that for every 1 m rise of fresh ground water table, there is 40 m fall of underlying sea water level. As the fresh ground water table increases and approaches the earth's surface by 1 m, the sea water level is pushed down by 40 m and vice-versa.

The interface is usually parabolic. Theoretically, it represents a stream line. Flow across it should not occur. However a narrow mixing zone between the fresh water and sea water interface takes place which is due to seasonal variation of water table and tides. Shape of the parabolic interface is given by

$$y = \sqrt{\left(\frac{80\,qx}{K}\right) + 880 \left(\frac{q}{K}\right)^2} \qquad (5.99)$$

where $x$ and $y$ are the coordinates originating from the point where sea level touches the ground surface, $x$ is positive away from sea and $y$ is positive down ward, Fig. 5.30, $q$ the seaward fresh water flow per unit width of ocean, $K$ the aquifer permeability. Total length of intrusion $l$ for an aquifer of height $h$ is given by

$$l = \frac{0.0125 Kh^2}{q} \quad \text{for } l > h \tag{5.100}$$

When sea water intrusion takes place, recharge wells are dug near the sea shore and recharging is done by fresh water so that sea water is pushed away. In Southern California, USA, such an experiment was carried out successfully.

## 5.14 METHODS OF GROUND WATER INVESTIGATION

The basic objective of ground water investigation is to locate suitable aquifers capable of yielding necessary quality and quantity of water at reasonable economy. Quality, quantity and cost are relative terms which depend on the purpose for which water to be supplied is intended. Ground water investigations are carried out by various methods. The important ones are discussed here. Figure 5.32 gives a broad picture of the various methods of ground water investigation.

### 5.14.1 Hydro-Geologic Investigation

In this investigation, hydrologic aspects of geology are investigated thoroughly. The method primarily relates the occurrence, storage and movement of ground water within a water bearing formation. Information gathered here are mainly on (a) physical and chemical properties of water bearing materials, (b) determination of coefficient of permeability and storage by conducting pumping test in the existing wells and (c) water table fluctuations.

*Remote Sensing*

Remote sensing has made rapid strides in the past two decades as an effective means of collecting information on earth without any physical contact. Satellite imageries taken from polar orbiting satellites or the aerial photographs taken from aeroplanes are interpreted, depending on the varying tonal expressions or signatures on the photos or imageries of the objects formed due to differences in their reflectance properties.

In remote sensing, a Multispectral Scanner System (MSS) records the scene in four bands of electromagnetic spectrum. Band number 4 is visible green, 5 is visible red, 6 and 7 are invisible reflected infra-red. MSS records the imagery in digital form which is received at the ground stations and converted back into pictures. It can be preserved and then processed in computer tapes. A number of features can be identified by such a system.

| | | |
|---|---|---|
| Moist grounds, water logged areas, springs | – | Dark tone |
| Saline soil | – | Light tone |
| Lack of vegetative cover, salt intrusion zones | – | White patches |

Other objects which can be identified from such imageries are drainage pattern, drainage density, vegetation, soil types, land slope, influent and effluent streams and irrigated areas.

Geological features like valley falls, abandoned rivers, back swamps, outcrop areas, weathered zones, joints, fractures and faults can also be easily identified. Advantages of this survey are (a) speed of operation, (b) possibility of survey for

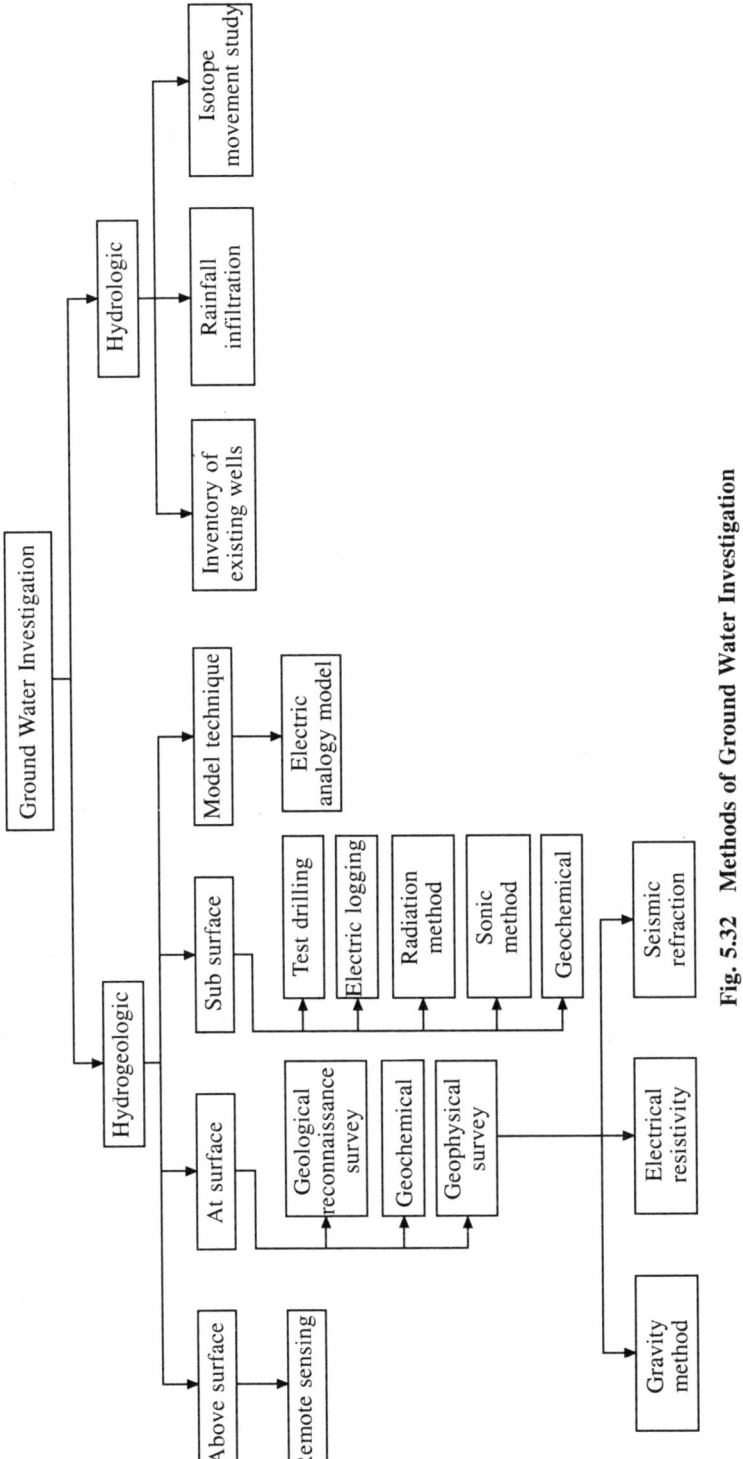

**Fig. 5.32 Methods of Ground Water Investigation**

inaccessible areas, (c) directly mapping the features, (d) checking the changes by repetitive mapping and (e) considerable reduction of the ground work.

For complete information, ground check is essential. Since the coverage is very wide, the resolution of the scanner imagery from satellites are usually poor. Therefore, remote sensing can be taken as an aiding tool in the quest of ground water survey.

### 5.14.2   At Surface

#### 1. Surface Geological Reconnaissance Survey
Before beginning any field investigation work, all information on geology of the area are required to be collected, compiled, analysed and interpreted in terms of water bearing properties. Existing geological maps, aerial photographs and other pertinent geological informations help to limit the areas to be further explored. By this the applicable method of exploration can be decided. Field reconnaissance help to cross check and supplement the information gathered. For this hydrologists utilise the knowledge on petrography, stratigraphy, structural geology, geomorphology and other geological specializations. Stratigraphy helps to locate the position and thickness of water bearing formations and continuity of confining beds whereas structural geology helps to locate water bearing formations which have been displaced by earths movements.

#### 2. Surface Geophysical Methods
Surface geophysical methods for prospecting ground water include (a) electrical resistivity method, (b) seismic method and (c) gravity surveys.

#### (a) Electrical Resistivity Method
This method is widely used for the determination of (1) position of water table, (2) thickness and depth of various formations and (3) zones of saline water and fresh water; and is simple, cheap and easy to interpret. The scheme consists of four equally spaced electrodes in a straight line (Fig. 5.33). Electric current is applied to the outside two electrodes $C_1$ and $C_2$. Voltage drop between the two inside electrodes $P_1$ and $P_2$ is measured. A series of measurements may be taken using different distances between electrodes, thus yielding series of values of apparent resistivity with depth.

Again a series of measurements at constant electrode spacing may be taken which give lateral geological changes at constant depth. It is assumed that the horizontal distance between electrodes is roughly equal to the depth of penetration. Various types of earth materials generally exhibit ranges of resistivity values. The fundamental factors that primarily affect are the water saturation, water quality and geological features.

#### (b) Seismic Refraction Method
The principle of this method is based on the earth's seismic waves or elastic shock wave propagation on earth materials. Denser is the material, faster is the travel of the wave. Depending on geology, portion of the wave is reflected back

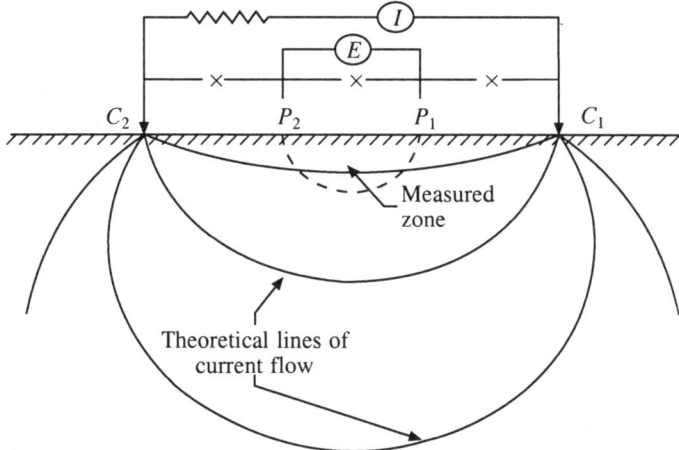

**Fig. 5.33  Four Electrode Electrical Resistivity Method**

at the structural or lithologic interface and returns back to land surface. These reflected waves generate earth motion causing seismometers to move, which is recorded in an oscillograph. Knowledge of wave velocities in different strata helps to interpret the geological formations at different depths. Waves while passing through earth materials of different densities, also get refracted and return back to the surface, which are picked up by a detector and is recorded by an oscillograph. A seismic travel time curve is produced from different arrival time of the waves at the seismometers from which depths and thickness of layers can be computed. This method gives results more accurately (~ 95%), if sound knowledge on the wave movement at different layers, proper interpretation and recording is made.

*(c) Gravity Method*
Gravitational field at various locations on the surface of the earth are measured with respect to an assumed datum. The variation of force of gravity is mainly due to rock densities. From this information, the location of water bearing zones can be delineated.

### 3. Surface Geochemical Methods
Chemical composition of ground water reflects composition of the aquifer material. Temperature of the water also indicates the depth of available ground water and the source. Hot water springs give an idea of the sulphur contamination of water drawn from great depths. Ground water from limestone formations contain good quantity of calcium, magnesium, carbonate and sulphate compounds. Therefore, geochemical methods of investigation may be used in conjuction with the surface-geologic methods to give more comprehensive knowledge of the geological formation of the area bearing ground water.

### 5.14.3  Sub-surface Geophysical Method
This gives more accurate result than the surface method and should be followed

when quality information is required. Various subsurface geophysical methods involve:

(a) Test drilling
(b) Electrical logging
(c) Radiation
(d) Sonic
(e) Geochemical

**(a) Test-drill Method:** It is the surest method of determining aquifer formations exceeding 100 m deep. A log of the rock samples is obtained while drilling is in progress. Over an area when good number of such drills are undertaken, a geological formation map of the area can be prepared which helps to reveal the character, depth and thickness of various strata, water levels and water quality. These test holes can be used as observation wells for recording water levels while conducting pumping tests.

**(b) Electrical Logging Method:** An electric logger comprising a recorder, a hoist unit, logging tools, electric generator, a mud resistivity measuring kit along with its accessories are mounted on a mobile van. The recorder has the components of oscillator, amplifier, servo motors, control panel, recorder channel, electronic potential meter, depth recorder, recording pens, chart paper and other materials. An electric logger may have single electrode or multielectrode logging system. At the bore hole, recording instrument is lowered and recordings on (i) resistivity of the aquifer zone and (ii) spontaneous potential logs are obtained by the instrument. Resistivity logs yield information on lithology, sequence and log formation boundaries, location of fresh and salt water bodies. Spontaneous potential log indicates permeable zones in terms of permeability. Information from both resistivity and potential logs give most of the data a hydrologist needs for any aquifer investigation.

**(c) Radiation Method:** Radiation loggers are essentially of two types (i) Gamma-ray and (ii) Neutron logger. The logger is lowered into a core hole. A Gamma-ray Sonde records the emission of gamma rays from rocks. The shale formations containing radioactive salts can be easily distinguished. In Neutron logger, neutrons are released from a source with high velocities into the formation and the numbers returning to the detector is recorded. More hydrogen concentration in the aquifer gives less counting of the neutrons. Porosity of the aquifer can be accurately measured by such a system. The loggers can be used in empty holes or dry wells resulting in better aquifer information of the area.

**(d) Sonic Method:** In this method the logger consists of a sonic sonde which emits sound waves between 2000–7000 m/sec. A single or double receiver system positioned at fixed distance in the sonde receives the sound waves. The sonic sonde is gradually lowered into a bore hole and the travel time of waves in different formations are recorded against the depth as it moves. This method gives reliable estimate of porosity of the formation. It is used in conjuction with other types of loggers discussed before to give the most acceptable result.

**(e) Geochemical Method:** Information gathered from various subsurface loggers are closely associated with geochemical methods as the chemical composition of the rock formations and the properties of ground water govern the resistivity, gamma ray radiation or neutron counting of the respective loggers. Many a times temperature logs are used along with other down-the-hole loggers, the information from which becomes very useful in the identification of water from different aquifers.

### 5.14.4   Model Study

On the basis of simiiarity between the Ohm's law governing the flow of current and the Darcy's law of ground water velocity, electric analogy models can be developed to simulate the ground water flow and recharge. The current I is taken to correspond the velocity $V$ of the ground water flow and the voltage difference E corresponds to the permeability $K$ of the prototype. A good knowledge on the aquifer properties help to build a prototype electric model of a basin. The model once validated for a particular area can give quick and reliable response of ground water position.

### 5.14.5   Hydrologic Investigation

Hydrologic information on: (a) depth of water table, (b) location of springs, (c) vegetation type, (d) base flow in rivers and (e) water available from precipitation and other sources recharging ground water provide valuable data on ground water studies. Investigation on: (i) information on existing wells, (ii) rainfall and infiltration relationship and (iii) tracer method are covered extensively under this study.

**(i) Information on Existing Wells:** On the area map, all the existing wells are located and the information on (i) strata log, (ii) yield, (iii) drawdown, (iv) depth of water table, (v) quality of water and (vi) amount of water withdrawn from ground water reservoir during various periods of the year are noted. For each information, contour maps are prepared which helps to guide a hydrologist to quantify ground water at various depths. Contours maps on porosity, storage coefficient and transmissibility of the area help to quantify the amount of ground water that can be exploited from a given area.

**(ii) Rainfall and Infiltration Relation:** Relation between rainfall and infiltration is very useful to access the annual ground water recharge of a basin. In India, rainfall period is limited to five months (June-October) and during the rest of the year there is no significant rainfall contributing to ground water recharge. Rainfall being a major source to ground water recharge, the percentage of this resource available to ground water potential at different locations helps to quantify ground water potential of a basin depending on the soil type and geological formations of the area.

**(iii) Tracer Method (Isotope Movement):** Radio isotopes like Tritium are used to investigate various flow and aquifer properties. A radio tracer is injected at the upstream flow point, i.e., higher level of water table and the time taken by the tracer to reach the location of the other measurement point is recorded. This gives information on the velocity of ground water flow, coefficient of permeability, flow rate and other aquifer parameters.

## PROBLEMS

5.1    Derive an expression for unsteady flow in unconfined aquifer clearly stating the assumptions made there in.

5.2    Explain the following terms:
(a) Porosity, (b) Specific yield, (c) Zone of aeration, (d) Perched aquifer, (e) Aquiclude, (f) Aquitard, (g) Transmissibility and (h) Leaky aquifer.

5.3    Briefly explain the terms:
(i) Radius of influence, (b) Cone of depression, (c) Partially penetrating well and (d) Drawdown.

5.4    Write notes on the following:
(i) Sea water intrusion, (b) Well loss and (c) Well interference.

5.5    Classify various methods of ground water investigation. Explain briefly the method as available under subsurface investigation.

5.6    Derive equilibrium flow equations for (a) Confined aquifer and (b) Unconfined aquifer.

5.7    During pumping of water from a 15 cm diameter well the following information are recorded.
Thickness of confined aquifer = 12 m.
Rate of pumping = 1000 lpm.
Drawdown at the well after 10 h of pumping = 2.00 m.
Drawdown at the well after 20 h of pumping = 3.00 m.
Determine the transmissibility, storage coefficient and coefficient of permeability of the aquifer.

5.8    From a fully penetrating well in a confined aquifer, water was pumped at the rate of 0.1 $m^3$/sec. The storage coefficient of the aquifer is $12 \times 10^{-4}$ and transmissivity 100 $m^2$/h. An observation well is located 20 m from the well. Find the drawdown at the observation well after 0.5, 5, 10 and 20 h of pumping.

5.9    A horizontal infiltration gallery 100 m long has water at an elevation of 373.0 m. The water table is at RL 379.0 m located 250 m away from the face of the gallery. If the permeability of the aquifer is 2 m/h, calculate the flow into the gallery.

5.10   How can you prepare the water table contours at a place? What are its uses?

5.11   During a month the average ground water table fluctuation is 12 m. If the area of this fluctuation covers 120.0 sq. km, calculate how many pumps discharging at the . rate of 0.5 $m^3$/sec are used for irrigation? Assume the working of the pump per day as 12 h and take the specific yield as 16%.

5.12   An undisturbed soil is oven dried. The sample weighs 614.5 g. Kerosene is added to it to saturate the sample and the saturation weight is 718.1 g. The saturated sample is lowered in a pot full with kerosene and it is found that 255 g of kerosene is displaced. Estimate porosity of the soil sample assuming no chemical reaction between the soil and kerosene.

5.13   Determine the storage coefficient if porosity is 25%, thickness of aquifer is 22 m, bulk modulus of water is 2.1 GN/ $m^2$ and that of soil skeleton is $2.9 \times 10^8$ N/$m^2$.

5.14   A river and a canal 1000 m apart receive 24 mm of rainfall per hour. Depth of water in river and canal taken from a common horizontal impervious strata is 12 m and

8.0 m. Compute the discharge/meter length if 50% of rain water infiltrates into the ground. Take permeability as 10 m/day. What will be the seepage during rain less period?

5.15 Two observation wells 15 and 30 m away from a pumping well of 20 cm diameter record drawdowns of 3.2 and 2.2 m respectively. If the well penetrates to a full static water table of 40 m, determine the transmissibility of the aquifer and the drawdown at the pumping well.

5.16 A 20 cm well discharges 200 lpm under a constant drawdown of 2.0 m. If the aquifer thickness is 30 m, calculate the yield from 30 and 40 cm wells from the same areas. Radius of influence can be taken as 300 m.

5.17 An artesian aquifer has a 15 cm diameter well. The strainer length of the well is 15 m. Assume the coefficient of permeability as 30 m/day and the radius of influence of the well as 300 m. What is the yield from the well if the drawdown is 2.5 m?

5.18 Describe the image well theory.

5.19 A 30 cm well is provided with a screen of 15 m out of 30 m thick water table aquifer. The well yields 2500 lpm with drawdown of 2.2 m. If the screen length is increased to 25 m, what will be the drawdown at the well?

5.20 An observation well 30 m away from the pumping well records a drawdown of 1.8 m after 20 min of pumping. Find the time when another well 50 m away will record the same drawdown.

5.22 A confined aquifer is composed of two distinct layers. The depths and coefficient of permeability are as follows:

| Layer | Depth | Coefficient of permeability |
|-------|-------|-----------------------------|
| A | 6.0 | 0.8 m/h |
| B | 8.0 | 1.2 m/h |

Calculate the equivalent permeability and transmissibility of the aquifer.

5.23 A river and canal run parallel to each other and are 2.0 km apart. Suddenly an industry located on the bank of the canal discharges its effluent in the water. The water carrying systems carry water at elevations of 79.9 and 75.2 m from an arbitrary datum. The elevation of the horizontal hard strata below is 65.0 m. If the permeability of the aquifer is 0.60 m/h and porosity is 30%, calculate the time taken by the pollution to reach the canal. Assume canal to be at higher elevation.

5.24 A well discharges at the rate of 2000 lpm. Calculate the transmissibility of the aquifer from the time-drawdown relation at an observation well 60.0 m away.

| Time (min) | 2 | 5 | 10 | 20 | 50 | 100 | 200 |
|------------|------|-----|-----|-----|-----|-----|------|
| Drawdown (m) | 0.32 | 1.2 | 2.7 | 4.4 | 6.0 | 7.5 | 10.5 |

3.25 Drawdown at a well of 20 cm diameter is to be limited to 4.50 m at 12 h of pumping. If transmissibility is 20 m²/h and storage coefficient is 0.004, calculate the discharge.

# Chapter Six

# Stream Flow

## 6.1 INTRODUCTION

A stream flow is that part of precipitation which appears in a stream as surface runoff and represents the total response of a basin undergoing frequent interactions between its various processes and storages. Precipitation falling on earth's surface may infiltrate into the ground or flow down on the surface to join the network of natural channels to finally drain out to large water bodies. Part of the water infiltrating into the ground may also join the stream (interflow) at a later stage as subsurface runoff and the rest may join the ground water table. Availability of ground water flow at the springs may also vary due to geological, topographical, natural and man made activities. Contribution from springs and the return flow from irrigation may join the stream forming the total runoff of a basin.

Stream flow constitutes a major phase in the hydrologic cycle. Some related terms are as follows:

## 6.2 TERMS USED

**Basin:** It is the area bounded by the highest contour called ridge line from where precipitated water is collected by surface and subsurface flows and drained out through the natural river. The ridge line forms a natural barrier dividing one basin from another. Figure 6.1 shows a watershed with land contours and streams of different orders. The term *catchment* or *watershed* or *drainage basin* is considered more or less synonymous to a basin. Watershed discharge $Q$ can be related to drainage area $A$ as

$$Q = xA^y \qquad (6.1)$$

where $x$ and $y$ are the parameters. Depending on the values of $x$ and $y$, $Q$ can be the peak flood, a minimum or mean flow from a catchment.

**Stream:** It is the natural flow channel in which water from a basin is collected and drained out to large water bodies.

**Overland flow:** Rain water which flows over the land surface in the form of a sheet of water to join the nearest stream is called overland flow. Thus, the surface runoff is considered as overland flow, so long as it does not join the nearest stream. The length of overland flow $L_{of}$ can approximately be calculated as

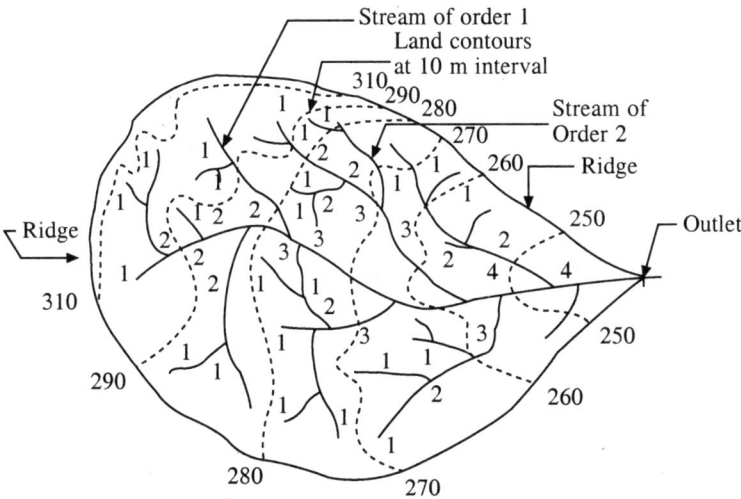

**Fig. 6.1   Watershed with Land Contours and Streams of Different Orders**

$$L_{of} = \frac{0.50}{D_d} \qquad (6.2)$$

where $D_d$ the drainage density.

**Surface Runoff:** Part of the precipitation and other drainage water of a basin which moves over the natural land surface and then through a network of channels of gradually larger sections, thus draining out water from the basin is called surface runoff.

**Subsurface Runoff:** Also known as *interflow* or *subsurface flow* is that part of precipitation which infiltrates into the ground and moves laterally or horizontally in the soil and meets the nearest stream before meeting the ground water table. Time taken for subsurface flow to reach a stream is always more than that of overland flow.

**Groundwater Flow:** Part of the water available in a stream from the depletion of groundwater or piezometric head is known as groundwater flow or groundwater runoff. Groundwater table nearly follows the profile of natural topography. Soon after rainy season in a mountainous regions, the fall of natural ground surface is more steeper than the fall of groundwater table. This gives rise to springs. Places where the ground-water table is higher than the river water level the stream is called effluent (Fig. 6.2a). The stream receives water from groundwater reservoir.

**Stream Flow:** The total runoff consisting of surface flow, subsurface flow, ground water or *base flow* and the precipitation falling directly on the stream is the stream flow or the total runoff of a basin. The division of flows are arbitrary. One type of flow at a place may change to another at a different location in the same stream. For example, surface flow may infiltrate to ground and becomes subsurface flow and may move underground to join either to a stream or to the ground water reservoir.

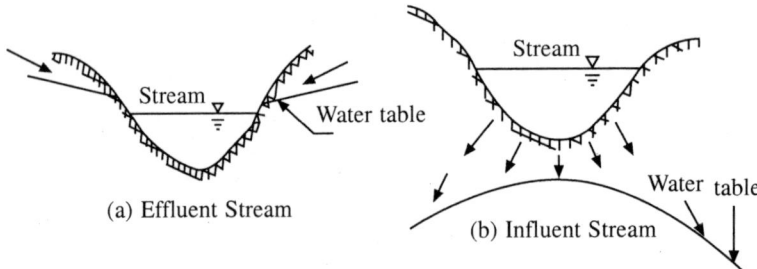

**Fig. 6.2   Effluent and Influent Streams**

**Influent Stream:** When the position of groundwater table is lower than the water level of a stream (Fig. 6.2b), such that water from the stream contributes to the groundwater storage, that stream is called influent. In early part of rainy season, this is the normal case for all rivers in India.

**Direct Runoff:** The direct or *storm runoff* is that part of stream flow occurring promptly as precipitation starts and contributes for an acceptable period after the storm ceases. Contribution from subsurface flow is considered nearly constant during the period.

**Base Flow:** It is that part of stream flow available mainly from ground water reservoir and delayed subsurface flow appearing during dry period (Fig. 6.3). For practical purposes the total flow in a stream is divided into two types viz. *direct runoff* and *base flow*, the distinction being mainly on time of arrival of flow in the stream.

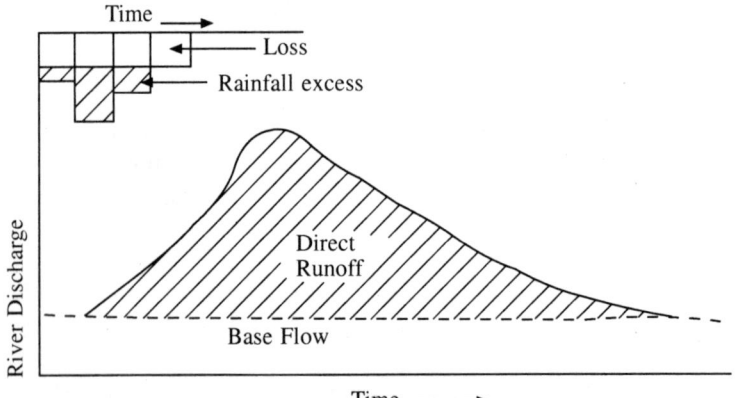

**Fig. 6.3   Catchment Rainfall and Runoff Process**

**Rainfall Excess:** The part or percentage of precipitation which is equal to the volume of direct runoff from a basin is called rainfall excess. Therefore, for a hydrologist dealing with water resources planning and management, all other terms, viz. interception, evapotranspiration, infiltration, depression storage and watershed leakages are called *losses*. Sometimes the term *effective rainfall* is used which is calculated from the volume of direct runoff and subsurface runoff. Effective rainfall is greater than rainfall excess by the quantity of subsurface flow, yet for all practical purposes they are considered the same.

**Channel Storage:** At any instant, the water content of a stream within its defined cross section is known as channel storage.

**Hydrologic Year:** The period of one year starting with the time when the ground and surface water storage of a basin is usually the minimum is called hydrologic or *water year*. The starting time varies from country to country. In India it is taken as the time just before the onset of the monsoon to the same period of next year (1st June to 31st May).

## 6.3 FACTORS AFFECTING RUNOFF

Factors influencing runoff of a basin from a storm are broadly classified into four major groups; viz. (i) climatic factors, (ii) basin characteristics, (iii) basin geology and (iv) basin infiltration characteristics.

### (i) Climatic Factors

The major climatic factors affecting runoff from rainfall are the precipitation characteristics which include (a) intensity, (b) duration, (c) aerial distribution, (d) direction of storm movement, (e) forms of precipitation and (f) evapotranspiration.

When precipitation is in the form of *snow*, the runoff from the basin is very less. Rise in temperature melts the snow and generates runoff from the basin at a later date.

*Intensity* of the storm is directly proportional to the magnitude of the flow appearing in the stream. A higher intense rainfall causes immediate rise in discharge of the stream and is found to be more pronounced in small catchments.

*Duration* of a storm has direct effect on the volume of runoff from a basin. Rainfall of given intensity occurring for longer duration gives rise to more runoff from a catchment. When the intensity is constant and the storm duration is longer, the peak discharge in the stream will rise gradually to a maximum and will continue at the maximum till the storm lasts.

*Aerial distribution* of the storm also affects the runoff of a stream both in magnitude and temporal distribution. Maximum runoff occurs when the entire area of a basin contributes to the runoff.

When a storm moves from upstream to downstream of a catchment at a rate equal to the movement of water in the channel then the effect will be a maximum flood peak at the outlet. Discharge in a stream is lower and can continue for a long time when the storm moves from the basin outlet to the upstream.

*Evapotranspiration* rate is inversely proportional to the runoff from a basin.

*(ii) Basin Characteristics*
The features affecting runoff from a storm of a basin are (a) size, (b) shape, (c) slope, (d) drainage density, (e) topography and (f) geology.

Total runoff from a catchment expressed in depth is independent of the area. Therefore, total volume of precipitation available at a river section is directly proportional to the catchment area and the peak flow can be approximated to the square root of drainage area. The peak flow per unit area decreases as area increases and the period of surface runoff increases with area.

The *shape* of the catchment affects the runoff peak in the following way. An elongated catchment has lesser peak than a fan-shaped catchment of same area (Fig. 6.4) and has long runoff durations than a fan-shaped catchment. As shown

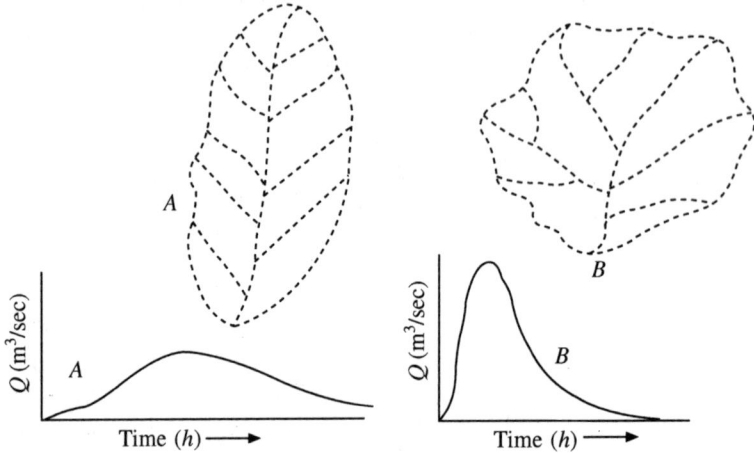

**Fig. 6.4    Effect of Catchment Shape on Runoff**

in Fig. 6.5a, a catchment area with a carrot-shape has peak flow occurring earlier than the catchment of type shown in Fig. 6.5b. This is because a larger catchment area in the latter case is contributing at the basin outlet. Depending on the shape, sometimes a catchment may have a multi-peak runoff pattern (Fig. 6.5c), even though all the three catchments may have the same area and characteristics.

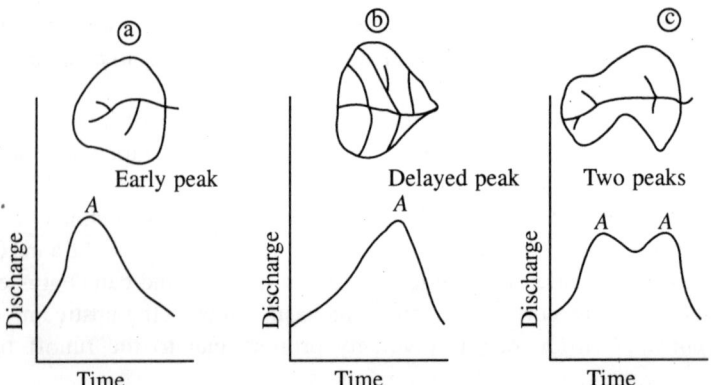

**Fig. 6.5    Affect of Concentration of Basin Area on Flood Peak.**

A catchment having extensive flat area gives rise to low peaks and less runoff whereas a catchment with steep slope produces high peak flood. Rate of *infiltration* from a flat catchment is more which affects the velocity of overland flow. Therefore, the time of arrival of peak at the outlet is late and so is the total time period of runoff for such flat shaped catchment. The *basin slope* plays an important role in urban hydrology where catchment is usually small. For a high intensity and long duration storm, the effect may be less pronounced.

The characteristics of a drainage net may be physically described by the following factors.

**(1) Drainage Density:** Drainage density, defined as the ratio of total length of all streams of the catchment divided by its area, indicates the drainage efficiency of the basin. The higher the value, quicker is the runoff and lesser is the infiltration and other losses. Thus, drainage density

$$D_d = \frac{L_s}{A} \qquad (6.3)$$

where $L_s$ is the total length (m) of all streams in a basin and $A$ the drainage area in sq. km.

Following the concept of stream orders proposed by Horton (1932), the drainage density of a basin can be quantitatively analyzed. $D_d$ is the ratio of total channel length to the total drainage area of a basin. In a topographic map drainage channels of all types are clearly visible. The smallest tributaries to which overland flow drains first (at the periphery of catchment) are designated as streams of *first order*. When two first order streams join, they give rise to a stream of *second order* and so on (Fig. 6.1). When a lower order stream joins a higher order stream, the order remains the same as that of higher order. A number of information about a drainage basin can be inferred by knowing the catchment area and lengths of different stream orders of a basin. If the drainage density is more, the disposal of runoff from a basin is quicker. A low drainage density catchment gives a larger period of surface runoff and more loss of rainfall. Drainage density between 2 and 750 per km have been reported.

**(2) Stream Density:** It is defined as the number of streams of given order per sq. km computed by dividing the total number of streams of the same order of the basin with their catchment area. Stream Density

$$D_s = \frac{N_s}{A} \qquad (6.4)$$

The shape of the basin is expressed by the following two factors.

**(3) Shape Factor ($B_s$):** U.S. Army Corps of Engineers (1954) proposed shape factor $B_s$ as the ratio of square of watershed length $L$ to the watershed area $A$, or

$$B_s = \frac{L^2}{A} \geq 1 \qquad (6.5)$$

For a square watershed $B_s \doteq 1$. If the watershed is long and narrow, then $B_s < 1$.

General *elevation of catchment* is an important feature in terms of its runoff producing capacity. Other things remaining the same (i) evapotranspiration is less from a higher elevation area, (ii) rainfall patterns may change with elevation from the mean sea-level, (iii) a higher elevation area usually has steeper slope than a low elevation area, (iv) higher altitude areas are less pervious and therefore infiltration is less in mountainous regions than the low lying alluvial zones and (v) ground water storage capacity is less for plateaus.

**(4) Channel Slope:** Channel slope affects the velocity and flow carrying capacity at any given location at its course or at the basin outlet. This factor is incorporated in Manning's and Chezy's equation. The bed slope is approximated to the energy slope under uniform flow conditions. However, the usual method to calculate the channel slope is to divide the fall of the channel by its map length. To accomplish this, the channel longitudinal section may be plotted on a graph paper. The height difference between the two points on the channel (plotted as ordinate) divided by the length of the channel (plotted in abscissa) gives the slope of the channel.

$$\text{Slope } S = \frac{\text{Elevation difference between two points of a channel}}{\text{Horizontal length between the points}} = \frac{h}{L} \quad (6.6)$$

**(5) Centroid of the Basin:** It represents the location of the point of weighted centre of a watershed and can be physically located on a cutout cardboard of the watershed map by hanging it at different corners from a overhanging thread. The point at which all the thread lines intersect is the centroid of the watershed. The distance between point nearest to the centroid in the channel and the basin outlet is nearly half the basin length. Snyder used this parameter to synthesize unit hydrograph for an ungauged basin.

**(6) Form Factor ($F_f$):** Horton (1932) expressed it as the ratio of watershed area $A$ to the square of its length $L^2$. It is also defined as the ratio of the width of the basin $W_b$ to its aerial length $L_b$ measured between the stream outlet to the most remote point on the basin. Value of form factor is always less than unity.

$$F_f = \frac{A}{L^2} = \frac{W_b}{L_b} < 1 \quad (6.7)$$

**(7) Compactness Coefficient:** Strahler (1964) defined it as the ratio of perimeter of the basin to the circumference of a circle with area equal to the area of the basin. It is expressed as

$$C_c = \frac{P}{\sqrt{(4\pi A)}} = 0.2821 \frac{P}{A^{0.5}} \geq 1 \quad (6.8)$$

where $A$ is the area of the basin in sq. km and $P$ the perimeter of the basin in km.

**(8) Elongation Ratio ($E_r$):** According to Schuman (1956), it is the ratio of the diameter of a circle of same area of the basin to the maximum length of the basin. $E_r$ values vary from 0.25 to 1.00.

$$E_r = \frac{\text{Diameter of circle of a watershed area}}{\text{Watershed length}} = \left(\frac{A}{0.786}\right)^{0.50} \frac{1}{L} \leq 1 \quad (6.9)$$

**(9) Circularity Ratio ($C_r$):** Miller (1959) proposed circularity ratio as the ratio of the basin area to the area of a circle having the same perimeter as the basin. Its value is always less than unity.

$$C_r = 12.57 \frac{A}{P^2} \leq 1 \quad\quad\quad (6.10)$$

---

**Example 6.1:** Find the drainage density, average overland flow length, form factor and the channel slope for a small watershed with the following data

    Area of the watershed = 122 km$^2$
    Distance between the outlet to the furthermost point = 18.35 km
    Elevation difference between the outlet and furthermost point = 1230 m
    Total length of channels of all orders = 618.8 km

**Solution**

$$\text{Channel Slope } S = \frac{1230 \times 10^{-3}}{18.35} = \frac{1}{14.92}$$

$$\text{Form factor} = \frac{\text{Watershed area}}{(\text{Watershed length})^2} = \frac{122}{(18.35)^2} = 0.362$$

$$\text{Drainage density} = \frac{\text{Total channel length}}{\text{Watershed area}} = \frac{618.8}{122} = 5.072 \text{ km/km}^2$$

$$\text{Average overland flow length } L_o = \frac{1}{2 \times \text{Drainage density}} = \frac{1}{2 \times 5.072} = 0.098 \text{ km}$$

$$= 98 \text{ m}$$

---

*(iii) Basin Geology*

Basin geology is responsible for the rate of infiltration during a storm. If good aquifer material forms the basin then surface runoff will be less due to increased infiltration, but for a basin composed of impervious materials the runoff will be highly peaked. Presence of faults, fissures and cracks in the geological formations results in the diversion of storm water to a new location where they terminate. Such formations may also divert water from a basin to an adjoining one as watershed leakage.

*(iv) Basin Infiltration Characteristics*

The main features are: (i) nature of the surface of the catchment and (ii) surface storage characteristics.

    *Vegetation* forces a good quantity of storm water to infiltrate and is primarily responsible for the reduction of surface runoff than a barren land. Cultivation activities like ploughing and digging increases the infiltration rate substantially. A thick grass cover is very effective in reducing the peak flow from a catchment.

Small depressions to large lakes reduce the amount of runoff from a basin to the extent that they hold water during a storm. Large reservoirs moderate a flood and so they attenuate the flood peak to the desired level.

## 6.4   STAGE MEASUREMENT

Discharge is defined as the volume of water flowing through a channel cross section per unit time. A river is a flow channel which collects water from the entire basin and drains out to large water bodies. It is possible to measure this part of precipitation more accurately. The technique of measurement of discharge belongs to a branch of science called *hydrometry*. For planning and management of any water resource project, accurate measurement of discharge of a basin is the first and the most essential requirement. In a hydrologic cycle, precipitation and river discharge are the only two components which can be measured accurately. Data from rainfall and the corresponding runoff can also be used to describe the rainfall-runoff-loss process adequately for a neighbouring ungauged basin. Therefore, adequate attention must be given to record stream flow at all possible stations of a stream. Discharge is measured in $m^3$/sec (cumec) or $ft^3$/sec (cusec). Measurements are recorded at least once in a day during normal periods and on a hourly basis during floods. It is difficult to measure river discharge continuously or at short intervals of say 1 hour. To overcome the difficulty, a *stage-discharge* (S-D) or *gauge-discharge* (G-D) relation at the site is established by careful observation of the entire range of stages and corresponding discharges. Once this relation, popularly called as *rating curve*, is established, the problem reduces to a simple reading of river stage and then finding the corresponding discharge from G-D graph or from G-D equation established at the site.

*Stage* is defined as the elevation of water surface at a location in a river or stream above a reference datum. In a river section, a number of devices can be used to measure stage.

Stage recorders popularly used at the various river gauging stations are described below.

### 6.4.1   Non-Recording Stage Recorders

*Staff Gauge*
A staff gauge is a scale graduated to meters and centimeters and is rigidly fixed at the river cross section or attached to a permanent bridge abutment, as shown in Fig. 6.6. Graduation of the staff should be clear, distinct and painted permanently. The level of water surface in contact with the gauge is measured by matching the reading of the staff and adding with it the reference datum level, which may be the mean sea level (MSL) or any other arbitrary level. For rivers with large fluctuations in water levels 3–4 numbers of staffs are required to be fixed at one side of the river cross section, and is known as *sectional gauge*. They are placed in position in the river section as shown in Fig. 6.7. Usually the staffs are made

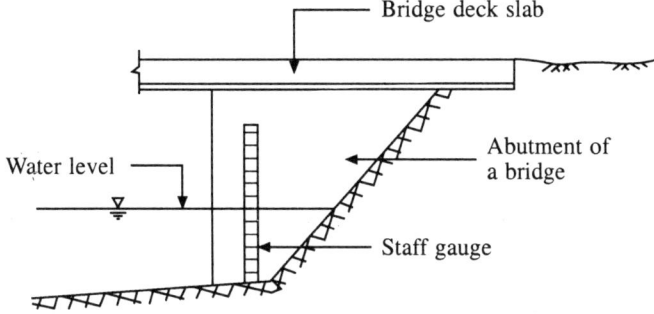

**Fig. 6.6   A Staff Gauge**

vertical and a overlap of at least 50 cm to the next immediate one is necessary. All the staff should refer to the same common datum and be calibrated.

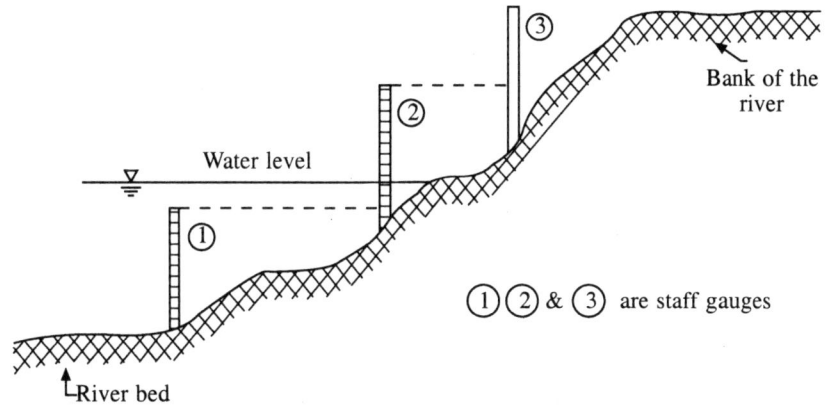

**Fig. 6.7   Sectional Staff Gauge**

The major problem with staff gauges is that they are likely to be affected by debris, boats, animals and water currents in floods. To overcome this difficulty, the following gauges may be used suitably.

*Tape or Wire Gauge*
A tape or wire is lowered from the structures like bridges or ropeways such that the weight attached at the end of the wire just touches the water surface. Measurements are made from the top datum from where the location of the wire-weight gauge is fixed (Fig. 6.8). The length of wire coming out of the drum of the gauge is read by a mechanical counter attached to a reel on which the wire or tape is wound, and subtracted from the reference datum. For better accuracy, an electric device may replace the weight which clicks a sound on contact with water surface. Thus water surface elevation can be determined with precision. Such type of gauges are called *Electric tape gauges*.

There are a number of other gauges like *crest stage indicator*, *hook-gauge* and *pressure gauge* that can be used to record the stage of a river. The crest stage indicator of US Geological Survey, records the highest stage of a river. These gauges have their own merits and limitations. Interested readers may refer to standard literatures for installation and use of such gauges.

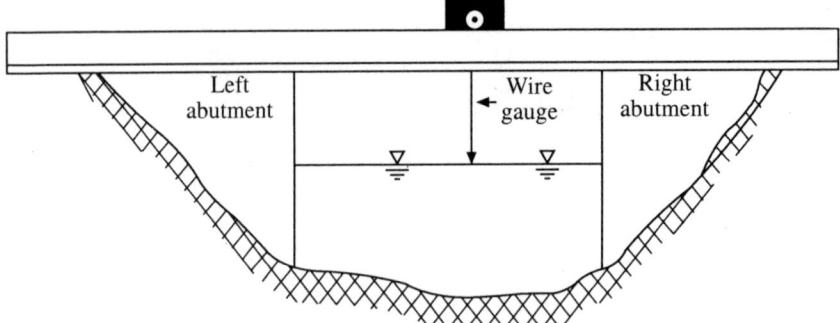

**Fig. 6.8    A Wire Gauge Suspended from Deck Slab of a Bridge**

### 6.4.2    Automatic Gauge Recorders

An automatic gauge records continuous stage of a river over time. There are a number of such gauges used at different situations for recording river gauges automatically. Some of the widely used gauges are discussed here.

**Water Stage Recorder:** A surface float is connected at one end of a wire which passes through a recorder, the other end of the rope is balanced by a suitable counter weight. Fluctuations in water surface level cause the float, rope and finally the wheel of the recorder to move. This causes the pen attached to the drum or the wheel of the recorder to move. The drum and the pen are connected with a rack and pinion arrangement which converts the angular movement of the drum to a linear one. The pen is attached to a clock mounted cylinder wound for 24 h or a week and its recording on the graph paper on the cylinder is continuous and reverses automatically when there is a fall of water level. The width of the graph may not accommodate the total fluctuation of the water level in a river. An arrangement to automatically reverse the pen to zero position is made in the instrument, when the pen reaches the outer edge.

Such an arrangement is protected from the main river by installing a stilling well. An arrangement to draw water from the main river at its lowest water level position is ensured through intake pipes at different levels between the river and intake well. The intake well should be high enough to take care of high floods. It may be located on the river bank adjacent to water or some distance away. Details of a water stage recorder are shown in Fig. 6.9.

Improvement to the automatic stage recorder can be made by adding instruments like (i) on board computer which records the stage, converts it into discharges continuously or at fixed time intervals, (ii) radio-transmitters can be attached to the system to transmit data signals (stages) to the controlling station either at certain intervals or on demand. These signals are converted to stages and flood warnings are issued, when water level exceeds a predefined danger level.

**Bubble Gauge:** A bubble gauge essentially consists of a small tube placed at the

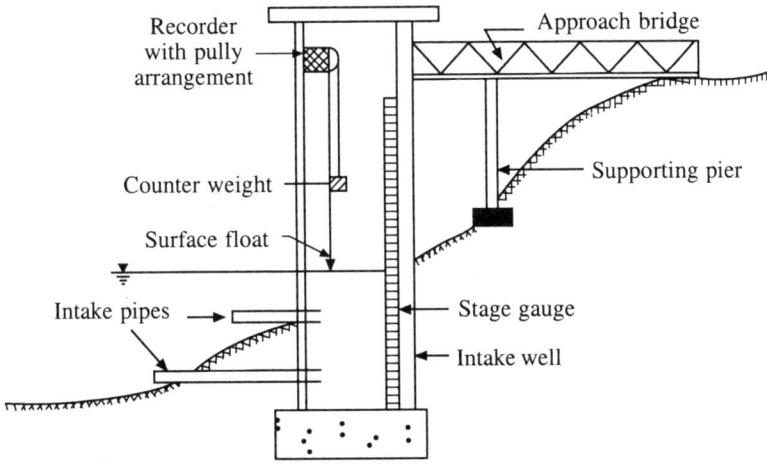

**Fig. 6.9    A Water Stage Recorder Installed in a Intake well**

lowest water level in a river through which compressed air (preferably Nitrogen gas) is slowly bubbled out (Fig. 6.10). The pressure in the tube equal to the water head above it is measured by a manometer connected to a recording device like pen and graph arrangement.

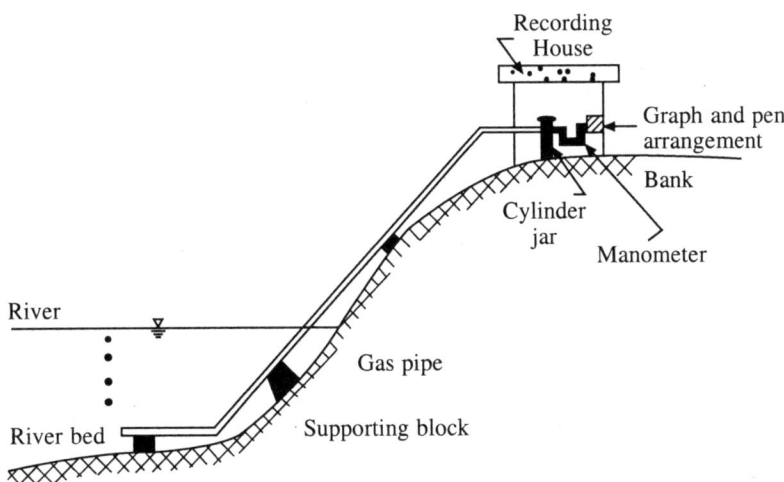

**Fig. 6.10    Arrangement of a Bubble Gauge Stage Recorder**

Gas from a high pressure cylindrical jar passes through a controlling unit to the small tube orifice. As the head of water in the river changes, the controlling unit automatically controls the gas pressure equal to the water head. The position of the pen connected to the free end of the manometer moves as the manometer head fluctuates, which is recorded on a graph paper. This device has a number of advantages (i) installation is cheap as intake well is not required here, (ii) high fluctuations in the river stage can be easily handled, (iii) the recording station can be located as much as 300 m away from the river section, (iv) there is no scope of clogging or choking of the system and (v) sensitivity of the instrument is better.

## 6.5   DISCHARAGE MEASUREMENT

Measurement of discharge is carried out by: (i) Discharge measuring structures, (ii) Approximate area-slope method, (iii) Slope method, (iv) Area-velocity method, (v) Radiotracer method, (vi) Dilution methods and (vii) Electromagnetic method.

### 6.5.1   Discharge Measuring Structures

Discharge measuring structures may be constructed for measuring discharges in small streams. They use indirect method of computing discharges from stages using standard equations. Various structures constructed across a stream for other purposes can be used for stream flow measurement. These are (i) Triangular or V-notch, (ii) Rectangular notch, (iii) Trapezoidal notch, (iv) Suppressed weir, (v) Broad crested weir, (vi) Hydraulic jump, (vii) Parshall flume, (viii) Venturi-flume and (ix) Drops. The equations used for such structures for calculating discharge are given in Table 6.1.

---

**Example 6.2:** At a suppressed weir, calculate discharge of a stream from the following data.

Water head over the weir = 2.30 m
Effective length of weir = 47 m
Thickness of the weir = 6.1 m

**Solution**

Ratio                                   $\dfrac{H}{l} = \dfrac{2.3}{6.10} = 0.377$

Since $H/l = 0.377$ ($< 0.40$), take coefficient of discharge over the weir as 0.864.

Discharge          $Q = 1.7 C_d L\, H^{3/2}$

$= 1.7 \times 0.864 \times 47 \times (2.3)^{3/2} = 240.8\ \text{m}^3/\text{sec}$

Discharge of the stream for the flow condition = 240.8 m³/sec

---

### 6.5.2   Approximate Area-Slope Method

During very high floods, a site may become inaccessible or the gauge-discharge setup may be fully inundated. Under such situations, discharge measurements can be accomplished by using area-slope method. The previous peak flood stages at two locations can be collected from the flood marks in the river course which gives the water surface slope of the peak flood. By knowing the distance between the points along the river, slope $S$ can be computed. Manning's equation can be used to calculate the discharge.

$$Q = \frac{1}{n} A R^{2/3} S_f^{1/2} \tag{6.11}$$

where $S_f$ is the slope of the energy line between the two points.

**Table 6.1  Discharge Over Various Types of Structures**

| Sl. No. | Structure name | Structure description | Equation used | Meaning of notations |
|---|---|---|---|---|
| (1) | (2) | (3) | (4) | (5) |
| 1. | Triangular notch (upto 1.5 m$^3$/sec) | Sharp crested V shaped triangular structure | $Q = (8/15)\sqrt{2g}\,c_d \tan(\theta/2)\,H^{5/2}$ head should be > 6 cm     (6.12) <br> $C_d = 0.60$ to $0.69$ | $H$ = head over notch(m) <br> $\theta$ = angle of notch <br> $C_d$ = coefficient of discharge |
| 2. | Rectangular notch | Sharp crested rectangular opening | | |
| | (i) Without end contraction | | $Q = (2/3)\,C_d\sqrt{2g}\,LH^{3/2}$     (6.13) <br> $C_d = 0.58$ to $0.7$ | $L$ = length of opening of the notch |
| | (ii) With end contraction | | $Q = 2/3\,C_d\sqrt{2g}\times(L - knH)H^{3/2}$     (6.14) | $k$ = coefficient of contraction (0.04-0.1) <br> $g$ = acceleration due to gravity 9.8 m/sec$^2$. <br> $n$ = twice the no. of piers |
| | (iii) With contraction and velocity of approach | | $Q = (2/3)\,C_d\sqrt{2g}\,L_e \cdot H_e^{3/2}$     (6.15) | $L$ = effective width of notch <br> $H_e$ = effective head of notch = $H + v_a^2/2g$ <br> $v_a$ = velocit of approach |
| 3. | Trapezoidal notch (Cipoletti notch) side slopes 1 H : 4V | Sharp crested trapezoidal | $Q = 1.86\,b\,H^{3/2}$     (6.16) | $L$ = $b$ = bottom width of notch. |
| 4. | Suppressed weir (weir without end contractions) | Weirs has considerable thickness in the direction of flow than notches | $Q = C_d\,Lh\sqrt{2g}\,(H - h)$     (6.17) <br> $Q_{max}$(at $h = 2/3H$) $= 1.7\,C_d\,LH^{3/2}$ <br> $C_d = 0.864$ to $1.0$ <br> $C_d$ depends on $H/l$ <br> $H/l < 0.4$, $C_d = 0.864$     (6.18) | $h$ = head of water at middle of weir <br> $l$ = thickness of weir |

*(Contd)*

| (1) | (2) | (3) | (4) | (5) | |
|---|---|---|---|---|---|
| 5. | Weir with end contractions | $L_e = L - 0.1\,nH$<br>$n$ = no. of ends or sides<br>$= 2 \times$ no. of sides | $Q = 1.7\,C_d\,L_e H^{3/2}$ | $n$ = no. of piers | (6.19) |
| 6. | Flumes with constricted width type. Standing wave used when critical velocity is developed at the constriction | Good for silt loaded streams. High discharge. | $Q = A C_f \sqrt{2gh}$<br>$h$ = head difference between U/S and throat.<br><br>$A = (a_1 a_2)/(a_1^2 - a_2^2)^{0.5}$ | $C_f = 0.95$ to1.0<br>$a_1$ = area at entrance<br>$a_2$ = area at throat<br>$C_f$ = coefficient of friction | (6.20) |
| 7. | Parshall flume | $Q = 0.001$ to 1000 m³/sec<br>$w = 7.5$ cm to 15 m | $Q = C \cdot w \cdot H_e^{2.58}$<br>$= 2.42\,w y_1^{2.58}$ | $H_e = y_1 + (V^2_{mean}/2g)$<br>$w$ = throat width in m<br>$y_1$= upstream gauge depth<br>$C$ (coefficient) = 2.2 | (6.21) |
| 8. | Ogee spillway | | $Q = C \cdot L_e H_e^{3/2}$ | $L_e$ = effective spillway length<br>$H_e$ = total head including velocity head<br>For high spillway neglect velocity head | (6.22) |
| 9. | Drops | Approach channel should be at least $20 \times$ depth of fall | $Q = \sqrt{2g}\,L H_c^{3/2}$ | $H_c$ = critical depth of water | (6.23) |

$R$ = Hydraulic mean radius (m) = $A/P$

$A$ = Area of cross section (m$^2$)

$P$ = Wetted perimeter of the channel cross section (m)

$n$ = roughness coefficient or rugosity coefficient

$Q$ = discharge (m$^3$/sec)

The area of cross section $A$ may vary between the two sections and also the slope of the energy line $S_f$ for various flood heights. Selection of $n$ for natural channels may vary from 0.02 to 0.10, depending on the roughness of the channel section and depth of flow. This method gives an approximate estimate of the discharge and should be used with caution.

The distance between the two river sections should be chosen preferably 100 times the flood depth in the channel because, the longer it is, the greater is the accuracy in the estimated value. The fall in the water head should be at least 20 cm. The river reach should be straight and uniform. Average values of channel parameters may be used in equation (6.11) for computation of river discharge.

---

**Example 6.3:** During the flood on 4/11/1990, all the gauge sites of the river Harbhangi (a tributary of Vansadhara) in Orissa were submerged. From the flood marks, the following data were collected. Compute the flood discharge.

1. River stages at two sites, 1.73 km apart are 380.38 and 400.2 m
2. Mean cross section of the channel = 335 m$^2$
3. Mean wetted perimeter of the area = 98 m (measured from the earlier river cross section).

**Solution**

Since it is a natural river, channel roughness coefficient $n$ may be taken as 0.035

Slope of the water surface $S_f = \dfrac{(400.2 - 380.38)}{1730} = 0.01146$

Using Manning's equation $Q = \dfrac{1}{n} A R^{1/3} S^{1/2}$; $\left( R = \dfrac{A}{P} = \dfrac{335}{98} = 3.418 \text{ m} \right)$

$Q = \dfrac{1}{0.035} \times 335 \times (3.418)^{2/3} \times (0.01146)^{1/2} = 2324 \, \text{m}^3/\text{sec}.$

---

### 6.5.3 Slope Method

The method discussed in Section 6.5.2 can be improved to give good results by incorporating the principle of conservation of energy between the two points of the selected reach. Considering reference datum as the channel bed at points (1) and (2) as shown in Fig. 6.11, Bernoulli's equation can be applied to calculate the head loss as

At section (1) energy head $\qquad H_1 = \dfrac{V_1^2}{2g} + Y_1$ $\qquad\qquad$ (6.24)

At section (2) energy head $\qquad H_2 = \dfrac{V_2^2}{2g} + Y_2$ $\qquad\qquad$ (6.25)

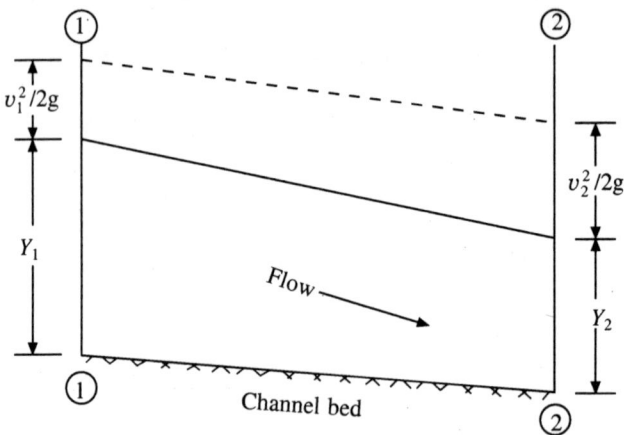

**Fig. 6.11   Slope Method of Calculation of Discharge**

Energy slope over length $l$ between the two sections

$$\left(\frac{H_1 - H_2}{l}\right) = S_f = \left(\frac{Q}{K}\right)^2 \tag{6.26}$$

where $Y_1$ and $Y_2$ are the depths of water in the channel at Sections (1) and (2), respectively, with velocities $V_1$ and $V_2$ and $K$ is the channel conveyance which can be obtained from Manning's equation as

$$Q = A \cdot \left(\frac{1}{n}\right) R^{2/3} S_f^{1/2} = K S_f^{1/2} \tag{6.27}$$

where
$$K = \frac{A \cdot R^{2/3}}{n} \tag{6.28}$$

$R$ is the hydraulic radius $= A/P$, $A$ the area of cross section of the channel in (m²), $P$ the wetted perimeter (m). From equation (6.27), the slope of energy line $S_f$ of equation (6.26) can be obtained. If two sections have different conveyance factors $K_1$ and $K_2$ with roughness $n_1$ and $n_2$ respectively, then the equivalent $K$ between those sections can be calculated as

$$K = \sqrt{(K_1 K_2)} \tag{6.29}$$

where $K_1 = \dfrac{A_1 R_1^{2/3}}{n_1}$ and $K_2 = \dfrac{A_2 R_2^{2/3}}{n_2}$ can be calculated for the two sections.
The procedure for using the area-slope method is:

1. Select the stream reach between two sections (1) and (2) as shown in Fig. 6.11.
2. From the cross sections at (1) and (2), find the depths of water during the flood at the two sections from the flood marks left by floating debris.
3. Compute cross sectional areas $A_1$ and $A_2$ and wetted perimeters $P_1$ and $P_2$ corresponding to the flood depths of step (2) of the particular flood.

4. Calculate the hydraulic radii $R_1$, $R_2$ and the conveyances $K_1$ and $K_2$ after selecting suitable roughness coefficient $n_1$ and $n_2$ for the sections (1) and (2) respectively.
5. Obtain the average or equivalent conveyance $K$ between Sections (1) and (2).
6. Assuming no sudden expansion or contraction between Sections (1) and (2), the loss in expansion and contraction between the sections can be neglected. To achieve this, the reach selected should be fairly uniform.
7. Calculate the discharge $Q$ from equation (6.27) by assuming a suitable value of $S_f$.
8. Calculate the velocities $V_1$ and $V_2$ at the two Sections (1) and (2) from the relation

$$Q = A_1 V_1 \quad \text{and} \quad Q = A_2 V_2.$$

9. From equations (6.24) and (6.25), calculate $H_1$, $H_2$ and the energy slope from equation (6.26).
10. The assumed value of $S_f$ of step (7) should be the same as the calculated value of $\dfrac{(H_1 - H_2)}{l}$ of step (9). If the two values differ then take the energy slope of step (9) and repeat steps (7) to (9) till the energy slope at the end of the iteration are the same.
11. The discharge calculated at the end of step (10) is the estimated flood discharge.

---

**Example 6.4:** During a high flood, a river reach of 1 km apart was having the following information.

| | | |
|---|---|---|
| Up stream: | Area of cross section | $A_1 = 180$ sq m |
| | Wetted perimeter | $P_1 = 50$ m |
| | Manning's roughness coefficient | $n_1 = 0.03$ |
| | Reduced level of water | $= 78.3$ m |
| Down stream: | Area of cross section | $A_2 = 183$ sq. km |
| | Wetted perimeter | $P_2 = 51$ m |
| | Manning's roughness coefficient | $n_2 = 0.025$ |
| | Reduced level of water | $= 78.0$ m |

Calculate the flood discharge. Neglect other losses.

**Solution**

Up stream (US) hydraulic radius $\quad R_1 = \dfrac{A_1}{P_1} = \dfrac{180}{50} = 3.6$ m

Down stream (DS) hydraulic radius $\quad R_2 = \dfrac{A_2}{P_2} = \dfrac{183}{51} = 3.59$ m

Conveyance of US $= K_1 = \left(\dfrac{1}{n_1}\right) A_1 R_1^{2/3} = \left(\dfrac{1}{0.03}\right) \times 180 \times 3.6^{2/3} = 14100$

Conveyance of DS $= K_2 = \left(\dfrac{1}{n_2}\right) A_2 R_2^{2/3} = \left(\dfrac{1}{0.025}\right) \times 183 \times 3.59^{23} = 17170$

Average conveyance for the reach $= \sqrt{(K_1 K_2)} = \sqrt{(14100 \times 17170)} = 15560$

Assume fall in energy head $78.30 - 78.0 = 0.30$ m between the reach of 1 km

$$S_f = \frac{0.30}{1000} = 0.0003$$

Discharge $Q = K\sqrt{S_f} = 15560 \times \sqrt{0.00030} = 269.5$ m$^3$/sec.

$V_1 = Q/A_1 = 269.5/180 = 1.497$ m/sec, and $V_2 = 269.5/183 = 1.473$ m/sec

Fall of head $= (h_1 - h_2) + \dfrac{(V_1^2 - V_2^2)}{2g} = (783 - 78) + \dfrac{(1.497^2 - 1.473^2)}{19.62}$

$$= 0.3036 = \text{say } 0.304 \text{ m}$$

**Iteration**

Taking fall of energy head now as 0.304/100

$$Q_1 = 15560 \sqrt{\left(\frac{0.304}{1000}\right)} = 271.43$$

$$V_1 = \frac{271.4}{180} = 1.508 \text{ m/sec}, \quad V_2 = \frac{271.4}{183} = 1.483 \text{ m/sec}$$

Fall of head $= (78.3 - 78.0) + \dfrac{(1.508^2 - 1.483^2)}{19.62} = 0.3038$

$$Q_1 = 15560 \sqrt{\left(\frac{0.3038}{1000}\right)} = 271.2 \text{ m}^3/\text{sec}$$

The flood discharge during the event is 271.2 m$^3$/sec.

---

### 6.5.4   Area-Velocity Method

This involves the measurement of velocity at the gauging site and the corresponding area to obtain river discharge directly in the field. Velocity of flowing water in a stream is measured by (a) current meter, (b) floats and (c) ultrasonic method. Velocity distribution at a section varies with depth of water and the location of the point in the section with respect to the banks. It is usually higher at the section where the depth of flow is the highest (Fig. 6.12a). A vertical profile of velocity distribution at a cross section of a stream is shown in Fig. 6.12b. The profile suggests that the velocity needs to be measured at number of locations at a section and also at different elevations to get the average velocity of the whole channel section. The following methods are used to measure the velocity.

### 6.5.4.1   Velocity Measurement by Current Meters

The following devices are used to measure velocity at a channel section.

*(a) Current Meters*

It is an instrument used for measuring velocity of water in a stream. Various kinds of current meters now in use are discussed next.

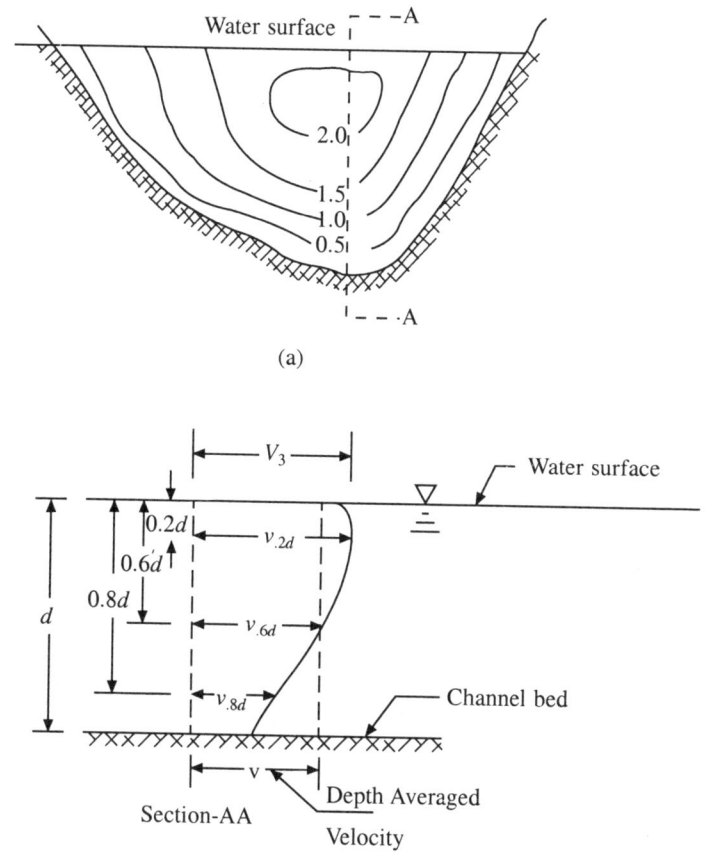

(a)

(b)

**Fig. 6.12   Velocity Distribution of a Channel Section**

*(i) Vertical Axis Type or Price Current Meter*

It is the most commonly used current meter consisting of a horizontal wheel carrying a series of conical cups around a vertical axis (Fig. 6.13). The other attachments of the current meter are the tail vans to keep it along the direction of flow of water and the balancing weight at the bottom. When the meter is lowered into water, the velocity of flowing water rotates the cups in a horizontal plane, the number of revolutions are recorded by an electromagnetic device. The rating formula for velocity of water is given by

$$V = (a + bN) \tag{6.30}$$

where $V$ is the velocity of flow in m/sec, $N$ the number of revolutions of the wheel per sec, $a$ and $b$ are constants supplied by the manufacturers. For standard current meters with 12.5 cm diameter cones, $a$ and $b$ are usually taken as 0.03 and 0.65 respectively. The meter is suspended by a cable to a depth of $0.6d$ from the surface of water, where $d$ is the depth of water at the location. At this depth the velocity recorded is taken as the mean velocity of the vertical section. For deeper streams mean of velocities recorded at 0.2$d$ and 0.8$d$ is taken as the average of the stream velocity at the section.

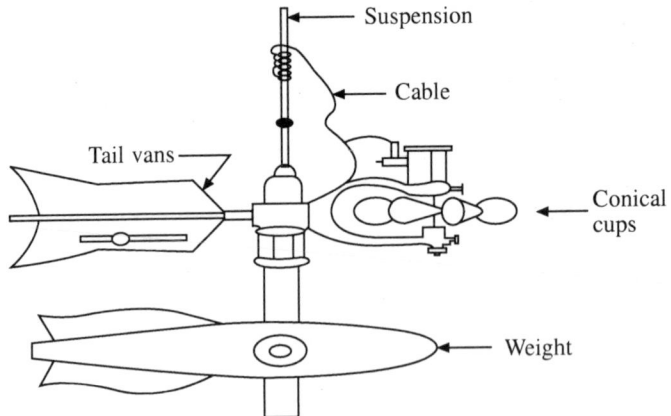

**Fig. 6.13   Vertical Axis type or Price Current Meter**

For river gorges a cable is tied to trees on both sides of the bank and the current meter is suspended from the cable into the stream at the desired depth. For wide rivers, a boat is anchored at a different locations by rope arrangement from the bank and the current meter is lowered from the boat. Usually such boats have projecting boom arrangements to avoid any disturbance arising due to position of the boat. The operating range of velocities of the stream water can vary from 0.15 to 4.0 m/sec. If the stream has strong vertical component of velocity then such type of current meters are unsuitable.

A small cup current meter known as *pigmy meter* is used for small velocity measurements as it runs faster for the same velocity of water. The values of $b$ and $a$ for pigmy type current meter are 0.3 and 0.008, respectively. Such current meters carry 5 cm diameter cones. When the velocity in a stream is high, the cable suspending the current meter may be deflected to form an arch-like segment which gives velocity reading at a different depth. Necessary correction may be applied for such a situation to bring the reading to the desired depth.

Current meters are *calibrated* in ponds or long channels where water is held stationary. A vehicle with cantilever arm projection to the channel helps to lower and move the current meter in the pond water. For each run, the current meter is moved at a predetermined speed, $v_i$ and the number of revolutions $N_i$, the cups make are counted. These readings plot a straight line in a graph paper from which the constants $a$ and $b$ can be determined. The coefficients are unique for each current meter. Current meters are to be serviced at regular intervals and whenever damage occurs, it should be repaired and calibrated.

*(ii) Propeller or Horizontal Axis Current Meter*

In this type, the propellers are fixed to one end of a horizontal shaft, the other end carries a set of fins. The fins help to stabilise the current meter against the velocity of flowing water. The centre of axis carry the sounding weight, as shown in Fig. 6.14. The current meter is suspended from a hoisting cable attached to the sounding weight. A mechanical gear and an electric counter arrangement records the number of revolutions of the shaft. An electric wire transmits the information to the receiver at the boat. Such type of current meters are rugged and can be used in the same way as the vertical axis type. A variety of propeller current meters are available in the market.

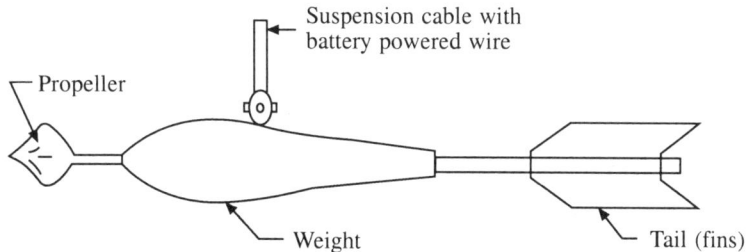

**Fig. 6.14   Propeller Type Current Meter**

*(iii) Optical Current Meter*
During very high floods, optical current meters are used as surface floats. A variable speed drive motor powered by dry cell battery is housed in a light weight float container. Optical devices like a set of rotating mirrors and a low power telescope installed in the current meter helps to locate it from an observation point at the upstream. As the meter moves down in the river water, the angle of the meter with respect to the upstream observation point increases. A relation between the velocity of water in the stream and the rotational speed of the mirror in the current meter is established. The velocity of stream so obtained can be multiplied by coefficients from 0.85 to 0.95 to give the average velocity of water.

### 6.5.4.2   Velocity Measurement by Floats
Floats are commonly used to measure the velocity of flowing water during high floods. Such floats give excellent results for small streams. Wind affects the measurement process to a great extent. The site selected for measurement of velocity by floats should be a straight reach, free from cross currents and eddies.

As shown in Fig. 6.15(a), a float is released at least 15 m upstream of the defined section A of the stream. A man standing at a downstream reach B 100 m away from the upstream section has a stop watch. Time taken by the float to travel between the stream reach of 100 m is noted. The distance travelled by the float divided by the time taken gives the surface velocity of the stream water. Average of three such readings multiplied by a reduction coefficient (0.85) gives the mean velocity for the section. Some relations have been proposed between the surface velocity measured by surface floats and mean velocity. Mysore Engineering Research Station proposed a relation of the type

$$V_m = 0.8529 V_s + 0.0085 \tag{6.31}$$

where $V_s$ is the velocity recorded by the surface float and $V_m$ the mean velocity for the vertical section. Any object that can float can be used as surface float. In the market a variety of specially designed floats are available. Some important types of floats are discussed here.

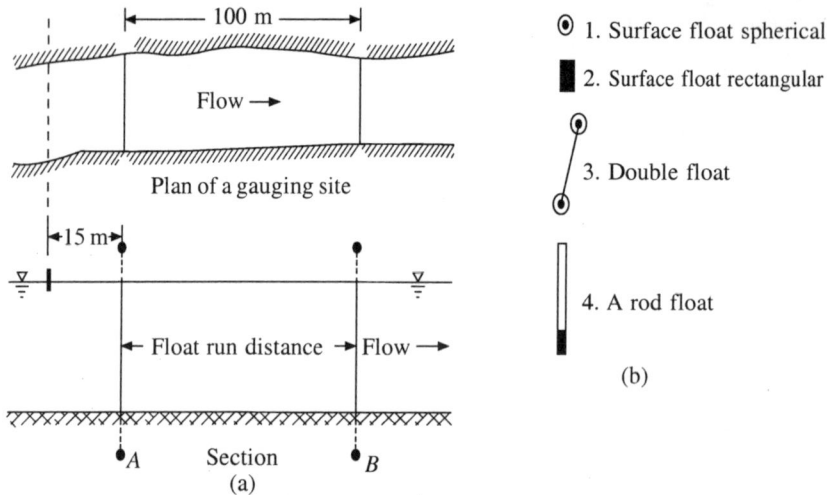

**Fig. 6.15   An Arrangement of Velocity Measurement by Floats**

**Types of Floats:** The essential requirements of a float are (i) stability while floating, (ii) distinct visibility and (iii) the float should not react with water. Four types of commonly used floats are shown in Fig. 6.15(b). A *surface float* may be either *wooden or metallic* and of various sizes and shapes. IS 3911 : 1966 recommends wooden surface floats for moderate velocities while metallic surface floats can be used for large turbulent rivers. A *subsurface* float usually has two floats tied together by a thin cord such that lower one is always at 0.2 h above the river bed, while the upper one floats at the water surface. Such a *double float system* gives average velocity for deep rivers directly even under considerable windy conditions. If the lower float is submerged to any other depth, then the system is termed as *twin float* and the velocity obtained is the mean of the velocity of the surface float and at the depth of lower float. A *rod float* which is long enough but does not touch the river bed can be used as a float. This float gives approximately the mean velocity at the section. When a shorter rod is used, then the mean velocity is calculated using the equation

$$V_m = V \left\{ 1.02 - 0.0116 \left( \frac{h_1}{h} \right)^{0.5} \right\} \tag{6.32}$$

where $V_m$ is the mean velocity in m/sec, $V$ the observed mean velocity of the rod float in m/sec, $h_1$ the distance between lower end of the rod float and bed level of river (m) and $h$ is the water depth of river (m). Lacy suggested the use of a single rod upto 0.80 m depth while for Parker, the depth of submergence of rod float varies between 0.91 to 0.97 times the water depth. Different investigators used different depths of rods, while care was taken to protect the rod bottom fouling against the river bed. Rod floats can have the maximum depth upto 4-5 m because of obvious reasons of handling and transportation.

### 6.5.4.3   Velocity Measurement by Ultrasonic Method

Another method to measure velocity of a stream is the use of ultrasonic sound waves moving in water from one side of the channel to other. Two transducers capable of emitting and recording sound waves are placed at the same reduced level on either side of the river banks, making an angle φ with respect to the river bank as shown in Fig. 6.16.

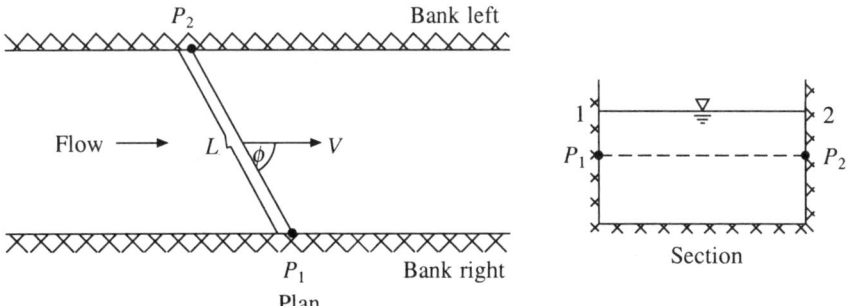

**Fig. 6.16   Velocity Measurement by Ultrasonic Method ($P_1$ and $P_2$ are Transducers)**

When transducer $P_2$ sends ultrasonic signals, it is received by $P_1$. The signals move with their own velocity in the water medium plus the velocity of water favouring it. When $P_1$ sends signals, it is received by $P_2$ with velocity of water opposing it. The velocity of water is computed from the relation

$$V = \frac{L \times (t_2 - t_1) \times \sec \phi}{2t_1 t_2} \qquad (6.33)$$

where $L$ is the distance between $P_1$ and $P_2$, $t_1$ and $t_2$ respectively represent the times taken by sound wave to reach from $P_2$ to $P_1$, and from $P_1$ to $P_2$. The method provides upto 3% accuracy. If velocity of flow at a different depth is required, then the transducers are provided at that depth and velocity measured. Thus a single or multiple point methods may be adopted to compute the average velocities. Wide and unstable rivers can easily be handled by this method. Thus the method is accurate, rapid and can be made automatic.

### 6.5.4.4   Measurement of Area

Figure 6.12 shows that the velocity of water at any river section varies in all the directions. The velocity is zero at the river periphery and changes rapidly as we move from the channel bed. A single area-velocity measurement for the entire section will give highly erroneous results. Therefore, the cross section of river at the gauging site is divided into a number of subsections by imaginary verticals. The number of such subsections for a stream are decided by the following criteria.

1. Discharge carried by any subsection should be upto 10% of the total discharge of the entire section.
2. The width of any subsection should be about 5% of the total width of the river at the site.

3. There should not be any large velocity difference between the adjacent subsections
4. The discharge variation between adjacent subsections should be between 5 and 10%.

Table 6.2 gives the number of subsections to be adopted for river channels for different widths, as per IS : 1192 specification.

**Table 6.2  Guidelines for Deciding Number of Subsections of a River Section for Discharge Measurement**

| River or channel width (m) | Number of observation sections | Maximum section width (m) |
|---|---|---|
| Upto 15 | 15 | 1.5 |
| Between 15–90 | 15 | 6 |
| Between 90–180 | 15 | 15 |
| Over 180 | 25 | To be fixed suitably |

For computation of area, determination of water depths at various subsections is carried out by any of the following methods.

**Wading or Sounding Rod:** A man walks across the river section with a graduated wading rod and the depth of water at each of the predefined subsection is measured.

**Lead-line or Cable Arrangement:** For river sections in plains, the lower end of a cable attached to a current meter with a sounding weight of 7-14 kg is lowered from a boat at the desired locations. By measuring the length of cable, the depth of water is obtained while velocity is recorded simultaneously by current meter.

**Echo-sounder:** For deep channels carrying the water at high velocities, an electroacoustic device called echo-sounder is often used. This instrument gives accurate results in the least time and is best suited for soft and mobile beds. A transducer is lowered at the surface of the water and high frequency sound waves are emitted from it. Reflected sound waves from river bed are received back by the instrument and the time lapse between transmission and reception is converted to depth of water at the location. A *recording type echo sounder* mounted on a boat graphs a continuous plot of stream bed on a moving paper chart drawn by stylus while the boat moves from one bank to the other. The instrument gives good result, a minimum height of 0.30 m of water in the river. The error of such mechanism is 1 cm/m of water depth.

### 6.5.4.5  *Measurement Procedure*
For area-velocity method, the following procedure called mid section method for measurement of discharge may be followed.

(i) Divide the cross-section of the river into $n$ number of vertical sections as per the guidelines given before.

(ii) Measure the depth of water at sections $h_1$, $h_2$, ..., $h_{n-1}$ by a wading rod, sounding rod or echo-sounder. For $n$ number of sections, there will be $(n+1)$ number of verticals but the depth $h$ at the beginning and end being zero, the number of verticals being $(n-1)$ as shown in Fig. 6.17.

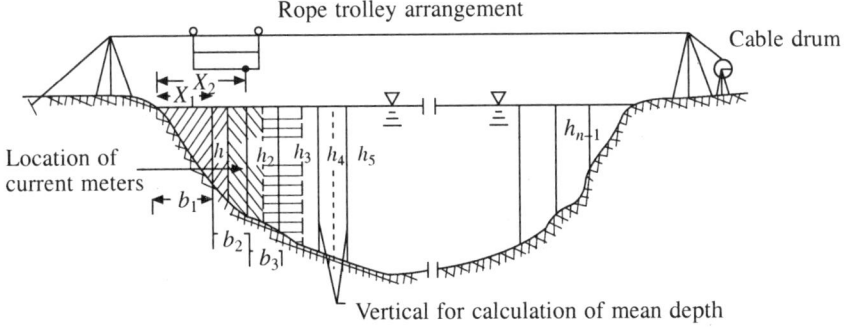

**Fig. 6.17   Discharge Calculation by Mid Section Method**

(iii) Measure the horizontal distances of the verticals $h_1$, $h_2$, ..., $h_{n-1}$ from the reference bank as $X_1$, $X_2$, ..., $X_{n-1}$.

(iv) Compute the width of each subsection by subtracting the distance of the vertical from the distance upto the previous vertical. Let $b_1 = x_1$, $b_2 = x_2 - x_1$, $b_3 = x_3 - x_2$, and so on.

(v) Compute the area at the subsections such that the depths $h_1$, $h_2$, $h_3$ ...,$h_n$ represent the mid points of the sections.

First and last triangular area $A_1 = \dfrac{h_2}{2}\left(\dfrac{b_2}{2} + b_1\right)$.

Intermediate areas $\qquad A_2 = h_2\left(\dfrac{b_2 + b_3}{2}\right)$.  (6.34)

(vi) Measure the velocity at locations $h_1$, $h_2$ ..., $h_n$ and compute the average velocity at each subsection by any of the following methods.

(a) **One point method:** Velocity observed at 0.60$h$ depth below the surface gives mean velocity within ± 5% accuracy $V_m = V_{0.6h}$.

(b) **Two point method:** Mean of two velocities recorded at 0.2$h$ and 0.8$h$ gives better results than method (a) with 2% accuracy. For this method the water depth should be greater than 0.6 m. Average velocity is calculated as $V_m = 0.5\,(V_{0.2h} + V_{0.8h})$.

(c) **By velocity profile:** A current meter is lowered and then raised from the bottom of channel at an uniform rate of 0.04 m/sec. The velocity of flow is recorded at close intervals. The velocity of flow is the average for the whole section. Two such complete cycles are observed and the average velocity should not vary more than 10%. Conversely velocity observations at close intervals are recorded and then a velocity profile is drawn. The average velocity is computed. Such procedure though time consuming, gives very accurate results. It is practised for calibration of stage-discharge relation or for experimental runs.

(d) **Surface velocity method:** This method is usually adopted during high floods. Here a surface float is used to get the surface velocity. This velocity when multiplied by a coefficient (0.85 to 0.95) gives average velocity of the sub section. The coefficient should be properly chosen.

(e) **Other methods:** Methods like

$$V_m = \frac{1}{3}(V_{0.2h} + V_{0.6h} + V_{0.8h}) \text{ or}$$

$$V_m = \frac{1}{6}(V_{0.1h} + V_{0.2h} + V_{0.4h} + V_{0.6h} + V_{0.8h} + V_{0.9h})$$

are sometimes used with the limitations that they are time consuming to measure. The percentage of accuracy achieved may not justify the approaches.

(vii) Multiply the area $(A_i)$ and respective average velocity for each subsection to find out the discharge for the subsection $(q_i)$ separately.

(viii) By mean section method, area and velocity is computed (Fig. 6.18) as

$$A_1 = b_1 \frac{(h_1 + 0)}{2}; \quad v_{a_1} = \frac{(0 + v_1)}{2}$$

$$A_2 = \frac{(h_1 + h_2)}{2}b_2; \quad v_{a_2} = \frac{(v_1 + v_2)}{2} \text{ and so on} \qquad (6.35)$$

and discharge is computed using the relations

$$q_1 = A_1 v_{a_1}; \quad q_2 = A_2 v_{a_2} \qquad (6.36)$$

For end sections the relation for triangles and for the middle section, rules for trapezoidal sections are used.

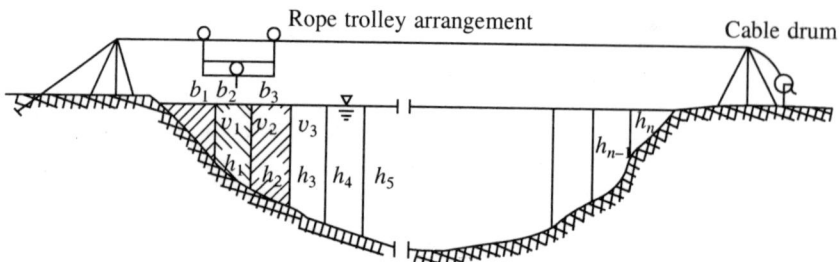

**Fig. 6.18   Discharge Calculation by Mean Section Method.**

(ix) Sum of discharges of all such subsections gives the total discharge for the section

$$Q = \Sigma A_i q_i. \qquad (6.37)$$

**Example 6.5:** A current meter was used to measure velocity of a river at a subsection 4.2 m deep. Calculate the average velocity of water at the sub-section. By surface float, the surface velocity was found to be 1.05 m/sec.

| Depth from surface | 0.42 | 0.84 | 1.68 | 2.54 | 3.36 | 3.78 |
|---|---|---|---|---|---|---|
| Velocity $V$(m/sec) | 1.1 | 1.3 | 1.25 | 0.95 | 0.60 | 0.55 |

**Solution**
Calculation of average velocity at the subsection is carried out in Table 6.3.

**Table 6.3    Average Velocity Computation for Example 6.5**

| Depth of record (m) | Depth (m) | Velocity (m/sec) | Calculation of Average velocity by | | | | | |
|---|---|---|---|---|---|---|---|---|
| | | | 1-point method | 2-point method | 3-point method | 6-point method | surface velocity | Velocity profile method |
| (1) | (2) | (3) | (4) | (5) | (6) | (7) | (8) | (9) |
| 0.42 | 0.1$d$ | 1.10 | | | | | | |
| 0.84 | 0.2$d$ | 1.30 | | | | | | |
| 1.68 | 0.4$d$ | 1.25 | | | | | | |
| 2.54 | 0.6$d$ | 0.95 | 0.95[a] | 0.95[b] | 0.95[c] | 0.942[d] | 0.958[e] | 0.957[f] |
| 3.36 | 0.8$d$ | 0.60 | | | | | | |
| 3.78 | 0.9$d$ | 0.55 | | | | | | |
| Average velocity at the section m/sec | | 0.95 | 0.95 | 0.95 | 0.95 | 0.942 | 0.958 | 0.957 |

a: At 0.6$d$; b: mean of 0.2$d$ and 0.8$d$; c: mean of 0.2$d$, 0.6$d$ and 0.8$d$; d: mean of all six observations; e: obtained by multiplying 0.9 to surface velocity (1.05); f: obtained by plotting the velocity profile and numerically integrating the velocity over depth.

**Example 6.6:** The rating curve of a current meter used for measuring velocity in a small river is given as $V = 0.62N + 0.032$ m/sec, where $N$ is the revolutions/sec. Calculate the discharge of the river from the following data. Velocity is measured at the mid of the sections.

| Distance from bank(m) | 0 | 2 | 5 | 8 | 12 | 15 | 18 | 21 | 23 | 24 |
|---|---|---|---|---|---|---|---|---|---|---|
| Depth(m) | 0 | 0.6 | 1.2 | 1.8 | 2.4 | 1.9 | 1.4 | 1.1 | 0.5 | 0 |
| $N$ at 0.6$d$ | 0 | 60 | 90 | 120 | 150 | 140 | 100 | 80 | 50 | 0 |
| Time (sec) | 0 | 150 | 140 | 140 | 160 | 140 | 140 | 140 | 140 | 0 |

**Solution**
The calculation is performed in Table 6.4 using equations (6.35) and (6.36).

**Table 6.4   Section Mean Velocity**

| Distance from bank(m) | N at 0.6 depth | Time (sec) | N/T (m/sec) | Velocity (m/sec) | Mean Velocity (m/sec) | Width of segment from col.(1) | Depth at b m away | Mean depth (m) | Segment discharge (m³/sec) |
|---|---|---|---|---|---|---|---|---|---|
| (1) | (2) | (3) | (4) | (5) | (6) | (7) | (8) | (9) | (10) |
| 0 | 0 | 0 | 0.0 | 0 | | | | | |
| | | | | | 0.14 | 2 | 0.6 | 0.3 | 0.084 |
| 2 | 60 | 150 | 0.4 | 0.28 | | | | | |
| | | | | | 0.356 | 3 | 0.6 | 0.9 | 0.961 |
| 5 | 90 | 140 | 0.643 | 0.431 | | | | | |
| | | | | | 0.497 | 3 | 1.2 | 1.5 | 2.237 |
| 8 | 120 | 140 | 0.857 | 0.563 | | | | | |
| | | | | | 0.589 | 4 | 1.8 | 2.1 | 4.948 |
| 12 | 150 | 160 | 0.938 | 0.614 | | | | | |
| | | | | | 0.633 | 3 | 2.4 | 2.15 | 4.083 |
| 15 | 140 | 140 | 1.000 | 0.652 | | | | | |
| | | | | | 0.564 | 3 | 1.9 | 1.65 | 2.992 |
| 18 | 100 | 140 | 0.714 | 0.475 | | | | | |
| | | | | | 0.431 | 3 | 1.4 | 1.25 | 1.616 |
| 21 | 50 | 140 | 0.571 | 0.386 | | | | | |
| | | | | | 0.310 | 2 | 1.1 | 0.80 | 0.496 |
| 23 | 50 | 140 | 0.357 | 0.235 | | | | | |
| | | | | | 0.127 | 1 | 0.5 | 0.25 | 0.029 |
| 24 | 0 | – | – | – | – | – | – | – | – |
| Total | | | | | | | | | 17.445 |

### 6.5.5 Radio-Tracer Method

A known quantity of suitable radio tracer is mixed to the flow at upstream of the gauge station. A *Geiger counter* at the station counts the number of clicks caused by the tracer in passing through the reach. Depending on the isotope activity the river may require 40 km distance of travel with a total of 750 cumecs of discharge to completely dispense the radio tracer. Discharge of the river can be calculated from the relation

$$Q = \frac{A \cdot K}{N} \tag{6.38}$$

where $Q$ is the discharge in m³/sec, $A$ the quantity of radio tracer in units decided in the laboratory, $K$ a constant for the isotope and the counter, $N$ the number of clicks generated by the tracer during its passage. Value of $K$ is determined in the laboratory.

### 6.5.6   Dilution Technique

In this method, a tracer like common salt or fluorescent dye of concentration $C_1$ is injected at constant rate of $q_1$ to a small stream of constant cross section. The flow in the stream is assumed to be steady. At the other station, sufficiently downstream of it, a pair of electrodes (for salts) or other suitable device is used to measure the tracer concentration. From the continuity equation, discharge $Q$ in the stream is calculated using

$$Q = \frac{q_1(C_1 - C_2)}{C_2 - C_0} \tag{6.39}$$

where $C_2$ is the final constant rate of the concentration of tracer in the downstream, $C_0$ the initial concentration of the tracer in the stream water present before the injection of $C_1$ into the stream. The mixing length depends on geometric dimension of the river cross section, discharge and flow conditions. It may vary from 1 km in small mountainous streams to more than 100 km in plain streams for large rivers. This method however gives reliable results both for mountainous streams and rivers in plains.

---

**Example 6.7:** Common salt solution of concentration 200 gm/l was added to a stream at a constant rate of 0.2 cm³/sec. Concentration of this salt in the stream already present was 0.01 ppm. At sufficiently downstream, the concentration of the salt in the stream water was measured as 0.05 ppm. Estimate the stream discharge.

**Solution**
Rate of injection of salt solution = 0.2 cm³/sec = $0.2 \times 10^{-6}$ m³/sec
Concentration of salt $C_1$ = 200 gm/l = 0.20 gm/gm
Final concentration of salt in water $C_2$ = 0.050 ppm = $0.050 \times 10^{-6}$ gm/gm
Initial concentration of salt in water already present $C_0$ = 0.01 ppm = $0.01 \times 10^{-6}$ gm/gm
Using equation (6.39), we get

$$Q = \frac{q_1(C_1 - C_2)}{C_2 - C_0} = \frac{0.20 \times 10^{-6} \times (0.20 - 0.5 \times 10^{-6})}{(0.050 \times 10^{-6} - 0.01 \times 10^{-6})} = 1.0 \text{ m}^3/\text{sec}$$

Discharge in the stream is 1.00 m³/sec.

---

### 6.5.7 Electromagnetic Induction Method
In this method a large coil is buried below a river bed which generates a vertical magnetic field across the full width of the river. When water of depth $h$ flows over it, a current of $I$ amperes in the coil produces a voltage $V$ in water which can be recorded by electrode probes at the banks. The voltage is proportional to the average velocity of water in the channel. Sophisticated equipments can measure discharge accurate upto 5%. The setup is costly and can be suited upto a river width of 70–80 m. The equation for discharge measurement can be expressed as

$$Q = \left[ C_1 \frac{(V \times h)}{1} + C \right]^K \tag{6.40}$$

where $h$ is the depth of water in the channel, $C_1$, $C$ and $K$ are constants. This method developed in U.K, is best suited in river sections which change due to sedimentation and weed growth. It is finding its increasing application. A setup of the scheme is shown schematically in Fig. 6.19.

### 6.6 STAGE-DISCHARGE RELATIONSHIP
The primary objective of measuring discharge directly at a gauging site is to

obtain a relation between the stage and discharge so that the stage can be converted to discharge from the developed relation. The discharge is plotted in abscissa and the stage as ordinate. This curve is widely known as *rating* or *G-Q* or *G-D curve*. It is used as reference curve for reading discharges from river stages.

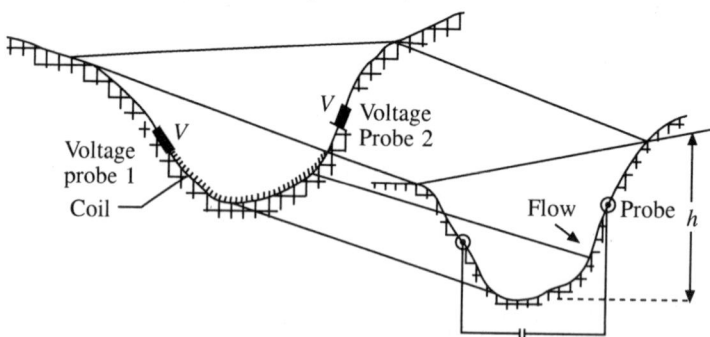

**Fig. 6.19   Discharge Measurement by Electro-Magnetic Induction Method**

Daily observation of discharge is expensive and time consuming. During high floods, it may not be possible to measure velocities at the gauging site at short intervals. The rating curve remains valid so long as the conditions at the gauging site remain stable. The combined effect of all the channel and flow parameters is termed as *control*. When the *G-D* curve remains unchanged with time, we call the site to be a *permanent control*, and when it changes with time, the site is known as *shifting control*.

At permanent control, the rating curve does not change with time but it may be necessary to check it periodically. A site may have a single control for the full stage or different controls each serving for different range of stages. A single stage-discharge relation assumes the form

$$Q = C (G - G_0)^n \qquad (6.41)$$

where $G$ is the gauge height (m) and $G_0$, the gauge height corresponding to zero discharge. It does not represent the river bed level but a value below it, $Q$ is the discharge in m³/sec, $C$ and $n$ are constants that can be evaluated using the method of least squares, however, the value of $G_0$ should be evaluated before.

**Evaluation of $G_0$:** There are a number of methods available to accomplish this. One of the simplest approach is trial-and-error search for $G_0$ which gives the best value of correlation coefficient $r$ and the lowest standard error for the observed set of stage and discharge values. A number of alternative methods are also available to determine the value of $G_0$ like graphical, semi-graphical or analytical approaches. One approach is to plot stage and discharge values on a plain graph paper and draw the best fit curve. It is extrapolated backwards to touch ordinate axis where the discharge is zero. Take $G$ as the value of $G_0$. Plot the discharge $Q$ and stage $(G - G_0)$ on a logarithmic scale which should be a straight line. If not, then chose another point, i.e., another value of $G_0$ close to it on trial to get a straight line plot. The curve is parabolic when $Q$ as abscissa is plotted against $G$ as ordinate.

In the other approach, three values of $Q$ are selected from a $G$-$D$ curve such that $\dfrac{Q_1}{Q_2} = \dfrac{Q_2}{Q_3}$. From a similar ratio of the right side of equation (6.41) we get

$$G_0 = \frac{(G_1 G_3 - G_2^2)}{(G_1 + G_3 - 2G_2)} \tag{6.42}$$

Using the above equation, value of $G_0$ can be evaluated. However, this method is less accurate than the previous one and should be used with caution.

**Looping of Rating Curve:** The curve should cover the full ranges of gauge and discharge observations. For most of the sites, the rating curve may exhibit some variations at low stages but for high stages it may remain almost fixed. For the same stage a rising flood may have increased velocity and discharge than during falling when the river has less velocity and discharge. This is true for all flood situations and for all rivers. The plot of stage and discharge relation for a complete flood period is not a single curve but forms a loop as shown in Fig. 6.20. This is because of the unsteady flow situations in a river during the rising and falling stages. For different floods the looping of the stage-discharge relation may be different. An average rating curve may be drawn for the site when the loop is not too wide apart. It is always desirable to draw separate rating curves for monsoon and non-monsoon months as the channel characteristics are different in the two distinct seasons.

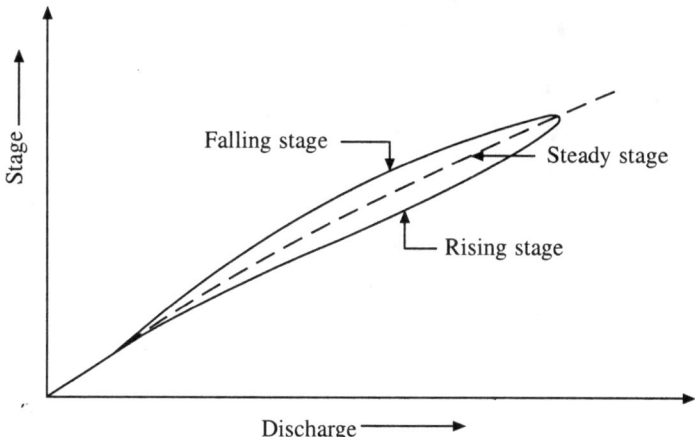

**Fig. 6.20  Stage-Discharge during a Complete Flood Period Showing a Loop Rating Curve**

During flood, correction may have to be applied to the rising and falling stage-discharge relations to get correct discharge from the observed stage. The correction can be carried out by correlating it with the stage of another secondary gauge located close to the main gauge, where the effect of wedge like storage is pronounced. The corrected discharge can be approximately computed from the equation

$$Q_a = Q_0 \sqrt{\left(\frac{\Delta Y_c}{\Delta Y}\right)} \qquad (6.43)$$

where $\Delta Y_c$ is the difference in the gauges between rising or falling stages of main and secondary gauge when the flow is $Q_a$, $\Delta Y$ is the difference in gauges between main and secondary gauge in normal flows for that stage when the flow is $Q_0$ (Fig. 6.21).

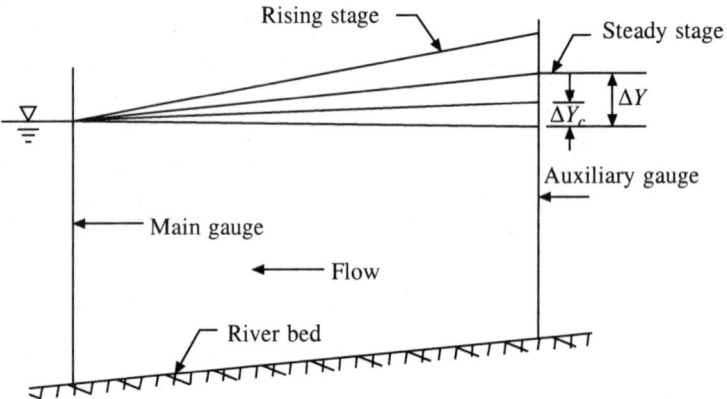

**Fig. 6.21 Correction to Flow Stages During Rising or Falling Periods**

### 6.6.1 Extension of Stage Discharge Relation

The stage-discharge relation established at a site is primarily used to interpolate discharges from the observed stages. In exceptional cases of heavy floods, it may be necessary to extrapolate the curve to read discharge from high stages. Such a situation does come in a river system. Extrapolation beyond the highest observed water level or lower than the lowest observed stage should be done with extra caution as the river control may change in those zones. It is necessary to check and examine the site for channel roughness, over bank spills and control of river at these highest and lowest zones. Such extrapolation may be subjected to risk and may lead to erroneous results and should be checked by more than one of the following methods.

*(a) Method I*
A stage-discharge relation (Fig. 6.22) when plotted on log-log paper takes the form

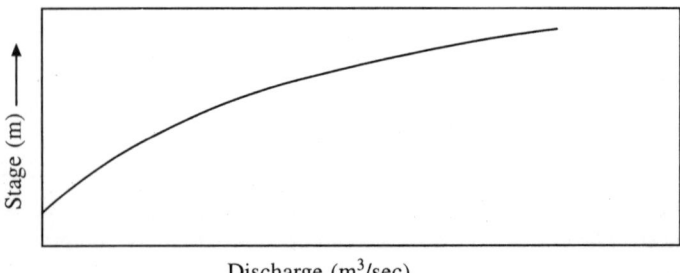

Discharge (m³/sec)

**Fig. 6.22   A Stage-Discharge Curve**

of a straight line (Fig. 6.23). The principle is used to extrapolate a *G-D* curve. Logarithmic of stage-discharge equation is used to fit a straight line which can be extrapolated easily. The procedure outlined for computation of zero gauge value $G_0$ may be followed. Taking logarithmic of equation (6.41) we get

$$\log Q = \log C + n \log (G - G_0) \qquad (6.44)$$

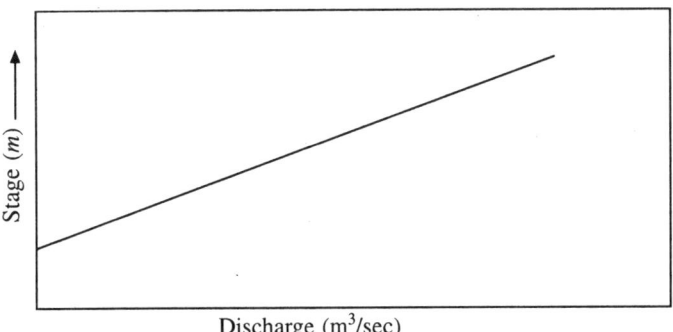

Fig. 6.23  **Stage Discharge Relation Log-Log Paper**

which is of the form of a straight line $y = a + bx$, where $y = \log Q$, $a = \log C$, representing the value of $Q$ where $(G - G_0)$ is unity, $n$ is the slope of the line and $x = \log (G - G_0)$. The straight line can be extended to compute the value of discharge for the highest and lowest stages. Value of $G_0$ must be determined before the use of this method.

*(b) Method II*

Manning's equation of uniform flow is represented as

$$Q = \frac{\sqrt{S}}{n} \times AR^{2/3} \qquad (6.45)$$

where $Q$ is the discharge in m³/sec, $S$ the river bed slope, $n$ the channel roughness, $A$ the area of cross section of the channel, $R$ the hydraulic mean radius (m). The procedure is to calculate values of $AR^{2/3}$ for the entire cross section at the river site for all the floods and plot a graph taking $AR^{2/3}$ in abscissa and stage as ordinate as shown in Fig. 6.24. Draw a mean curve passing through the data points. Assuming the highest observed $Q$ gives the value of $\sqrt{S}/n$ which is fairly constant for all stage and discharges above it, the value of $\sqrt{S}/n$ is calculated from equation (6.45) for the highest observed $Q$. To accomplish this, the value of $AR^{2/3}$ must be read from the graph of $AR^{2/3}$ vs. stage. Now, for the highest flood corresponding to the new stage for which extrapolation is required, $AR^{2/3}$ is read from the graph for the new stage and $\sqrt{S}/n$ is already known. Value of $Q$ can be easily computed from equation (6.45).

*(c) Method III*

Chezy's equation can also be used successfully to extend the stage discharge relation. Chezy's equation for discharge can be written as

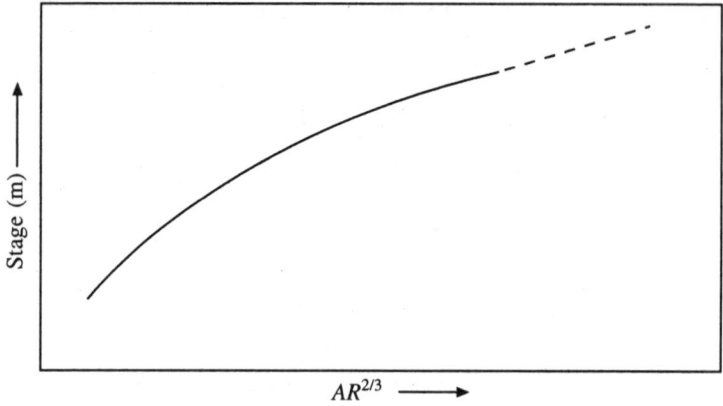

**Fig. 6.24   Plot between $AR^{2/3}$ and Stage for Extrapolation of Rating Curve**

$$Q = AV = AC \sqrt{RS} \qquad (6.46)$$

where $A$ is the area of the channel cross section, $C$ the coefficient, $S$ the slope of the energy line and $R$ is the hydraulic radius. Assuming $C\sqrt{S}$ as constant for the section for any stage and approximating $R$ as the depth of flow $d$, a plot between $Q$ vs. $A\sqrt{d}$ will be a straight line. Therefore, plot a graph between $Q$ vs. $A\sqrt{d}$ for all observed records and extrapolate the straight-line plot to get $Q$ for the $A\sqrt{d}$ of the high stage.

**Effeciency of system:** After fitting the stage-discharge curve, the standard error of estimate is calculated from the equation

$$S_e = \sqrt{\frac{1}{n} \Sigma (Q_0 - Q_c)^2} \qquad (6.47)$$

where $S_e$ is the standard error of estimate, $n$ the number of data, $Q_0$ the observed discharge, $Q_c$ the computed values of discharges from stages. A pair of curves drawn through points at 1.96 times $S_e$ distance on either side of the fitted stage-discharge curve is known as 95% confidence limit curve, which means that 19 out of 20 computed discharge values from the stages will be lying in this zone. The efficiency of the fitted curve is calculated as

$$\text{Efficiency } (\eta) = \frac{\text{Variance of observed discharge} - S_e^2}{\text{Variance of observed discharges}} \times 100 \qquad (6.48)$$

Closer is the value of $\eta$ to 100%, better is the efficiency of the fitted curve.

---

**Example 6.8:** Stage and corresponding discharge values of a river site are given as follows. Determine stage for zero discharge. Calculate discharge corresponding to stage of 226.85 m.

| Stage (m) | 221.3 | 221.65 | 221.95 | 222.8 | 223.4 | 223.85 | 224.55 | 225.4 | 225.75 |
|---|---|---|---|---|---|---|---|---|---|
| Discharge ($m^3$/sec) | 12 | 55 | 100 | 295 | 490 | 640 | 1010 | 1300 | 1550 |

**Solution**

The stage discharge relation tried is of the form $Q = C(G - G_0)^n$

Three $G_0$ elevations have been tried as shown in Table 6.5. The standard error of estimate $S_e$ is minimum for the assumed $G_0 = 221.1$ m. Therefore $G_0$ taken for the study is 221.10 m.

**Table 6.5   Trial value of $G_0$ at 221.0 m**

| Stage G (m) | Discharge $Q_0$ m³/sec | G − 221.0 | $x = \log$ (col.3) | $y = \log$ (col.2) | $x^2$ | $y^2$ | $xy$ | $Q_c = 106$ × $(G-221)^{1.735}$ | $(Q_0 - Q_c)^2$ |
|---|---|---|---|---|---|---|---|---|---|
| (1) | (2) | (3) | (4) | (5) | (6) | (7) | (8) | (9) | (10) |
| 221.3 | 12 | 0.30 | −0.523 | 1.079 | 0.273 | 1.164 | −0.564 | 13.1 | 1.2 |
| 221.65 | 55 | 0.65 | −0.187 | 1.740 | 0.035 | 3.029 | −0.326 | 50.1 | 23.1 |
| 221.95 | 100 | 0.95 | −0.022 | 2.000 | 0.000 | 4.000 | −0.044 | 96.9 | 9.1 |
| 222.80 | 295 | 1.80 | 0.255 | 2.470 | 0.651 | 6.100 | 0.630 | 294.0 | 1.0 |
| 223.40 | 490 | 2.40 | 0.380 | 2.690 | 0.144 | 7.237 | 1.023 | 484.4 | 31.9 |
| 223.85 | 640 | 2.85 | 0.455 | 2.806 | 0.207 | 7.875 | 1.276 | 652.6 | 160.4 |
| 224.55 | 1010 | 3.35 | 0.550 | 3.004 | 0.302 | 9.026 | 1.653 | 955.5 | 2971.0 |
| 225.40 | 1300 | 4.40 | 0.643 | 3.114 | 0.414 | 9.700 | 2.004 | 1386.8 | 7536.5 |
| 225.75 | 1550 | 4.75 | 0.677 | 3.190 | 0.458 | 10.178 | 2.159 | 1583.8 | 1144.2 |
| Sum | | | 2.228 | 22.083 | 2.464 | 58.309 | 7.811 | | 11878.4 |

Now
$$\Sigma x = 2.228 \qquad \Sigma y = 22.094 \qquad \Sigma xy = 7.811$$
$$\Sigma x^2 = 1.900 \qquad \Sigma y^2 = 58.306 \qquad (\Sigma x)^2 = 4.966 \; (\Sigma y)^2 = 488.16$$

Taking logarithmic of equation (6.41) we get

$$\log Q = \log C + n \log (G - G_0), \text{ which is the form of } y = a + bx$$

The coefficients are calculated from the relation

$$n = \frac{N(\Sigma\, xy) - (\Sigma\, x)(\Sigma\, y)}{N(\Sigma\, x^2) - (\Sigma\, x)^2} \quad \text{and} \quad b = \log C = \frac{\{(\Sigma\, y) - n(\Sigma\, x)\}}{N}$$

or
$$n = \frac{(9 \times 7.811 - 2.228 \times 22.094)}{(9 \times 1.90 - 2.228^2)} = 1.735$$

$$\log C = \frac{(22.094 - 1.735 \times 2.228)}{9} = 2.025208 \text{ or } C = 10^{2.025208} = 1.06.0$$

The equation becomes $Q = 106.0\,(G - 221.0)^{1.735}$.

Now substituting the values of $G$ from col. (1), the computed values of $Q$ are obtained. These values are given in col. (9) of Table 6.5. Standard error of estimate is obtained from equation (6.47) as

$$S_e = \left\{ \frac{1}{n} \Sigma\, (Q_0 - Q_e)^2 \right\}^{1/2} = \left\{ \frac{11878.4}{9} \right\}^{1/2} = 36.3$$

Correlation coefficient between the stage and discharge in cols. (4) and (5) is obtained as

$$r = \frac{N(\Sigma\, xy) - (\Sigma\, x)(\Sigma\, y)}{\sqrt{N(\Sigma\, x^2) - (\Sigma\, x^2)} \; \sqrt{N(\Sigma\, y^2) - (\Sigma\, y^2)}}$$

or
$$r = \frac{9 \times 7.81 - 2.228 \times 22.094}{\sqrt{9 \times 1.9 - 2.228^2} \; \sqrt{9 \times 58.306 - 22.094^2}} = 0.999$$

This means the correlation between stage and discharge are very good.

For Second and third trials refer Tables 6.6 and 6.7, respectively.

**Table 6.6    Trial Value of $G_0$ at 221.1 m**

| Stage $G$ (m) | Discharge $Q_0$ m³/sec | $G-221.1$ | $x = \log$ (col. 3) | $y = \log$ (col. 2) | $x^2$ | $y^2$ | $xy$ | $Q_c = 137.5$ $(G-221.1)^{1.547}$ | $(Q_0-Q_c)^2$ |
|---|---|---|---|---|---|---|---|---|---|
| (1) | (2) | (3) | (4) | (5) | (6) | (7) | (8) | (9) | (10) |
| 221.3 | 12 | 0.30 | −0.523 | 1.079 | 0.273 | 1.164 | −0.564 | 13.1 | 1.2 |
| 221.3 | 12 | 0.20 | −0.700 | 1.079 | 0.488 | 1.164 | −0.754 | 11.4 | 0.4 |
| 221.65 | 55 | 0.55 | −0.259 | 1.740 | 0.067 | 3.029 | −0.452 | 54.5 | 0.2 |
| 221.95 | 100 | 0.85 | −0.071 | 2.000 | 0.005 | 4.000 | −0.141 | 106.9 | 48.0 |
| 222.80 | 295 | 1.70 | 0.230 | 2.470 | 0.053 | 6.100 | 0.569 | 312.4 | 305.7 |
| 223.40 | 490 | 2.30 | 0.362 | 2.690 | 0.131 | 7.237 | 0.973 | 498.8 | 77.5 |
| 223.85 | 640 | 2.75 | 0.439 | 2.806 | 0.193 | 7.875 | 1.232 | 657.6 | 311.5 |
| 224.55 | 1010 | 3.45 | 0.538 | 3.004 | 0.289 | 9.026 | 1.616 | 934.0 | 5770.9 |
| 225.40 | 1300 | 4.30 | 0.633 | 3.114 | 0.401 | 9.700 | 1.972 | 1313.2 | 175.1 |
| 225.75 | 1550 | 4.65 | 0.667 | 3.190 | 0.445 | 10.178 | 2.129 | 1482.2 | 4591.1 |
| Total | | 1.841 | | 22.094 | 2.074 | 58.306 | 7.146 | | 11281.6 |

Solving for $n$ and $C$ as done above, we get $n = 1.547$, log $C = 2.138455$ or $C = 137.5$

The equation becomes $Q = 137.5(G - 221.1)^{1.547}$

Correlation coefficient $r = 0.9995$

Standard error of estimate $S_e = 35.4$

**Table 6.7    Trial Value of $G_0$ at 221.2 m**

| Stage $G$ (m) | Discharge $Q_0$ m³/sec | $G-221.2$ | $x = \log$ (col. 3) | $y = \log$ (col. 2) | $x^2$ | $y^2$ | $xy$ | $Q_c = 184.7 \times$ $(G-221.2)^{1.30}$ | $(Q_0-Q_c)^2$ |
|---|---|---|---|---|---|---|---|---|---|
| (1) | (2) | (3) | (4) | (5) | (6) | (7) | (8) | (9) | (10) |
| 221.3 | 12 | 0.10 | −1.000 | 1.079 | 1.000 | 1.164 | −1.079 | 9.25 | 7.5 |
| 221.65 | 55 | 0.45 | −0.365 | 1.740 | 0.120 | 3.029 | −0.603 | 65.4 | 108.4 |
| 221.95 | 100 | 0.75 | −0.125 | 2.000 | 0.156 | 4.000 | −0.250 | 127.1 | 732.8 |
| 222.80 | 295 | 1.60 | 0.204 | 2.470 | 0.042 | 6.100 | 0.504 | 340.2 | 2049.3 |
| 223.40 | 490 | 2.20 | 0.342 | 2.690 | 0.117 | 7.237 | 0.921 | 514.8 | 613.7 |
| 223.85 | 640 | 2.65 | 0.423 | 2.806 | 0.179 | 7.875 | 1.188 | 655.7 | 245.6 |
| 224.55 | 1010 | 3.35 | 0.525 | 3.004 | 0.276 | 9.026 | 1.577 | 889.2 | 14579.9 |
| 225.40 | 1300 | 4.20 | 0.623 | 3.114 | 0.388 | 9.700 | 1.941 | 1193.1 | 11419.3 |
| 225.75 | 1550 | 4.55 | 0.658 | 3.190 | 0.433 | 10.178 | 2.099 | 1324.0 | 51084.6 |
| Total | | 1.304 | | 22.094 | 2.571 | 58.309 | 6.298 | | 80841.1 |

Solving for $n$ and $C$ as done above, we get $n = 1.30$, log $C = 2.266567$ or $C = 184.7$

The equation becomes $Q = 184.7(G - 221.2)^{1.30}$
Correlation coefficient $r = 0.995$
Standard error of estimate $S_e = 94.77$

Comparing the results of Tables 6.5, 6.6 and 6.7, the minimum standard error is obtained for $G_0$ at 221.10 m.

Therefore the final equation for the problem becomes

$$Q = 137.548 \, (G - 221.10)^{1.547}$$

Corresponding to $G_0 = 226.85$,

$$Q(\text{m}^3/\text{sec}) = 137.548 \, (226.85 - 221.10)^{1.547} = 2059.3 \text{ m}^3/\text{sec}.$$

**Example 6.9:** The following data are observed from a stream gauging station

| | | |
|---|---|---|
| Main gauge reading (m) | 19.3 | 19.3 |
| Auxiliary gauge reading (m) | 19.12 | 18.72 |
| Discharge at main gauge (m³/sec) | 6.80 | 10.50 |

Calculate the discharge when the main gauge is still 19.3 m and auxiliary gauge is 19.0 m. Auxiliary gauge is at the downstream of the main gauge.

**Solution**

To solve the problem, a relation between change in gauge of main and auxiliary station can be correlated to the observed discharge in the following way.

$$S_1 = 19.3 - 19.12 = 0.18 \text{ corresponds to } Q_1 = 6.8 \text{ m}^3/\text{sec}$$

$$S_2 = 19.3 - 18.72 = 0.58 \text{ corresponds to } Q_2 = 10.5 \text{ m}^3/\text{sec}$$

$\therefore$ The problem is to calculate $Q_3$ when $S_3 = 19.3 - 19 = 0.30$

The relation is essentially nonlinear. Let the form of equation be $S = Q^k$

For the given data $0.18 = 6.8^m$ and $0.58 = 10.5^n$

The value of $k$ should satisfy both the above relations. To do this let us reduce them to the following form

$$\left( \frac{0.18}{0.58} \right) = \frac{(6.8)^m}{(10.5)^n} = \left( \frac{6.8}{10.5} \right)^k, \quad \text{where } k = m/n$$

or

$$0.3103 = (0.6476)^k$$

Taking logarithmic to both sides, $\log (0.3103) = k \log (0.6476)$

or $\quad -0.508155 = -k \times 0.18868$ which gives $k = 2.6932$

Therefore the relation becomes $\dfrac{S_2}{S_3} = \left( \dfrac{Q_2}{Q_3} \right)^{2.6932}$

Now for $S_2 = 0.18$, $S_3 = 0.30$ and $Q_2 = 6.8$, $Q_3 = ?$

Applying the relation we get $\left( \dfrac{0.18}{0.3} \right) = \left( \dfrac{6.8}{Q} \right)^{2.6932}$

which gives $Q = 8.22 \text{ m}^3/\text{sec}$.

## 6.7 REQUIREMENT OF A GOOD GAUGE-DISCHARGE SITE

Gauge and discharge measurement should preferably be located at the same site (since good correlation always exists between the two) fulfilling the following requirements:

1. The river reach must be stable and fairly straight on both upstream and downstream for a length of 0.75 to 1.00 km.

2. Elevation and discharge relation should always be uniform, i.e., site is not subjected to shifting control.
3. Site should be easily accessible during all times in a year.
4. Site should be sufficiently upstream to the flood forecasting area so that flood warnings can be given in advance.
5. The site should be sensitive to all stage and discharges, i.e., for a small change in discharge, measurable change in stage should occur.
6. Back water or tidal effect should be the minimum.
7. Site should be away from bridges. If bridge sites cannot be avoided then it should be located upstream to the bridge site at a distance equal to 4 times the width of the bridge.
8. When a tributary joins, then the site should be located 0.8 km upstream or downstream of their confluence.
9. At a site, wind action and disturbance due to animals should be minimum.
10. Site should have stable and high banks to contain floods.
11. Rock outcrops and vegetal growth at the reach should be the minimum.
12. Islands should not be present at the gauging section.
13. Cross section of the entire reach of the river should be fairly uniform.
14. Cross currents, vortex and eddies formation, reverse slope in parts of the channel bed should be absent at the *G-D* site.

### 6.7.1  Network Design

No definite relation can be established to fix the number of stations to gauge completely a river system. Adequacy of the gauge network is sometimes related to the population density, industrial, mineral and other important features of the basin in question. A basin should have two types of stations.

(i) *The base or permanent stations:* For which a long term data for the important tributaries and the main river are always collected.
(ii) *Auxiliary or Secondary stations:* Short term data for other locations in the river and tributaries are always needed to form a network of gauging sites. The stations should be located at the existing or potential dam sites and other strategic places for flood forecasting purposes. Data from secondary stations are correlated with the primary stations to give an extended long term series at the secondary site.

The network (based on World Meteorological Organization guidelines) that can be followed are:

1. In principal streams, the first set of stations should be located from the upstream, draining an area of 1300 sq. km.
2. Subsequent stations should be added where the basin area doubles approximately.
3. Whenever a major tributary join, a station should be fixed downstream of their confluence.
4. Guidelines for the requirement of good *G-D* site for a station should be strictly followed.

## 6.8 RUNOFF COMPUTATION

Stream flow measures the part of precipitation excess from a watershed over a period of time. A hydrologist or an engineer-in-charge of water resources project planning and management may be interested to know the total quantity of runoff from the watershed at the project site over the period of years. Thus the total runoff from a project catchment over the period of one year is called the *annual yield*. It represents the total runoff volume from the catchment. At any potential project site, availability of a long term runoff data covering a period of 30 years or more is essential to study the economics of the project and to fix the pattern of demands from the project. Yield at a site for longer years help to fix the design parameters of the project like the height, storage capacity, release patterns for irrigation, power generation, municipal demands and other requirements. In a developing country like India, such requirements are hardly available at the potential sites. Therefore reliable methods of generating the stream flow data should be used to take care of the complex rainfall-runoff process. The method should effectively combine the precipitation, interception, depression storage, infiltration and evapotranspiration variables to produce a net output at the desired location in a basin. In hydrology two watersheds cannot be alike and the same watershed may not behave in the same way under two similar circumstances. Therefore, a broad approach to get fairly acceptable results should be used to solve the problems. Various methods used for the runoff analysis are: (a) Extension of flow data, (b) Rainfall-runoff correlation, (c) Empirical relations, (d) Runoff simulation models.

### 6.8.1 Extension of Runoff Records

When a runoff series at the site of interest is available for a period of say 12 years or more, the runoff volumes can be synthetically generated by the following procedures.

**(1) Thomas-Fiering Model:** The model has been applied successfully to generate sequentially the monthly, 10-daily or weekly volumes of discharges from a serially dependent series. The model is an extension of the *Markov model* which assumes that a monthly or 10-daily variable is dependent only on the just recent one or two variables involving non-stationarity both in mean and standard deviation. Any monthly runoff volume is dependent on the just preceding month to some degree and the monthly means and standard deviations are non-stationary due to solar cycles. For example, flow volumes of July are dependent on the volumes of June to some extent, the mean and standard deviation of all June months are different to that of July. The model which is described in detail in Chapter 2 is again reproduced here.

$$q_{j+1} = q_{avj+1} + b_{j,j+1}(q_i - q_{av}) + Z_i S_{j+1} \sqrt{1 - r_{j,j+1}^2} \qquad (6.49)$$

where $q_{j+1}$, $q_j$ are the discharge volumes during $(j+1)$th and $j$th month; $q_{avj+1}$, and $q_{avj}$ the mean monthly discharge volumes for $(j+1)$th and $j$th months; $S_{j+1}$, $S_j$ the standard deviations for $(j+1)$th and $j$th months, $r_{j,j+1}$ the correlation coefficient between the months $j$ and $(j + 1)$, $Z_j$ the random independent variate with zero mean and unit variance, $b_{j,j+1} = r_{j,j+1} (S_{j+1}/S_j)$. When $j$ represents June then $(j + 1)$ should mean July, and so on. Example on Thomas-Fiering model is given in Chapter 2.

**(2) Harmonic Analysis Model:** When the period of analysis of a time series is less than a year then the series is affected by seasonality. Such a time series is non-stationary in nature. The series free from deterministic components like trend (if present) and periodicity are modeled by fitting upto 6 numbers of harmonics as outlined in Chapter 2. The residuals after removing periodicity are quantified as stationary stochastic component or component due to randomness. The residual series may be having dependence which is tested by serial correlation coefficients of different lags. If dependence in the data exists then a moving average or linear autoregressive model is used. If series are independent, i.e., serial correlation coefficient of different lags are zero, then the statistics of residual stochastic components are calculated and a random number series is reduced to the properties of the above statistics. Using the reduced random number with the properties of the time series, the process is reversed to synthesise the probable sequence of the series.

**(3) Correlation with the Adjoining Station:** When a long-term runoff data at an adjacent base station is available then a correlation of the following forms may be used to calculate the flow of the required station

$$Q_R = Q_B \left( \frac{A_R}{A_B} \right)^{0.75} \tag{6.50}$$

or
$$Q_R = Q_B \left( \frac{A_R}{A_B} \right)^{0.67} \tag{6.51}$$

where $Q_R$ and $Q_B$ are discharges in m$^3$/sec of regressed and the base stations, $A_R$ and $A_B$ are the corresponding catchment area in sq. km. Selection of equation (6.50) or (6.51) is made depending on Dicken's or Ryve's formula used for computation of discharge in that area. Normally for north India, equation (6.50) may be used and for south India, equation (6.51) is preferred. The relation is applicable only where discharge and catchment variation is within 10% of the base station.

---

**Example 6.10:** Monthly runoff data of river Mahanadi for a base station A from 1975 to 1991 is given below along with an auxiliary station B data from 1975 to 1991. Extrapolate the data of the auxiliary station from 1975 to 1991. Catchment area of station A is 7890 sq. km and that of B is 7280 sq. km

| Year (19 ) | 75 | 76 | 77 | 78 | 79 | 80 | 81 | 82 | 83 | 84 | 89 | 86 | 87 | 88 | 89 |
|---|---|---|---|---|---|---|---|---|---|---|---|---|---|---|---|
| | 90 | 91 | | | | | | | | | | | | | |
| $Q_B$ m$^3$/sec | 522 | 659 | 1196 | 604 | 586 | 616 | 464 | 769 | 523 | 527 | 537 | 650 | 684 | 557 | 765 |
| | 580 | 1153 | | | | | | | | | | | | | |
| $Q_A$ m$^3$/sec | – | – | – | – | – | – | – | 271 | 210 | 211 | 154 | 182 | 246 | 153 | 244 |
| | 183 | 322 | | | | | | | | | | | | | |

**Solution**

The location of both the catchments are in Orissa, where Dicken's equation is used to calculate flood discharge. Using equation (6.50)

$$Q_A = Q_B \left( \frac{A_A}{A_B} \right)^{0.75} = Q_B \left( \frac{7280}{7890} \right)^{0.75} = 0.9414 \, Q_B$$

Extrapolation of data of station A from 1981 to 1975 is carried out using the above equation

| Year | 1975 | 1976 | 1977 | 1978 | 1979 | 1980 | 1981 |
|---|---|---|---|---|---|---|---|
| Discharge of A, $Q_A$ | 522 | 659 | 1196 | 604 | 586 | 616 | 464 |
| Discharge of B, $Q_B$ | 491.4 | 620.4 | 1126.9 | 568.8 | 551.7 | 579.9 | 436.8 |

**(4) Regression Techniques:** Various regression equations discussed in Chapter 2 correlate two sets of data, one being dependent and the other independent in a way the regression coefficients describe them. In the present problem, the base station for which a long term data is available is taken as dependent variable ($q_b$) and the station for which runoff is to be computed is taken as independent variable ($q_r$). The equation of a straight line can be represented as

$$q_r = a + b \, q_b \tag{6.52}$$

where $a$ and $b$ are regression coefficients. The coefficients can be computed from the relations.

$$b = \frac{N \Sigma q_b q_r - \Sigma q_r \cdot \Sigma q_b}{N \Sigma q_b^2 - (\Sigma q_b)^2} \tag{6.53}$$

$$a = \frac{\Sigma q_r \cdot \Sigma q_b^2 - \Sigma q_b \cdot \Sigma q_b q_r}{N \Sigma q_b^2 - (\Sigma q_b)^2} \tag{6.54}$$

and the correlation between the two sets of variables is represented by the coefficient

$$r = \frac{N \Sigma q_b q_r - \Sigma q_r \cdot \Sigma q_b}{\sqrt{N \Sigma q_b^2 - (\Sigma q_b)^2} \, \sqrt{N \Sigma q_r^2 - (\Sigma q_r)^2}} \tag{6.55}$$

in which $r$ lies between 0 to +1 as it is positively correlated. A value of 0.6 to 1.00 indicates good correlation. Regression analysis of concurrent 10-daily or monthly to monthly flow should be established.

---

**Example 6.11:** Solve example 6.10 by regression method.

**Solution**

Solution is carried out in Table 6.8.

Using equations (6.52) through (6.55), the values of coefficients $a$, $b$ and the correlation coefficient $r$ are calculated by least square method by fitting an equation of straight line. The parameters are

$$b = 0.328394, \, a = 1.762070 \text{ and } r = 0.939038$$

**Table 6.8   Regression Method of Obtaining Flow Sequence from the Available Data of Neighbouring Station**

| Year | Discharge $Q_B = x$ (m³/sec) | Discharge $Q_A = y$ (m³/sec) | $x \times x$ (2) × (2) | $y \times y$ (3) × (3) | $x \times y$ (2) × (3) | Year | Discharge $Q_B$ (cumec) | Discharge $Q_A = 1.762$ $+ 0.3284Q_B$ |
|------|------|------|------|------|------|------|------|------|
| (1) | (2) | (3) | (4) | (5) | (6) | (7) | (8) | (9) |
| 1982 | 769 | 271 | 591361 | 73441 | 208399 | 1975 | 522 | 173.1837 |
| 1983 | 523 | 210 | 273529 | 44100 | 109830 | 1976 | 659 | 218.1737 |
| 1984 | 527 | 211 | 277729 | 44521 | 111197 | 1977 | 1196 | 394.5212 |
| 1985 | 537 | 154 | 288369 | 23716 | 82698 | 1978 | 604 | 200.1120 |
| 1986 | 650 | 182 | 422500 | 33124 | 118300 | 1979 | 586 | 194.2009 |
| 1987 | 684 | 246 | 467856 | 60516 | 168264 | 1980 | 616 | 204.0527 |
| 1988 | 557 | 153 | 310249 | 23409 | 85221 | 1981 | 464 | 154.1368 |
| 1989 | 765 | 244 | 585225 | 59536 | 186660 | | | |
| 1990 | 580 | 183 | 336400 | 33489 | 106140 | | | |
| 1991 | 1153 | 322 | 1329409 | 103684 | 371266 | | | |
| Sum | 6745 | 2176 | 4882627 | 40536 | 1547975 | | | |

The equation developed is of the form : $Q_A = 1.76207 + 0.328394\ Q_B$
Using the equation, the flow series from 1975 to 1981 are generated as shown in col. (9) of Table 6.8.

**(5) Longbeins Method:** This method is preferable when short term runoff data of station $r$ is in the neighbourhood of a base station $b$, where long term data is available. The two stations should preferably be of the same basin. A 10-daily to 10-daily or monthly to monthly data for the two stations is correlated as follows:

1. Prepare a table in which the first column is the common years for which data for both stations are available.
2. In the second column enter the 10 daily or monthly values of the base station against the year of col. (1).
3. In the third column enter the logarithmic values of col. (2). Sum the logarithmic values for the column and find mean of the series log $q_{avb}$.
4. In the fourth column enter the value of (log $q_b$ – log $q_{avb}$).
5. Columns 5, 6, and 7 should repeat the procedures of steps (2), (3) and (4) respectively with data of station $r$.
6. In col. (8), enter the square of values of col. (4).
7. In the next two columns enter the values of cols. (7) × (4) and (7) × (7).
8. A linear relation of the type $y = a + bx$ is to be developed taking $x_i$ as the data of col. (4) and $y_i$ as the data of col. (7). Since the variables are the deviations from the mean, the intercept $a$ of the equation will be zero and the slope $b$ is computed as

$$b = \frac{\Sigma\, xy}{\Sigma\, x^2} \tag{6.56}$$

The correlation coefficient $r$ is computed as

$$r = \frac{\Sigma\, xy}{\Sigma\, x^2 \cdot \Sigma\, y^2} \tag{6.57}$$

9. In col. (11) generate the series taking the equation $y = bx$.
10. The logarithm of discharge for station $r$ is obtained in col. (12) by adding $y_i$ series of col. (10) with the logarithm mean of series $r$.
11. Discharge of station $r$ is obtained by taking antilog of col. (12) and the values are entered in col. (13).

The details of the procedure are outlined in the following example.

---

**Example 6.12:** Solve example 6.10 by Longbein's method.

**Solution**
Computations are carried out in Table 6.9.
From step (8), the slope of the fitted line $b$ and the correlation coefficient are computed as

$$\text{Slope } b = \frac{0.0835}{0.1037} = 0.804711$$

$$\text{Correlation coefficient} = \frac{0.0835}{(0.1037 \times 0.10)^{1/2}} = 0.81941$$

The equation becomes $y_i = 0.8047\, x_i$ or $Q_r = 0.8047 Q_B$

In col. (11) the flow series from 1975 to 1981 are generated by using the equation $Q_r = 0.8047 Q_B$. To this series when the mean flow of 2.326038 of station $r$ is added, it gives the logarithmic series of station $r$ from 1975–81. Taking antilogarithms of the series of col. (12), the generated flow series for the period from 1975 to 1981 are computed in col. (13) of Table 6.9.

---

**(6) Flow Duration Curve Method:** A flow duration curve is obtained by plotting the percent of time the flow has equalled or exceeded in abscissa against the magnitude of flow as ordinate. It is also known as *frequency-discharge curve*. The following procedure can be followed to obtain the curve.

1. Arrange the given series of river discharges in descending order of magnitude. The data can be daily, weekly, 10-daily or monthly volumes.
2. Assign serial numbers to the orders of step (1). For the highest value take $m = 1$, and for the lowest value $m = N$, where $N$ is the number of data used in the analysis.
3. Calculate the percentage probability of the magnitude being equalled or exceeded as

$$P = \left[ \frac{m}{(N + 1)} \right] \times 100 \qquad (6.58)$$

4. Plot a graph with $P$ in abscissa and $Q$ as ordinate. The resulting graph called flow-duration curve (Fig. 6.25) represents the cumulative frequency distribution of stream flow data providing following information

(i) The *dependability of the flow* at any percent of time can be read from the curve. For example, the flow $Q_i$ corresponding to $P = 50$ or 75% gives 50% or the 75% dependable flow magnitudes for the stream.

## Table 6.9  Longbein's Method of Extension of Flow Records

| Year | Discharge $Q_b$ (cumec) | Log (2) | (3)−2.815 $x_i$ | Discharge $Q_r$ (cumec) | Log (5) | (3)−2.326 $y_i$ | $x_i x_i$ | $y_i y_i$ | $x_i y_i$ | $y_i = (4)$ × 0.8047 | Col. (11) +2.326 | Discharge $Q_r$ |
|---|---|---|---|---|---|---|---|---|---|---|---|---|
| (1) | (2) | (3) | (4) | (5) | (6) | (7) | (8) | (9) | (10) | (11) | (12) | (13) |
| 1975 | 522 | 2.717 | −0.0980 | | | | | | | −0.079 | 2.2469 | 176.6 |
| 1976 | 659 | .818 | 0.0029 | | | | | | | 0.0024 | 2.3284 | 213.0 |
| 1977 | 1196 | 3.077 | 0.2618 | | | | | | | 0.2107 | 2.5367 | 344.1 |
| 1978 | 604 | 2.781 | −0.0348 | | | | | | | −0.0281 | 2.2979 | 198.6 |
| 1979 | 586 | 2.767 | −0.0479 | | | | | | | −0.038 | 2.2874 | 193.8 |
| 1980 | 616 | 2.789 | −0.0263 | | | | | | | −0.0212 | 2.3048 | 201.7 |
| 1981 | 464 | 2.666 | −0.1493 | | | | | | | −0.1202 | 2.2058 | 160.6 |
| 1982 | 769 | 2.885 | 0.0700 | 271 | 2.4329 | 0.1069 | 0.0049 | 0.0114 | 0.0074 | | | |
| 1983 | 523 | 2.718 | −0.0973 | 210 | 2.3222 | −0.003 | 0.0094 | 0.0000 | 0.0004 | | | |
| 1984 | 527 | 2.721 | −0.0940 | 211 | 2.3242 | −0.001 | 0.0088 | 0.0000 | 0.0001 | | | |
| 1985 | 537 | 2.729 | −0.0859 | 154 | 2.1875 | −0.138 | 0.0073 | 0.0191 | 0.0119 | | | |
| 1986 | 650 | 2.819 | −0.0029 | 182 | 2.2601 | −0.065 | 0.0000 | 0.0043 | 0.0002 | | | |
| 1987 | 684 | 2.835 | 0.0191 | 264 | 2.3909 | 0.064 | 0.0004 | 0.0042 | 0.0012 | | | |
| 1988 | 557 | 2.745 | −0.0700 | 153 | 2.1847 | −0.141 | 0.0049 | 0.0199 | 0.0099 | | | |
| 1989 | 765 | 2.883 | 0.0677 | 244 | 2.3874 | 0.061 | 0.0045 | 0.0037 | 0.0041 | | | |
| 1990 | 580 | 2.763 | −0.0524 | 183 | 2.2624 | −0.063 | 0.0027 | 0.0040 | 0.0033 | | | |
| 1991 | 1153 | 3.061 | 0.2459 | 322 | 2.5078 | 0.182 | 0.0604 | 0.0331 | 0.0447 | | | |
| Sum = | | 28.1589* | | | 23.2601 | | 0.1033 | 0.0997 | 0.0832 | | | |
| Mean = | | 2.815 | | | 2.32601 | | | | | | | |

*Summation from 1982 to 1991 only.

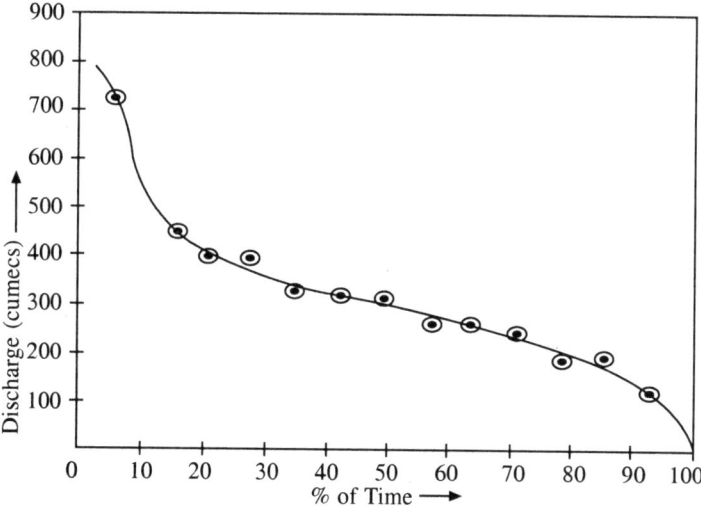

**Fig. 6.25 Flow Duration Curve**

(ii) It is used to extend the stream flow data in the following way. Let there be two stations $A$ and $B$ with long term (say period $Y$) runoff data available for station $A$ and short term (say period $X$) data available for station $B$. To extrapolate the data of station $B$ equal to the length of $A$, plot a flow duration curve of $B$ for length of data $X$. Plot two more flow duration curves for station $A$ for periods $X$ and $Y$, as shown in Fig. 6.26. The problem is to constitute a flow duration curve of station $B$ for the period $Y$. Let the graphs be designated as $A_x$, $A_y$, $B_x$ and $B_y$.

To constitute the curve $B_y$ from the three known curves $A_x$, $A_y$ and $B_x$, take a point $P_1$ on the curve $B_x$ and draw a line parallel to discharge ordinate till it touches line $A_x$ curve at $P_2$. Draw a line parallel to the time scale to reach $A_y$ curve at $P_3$ and then locate the point $P_4$ such that points $P_1$, $P_2$, $P_3$ and $P_4$ form corners of a rectangle. The lines $P_1P_2$, $P_2P_3$, $P_3P_4$ and $P_1P_4$ should be parallel to their respective coordinates. $P_4$ is a point on the curve $B_y$. Likewise take several points on curve $B_x$ and get a series of points $P_4$ which should complete the curve drawing of $B_y$. The values for different years of discharge can be found from the plotted curve $B_y$ from the $A_y$ time percentage-discharges of station $A$.

(iii) Flow duration curve on a log-log paper is an useful tool, from which the variability of stream flow can be compared with the data of neighbouring stream records.

(iv) The flow-duration curve is also used for flood-control studies, hydropower potential studies, sediment studies and drainage character studies of a basin.

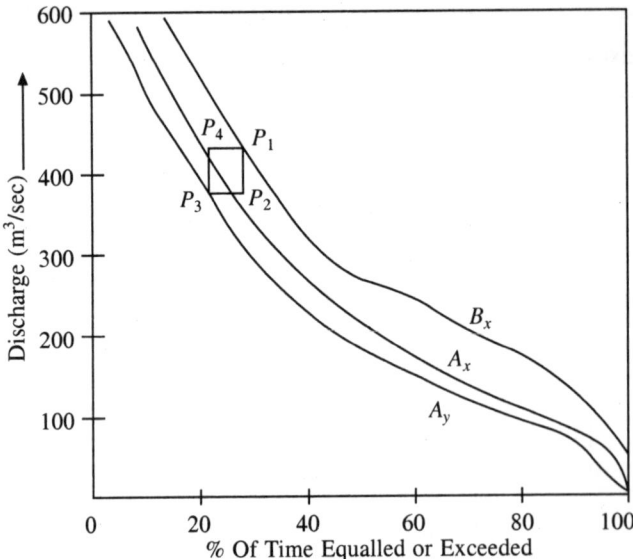

**Fig. 6.26    Extension of Stream Flow Data Using Flow-Duration Curve**

**Example 6.13:** Draw a flow duration curve from the annual peak flow discharges at a gauge site of a river in Orissa using the following data .

| Year | 1980 | 1981 | 1982 | 1983 | 1984 | 1985 | 1986 | 1987 | 1988 | 1989 | 1990 | 1991 | 1992 |
|---|---|---|---|---|---|---|---|---|---|---|---|---|---|
| Discharge (m³/sec) | 311 | 720 | 430 | 330 | 248 | 125 | 405 | 400 | 233 | 178 | 248 | 321 | 186 |

**Solution**
The information required for plotting flow-duration curve are calculated in Table 6.10 Plot of the flow duration curve is shown in Fig. 6.25.

**Table 6.10    Data for Construction of Flow Duration Curve for Example 6.13**

| Year | Peak flow $Q_i$ (m³/sec) | Peak flow in decreasing order | No. of times flood exceeded or equalled | % times flood equalled or exceeded $100m/(N + 1)$ |
|---|---|---|---|---|
| (1) | (2) | (3) | (4) | (5) |
| 1980 | 311 | 720 | 1 | 7.1 |
| 1981 | 720 | 430 | 2 | 14.3 |
| 1982 | 430 | 405 | 3 | 21.4 |
| 1983 | 330 | 400 | 4 | 28.6 |
| 1984 | 248 | 330 | 5 | 35.7 |
| 1985 | 125 | 321 | 6 | 42.9 |
| 1986 | 405 | 311 | 7 | 50.0 |
| 1987 | 400 | 248 | 8 | 57.1 |
| 1988 | 233 | 248 | 9 | 64.3 |
| 1989 | 178 | 233 | 10 | 71.4 |
| 1990 | 248 | 186 | 11 | 78.6 |
| 1991 | 321 | 178 | 12 | 85.7 |
| 1992 | 186 | 125 | 13 | 92.9 |

## 6.9 RUNOFF FROM RAINFALL RECORDS

At a potential water resources project site a hydrologist should have long term runoff volumes as the primary data for its planning and management. In a developing country like India, seldom does one get such a situation where discharge observation at the required site for more than 30 years or so is available.

However, rainfall is gauged extensively and rather uniformly over the country from which reliable runoff informations can easily be computed. The following situations are generally encountered at any site.

### 6.9.1 Use of Rainfall-Runoff Data at the Site

At the project site a short term runoff data (of 10 years or more) and a long term rainfall data covering the entire catchment upto a project site is available. Steps to compute yield series for such a situation are:

1. Prepare a map of the basin upto the project site and mark the rain gauge stations.
2. Check the consistency of the data for each station and correct them whenever the data is found inconsistent.
3. Fill the gaps in data (if any) by a suitable methods.
4. Compute the average rainfall over the catchment by any of the methods outlined in the precipitation chapter. It is better if monthly total rainfalls are computed.
5. Correlate the monthly rainfall depths with the corresponding runoff depths over the catchment say for the month of June and develop a regression relationship between them. When 10 years data are available, then ten numbers of June rainfall-runoff data are used to develop the relation.
6. Similarly develop relationships between rainfall and runoff for other months of the year. If the rainfall is predominant in the monsoon months only then develop relationships only for those months (say from June to October = 5 regression relations to be developed). It is better if the best fit straight line relationship can be developed for each month than going for higher order equations which may not justify the rainfall-runoff process.
7. Compute the correlation coefficient $r$ for each month. The monthly correlation coefficients should preferably be close to +1. Under no case should it be less than 0.6, because the variance explained by fitting such a line is taken as $0.6^2 = 0.36$ or 36%.
8. Using the monthly relation (say, for June), compute the runoff for June for the years for which only average rainfall is available but not the concurrent runoff. Likewise extend the runoff series for all other months of the year using the corresponding monthly regression equations.
9. Sum the runoff depths for all the monsoon months thus extended which should give the total runoff depth for the monsoon periods of the respective years.
10. Prepare the monsoon total of runoff volumes for all the years on record by multiplying the catchment area. Keep the observed runoff volumes undisturbed in the series for the years of the available data.

11. If the non-monsoon runoff are mostly contributed from ground water source, i.e., when rainfall in these months are very less, then the flows from say November to May is computed on percentage basis with respect to the monsoon flows for each month. The percentages for the non-monsoon months are fixed on the basis of observed data.

12. The non-monsoon volumes thus computed are added to give total non-monsoon yield at the site.

13. The annual series is computed adding the monsoon and non-monsoon yields.

14. The annual series are arranged in decreasing order of magnitude and are serially numbered.

15. Percent dependable flows are marked on the basis of relation of probability of exceedence, i.e., $P = [(m/N+1) \times 100]$, where $P = 0.5$ means 50% dependable flow, $P = 0.75$ means 75% dependable and so on, $m$ stands for the rank of the data at which dependable flow occurs and $N$ the total data length.

---

**Example 6.14:** Compute the yield series for Manjore irrigation project (Orissa) from the data given in the following tables. The columns for $x$ represent the rainfall depths and $y$ runoff depths averaged over the catchment. Free catchment of the project is 285.4 sq. km. The years of available data are entered in the first columns of the respective tables.

**Solution**
Monthwise rainfall-runoff analysis are carried out in the following tables, where $X$ is the average catchment rainfall in mm, $Y$ the runoff averaged over the basin in mm, $S_E$ the standard error and $r_j$ the correlation coefficient between $X$ and $Y$ series for the months $j$.

**Table 6.11    Rainfall-Runoff Correlation for the Month of June**

| Year | $X_i$ | $Y_i$ | $X_i^2$ | $Y_i^2$ | $X_i Y_i$ |
|------|-------|-------|---------|---------|-----------|
| (1)  | (2)   | (3)   | (4)     | (5)     | (6)       |
| 1984 | 435.5 | 32.5  | 189660  | 1056    | 14154     |
| 1985 | 143.4 | 1.4   | 20544   | 2       | 201       |
| 1986 | 550.6 | 119.5 | 303160  | 14280   | 65797     |
| 1987 | 116.8 | 16.2  | 13642   | 262     | 1891      |

Most acceptable equations developed are

1. $Y = -23.98 + 0.211X : S_E = 47.57 :$    $r = 0.855$ accepted for the project

2. $Y = 0.000707\ X^{1.827} : S_E = 2.094 :$    $r = 0.78$ (by discarding more values).

Most acceptable equations developed for July (Table 6.12) are

$Y = -31.998 + 0.44\ X$  $: S_E = 148.56$  $: r = 0.611$

$Y = 0.0361\ X^{1.3667}$    $: S_E = 1.127$    $: r = 0.803$ accepted for the project.

**Table 6.12 Rainfall-Runoff Correlation for the Month of July**

| Year | $X_i$ | $Y_i$ | $X_i^2$ | $Y_i^2$ | $X_i Y_i$ |
|------|-------|-------|---------|---------|-----------|
| (1) | (2) | (3) | (4) | (5) | (6) |
| 1980 | 283.1 | 63.9 | 80146 | 4085 | 18094 |
| 1981 | 106.0 | 23.7 | 11236 | 563 | 2514 |
| 1982 | 260.3 | 93.3 | 67756 | 8695 | 24272 |
| 1983 | 302.8 | 97.3 | 91688 | 9459 | 29450 |
| 1984 | 242.7 | 57.1 | 58903 | 3259 | 13854 |
| 1985 | 251.5 | 38.7 | 63252 | 1500 | 9741 |
| 1986 | 355.6 | 251.2 | 126451 | 63085 | 89315 |
| 1987 | 449.2 | 110.2 | 201780 | 12144 | 49500 |

**Table 6.13 Rainfall-Runoff Correlation for the Month of August**

| Year | $X_i$ | $Y_i$ | $X_i^2$ | $Y_i^2$ | $X_i Y_i$ |
|------|-------|-------|---------|---------|-----------|
| (1) | (2) | (3) | (4) | (5) | (6) |
| 1980 | 116.1 | 49.1 | 13479 | 2410 | 5699 |
| 1981 | 484.0 | 238.4 | 234256 | 56843 | 15394 |
| 1982 | 923.1 | 477.2 | 852114 | 227702 | 440486 |
| 1983 | 438.8 | 229.8 | 192545 | 52811 | 440486 |
| 1984 | 765.7 | 247.4 | 586297 | 61202 | 189427 |
| 1985 | 464.5 | 590.9 | 417962 | 349208 | 382041 |
| 1986 | 238.0 | 194.7 | 56644 | 37929 | 46352 |
| 1987 | 115.5 | 34.8 | 133 | 1214 | 4024 |

Most acceptable equations developed are

$Y = 18.592 + 0.513 X$    $S_E = 302.7 : r = 0.803$

$Y = 0.2301 X^{1.13268}$    $S_E = 1.001 : r = 0.926$ accepted for the project.

**Table 6.14 Rainfall-Runoff Correlation for the Month of September**

| Year | $X_i$ | $Y_i$ | $X_i^2$ | $Y_i^2$ | $X_i Y_i$ |
|------|-------|-------|---------|---------|-----------|
| (1) | (2) | (3) | (4) | (5) | (6) |
| 1980 | 238.6 | 69.3 | 56930 | 4813 | 16552 |
| 1981 | 136.0 | 33.6 | 18496 | 1127 | 4566 |
| 1982 | 75.9 | 19.7 | 3761 | 388 | 1496 |
| 1983 | 238.7 | 91.1 | 80486 | 8301 | 25848 |
| 1984 | 110.3 | 66.1 | 12166 | 4370 | 7291 |
| 1985 | 317.2 | 314.1 | 100616 | 98646 | 99626 |
| 1986 | 264.5 | 147.1 | 69960 | 21663 | 38930 |
| 1987 | 143.5 | 49.3 | 20592 | 2413 | 7075 |

Most acceptable equations developed are

$Y = -60.511 + 0.812\ X$    $S_E = 161.4 : r = 0.769$

$Y = 0.04427\ X^{1.4265}$      $S_E = 1.161 : r = 0.859$ accepted for the project.

**Table 6.15 Rainfall-Runoff Correlation for the Month of October**

| Year | $X_i$ | $Y_i$ | $X_i^2$ | $Y_i^2$ | $X_i\,Y_i$ |
|------|-------|-------|---------|---------|------------|
| (1) | (2) | (3) | (4) | (5) | (6) |
| 1980 | 32.5 | 10.3 | 1056 | 105 | 233 |
| 1981 | 2.5 | 20.6 | 6 | 423 | 51 |
| 1982 | 0.0 | 2.6 | 0 | 7 | 0 |
| 1983 | 51.5 | 52.7 | 2652 | 2779 | 2715 |
| 1984 | 49.1 | 27.3 | 2421 | 746 | 1343 |
| 1985 | 89.4 | 103.2 | 7992 | 10761 | 9274 |
| 1986 | 22.8 | 82.0 | 520 | 6895 | 1893 |
| 1987 | 63.0 | 68.5 | 3969 | 4696 | 4317 |

Most acceptable equations developed are

$Y = 12.821 + 0.856X$    $: S_E = 67.9 : r = 0.714$

$Y = 21.486\ X^{0.18816}$     $: S_E = 1.878 : r = 0.82$ accepted for the project.

 Two values of correction coefficient $r$ for the moments from June to October (Tables 6.11 to 6.15) are obtained when one or more sets of data are discarded from the analysis by considering them as freak events. Moonsoon yield series for the project is calculated in Tables 6.16 and 6.17.

### 6.9.2   Use of Rainfall-Runoff Relation of Neighbouring Sites
A situation may arise when only the average catchment rainfall upto the required site (say A) is available. To solve the problem the average catchment rainfall and the corresponding runoff at another location either upstream or downstream of the present site (say B) may be collected and a rainfall-runoff relationship for all the months may be established. When such sites at upstream or downstream locations are not available, a neighboring basin which is considered hydrometeorologically homogeneous may be searched for concurrent rainfall and runoff data at its site $B_1$. Under both the circumstances, monthly regression equations developed either at B or $B_1$ are considered to hold good at the present site A. The yield series is synthesised using the regression equation developed at the other site with the average rainfall of the catchment at the required site at A. Similar procedure outlined in example 6.14 may be followed.

### 6.9.3   Empirical Relations
The method is used when rainfall runoff relation at the neighbouring station is not available. Various investigators proposed empirical relations to compute runoff from a watershed on the basis of the available rainfall data, catchment and climatic characteristics. These relations are purely regional in nature and should be used with caution while using at the other regions. Some of the relations developed are discussed as follows.

**Table 6.16 Calculation of Runoff Series for Manjore Project**

| Year | June | | July | | August | | September | | October | | Total |
|---|---|---|---|---|---|---|---|---|---|---|---|
| | RF | ROF | RF | ROF | RF | ROF | RF | ROF | RF | ROF | Σ Runoff |
| (1) | (2) | (3) | (4) | (5) | (6) | (7) | (8) | (9) | (10) | (11) | (12) |
| 1968 | 112.7 | 0.6 | 379.7 | 120.9 | 2595.3 | 144.7 | 160.3 | 61.3 | 73.6 | 48.2 | 375.7 |
| 1969 | 146.4 | 7.7 | 572.2 | 211.8 | 434.3 | 224.0 | 121.4 | 41.3 | 9.2 | 32.6 | 517.4 |
| 1970 | 546.5 | 92.1 | 121.9 | 186.7 | 411.2 | 210.6 | 249.0 | 114.9 | 86.5 | 49.7 | 654.0 |
| 1971 | 224.2 | 24.1 | 442.2 | 148.9 | 618.4 | 334.4 | 216.8 | 94.4 | 266.1 | 61.4 | 663.2 |
| 1972 | 73.5 | 0.0 | 505.4 | 171.0 | 339.4 | 169.4 | 160.9 | 61.7 | 31.0 | 41.0 | 443.1 |
| 1973 | 87.0 | 0.0 | 544.6 | 193.1 | 433.4 | 223.5 | 278.8 | 135.1 | 134.5 | 54.0 | 605.7 |
| 1974 | 40.1 | 0.0 | 266.4 | 59.8 | 383.8 | 194.8 | 416.0 | 238.9 | 94.3 | 50.5 | 540.0 |
| 1975 | 113.3 | 0.7 | 302.1 | 88.5 | 537.4 | 285.2 | 188.1 | 77.1 | 95.1 | 50.6 | 502.1 |
| 1976 | 132.4 | 4.7 | 300.7 | 87.9 | 483.8 | 253.2 | 172.6 | 68.2 | 12.7 | 34.7 | 448.7 |
| 1977 | 170.0 | 12.0 | 303.7 | 89.1 | 367.7 | 185.6 | 355.5 | 191.0 | 41.3 | 43.2 | 520.9 |
| 1978 | 204.1 | 19.9 | 255.5 | 70.4 | 372.1 | 188.0 | 166.6 | 64.8 | 103.9 | 51.4 | 394.5 |
| 1979 | 103.5 | 0.0 | 237.9 | 61.8 | 385.8 | 180.5 | 122.6 | 41.8 | 24.0 | 39.0 | 323.1 |
| 1980 | 426.4 | 66.8 | — | 63.9 | — | 49.1 | — | 69.3 | — | 10.3 | 259.4 |
| 1981 | 157.3 | 10.0 | — | 23.7 | — | 238.4 | — | 49.4 | — | 20.6 | 342.1 |
| 1982 | 113.9 | 0.8 | — | 93.2 | — | 477.1 | — | 33.6 | — | 7.7 | 607.4 |
| 1983 | 81.5 | 0.0 | — | 97.3 | — | 229.8 | — | 19.1 | — | 52.7 | 398.9 |
| 1984 | — | 32.5 | — | 57.1 | — | 247.4 | — | 66.1 | — | 27.3 | 430.4 |
| 1985 | — | 1.4 | — | 38.7 | — | 590.9 | — | 314.1 | — | 103.7 | 1048.8 |
| 1986 | — | 119.5 | — | 151.2 | — | 194.8 | — | 147.2 | — | 83.0 | 695.7 |
| 1987 | — | 16.2 | — | 110.2 | — | 34.8 | — | 49.3 | — | 68.5 | 279.0 |

RF = Rainfall in mm; ROF = Runoff in mm.

### Table 6.17   Calculation of Yield Series or the Runoff Volumes

| Year | Monsoon runoff (mm) | Monsoon runoff (Ha.m) | Non-Monsoon runoff taken as 7.6% of monsoon yield as per actual observation | Total runoff (Ha.m) | Runoff in decreasing order (Ha.m) | Rank |
|------|------|------|------|------|------|------|
| (1) | (2) | (3) | (4) | (5) | (6) | (7) |
| 1968 | 375.7 | 10723 | 815 | 11538 | 32206 | 1 |
| 1969 | 517.3 | 14764 | 1122 | 15886 | 21363 | 2 |
| 1970 | 654.0 | 18666 | 1419 | 20085 | 20085 | 3 |
| 1971 | 663.1 | 18924 | 1438 | 20362 | 20362 | 4 |
| 1972 | 443.1 | 12644 | 961 | 13604 | 18651 | 5 |
| 1973 | 605.7 | 17286 | 1314 | 18600 | 18600 | 6 |
| 1974 | 543.1 | 15572 | 1180 | 16702 | 16702 | 7 |
| 1975 | 502.0 | 14326 | 1089 | 15415 | 15995 | 8 |
| 1976 | 448.6 | 12803 | 973 | 13776 | 15886 | 9 |
| 1977 | 520.8 | 14865 | 1130 | 15995 | 13415 | 10 |
| 1978 | 394.5 | 11258 | 856 | 12114 | 13776 | 11 |
| 1979 | 523.7 | 9240 | 702 | 9942 | 13604 | 12 |
| 1980 | 254.4 | 7260 | 552 | 7812 | 13216 | 13 |
| 1981 | 242.1 | 9763 | 742 | 10505 | 12249 | 14 |
| 1982 | 607.4 | 17334 | 1317 | 18651 | 12114 | 15 |
| 1983 | 398.9 | 11384 | 865 | 12249 | 11538 | 16 |
| 1984 | 430.4 | 12283 | 933 | 13216 | 10505 | 17 |
| 1985 | 1048.8 | 29931 | 2275 | 32206 | 9942 | 18 |
| 1986 | 695.7 | 19854 | 1509 | 21363 | 8567 | 19 |
| 1987 | 279.0 | 7962 | 605 | 8567 | 7812 | 20 |

Runoff in Ha.m of col. (3) is obtained by multiplying catchment area in Ha. to the runoff depths in m of col. (2).

For 50% dependable yield $m = \dfrac{50}{100} \times (20 + 1) = 10.5$th value or say the 11th value

The 11th value from the above table is read as 13776 Ha.m

For 75% dependable yield $m = \dfrac{75}{100} \times (20 + 1) = 15.75$th value or say 16th value

The 16th value from the above table is read as 11538 Ha.m.

For 90% dependable yield $m = \dfrac{90}{100} \times (20 + 1) = 18.9$th value or say 19th.

The 19th value from the above table is read as 8567 Ha.m.

#### 6.9.3.1   Strange Table

Strange (1928) studied some rainfall and runoff process in Konkan region (India) and proposed a table of coefficients for the computation of runoff from the observed rainfalls. The catchment character was classified as *good, average* or *bad* and surface condition before the rainfall as *dry, damp* or *wet*. The proposed coefficients are given in Table 6.18.

**Table 6.18   Strange's Table**

| Daily rainfall (mm) | Surface condition and runoff as% of rainfall | | | Remarks |
|---|---|---|---|---|
| | Dry | Damp | Wet | |
| (1) | (2) | (3) | (4) | (5) |
| 5 | – | 4 | 7 | |
| 6.25 | – | 4.2 | 8 | |
| 7.5 | – | 4.5 | 8.5 | |
| 10 | 1 | 5 | 10 | |
| 12.5 | 1.2 | 6 | 12 | |
| 15 | 1.5 | 7 | 13.0 | |
| 18.75 | 1.7 | 8 | 14.8 | |
| 20 | 2 | 9 | 15 | For good catchment |
| 25 | 3 | 11 | 18 | add upto 25% of |
| 30 | 4 | 13 | 20 | yield to this value. |
| 35 | 5.7 | 15.5 | 24 | |
| 40 | 7 | 18 | 28 | For bad catchment |
| 45 | 8.5 | 20 | 31 | deduct upto 25% of |
| 50 | 10 | 22 | 34 | yield to this value. |
| 60 | 14 | 28 | 41 | |
| 70 | 18 | 33 | 48 | |
| 80 | 22 | 39 | 55 | |
| 90 | 25 | 44 | 62 | |
| 100 | 30 | 50 | 70 | |

Using Table 6.18, daily runoff from the catchment can be computed from which annual yield series can be reconstructed. Relation in the form of total monsoon rainfall in cm against the catchment yield is given in Table 6.19, which can be used as check to the sum of the runoff calculations made over the monsoon period.

**Table 6.19   Strange's Table for Total Monsoon Period**

| Total monsoon rainfall (mm) | Total monsoon yield in (mm) | Remarks |
|---|---|---|
| 250 | 44 | The figures of runoff are applicable for |
| 500 | 150 | good catchment. For average catchment |
| 750 | 260 | take 75% of the runoff and for a bad |
| 1000 | 375 | catchment take upto 50% of the above |
| 1250 | 475 | values of the total runoff (for monsoon). |
| 1500 | 590 | |
| 1750 | 700 | |

### 6.9.3.2   Binnie's Runoff Percentages

Sir Alexander Binnie (1870) studied more than 40 catchments in India and proposed the following percentage of runoff to be used in the regions of Madhya Pradesh and Maharashtra. From Table 6.20, the annual runoff volumes can be calculated by knowing the annual rainfall and the catchment area.

**Table 6.20   Binnie's Annual Runoff Percentages**

| Annual rainfall (mm) | 500 | 600 | 700 | 800 | 900 | 1000 | 1100 |
|---|---|---|---|---|---|---|---|
| Annual runoff % | 15 | 21 | 25 | 29 | 34 | 38 | 40 |

### 6.9.3.3   Barlow's Runoff Percentage

Barlow (1912) carried out extensive studies on the catchment of Uttar Pradesh and proposed the following percentages of runoff for various types of catchments (Table 6.21). These percentages may be multiplied to the average catchment rainfall upto 130 sq. km in size to get the runoff volumes.

**Table 6.21   Barlow's Percentage Coefficients for Runoff Computation from Rainfall**

| Nature of rainfall | Catchment types | | | | |
|---|---|---|---|---|---|
| | A | B | C | D | E |
| 1.   Light rain upto 25 mm/day | 7 | 12 | 16 | 28 | 36 |
| 2.   Medium rainfall of varying intensity 25–75 mm/day | 10 | 15 | 20 | 35 | 45 |
| 3.   Continuous down pour greater than 75 mm/day | 15 | 18 | 32 | 60 | 81 |

A: flat, cultivated and black cotton soil, B : flat, partly cultivated soil, C: average catchment, D: hills and plains with little cultivation, E: hilly and steep regions with hardly any cultivation.

### 6.9.3.4   Runoff Equations

For different regions of the country equations have been proposed for computing runoff from the rainfall and other catchment data. An abstract of the widely used equations in India is given in Table 6.22.

There are many such equations developed for different catchments of the country. Presentation of all the equations are beyond the scope of this book. The values of $P$ and $R$ can be taken as cms except in equation (6.65) which is in million m$^3$.

---

**Example 6.15:** In Orissa, Jeera project is proposed at a site draining a catchment of 132 sq. km. Monthly average rainfall over the catchment for July is given below. Calculate yield at the project site for the month if catchment is in hilly region. There is no other runoff information available from the project.

| Dates of 7/1996 | 2 | 5 | 7 | 12 | 14 | 22 | 25 | 26 | 28 | 31 |
|---|---|---|---|---|---|---|---|---|---|---|
| Rainfall (mm) | 12.5 | 25 | 20 | 40 | 7.5 | 35 | 10 | 60 | 12.5 | 22 |

**Table 6.22  Various Runoff Equations**

| Sl. No. (1) | Name of the Investigator and the Formula (2) | Region of investigation (3) | Symbol meanings (4) | Remarks (5) |
|---|---|---|---|---|
| 1. | Khosla : $R_m = P_m - 0.48T_m$ (1960) | India and USA | $R_m$ = Monthly runoff in cm $T_m > 4.5°C$ $P_m$ = precipitation in cm | (6.59) |
| | If $T_m < 4.5°C$ than the term $(0.48T_m)$ is replaced by the following figure | | | |
| 2. | Inglish:$R = 0.85P - 30.5$ & Desouza: | Ghat regions (western ghats) | $P$ = Annual precipitation in cm | (6.60) |
| | $R = (0.00394P^2 - 0.0701P)$ | Plain regions (Deccan plate) | $R$ = Runoff in cm | (6.61) |
| 3. | ICAR: $R = 1.511(P^{1.44})(T_m)^{-1.34}(A^{-.0613})$ | Nilgir hills | $A$ is watershed area km² $P$ and $R$ in cms | Annual (6.62) |
| 4. | C. Vermuel: $R = P - (0.16P + 39.44)(0.09T + 0.12)$ | | $P$ and $R$ in cms | Annual (6.63) |
| 5. | I.G. Justin: $R = 0.284\ S^{0.155}\ P^2\ (1.8T + 32)^{-1}$ | | $S$ = Slope of the catchment | (6.64) |

| $T_m$ | 4.5°C | -1°C | -12°C | -18°C |
|---|---|---|---|---|
| Replacement of $(0.48T_m)$ | 2.17 | 1.80 | 1.25 | 1.00 |

*(Contd)*

| (1) | (2) | (3) | (4) | (5) |
|---|---|---|---|---|
| 6. | Dhir, Ahuja and Majumdar: | | | |
| | $R = 13.18\ P + 86.5$ | Machakund (A.P.)(2220 km$^2$) | | (6.65) |
| | $R = 120\ P - 4945$ | Chambal (Rajasthan) 22500 km$^2$ | | (6.66) |
| | $R = 34.6\ P - 1510$ | Tawa (M.P.) 5950 km$^2$ | | (6.67) |
| | $R = 90.5\ P - 4800$ | Manjara (A.P.) 21700 km$^2$ | | (6.68) |
| | $R = 435\ P - 17200$ | Tapi (Gujarat) 64400 km$^2$ | | (6.69) |
| | $R = 13400\ P - 575000$ | Damodar (W.B.) 19900 km$^2$ | | (6.70) |
| | $R = 4270\ P - 254000$ | Maniari (805 km$^2$) | | (6.71) |
| 7. | UP Irrigation Research Institute: | | | |
| | $R = 5.45\ P^{0.6}$ | Ganga 23400 km$^2$(Haridwar) | | (6.72) |
| | $R = 0.35\ P^{1.1}$ | Yamuna at Tajewala 11150 km$^2$ | | (6.73) |
| | $R = 2.7\ P^{0.8}$ | Sarada at Bana basa 14960 km$^2$ | | (6.74) |

$P$ = Annual rainfall (cm); $R$ = Annual runoff (cm).

**Solution**

The following table is prepared to calculate the yield at the project site for July, 1996. For this period we can assume the soil to be wet for the region of Orissa as the onset of monsoon here begins around 10th of June.

**Table 6.23   Runoff Computation Using Strange's Coefficients**

| Date | Rainfall (mm) | Runoff % from Strange table | Runoff (mm) |
|------|------|------|------|
| (1) | (2) | (3) | (4) |
| 2/7/96 | 12.5 | 12 | 1.5 |
| 5/7/96 | 25 | 18 | 4.5 |
| 7/7/96 | 20 | 15 | 3.0 |
| 12/7/96 | 40 | 28 | 11.2 |
| 14/7/96 | 7.5 | 8.5 | 0.64 |
| 22/7/96 | 35 | 24 | 8.4 |
| 25/7/96 | 10 | 10 | 1.0 |
| 26/7/96 | 60 | 41 | 24.6 |
| 28/7/96 | 12.5 | 12 | 1.5 |
| 31/7/96 | 22 | 16.5 | 3.63 |
| Total | 244.5 | | 59.97 |

Since catchment is in hilly region, it can be considered as good, 25% yield can be added to get the total yield for July .

$$\text{Total yield for July} = 59.97 \times \frac{125}{100} = 74.96 \text{ mm}$$

$$\text{Yield from the whole catchment} = \frac{74.96 \times 132 \times 10^6}{1000} \text{ m}^3 = 9.90 \text{ million m}^3.$$

Yield for July = 9.90 M.cum.

## 6.10   RUNOFF SIMULATION MODELS

*Simulation* is defined as the mathematical representation of the response of a hydrologic system to a series of events during a given time. On the other hand *synthesis* means the generation of (say flow) sequences preserving the statistical properties of the observed record. Thus synthesis helps to overcome data inadequacies.

A simulation technique in which the mathematical relationships describing the interdependence of various parameters in the system are formulated is called a model. A model is essentially represented as shown in Fig. 6.27.

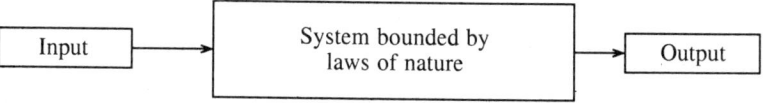

**Fig. 6.27   Concept of a Model**

Simulation models are classified into the following groups.

1. *Physical Analog and Mathematical Model:* In physical modeling, the structure is rebuilt to a scale of say 1:100 and the behaviour of the structure to the possible natural process in the same scale is tested. In mathematical model, a simple mathematical statement say $y = a + bx$ can take care of the complex rainfall-runoff process of the catchment.

2. *Continuous and Discrete Model:* A model which takes care of the continuous inputs and outputs of a system is called continuous. When time step is involved in the model (say at 3 h intervals of rainfall or runoff data) then the model becomes discrete.

3. *Static and Dynamic:* When a model is independent of time then it is called static model. A unit hydrograph which represents the response of unit rainfall excess is a static model whereas the infiltration process or the evapotranspiration process is a dynamic process.

4. *Descriptive and Conceptual:* Descriptive models describe the natural processes through basic fundamental equations, whereas the conceptual models describe the system through the concept of probabilities or similarities.

5. *Lumped and Distributed:* When spatial variation is considered for all the parameters, it is called distributed model and when such variations are not considered, the model is lumped.

6. *Black box and Structural:* In black box models, the process of transformation of the input to output are not understood. In structural models, definite laws and principles are employed for the transformation.

7. *Deterministic and Stochastic:* When a random parameter is not included, the model is deterministic and a stochastic model includes the randomness in some sequence. The models are described in Chapter 2.

8. *Event and Sequential:* An event model considers all parameters of a single event only whereas a sequential model considers the sequence or group of events over a length of time.

9. *General and Catchment Specific:* Some models are capable of incorporating wide range of parameters. These models can be used for simulating the input and output processes for any basin. The catchment specific models are developed to serve the catchment for which it is developed.

### 6.10.1  Steps in Modelling

(A) **Identification of the system:** The causes of underlying processes and their effects are identified.

(B) **Conceptualisation of the system:** It is done in three steps. The techniques that are to be used for the system are selected, mathematical equations are formulated and in the last the mathematics is to be translated to computer programs as algorithms.

(C) **Verification of the model:** With the series of known inputs and outputs, the model should be verified critically for all ranges of informations.

(D) **Implementation of the model:** Improvements to the model may be carried out by changing algorithms and the accuracy of the reproduction of the model may be tested with several sets of inputs and outputs of all ranges before finally accepting it as a tool for simulation.

Fitting a simple rainfall – runoff regression equation for a catchment can be considered as a model in its simplest term, if it can reproduce to an acceptable accuracy, the runoff's for all ranges of rainfall – depths that can occur over the basin.

A review of literatures indicate a large number of rainfall – runoff models that have been developed and used by various agencies. *Crawford and Linsley* (1959) gave a new dimension to the modeling approach and proposed an *Explicit Soil and Moisture Accounting* (ESMA) model known as *Stanford Watershed Model* (SWM). Since then the model has undergone many improvements and the present model in FORTRAN was translated from the original ALGOL by James. The SWM-IV is a lumped model requiring data on hourly precipitation, daily evapotranspiration and the catchment characteristics. The fundamental continuity equation used by the model is

Precipitation = Evapotranspiration + Surface runoff + Change in storage  (6.75)

where all the units are considered in cm or mm. A typical flow chart of SWM-IV model is given in Fig. 6.28. Actual computation proceeds from one function to another as encircled in the figure. The soil moisture is modeled with three reservoirs. The (i) upper zone representing rapid runoff response, (ii) lower zone representing long term infiltration and (iii) deep ground water storage zone representing base flow in the stream. It requires 3 to 6 years data for calibration to adjust several default parameters until the model output graph matches exactly with the recorded stream flow data. The model gives better results for yield series than for flow hydrographs and has since been improved considerably. Following models are the changed version of the SWM-IV: (i) *Kentucky Watershed Model* (KWM), 1970 developed by Liou for a catchment upto 5000 sq. km, (ii) *Kentucky self-calibrating version model with an optimization program* (OPSET) to optimally calibrate the parameters, (iii) *Hydrocomp Simulation Program* (HSP-1966) to solve a large number of catchment simulation problems, (iv) *Stream-Flow Synthesis and Reservoir Regulation Model* (SSARR-1968) for dealing with large watersheds. This model has been tested and verified in USA and (v) *National Weather Service River Forecasting System* (NWSRFS) model.

### 6.10.2  HEC Model

Hydrologic engineering centre of US Army corps of Engineers has developed a series of computer packages for solving most of the hydrological problem encountered in river valley system. The packages are named HEC-1 to HEC-6. They can be used for flood analysis, stream flow synthesis, reservoir operation, flood mapping and solving variety of other problems. HEC-1 with six subroutings is capable of solving problems like determining an optimal unit hydrograph, snowmelt computation, hydrograph routing and balancing of hydrographs. Detail discussion on HEC packages is beyond the scope of this book. Some of the

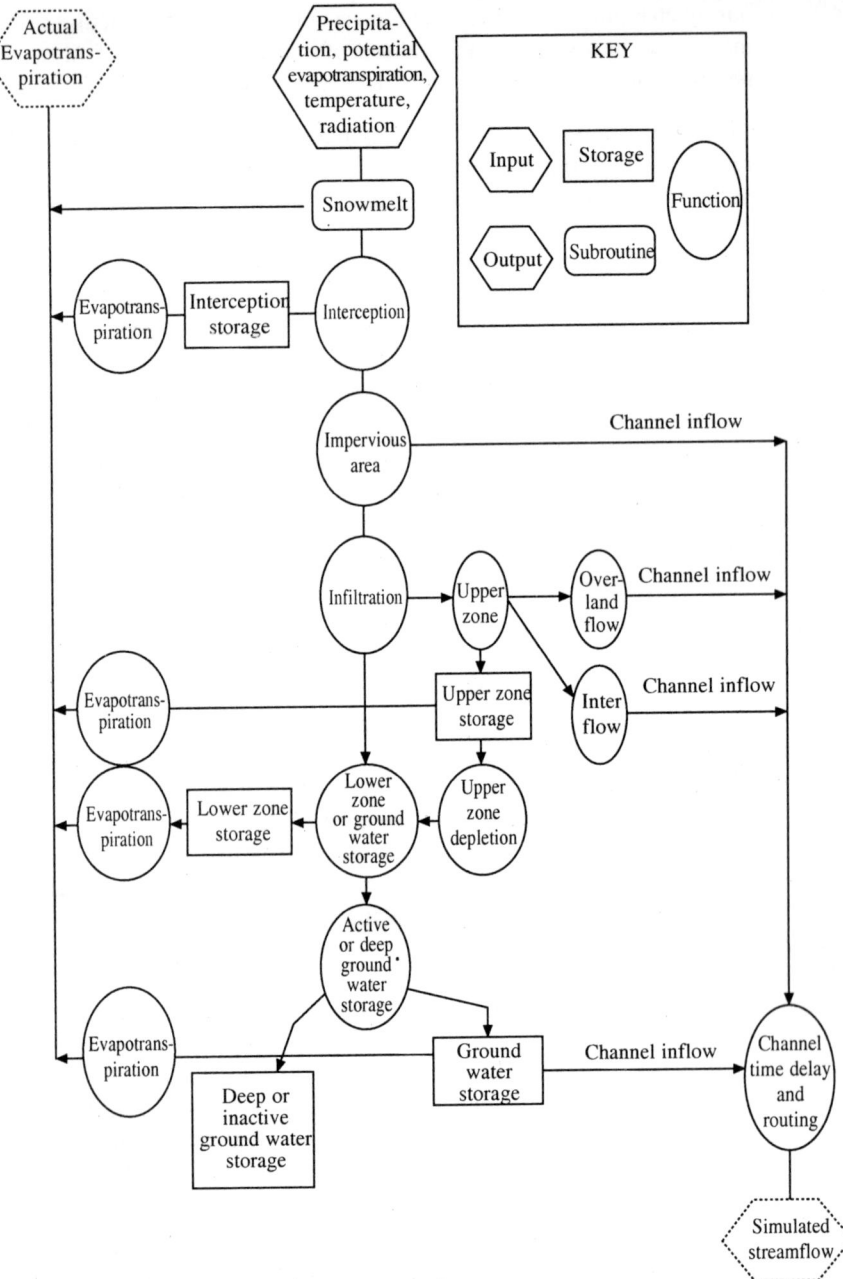

**Fig. 6.28   Flow Chart of Stanford Watershed Model**

models and their addresses are given in Table 6.24. Interested readers may contact at the addresses given in the Table for detailed information about the models.

**Table 6.24** Information on some Watershed Models

| Sl. No. | Name of model | Computer language | Hardware requirement | Inputs | Outputs | Purpose | Address of the developer |
|---------|---------------|-------------------|----------------------|--------|---------|---------|--------------------------|
| 1. | Stream flow synthesis reservoir regulation (SSARR) | FORTAN IV | IBM-360-50 Basin data | Rainfall runoff PET, Temp. | Hydrograph | Flood for operation | Corps of Engineers, Portland, Oregor, USA |
| 2. | National Weather Service Hydrology Model (NWSHM) | FORTAN IV | IBM-486 or higher | -do- hydrograph | Continuous | Flood forecosting | National weather Service, Silver Spring, Mary Land, USA |
| 3. | Serial storage type model (Tank Model I & II) | FORTAN IV | PC/AT 486 or higher | -do- | Daily discharge & hydrograph | River discharge forecasting | National research centre for disaster prevention 1-Ginza Higats-6, Chuo-ku, Tokyo, Japan |
| 4. | HSP | PL/I | PC/AT 486 or Higher | -do- | -do- | Hydropower system operation, Flood forecasting. | Hydrocomp Institute, Inc. 591 Lytton Av., Palo, Alto, California 94301, USA. |
| 5. | CBM-Common wealth Bureau of Meteorology | FORTAN | -do- | -do- | discharge | Short term flood forecasting | Commonwealth Bureau of Meteorlogy, P. Box 1289K Melbourn Vic-3001 Australia |
| 6. | R.R model of Hydro Meteorological Centre (HMC-RR) | ALGOL-60 BASM | | Rainfall Runoff | -do- | -do- | H.M. Centre of USSR, Bolsevists Kaja-13, Moscow-123376, USSR |

*(Contd)*

| Sl. No. | Name of model | Computer Language | Hardware requirement | Inputs | Outputs | Purpose | Address of the developer |
|---|---|---|---|---|---|---|---|
| 7. | Girard-I | FORTAN 4G | IBM 350–50 | Rainfall runoff and Topographical Data | Discharge daily | -do- | Orstorm, 19rue E, Carrier 75018 Paris, France |
| 8. | DISPRIN | FORTRAN | POP-85 | | -do- | Hydrograph for flood forecasting sub-basin wise and Res. operation. | W.R. Board, Reading bridge house, Reading U.K. |
| 9. | BIDM | FORTRAN -IV | IBM 360-40 | R.R and Channel routing parameters | -do- | Forecasting | Societe Grenobloise d' Etudes et Application, Hydrauliques (SUGREAH) 84–86, Av. Leon Blun 38-Grenoble. France |

## PROBLEMS

6.1 What are the essential requirements of an ideal gauge and discharge site?

6.2 Briefly explain the factors that affect the runoff from a basin.

6.3 Discuss the various water stage recorders.

6.4 In a small stream a discharge measuring structure is to be constructed. What are the various types of structures you can suggest and which you will select finally?

6.5 Describe the types of current meters normally used to measure mean velocity of water in rivers.

6.6 What are the methods available to measure velocity of water at a river section ?

6.7 Explain dilution method of flow measurement.

6.8 What is a rating curve? Name different methods which can be applied for extending rating curve.

6.9 Briefly discuss the various methods normally used to extend runoff data.

6.10 For a catchment only rainfall data is available. How you can compute runoff from the basin?

6.11 What do you understand by simulation? How many types of simulation models you know? Write the steps you will consider to model catchment.

6.12 The following data are observed in a stream by a Price current meter.

| Distance from bank (m) | 0 | 3 | 5 | 7 | 9 | 12 | 15 | 18 | 21 | 23 | 25 | 27 |
|---|---|---|---|---|---|---|---|---|---|---|---|---|
| Depth (m) | 0 | 0.6 | 1.2 | 2.05 | 2.35 | 2.1 | 1.9 | 1.6 | 1.4 | 1.0 | 0.4 | 0 |
| No. of revolutions at 0.6d | 0 | 90 | 95 | 135 | 142 | 125 | 115 | 110 | 95 | 90 | 76 | 0 |
| Time (seconds) | 0 | 184 | 125 | 125 | 125 | 125 | 125 | 125 | 125 | 1254 | 125 | 0 |

Current meter rating equation is given as $V = 0.33 + 0.03N$ m/sec, where $N$ is the number of revolutions/sec. Calculate river discharge.

6.13 Following data are obtained while gauging a stream

| Main gauge reading (m) | 20.10 | 20.10 |
|---|---|---|
| Auxiliary gauge reading (m) | 19.82 | 19.13 |
| Discharge (cumec) | 5.40 | 9.35 |

Calculate discharge when the main gauge is 20.1 m and the auxiliary gauge is 19.52 m.

6.14 Stage-discharge data at a site are given below. Find stage corresponding to zero-discharge.

| Stage(m) | 25.9 | 26.89 | 27.91 | 29.06 | 31.0 | 33.52 | 34.4 | 35.38 | 36.42 | 38.01 | 39.05 | 39.55 |
|---|---|---|---|---|---|---|---|---|---|---|---|---|
| Discharge | 8.9 | 23 | 36 | 47 | 120.8 | 285.3 | 380 | 456 | 533 | 590 | 680 | 690 |

Calculate discharge for 32.25 and 39.4 m stages.

6.15 Two river sections 7.50 km away along the river course have the following data during a flood.

| | Elevation (m) | Area of cross section (m) | Hydraulic radius (m) |
|---|---|---|---|
| U/S section X | 205.87 | 75.30 | 2.75 |
| D/S section Y | 203.50 | 95.38 | 3.10 |

Estimate discharge in the river if Manning's roughness coefficient is 0.025 for both the sections. Assume loss of head between the two sections as negligible.

6.16  Estimate the discharge in a stream when 250 g/l of common salt was discharged into it at constant rate of 32 l/sec. Salt concentration already present in the stream water was 8 parts per million (PPM) and after the dilution its concentration increased to 50 PPM measured after through mixing at the downstream of the station.

6.17  What do you understand by a flow duration curve? In what way it is going to help an engineer in water resource planning.

6.18  Compute runoff volumes from a catchment of 80 sq. km from the following data.

| Months | J | F | M | A | M | J | J | A | S | O | N | D |
|---|---|---|---|---|---|---|---|---|---|---|---|---|
| Rainfall (mm) | 7.5 | 9 | 12 | 42 | 70 | 110 | 230 | 280 | 150 | 100 | 60 | 25 |
| Temp. $^0$C | 22 | 24 | 34 | 36 | 37 | 33 | 32 | 32 | 28 | 25 | 22 | 20 |

Use Khosla's runoff formula and compare its result with Strange's approach. The catchment is located in Andhra Pradesh.

6.19  Annual rainfall and runoff in mm over a catchment are given below.

| year (19-) | 77 | 78 | 79 | 80 | 81 | 82 | 83 | 84 | 85 | 86 | 87 | 88 |
|---|---|---|---|---|---|---|---|---|---|---|---|---|
| Rainfall (mm) | 910 | 1110 | 605 | 1300 | 1470 | 990 | 1480 | 520 | 1195 | 900 | 660 | 750 |
| Runoff (mm) | 305 | 515 | 245 | 620 | 750 | 403 | 654 | 165 | 472 | 390 | 275 | 230 |

Develop a rainfall-runoff relation by the method of least square. Find the degree of association between them. If the rainfall for a year is 1210 mm, what will be the runoff for that year ?

6.20  Is it necessary to extend the stream flow records for a river basins in India ? Describe Longbein's method for extending stream flow data.

6.21  During a high flood, the surface velocity recorded by a float at a place are given below. Compute the total discharge at the site and plot the river cross-section from the following data.

| Dist. from left bank (m) | 10 | 15 | 20 | 25 | 30 | 35 | 40 | 45 | 50 | 56 |
|---|---|---|---|---|---|---|---|---|---|---|
| Depth of flow (m) | 0 | 1.8 | 2.7 | 3.05 | 3.7 | 6.1 | 4.2 | 2.5 | 1.4 | 0 |
| Time taken for float for 100 m run (sec) | – | 110 | 80 | 50 | 35 | 25 | 32 | 85 | 95 | 0 |

6.22  The following data are available at a *G-D* site

| Stage (m) | 2.4 | 3.3 | 4.1 | 4.5 | 5.2 | 5.5 | 5.75 | 6.1 | 6.5 |
|---|---|---|---|---|---|---|---|---|---|
| Discharge (m$^3$/sec) | 220 | 356 | 483 | 551 | 721 | 806 | 885 | 998 | 1195 |

Plot the rating curve and obtain the stage for zero discharge using arithmetic and log-log paper.

6.23  From the rating curve at a site the following information is obtained:

| Depth of flow (m) | 0.5 | 0.7 | 1.0 | 1.4 | 1.6 | 1.8 | 1.9 | 2.0 | 2.8 |
|---|---|---|---|---|---|---|---|---|---|
| Area (sq.m) | 30 | 120 | 170 | 222 | 310 | 380 | 475 | 495 | 780 |
| Discharge (cumecs) | 32 | 145 | 220 | 303 | 430 | 550 | 730 | 780 | – |

Extend the curve and estimate the flow for the channel depth of 2.80 m.

6.24  How can you carry out discharge measurement by (a) electromagnetic induction method and (b) ultrasonic method?

6.25 Explain how discharge can be computed by slope-area method. What data are required for the calculation ?

6.26 Using current meter how you will proceed to measure section mean velocity at a site.

6.27 Daily flows recorded at a gauging site for a period of 10 days are 15, 27, 65, 120, 100, 86, 72, 61, 45 and 33 m$^3$/secrespectively. Obtain the volume of flow in 4 units. If the catchment area is 780 sq. km, what is the depth of flow over the area?

6.28 (a) Explain a method of extension of stage – discharge data.

　　 (b) What do you understand by isovel? At a river cross section, the distribution isovel are shown. How can you compute the discharge at the site.

6.29 Salt solution with concentration of 75 gm/lit is mixed into a stream at a rate of 20 lit/sec. Taking the initial salt concentration of the stream as 0.3 ppm, determine the discharge in the stream if the recorded concentration of salt at the downstream is found to be 2.2 ppm.

6.30 At a gauging station, the rating curve is established as $Q = (G - 52)^{1.35}$, where $Q$ is the discharge in m$^3$/sec and $G$ corresponds to the river gauge. The observed stage at the site on a day are as follows.

| Time(h) | 0 | 2 | 4 | 6 | 8 | 10 | 12 | 14 | 16 | 18 | 20 | 22 |
|---------|------|-------|------|------|------|------|------|-------|------|-------|------|------|
| Stage(m) | 54.8 | 54.95 | 55.2 | 55.7 | 56.3 | 55.9 | 55.2 | 54.85 | 54.4 | 54.15 | 53.9 | 53.6 |

Plot the stage-discharge curve. Calculate discharge for 52.25 and 53.4 m stages. What is the mean flow during the observation period?

6.31 Differentiate between (a) Permanent and shifting control, (b) Stream flow and runoff (c) horizontal and vertical axis currentmeter.

6.32 What factors are considered while locating a gauge-discharge site.

# Chapter Seven

# Hydrograph

## 7.1 INTRODUCTION

Hydrograph is a graphical representation of flow of a river at a location with time. It is the total response or the output of a watershed beginning with precipitation as the hydrological exciting agent or input. The system representing the catchment's physiographic, geologic and hydrometeorologic effects is complex and is difficult to model it accurately due to high variability of these parameters in space and time. A hydrograph comprises three phases, namely, surface, subsurface and base flow. A schematic representation of the runoff process showing the components are presented in Fig. 7.1. The factors affecting the hygrograph at a place being complex and interrelated, a basin may not produce two exact flood hydrographs with two similar precipitations as input, nor can two basins of the same drainage area produce the same flood hydrographs with similar precipitation. When large number of hydrographs of a basin at a location are analysed, one can notice their irregular shapes representing the complexity of the storm and the catchment character. To begin with, a hydrograph resulting from a single storm is taken and its components studied.

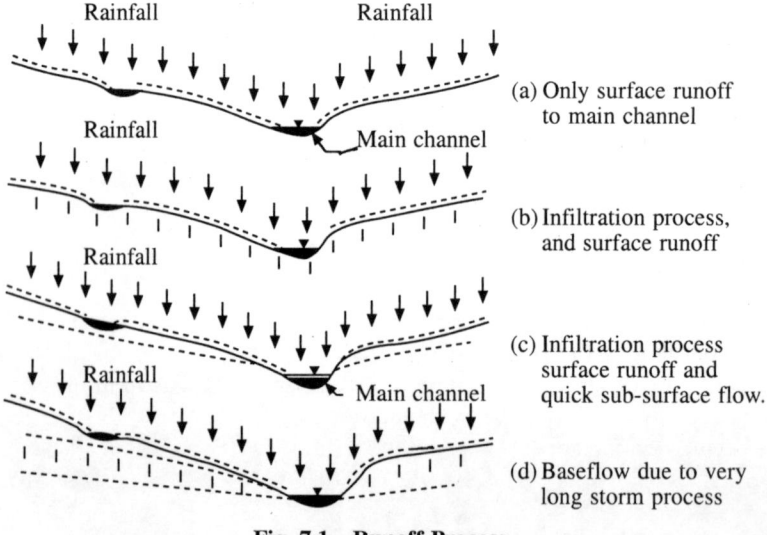

**Fig. 7.1  Runoff Process**

## 7.2  HYDROGRAPH CONCEPT

Let us assume a basin area (Fig. 7.2(a)) having uniform basin characteristics with the entire area subjected to uniform rainfall and the conditions of the catchment producing runoff are also uniform. The areas $OA_1A_1, A_1A_1A_2A_2, A_2A_2A_3A_3$ are so divided that water takes 1 h to reach from one boundary to the next below. Let us examine the hydrograph resulting from the first hour of the storm. Rainfall over the area $OA_1A_1$ is occurring for 1 h. The runoff from line $A_1A_1$ takes 1 h to reach at $O$. The hydrograph from the catchment area $OA_1A_1$ (Fig. 7.2(a)) is shown as triangle $01'2$ in Fig. 7.2(b). At the end of the first hour when the runoff from the point $O$ is still continuing at the catchment outlet, contribution from the boundary line $A_1A_1$ has started arriving. At this time all the area lying in the boundary $OA_1A_1$ is contributing at $O$ due to 1 h rainfall. Therefore, the peak runoff resulting from the area $OA_1A_1$ due to 1 h rainfall occurs at the end of the first hour as shown in triangular diagram. The last drop of rain occurring at the end of 1 h storm from the line $A_1A_1$ takes one more hour to reach the outlet $O$, i.e., at the end of second hour, when the last drop of water from the boundary line $A_1A_1$ reaches at $O$ the runoff hydrograph becomes zero. It may be noted that the whole area contributed for 1 h only at the outlet due to 1 h rainfall. Runoff from boundary area $A_1A_1$ contributed at $O$ from end of first hour to end of second hour and the runoff from $O$ at $O$ is available for 1 h beginning and ending with storm period from zero to one hour.

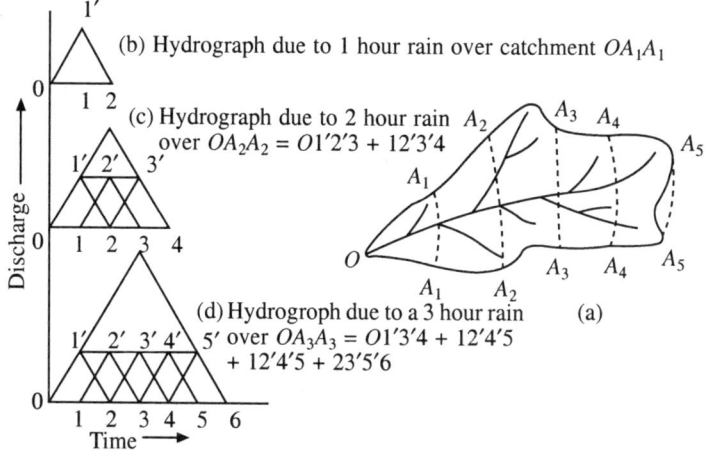

Fig. 7.2  Concept of Hydrograph Formation

Imagine that rainfall for 1 h occurred only in the area $A_1A_1A_2A_2$ which has started contributing to the outlet $O$ after end of 1 h and continued till the end of third hour. The resulting hydrograph is a triangle $12'3$ as shown in Fig. 7.2(c), with peak of flow at end of second hour. Similarly if a $2h$ rainfall is imagined to have occurred in the area $OA_2A_2$ then the resulting hydrograph at $O$ is the large triangle shown in Fig. 7.2(c). Similarly a 3 h rainfall concentration over the area $OA_3A_3$ will produce a hydrograph of shape shown in Fig. 7.2(d).

Thus the hydrographs produced by 1 h rainfall in the area are triangles at lags

. 7.2. The combined effect will be when the total area is subjected
.m rainfall. The resulting hydrograph is the observed discharge from
.nent at outlet $O$. Any rainfall exceeding this time with the same intensity
.duce a larger time base hydrograph. However, the peak of the hydrograph
.e of same magnitude and of longer time equal to the total rainfall period
n. .us the time of travel of the drop of water from extreme boundary point to the
basin outlet.

## 7.3 COMPONENTS OF A HYDROGRAPH

The essential components are discussed below.

**(i) Rising Limb:** It is the portion from $A$ to $B$ (Fig. 7.3) of the hydrograph
representing the ascending portion of the curve. It is mainly influenced by the
storm and basin characteristics. In general, the rising limb is concave, rising
slowly in the early stage of flood but more rapidly towards the end portion. This
is due to the fact that at the initial stage of a storm, losses are variable and high.
The flow begins to build up in the channel as the storm duration increases. The
rising limb is attributed due to the contribution of more and more area at the
gauging site over time, beginning with the onset of the storm. It gradually
reaches the peak when maximum area contribute their runoff at the given outlet.

**(ii) The Peak or Crest Segment:** It is represented by the portion from $B$ to $C$ of
the curve showing points of inflections of the rising and the falling limb. Peak
of the hydrograph occurs when all portions of the basin contributes at the outlet
simultaneously at the maximum rate. For large watersheds, it represents the time
when highest concentration of area-depth (cm) of runoff is received at the outlet.
The points of inflections represent the change point of contribution of the catchment
to the channel system. Depending on the rainfall-catchment characteristics, the
peak may be sharp, flat or may have several well defined peaks, but the highest
discharge is the point of special interest to all hydrologists.

**(iii) The Recession Limb:** This limb is the convex curve (from point $C$ to $D$ in
Fig. 7.3) representing the withdrawal of water from the storage build up in the
basin with maximum at point $C$. Point $D$ represents where the contribution to the
channel is purely from ground water. By this time, contribution from the storm
is absent and the channel characteristic mostly determine the type of recession.
When a storm concentrates more near the outlet, then the length of this curve is
shorter, whereas if rainfall concentrates at the far end of the catchment, the
recession curve is longer. The slope of recession curve indicates the rate of
withdrawal of water from the channel storage. The curve is mathematically
represented as

$$Q_t = Q_o \, K_r^t \qquad\qquad (7.1)$$

where $Q_t$ is the discharge $t$ days after the initial discharge $Q_o$ which may be taken
as the discharge at point $C$ in the curve and $K_r^t$ the recession constant with a
value less than unity. Barnes (1940), based on the study of large number of

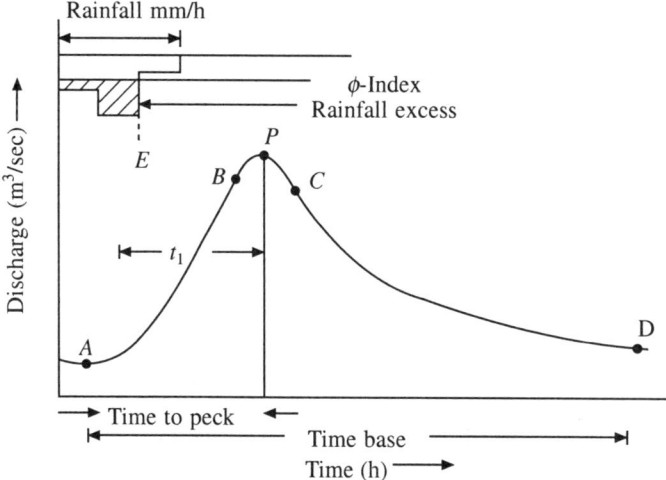

Fig. 7.3  Separation of Base Flow

actual hydrographs found that equation (7.1) holds good for the lower portion of the curve where the contribution is mainly from ground water. In the upper part of the curve, the contribution from all the three phases, i.e., surface storage, subsurface (inter) flow and ground water flow exists. Therefore, the recession constant $K_r^t$ is taken as product of all three components as

$$K_r^t = K_s \cdot K_i \cdot K_g \tag{7.2}$$

where $K_s$ represents the recession constant for surface storage, $K_i$ the recession constant for interflow and $K_g$ is the recession constant for ground water storage. Plot of equation (7.1) on a semilogarithmic graph paper should plot a straight line. The plot of the recession curve on a semilog paper with $Q$ on log-scale is not a straight line, but consists of three segmental curves. This is due to the effect of three components making the recession curve. When the line deviates, it represents the contribution from a different recession. Studies on large number of recession curves indicates that $K_g$ varies between 0.9 and 1.0, $K_i$ between 0.55 and 0.9 and $K_s$ from 0.08 to 0.4 and will be further discussed. The following example describe the computation of three coefficients.

---

**Example 7.1:** Flow observation by Central Water Commission from 6/10/1998 to 9/10/1998 for Ib river, a tributary of Mahanadi at Sundargarh gauge-discharge site are given below. Compute the recession coefficients for the channel storage, interflow and base flow.

| Date: | 6/10/98 | | | | 7/10/98 | | | | 8/10/98 | | | | 9/10/98 |
|---|---|---|---|---|---|---|---|---|---|---|---|---|---|
| Time (h) | 24 | 6 | 12 | 18 | 24 | 6 | 12 | 18 | 24 | 6 | 12 | 18 | 24 | 6 |
| Q (Cumec) | 100 | 610 | 1860 | 1570 | 970 | 700 | 540 | 630 | 330 | 260 | 217 | 192 | 170 | 150 |

**Solution**

Table 7.1 shows the calculation of recession coefficients.

### Table 7.1   Computation of Recession Coefficients for Example 7.1

| Date from the end of recession | Discharge (m³/sec) | $K_g$ | Base flow (m³/sec) | Residual col. (2)–(4) | $K_i$ | Interflow | Surface storage col. (5)–(7) | $K_s$ |
|---|---|---|---|---|---|---|---|---|
| (1) | (2) | (3) | (4) | (5) | (6) | (7) | (8) | (9) |
| 9/10/98 | | | | | | | | |
| 6 | 150 | – | 150 | 0 | – | | | |
| 24 | 170 | 0.88* | 170 | 0 | – | | | |
| 8/10/98 | | | | | | | | |
| 18 | 192 | 0.88* | 192 | 0 | – | | | |
| 12 | 217 | 0.88 | 217 | 0 | – | | | |
| 6 | 260 | 0.83 | 247† | 13 | – | 13 | 0 | |
| 24 | 330 | 0.781 | 280† | 50 | 0.26** | 50 | 0 | |
| 7/10/98 | | | | | | | | |
| 18 | 430 | 0.77 | 318 | 112 | 0.45** | 112 | 0 | |
| 12 | 540 | 0.80 | 362 | 178 | 0.63 | 178 | 0 | |
| 6 | 700 | 0.77 | 411 | 289 | 0.62 | 283†† | 6 | |
| 24 | 970 | 0.72 | 467 | 503 | 0.58 | 448†† | 54 | 0.12 |
| 6/10/98 | | | | | | | | |
| 18 | 1570 | 0.62 | 531 | 1039 | 0.48 | 712 | 327 | 0.17 |

$$*\frac{150}{170} = 0.88, \quad \frac{192}{217} = 0.88; \quad **\frac{13}{50} = 0.26, \quad \frac{50}{112} = 0.45$$

The recession curve from discharge of 1570 m³/sec to 150 m³/sec is entered in the table in reverse order. The plot of the graph in the semilog paper is shown in Fig. 7.4. Change in the plot can be observed at 260 m³/sec for curve $K_g$ in col. (2) and 289 m³/sec for curve $K_i$ in col. (5). The flows beyond them are calculated as follows.

$$† 247 = \frac{217}{0.88}, \quad 280 = \frac{247}{0.88} \quad \text{in col. (4)}$$

$$†† 283 = \frac{178}{0.63}, \quad 448 = \frac{283}{0.63} \quad \text{in col. (7)}$$

The recession constants taken for the storm are :

$$K_g = 0.88 \qquad K_i = 0.63 \qquad K_s = 0.17$$

$$K_r = K_g K_i K_s = 0.88 \times 0.63 \times 0.17 = 0.094$$

*Time to Peak $t_p$*

It is the time lapse between the starting of the rising limb *A* to the peak of the hydrograph *P* (Fig. 7.3). It is represented in days for large basins and in hours for small basins. Factors like distribution of rainfall over the basin, duration of storm, travel time of water in the channel and other catchment characteristics govern the time to peak of a hydrograph.

*Time Lag $t_l$*

Time lag is the time interval between centre of mass of rainfall hyetograph to the centre of mass of runoff hydrograph, from the same axis. Many hydrologists

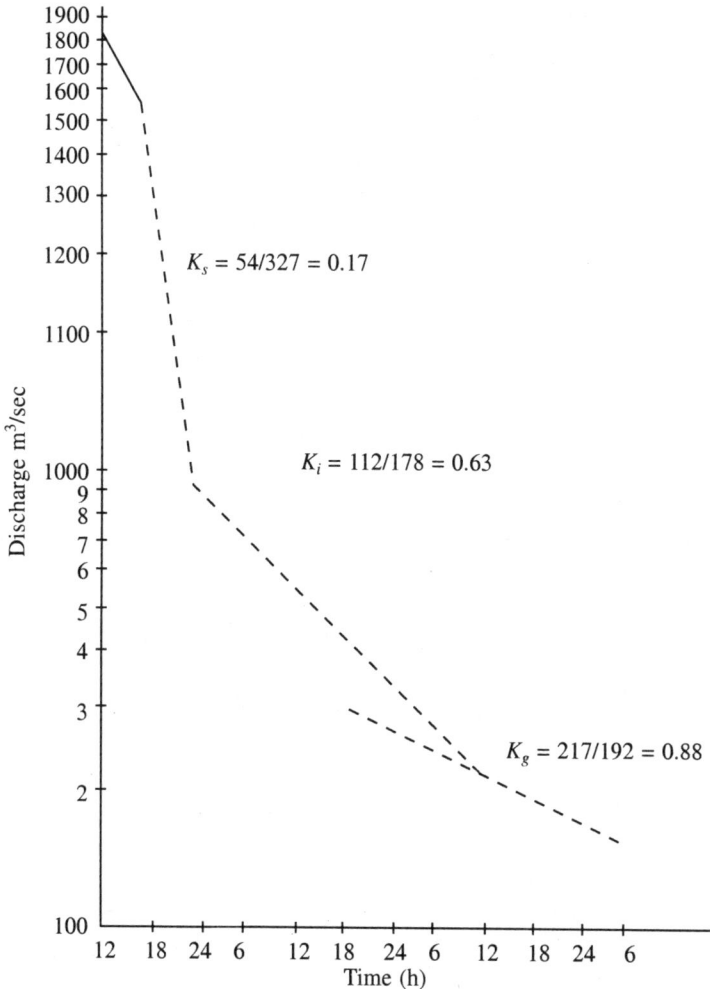

**Fig. 7.4  Plot of Recession Limb for Example 7.1**

also refer to it as the time lapse between the centre of mass of the effective rainfall and the peak of the hydrograph as shown in the Fig. 7.3.

*Time of Concentration ($t_c$)*

It is the longest time taken by a drop of water to reach the outlet or the point of consideration in the channel from the furthest point of the catchment, it can be estimated using the following Kirpich (1940) formula

$$t_c = 0.000323 L^{0.77} S^{-0.385} \tag{7.3}$$

where $t_c$ is the time of concentration (h), $L$ the maximum length of travel of water (m), $S$ the slope of the channel given by $H/L$ ratio and $H$ is the elevation difference between the remote point in the channel and the outlet point. Time to peak plus the storm duration represents approximately the time of concentration of a basin. The lag time $t_l$ and time of concentration $t_c$ can be related in a simple way as

$$t_c = 1.42t_l \qquad (7.4)$$

where $t_l$ relates to peak discharge of the hydrograph.

### Time Base of Hydrograph ($T_b$)

It is considered as the time between the starting of the runoff hydrograph (point A) to the end of direct runoff due to storm (point D)

$$T_b = t_c + t_r \qquad (7.5)$$

where $t_r$ is the storm duration (h) and $t_c$ is time of concentration in hours.

### Direct Runoff

It is that part of the precipitation which appears quickly as flow in the river. It includes the surface runoff and the subsurface runoff (inter flow) of a basin at its outlet.

### Base Flow

The part of runoff represented by a hydrograph which receives water from the ground water storage is called base flow. Natural springs and channel bank storages contributes to this flow during non-rainy periods. The total runoff of a storm is usually divided into two components, viz., (i) base flow and (ii) direct runoff.

When hydrograph resulting from a storm is available at a location, first separate the base flow, the residual of which is the *direct runoff hydrograph* (DRH). Of the procedures available for base flow separation, four are discussed as follows:

**Procedure I:** It is the simplest form of base flow separation which is obtained by joining the points A and D of the hydrograph by a straight line as shown in Fig. 7.5. Point A is clearly distinguishable in the plot of the hydrograph. Point D is obtained by locating it N days away from the peak time of the flood. Linsly proposed the following equation for estimating N:

$$N = 0.83A^{0.20} \qquad (7.6)$$

where A is the drainage area of the basin in km². For small watersheds this method of separation of base flow from the storm hydrograph gives erroneous results. Experience and inspection of good number of hydrographs of a basin help to reasonably locate point D such that the time base of the hydrograph is not too long and ground water contribution does not affect the direct runoff resulting from the storm.

**Procedure II:** In this procedure, the curve of the hydrograph representing base flow before the onset of storm ($A'A$ in Fig. 7.5) is extended till it meets at point E vertically below the peak of the hydrograph. This curve would represent the hydrograph by base flow in the absence of storm. The points, E and D are joined to give a base flow curve *AED* (Fig. 7.5). This method is used frequently by hydrologists in the field. Location of point *D* in the hydrograph is obtained as discussed earlier.

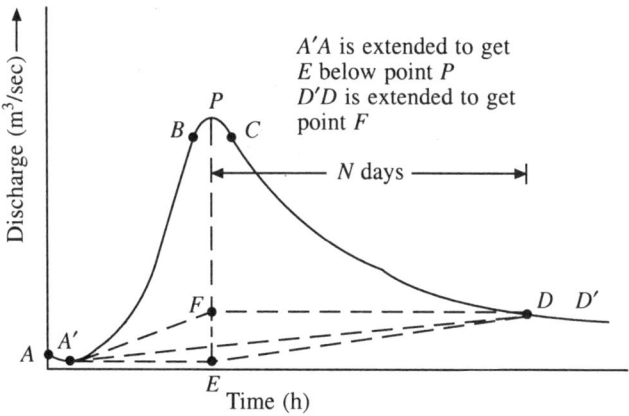

**Fig. 7.5 Separation of base flow**

**Procedure III:** This is used where significant quantity of stream flow is derived from the ground water storage and it appears at the stream outlet quickly along with the direct runoff. The procedure is to locate a point *D'* in the recession curve away from point *D* and extend the curve *D'D* backward so that a point vertically below the peak flow *F* is marked. Join point *F* with *A* so that the curve *AFD* gives the base flow separation line (Fig. 7.5).

For flood studies, base flow contribution is usually small and any of the above methods may not affect the magnitude of direct runoff considerably from a storm. Therefore, inaccuracies in the estimation and separation of base flow is not very important in flood studies.

**Procedure IV:** This procedure, practised by most of the field engineers, involves visual inspection of the hydrograph and selection of a point *D* in the recession limb such that it almost represents the end of the drainage of the channel storage from a storm. *A* straight line joining *A* and *D* gives a reasonable estimate of ground water contribution to the storm hydrograph (Example 7.2). Procedures I and IV are essentially the same if selection of point *D* is made on the basis of experience and judgement.

*Rainfall Excess*
The area of direct runoff hydrograph is the volume of direct runoff from a watershed from a given storm. The runoff volume is the result of the precipitation in excess of the losses like interception, depression storage, evaporation, transpiration and infiltration during the storm rainfall–runoff process. Therefore, for analysis of any storm, a rainfall hyetograph is to be prepared and the total loss is to be subtracted from it to give the *effective rainfall hyetograph* (ERH) which should be equal to the depth of direct runoff. Stepwise procedure to compute ERH from the analysis of rainfall and runoff are as follows:

(1) Plot the flood hydrograph to scale on a graph paper.
(2) Separate the base flow by any of the methods.
(3) The ordinates of direct runoff hydrograph are readout from the hydrograph (Fig. 7.6a).

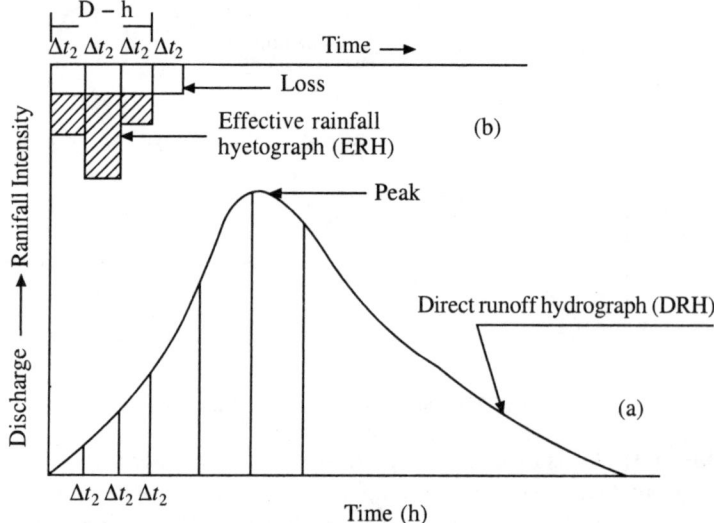

Fig. 7.6  **Direct Runoff Hydrograph and the Corresponding ERH.**

(4) Compute the area of DRH by using trapezoidal formula or by planimetering. Using trapezoidal formula the volume of DRH is obtained as

$$\text{Volume of DRH} = \Delta t_1 \times 60 \times 60 \; [h_1 + h_2 + h_3 + \dots + h_n] \qquad (7.7)$$

where, $h_1, h_2, \dots, h_n$ are the ordinates of DRH at time interval of $\Delta t_1$ hours.

(5) Compute the depth of direct runoff. To achieve this, divide the volume of runoff by the catchment area, taking due consideration of the units used.
(6) Compute the average storm rainfall over the basin from the rain gauge stations in and around the basin.
(7) Compute the rainfall intensities of the storm in mm/h at $\Delta t_2$ interval by using the data of a self-recording rain gauge from the catchment producing the above flood in the time interval of $\Delta t_2$ hours.
(8) Plot rainfall hyetograph in the same graph paper having the hydrograph, by taking time in abscissa in the opposite side of DRH and intensity of rainfall in the ordinate adjacent to the time axis of the DRH as shown in Fig. 7.6b.
(9) The initial losses may be subtracted from the first block of the rainfall histogram (hyetograph).
(10) Assume a trial $\phi$-index and compute the total depth of rainfall excess. (Note that the excess rainfall being the cause of the hydrograph, both the hydrograph and the hyetograph should start at the same time.) This depth should be the same as the DRH depth computed under step (5) above, if not, change the level of $\phi$-index on trial and error so that they match exactly. The procedure is explained through Example 7.2.

**Example 7.2:** A storm event of October, 1998 of Ib river at Sundargarh site is as follows.

| Average catchment rainfall (mm) | 7.76 | 33.72 | 8.16 | 0 | 0 | 4.26 | 4.94 |
|---|---|---|---|---|---|---|---|
| Time (6/10/98-7/10/98) | 21 | 24 | 03 | 06 | 9 | 12 | 15 |

Catchment area of the basin is 5870 sq. km. The observed hydrograph is given in example 7.1. Calculate $\phi$-index and excess rainfall.

**Solution**

The hydrograph ordinates are entered in col. (2) of Table 7.2. The base flow at the beginning of the storm is 100 m³/sec and at the end 150 m³/sec. Therefore, a linear variation between 100 and 150 of the base flow is assumed as shown in col. (3). The magnitudes of direct runoff hydrograph are obtained as in col. (4) = col. (2) – col. (3).

**Table 7.2   Rainfall Excess Estimation from Observed Flood Record**

| Time | Discharge (m³/sec) | Baseflow (m³/sec) | DRH (m³/sec) | Rainfall (mm) | Rainfall intensity mm/h | $\phi$-index (mm/h) | Rainfall excess (mm) |
|---|---|---|---|---|---|---|---|
| (1) | (2) | (3) | (4) | (5) | (6) | (7) | (8) |
| 21 | | | | 7.76 | 2.59 | 3.54 | – |
| 24 | 100 | 100 | 0 | 33.72 | 11.24 | 3.54 | 7.7 |
| 3 | | | | 8.16 | 2.72 | 3.54 | – |
| 6 | 610 | 105 | 505 | 0 | 0 | – | – |
| 9 | | | | 0 | 0 | | |
| 12 | 1860 | 110 | 1750 | 4.26 | 1.42 | 3.54 | – |
| 15 | | | | 4.94 | 1.65 | 3.54 | – |
| 18 | 1570 | 115 | 1455 | Total depth of rainfall excess = 7.7 × 3 = 23. 1 mm | | | |
| 24 | 970 | 130 | 850 | | | | |
| 6 | 700 | 125 | 575 | | | | |
| 12 | 540 | 130 | 410 | | | | |
| 18 | 430 | 135 | 295 | | | | |
| 24 | 330 | 140 | 190 | | | | |
| 6 | 260 | 145 | 115 | | | | |
| 12 | 217 | 150 | 67 | | | | |
| 18 | 192 | 150 | 42 | | | | |
| 24 | 170 | 150 | 20 | | | | |
| 6 | 150 | 150 | 0 | | | | |

$\Delta t_1$ of DRH ordinates = 6 h = 6 × 60 × 60 sec

$\Delta t_2$ of ERH ordinates = 3 h

Area of the DRH which is equal to the volume of direct runoff from the area is calculated below.

$$\text{Area} = 6 \times 60 \times 60 \{0 + 505 + 1750 + 1455 + 850 + 575 + 410 + 295$$
$$+ 190 + 115 + 67 + 42 + 20 + 0\}$$
$$= 6 \times 3600 \times \{6274\} = 135.5184 \times 10^6 \, m^3$$

$$\text{Depth of direct runoff} = \frac{(135.5184 \times 10^6 \times 1000)}{(5870 \times 10^6)} = 23.10 \text{ mm.}$$

Taking $\phi$-index as 3.54 mm/h, the depth of rainfall excess is 7.7 mm/h × 3 h = 23.1 mm which equals to the direct runoff depth from the storm-rainfall for the catchment. Plot of DRH and ERH are shown in Fig. 7.7.

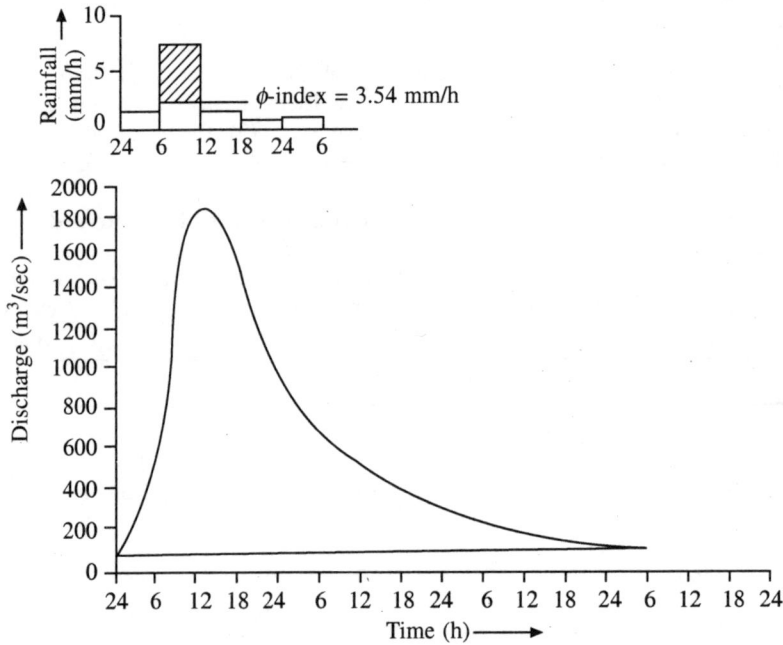

**Fig. 7.7 DRH, ERH and $\phi$-index for Example 7.2**

## 7.4 UNIT HYDROGRAPH (UH)

An unit hydrograph or *unit graph* is the hydrograph of direct runoff resulting from unit depth of 1 cm of rainfall excess generated uniformly over the basin for a specified duration (*D*-hours). The term unit depth of rainfall excess means excess rainfall over and above all losses in the basin under consideration (see Chapter 4 for details). The duration is the period of rainfall excess which is assumed to be uniformly distributed over the basin area. The specified duration is important as the shape and the peak of unit hydrograph of a basin depends on it. Thus for a period of 3 h rainfall excess over the basin, the unit hydrograph is named as 3-h unit hydrograph. Sketch of an UH is shown in Fig. 7.8. A watershed can have as many unit-hydrographs depending on the number of corresponding periods of rainfall excess. An unit hydrographs of 2, 6 or 12 h indicates that it is the duration of rainfall excess giving rise to the unit hydrograph. It never means the duration of the occurrence of unit hydrograph.

The concept of unit hydrograph was originally put forth by Sherman in 1932. The present definition is more refined and an universally accepted one based on the following principles and assumptions.

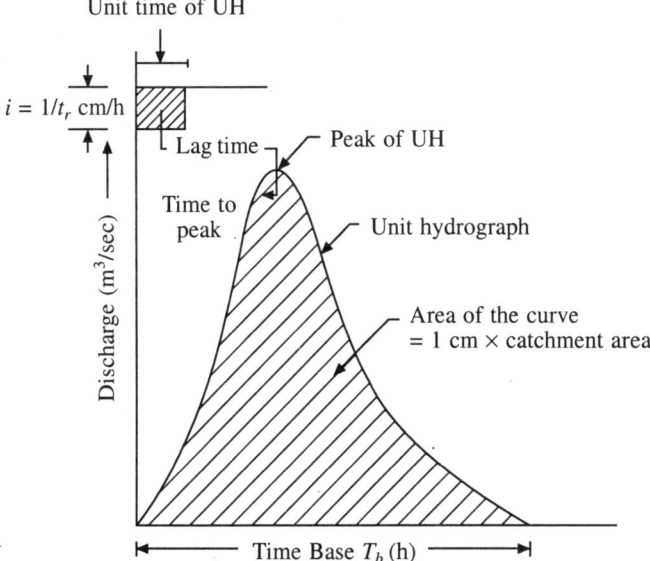

Unit time of UH

$i = 1/t_r$ cm/h

Lag time — Peak of UH

Time to peak

Unit hydrograph

Area of the curve
= 1 cm × catchment area

Discharge (m³/sec)

Time Base $T_b$ (h)

**Fig. 7.8   Sketch of an Unit Hydrograph**

### 7.4.1   Assumptions and Conditions in Unit Hydrograph

1. *The principle of time invariance* applies to the unit hydrograph, which means that a given rainfall excess will produce the same direct runoff hydrograph whatever may be the season of the year.
2. *The principle of linearity* is that when a *D*-hour rainfall excess of 1 cm depth over the basin produces a direct runoff hydrograph of 1 cm, then *x* cm depth of rainfall excess of the same duration (*D*-hour) over the same basin will produce a runoff of *x* cm depth. Conversely, the ordinates of the direct runoff hydrograph from *x* cm rainfall excess of *D*-hour duration may be brought to unit hydrograph of *D*-hour by dividing the ordinates of DRH by $(x/1)$. Similarly, if $x_1$ cm of rainfall excess produces $y_1$ cm of runoff and $x_2$ cm produces $y_2$ cm, then $(x_1 + x_2)$ cm of rainfall excess will produce $(y_1 + y_2)$ cm of DRH.
3. A storm of duration of rainfall excess $t_r$ hours will always produce a surface runoff hydrograph of base $T_b$-hours regardless the intensity of rainfall in the period $t_r$. Time base $T_b$ of runoff is sum of $t_c$ and $t_r$ where $t_c$ is the time of concentration of the watershed upto the desired outlet.
4. An unit hydrograph is a lumped response of the catchment at the basin outlet.
5. The area of DRH due to unit hydrograph is the area of the basin multiplied by 1 cm.
6. When the duration of rainfall excess is *D*-h, the excess rainfall intensity is 1/*D*-cm/h. This is because the total rainfall producing a UH is 1 cm. For 1 h duration of rainfall excess, the intensity of rainfall excess is 1 cm/h.
7. Since the consideration for UH is the rainfall excess, ·the antecedent or

subsequent storm conditions have no action to the present derivation of UH.

### 7.4.2   Limitations of Unit Hydrograph

1. The maximum catchment limitation for derivation or application of unit hydrograph theory is upto 5000 $km^2$. For catchment exceeding this area, the basic assumption of uniform rainfall distribution over the basin due to the storm may be violated. However, the limitation on size of catchment area should also take into account the orographic features of the basin.
2. The application of UH is not suitable for very long basins.
3. The lower limit on the size of a basin to which the UH concept apply should preferably be more than 2 $km^2$.
4. When a large portion of a basin is covered with snow, the UH principle should not be applied.
5. The duration of rainfall excess should preferably be 1/3 to 1/5 of basin lag.
6. UH should not be derived from a catchment where large storage exists.
7. If there is high variation in the rainfall intensity over the basin then for such storms UH should not be derived.

Fluctuations of intensities of rainfall in *D*-hour is taken care of by the catchment characteristics producing runoff. Total affect of variation of intensity on unit hydrograph is usually neglected. However it is not possible to quantify this limitation of variation in UH.

### 7.4.3   Uses of Unit Hydrograph
The important purposes for which a unit hydrograph can be used are :

1. *Computation of flood hydrograph for design of structure*: When the probable maximum precipitation (PMP) or the standard project storm (SPS) for a basin is known, the UH is convoluted over the excess rainfall of the histogram blocks of PMP or SPS to obtain the design flood at the project site (see Chapter-8 for details).
2. *Extension of the flow records at a site*: The method is more or less the same as (1). All storm precipitation depths, i.e., rainfall excess for the entire period under consideration are multiplied successively by UH ordinates and added up to compute the runoff volumes. The work is enormous and cannot be handled manually. Use of a digital computing machine is necessary to calculate the runoff values for each storm.
3. *Flood forecasting models*: Unit hydrographs developed for a basin are stored in a computer. Knowing the excess rainfall depths from telemeter-gauges, flood can be forecasted for the basin by convoluting the UH over the excess storm rainfall and carrying out the channel routing if necessary.
4. *Comparing the catchment characteristics*: Two unit hydrographs of the same unit durations derived from two adjoining basins can be used to compare the hydro-meteorological characteristics of the basins.

## 7.5   DERIVATION OF UNIT HYDROGRAPH

### 7.5.1   From Simple Storm Hydrograph

A basin under study should have a stream gauging station at the desired site with discharge records available at hourly or short intervals. It should also have good coverage of rain gauge stations with at least one self recording rain gauge (SRRG). With the type of rainfall and runoff data available from these stations, the following is the procedure to derive UH from a basin.

1. Collect the rainfall and corresponding runoff data for all isolated storms. The selected storms should preferably have nearly equal durations of rainfall.
2. Take a storm hydrograph and read its ordinates at $\Delta t_1$ interval. Compute the depth ($h_r$) of surface runoff from the basin after separating the base flow. This is discussed in section 7.3.
3. Take the corresponding storm rainfall for all stations of the basin and compute the average basin rainfall by Thiessen polygon or other suitable method. Convert daily rainfall to convenient $\Delta t_2$-hour intensities depending on the total duration of the storm. Total storm period may be divided into 2 to 6 h intervals. The breakup of the daily average precipitation to $\Delta t_2$-hour should be done on the basis of the percentage of hourly rainfall records of automatic raingauges (SRRG) of the watershed.
4. Plot the rainfall hyetograph of the average storm precipitation of the basin. Separate the initial losses and calculate the $\phi$-index or the infiltration losses such that the depth of rainfall excess ($h_e$) is equal to the depth of runoff ($h_r$) obtained from the storm DRH. This step is outlined in section 7.3.
5. The total duration of rainfall excess obtained from the rainfall hyetograph is the unit duration (D-hour) of unit hydrograph. It may be noted here that breaking the storm hyetograph to $\Delta t_2$-hours has nothing to do with the unit duration of the unit hydrograph. Usually D is an integral multiple of $\Delta t_2$ hours.
6. Divide all the ordinates of the DRH by the factor $h(=h_e)$ cm/1cm to give 1 cm rainfall excess over the basin. This resulting hydrograph is the unit hydrograph of D-hour called D-hour unit hydrograph for the basin.
7. Repeat the procedure from steps (2) to (6) for all other storms selected under step (1) for the basin.
8. Plot all the Unit hydrographs on a graph paper from the common zero–zero axis. The various storms may not give identical unit hydrographs.
9. Average all the unit hydrographs. For this the peak of all unit hydrographs are averaged first to give $Q_p$ and the time to peak $t_p$ are averaged to give $t_p$. All other ordinates of the hydrographs are averaged. The time base $T_b$ of the average unit hydrograph is obtained by averaging $T_b$ of all UH's.
10. The total depth of runoff for the average UH should be 1 cm. For this, plot the average unit hydrograph and compute its area by trapezoidal method or by planimetering the hydrograph. Sum of all ordinates of the unit hydrograph multiplied by the time interval of ordinates in seconds

gives the volume of UH, which when divided by the catchment area should yield 1 cm. If the depth of UH is having slight departure from unit then the unit hydrograph boundaries are to be adjusted and procedure of this step repeated to yield 1cm rainfall excess over the basin. The following example tries to explain the above method.

---

**Example 7.3:** Catchment average rainfall and the corresponding observed runoff for two flood events for an irrigation project are given below. Compute UH for each and an average unit hydrograph for the catchment. What is the duration of the unit hydrograph derived by you?

**Table 7.3    Rainfall and Runoff Data for Example 7.3**

| Time (h) | Event-I Flood magnitude (m³/sec) | Event-I Average rainfall (mm) | Event-II Flood magnitude (m³/sec) | Event-II Average rainfall (mm) |
|---|---|---|---|---|
| (1) | (2) | (3) | (4) | (5) |
| 03 | 87 | 14.53 | 113 | 29.36 |
| 06 | | 22.31 | | 13.42 |
| 09 | 155 | 30.97 | 113 | 33.72 |
| 12 | | 21.28 | | 37.41 |
| 15 | 2912 | | 224 | |
| 21 | 3774 | | 1565 | |
| 3 | 4283 | | 5727 | |
| 9 | 2467 | | 4102 | |
| 15 | 1492 | | 3152 | |
| 21 | 791 | | 2202 | |
| 3 | 613 | | 1805 | |
| 9 | 486 | | 1021 | |
| 15 | 370 | | 815 | |
| 21 | 285 | | 644 | |
| 3 | 204 | | 543 | |
| 9 | 165 | | 441 | |
| 15 | 130 | | 351 | |
| 21 | 102 | | 328 | |
| 3 | | | 240 | |
| 9 | | | 160 | |

Take catchment area at the project site as 5870 sq. km.

**Solution**

Calculations for runoff analysis are carried out in Table 7.4.

Here $\Delta t_1 = 6$ h and $\Delta t_2 = 3$ h.

$$\text{Depth of runoff for event- I} = \frac{(6 \times 3600 \times 13053)}{(5870 \times 10^6)} = 0.048 \text{ m} = 48.0 \text{ mm} = 4.8 \text{ cm}$$

where 13053 is the sum of flood magnitude of event-I.

**Table 7.4  Computation of Runoff Depth over the Catchment for Example 7.3**

| Time (h) | Event-I Discharge (m³/sec) | Event-I Baseflow (m³/sec) | Event-I DRH (m³/sec) | Event-II Discharge (m³/sec) | Event-II Baseflow (m³/sec) | Event-II DRH (m³/sec) | UH ordinates Event-I col.(4)/4.80 | UH ordinates Event-II col.(7)/7.38 | UH ordinates Average | UH ordinates Adjusted (UH) |
|---|---|---|---|---|---|---|---|---|---|---|
| (1) | (2) | (3) | (4) | (5) | (6) | (7) | (8) | (9) | (10) | (11) |
| 3 | 87 | 87 | 0 | 113 | 113 | 0 | 0.0 | 0.0 | 0 | 0.0 |
| 9 | 155 | 88 | 67 | 113 | 113 | 0 | 14.0 | 0.0 | 7.0 | 301.4 |
| 15 | 2912 | 89 | 2823 | 224 | 113 | 111 | 587.8 | 15.0 | 301.4 | 816.7 |
| 21 | 3774 | 90 | 3684 | 1565 | 113 | 1452 | 767.5 | 196.7 | 482.1 | 517.5 |
| 3 | 4283 | 91 | 4192 | 5727 | 113 | 5614 | 872.7 | 760.5 | 816.7 | 351.5 |
| 9 | 2467 | 92 | 2375 | 4102 | 113 | 3989 | 494.5 | 540.4 | 517.5 | 214.0 |
| 15 | 1492 | 93 | 1399 | 3152 | 113 | 3039 | 291.3 | 411.7 | 351.5 | 168.5 |
| 21 | 791 | 94 | 697 | 2202 | 113 | 2089 | 145.1 | 283.0 | 214.0 | 102.1 |
| 3 | 613 | 95 | 518 | 1805 | 113 | 1692 | 107.8 | 229.2 | 168.5 | 76.0 |
| 9 | 486 | 96 | 390 | 1021 | 113 | 908 | 81.2 | 123.0 | 102.1 | 35.5 |
| 15 | 370 | 97 | 273 | 815 | 113 | 702 | 56.8 | 95.1 | 76.0 | 40.0 |
| 21 | 285 | 98 | 187 | 644 | 113 | 531 | 38.9 | 72.0 | 55.5 | 29.0 |
| 3 | 204 | 99 | 105 | 543 | 113 | 430 | 21.9 | 58.2 | 40.0 | 19.1 |
| 9 | 165 | 100 | 65 | 441 | 113 | 328 | 13.5 | 44.4 | 29.0 | 14.6 |
| 15 | 130 | 101 | 29 | 351 | 113 | 238 | 6.0 | 32.2 | 19.1 | 8.6 |
| 21 | 102 | 102 | 0 | 328 | 113 | 215 | 0.0 | 29.1 | 14.6 | 3.2 |
| 3 | – | – | – | 240 | 113 | 127 | – | 17.2 | 8.6 | 0 |
| 9 | – | – | – | 160 | 113 | 47 | – | 6.4 | 3.2 | 0 |
| Total | | | | 13053 | | 20060 | | | | |

$$\text{Depth of runoff for event- II} = \frac{(6 \times 3600 \times 20060)}{(5870 \times 10^6)} = 0.0738 \text{ m}$$

$$= 73.8 \text{ mm} = 7.38 \text{ cm}$$

Analysis of rainfall is carried out in Table 7.5.

**Table 7.5   Analysis of Rainfall for Obtaining Unit Duration of UH for Example 7.3**

| | Event - I | | | | Event - II | | | |
|---|---|---|---|---|---|---|---|---|
| Time (h) | Rainfall (mm) | Rainfall intensity (mm/h) | $\phi$–index (mm/h) | Rainfall excess ( mm) | Rainfall (mm) | Rainfall intensity (mm/h) | $\phi$–index (mm/h) | Rainfall excess (mm) |
| (1) | (2) | (3) | (4) | (5) | (6) | (7) | (8) | (9) |
| 3–6 | 14.53 | 4.84 | 3.42 | 1.42 | 29.36 | 9.79 | 3.34 | 6.45 |
| 6–9 | 22.31 | 7.44 | 3.42 | 4.02 | 13.42 | 4.47 | 3.34 | 1.13 |
| 9–12 | 30.97 | 10.32 | 3.42 | 6.90 | 33.72 | 11.24 | 3.34 | 7.90 |
| 12–15 | 21.28 | 7.09 | 3.42 | 3.67 | 37.41 | 12.47 | 3.34 | 9.13 |
| Total | | | | 16.01 | | | | 24.61 |

Rainfall excess = 16.01 × 3 = 48.03 mm      Rainfall excess = 24.61 × 3 = 73.83 mm

Rainfall excess period for both the storms are 12 h. Therefore, the unit duration of unit hydrograph is 12 h. Rainfall excess depth for event-I is 4.8 cm. All the ordinates of DRH of event-I are divided by 4.8 to get the UH ordinates. The DRH and UH are plotted in Fig. 7.9a. Similarly the DRH ordinates of event-II are divided by the corresponding rainfall excess depth of 7.38 to get the ordinates of UH. The DRH and UH for the second event is plotted in Fig. 7.9b. The two unit hydrographs derived are composited in Fig. 7.9c. The average UH is obtained as outlined in steps (9) and (10). The ordinates of 12-hour unit hydrographs are given in col. (11) of Table 7.4 from which we get

$$\frac{6 \times 3600 \times \Sigma(\text{col. } 11) \times 100}{5870 \times 10^6} = 1.0 \text{ cm}$$

**Example 7.4:** Rainfall of 45 mm and 58 mm occurred successively over the basin of example 7.3 continuously for 12 h each. Assume a loss rate of 0.1 cm/h. Compute runoff due to the storm assuming a constant base flow of 100 m³/sec. Use the UH of example 7.3.

**Solution**
The loss rate is 0.10 cm/h (refer Table 7.6). Over 12 h the loss is 1.2 cm. The rainfall of 45 mm or 4.5 cm reduces to (4.5 – 1.2) = 3.3 cm and the rainfall of 5.8 cm reduces to (5.8 – 1.2) = 4.6 cm in successive 12 h blocks. The resulting ordinates of DRH due to 3.3 cm of rainfall excess occurring in 12 h can be calculated by multiplying the ordinates of UH by 3.3. The DRH due to 4.6 cm rainfall excess occurs after 12 h of the beginning of the storm. Therefore, the ordinates of DRH due to 4.6 cm of rainfall excess is lagged by 12 h as given in col. (4) of Table 7.6. The method of superposition of unit hydrograph allows us to combine the effects of 3.3 and 4.6 cm hydrographs as shown in col. (5) of Table 7.6. The flood hydrograph is obtained by adding the base flow of 100 m³/sec as given in col. (5). The resulting flood hydrograph is given in col. (7).

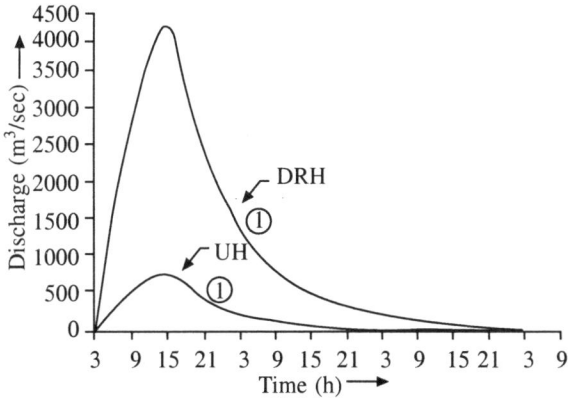

**Fig. 7.9a    DRH and UH of Event-I of Example 7.3**

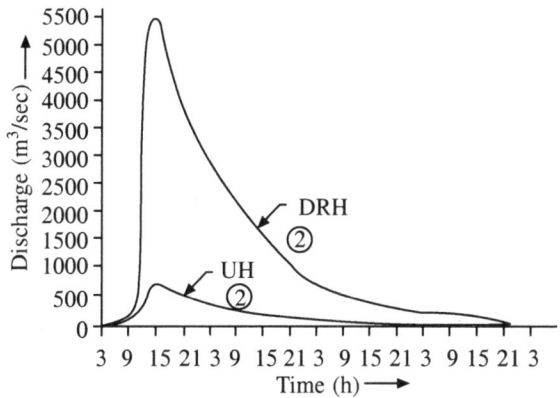

**Fig. 7.9b    DRH and UH of Event-II of Example 7.3**

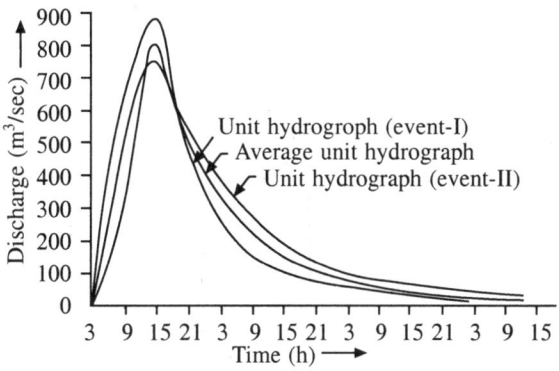

**Fig. 7.9c    Average UH of Example 7.3**

### 7.5.2    From Complex Storms

On the event of non-availability of isolated storms producing single peaked hydrograph of a basin, the available complex hydrographs are to be utilised for the derivation of unit hydrograph. A complex hydrograph is the overall effect of

Table 7.6    Calculation of Flood Hydrograph for Example 7.4

| Time (h) | Ordinates of UH (m³/sec) | DRH due to 3.3 cm RE (2) × 3.3 (m³/sec) | DRH due to 4.6 cm RE (2) × 4.6 (m³/sec) | DRH due to both RE blocks (3) + (4) (m³/sec) | Base flow 100 (m³/sec) | Flood hydrograph (m³/sec) |
|---|---|---|---|---|---|---|
| (1) | (2) | (3) | (4) | (5) | (6) | (7) |
| 3 | 0.0 | 0.0 | 0.0 | 0.0 | 100 | 100.0 |
| 9 | 301.4 | 994.6 | 0.0 | 994.6 | 100 | 1094.6 |
| 15 | 816.7 | 2695.1 | 0.0 | 2695.1 | 100 | 2795.1 |
| 21 | 517.5 | 1707.7 | 1386.4 | 3094.2 | 100 | 3194.2 |
| 3 | 351.5 | 1160.0 | 2756.8 | 3916.8 | 100 | 4016.8 |
| 9 | 214.0 | 706.2 | 2380.5 | 3086.7 | 100 | 3186.7 |
| 15 | 168.5 | 556.0 | 1616.9 | 2172.9 | 100 | 2272.9 |
| 21 | 102.1 | 337.0 | 984.4 | 1321.4 | 100 | 1421.4 |
| 3 | 76.0 | 250.8 | 775.1 | 1025.9 | 100 | 1125.9 |
| 9 | 55.5 | 183.3 | 469.7 | 652.9 | 100 | 752.9 |
| 15 | 40.0 | 132.0 | 349.6 | 481.6 | 100 | 581.6 |
| 21 | 29.0 | 95.7 | 255.3 | 351.0 | 100 | 451.0 |
| 3 | 19.1 | 63.0 | 184.0 | 247.0 | 100 | 347.0 |
| 9 | 14.6 | 48.2 | 133.4 | 181.6 | 100 | 281.6 |
| 15 | 8.6 | 28.4 | 87.9 | 116.3 | 100 | 216.3 |
| 21 | 3.2 | 10.6 | 67.2 | 77.8 | 100 | 177.8 |
| 3 | 0.0 | 0.0 | 39.6 | 39.6 | 100 | 139.6 |
| 9 | 0.0 | 0.0 | 14.7 | 14.7 | 100 | 114.7 |
| 15 | 0.0 | 0.0 | 0.0 | 0.0 | 100 | 100.0 |

several overlapped storms of varying intensities giving rise to multi-peaked hydrograph. It is difficult to separate the effect of the individual isolated periods of rainfalls producing DRH. Of the several methods available to derive UH from such complex hydrographs, two are described as follows:

*Method I (Inverse of Flood Computation Method)*

1. Let the complex DRH be the result of rainfall excess of four consecutive rainfall blocks of magnitudes $R_1, R_2, R_3$ and $R_4$ cm with varying intensities but durations of D-hours each. The rainfall hyetograph is shown in Fig. 7.10a. From the rainfall records numerical value of D-hour and the magnitudes $R_1, R_2, R_3$, and $R_4$ are known.
2. Take the complex hydrograph and plot it on a graph paper. Separate the base flow. Find the DRH and replot it along with ERH in the same paper as shown in Fig. 7.10b.
3. The ordinates of the complex DRH are read at D-hour intervals. Let the ordinates be $Q_1, Q_2, Q_3, ... , Q_n$ m³/sec.
4. Let the unit hydrograph ordinates be at $U_1, U_2, U_3, ... , U_n$ m³/sec at the same interval of D-hour.
5. The first ordinate $Q_1$ of the complex DRH is the result of the first ordinate of unit hydrograph $U_1$ and the rainfall excess $R_1$ only, we can write it as

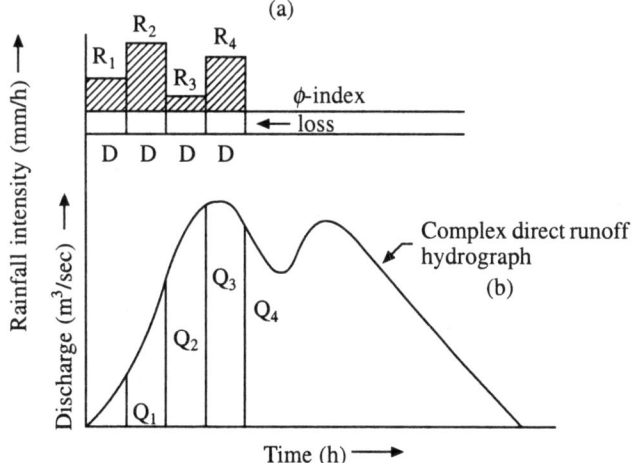

Fig. 7.10 **A Complex Direct Runoff Hydrograph Resulting from 4-Blocks of Rainfall Excess**

$$Q_1 = R_1 U_1 \tag{7.8}$$

From equation (7.8) $U_1$ can be computed. The second ordinate $Q_2$ is the result of the $R_1$ and $R_2$, expressed as

$$Q_2 = R_1 U_2 + R_2 U_1 \tag{7.9}$$

This is because we have read the ordinates of DRH at $D$-hour intervals. In equation (7.9), all except $U_2$ are known from which $U_2$ can be calculated. The third ordinate of complex DRH is expressed as

$$Q_3 = R_1 U_3 + R_2 U_2 + R_3 U_1 \tag{7.10}$$

from which the only unknown $U_3$ is computed. The fourth, fifth and other higher ordinates of DRH are expressed as

$$Q_4 = R_1 U_4 + R_2 U_3 + R_3 U_2 + R_4 U_1$$

$$Q_5 = R_1 U_5 + R_2 U_4 + R_3 U_3 + R_4 U_2 \tag{7.11}$$

$$Q_6 = R_1 U_6 + R_2 U_5 + R_3 U_4 + R_4 U_3$$

$$Q_7 = R_1 U_7 + R_2 U_6 + R_3 U_5 + R_4 U_4$$

$$\cdots$$

In the equation involving $Q_4$, the only unknown is $U_4$ which can be computed. In this way all the ordinates of the UH can be computed. From equation (7.11), all the subsequent ordinates of the unit hydrograph are computed. The system of equations can also be solved by matrix inversion and transposition as

$$\{U\} = \{R^{-1}\}\{Q\} = \{R^T R\}^{-1} R^T Q \tag{7.12}$$

This method becomes handicap in the process of computations when error in $R_1$, $R_2$, $R_3$ or $Q_1$, $Q_2$, $Q_3$ are inheritant in the data. This happens because various intensities of rainfall are grouped to form $R_1, R_2, \ldots, R_n$ or the base flow separation may be erroneous. The error in $U_1$ is propagated to $U_2$ and finally in the recession limb of complex UH, negative values may appear in the calculations as it proceeds. Care may be taken to calculate $R_i$ values again or an optimisation procedure may give a correct UH.

---

**Example 7.5:** Average catchment rainfall excess and the corresponding runoff from a river basin in Orissa from 6/8/85 to 10/8/85 are give below. Derive an UH for the basin. Also mention the unit duration of the UH.

| Time (h) from 06/08/85 | 18 | 24 | 6 | 12 | 18 | 24 | 6 | 12 | 18 | 24 | 6 | 12 | 18 | 24 | 6 |
|---|---|---|---|---|---|---|---|---|---|---|---|---|---|---|---|
| Observed discharge (m³/sec) | 230 | 940 | 2990 | 3260 | 2350 | 2900 | 1760 | 1514 | 1240 | 910 | 713 | 527 | 430 | 393 | 280 |
| Avg. catchment rainfall (mm) | 20.1 | 23.15 | 1.75 | 15.6 | | | | | | | | | | | |

Catchment area at the project site = 5928 sq.km.

**Solution**
On a straight line assumption between the initial 230 m³/sec and final 280 m³/sec, the base flow is separated and the ordinates of the direct runoff hydrograph are computed as given below.

---

| Time (h) from 6/08/85 | 18 | 24 | 6 | 12 | 18 | 24 | 6 | 12 | 18 | 24 | 6 | 12 | 18 | 24 | 6 |
|---|---|---|---|---|---|---|---|---|---|---|---|---|---|---|---|
| Observed discharge (m³/sec) | 230 | 940 | 2990 | 3260 | 2350 | 2900 | 1760 | 1514 | 1240 | 910 | 713 | 527 | 430 | 393 | 280 |
| Baseflow (m³/sec) | 230 | 233 | 236 | 239 | 242 | 245 | 251 | 254 | 257 | 260 | 263 | 266 | 270 | 280 | 280 |
| DRH(m³/sec) | 0 | 707 | 2754 | 3021 | 2108 | 2655 | 1509 | 1260 | 983 | 650 | 450 | 261 | 160 | 113 | 0 |

---

Let the unit duration of the UH be 6 h.
Sum of DRH ordinates is 16631 m³/sec

$$\text{Depth of total runoff} = \frac{(6 \times 3600 \times 16631)}{(5928 \times 10^6)} = 0.0606 \text{ m} = 60.6 \text{ mm}$$

Total effective rainfall = 20.1 + 23.15 + 1.75 + 15.6 = 60.6 mm (given)
The depth of rainfall at 6-h interval are given as

$$R_1 = 2.01 \text{ cm}$$
$$R_2 = 2.315 \text{ cm}$$
$$R_3 = 0.175 \text{ cm}$$
$$R_4 = 1.56 \text{ cm}$$

Using equation (7.8) $Q_1 = R_1 U_1$ or $707 = 2.01 \times U_1$, which gives $U_1 = 351.7$ m³/sec

Equation (7.9 ) gives $2754 = 2.01 \times U_2 + 2.315 \times 351.7$

From which we get $U_2 = 965.1$ m³/sec

Equation (7.10) gives $3021 = 2.01 \times U_3 + 2.315 \times 965.1 + 0.175 \times 351.7$

Which gives $U_3 = 360.8$ m³/sec.

Again $2108 = 2.01 \times U_4 + 2.315 \times 360.8 + 0.175 \times 965.1 + 1.56 \times 351.7$

or $U_4 = 276.2$ m$^3$/sec

$2655 = 2.01 U_5 + 2.315 \times 276.2 + 0.175 \times 360.8 + 1.56 \times 965.1$ which gives

$U_5 = 222.3$ m$^3$/sec

$1509 = 2.01 U_6 + 2.315 \times 222.3 + 0.175 \times 276.2 + 1.56 \times 360.8$ which gives

$U_6 = 190.6$ m$^3$/sec

$1260 = 2.01 U_7 + 2.315 \times 190.6 + 0.175 \times 222.3 + 1.56 \times 276.2$ which gives

$U_7 = 173.6$ m$^3$/sec

Likewise the unit hydrograph ordinates are calculated as follows:

$983 = 2.01 U_8 + 2.315 \times 173.6 + 0.175 \times 190.6 + 1.56 \times 222.1 \Rightarrow U_8 = 100.00$ m$^3$/sec

$650 = 2.01 U_9 + 2.315 \times 100 + 0.175 \times 173.6 + 1.56 \times 190.6 \Rightarrow U_9 = 45.2$ m$^3$/sec

$450 = 2.01 U_{10} + 2.315 \times 45.2 + 0.175 \times 100 + 1.56 \times 173.6 \Rightarrow U_{10} = 28.4$ m$^3$/sec

$261 = 2.01 U_{11} + 2.315 \times 28.4 + 0.175 \times 45.2 + 1.56 \times 100 \Rightarrow U_{11} = 15.6$ m$^3$/sec

$111 = 2.01 U_{12} + 2.315 \times 15.6 + 0.175 \times 28.4 + 1.56 \times 45.2 \Rightarrow U_{12} = 0$

---

*Method II (Collins' Method)*
This method uses trial and error approximations to compute unit hydrograph from a complex DRH. The steps involved are

1. The effective rainfall hyetograph ERH is obtained, using the same process as outlined in the computation of rainfall excess.
2. Assume an unit hydrograph. One approximation is by reducing the DRH ordinates from the complex storm to 1 cm by dividing the ordinates of ERH magnitude.
3. Apply the unit hydrograph to all effective rainfall blocks as discussed in Example 7.4 except the largest block of rainfall excess.
4. The resulting hydrograph is subtracted from the actual DRH of the complex storm giving rise to a residual hydrograph.
5. The resulting residual hydrograph is due to the largest block of excess rainfall hyetograph. Reduce the residual hydrograph to unit hydrograph by dividing the ordinates of residual hydrograph by the RE depth of the largest block. This gives the first trial of UH.
6. Find a weighted average unit hydrograph between the first trial unit hydrograph and the assumed unit hydrograph. Take this unit hydrograph as the base for the second trial iteration and repeat the steps to get the second unit hydrograph under step (5).
7. Compare the second residual UH under step (6) with that of the UH under step (5). If the ordinates of the two UH are nearly comparable then the second UH is taken as the acceptable for the basin. If they do not, then repeat the steps till the last UH and the UH just proceeding to it are comparable.

The correctness of the method can be checked by the method of least square. A discharge hydrograph is computed from the derived unit hydrograph from a known rainfall histogram by multiplying the ordinates of UH by each block of rainfall excess and lagging the resulting hydrographs by the time period of the UH and adding the ordinates. The derived (computed) hydrograph is used to find out the standard error of estimate with the observed one and if it is found too large then the unit hydrograph ordinates are readjusted.

## 7.6   S-HYDROGRAPH

S-hydrograph or S-curve is produced when a continuous effective rainfall at a constant rate occurs for an infinite time over the basin. It is a theoretical concept. The curve of this hydrograph looks like a deformed S-shape. The curve can be obtained when a series of $D$-hour unit hydrographs spaced at $D$-hour apart are summed up. Fig. 7.11 shows a S-hydrograph. This is the result of a series of 1 cm rainfall excess occurring over $D$-hour duration, i.e., $D$-hour unit hydrographs spaced at $D$-hour interval and summed for infinite time. The curve is very steep initially and reaches an equilibrium discharge $Q$ (m$^3$/sec) $= 2.78 A \cdot I = 2.78A/D$, after a time equal to the base of unit hydrograph. Here $A$ is the drainage area of the basin in sq. km, from which UH is derived and $I$ is the intensity of rainfall ($= 1$ cm/$D$-hour) and $D$ in hour. If the time of concentration of the basin is $t_c$ and a $D$-hour UH is used for derivation of S-hydrograph for the catchment, then ideally $t_c/D$ numbers of UH's are sufficient to get the equilibrium discharge from the catchment. The equilibrium discharge is that stage when the rainfall intensity continues upto $t_c$, i.e., when all the catchment area is contributing at the basin outlet. The S-hydrograph reaches equilibrium at this stage. Any more addition of $D$-hour UH will not change the shape of the curve. While constructing S-hydrograph, it can be noticed that undulations are formed for some more time

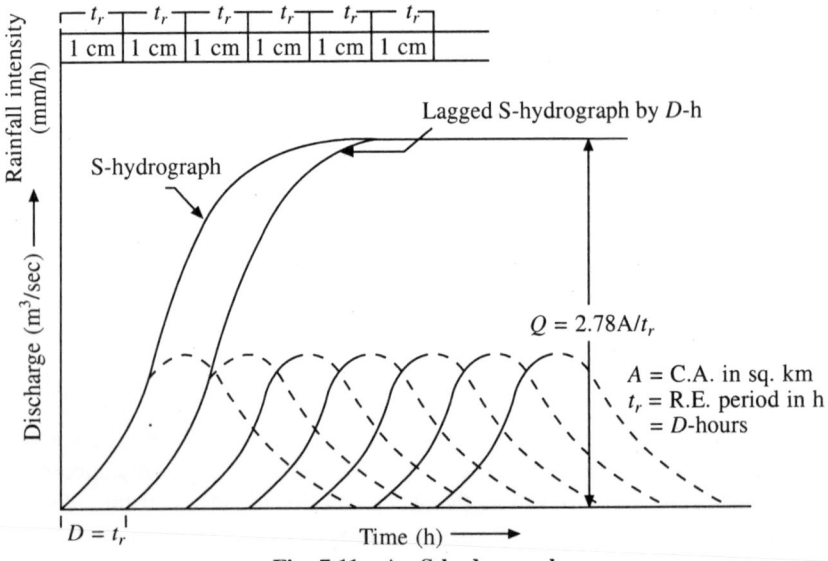

Fig. 7.11   An S-hydrograph

away from the point of equilibrium discharge. These are the results of small errors in the UH ordinates which magnify beyond the equilibrium stage of the summation hydrograph. A smooth curve beyond the time base distance of the UH should be drawn under such a situation.

An important application of S-hydrograph is to derive unit hydrograph of other ($D'$-hour) durations from a unit hydrograph of $D$-hour duration. Let a $D$-hour UH be used to derive S-hydrograph. The intensity of the rainfall $i = 1/D$ cm/h. Let the ordinate of S-curve at any instant $t$ is denoted as $S_1$. Let the S-curve is shifted by time $dt$ hours. The ordinates of this S-curve at the same instant $t$ is say $S_2$ as shown in Fig. 7.12. The ordinate difference $(S_1 - S_2)$ is the DRH arising

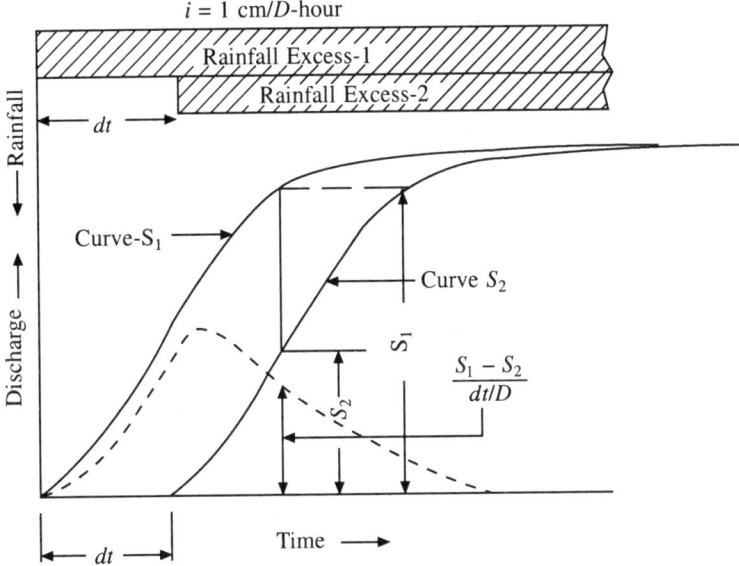

**Fig. 7.12   Derivation of D-hour UH from S-Hydrograph**

due to rainfall period from zero to $dt$ hours with rainfall intensity of 1cm/$D$ hour. The rainfall depth during $dt$ hours = $(dt/D)$ cm. Therefore $(S_1 - S_2)$ divided by $(dt/D)$ cm results a DRH due to R.E. of 1cm in $dt$ hours, i.e., $dt$ unit hydrograph.

$$\text{Therefore } dt \text{ hour UH ordinate} = \frac{(S_1 - S_2)}{(dt/D)} = \frac{(S_1 - S_2)}{i \times dt} \qquad (7.13)$$

where $i = 1/D$ cm/h. From the above relation, steps to derive UH of other durations $D'$ hour from the given $D$-hour UH are given.

1. Construct a S-hydrograph from a $D$-hour unit hydrograph.
2. Write the ordinates of S-hydrograph in a column against time. Let it be denoted by $S_x$.
3. Let all the S-hydrograph ordinates in the next column is lagged by time equal to the desired $D'$ hour duration for which new UH is to be constructed. Let the S-hydrograph be denoted by $S_y$.
4. Subtract the ordinates of the two S-hydrographs $(S_x - S_y)$.

5. Divide the ordinates of the column of step (4) by ($D'/D$). The resulting hydrograph is an UH due to 1 cm depth of rainfall excess or 1 cm rainfall excess in $D'$-hour or intensity of $1/D'$ cm/h rainfall excess.

---

**Example 7.6:** A 4 h hydrograph for a project site in Mahanadi Basin is given below. Calculate (i) A 12-h unit hydrograph and (ii) 2-h UH by S-hydrograph approach.

| Time (h) | 0 | 2 | 4 | 6 | 8 | 10 | 12 | 14 | 16 | 18 | 20 | 22 | 24 | 2 |
|---|---|---|---|---|---|---|---|---|---|---|---|---|---|---|
| UH ordinates (m³/sec). | 0 | 30 | 110 | 170 | 210 | 180 | 120 | 80 | 40 | 35 | 20 | 15 | 5 | 0 |

**Solution**
Required S- hydrograph from the given 4-h UH is calculated in Table 7.7.

**Table 7.7   Unit Hydrograph of Other Durations by S-curve Approach**

| Time<br>(h) | Ordinate<br>of UH<br>(m³/sec) | S-hydro-<br>graph<br>ordinate<br>(m³/sec) | Col. (3)<br>lag by<br>12h<br>(m³/sec) | Col.(3)<br>– Col.(4)<br>(m³/sec) | 12-h UH<br>Col.(5)/<br>(12/4)<br>(m³/sec) | Col.(3)<br>lag by<br>2h<br>(m³/sec) | Col.(3)<br>– Col.(7)<br>(m³/sec) | 2-hUH=<br>col.(8)/<br>(2/4)<br>(m³/sec) |
|---|---|---|---|---|---|---|---|---|
| (1) | (2) | (3) | (4) | (5) | (6) | (7) | (8) | (9) |
| 0 | 0 | 0 | – | 0 | 0 | – | 0 | 0 |
| 2 | 30 | 30 | – | 30 | 10 | 0 | 30 | 60 |
| 4 | 110 | 110 | – | 110 | 36.7 | 30 | 80 | 160 |
| 6 | 170 | 200* | – | 200 | 66.7 | 110 | 90 | 180 |
| 8 | 210 | 320 | – | 320 | 106.7 | 200 | 120 | 240 |
| 10 | 180 | 380** | – | 380 | 126.7 | 320 | 60 | 120 |
| 12 | 110 | 430 | 0 | 440 | 146.7 | 380 | 50 | 100 |
| 14 | 80 | 460 | 30 | 430 | 143.3 | 430 | 30 | 60 |
| 16 | 40 | 470 | 110 | 360 | 120.0 | 460 | 10 | 20(35) |
| 18 | 35 | 495 | 200 | 295 | 98.3 | 470 | 25 | 50(25) |
| 20 | 30 | 500 | 320 | 180 | 60.0 | 495 | 5 | 10 |
| 22 | 15 | 510 | 380 | 130 | 43.3 | 510 | 0 | 0 |
| 24 | 5 | 505(510) | 430 | 75 | 25.0 | 510 | | |
| 2 | 0 | 510 | 460 | 50 | 16.7 | 515 | | |
| 4 | | 510 | 470 | 40 | 13.3 | | | |
| 6 | | 510 | 495 | 15 | 5.0 | | | |
| 8 | | 510 | 500 | 10 | 3.3 | | | |
| 10 | | 510 | 510 | 0 | 0.0 | | | |

\* $200 = 30 + 170$; \*\* $380 = 200 + 180$, and so on.

Figures in brackets show adjusted values. The S-hydrograph ordinates are calculated by successively adding the ordinates of UH at interval equal to unit duration of UH (4-h here). Thus the ordinate of 0 h + 4 h gives the 4th hour S-hydrograph ordinate. The 8th h S-hydrograph ordinate is obtained by adding the 4th and 8th h UH ordinates. Similarly the 10th h S-hydrograph is obtained by adding 2nd, 6th and 10th h UH ordinates together. The process is repeated to cover the entire hydrograph.

For derivation of 12 h UH, the S-hydrograph of col. (3) is lagged by 12-h in col. (4). The difference between cols. (3) and (4) is entered in col. (5). The ordinates of 12-h UH of col.(6) is obtained by dividing col. (5) by (12/4), here 4 is the duration of UH from which S-hydrograph is obtained.

The rest columns showing the computation of 2-h unit hydrograph are self explanatory.

## 7.7 CHANGE OF UNIT DURATION OF UNIT HYDROGRAPH

The duration of a unit hydrograph is fixed on the basis of isolated rainfall excess period producing DRH. If good number of unit hydrographs can be derived from the observed events then the unit durations of these UH's may be different. They may be averaged to obtain a D-hour UH, when their durations vary between 0.9 D and 1.1 D. While dealing with field problems it may be required to produce unit hydrograph of other durations from a $D$-hour unit hydrograph, such as when 3-h rainfall excess blocks are required to be convoluted to produce flood hydrograph using a 4-h UH for the basin. For such a situation, the 4-h UH is to be changed to a 3-h UH. Depending on the requirement either of the following two methods discussed below with examples, can be used to change the unit duration of UH.

(i) When the required duration is an integer multiple of $D$-hour.
(ii) When the required duration is a real multiple of $D$-hour.

### 7.7.1 Required Duration is an Integer Multiple of $D$-hour

For such type of problem the *principle of superposition* is used. If the given unit hydrograph has unit duration of say 6-h and it is required to compute unit hydrograph of say 18 h duration, then the procedure is:

1. Prepare a table in which the col. (1) contains the time $(t_i)$ intervals of the ordinates of the 6-h unit hydrograph.
2. In col. (2), the corresponding ordinates of the UH against the $(t_i)$ are entered.
3. In the next col., the ordinates of the UH as in col. (2) are entered with lag of duration of the unit hydrograph, i.e., 6 h here.
4. In the next col., the data of col. (2) is again entered with a further lag of $2 \times D$-hours, i.e., 12 h here.
5. In col. (5), the values of cols. (2), (3) and (4) are added row-wise and entered.
6. In col. (6), the ordinates of col. (5) are divided by (18 h/6 h) 3 and entered. This gives the desired 18-h unit hydrograph.
7. When the shifting of the unit duration of the UH is too large then smoothening of the ordinates of the new UH may be necessary.

---

**Example 7.7:** Calculate a 12-h unit hydrograph from the UH of example 7.6 using the principle of superposition.

**Solution**

Computations are carried out in Table 7.8. The 4 h UH ordinates are entered against time in cols. (1) and (2). In col. (3), the UH ordinates are lagged by 4 h and in col. (4) it is lagged by 8 h. In col. (5), the ordinates of 3 cm DRH are obtained by summing the cols. (2), (3) and (4) row-wise. The 12-h UH ordinates are obtained by dividing the elements of col. (5) by 3 (12/4).

**Table 7.8   Conversion of 4 h Unit Hydrograph to 12 h Unit Hydrograph**

| Time (h) | UH ordinates (m³/sec) | UH lagged by 4h (m³/sec) | UH lagged by 8h (m³/sec) | Total 3 cm R.E. in 12 h (m³/sec) cols. (2) + (3) + (4) | 12-h UH ordinates (m³/sec) col. (5) /(12/4) |
|---|---|---|---|---|---|
| (1) | (2) | (3) | (4) | (5) | (6) |
| 0 | 0 | – | – | 0 | 0.0 |
| 2 | 30 | – | – | 30 | 10.0 |
| 4 | 110 | 0 | – | 110 | 36.7 |
| 6 | 170 | 30 | – | 200 | 66.7 |
| 8 | 210 | 110 | 0 | 320 | 106.7 |
| 10 | 180 | 170 | 30 | 380 | 126.7 |
| 12 | 110 | 210 | 110 | 430 | 143.3 |
| 14 | 80 | 180 | 170 | 430 | 143.3 |
| 16 | 40 | 110 | 210 | 360 | 120.0 |
| 18 | 35 | 80 | 180 | 295 | 98.3 |
| 20 | 30 | 40 | 110 | 180 | 60.0 |
| 22 | 15 | 35 | 80 | 130 | 43.3 |
| 24 | 5 | 30 | 40 | 75 | 25.0 |
| 2 | 0 | 15 | 35 | 50 | 16.7 |
| 4 | | 5 | 30 | 35 | 11.7 |
| 6 | | 0 | 15 | 15 | 5.0 |
| 8 | | | 5 | 5 | 1·7 |
| 10 | | | 0 | 0 | – |

### 7.7.2   Required Duration is a Real Multiple of *D*-hour

The problem can be solved by the help of S-hydrograph. As an example consider an unit hydrograph of 4-h duration. The requirement is to compute another UH of say 10-hour duration. The method of superposition as discussed earlier cannot be used here as 10 is not an integer multiple of 4. The method suggested by Morgam and Hullinghors by S-hydrograph is described. For details of this method example 7.6 may be followed. The S-hydrograph of col. 4 of Table 7.7 is lagged by 10 h instead of 12 h. Then in col. 6, the 10 h UH is obtained by dividing the values of col. 5 by (10/4).

### 7.8   INSTANTANEOUS UNIT HYDROGRAPH (IUH)

Unit hydrographs are named according to their duration of rainfall excess. With decrease in the unit duration, an unit hydrograph will show a marked shifting of its peak and geometry towards the left axis. In a limiting case, when the duration of the rainfall excess becomes infinitesimally small, i.e., 1 cm of rainfall excess is spread over the catchment uniformly and instantaneously, the resulting DRH is called an Instantaneous Unit Hydrograph or IUH. The notation of IUH is U (0, *t*). Note that it is impossible for a basin to get rainfall excess of 1 cm in zero time. IUH is only a concept used to investigate rainfall-runoff process of a basin theoretically and is defined as a fictitious unit hydrograph representing the surface hydrograph from a basin resulting from instantaneous rainfall excess volume of 1 cm over the basin. The advantage of IUH is the elimination of a major parameter "the duration of effective rainfall" from the unit hydrograph. An IUH can provide many important information about the catchment characteristics. The ordinate of DRH at time $t$, derived from an IUH is given as

$$Q_t = \int_0^{t'} U(t - \tau)I(\tau)d\tau \qquad (7.14)$$

where $U(t - \tau)$ is the IUH ordinates at the time $\tau$, $I(\tau)$ the rainfall excess function of duration $t_0$ at the time $\tau$ and duration $d\tau$, $t'$ the time $t$ when $t \le t_0$ and $t' = t_0$ when $t > t_0$. The above integral is called the *convolution integral* or *Duhamel integral* in which the IUH, $U(t - \tau)$ is called the *Kernel function*. Convolution of IUH and $I(\tau)$ is shown in Fig. 7.13. Shape of an IUH resembles a single peaked hydrograph.

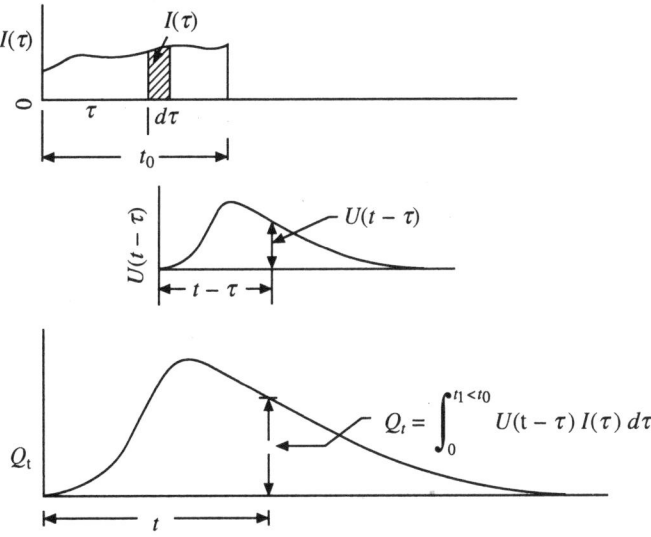

**Fig. 7.13  Convolution of *I* (*t*) and IUH**

The process is the same as we compute flood hydrograph from a rainfall excess histogram. Here an instantaneous rainfall function $I(t)$ of duration $t_0$ is applied to the IUH to get DRH from a catchment whose discharge at any time $t$ is given by the integral. If unit hydrograph is represented as $U(D, t)$, where $D$ is the unit duration or simply the duration of the rainfall excess and $t$ represents the time, then the ordinate of the UH is represented simply as $U$ or $U(D, t)$. For an IUH, since $D$ is zero, the ordinate of IUH at time $t$ is $U(0, t)$ or simply $U(t)$. The properties of IUH are

   (i) $U(t)$ is always positive for $t > 0$
   (ii) $U(t)$ is zero for $t \le 0$
   (iii) When $t$ goes to $\infty$, $U(t)$ tends to be zero

(iv) Integration of $U(t)dt$ between the limits 0 to $\infty$ is the unit depth of the catchment.

(v) Integration of $U(t)$ $tdt$ between the limits 0 to $\infty$ is the time lag of IUH (it is the interval between centroid of ERH and DRH).

(vi) Time to peak of IUH is always less than the time of the centroid of the curve.

Many conceptual models are available in literature to derive instantaneous unit hydrograph. Some important models are discussed in section 7.10. Before we discuss these models, the procedure to derive UH from IUH should be known.

### 7.8.1   Derivation of UH from IUH

Instantaneous unit hydrograph derived by any of the conceptual models can be used to derive unit hydrograph of required durations, the stepwise procedure is

1. Prepare a table in which the col. (1) contains time and col. (2) contains IUH ordinates.
2. In col. (3) enter the IUH ordinates of step-2 by lagging $D$-hour, the required duration of UH.
3. In the next column add the two IUH ordinates (of col. (2) and col. (3)) row-wise.
4. Divide the ordinates of col. (4) by 2 and enter them in col. (5). These ordinates are the required UH of $D$-hour duration.
5. It is to be noted that the $D$-hour UH obtained by the above process should not be more than 3 h. If UH of larger duration is required, then the process of S-hydrograph or the method of superposition described earlier may be adopted.

The background of the above procedure is discussed here. The ordinates of a $dt$ unit hydrograph at any instant $t$, presented in equation (7.13) is reproduced here.

$$\frac{(S_1 - S_2)}{i \, dt} \tag{7.15}$$

where $S_1$ and $S_2$ are the ordinates of the S-hydrograph at lag of $dt$ hours. If the intensity of rainfall for the UH from which S-hydrograph is derived is $i = 1$cm/ h, i.e., the S-hydrograph is derived from a 1 h UH then

$$U(t) = \frac{\Delta S}{dt} \tag{7.16}$$

is approximately the slope of S-hydrograph at any instant. In a limiting case where $dt \to 0$, the time of rainfall excess becomes infinitesimally small. As $\Delta S$ also becomes very small, the slope of S-curve at any instant $t$ is the ordinate of IUH.

$$U(t) = \frac{ds}{dt} \text{ as } dt \to 0 \tag{7.17}$$

or $$ds = U(t) \, dt \tag{7.18}$$

Integrating between limits 1 and 2 of the two curves $S_1 - S_2$

$$S_2 - S_1 = \int_1^2 U(t)\, dt \approx \frac{1}{2} [U(t_2) + U(t_1)] (t_2 - t_1) \qquad (7.19)$$

as between the integration points 1 and 2 we can always take

$$U(t) \approx \frac{1}{2} [U(t_2) + U(t_1)]$$

or we can take $\qquad \dfrac{(S_2 - S_1)}{(t_2 - t_1)} \approx \dfrac{1}{2} [U(t_2) + U(t_1)] \qquad (7.20)$

But $\dfrac{(S_2 - S_1)}{(t_2 - t_1)}$ is the UH ordinate of $(t_2 - t_1)$ hour duration. Thus by taking average of IUH ordinates at $dt = (t_2 - t_1)$ interval gives a $dt$-hour UH. It is desirable to keep $dt$ as 1 hour or less because between $t_2$ and $t_1$, the profile of the hydrograph is usually curvilinear and for large difference of time it may lead to error in its average value.

## 7.9 DERIVATION OF IUH

Instantaneous unit hydrograph can be derived by either of the following approaches.

1. From S-Hydrograph
2. From Conceptual models
3. By fitting Harmonic series to DRH and ERH
4. Theoretically from Laplace transform function

Since approaches (1) and (2) are used widely by field engineers for determination of IUH, we limit our discussion to these two methods only.

### 7.9.1 From S-Hydrograph

As already outlined, the two S-hydrographs (say $S_1$ and $S_2$) when separated at duration $dt$ hours and their ordinates subtracted and divided by ratio $dt/D$, where $D$ is the unit duration of the unit hydrograph from which S-hydrograph is derived, gives a $dt$-hour unit hydrograph for the basin. The principle is extended to derive IUH from S-hydrograph. From equation (7.13), we know that $dt$-hour UH is given as $\dfrac{(S_1 - S_2)}{i\, dt}$. In a limiting case where $dt \to 0$, the time of rainfall excess becomes infinitesimally small and the result is the IUH. The IUH ordinates at any instant $t$ is given as

$$u(t) = \frac{\Delta S}{i \Delta t} = \frac{\Delta S}{\Delta t} = \frac{ds}{dt} \qquad (7.21)$$

where $i$ is 1cm/h. This means that when a S-hydrograph derived from an UH of 1 h rainfall excess (1 cm/h), the slope of the S-hydrograph gives IUH ordinate at any time $t$ from the origin of the S-hydrograph.

### 7.9.2 From Conceptual Models

Conceptual models have made rapid strides in hydrology and are increasingly used for derivation of IUH since Zoch (1934) proposed his work simulating a linear channel in series with a linear reservoir. The catchment action is considered as analogous to the response of linear reservoirs or linear channels. Discussion on all the available models on the derivation of IUH are beyond the scope of this book, however the following models deserves special attention.

### 7.9.2.1 Clark Model

Clark (1945) proposed an IUH due to an instantaneous rainfall excess over a drainage area by considering a linear channel in series with a linear reservoir. Study of unit hydrograph showed that

(i)   the peak of DRH occurs sometime after ERH. This is known as *translation.*
(ii)  depth of highest ERH block is always higher than the depth of peak DRH. The effect is called *attenuation.*

This means that the flood peak is translated to the basin outlet which is the affect of time. *Attenuation* is achieved due to storage affect of the catchment. In Clark's model, the translation is first achieved by channel travel time of the time area histogram and the attenuation by routing the output through a linear reservoir at watershed outlet. The terms used in the modelling approach are described next.

**Linear Channel:** It is a fictitious channel in which the time required to translate a given discharge $Q$ of any magnitude through a given reach is constant. Thus the linear channel helps in delaying the arrival of discharge of any magnitude at the outlet without affecting peak. It does translation of the inflow to the outflow without attenuation. Once the channel reach is given the velocity is taken as constant for any flood magnitude.

**Linear Reservoir:** It is a fictitious reservoir in which the storage is directly proportional to the discharge. Consider a simple sharp crested weir. The equation over this crest is given as

$$Q = \frac{2}{3} C_d \sqrt{2g} \, h^{3/2} \tag{7.22}$$

or $$Q = K_1 h^{3/2} \tag{7.23}$$

Since storage is the product of area and the height or $S = A \times h$, $h = S/A$,

$$Q = K_1 \left(\frac{S}{A}\right)^{3/2} = \frac{K_1 S^{3/2}}{A^{3/2}} \tag{7.24a}$$

$$S = \left(\frac{A^{3/2}}{K_1} Q\right)^{2/3} \tag{7.24b}$$

or $$S = KQ^a \tag{7.25}$$

For a linear reservoir, let us assume $S = KQ$, where $K = A^{3/2}/K_1$, $K$ is the characteristics of the reservoir called *storage coefficient.* The storage and discharge relation is always nonlinear as shown in equation (7.25). For a linear reservoir the assumption of $S = KQ$ is conceptual, $K$ has the dimension of time. The linear reservoir incorporates the attenuation of the flood wave.

**Routing:** It is the technique of computing the output of a system at a downstream location by knowing the parameters at an upstream location.

**Time-Area Histogram:** First we must understand the concept of channel travel time before preparing time-area histogram of a basin. The *channel travel time* is the time taken by water to move along the channel from one point to another along it and is less than the time of concentration due to the combined effect of wave movement and translation of water in channel. The travel time is physically taken as the period between the end of the rainfall excess and the point of inflection of the runoff hydrograph at the end of peak segment. In absence of any data, travel time can be approximated to the time of concentration given by Kirpich (1940). The total travel time of the basin is divided into say $n$ equal parts. The total catchment area is divided into the same $n$ number of subareas by lines of equal travel time called isochrone. Thus an *isochrone* is defined as the line joining places in the watershed that are away from the common outlet by the same time of travel of water. For making time area histogram for a given watershed, and subsequently to obtain IUH for the basin using Clark model the following steps may be followed.

1. Find the largest path of the river and plot its profile as longitudinal section. Let $L$ be the total length.
2. Divide $L$ into number of equal segments $n$($n$ being arbitrary) as the number of blocks of time-area histogram to be prepared. If a watershed is to be divided into $n$ time-area histograms, then

$$l = \frac{l_1}{(S_1)^{0.5}} = \frac{l_2}{(S_2)^{0.5}} = \dots = \frac{l_n}{(S_n)^{0.5}} \qquad (7.26)$$

   where $l_1, l_2, \dots, l_n$ are the length segments with their sum equal to $l$ and $S_1, S_2, \dots, S_n$ are the slope of the segments. The basic assumption made here is that the time of travel of any elemental area is proportional to $\frac{l_i}{(S_i)^{0.5}}$.

3. Find time of travel $T_t$. This is the distance between end of ERH and point of contraflexure of recession limb of DRH.
4. Repeat the exercise of step (2) along the paths for the tributaries.
5. Draw the isochrones passing through these points at distance $l$ apart (Fig. 7.14a)
6. Measure the areas enclosed by successive isochrones. Let them be denoted as $A_1, A_2, \dots, A_n$.
7. Prepare a bar graph by plotting the time $\Delta t$ ($\Delta t = T_t/n$, where $n$ is the number of such subareas formed by isochrones) in abscissa and area $A_1$, $A_2 \dots A_n$ as ordinate. The resulting graph is the time-area histogram (Fig. 7.14b).

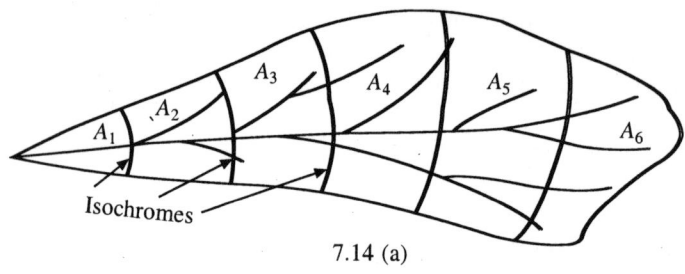

7.14 (a)

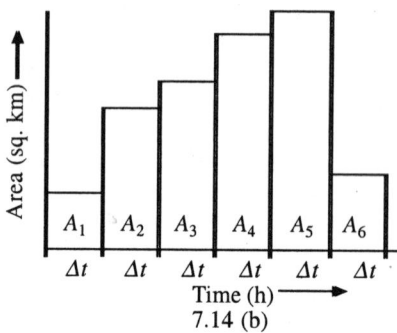

Time (h) ⟶
7.14 (b)

**Fig 7.14   Time area histogram**

8. Each area represents a runoff volume of $(A_r \times R_e)$ km$^2$-cm moving out in time D$t$ hours, where $A_r$ is the area of the interisochronal area in km$^2$, $R_e$ the rainfall excess in cm. The time $\Delta t (= \Delta T_t)$ is travel time between isochrones in hours. The outflow hydrograph is obtained as the sequence of $A_r \times R_e$ of the subcatchment-rainfall excess moves. This does not provide scope for storage properties. To overcome this, the $A_r \times R_e$ volumes are routed at the channel outlet by a linear reservoir.

9. The routing is done as follows.

The general continuity equation of flow is given as

$$\text{Inflow} - \text{Outflow} = \text{Change in storage}$$

or $\qquad \dfrac{I_1 + I_2}{2} \Delta T_t - \dfrac{Q_1 + Q_2}{2} \Delta T_t = S_2 - S_1 \qquad (7.27)$

Taking $S_2 = KQ_2$ and $S_1 = KQ_1$ equation (7.27) can be written as

$$0.5 \times (I_1 + I_2) \Delta T_t - 0.5 (Q_1 + Q_2) \Delta T_t = K(Q_2 - Q_1) \qquad (7.28)$$

where $\Delta T_t$ is the time of travel of water from the end of one isochrone to the other, $Q_1$ and $Q_2$ are the outflows at the beginning and end of $\Delta T_t$, $I_1$ and $I_2$ are the time area histograms at the beginning and end of the time interval and $K$ is the storage coefficient. From equation (7.28) we can derive $Q_2$ (by eliminating $S_1$ and $S_2$ : $S_1 = KQ_1$ and $S_2 = KQ_2$) as

$$Q_2 = C_0 I_2 + C_1 I_1 + C_2 Q_1 \qquad (7.29)$$

where

$$C_0 = \frac{0.5\Delta T_t}{K + 0.5\Delta T_t}, \; C_1 = \frac{0.5\Delta T_t}{K + 0.5\Delta T_t}, \text{ and } C_2 = \frac{K - 0.5\Delta T_t}{K + 0.5\Delta T_t} \qquad (7.30)$$

The routed outflow from equation (7.29) of the time area histogram gives the required IUH for the basin, $I_1$ and $I_2$ are the inflow hydrograph ordinates. For routing time-area-histogram $C_0$ is taken as $C_1$, as the inflow at $\Delta T_t$ interval is taken as the block of time area histogram of same travel time $\Delta T_t$ in hours, the values of $I_1$ and $I_2$ can be taken as equal. The inflow rate due to the inter-isochrone with area $A_r$ in sq. km and 1 cm rainfall excess during time interval $\Delta T_t$ is given as

$$I = \frac{A_r \times 10^4}{3600 \, \Delta T_t} = \frac{2.78 \, A_r}{\Delta T_t} \qquad (7.31)$$

Therefore

$$Q_2 = 2C_1 \, I_1 + C_2 \, Q_1 \qquad (7.32)$$

and

$$C_0 + C_1 + C_2 = 1.0$$

*Determination of Coefficients K and $\Delta T_t$*
The recession limb of the DRH is expressed as

$$Q_t = Q_0 \, e^{-t/K} \qquad (7.33)$$

where $Q_t$ is the outflow at any other time $t$, $Q_0$ is the outflow at the beginning and $K$ is the storage constant. Taking logarithm of equation (7.33)

$$\ln Q_t = \frac{-t}{K} \ln Q_0$$

or

$$K = -t \frac{\ln Q_0}{\ln Q_t} = t \frac{\ln Q_t}{\ln Q_0}$$

or

$$K = t \ln \left( \frac{Q_0}{Q_t} \right) \qquad (7.34)$$

Taking two points in the recession curve $(t_1, Q_1)$ and $(t_2, Q_2)$, two equations of (7.34) are formed. Subtracting one from the other, the only unknown $Q_0$ is computed. Resubstituting the values we get $K$ in hours. It can also be determined from equation (7.33) by taking two points in recession curve and substituting the values of $Q_t$ and $t$ in it and solving for $K$.

$T_t$ is determined from the plot of the observed DRH and ERH. It is the distance along $X$ axis between the end of ERH and point of contraflexure (inflection) of the recession limb of DRH in hours as shown in Fig. 7.3 as *EC*, where point $C$ represents end of running of the surface water flow into the channel.

Travel time can be approximately computed from equation (7.3).

**Assumptions in Clark Model:** The following assumptions are made while using Clark's model.

1. Input is instantaneous.
2. Input is unit.

3. Rainfall excess first undergoes translation and then attenuation.
4. Translation is achieved by travel time area histogram.
5. Attenuation is achieved by routing the travel time area histogram through a linear reservoir at the catchment outlet.
6. Model is lumped at the outlet.
7. Model is time invariant.

---

**Example 7.8:** The observed hydrograph for a catchment area of 900 sq. km is:

| Time (h) | 0 | 3 | 6 | 9 | 12 | 15 | 18 | 21 | 24 |
|---|---|---|---|---|---|---|---|---|---|
| $Q$(m³/sec) | | 100 | 200 | 800 | 1200 | 1800 | 1600 | 1000 | 400 | 100 |

The rainfall hyetograph is:

| Time (h) | 0 | 3 | 6 | 9 |
|---|---|---|---|---|
| Rainfall Intensity (mm/h) | | 16 | 12 | 22 |

(a) Apply Clark's model to evaluate (i) IUH and (ii) UH of 3 h duration.
(b) What will be the affect on the peak if the catchment is divided into more subareas by more isochrones?

**Solution**
(a) From the observed hydrograph base flow at the rate of 100 m³/sec is subtracted to get the DRH. Rainfall hyetograph and DRH are plotted in the same graph paper as shown in Fig. 7.15.

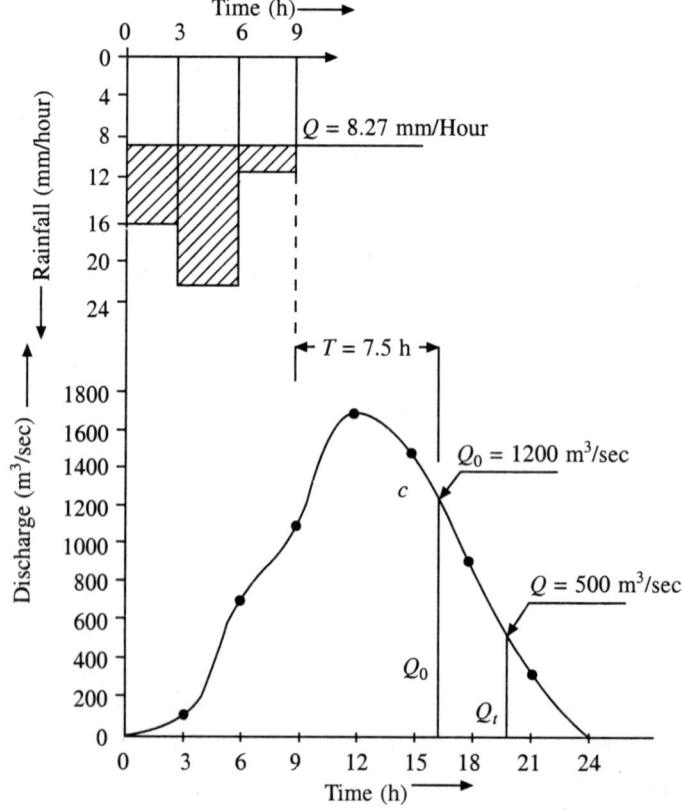

**Fig. 7.15   DRH and ERH for Example 7.8**

Time of translation = Time to the point of contraflexure

− Time of the end of rainfall excess function

or $T_t = 16.5 − 9.0 = 7.5$ h (Fig. 7.15).

The point C is selected by looking at the DRH.

DRH ordinate at $T_{16.5} = 1200$ m$^3$/sec

For the recession limb, the DRH ordinates are given below.

| Time (h). | 16.5 | 18.0 | 19.5 | 21.0 | 22.5 | 24.0 |
|---|---|---|---|---|---|---|
| DRH ordinates (m$^3$/sec) | 1200.0 | 900.0 | 600.0 | 300.0 | 150.0 | 0.0 |

Time versus recession limb of DRH are plotted in a semi-log paper as shown in Fig. 7.16.

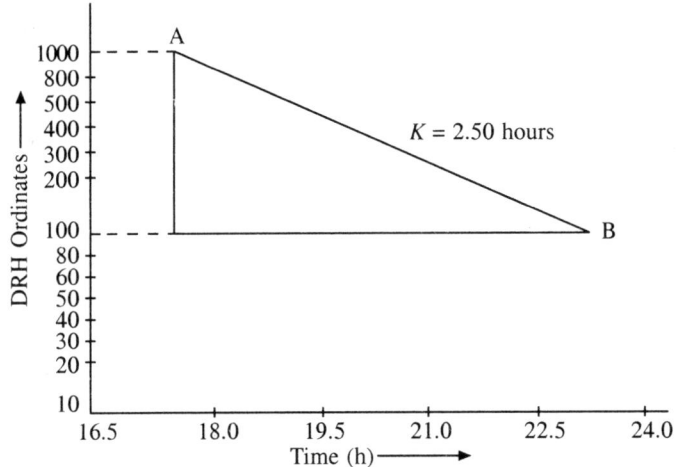

**Fig. 7.16 Plot of the Recession Limb of DRH on Semilog Paper to Determine $K$ of Example 7.8**

Two points A and B are marked in the graph and the following $Q_t$ and $t$ values are read:

$$Q_{t1} = 1000 \qquad : \qquad t_1 = 17.7$$

$$Q_{t2} = 100 \qquad : \qquad t_2 = 23.5$$

The equation between $Q_t$ and $Q_0$ is given by $Q_t = Q_0\, e^{-t/K}$.

Therefore $1000 = 1200\, e^{-17.7/K}$ ... $\qquad$ (1)

and $100 = 1200\, e^{-23.5/K}$ ... $\qquad$ (2)

Dividing (1) by (2) we get $10 = e^{(-17.7+23.5)/K}$ or taking logarithmic with base 10

$$\log_{10} 10 = \{(23.5 − 17.7)\, \log_{10} e\}/K$$

or $K = (23.5 − 17.7)\,(\log_{10} e/\log_{10} 10) = 5.8 \times 0.4343 = 2.50$ h

The coefficients $C_0$, $C_1$ and $C_2$ are calculated below. Take $\Delta T_t = 1.5$ h

$$C_0 = C_1 = \frac{0.5\Delta T_t}{(K + 0.5\Delta T_t)} = \frac{(0.5 \times 1.5)}{(2.50 + 0.50 \times 1.5)} = 0.2305$$

and $$C_2 = \frac{K - 0.5\Delta T_t}{(K + 0.5\Delta T_t)} = \frac{(2.50 + 0.50 \times 1.5)}{(2.50 + 0.50 \times 1.5)} = 0.539$$

Check $$C_0 + C_1 + C_2 = 0.2305 + 0.2305 + 0.5390 \doteq 1.00$$

Let the routing interval = 1.50 h. The catchment area is shown in Fig. 7.17. The area is divided into time area blocks by drawing isochrones. This is carried out by taking the

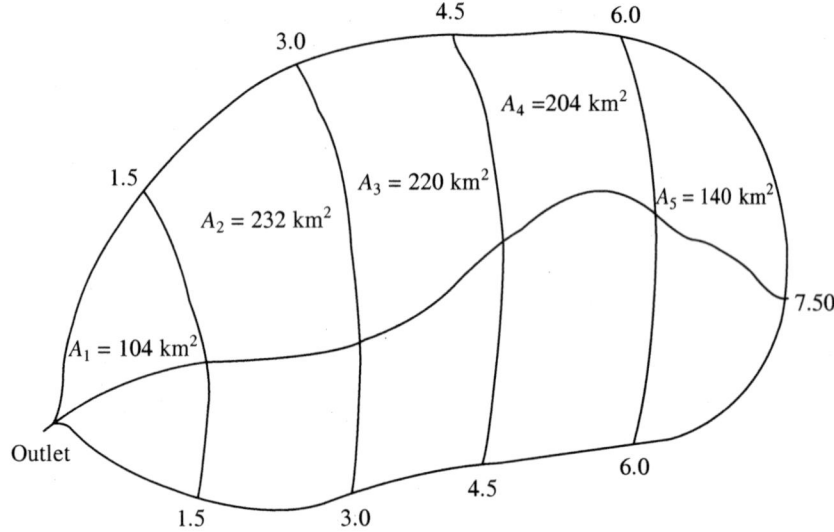

**Fig. 7.17    Drawing of Isochrones for the Catchment of Example 7.8 (Time-Area Histogram Blocks as Shown in the Figure).**

travel time or translation time as $T = 7.5$ h. The stream length is measured by thread and divided into 5 equal parts. At each such point isochronal lines are drawn by hand. Area between the isochrones are measured by a transparent graph sheet by counting squares. The time area histograms are plotted in Fig. 7.18 by taking $\Delta T_t$ as 1.5 h in abscissa and the inter isochronal areas as ordinate. Each subarea $A_r$ represents a volume equal to $A_r$ km²-cm = $A_r \times 10^4$ m³.

This volume moves in time $\Delta T_t = 1.5$ h.

Therefore $$I = \frac{A_r \times 10^4}{(3600 \times \Delta T_t)} \text{ where } A_r = A_1, A_2, A_3 \ldots \text{ are in km}^2.$$

or $$I = t \frac{10^4}{(3600 \times 1.5)} A_r \text{ m}^3/\text{sec} = 1.852 A_r \text{ m}^3/\text{sec}.$$

Outflow is calculated from the relation

$$Q = C_0 I_1 + C_1 I_2 + C_2 Q_1 = 2 \times 0.2305 I + 0.539 Q_1$$

and the results are given in Table 7.9.

At the beginning of routing $I_1 = 0$, $I_2 = 0$ and $Q_1 = 0$. Therefore, IUH ordinate at the end of first time step = 0 m³/sec. For the second time step: $C_0 I_1 = 45$, $C_1 I_2 = 45$ and $C_2 Q_1 = 0$, IUH ordinate = 90 m³/sec. Likewise the routing is carried out. Since it is required to calculate a 3h UH from the IUH, the ordinates of IUH are lagged by 3 h in col. (8). The IUH ordinates of cols. (7) and (8) are summed up in col. (9). The elements of col. (9) when divided by 2 give the ordinates of the required 3 h UH.

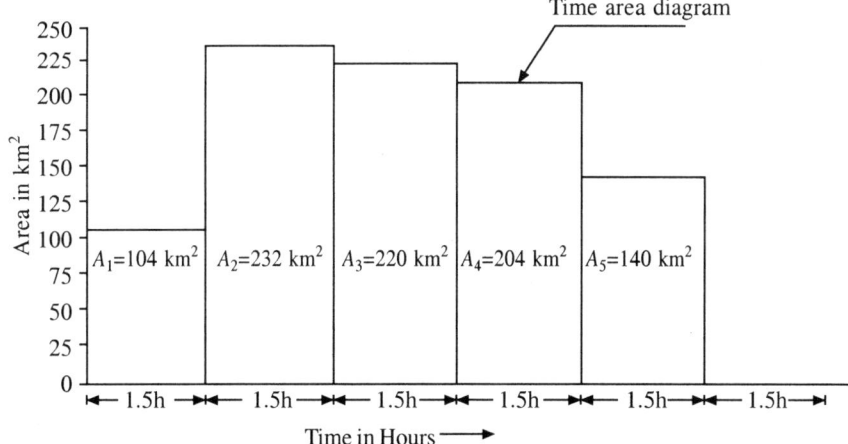

Fig. 7.18   Time-Area Histogram Blocks of Example 7.8

Table 7.9   Clark's Model of Computing IUH for Example 7.8

| Time (h) | Area (km²) | $I$ (m³/sec) 1.852$A_r$ | 0.2305 $\times I_1$ (m³/sec) | 0.2305 $\times I_2$ (m³/sec) | 0.539$Q$ (m³/sec) | Total IUH (m³/sec) | Lag of col.(7) by 3h | Col (7)+ col(8) | Col. (9)/2= 3h UH |
|---|---|---|---|---|---|---|---|---|---|
| (1) | (2) | (3) | (4) | (5) | (6) | (7) | (8) | (9) | (10) |
| 0.0 | 0 | 0 | 0 | 0 | 0 | 0 | 0 | 0 | 0 |
| 1.5 | 104 | 193 | 45 | 45 | 0 | 90 | 0 | 90 | 45 |
| 3.0 | 232 | 430 | 99 | 99 | 49 | 247 | 0 | 247 | 124 |
| 4.5 | 220 | 408 | 94 | 94 | 133 | 321 | 90 | 411 | 206 |
| 6.0 | 204 | 378 | 87 | 87 | 173 | 347 | 247 | 594 | 297 |
| 7.5 | 140 | 259 | 60 | 60 | 187 | 307 | 321 | 628 | 314 |
| 9.0 | 0 | 0 | 0 | 0 | 165 | 165 | 347 | 512 | 256 |
| 10.5 | 0 | 0 | 0 | 0 | 89 | 89 | 307 | 396 | 198 |
| 12.0 | 0 | 0 | 0 | 0 | 48 | 48 | 165 | 213 | 107 |
| 13.5 | 0 | 0 | 0 | 0 | 26 | 26 | 89 | 115 | 58 |
| 15.0 | 0 | 0 | 0 | 0 | 14 | 14 | 48 | 62 | 31 |
| 16.5 | 0 | 0 | 0 | 0 | 8 | 8 | 26 | 34 | 17 |
| 18.0 | 0 | 0 | 0 | 0 | 4 | 4 | 14 | 18 | 9 |
| 19.5 | 0 | 0 | 0 | 0 | 2 | 2 | 8 | ·10 | 5 |
| 21.0 | 0 | 0 | 0 | 0 | 1 | 1 | 4 | 5 | 3 |

(b) The behaviour of the system by drawing more number of isochrones is tested here. The inter-isochrone travel time is taken as 1 h and translation time is approximated to 7 h. Computations are carried out as follows.

Let the translation time $T_t$ = 7.00 h

Let the inter isochronal time $\Delta t$ = 1 h

The channel length is divided into 7 parts and area bounded between the isochrones are measured after redrawing the isochrones

$$A_1 = 55 \text{ km}^2, A_2 = 119 \text{ km}^2\ A_3 = 161 \text{ km}^2\ A_4 = 170 \text{ km}^2,$$

$$A_5 = 160 \text{ km}^2, \; A_6 = 137 \text{ km}^2 \; A_7 = 98 \text{ km}^2$$

$$C_0 = C_1 = \frac{(0.5 \times \Delta T_t)}{(K + 0.5\Delta T_t)} = \frac{(0.5 \times 1.0)}{(2.5 + 0.5)} = 0.167$$

$$C_2 = \frac{(2.50 - 0.5 \times 1.0)}{(2.50 + 0.50 \times 1)} = \frac{2.0}{3.0} = 0.667$$

$$I = \left\{ \frac{10^4}{(3600 \times 1.0)} \right\} A_r = 2.778 A_r \text{ m}^3/\text{sec}.$$

**Table 7.10    Computation of IUH by Clark's Model ($T_t = 7.0$ h and $\Delta T_t = 1$ h)**

| Time (h) | Area (km²) | $I$ (m³/sec) =2.778$A_r$ | 0.167$I_1$ (m³/sec) | 0.167$I_2$ (m³/sec) | 0.667$Q$ (m³/sec) | Total IUH (m³/sec) | col. (7) lag by 3 h | Sum of cols. (7 + 8) | Sum/2 = 3 h UH |
|---|---|---|---|---|---|---|---|---|---|
| (1) | (2) | (3) | (4) | (5) | (6) | (7) | (8) | (9) | (10) |
| 0 | 0 | 0 | 0 | 0 | 0 | 0 | – | 0 | 0 |
| 1 | 55 | 153 | 26 | 26 | 0 | 52 | – | 52 | 26 |
| 2 | 119 | 331 | 55 | 55 | 35 | 145 | – | 145 | 73 |
| 3 | 161 | 448 | 75 | 75 | 97 | 247 | 0 | 247 | 124 |
| 4 | 170 | 472 | 79 | 79 | 165 | 322 | 52 | 374 | 187 |
| 5 | 160 | 445 | 74 | 74 | 215 | 363 | 145 | 508 | 254 |
| 6 | 137 | 381 | 64 | 64 | 242 | 370 | 247 | 617 | 309 |
| 7 | 98 | 272 | 45 | 45 | 247 | 337 | 322 | 659 | 330 |
| 8 | 0 | 0 | 0 | 0 | 225 | 225 | 363 | 588 | 294 |
| 9 | 0 | 0 | 0 | 0 | 150 | 150 | 370 | 520 | 260 |
| 10 | 0 | 0 | 0 | 0 | 100 | 100 | 337 | 437 | 219 |
| 11 | 0 | 0 | 0 | 0 | 67 | 67 | 225 | 292 | 146 |
| 12 | 0 | 0 | 0 | 0 | 45 | 45 | 150 | 195 | 98 |
| 13 | 0 | 0 | 0 | 0 | 30 | 30 | 100 | 130 | 65 |
| 14 | 0 | 0 | 0 | 0 | 20 | 20 | 67 | 87 | 43 |
| 15 | 0 | 0 | 0 | 0 | 13 | 13 | 45 | 58 | 29 |
| 16 | 0 | 0 | 0 | 0 | 9 | 9 | 30 | 39 | 19 |
| 17 | 0 | 0 | 0 | 0 | 6 | 6 | 20 | 26 | 13 |
| 18 | 0 | 0 | 0 | 0 | 4 | 4 | 13 | 17 | 9 |

It can be seen that by increasing the blocks of the time area histogram from 5 to 7, the peak of the 3 h UH changed from 314 to 330 m³/sec. The more we increase the time area histogram blocks, the higer value of peak the system yields. Plot of the IUH by Clark's method for $K = 2.5$ h is shown in Fig. 7.19.

### 7.9.2.2    *Nash Model*

Nash (1957) proposed a cascade of linear reservoirs of equal sizes to represent the IUH for a catchment. All the reservoirs have the same storage constant $K$. Number of identical reservoirs required to model a catchment is computed from an observed event of DRH and the corresponding ERH. The model is lumped and time invariant. An input of 1 cm of excess rainfall over the catchment is applied to the first reservoir instantaneously. The routed outflow from the first

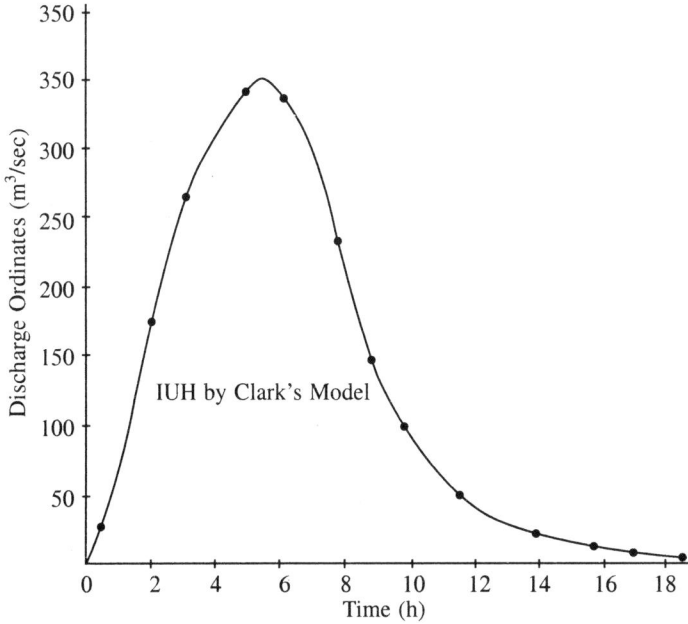

**Fig. 7.19 IUH by Clark's Model for *K* = 2.5 (Example 7.8)**

reservoir becomes the input to the second reservoir in series and the second reservoir output becomes the input to the third, and so on. Output from the last (*n*th) reservoir is the output from the system representing an IUH for the basin. The cascade of *n* number of identical linear reservoirs along with the shapes of the routed outflow hydrographs are shown in Fig. 7.20.

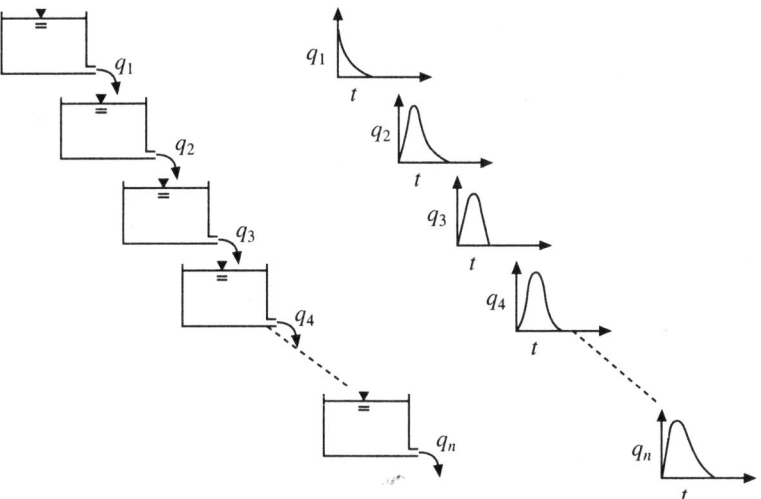

**Fig. 7.20 A Series of Linear Reservoirs Used in Nash Model**

### Working of the Model

*First Linear Reservoir*

Let the duration of input is $t_0$. Continuity equation for a reservoir is given as

$$\text{Inflow} - \text{Outlfow} = \text{Change in storage}$$

or
$$I - Q = \frac{ds}{dt} \qquad (7.35)$$

The characteristic equation of a linear reservoir (equation 7.25) is taken as

$$S = KQ$$

As inflow is instantaneous, i.e., by the time the outflow begins the inflow $I = 0$. For an instantaneous inflow for $t = 0$ and $I = 0$, equation (7.35) at any instant $dt$ can be written as

$$- Q = \frac{d\,(S)}{dt} = \frac{d\,(K \cdot Q)}{dt} = K\frac{dQ}{dt}$$

or
$$\frac{dt}{K} = \frac{-\,dQ}{Q} \quad \text{or} \quad \frac{-t}{K} = \ln Q + C \qquad (7.36)$$

for $t = 0 \rightarrow Q = Q_0$, i.e., the inflow into the reservoir is complete and the outflow takes place at the maximum rate of $Q_0$.

The constant $C = -\ln Q_0$. Therefore equation (7.36) becomes

$$\frac{-t}{K} = \ln Q - \ln Q_0 = \ln\left(\frac{Q}{Q_0}\right)$$

Taking antilogarithm we get $Q_0 e^{-t/K} = Q$

or
$$Q = Q_0\, e^{-t/K} = \frac{1}{K}\, S_0\, e^{-t/K} = (1/K)e^{-t/K}$$

(as $S_0 = Q_0 K$, $Q_0 = \dfrac{S_0}{K} = \dfrac{1}{K}$ as when the input is unit the storage $S_0 = 1$)

$\therefore$
$$Q_1 = \left(\frac{1}{K}\right) e^{-t/K} \qquad (7.37)$$

which is the output from the first linear reservoir with instantaneous unit input. This represents the IUH for the catchment considering only one linear reservoir.

*Second Linear Reservoir*

Outflow from the first linear reservoir is the inflow into the second linear reservoir. Continuity equation for the second linear reservoir is written as

$$I_2 - Q_2 = \frac{dS_2}{dt} \qquad (7.38)$$

where
$$I_2 = Q_1 = \left(\frac{1}{K}\right) e^{-t/K} \qquad (7.39)$$

for this linear reservoir, the characteristic equation is

$$S_2 = K Q_2 \qquad (7.40)$$

Combining equations (7.38), (7.39) and (7.40) we get

$$\frac{dQ_2}{dt} + \frac{Q_2}{K} = \frac{I_2}{K} = \frac{I}{K^2}e^{-t/K} \quad \text{or} \quad dQ_2 = \frac{1}{K^2} \cdot e^{-t/K} \cdot dt \qquad (7.41)$$

Solution to this equation is obtained by integration. By applying the standard form of integration to equation (7.41)

$$Q_2 e^{t/K} = \int \left(\frac{1}{K^2}\right) e^{-t/K} \cdot e^{-t/K} \, dt + C = \int \left(\frac{1}{K^2}\right) dt + C = t/K^2 + C$$

for $t = 0$, the outflow from the second reservoir $Q_2 = 0$, Therefore $C = 0$.

$$\therefore \qquad Q_2 = \left(\frac{1}{e^{t/K}}\right)\left(\frac{t}{K^2}\right) = \left(\frac{t}{K^2}\right)\left(\frac{1}{e^{-t/K}}\right) = \left(\frac{t}{K^2}\right) e^{-t/K}$$

or $\qquad Q_2 = \left(\frac{t}{K^2}\right) e^{-t/K} \qquad (7.42)$

*Third Linear Reservoir*
Input to this reservoir is the output from the second linear reservoir. The routed outflow through this reservoir is computed as

$$I_3 = Q_2 = \left(\frac{t}{K^2}\right) e^{-t/K}$$

Now $\qquad \dfrac{dQ_3}{dt} + \dfrac{Q_3}{K} = \dfrac{I_3}{K} = \dfrac{1}{K^2} \cdot \dfrac{1}{K^2} e^{-t/K} = \dfrac{t}{K^3} e^{-t/K}$

Integrating we get

$$Q_3 \cdot e^{t/K} = \int \left(\frac{t}{K^3}\right) e^{-t/K} \cdot e^{t/K} \cdot dt = \int \left(\frac{t}{K^3}\right) dt = \frac{t^2}{2K^3} \qquad (7.43)$$

or $\qquad Q_3 = \left(\dfrac{t^2}{2K^3}\right) e^{-t/K} = \left(\dfrac{1}{2.1}\right)\left(\dfrac{1}{K}\right)\left(\dfrac{t}{K}\right)^2 e^{-t/K} \qquad (7.44)$

Similarly for the fourth linear reservoir it can be shown that

$$Q_4 = \left(\frac{1}{3.2.1}\right)\left(\frac{1}{K}\right)\left(\frac{t}{K}\right)^3 e^{-t/K} \qquad (7.45)$$

For *n*th liner reservoir we can write the equation as

$$Q_n = \frac{1}{n-1!}\frac{1}{K}\left(\frac{t}{K}\right)^{n-1} e^{-t/K} \qquad (7.46)$$

which is the IUH represented by $U(0, t)$ or $U(t)$. The equation

$$Q(t) = \frac{1}{(n-1)!} \frac{1}{K} \left(\frac{t}{K}\right)^{n-1} e^{-t/K} \tag{7.47}$$

represents the IUH ordinate at any time $t$ from the Nash model. This is because 1 cm of rainfall excess is taken as an instantaneous input to the system. The ordinates of IUH given by equation (7.47) can be evaluated when the values of $n$ and $K$ are known.

*Determination of Parameters* n *and* K *for Nash Model*
If a catchment has a total lag time $= T$, then each reservoir has a lag time $(K)$ of $T/n$. We can study the effect of outflow hydrograph with changing $n$. The number of linear reservoirs required to model a catchment can be estimated by method of moments. It should preferably be an integer. This can be evaluated by taking the first moment of IUH about origin, i.e., when $t = 0$

$$M_1 = n\,K \tag{7.48}$$

The first moment $M_1$ represents the lag time of the centroid of IUH. This is same as the difference of the moments of DRH and ERH from the centroid.

$\therefore$
$$M_{DRH1} - M_{ERH1} = nK \tag{7.49}$$

The second moment of IUH about the origin, i.e., $t = 0$ is given as

$$M_2 = n\,(n+1)K^2 \tag{7.50}$$

which is equated to the second moment of DRH and ERH about the origin, i.e. $t = 0$. Further it was shown that

$$M_{DRH2} - M_{ERH2} = n(n+1)K^2 + 2nK\,M_{ERH1} \tag{7.51}$$

where $M_{DRH2}$ and $M_{ERH2}$ are the second moment of the DRH and ERH respectively about the origin.
    Solving equations (7.49) and (7.51), the values of $n$ and $K$ can be evaluated.

---

**Example 7.9:** The drainage area of a river is 2200 sq. km. The effective time lag was found to be 9.6 h. Compute IUH for the drainage basin if the following values of number of linear reservoirs are selected:

(i) $n = 3$,   (ii) $n = 4$,   (iii) $n = 5$

Use Nash's equation and plot all the 3 IUH's on a single plot for comparison.

**Solution**
    Here catchment area = 2200 sq. km
    Effective time lag = 9.6 h.
    Number of linear reseviors can be = 3, 4 or 5

**Case 1:** Take three number of linear reservoirs to represent the catchment in the first case.

Time lag $= n \times K$

$K$ = Time lag / Number of linear reservoirs = 9.6/3 = 3.2 h

Unit volume = Catchment area $\times$ 1 cm = $2200 \times 10^6/100$ m$^3$ = $22 \times 10^6$ m$^3$

This value is to be multiplied with $U(0, t)$ ordinates to get the IUH for the drainage basin. IUH ordinates for the basin are given in the col. (6) of table 7.11, for $n = 3$, $K = 3.2$ hours.

$$\frac{1}{(n-1)!}\left(\frac{1}{K}\right) = \frac{1}{(2! \times 3.2 \times 3600)} = 4.34027 \times 10^{-5}/\text{sec}$$

Computation of IUH ordinates are carried out in Table 7.11.

**Table 7.11  Computation of IUH using Nash Model for Example 7.9**
**($n = 3$, $K = 3.2$ h)**

| Time (h) | $t/K$ col. (1)/3.2 | $(t/K)^{n-1}$ | $e^{-t/K}$ | $U(0, t)$ = cols. (3) × (4) × 4.3403 × 10⁻⁵ (×10⁻⁶) | $U(0, t) \times 22$ × 10⁶ = col. (5) × 22 × 10⁶ |
|---|---|---|---|---|---|
| (1) | (2) | (3) | (4) | (5) | (6) |
| 0 | 0.000 | 0.000 | 1.000 | 0.000 | 0 |
| 0.5 | 0.15625 | 0.02441 | 0.85535 | 0.9062 | 20 |
| 1.0 | 0.3125 | 0.09766 | 0.73162 | 3.1011 | 68 |
| 1.5 | 0.46875 | 0.21973 | 0.62578 | 5.9779 | 131 |
| 2.0 | 0.625 | 0.39063 | 0.53526 | 9.0749 | 200 |
| 2.5 | 0.78125 | 0.61035 | 0.45783 | 12.1284 | 267 |
| 3.0 | 0.9735 | 0.87891 | 0.39161 | 14.9385 | 329 |
| 3.5 | 1.09375 | 1.19629 | 0.33496 | 17.3917 | 383 |
| 4.0 | 1.250 | 1.5625 | 0.28651 | 19.4298 | 428 |
| 4.5 | 1.40625 | 1.97754 | 0.24506 | 21.0337 | 463 |
| 5.0 | 1.5625 | 2.44161 | 0.20961 | 22.2112 | 489 |
| 6.0 | 1.8750 | 3.51563 | 0.15336 | 23.400 | 515 |
| 7.0 | 2.1875 | 4.78516 | 0.112197 | 23.302 | 513 |
| 8.0 | 2.500 | 6.2500 | 0.08208 | 22.2669 | 490 |
| 10.0 | 3.125 | 9.7656 | 0.04394 | 18.6229 | 410 |
| 12.0 | 3.750 | 14.0625 | 0.02352 | 14.354 | 316 |
| 15.0 | 4.6875 | 21.9726 | 0.00921 | 8.7830 | 93 |
| 18.0 | 5.6250 | 31.6406 | 0.003606 | 4.9526 | 109 |
| 21.0 | 6.5625 | 43.0664 | 0.001412 | 2.6400 | 58 |
| 25.0 | 7.825 | 61.0350 | 0.000405 | 1.0719 | 24 |
| 30.0 | 9.375 | 87.891 | 0.0000848 | 0.3236 | 7 |
| 35.0 | 10.9375 | 119.629 | 0.00001778 | 0.0092 | 2 |
| 40.0 | 12.50 | 156.25 | $3.72605 \times 16^{-6}$ | 0.0000 | 0.6 |

**Case II:**
$$n = 4, K = \frac{9.6}{4} = 2.40 \text{ h}$$

$$\frac{1}{(n-1)!} \times \frac{1}{K} = \frac{1}{(6 \times 2.40 \times 60 \times 60)} = 1.92901 \times 10^{-5}/\text{sec}$$

Details of the calculations of IUH and UH for $n = 4$ and $K = 2.4$ h are given in Table 7.12.

**Case III:**
$$n = 5, K = \frac{9.6}{5} = 1.92 \text{ hours}$$

$$\frac{1}{(n-1)!} \times \frac{1}{K} = \frac{1}{(4! \times 1.92 \times 60 \times 60)} = 6.0282 \times 10^{-6}/\text{sec}.$$

IUH and UH for $n = 5$ and $K = 1.92$ h are calculated in Table 7.13.

**Table 7.12   Computation of IUH using Nash Model for Example 7.9**
**($n = 4$ and $K = 2.4$ h)**

| Time (h) | $t/K$ col. (1)/2.4 | $(t/K)^{n-l}$ | $e^{-t/K}$ | $U(0, t) = $ cols. (3) $\times 4 \times 1.929 \times 10^{-5}$ $(\times 10^{-6})$ | $U(0, t) \times 22$ $\times 10^6 = $ col. (5) $\times 22 \times 10^6$ |
|---|---|---|---|---|---|
| (1) | (2) | (3) | (4) | (5) | (6) |
| 0 | 0.000 | 0.000 | 1.000 | 0.000 | 0 |
| 0.5 | 0.2083 | 0.0091 | 0.8119 | 0.1416 | 4 |
| 1.0 | 0.4167 | 0.0723 | 0.6592 | 0.9198 | 20 |
| 1.5 | 0.6250 | 0.2441 | 0.5353 | 2.5208 | 56 |
| 2.0 | 0.8333 | 0.5787 | 0.4347 | 4.8531 | 107 |
| 2.5 | 1.0462 | 1.1303 | 0.3528 | 7.6936 | 169 |
| 3.0 | 1.2500 | 1.9531 | 0.2865 | 10.7943 | 238 |
| 3.5 | 1.4583 | 3.1015 | 0.2326 | 13.9175 | 306 |
| 4.0 | 1.6667 | 4.6296 | 0.1888 | 16.8676 | 371 |
| 4.5 | 1.8750 | 6.5918 | 0.1534 | 19.5000 | 429 |
| 5.0 | 2.0833 | 9.0422 | 0.1245 | 21.7186 | 478 |
| 6.0 | 2.500 | 15.625 | 0.0821 | 24.7410 | 544 |
| 7.0 | 2.9167 | 24.812 | 0.0541 | 25.9001 | 570 |
| 8.0 | 3.3333 | 37.037 | 0.0357 | 25.4880 | 561 |
| 10.0 | 4.1667 | 72.338 | 0.0155 | 21.6336 | 476 |
| 12.0 | 5.000 | 125.000 | 0.0067 | 16.2470 | 357 |
| 15.0 | 6.250 | 244.140 | 0.0019 | 9.0915 | 200 |
| 18.0 | 7.500 | 421.87 | 0.0006 | 4.5010 | 99 |
| 21.0 | 9.167 | 770.257 | 0.0000105 | 1.5516 | 34 |
| 25.0 | 10.833 | 1271.41 | $1.9737 \times 10^{-5}$ | 0.484 | 34 |
| 30.0 | 12.508 | 1953.12 | $3.7267 \times 10^{-6}$ | 0.1404 | 3 |
| 35.0 | 14.580 | 3402.00 | $4.6551 \times 10^{-7}$ | 0.0028 | 1 |

The three IUH's obtained by taking $n$ as 3, 4 and 5 are plotted in Fig. 7.21 for comparison. It can be seen that when $n$ increases, the peak of IUH also increase. Time to peak increases as $n$ increases.

**Example 7.10:** The observed hydrograph ordinates for a catchment area are given below

| Time (h) | 0 | 3 | 6 | 9 | 12 | 15 | 18 | 21 | 24 |
|---|---|---|---|---|---|---|---|---|---|
| $Q$(m³/sec) | 100 | 200 | 800 | 1200 | 1800 | 1600 | 1000 | 400 | 100 |

Catchment area = 900 sq. km.
Precipitation data for the hydrograph are

| Time (h) | 0–3 | 3–6 | 6–9 |
|---|---|---|---|
| Rainfall intensity (mm/h) | 16 | 22 | 12 |

Compute (i) $\phi$-index, (ii) ERH, (iii) DRH, (iv) $nK$, (v) $n$ and $K$, (vi) IUH, (vii) 3-h UH and (viii) 3 h UH from Sherman's approach and compare with result of the UH obtained in Example 7.9.

**Solution**

(i) *$\phi$-index*
The ordinates of DRH are obtained by subtracting the base flow from the observed hydrograph at the rate of 100 m³/sec. Calculations are carried out as follows:

**Table 7.13** **Computation of IUH using Nash Model for Example 7.9**
**(n = 5 and K = 1.92 h)**

| Time (h) | $t/K$ col. (1)/1.92 | $(t/K)^{n-1}$ | $e^{-t/K}$ | $U(0, t)$ = cols. (3) × (4) × 6.0282 × $10^{-6}$ ($\times 10^{-6}$) | $U(0, t) \times 22$ × $10^6$ = col. (5) × 22 × $10^6$ |
|---|---|---|---|---|---|
| (1) | (2) | (3) | (4) | (5) | (6) |
| 0 | 0.000 | 0.000 | 1.000 | 0.000 | 0 |
| 0.5 | 0.2604 | 0.0046 | 0.770 | 0.021 | 0.5 |
| 1.0 | 0.5208 | 0.0736 | 0.594 | 0.2635 | 6 |
| 1.5 | 0.7813 | 0.3725 | 0.4578 | 1.0581 | 33 |
| 2.0 | 1.0417 | 1.1774 | 0.3529 | 2.5044 | 55 |
| 2.5 | 1.3021 | 2.8745 | 0.2720 | 4.7120 | 104 |
| 3.0 | 1.5625 | 5.9605 | 0.2096 | 7.5315 | 166 |
| 3.5 | 1.8229 | 11.0425 | 0.1615 | 10.7535 | 237 |
| 4.0 | 2.0833 | 18.8380 | 0.1245 | 14.140 | 311 |
| 4.5 | 2.3437 | 30.1748 | 0.0959 | 17.4563 | 384 |
| 5.0 | 2.6042 | 45.991 | 0.0739 | 20.506 | 451 |
| 6.0 | 3.1250 | 95.3674 | 0.0439 | 25.2589 | 556 |
| 7.0 | 3.6458 | 176.680 | 0.0261 | 27.7976 | 612 |
| 8.0 | 4.1667 | 301.408 | 0.0155 | 28.1694 | 620 |
| 10.0 | 5.2083 | 735.86 | 0.0055 | 24.2685 | 534 |
| 12.0 | 6.250 | 1525.8 | 0.0019 | 17.7558 | 391 |
| 15.0 | 7.805 | 3725.3 | $4.046 \times 10^{-4}$ | 9.0870 | 200 |
| 18.0 | 9.375 | 7725.0 | $8.4818 \times 10^{-5}$ | 3.9498 | 81 |
| 22.0 | 11.4583 | 17238 | $1.0561 \times 10^{-4}$ | 1.0979 | 24 |
| 26.0 | 13.5417 | 33627 | $1.315 \times 10^{-6}$ | 0.2666 | 6 |
| 30.0 | 15.625 | 59605 | $1.6373 \times 10^{-7}$ | 0.0588 | 1 |

| Time (h) | 0 | 3 | 6 | 9 | 12 | 15 | 18 | 21 | 24 |
|---|---|---|---|---|---|---|---|---|---|
| $Q$ (m³/sec) | 100 | 200 | 800 | 1200 | 1800 | 1600 | 1000 | 400 | 100 |
| Base flow | 100 | 100 | 100 | 100 | 100 | 100 | 100 | 100 | 100 |
| DRH ordinates (m³/sec) | 0 | 100 | 700 | 1100 | 1700 | 1500 | 900 | 300 | 0 |

Volume of DRH = 3 × 3600{0 + 100 + 700 + 1100 + 1700 + 1500 + 900 + 300 + 0)

= 3 × 3600 × 6300 m³

Depth of runoff $d = \dfrac{(3 \times 3600 \times 6300)}{(900 \times 10^6)}$ m = 0.0756 m = 75.6 mm = 7.56 cm

Depth of rainfall at end of 3rd, 6th and 9th hour are 48, 66 and 36 mm, respectively.

| | | | | |
|---|---|---|---|---|
| Rainfall depths (mm) | : | 48.00 | 66.00 | 36.00 |
| $\phi$-index (mm) | : | 24.80 | 24.80 | 24.80 |
| Runoff depth (mm) | : | 23.20 | 41.20 | 11.20 |
| Effective rainfall intensity | : | 7.73 | 13.73 | 3.73 |

Total runoff depth = 23.20 + 41.20 + 11.20 = 75.60 mm, which is same as the depth of rainfall excess obtained previously from DRH. Therefore $\phi$-index = 24.8/3 = 8.27 mm/h (Fig. 7.22).

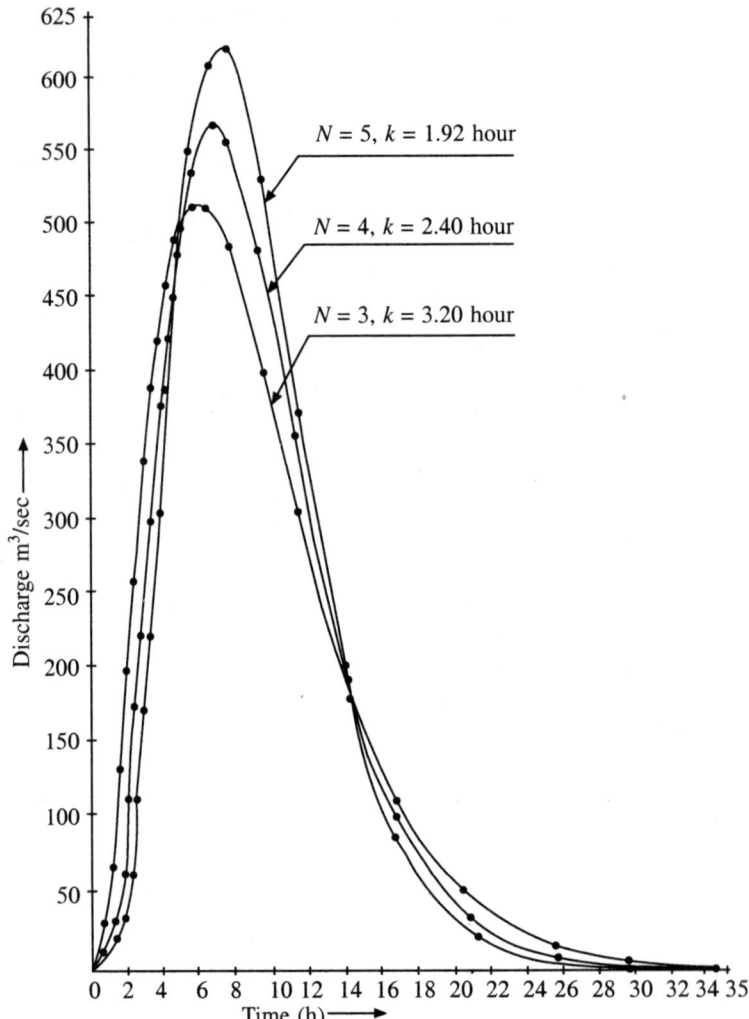

**Fig. 7.21    Comparison of IUH's of Nash Model for *N* = 3, 4 and 5 for Example 7.9**

**(ii) *ERH***
Effective rainfall hyetograph is obtained after subtracting the losses ($\phi$-index) from the rainfall hyetograph, shown in Fig. 7.22. Intensity of effective rainfall given above is shown in Fig. 7.22.

**(iii) *DRH***
Direct runoff hydrograph is obtained after subtracting base flow from the observed hydrograph. The base flow taken as 100 m³/sec is assumed as constant throughout. This is shown in Fig. 7.23.

**(iv) *nK***

Taking first moment about origin $M_{\text{DRH1}} - M_{\text{ERH1}} = nK$                 (1)

where $M_{\text{DRH1}}$ is the first moment of the DRH about origin and $ME_{\text{ERH1}}$ is the first moment

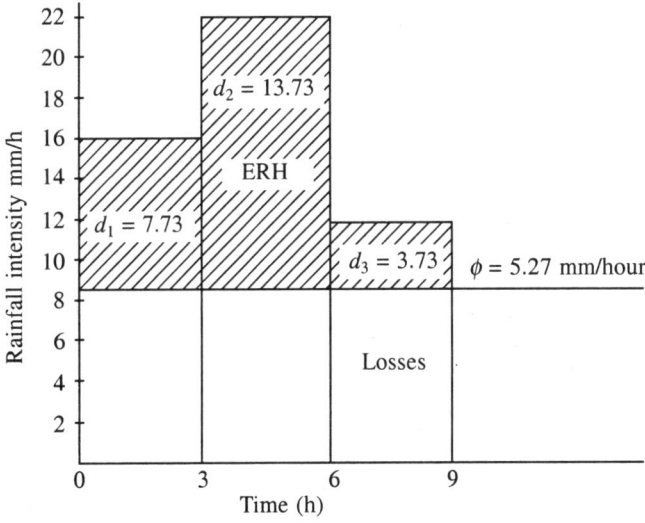

**Fig. 7.22   ERH and $\phi$-Index for Example 7.10**

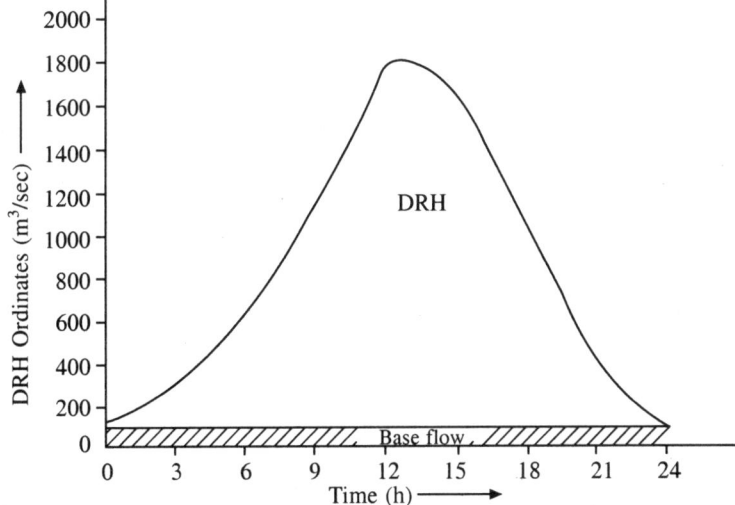

**Fig. 7.23   Observed Hydrograph and Base Flow Separation to obtain DRH**

of the ERH about origin. If $t_i$ is the distance of centroids of the hyetograph blocks from the origin, then

$$M_{ERH1} = \frac{(I_1t_1 + I_2t_2 + I_3t_3)}{(I_1 + I_2 + I_3)} = \frac{\Sigma I_i t_i}{\Sigma I_i}$$

Substituting the values

$$M_{ERH1} = \frac{(7.73 \times 1.5 + 13.73 \times 4.5 + 3.73 \times 7.5)}{(7.73 \times 13.73 + 3.73)} = \frac{101.35}{25.19} = 4.024 \text{ h}$$

$$M_{DRH1} = \frac{(R_1 t_1 + R_2 t_2 + \ldots + R_7 t_7)}{(R_1 + R_2 + \ldots + R_7)} = \frac{\Sigma R_i t_i}{\Sigma R_i}$$

The DRH curve is approximated to a series of rectangles. The ordinates of DRH and the distance of these ordinates from axis are multiplied to give $M_{DRH1}$.

$$M_{DRH1} = \frac{(100 \times 3 + 700 \times 6 + 1100 \times 9 + 1700 \times 12 + 1500 \times 15 + 900 \times 18 + 300 \times 21)}{(100 + 700 + 1100 + 1700 + 1500 + 900 + 300)}$$

$$= \frac{79800}{6300} = 12.67 \text{ h}$$

Substituting in equation (1), we get the total lag $nK = 12.67 - 4.024 = 8.646$.

(v) *Values of n and K*

Taking second moments of DRH and ERH about origin we get

$$M_{DRH2} = \frac{(R_1 t_1^2 + R_2 t_2^2 + \ldots + R_7 t_7^2)}{(R_1 + R_2 + \ldots + R_7)}$$

$$= \frac{(100 \times 3^2 + 700 \times 6^2 + 1100 \times 9^2 + 1700 \times 12^2 + 1500 \times 15^2 + 900 \times 18^2 + 300 \times 21^2)}{(100 + 700 + 1100 + 1700 + 1500 + 900 + 300)}$$

$$= \frac{1121400}{6300} = 178 \text{ h}^2$$

$$M_{ERH2} = \frac{(I_1 t_1^2 + I_2 t_2^2 + I_3 t_3^2)}{(I_1 + I_2 + I_7)}$$

$$= \frac{(7.73 \times 1.5^2 + 13.73 \times 4.5^2 + 3.73 \times 7.5^2)}{(7.73 + 13.73 + 3.73)}$$

or
$$M_{ERH2} = \frac{505.237}{25.20} = 20.06 \text{ h}^2$$

But the second moment about origin of the system is given as

$$M_{DRH2} - M_{ERH2} = n(n+1)K^2 + 2nK \times M_{ERH1}$$

Substituting the values,

$$178 - 20.06 = n^2 K^2 + nK \times K + 2 \, nK \, M_{ERH1} \tag{2}$$

or
$$157.94 = (8.646)^2 + 8.646 \times K + 2 \times 4.024 \times 8.646$$

or
$$8.646 \, K = 157.94 - (8.646)^2 - 2 \times 4.024 \times 8.646$$

or
$$K = \frac{13.604}{8.646} = 1.573 \text{ h}$$

Since
$$nK = 8.646, \, n = \frac{8.646}{K} = \frac{8.646}{1.573} = 5.50$$

Take
$$n = 5$$

Therefore
$$K = \frac{8.646}{5} = 1.73 \text{ h}$$

(vi) *IUH*
For a system of 5 linear reservoirs, the IUH is given by

$$Q = \frac{1}{(n-1)!} \times \left(\frac{1}{K}\right) \left(\frac{t}{K}\right)^{n-1} e^{-t/K}$$

Now $\quad \dfrac{1}{(n-1)!} \times \dfrac{1}{K} = \dfrac{1}{(4 \times 3 \times 2 \times 1)} \times \dfrac{1}{(1.73 \times 3600)} = 6.6902 \times 10^{-6}$

The calculation of IUH is carried out in Table 7.14. IUH ordinates for the basin are given in col. (6) of Table 7.14.

**Table 7.14   IUH by Nash Model for Example 7.10**

| Time (h) | $t/K$ col. (1)/1.73 | $(t/K)^{n-1}$ | $e^{-t/K}$ | $U(0, t)$ = col. (3) $\times$ (4) $\times$ 6.6902 $10^{-6}$ ($\times 10^{-6}$) | $U(0, t) \times 9$ $\times 10^6$ = (5) $\times 9 \times 10^6$ |
|---|---|---|---|---|---|
| (1) | (2) | (3) | (4) | (5) | (6) |
| 0 | 0.000 | 0.000 | 1.000 | 0.000 | 0 |
| 1.0 | 0.678 | 0.1116 | 0.561 | 0.41886 | 4 |
| 2.0 | 1.156 | 1.786 | 0.3147 | 3.76025 | 34 |
| 3.0 | 1.734 | 9.043 | 0.1766 | 10.6842 | 96 |
| 4.0 | 2.3121 | 28.579 | 0.0991 | 18.9478 | 171 |
| 5.0 | 2.890 | 69.77 | 0.0556 | 25.6416 | 234 |
| 6.0 | 3.468 | 144.70 | 0.0312 | 30.1838 | 272 |
| 7.0 | 4.0462 | 268.05 | 0.0175 | 31.3626 | 282 |
| 8.0 | 4.6243 | 457.27 | 0.0098 | 30.0156 | 270 |
| 9.0 | 5.2023 | 732.46 | 0.0055 | 26.9707 | 243 |
| 10.0 | 5.7803 | 1116.39 | 0.00309 | 23.0625 | 208 |
| 12.0 | 6.9364 | 2315.0 | $9.7176 \times 10^{-4}$ | 15.0504 | 136 |
| 15.0 | 8.6705 | 5652.0 | $1.71573 \times 10^{-4}$ | 6.4877 | 58 |
| 18.0 | 10.4046 | 11719.0 | $3.02928 \times 10^{-5}$ | 2.37503 | 21 |
| 21.0 | 12.1387 | 21712.0 | $5.34847 \times 10^{-7}$ | 0.7769 | 7 |
| 24.0 | 13.8728 | 37039.0 | $0.9443 \times 10^{-6}$ | 0.2340 | 2 |
| 27.0 | 14.4510 | 43609.0 | $0.52967 \times 10^{-6}$ | 0.1545 | 1 |

(vii) *Derivation of 3 h UH from IUH*
The procedure to obtain 3-h unit hydrograph from the IUH is to lag two IUH's by 3 h, the corresponding ordinates are summed up and divided by 2. The resulting hydrograph is a 3-h UH, with a peak less than the corresponding IUH and time base larger than the IUH by 3 h. Calculation of 3-h UH from the IUH of Nash model is carried out in Table 7.15.

(viii) *Derivation of a 3-hour UH from S-curve and Sherman's approach*
Since rainfall intensity affecting the runoff from 6th to 9th h is of 3.73 mm/h, the unit hydrograph can be taken as 6 h duration only. This is because the ERH from 3-6 and 0-3 h are 13.73 mm/hour and 7.73 mm/h respectively. Therefore, the problem is to find a UH of 3h duration from 6 *h* duration UH using S-curve. The method is illustrated in Table 7.16. Depth of runoff = 7.56 cm.

### Table 7.15   Deriving UH Ordinates from Nash IUH

| Time (h) | IUH ordinates (m³/sec) | Lagging of IUH by 3 h (m³/sec) | Sum of two hydrographs cols. (3) + (2) (m³/sec) | Ordinate of 3 h UH (m³/sec) col.(4)/2 |
|---|---|---|---|---|
| (1) | (2) | (3) | (4) | (5) |
| 0 | 0.0 | – | 0.0 | 0.0 |
| 1.0 | 4.0 | – | 4.0 | 2.0 |
| 2.0 | 34.0 | – | 34.0 | 17.0 |
| 3.0 | 96.0 | 0.0 | 96.0 | 48.0 |
| 4.0 | 171.0 | 4.0 | 175.0 | 88.0 |
| 5.0 | 234.0 | 34.0 | 268.0 | 134.0 |
| 6.0 | 272.0 | 96.0 | 368.0 | 184.0 |
| 7.0 | 282.0 | 171.0 | 453.0 | 227.0 |
| 9.0 | 243.0 | 272.0 | 515.0 | 258.0 |
| 10.0 | 208.0 | 282.0 | 490.0 | 245.0 |
| 12.0 | 136.0 | 243.0 | 379.0 | 190.0 |
| 15.0 | 58.0 | 136.0 | 194.0 | 97.0 |
| 18.0 | 21.0 | 58.0 | 79.0 | 40.0 |
| 21.0 | 7.0 | 21.0 | 28.0 | 14.0 |
| 24.0 | 2.0 | 7.0 | 9.0 | 5.0 |
| 27.0 | 1.0 | 2.0 | 3.0 | 2 |
| 30.0 | 0.0 | 1.0 | 1.0 | 0 |

Ordinate of S-hydrograph beyond $T_c = A \times 1$ cm/$D$ hour.

$$= 900 \times 10^6 \left( \frac{1}{100} \right) \times \frac{1}{(6 \times 3600)} = 416.7 \text{ say } 417 \text{ m}^3/\text{sec}$$

Time base of the UH $= T_c$ + RE duration

or $\qquad T_c$ = Time base – RE duration = 24 – 6 = 18 h.

As can be seen from the ordinates of 3-h UH derived from IUH and DRH (ref. cols. 10 and 11 of Table 7.16), the value of UH from IUH show:

(i)   Steep rise at the beginning and the base extends over a longer period.

(ii)  Peak is more than the 3 h UH obtained from DRH analysis

(iii) Peak of UH occurs earlier than DRH-analysis approach of obtaining UH.

For comparison, the 3 h UH obtained by Clark model in Example 7.8 and the Nash model and by Sherman's approach (DRH and ERH analysis) are plotted in the same paper (Fig. 7.24). The peak of the 3-hour unit hydrograph obtained by Sherman's approach is observed at 12th hour with ordinate of 225 m³/sec. The corresponding peak discharge of the unit hydrograph obtained by Nash model and Clark's model are given in Table 7.17.

Since Sherman's UH (3 h) gives lesser peak and the occurrence of the peak is later than others, it means that the storage effect is more here. Therefore, this model is suitable for flat areas. Clark's model gives the highest peak. It also occurs earlier than the other two. Therefore this model is suitable for hilly areas where soil slope is more and obstructions are less. Nash model gives an intermediate value. Therefore, for an intermediate type of catchment, this model is suitable.

**Table 7.16  Comparison of UH Derived by Conceptual and Conventional Approaches (Sherman)**

(Units: m³/sec)

| Time (h) | Hydrograph ordinates | Base flow | DRH cols. (2)–(3) | 6 h UH col. (4) /7.56 | S-curve ordinates | S-curve lag by 3 h | Diff. cols (6)–(7) | 3h UH col (8) /(3/6) | UH ordinates from | |
|---|---|---|---|---|---|---|---|---|---|---|
| | | | | | | | | | IUH | Shermans |
| (1) | (2) | (3) | (4) | (5) | (6) | (7) | (8) | (9) | (10) | (11) |
| 0 | 100 | 100 | 0 | 0 | 0 | 0 | 0 | 0 | 0 | 0 |
| 3 | 200 | 100 | 100 | 13 | 13 | 0 | 13 | 26 | 48 | 13 |
| 6 | 800 | 100 | 700 | 93 | 93 | 13 | 80 | 160 | 184 | 93 |
| 9 | 1200 | 100 | 1100 | 146 | 159 | 93 | 66 | 132** | 258 | 146 |
| 12 | 1800 | 100 | 1700 | 225 | 318 | 159 | 159 | 318 | 190 | 225 |
| 15 | 1600 | 100 | 1500 | 198 | 357 | 318 | 39 | 78** | 97 | 198 |
| 18 | 1000 | 100 | 900 | 119 | 437* | 357 | 60 | 120 | 40 | 119 |
| 21 | 400 | 100 | 300 | 40 | 397* | 417 | 0 | 0 | 14 | 40 |
| 24 | 100 | 100 | 0 | 0 | 437* | – | – | – | 5 | 0 |

\* The values need to be adjusted to the catchment area times 1 cm rainfall depth = 417 m³/sec.

\*\* Shows that the unit duration of the unit hydrograph as obtained directly from DRH and ERH analysis does not give satisfactory results.

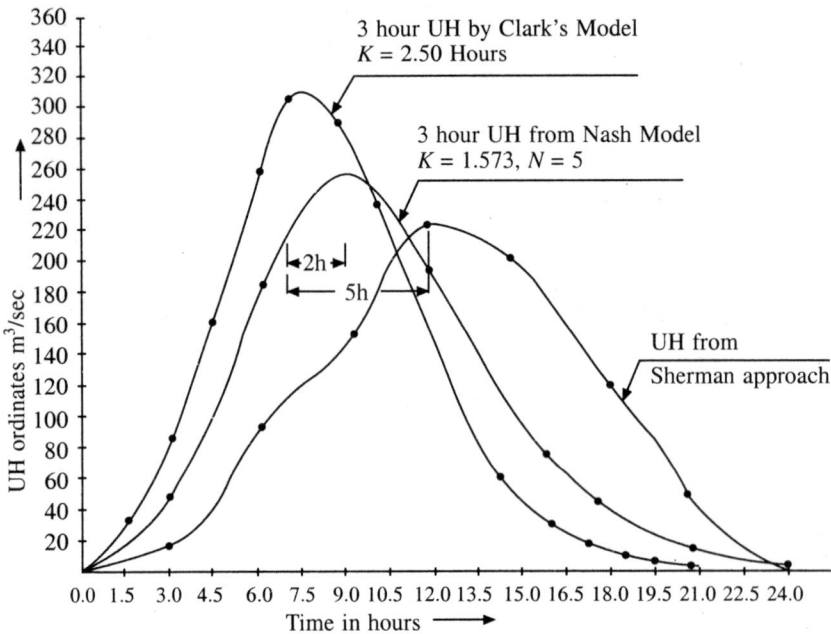

**Fig. 7.24** **Comparison of Unit Hydrographs Derived by Clark, Nash and Sherman's Approach**

**Table 7.17** **Comparison of UH by Nash, Clark and Sherman's Approach**

| Serial no. | Model type | Peak ordinate of UH (m³/sec) | Time to peak ordinate |
|------------|------------|------------------------------|-----------------------|
| (1) | (2) | (3) | (4) |
| 1. | Sherman's Approach | 225.0 | 12th h |
| 2. | Nash model: $n = 5$ | | |
| | $K = 1.573$ h | 258.0 | 9th h |
| 3. | Clarks model $K = 2.50$ h | 330.0 | 7th h |

### 7.9.2.3  Chow-Kulandaiswamy Model

Chow and Kulandaiswamy (1964) proposed a general hydrologic system model in the form of storage equation in nonlinear form. The rainfall and runoff relationship is solved by systems approach. A combination of linear reservoirs are used to model the system. The continuity equation for movement of water can be written as

$$I - Q = \left( \frac{ds}{dt} \right) = \Delta S \tag{7.52}$$

where $I$ and $Q$ are in the form of hydrographs. Since slopes of the hydrograph are changing, the storage is a function of $I$, $Q$ and their higher derivatives. Therefore, the storage equation in the most generalised form was proposed as

$$S = \sum_{m=0}^{M} a_m (I, Q) \frac{d^m I}{dt^m} + \sum_{n=0}^{N} b_n (I, Q) \frac{d^n I}{dt^n} \qquad (7.53)$$

where the coefficients $a_1, a_2, \ldots, a_m$ and $b_1, b_2, \ldots, b_n$ are the functions of $I$ and $Q$. For simplification, the coefficients are assumed to be the functions of average inflow and outflow. The approximation so made does not affect the overall result. The storage $S$ is a function of inflow $I$ and outflow $Q$. Substituting equation (7.53) in (7.52), expanding in Taylor's series and dropping the higher insignificant terms, the resulting differential equation for outflow from the group of nonlinear reservoirs can be written as

$$Q(t) = -\left\{ \frac{D + a_1 D^2 + \ldots + a_{m-1} D^m + a_m D^{m+1} - 1}{b_0 D + b_1 D^2 + \ldots + b_{n-1} D^n + b_n D^{n+1} + 1} \right\} I \qquad (7.54)$$

where $D = \dfrac{d}{dt}$ is the differential operator. The transfer function of (7.54) can be written as

$$\psi = -\frac{M(D)}{N(D)} \qquad (7.55)$$

where $M(D)$ and $N(D)$ are polynomials defined as the numerator and denominator of equation (7.55). It can be shown that a linear reservoir is a particular case of the general transfer function of Chow-Kulandaiswamy's equation (7.55). Linear reservoir equation is given as $S = K \cdot Q$

or
$$\left( \frac{dS}{dt} \right) = K \left( \frac{dQ}{dt} \right) = K \times D(Q)$$

Again
$$I - Q = \left( \frac{dS}{dt} \right) = K \cdot D(Q)$$

or
$$I = (1 + KD) Q \text{ or } Q = \left\{ \frac{1}{(1 + KD)} \right\} I \qquad (7.56)$$

Take the first term of the numerator and the first two terms of the denominator in the given general equation (7.54), we get

$$Q = -\left\{ \frac{-1}{1 + b_0 D} \right\} I = \frac{1}{1 + b_0 D} I \qquad (7.57)$$

Taking coefficient $b_0 = K$, equations (7.56) and (7.57) can be shown to be the same. It can also be proved that the Muskingum's channel routing equation can be easily derived by taking the first two terms of numerator and the first two terms of denominator of Chow-Kulandaiswamy's general equation (For Muskingum's channel routing equation see Chapter 9).

*Assumptions in the Derivation of the General Equation*
Following assumptions are made in the derivation of Chow-Kulandaiswamy general equation:

1. The system is lumped at the outlet.
2. The system is nonlinear. It can be made linear by taking inflow and outflow as function of maximum or steady stage.
3. Storage is a function of $I$, $Q$ and their higher order derivatives
4. Multiplication of derivatives are small and therefore they are neglected. Similarly higher order derivatives are small. They are also assumed to be negligible.
5. The system is time variant.
6. Studies show that the coefficients $a_0$, $a_1$, ..., $a_m$ decrease exponentially with increase in the peak discharge, whereas $b_0$, $b_1$, ..., $b_n$ do not change in any definite trend.
7. IUH derived from the general equation vary from storm to storm indicating non-linearity.

### 7.9.2.4  Dooge's Model

In this model, the watershed area is divided into $n$ number of subareas by drawing isochrones. A linear channel in series with a linear reservoir, model the rainfall – runoff process for each subarea. For the first subarea, the time area block represents the outflow from the linear channel which is taken as the input to the linear reservoir in series for the subarea. For the second subarea, the outflow from the first area along with the outflow from the linear channel of the second area is taken as the input to the linear reservoir. The output IUH from the basin is represented by the following equation.

$$Q_i = \frac{1}{t_c} \int_0^{t' \leq t_c} \frac{\delta(t - \tau)}{\sum_{i=1}^{\tau}(1 + K_i D)} w(\tau/t_c)d\tau \qquad (7.58)$$

where $t_c$ is the time of concentration, $i$ the order of reservoir, $K_i$ the storage constant, $D = \dfrac{d}{dt}$, $\delta(t - \tau)$ is the *dirac-delta function*, $w\left(\dfrac{\tau}{t_c}\right)$ the ordinate of dimensionless time-area histogram, $t$ is time elapsed and $t$, the time of translation between the subarea and outlet. Solution of equation (7.58) for computation of IUH is difficult for practical applications.

### 7.9.2.5  Singh's Model

Singh (1962) improved the IUH model proposed by Dooge. It consists of a linear channel and two linear reservoirs of storage coefficients $K_1$ and $K_2$ in series. Singh's model is modified to suite field problems. IUH for the model is given by

$$U_t = \frac{1}{(K_2 - K_1)} \int_0^{t' \leq t_c} \left[ e^{-\frac{(t - \tau)}{K_2}} - e^{-\frac{(t - \tau)}{K_1}} \right] w(\tau) d\tau \qquad (7.59)$$

where $K_2$ and $K_1$ are storage coefficients in series of the two linear reservoirs.

### 7.9.2.6  Diskin Model

Diskin (1964) proposed a model for IUH with two branches of linear reservoir in parallel to each other. One branch has $n$ linear identical reservoirs of storage coefficient $K_1$ in series and the other branch with $m$ identical linear reservoirs of storage coefficient $K_2$ in series. With $A$ and $B$ as input to the first and second series of the branches where $A + B = 1$, the equation of IUH can be written as

$$U_t = \frac{A}{K_1 (n-1)!} \left( \frac{t}{K_1} \right)^{n-1} e^{t/K_1} + \frac{B}{K_2 (M-1)!} \left( \frac{t}{K_2} \right)^{m-1} e^{t/K_2} \qquad (7.60)$$

The centroid of the IUH has the lag time equal to $t_{cl} = AnK_1 + BmK_2$. The linear channel is used to produce a time area diagram for the drainage basin.

Many other investigators have contributed significantly to the concept of IUH modelling since 1937, the details are beyond the scope of discussion in this book. The investigators O' Kelly (1955), Sugawara and Maruyama (1956), Rose et al. (1961) have used linear reservoir and or channels in series or parallel to simulate the catchment effect, while deriving IUH.

## 7.10 SYNTHETIC UNIT HYDROGRAPH

All the forgoing models use ERH and DRH to derive an unit hydrograph for the basin. When a catchment is ungauged, the established empirical relationship between the unit hydrograph parameters and catchment characteristics may be used to synthesize unit hydrographs for a basin. Various authors have proposed their synthetic models to derive an unit hydrograph. Some important and popular methods are discussed as follows.

### 7.10.1 Snyder's Approach

The model proposed by Snyder (1938) is accepted as standard practice for derivation of unit hydrograph for a catchment, where rainfall and runoff data are not available. Though empirical relations were originally developed for their appalachian in the catchment of eastern USA, the modified equations suiting to Indian conditions are discussed below.

*Basin lag* or *lag time* or simply the *lag* is approximately taken as the time difference between the centroid of ERH and centroid of DRH representing physically the mean travel time of water from all parts of catchment to the basin outlet for the storm. It is also taken as mid point of unit rainfall excess to peak of unit DRH (Fig. 7.25).

Lag time (h) $\qquad\qquad\qquad t_p = C_t (l.l_c)^{0.3}$ $\qquad\qquad\qquad$ (7.61)

Duration of rainfall excess for standard storms $t_{re} = \dfrac{t_p}{5.5}$ $\qquad\qquad$ (7.62)

Peak discharge in (m³/sec) $\qquad Q_{pr} = \dfrac{2.78\, C_p A}{t_p}$ $\qquad\qquad$ (7.63)

Time base of the hydrograph (h) $\qquad T_b = (72 + 3\, t_p)$ $\qquad\qquad$ (7.64)

where $C_p$ is a coefficient of the area varying between 0.56 to 0.69. Miller (1983) found the coefficient varying from 0.23 to 0.67 for USA catchments. For Texas catchment, Clark (1969) calculated the coefficient varying from 0.31 to 1.22. A is the drainage area in km$^2$, $l$ the length of the river measured from the furthest point of the basin to the gauging site, $l_c$ the distance of the centroid of the area from the gauging site along the river. If centroid is not located on the river course, then it is projected to the nearest river and length $l_c$ is measured. $C_t$ is the coefficient varying between 1.35 and 1.60 reported by Snyder, but $C_t$ varying from 0.30 to 6.0 have been reported. Equation (7.61) has been modified by Linsly et al (1958) and proposed the equation in the following form.

$$t_p = C\left(\frac{l \cdot l_c}{\sqrt{S}}\right)^n$$    (7.65)

where $n = 0.38$, $C = 1.72$ for mountainous zones, 1.0 for foot hill and 0.50 for valley may be taken, $S$ is the general slope of the basin. When the duration of rainfall excess is any nonstandard value, i.e., $t_{re}$ is not taken as per equation (7.62) then let it be denoted as $t_r$. The duration of basin lag approximated between mid point of RE to peak of UH, $t_{np}$ in hour is given as

$$t_{np} = t_p + 0.25 \ (t_r - t_{re})$$    (7.66)

where $t_{re}$ is calculated by equation (7.62), $t_r$ is assumed on the basis of experience and $t_p$ is calculated by equation (7.61). Value of $t_p$ should be replaced by $t_{np}$ in equation (7.63) to get $Q_{pr}$. Equation (7.64) holds good for large catchments but for small catchments the following equation proposed by Taylor and Schwartz (1952) may be followed

$$T_b = 5 \ (t_{np} + 0.5t_r)$$    (7.67)

where $T_b$ is in hours and $t_{np}$ is computed from equation (7.66). It is preferable if $T_b$ is taken between 4 and 5 times the time to peak. A synthetic UH is plotted in Fig. 7.25. Stepwise procedure to derive UH by Snyder's approach is given below.

1.  Take data from an adjacent basin with nearly the same catchment area as the basin in question. Derive unit hydrograph for the adjoining basin from the observed rainfall and runoff data. Measure Snyder's coefficients

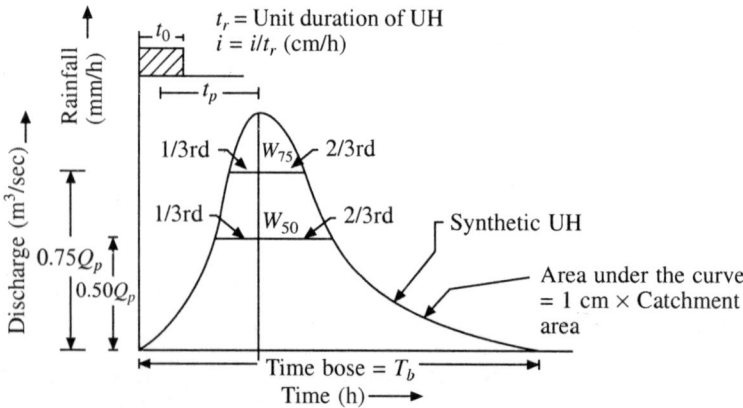

**Fig. 7.25   A Synthetic UH proposed by Snyder**

for the adjacent basin by comparing the values of $t_p$, $Q_{pr}$, $T_b$ and other parameters with that of the derived UH. The coefficients so derived help to compare the UH for the ungauged basin in question.

2. Compute the Snyder's parameters of the basin in question using equations (7.61) through (7.67). The values should be comparable with the adjacent basin.

3. Sketch the tentative unit hydrograph by taking the following two additional features proposed by US Army Corps of Engineers.

$$W_{50} = \frac{5.87}{q_{pru}^{1.08}} \qquad (7.68)$$

$$W_{75} = \frac{3.354}{q_{pru}^{1.08}} \qquad (7.69)$$

where $q_{pru}$ is the peak discharge of the UH per unit drainage area = $Q_{pr}/A$ in (m³/sec)/km², $W_{50}$ the width of the UH at 50% of the peak discharge, $W_{75}$ the width of the UH at 75% of the peak discharge. The values of $W_{50}$ and $W_{75}$ are in time units and therefore it is easy to plot them.

4. Care may be taken to plot the widths of $W_{75}$ and $W_{50}$ in the ratio of 1: 2 from the peak i.e., the widths of $W_{50}$ and $W_{75}$ may be divided into three parts and one part may be provided to the left of the peak and two parts to right of peak (Fig. 7.25). Hudlow and Clark (1969) proposed 40% of the widths to the left of peak and 60% to the right for optimum results.

5. The volume of the unit hydrograph should be 1 cm × area of the catchment in km². Adjust the ordinates of the UH if it does not.

6. Since $T_b$ is tentative, a S-hydrograph is plotted from the UH. The S-hydrograph is smoothened with equilibrium at 1 cm × Area.

7. From the S-hydrograph, the UH of desired duration may be computed back.

### 7.10.2 Synthetic UH for Indian Catchment

Central Water Commission of India recommends the following empirical relations for developing a synthetic unit hydrograph.

$$Q_{pr} = 4.44 \, A^{0.75} \text{ for } S < 0.0028$$

or $\qquad Q_{pr} = 222 \, A^{0.75} \, S^{0.67} \text{ for } S > 0.0028 \qquad (7.70)$

where $S$ is the mean slope of the catchment, $A$ the catchment area in km².

$$\text{Lag time} = t_{lp} = 3.95 \left( \frac{Q_p}{A} \right)^{-0.9} \qquad (7.71)$$

Duration of rainfall excess is given as $D = 1.10 \, t_{lp}$ $\qquad (7.72)$

Time to peak is derived as per the equation (7.61). It will be preferable if the lag time is calculated by considering 1 h UH, i.e., the time lapse between centre of one hour rainfall excess and peak of UH. The mean slope is calculated by dividing the length of the total longitudinal section of the river to small lengths

of say $l_1, l_2, ..., l_n$ with slopes $s_1, s_2, ..., s_n$ and using the relation $\left[ l_e / \Sigma \left( \dfrac{l_i}{S_i^{0.5}} \right) \right]^2$

where $l_e$ is length from centroid of catchment to the outlet of the gauging site. Other parameters can be computed using Snyder's approach.

### 7.10.3　Goel et al. Method

A synthetic hydrograph proposed by Goel, Hussian and Das (1977) for the catchment in India has the following parameters.

$t_p$ can be derived from equation (7.65), where $C = 1.13$ and $n = 0.277$

$$Q_{pr} = 0.315\ A^{0.93}\ S^{0.53} \tag{7.73}$$

$$T_b = 4.3 \left( \frac{l \cdot l_c}{\sqrt{S}} \right)^{0.28} \tag{7.74}$$

$$W_{50} = 2.18(q)^{-1.12} \tag{7.75}$$

$$W_{75} = 0.81(W_{50})^{0.72} \tag{7.76}$$

$$W_{0.50} = 0.69(W_{50})^{0.69} \tag{7.77}$$

$$W_{0.75} = 0.6(W_{0.50})^{0.95} \tag{7.78}$$

where $q$ is the peak discharge per unit catchment area, ( $q = \dfrac{Q_{pr}}{A}$, where $Q_{pr}$ is the peak discharge in m³/sec given in equation (7.73) and $A$ is the catchment area in km²) and $W_{0.50}$ and $W_{0.75}$ are the widths of the rising side of the UH for discharges equal to $0.5Q_{pr}$ and $0.75Q_{pr}$ heights, respectively.

---

**Example 7.11:** An unit hydrograph is to be developed for an ungauged catchment for which there is no information of any kind. An adjoining catchment is thoroughly gauged. It has 3-h UH with peak of 140 m³/sec appearing 37 h from the start of rainfall excess. Determine the Snyder's coefficients of the hydrograph to be used for the adjoining ungauged catchment for formulation of 3 h UH.

|  | Gauged catchment |
|---|---|
| Area (*A*) | 2718 sq. km |
| Basin length along the river (*l*) | 148 km |
| Centroid length along the river (*l_c*) | 76 km |

**Solution**

Rainfall excess duration $t_r = 3$ h.

Time to peak from beginning of storm = 37 h. This equals half of the rainfall excess duration plus $t_{np}$. Or

$$37 = \frac{t_r}{2} + t_{np} = \frac{3}{2} + t_{np}$$

Therefore, $t_{np} = 35.5$ h

The modified basin lag $t_{np}$ can be obtained from equation (7.66) as

$$t_{np} = t_p + 0.25 \, (t_r - t_{re})$$

$$= t_p + 0.25 \left( 3 - \frac{t_p}{5.5} \right) = \frac{21 t_p}{22} + 0.75$$

or

$$35.5 = \left( \frac{21}{22} \right) t_p + 0.75$$

Therefore

$$t_p = (35.5 - 0.75) \times \left( \frac{22}{21} \right) = 36.4 \text{ h}$$

but

$$t_p = C_t \, (l \times l_c)^{0.3} \text{ or } 36.4 = C_t \, (148 \times 76)^{0.3}$$

Therefore

$$C_t = \frac{36.4}{(148 \times 76)^{0.3}} = 2.22$$

Peak discharge

$$Q_{pr} = 2.78 \, C_p.A \, /t_{np} \text{ or } 140 \doteq (2.78 \times C_p \times 2718)/35.5$$

or

$$C_p = \frac{(35.5 \times 140)}{(2.78 \times 2718)} = 0.658.$$

Coefficients $C_t = 2.22$ and $C_p = 0.658$ can be taken for the adjoining ungauged catchment to derive 3 h synthetic unit hydrograph.

**Example 7.12:** For a basin of 198 sq. km, construct a 4 h unit hydrograph from the following data. The length of the main channel is 21.6 km. Length of centroid from the outlet along the river is 11.2 km. Use Snyder's method. The coefficient $C_t$ for the neighbouring catchment is found to be 1.5. Take $C_p$ as 0.59.

**Solution**

Here catchment area $A$ = 198 sq. km

Length of channel $l$ = 21.6 km

Length of centroid from basin outlet $l_c$ = 11.2 km

Desired duration of rainfall excess period = 4 h

Lag time $t_p = C_t \, (l \times l_c)^{0.3} = 1.5 \, (21.6 \times 11.2)^{0.3} = 7.78$ h

Duration of rainfall excess $t_{re} = t_p/5.5 = 7.78/5.5 = 1.41$ h $\neq 4$ h

Since the desired duration of rainfall excess is 4 h the value of $t_p$ needs to be modified

$$t_{np} = t_p + 0.25 \, (t_r - t_{re})$$

$$= 7.78 + 0.25 \, (4 - 1.41) = 8.42 \text{ h}$$

Peak discharge $a$ $\quad Q_{pr} = 2.78 \, C_p \, A/t_p = (2.78 \times 0.59 \times 198)/8.42 = 38.57 \text{ m}^3/\text{sec/km}^2$

$$q_{pru} = \frac{38.57}{198} = 0.1948 \text{ m}^3/\text{sec/sq km.}$$

Time base of the hydrograph $T_b = (72 + 3t_p) = (72 + 3 \times 8.42) = 97.3$ say 97 h

which is very large for the basin. Therefore the equation for small watershed should be applied.

For small catchment $T_b = 5 \, (t_{np} + 0.5 \, t_r) = 5 \, (8.42 + 0.5 \times 4 \,) = 52$ h

$$W_{50} = \frac{5.87}{q_{pru}^{1.08}} = \frac{5.87}{(0.1948)^{1.08}} = 34.35 \text{ h}$$

$$W_{75} = \frac{3.354}{(0.1948)^{1.08}} = 19.63 \text{ h}$$

The unit hydrograph is plotted in Fig. 7.26.

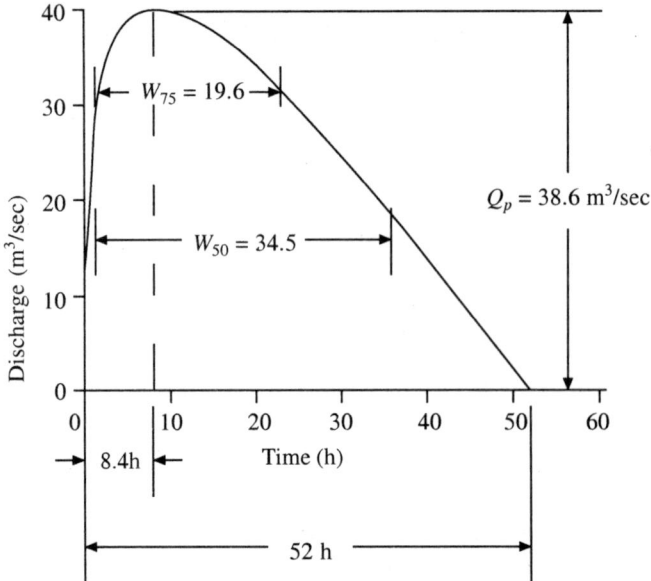

**Fig. 7.26   Synthetic Unit Hydrograph for Example 7.12**

## 7.11   DIMENSIONLESS UNIT HYDROGRAPH

U.S. Soil Conservation Service proposed a dimensionless UH on the basis of analysis of a large number of unit hydrographs from different areas of various sizes. The method is to express the discharge ordinates of UH as ratio $(Q_t/Q_p)$ and abscissa as $(t/t_p)$, where $Q_t$ are the ordinates of unit hydrograph with $Q_p$ as the peak ordinate, $t$ the time from the beginning of UH and $t_p$ the time to peak, i.e., time when $Q_t = Q_p$. Thus if $Q_p$ and $t_p$ are known, the entire UH can be plotted using the standard ratio between $Q_t/Q_p$ and $t/t_p$ from Table 7.18.

**Table 7.18   Parameters of Dimensionless Unit Hydrograph**

| $t/t_p$ | 0.0 | 0.20 | 0.40 | 0.6 | 0.80 | 1.0 | 1.2 | 1.4 | 1.6 | 1.8 | 2.0 | 2.5 | 2.75 | 3.0 | 3.5 | 4.0 | 4.5 | 5.0 |
|---|---|---|---|---|---|---|---|---|---|---|---|---|---|---|---|---|---|---|
| $Q_t/Q_p$ | 0.0 | 0.075 | 0.28 | 0.6 | 0.90 | 1.0 | 0.92 | 0.75 | 0.56 | 0.42 | 0.32 | 0.22 | 0.155 | 0.075 | 0.036 | 0.18 | 0.009 | 0.004 |

For an ungauged basin the above table can be used to compute UH. For small catchments the following equations can be used.

Time to peak can be calculated from the relation

$$t_p = 0.50\, t_{re} + t_l \tag{7.79}$$

$$t_{re} = 0.133 t_c \tag{7.80}$$

where $t_c$ is the time of concentration, $t_{re}$ the duration of rainfall excess, $t_p$ the time to peak in hours from beginnings of rainfall to peak discharge, $t_1$ is the time lag between centroid of rainfall excess to peak discharge. Time lag or time to peak can be calculated using

$$t_1 = 0.60 t_e \qquad (7.81)$$

The peak discharge $Q_p$ can be calculated from the following equation.

$$Q_p = \frac{2.08\,A}{t_p} \qquad (7.82)$$

where $A$ is the area (sq. km). Knowing $Q_p$ and $t_p$, the UH for a basin can be computed. The plot of a dimensionless UH is shown in Fig. 7.27. The utilities of dimensionless UH are

1. Dimensionless UH is an excellent tool to compare the UH of various catchments of a region.
2. The UH derived for the neighboring basin can be made dimensionless and transposed to the ungauged basin.
3. It can be used effectively to average the UH's derived for various catchment of the region.

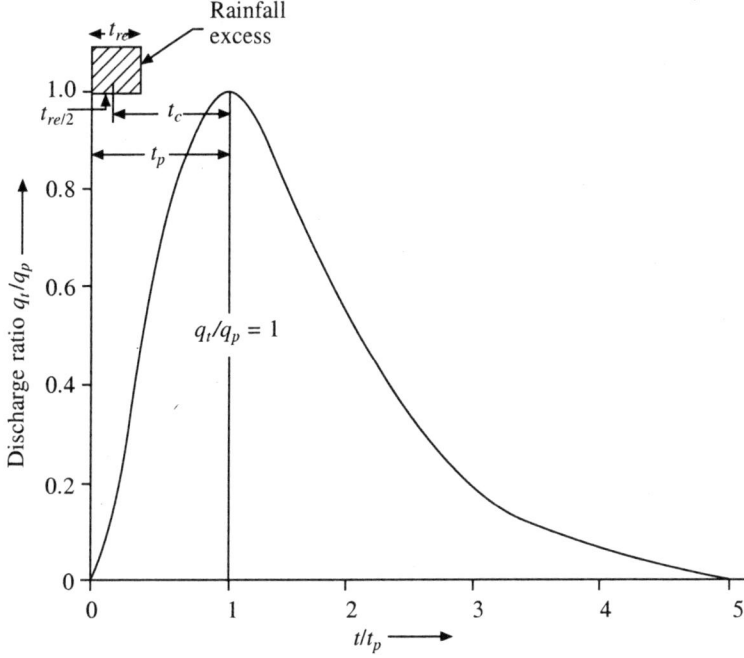

**Fig. 7.27   A Dimensionless Unit Hydrograph**

**Example 7.13:** Obtain a 30 min UH for a basin of 12.5 sq. km, the time of concentration can be taken as 2.5 h.

**Solution**

Here time of rainfall excess = duration of UH = 30 min = 0.50 h

Lag time $t_1 = 0.60 \times t_c = 0.60 \times 2.5 = 1.5$ h

Time to peak $\qquad t_p = \dfrac{t_r}{2} + t_1 = \dfrac{0.50}{2} + 1.5 = 1.75$ h

Now $\qquad\qquad Q_p = \dfrac{2.08 \times A}{t_p} = \dfrac{2.08 \times 12.5}{1.75} = 14.84$ cumecs

Using Table 7.18 the UH can be obtained for the basin. Since $t_p$ and $q_p$ are known, the rows of the $t/t_p$ can be multiplied by $t_p$ (=1.75) to get the time, corresponding to which the row $q/q_p$ can be multiplied with $Q_p$ (=14.86) to get the values of the complete unit hydrograph.

---

## 7.12   DISTRIBUTION GRAPH

Another modification to the presentation of UH was proposed by Bernard (1935). The graph is known as *distribution* or *percentage graph*, and is basically a unit hydrograph whose ordinates $d_i$ at fixed intervals of $\Delta t$ are expressed as percentage of the total runoff. The blocks of the distribution graphs forming the histogram are at the time interval $\Delta t = D - h$, i.e., unit duration of the unit hydrograph. To facilitate the distribution, the entire time base $(T_b)$ of the unit hydrograph is converted to $n$ blocks such that $n = T_b/D$. Thus the sum of all block areas of the distribution graph is equal to 100%. A typical percentage unit hydrograph looks like the one given in Fig. 7.28 where

$$A_1 + A_2 + \dots + A_{10} = 100\% \text{ of } A$$

$$\frac{T_b}{D} = n \ (= 10 \text{ in the figure})$$

The graph is useful for comparing the UH of different basins with similar hydrometeorological conditions. Runoff characteristics of different basins can be compared by such a unit graph. A given ERH can be superimposed on the distribution graph in the same way the flood is calculated from a known ERH and UH to produce a DRH. The following example illustrates the procedure.

---

**Example 7.14:** The ordinates of an 3 h UH are given below. Prepare a distribution graph for the unit hydrograph.

| Time (h) from 1/3/98 | 5 | 8 | 11 | 14 | 17 | 20 | 23 | 2 | 5 | 8 | 11 | 14 |
|---|---|---|---|---|---|---|---|---|---|---|---|---|
| UH ordinates (m³/sec) | 0 | 10.9 | 28.2 | 67.7 | 92.3 | 84.4 | 36.5 | 22.6 | 13.9 | 9.8 | 4.9 | 0 |

**Solution**
Calculation are carried in Table 7.19. The unit hydrograph is plotted in Fig. 7.28. The UH is divided to 3 h blocks to convert it to the distribution graph. The area blocks of UH are given in Table 7.19.

The distribution graph for the example is plotted in Fig. 7.29.

**Example 7.15:** The distribution graph of example 7.14 is applied to a neighbouring catchment of 300 sq. km with rainfall excess of 6.0, 2.5 and 4 cm, respectively, for 3 h blocks. Calculate ordinates of DRH from the rainfall data.

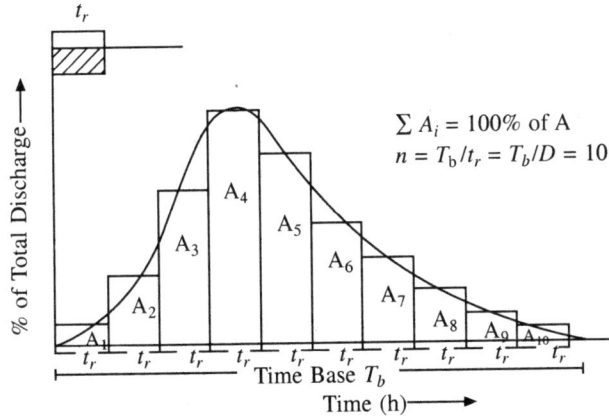

**Fig. 7.28   A Distribution Graph (After Bernard 1935)**

$$\Sigma A_i = 100\% \text{ of } A$$
$$n = T_b/t_r = T_b/D = 10$$

**Table 7.19   Preparation of a Distribution Graph**

| Time (h) | Time interval $\Delta t$ (h) | UH ordinates $d_i$ (m³/sec) | Area of blocks cumec-h cols.(2) × (3) | % of blocks of histogram =100 × col. (4)/1113.6 | |
|---|---|---|---|---|---|
| (1) | (2) | (3) | (4) | (5) | |
| 5 | 0 | 0 | 0 | 0 | |
| 8 | 3 | 10.9 | 32.7 | 3.0 | ($A_1$) |
| 11 | 3 | 28.2 | 84.6 | 7.6 | ($A_2$) |
| 14 | 3 | 67.7 | 203.1 | 18.3 | ($A_3$) |
| 17 | 3 | 92.3 | 276.9 | 24.9 | ($A_4$) |
| 20 | 3 | 84.4 | 253.2 | 22.7 | ($A_5$) |
| 23 | 3 | 36.5 | 109.5 | 9.8 | ($A_6$) |
| 2 | 3 | 22.6 | 67.8 | 6.1 | ($A_7$) |
| 5 | 3 | 13.9 | 41.7 | 3.7 | ($A_8$) |
| 8 | 3 | 9.8 | 29.4 | 2.6 | ($A_9$) |
| 11 | 3 | 4.9 | 14.7 | 1.3 | ($A_{10}$) |
| 14 | 3 | 0 | 0 | 0 | |
| Total | | | 1113.6 | 100% | |

**Solution**

Catchment area = 300 sq. km = $300 \times 10^6 \text{ m}^2$

Runoff due to 1 cm of rainfall excess = $\left\{ \dfrac{300 \times 10^6}{(3 \times 60 \times 60)} \right\} \times \left( \dfrac{1}{100} \right) = 277.8 \text{ m}^3/\text{sec}$

The calculation are carried out in Table 7.20.

The runoff coordinates given in col.(8) of the Table 7.20 represent the centre points of the blocks of the DRH graph. For example for plotting the DRH, take 50 m³/sec against time abscissa of 1.5 h, 147.2 m³/sec at 4.5 h, and so on, till the entire hydrograph is covered.

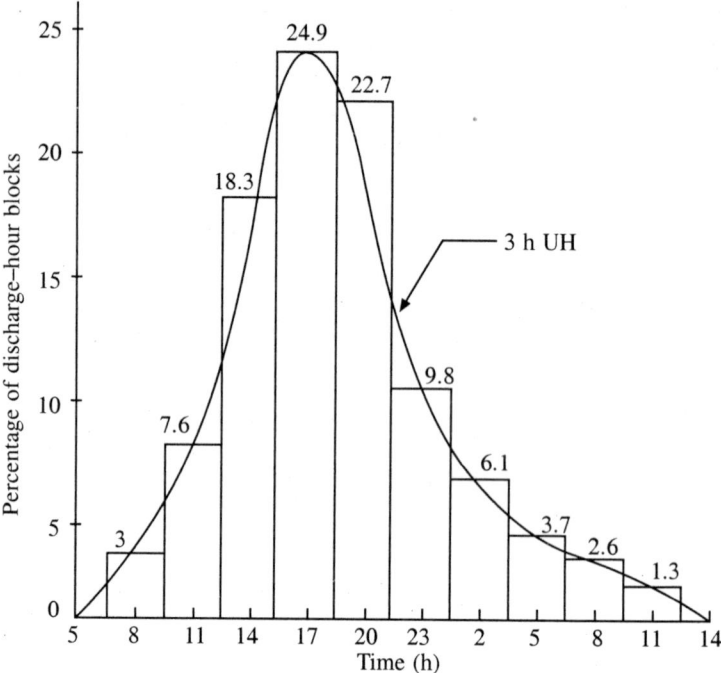

**Fig. 7.29    Area-Blocks of UH to Convert to Distribution Graph for Example 7.14**

**Table 7.20    Computation of DRH from rainfall excess data for Example 7.15**

| Time (h) say from 0 h | Rainfall excess (cm) | % of distribution graph | Rainfall excess as per distribution graph due to | | | Runoff (cm) in 3h cols{(4) +(5)+(6)} | Runoff= (m³/sec) col(7)× 277.8 |
|---|---|---|---|---|---|---|---|
| | | | 6 cm | 2.5 cm | 4 cm | | |
| (1) | (2) | (3) | (4) | (5) | (6) | (7) | (8) |
| 0–3 | 6 | 3 | 0.18 | 0.00 | 0.00 | 0.180 | 50.0 |
| 3–6 | 2.5 | 7.6 | 0.456 | 0.075 | 0.00 | 0.531 | 147.2 |
| 6–9 | 4 | 18.3 | 1.098 | 0.190 | 0.12 | 1.408 | 391.1 |
| 9–12 | | 24.9 | 1.494 | 0.456 | 0.304 | 2.254 | 626.1 |
| 12–15 | | 22.7 | 1.362 | 0.623 | 0.732 | 2.717 | 754.7 |
| 15–18 | | 9.8 | 0.588 | 0.568 | 0.996 | 2.152 | 597.8 |
| 18–21 | | 6.1 | 0.366 | 0.245 | 0.908 | 1.519 | 421.9 |
| 21–24 | | 3.7 | 0.222 | 0.153 | 0.392 | 0.767 | 213.1 |
| 24–3 | | 2.6 | 0.156 | 0.093 | 0.244 | 0.493 | 136.9 |
| 3–6 | | 1.3 | 0.078 | 0.065 | 0.148 | 0.291 | 80.8 |
| 6–9 | | | | 0.033 | 0.104 | 0.137 | 38.1 |
| 0–12 | | | | | 0.052 | 0.052 | 14.4 |

## PROBLEMS

7.1.    (a) Draw a unit hydrograph and explain the salient features.

(b) Define a unit hydrograph. Write all its assumptions and limitations.

7.2 For an ungauged catchment how can you derive a unit hydrograph?

7.3 Write the steps involved in computing a *D*-h UH from rainfall- runoff data.

7.4. (a) What is IUH? What are its advantages over UH?

(b) Define a S-hydrograph. Mention its uses.

7.5 Explain the terms:

(a) Isochrone, (b) Linear channel, (c) Linear reservoir, (d) Hyetograph,

(e) Hydrograph, (f) Time of concentration.

7.6 How you will determine an IUH by Clark's model?

7.7 Derive Nash's equation for *n* number of reservoirs.

7.8 Explain the procedure to derive effective rainfall hyetograph from observed rainfall-runoff data of a basin.

7.9 Average daily discharge for a basin draining 1640 sq. km are given below.

| Date | 7 | 8 | 9 | 10 | 11 | 12 | 13 | 14 | 15 | 16 |
|---|---|---|---|---|---|---|---|---|---|---|
| Discharge (m³/sec) | 248 | 202 | 164 | 232 | 367 | 395 | 367 | 294 | 186 | 185 |

Determine (a) Total volume of runoff, (b) Mean daily discharge and (c) Runoff depth over the basin.

7.10 The ordinates of a 1-h UH are given below. Compute S-hydrograph and derive a 3-h UH from the data

| Time (h) | 0 | 1.5 | 3 | 4.5 | 6 | 7.5 | 9 | 10.5 | 12 |
|---|---|---|---|---|---|---|---|---|---|
| 1-h UH ordinates (m³/sec) | 0.3 | 1.2 | 15 | 4 | 1.4 | 0.9 | 0.5 | 0.15 | 0.0 |

7.11 The ordinates of a 4-h UH are given below. If there is 4 cm effective rainfall occurring uniformly for 4-h, calculate the DRH resulting from the storm.

| Time (h) | 0 | 2 | 4 | 6 | 8 | 10 | 12 | 14 | 16 | 18 | 20 | 22 | 24 |
|---|---|---|---|---|---|---|---|---|---|---|---|---|---|
| 4 h UH ordinate (m³/sec) | 0 | 9 | 24 | 50 | 66 | 72 | 65 | 53 | 28 | 20 | 11 | 6 | 0 |

7.12 A watershed of 3130 sq. km was subjected to a storm of 4-h duration from which the following hydrograph resulted.

| Time (h) | 3 | 6 | 9 | 12 | 15 | 18 | 21 | 24 | 3 | 6 | 9 | 12 | 15 | 18 | 21 |
|---|---|---|---|---|---|---|---|---|---|---|---|---|---|---|---|
| Discharge (m³/sec) | 20 | 16 | 175 | 270 | 230 | 200 | 170 | 150 | 130 | 115 | 100 | 90 | 80 | 70 | 60 |

What is the rainfall excess for the storm? Obtain an UH for the watershed.

7.13 Results of three numbers of 6-h storm hydrographs over a basin of 180 sq. km are given below. Obtain the representative UH for the basin.

Time (h)  1.5  3   4.5   6   7.5   9   10.5   12   13.5   15   16.5   18   19.5  21.0

Hydrograph ordinates are (m³/sec)

Event I   165 1080 2250 2305 1560 970  810  600  480  390 300 210 150 60

Event II   120 720  1500 2070 3210 2070 1730 1290 1040 750 510 335 210 70

Event III  90  510  1200 1990 2880 2340 1980 1470 1200 920 720 480 330 180

7.14 A basin has an area of 360 sq. km. Length of the longest stream is 26.0 km and length of the centroid along the stream from the outlet 9.80 km. Determine a synthetic UH for the basin.

7.15 Generate an IUH using Nash model for various values of $n$ keeping $K$ as constant. Discuss the results.

7.16 A drainage basin has an area of 160 sq. km. Storage constant $K$ is 8 h and time of concentration is 6.0 h. The inter-isochronal area are given below. Obtain an UH for the basin using Clark model.

| Time (h) | 0-1 | 1-2 | 2-3 | 3-4 | 4-5 | 5-6 |
|---|---|---|---|---|---|---|
| Isochrone area (km$^2$) | 8 | 40 | 23 | 47 | 30 | 12 |

7.17 The values of Nash model coefficients are $K = 6$, $n = 3$. Determine the ordinates of IUH in m$^3$/sec, if the catchment area is 350 km$^2$. Compute the ordinates at 2 h interval.

7.18 ERH and DRH resulting from a storm are given below

| Time (h) | 0 | 6 | 12 | 18 | 24 | | 30 | 36 | 42 | 48 | 54 | 60 | 66 |
|---|---|---|---|---|---|---|---|---|---|---|---|---|---|
| ERH (cm) | | 4.5 | 3.0 | 3.6 | 2.5 | | | | | | | | |
| DRH (m$^3$/sec) | 0 | 18 | 145 | 370 | 400 | | 270 | 170 | 80 | 40 | 20 | 10 | 0 |

Determine the values of Nash coefficients $n$ and $K$.

7.19 Mean Daily discharge in cumecs observed at a stream site are given below.

| Date | 6 | 7 | 8 | 9 | 10 | 11 | 12 | 13 | 14 | 15 | 16 |
|---|---|---|---|---|---|---|---|---|---|---|---|
| Discharge | 180 | 1270 | 5350 | 8150 | 6580 | 1540 | 510 | 280 | 220 | 200 | 180 |

Obtain the recession coefficients for the hydrograph.

7.20 From the data of the recession curve of a hydrograph, it is observed that on a certain day the peak recession is 800 m$^3$/sec. After 24 h the recession constant $K_r$ is found to be 0.85. Obtain the flow in the recession on the 7th day in the channel.

7.21 From the toposheet it is observed that the fall of elevation from the remote point in the watershed to the outlet is 1600 m between a length of 130 km in the channel. Obtain the time of concentration of the basin.

7.22 On a recession curve of a hydrograph, two points at 4 days apart records discharge of 210 and 120 m$^3$/sec respectively. What is the recession constant for this curve?

7.23 Ordinates of 4 h UH in m$^3$/sec are given below. Obtain 3 h and 6 h UH by S-hydrograph approach.

| Time (h) | 0 | 2 | 4 | 6 | 8 | 10 | 12 | 14 | 16 | 18 | 20 | 22 |
|---|---|---|---|---|---|---|---|---|---|---|---|---|
| Ordinates of 4h UH (m$^3$/sec) | 0 | 40 | 160 | 260 | 300 | 260 | 180 | 105 | 55 | 30 | 10 | 0 |

7.24 Consider a 3-h UH to be a triangle whose base is 8 h and the peak of 4 m$^3$/sec occurs at 3-h from the origin of the time. If rainfall of intensity 25 mm/h occurred for 3-h, compute the DRH and peak discharge.

7.25 Construct a 2-h UH from the UH of the exercise 7.24.

7.26 Derive a 4 h UH by Snyder's method from the following data.
Length of the main stream = 12 km.
Catchment area = 60 sq. km.
Centroid of the watershed along the channel = 4.5 km.
Take $C_t = 2.2$, $C_p = 0.70$ and plot the unit hydrograph.

7.27 Ordinates of direct runoff hydrograph resulting from 4 h storm excess of 5 cm are given below. Determine the catchment area and the ordinates of UH in m$^3$/sec. Obtain 3 h UH by S-hydrograph approach.

| Time (h) | 0 | 4 | 8 | 12 | 16 | 20 | 24 | 28 | 32 | 36 | 40 | 44 |
|---|---|---|---|---|---|---|---|---|---|---|---|---|
| Ordinates of 4h UH (m$^3$/sec) | 0 | 60 | 280 | 400 | 320 | 280 | 210 | 155 | 110 | 70 | 30 | 0 |

7.28 Discuss a method to obtain UH from complex storms. What do you understand by the the principle of linearity and time invariance in unit hydrograph?

7.29 Draw a neat diagram of a flood hydrograph resulting from an isolated storm and briefly discuss the salient features.

7.30 Determine the recession constants of the hydrograph of exercise 7.27 if a constant base flow of 25 m³/sec is to be added to the DRH ordinates.

7.31 Ordinates of 3-h UH are given below. If rainfall intensity of 25 mm/h occurs for a period of 6 h followed by another 3 h rainfall of intensity of 10 mm/h with a gap of 3 h, what peak flow the system will produce ?

| Time (h) | 0 | 3 | 6 | 9 | 12 | 15 | 18 | 21 | 24 | 27 | 30 | 33 |
|---|---|---|---|---|---|---|---|---|---|---|---|---|
| Ordinates of 3-h UH (m³/sec) | 0 | 35 | 160 | 200 | 180 | 155 | 125 | 105 | 80 | 50 | 20 | 0 |

# Chapter Eight

# Design Flood

## 8.1 INTRODUCTION

A flood is an unusual high stage of a river that overflows the natural or man made banks spreading water to its flood plains that are thickly populated due to the obvious advantage of water supply and irrigation. It is possible to predict and contain a flood to a reasonable extent with the forecasting models and technologies. The first information required to predict a flood at a particular place and time is the measurement of all the floods for maintaining a good record. Analysis of flood records gives an in-depth knowledge based on which flood prediction and protection measures can be carried out. However, for urban areas where catchments sizes are small, flood prediction may be carried out on empirical relations. For design of culverts, bridges, barrages, small dams, embankments, protective works and water supply scheme, the peak flood discharge is the greatest concern on the basis of which the sizes, capacities, locations and outlets of these structures are fixed. The magnitudes of floods are described by flood-discharge, elevation and volume, each of which is important for specific design of a structure. For design of important structures, complete flood hydrograph at the site is an important requirement. A design flood to be considered for a structure is dependent on a large number of factors, but the importance of the structure vis-a-vis its desired objective has to be kept in mind.

A flood accepted for the design of a structure is based on (i) importance of the structure, (ii) economy, (iii) probable effect at its downstream due to its sudden damage, (iv) life expectancy of the structure, (v) inconvenience it can cause to traffic, (vi) population density of the downstream area, (vii) submergence of mineral, industrial and other strategic areas and (viii) economic condition of the people of the affected area. Damages due to small structures like minor irrigation projects, small causeways or bridges create temporary disruptions of the area. Loss to life and property from such damages are small. However, for large dams and important bridges, no risk can be taken while designing them.

## 8.2 DESIGN FLOOD

A flood used for the design of a structure on considerations of its safety, economy, life expectancy and probable damage considerations is called the *design flood*. For important structures located at strategic locations, virtually no-risk can be taken for its failure. The flood selected for design of such structures should probably be the highest one. For other structures, some probability of failure can be allowed. Depending on the size of a water resources project, any of the following types of flood can be estimated.

*Frequency Based Flood (FBF)*
A design flood estimated using flood-frequency analysis for an accepted return period (say 100 or 1000 years ) is called FBF. Sometimes frequency analysis of rainfall data is carried out and a suitable rainfall-runoff model generates the required FBF.

*Probable Maximum Flood (PMF)*
US Army Corps of Engineers (1979) defined it as the extreme flood, which is physically possible in a region due to the most severe combination of critical meteorological and hydrological factors that are reasonably possible over the region under consideration. PMF is used for the design of all important structures with virtually no-risk criterion. At the maximum reservoir level when a dam fails, the flood water can completely wash away life and property of the area down below. Failure of such dams can create immense loss to a nation. Virtually 100% safety against such failures should be ensured. A PMF therefore cannot be assigned a specific return period on the basis of flood frequency studies. PMF is associated with probable maximum precipitation (PMP) and is discussed in Chapter 3.

*Standard Project Flood (SPF)*
A flood computed from the standard project storm (SPS) that have occurred over the project area under consideration or on the adjoining areas with similar hydrometeorological and basin characteristics without its maximization as in PMP is called standard project flood. The flood is considered reasonably characteristic of the region. It usually varies between 40 and 60% of the probable maximum flood.

*Regression Formulae*
A regression relation of the form $Q = CA^n$ can be developed for the region under consideration, where $Q$ is the discharge rate in $m^3$/sec, $A$ the drainage area in sq. km, $C$ and $n$ are constants of regression which are to be evaluated for the region. Various investigators have proposed regression type equations to suit the specific area under consideration. Before we discuss the above floods, guidelines by Central Water Commission, India (1972) for selection of design flood of a project depending on its size and the importance are given below.

U.S. Army Crops of Engineers (1979) classified water resources structures as low, significant or high hazards potential depending on the following criterion.

**Low Hazard** : Almost no loss of life and minimal economic loss. There is no permanent structure of human habitation and the region is considered almost undeveloped.

**Significant** : Loss of life for a few may be possible. There is no urban development in the region. The economic loss due to agriculture, industry or structures is appreciable.

**High** : Loss of life of the people are more than a few. The economic loss is excessive.

The hazard potential classification of the structures (Table 8.1) are directly related to the hydraulic head of the dam, it's storage capacity and the catchment area draining upto the project site. Therefore, the selection of design floods for low, medium or high structures are based on the potential hazards.

**Table 8.1   Classification of a Project for Selection of Design Flood**

| Structure | Storage in M.m³ | Head in m | Type of flood |
|---|---|---|---|
| **On the basis of size** | | | |
| (i) Dams/Barrages (Large projects) | > 60 | >30 | (a) PMF from PMP or (b) Flood with return period of 1000 years or more. |
| (ii) Permanent Barrage/ Small dams | < 60 and > 10 | 12–30 | (a) SPF from SPS (b) Flood with return period 100 years (c) a or b whichever is higher |
| (iii) Small structures (minor projects) | C.D. works culverts/bridges Dams upto 10 M m³ storage, Pickup barrages | 7.5–12 | (a) Flood of return period of 50 or 100 years depending on the importance of the structure |
| **On the basis of Catchment** | | | |
| (i) Small size catchment | Upto 250 km² | | – Rational method – Flood frequency studies – UH approach – Flood peak versus catchment area |
| (ii) Medium size catchment | 250–5000 km² | | – UH approach – Flood peak vs. Catchment area |
| (iii) Large size catchment | > 5000 km² | | – UH approach and flood routing – Flood frequency studies – Flood peak versus catchment area |

The design flood is considered as inflow at the upstream of the structure which the structure should pass safely. The problem of computation of design flood for a structure is carried out under the following methods:

(1) For ungauged catchment
  (a) Rational approach
  (b) Empirical equations
  (c) Envelope curves

(2) For gauged catchment
  (a) Unit hydrograph approach
  (b) Flood frequency analysis

## 8.3 FLOOD PEAK ESTIMATION FOR UNGAUGED CATCHMENTS

### 8.3.1 Rational Method

Among various types of empirical relations, *rational formula* is the most rational method of calculating peak discharge for small catchments. It can be considered as the representative of all other empirical relations. The formula is called rational because of the units of the quantities considered being numerically consistent. The original formula was in FPS unit in which rainfall was measured in inches and area in acres (one acre-inch of rainfall for one hour produces one cusec of flow from the area). The maximum rate of runoff from the watershed appears when the entire area contributes at the basin outlet. The runoff gradually increases from zero to peak when rainfall duration reaches the time of concentration $t_c$ and thereafter it becomes constant for the period of rainfall excess ( $t - t_c$), i.e., from time $t_c$ onwards. After the cessation of rain, the runoff recedes gradually to become zero at time $t_c$ from the end of the peak. In rational formula, a certain percentage of rainfall is considered as runoff. The peak discharge in the original FPS unit is given by

$$Q_p = CIA \tag{8.1}$$

or in SI units

$$Q_p = 0.278 \, CIA \tag{8.2}$$

where $Q_p$ is the peak discharge in m$^3$/sec , $C$ the runoff coefficient representing a ratio of runoff to rainfall, $A$ the catchment area in km$^2$ and $I$ the rainfall intensity in mm/h whose duration should atleast be equal to the time of concentration of the basin. The value of coefficient $C$ varies from 0.05 to 0.95. It represents the total cumulative effect of the watershed losses. The factors affecting the value of $C$ are (i) initial losses, (ii) depression storage, (iii) nature of the soil, (iv) surface slope, (v) degree of saturation, (vi) rainfall intensity, (vii) geology of the catchment and (viii) geo-hydrological characteristics of the basin. Table 8.2 gives the value of $C$ for different types of catchment.

When a watershed of area $A$ has distinct zones of $C$ then divide the area into sub zones of $A_1$, $A_2$ ,..., $A_n$, with their runoff coefficients $C_1$, $C_2$, ..., $C_n$ and a weighted $C$ is computed as $C_W = \Sigma(C_i A_i)/A$. The value of $C_w$ is to be used in equation 8.2 instead of $C$.

To find intensity of rainfall $I$ at a place, any of the following formulae can be used depending on their suitability. The general form correlating intensity - duration-return period is given as

$$I = \frac{K \cdot T^a}{(t_c + b)^n} \tag{8.3}$$

where $I$ is the intensity of rainfall (cm/h), $T$ the return period (years), $t_c$ the time of concentration of the watershed in (h), $K$, $a$, $b$ and $n$ are constants. For different places of India these constants can be evaluated and used (Table 8.3). However, some specific equations correlating the maximum intensity of rainfall with time of storm are (refer equations 8.4 through 8.11):

**Table 8.2   Values of Runoff Coefficient *C***

| Type of Area | Value of *C* |
|---|---|
| *Urban area* | |
| Lawns | |
| Sandy soil flat 2% | 0.05–0.10 |
| Sandy soil average 2.7% | 0.1–0.15 |
| Sandy soil steep < 7% | 0.15–0.2 |
| Heavy soil flat to steep | 0.13–0.35 |
| Heavy soil average | 0.18–0.22 |
| Heavy soil steep | 0.25–0.35 |
| *Business area* | 0.5–0.95 |
| *Residential area* | |
| Single house area | 0.3–0.5 |
| Flats | 0.5–0.7 |
| Suburban | 0.25–0.4 |
| Apartments | 0.50–0.7 |
| *Industrial Area* | |
| Unimproved | 0.1–0.3 |
| Light areas | 0.5–0.8 |
| Heavy areas | 0.6–0.9 |
| Rail yards | 0.2–0.4 |
| *Streets* | 0.7–0.95 |
| *Agriculture area* | |
| Flat zones | 0.1–0.5 |
| Hilly | 0.3–0.7 |

## Equations for Competation of Maximum Intensity of a Storm

| Equation | Condition of its use | Proposed by | |
|---|---|---|---|
| $I = 30.48/t^{0.5}$ | Applicable for ordinary to sever storms | C.E. Gregory | (8.4) |
| $I = 15.24/t^{0.5}$ | Applicable to winter storms | C.E. Gregory | (8.5) |
| $I = 81.25/t^{0.8}$ | Applicable to storm of maximum intensity | C.E. Gregory | (8.6) |
| $I = (348/t)^{0.5}$ | For storms expected each year | E.W. Clark | (8.7) |
| $I = (1045/t)^{0.5}$ | For storms expected once in 8 years | E.W. Clark | (8.8) |
| $I = (2090/t)^{0.5}$ | For storms expected once in 15 years | E.W. Clark | (8.9) |
| $I = 98.4/t^{0.687}$ | Storms of maximum intensity | Sherman | (8.10) |
| $I = 63.8/t^{0.687}$ | Applicable to ordinary storms | Sherman | (8.11) |

where $I$ is the intensity of rainfall (mm/h), $t$ the time (min). If there is no self recording rain gauge, then the following formula can be used to obtain the maximum intensity of the storm that is likely to occur during an interval of any one hour within the storm duration.

$$I = \frac{F}{2} + \frac{1}{t} \qquad (8.12)$$

where $I$ is the maximum intensity of rainfall in mm/h, $F$ the total rainfall of the storm in mm and $t$ the duration of storm in h. However, the term maximum is not clearly defined and is subjective. The concept of frequency can be incorporated to this.

*Limitations of Rational Formula*

1. The formula is applicable to small catchments. The watershed area can be maximum upto 50 km$^2$.
2. Duration of rainfall intensity should be more than the time of concentration of the basin.
3. It gives the peak of the hydrograph but does not provide the complete hydrograph.
4. It plots a straight line relation between $Q_p$ and $I$ with intercept zero whereas nature does not follow such a simple equation.
5. Rainfall intensity must be constant over the entire watershed during the time of concentration.
6. Coefficient $C$ is assumed to be same for all storms which means the losses are constant for all storms.

The first step in using the rational formula is to compute $t_c$ for the watershed. Once $t_c$ is known, all previous rainfall records can be checked and the rainfall durations exceeding $t_c$ are analysed to get rainfall intensities $I$ for which return period $T$ is computed as discussed in Chapter 2. Coefficient $C$ can be read from Table 8.2. Equation 8.2 is used to give the peak discharge $Q_p$. The rational formula is widely applied for sewer design in urban areas.

---

**Example 8.1:** An outlet is to be designed for a small town covering 12 km$^2$, of which road area is 30%, residential area is 50% and the rest is industrial area. The slope of the catchment is 0.005 and the maximum length of the town measured on the map is 1.6 km. From depth duration analysis for the catchment, the following informations are obtained.

| Rainfall duration (min) | 30 | 40 | 50 |
|---|---|---|---|
| Rainfall depth (mm) | 30 | 40 | 44 |

Calculate the peak discharge.

**Solution**

Time of concentration can be calculated from Kirpich equation (7.3)

$t_c = 0.000323 \, L^{0.77} \, S^{-0.385} = 0.000323 \times 1600^{0.77} \times 0.005^{-0.385} = 0.728$ h $= 43.7$ min.

Rainfall for $t_c = 43.7$ min is computed by interpolating data given in the problem.

$$\text{Depth of rainfall} = 40 + \frac{40 - 40}{10} \times 3.7 = 41.48 \text{ mm}$$

$\therefore$ $\qquad\qquad\qquad I = 41.48$ mm in 43.7 min $= 56.95$ mm/h

From Table 8.2, $C$ for road is 0.80, for residential area is 0.40 and for industrial area is 0.20.

**Table 8.3  Intensity-Duration-Return Period Coefficients of Equation (8.3)**

| Zone | Station | $K$ | $a$ | $b$ | $n$ |
|------|---------|-----|-----|-----|-----|
| (1) | (2) | (3) | (4) | (5) | (6) |
| Southern Zone | Bangalore | 6.275 | 0.1262 | 0.50 | 1.1280 |
| | Hyderabad | 5.250 | 0.1354 | 0.50 | 1.0295 |
| | Kodainand | 5.914 | 0.1711 | 0.50 | 1.0086 |
| | Madras | 6.126 | 0.1664 | 0.50 | 0.8207 |
| | Mangalore | 6.744 | 0.1395 | 0.50 | 0.9374 |
| | Tiruchirapalli | 7.135 | 0.1638 | 0.50 | 0.9624 |
| | Trivanddrum | 6.762 | 0.1536 | 0.50 | 0.8158 |
| | Visakhapatanam | 6.646 | 0.1692 | 0.50 | 0.9963 |
| | Southern Zone | 6.311 | 0.1523 | 0.50 | 0.9465 |
| Northern Zone | Agra | 4.911 | 0.1667 | 0.25 | 0.6293 |
| | Allahabad | 8.570 | 0.1692 | 0.50 | 1.0490 |
| | Amritsar | 14.41 | 0.1304 | 1.40 | 1.2963 |
| | Dehradun | 6.00 | 0.220 | 0.50 | 0.8000 |
| | Jaipur | 6.219 | 0.1026 | 0.50 | 1.1172 |
| | Jodhpur | 4.098 | 0.1677 | 0.50 | 1.0369 |
| | Lucknow | 6.0674 | 0.1813 | 0.50 | 1.0331 |
| | New Delhi | 5.208 | 0.1574 | 0.50 | 1.1072 |
| | Srinagar | 1.503 | 0.2730 | 0.25 | 1.0636 |
| | Northern zone | 5.914 | 0.1623 | 0.50 | 1.0127 |
| Central Zone | Bagratawa | 8.5704 | 0.2214 | 1.25 | 0.9331 |
| | Bhopal | 6.9296 | 0.1892 | 0.50 | 0.8767 |
| | Indore | 6.9280 | 0.1394 | 0.50 | 1.0651 |
| | Jabalpur | 11.379 | 0.1746 | 1.25 | 1.1206 |
| | Jagdalpur | 4.7065 | 0.1084 | 0.25 | 0.9902 |
| | Nagpur | 11.45 | 0.1560 | 1.25 | 1.0324 |
| | Punasa | 4.7011 | 0.2608 | 0.50 | 0.8653 |
| | Raipur | 4.6830 | 0.1389 | 0.15 | 0.9284 |
| | Thikri | 6.088 | 0.1747 | 1.00 | 0.8587 |
| | Central zone | 7.4645 | 0.1712 | 0.75 | 0.9599 |
| Western Zone | Aurangabad | 6.081 | 0.1459 | 0.50 | 1.0923 |
| | Bhuj | 3.823 | 0.1919 | 0.25 | 0.9902 |
| | Mahabaleswan | 3.483 | 0.1267 | 0.00 | 0.4853 |
| | Nandurbar | 4.254 | 0.2070 | 0.25 | 0.7704 |
| | Vengusta | 6.863 | 0.1670 | 0.75 | 0.8683 |
| | Veraval | 7.787 | 0.2087 | 0.50 | 0.8908 |
| | Western zone | 3.974 | 0.1647 | 0.15 | 0.7327 |
| Easter Zone | Agartala | 8.097 | 0.1177 | 0.50 | 0.8191 |
| | Dumdum | 5.940 | 0.1150 | 0.15 | 0.9241 |
| | Guwahati | 7.206 | 0.1557 | 0.75 | 0.9401 |
| | Gaya | 7.176 | 0.1483 | 0.50 | 0.9459 |
| | Imphal | 4.939 | 0.1340 | 0.50 | 0.9719 |
| | Jamshedpur | 6.930 | 0.1307 | 0.50 | 0.8737 |
| | Jharsuguda north | 8.596 | 0.1392 | 0.75 | 0.8740 |
| | N Lakmipur | 14.070 | 0.1256 | 1.25 | 1.0730 |
| | Sagar Island | 16.524 | 0.1402 | 1.50 | 0.9635 |
| | Shillong | 6.728 | 0.1502 | 0.75 | 0.9575 |
| | Eastern Zone | 6.933 | 0.1353 | 0.50 | 0.8801 |

$\therefore$    Composite $C = \dfrac{0.8(0.3 \times 12) + 0.4(05 \times 12) + 0.2(0.2 \times 12)}{12}$

$$= 0.24 + 0.20 + 0.04 = 0.48$$

$\therefore$    $Q_p = 0.278 \times 0.48 \times 56.95 \times 12 = 91.2 \ \text{m}^3/\text{sec}$

## 8.3.2 Empirical Equations

Various regression type equations relating catchment area, river bed-slope and return periods to the flood peak discharge are available in the literature. Such relations are purely regional in nature. The formulae hold good for the region only where the investigation was carried out and sufficient care be taken to use them elsewhere. Empirical equations are employed for structures of less importance. Following approaches are usually carried out under this category of analysis.

### 8.3.2.1 Preliminary Investigation

For preliminary investigation a rough estimate of the flood peak is made from the information of the flood marks at the river reach for all high flood events occurred earlier. Using Manning's equation, the flood discharge is computed. A factor of safety of 1.5 is normally applied to get the necessary design flood peak at the site. Care should be taken to collect data from all the previous flood marks. The river slope $S$ is computed from the knowledge of two flood marks at distance $L$ apart along the river. The area of river cross-section $A$ can be surveyed corresponding to the river stage of the flood marks. A suitable rugosity coefficient $n$ is selected depending on the channel roughness. Such a method is used only for preliminary investigation of flood. The Manning's equation

$$Q_p = \frac{1}{n}A \cdot \left(\frac{A}{P}\right)^{2/3} S^{1/2} \tag{8.13}$$

can be used to calculate the flood peak $(Q_p)$ in $\text{m}^3/\text{sec}$ where, $P$ is the wetted perimeter of the river in m.

### 8.3.2.2 Flood Peak Formulae

Most of the empirical equations are derived by correlating the catchment area with the observed flood peaks in the following ways.

**(1) Dicken's Formula (1865):** Dicken proposed the following form of the catchment flood peak relation for the regions of Central and North India catchments.

$$Q_p = CA^{3/4} \tag{8.14}$$

where $A$ is the catchment area in $\text{km}^2$, $Q_p$ the maximum flood discharge ($\text{m}^3/\text{sec}$) and $C$ a coefficient varying from 2.8 to 28 depending on the location of the place. The original formula has been tested by various hydrologists at other locations of India and the coefficient $C$ up to 40 for the Western Ghat hills is recommended. The range of $C$ for various regions of India are given in Table 8.4.

Selection of correct value of $C$ depends on the experience of a hydrologist, his knowledge of local catchment and rainfall characteristics. For mountainous

**Table 8.4   Dicken's Coefficient for Various Regions of India**

| Sl. No. | Area description | Value of $C$ |
|:---:|:---|:---:|
| (1) | (2) | (3) |
| 1. | For flat, sandy or cultivated plains of north India | 2.8 to 6.0 |
| 2. | Undulating area with hard soil for north Indian hills | 11.0 to 14.0 |
| 3. | Undulating country for central India region | 14.0 to 28.0 |
| 4. | Catchment covered with precipitous hills as of Orissa, Andhra | 22.0 to 28.0 |
| 5. | Western ghat regions of India | 20.0 to 40.0 |

regions, the values of $C$ in the range of 14–28 are usually selected. For the estimation of flood peaks for small structures, Dicken's equation with the coefficient $C$ varying between 25 and 28 is widely used for Orissa catchments.

**(2) Ryve's Formula (1884):** Ryve modified Dicken's equation to suite the catchments of Tamil Nadu. The proposed relation takes the following form.

$$Q_p = CA^{2/3} \qquad (8.15)$$

where $Q_p$ is peak discharge in m$^3$/sec and $A$ is the area in sq.km. Ryve suggested the following values:

   (i)   $C = 6.8$ for areas within 75 km from the coast.
   (ii)  $C = 8.5$ from 75 to 175 km from the coast.
   (iii) $C = 10.2$ for a limited areas near hills.

However, the coefficient has been modified to suit parts of Karnataka and Kerala and its value exceeding 40 have been reported for certain river projects built earlier.

**(3) Inglis Formula (1940):** Inglis studied a good number of catchments in the earlier Bombay state, parts of west Deccan plateau and suggested the following relation between catchment area and peak discharge

$$Q_p = \frac{124\,A}{\sqrt{(A + 10.4)}} \qquad (8.16)$$

Other investigators suggested the following formulae correlating the peak discharge $Q_p$ in m$^3$/sec and watershed area $A$ in sq. km.

**(4) Ali Nawaz Jung Bahadur Formula**

$$Q_p = C\,(0.3906A)^{0.925-(1/14)\,\log 0.3906A} \qquad (8.17)$$

where $C$ varies from 49 to 60.

**(5) Coutagne Formula:** This is used in France

$$Q_p = 150(A)^{0.5} \qquad (8.18)$$

and is applicable for area between 400 and 3000 km$^2$.

**(6) Meyer Formula (1926):** The formula is extensively used in USA

$$Q_p = 175 \ (A)^{1/2} \qquad (8.19)$$

For use of this equation, the area should be higher than 10 km$^2$.

**(7) Fanning (USA) Formula**

$$Q_p = 2.64 \ (A)^{0.8} \qquad (8.20)$$

**(8) Creager (USA) Formula**

$$Q_p = 130(0.386A)^{0.894(0.386A)^{-0.08}} \qquad (8.21)$$

**(9) Boston Society of Civil Engineers Formula**

$$Q_p = C \ (A)^{1/2} \qquad (8.22)$$

where $C = 3.5$ for rainfall $< 500$ mm, 8.4 for rainfall between 500 and 750 mm and 35 for rainfall between 750 and 1000 mm or more.

**(10) Rhind's Formula:**

$$Q_p = 0.098 \ \text{CSR} \ (0.386A)^P \qquad (8.23)$$

where $Q_p$ is flood peak in m$^3$/sec, $S$ the average slope above the point of interest taken for a length of 5 km along the river, $R$ the highest annual rainfall recorded in the area in cm, $C$ and $P$ the coefficients.

### 8.3.2.3 *Empirical Relations with Return Periods*
Some important empirical equations correlating the peak discharge with the return periods are given below.

**(1) Horton's Equation**

$$q_p = 71.2 \ ( \ T)^{1/4} \ (A)^{-1/2} \qquad (8.24)$$

where $q_p$ is the flood peak in m$^3$/sec per sq. km, $A$ a drainage area in km$^2$, $T$ the return period in years.

**(2) US Geological Survey (1955):** The mean annual average flood ($Q_{2.33}$ m$^3$/s) with a return period of 2.33 years can be computed from the relation

$$Q_{2.33} = 0.0147 \ CA^{0.7} \qquad (8.25)$$

where $C$ is a constant whose value varies widely from 1 to 100 and $A$ is the catchment area in sq. km. The equation follows Gumbel's extreme value distribution which has the property of $T = 2.33$ years for average value of the series when data length $N$ is large.

**(3) Fuller's Equation (1914)**

$$Q_{PT} = Q_{av} \ (1 + 0.8 \log T)$$

or $\qquad Q_{PT} = C\,A^{0.8}\,(1 + 0.3474\,\ln T)$ $\qquad\qquad$ (8.26)

and $\qquad Q_{max} = Q_{PT}\,(1 + 2.66A^{-0.3})$

where $Q_{pT}$ is the maximum 24 h flood in $m^3$/sec with return period $T$ years, $A$ the catchment area in $km^2$, $C$ a constant varying between 0.03 and 2.8, $Q_{av}$ the yearly average flood, $Q_{max}$ the maximum instantaneous discharge.

**(4) Pettis Formula**

$$Q_p = C(P \cdot B)^{5/4} \qquad\qquad (8.27)$$

where $Q_p$ is the peak flood discharge of 100 year return period in $m^3$/sec, $P$ the one day rainfall of 100 year return period in cm, $B$ the average width of the basin and $C$ a coefficient (0.195 for desert and 1.51 for humid regions).

**(5) UP Irrigation Research Institute:** Dicken's coefficient $C$ can be computed for a given return period $T$ as

$$C = 0.442\,\ln(0.6T)\,\ln\!\left(\frac{1185}{P}\right) + 4 \qquad\qquad (8.28)$$

where the percentage factor $p = \left\{\dfrac{(a + 6)}{(A + a)}\right\} \times 100$, $a$ is snow covered area in $km^2$ of the basin and $A$ the remaining area of the basin.

### 8.3.2.4   SCS Curve Number Method

This method also known as *hydrologic soil cover complex number method*, was developed by Ogrosky and Mockus (1957) for determining peak rate of runoff from small watersheds. A runoff *curve number* (CN) is developed through field studies by measuring runoff from different soils at various locations. The antecedent moisture condition and the physical characteristics of the watershed are correlated to give hydrologic soil groups. Soil of any watershed can be classified into the following four hydrologic groups.

**Group A:**  A low runoff potential group with very high infiltration rate. From such soils, even under wet condition, the runoff expectations are low. Infiltration rate is 8-12 mm/h. Transmission rate is very high for such soils.

**Group B:**  Moderately low runoff potential soil groups with moderate rate of water transmission. Soil textures vary from fine to moderately course. Final infiltration rate is 4-8 mm/h.

**Group C:**  Moderately high runoff potential with low infiltration rates with moderately good to well drained soils. Texture is moderately fine to moderately coarse with slow rate of water transmission. Final rate of infiltration is 1-4 mm/h.

**Group D:**  Very slow infiltration rate when thoroughly wet. Clay soils form such groups. The final infiltration rate for such soils vary from 0 to 1 mm/h. Such soils have very low rate of transmission.

The above classification of soils based on effective depth, average clay content, infiltration and probability. Fig. 8.1 shows the hydrologic soil group. Table 8.5 gives runoff curve numbers for hydrologic soil cover complexes (antecedent moisture condition, AMC-II, $I_a = 0.25$) and Table 8.6 gives conversion from type AMC-II to AMC-I or AMC-III.

AMC-I    =    Lowest runoff potential. The watershed soils are dry enough for satisfactory cultivation to take place.

AMC-II   =    Average condition.

AMC-III  =    Highest runoff potential. The watershed is practically saturated from antecedent rain. The AMC group is determined using five day antecedent rainfall.

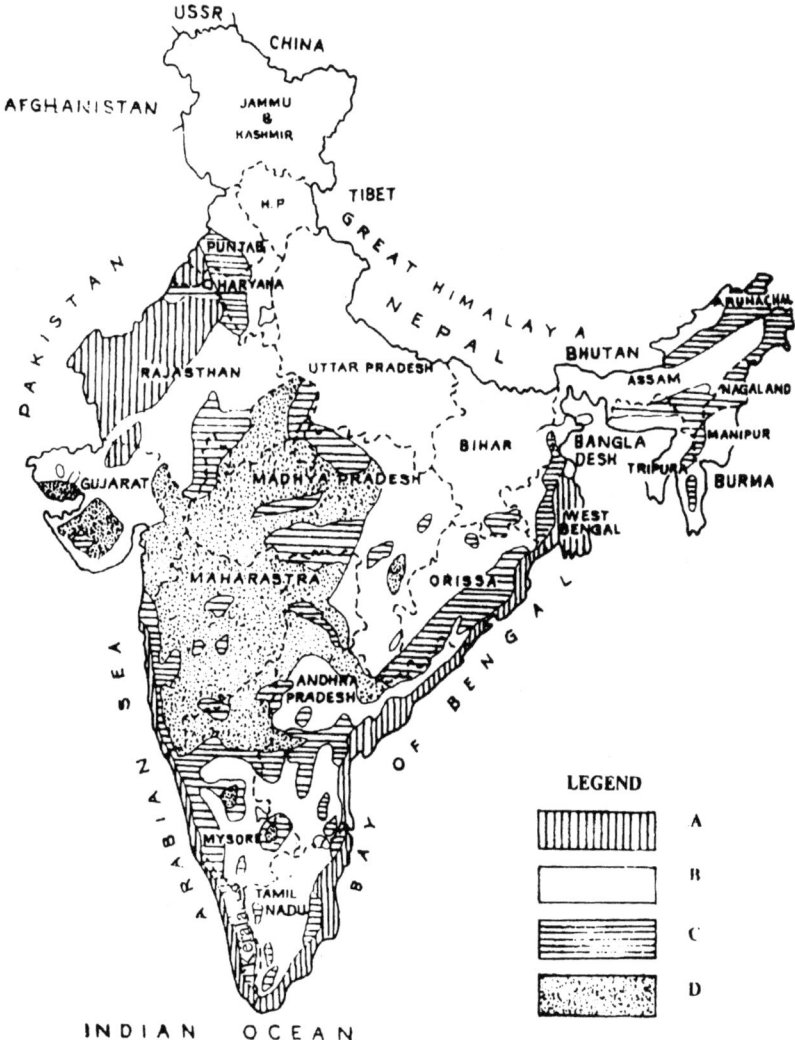

Fig. 8.1   The hydrologic soil groups of India.

**Table 8.5    Runoff Curve Numbers for Hydrologic Soil Cover Complexes**
(after Soil Conservation Service, 1969; for watershed condition II, $I_a = 0.2\ S$)

| Land use | Cover treatment practice | Hydrologic condition | Hydrologic soil group | | | |
|---|---|---|---|---|---|---|
| | | | A | B | C | D |
| Fallow row crops | Straight row | – | 77 | 86 | 91 | 94 |
| | Straight row | poor | 72 | 81 | 88 | 91 |
| | Straight row | good | 67 | 78 | 85 | 89 |
| | Contoured | poor | 70 | 79 | 84 | 88 |
| | Contoured | good | 65 | 75 | 82 | 86 |
| | Contoured | poor | 66 | 74 | 80 | 82 |
| | Contoured | good | 62 | 71 | 78 | 81 |
| Small grain | Straight row | – | 65 | 76 | 84 | 88 |
| | Straight row | poor | 63 | 78 | 83 | 87 |
| | Straight row | good | 63 | 74 | 82 | 85 |
| | Contoured | poor | 61 | 83 | 81 | 84 |
| Close seeded | Straight row | – | 66 | 77 | 85 | 89 |
| legumes or rotation | Straight row | good | 58 | 72 | 81 | 85 |
| meadow | Contoured | good | 64 | 75 | 83 | 85 |
| | Contoured | poor | 55 | 69 | 78 | 83 |
| | Contoured | good | 63 | 73 | 80 | 83 |
| | Contoured | good | 51 | 67 | 76 | 80 |
| Pasture | Straight row | poor | 68 | 79 | 86 | 89 |
| or range | Straight row | fair | 49 | 69 | 79 | 84 |
| | Contoured | good | 39 | 61 | 79 | 80 |
| | Contoured | good | 42 | 67 | 81 | 88 |
| | Contoured | poor | 25 | 59 | 75 | 83 |
| | Contoured | good | 6 | 35 | 70 | 79 |
| Meadow (permanent) | | good | 30 | 58 | 71 | 78 |
| Wood lands (farm wood lots) | | poor | 45 | 66 | 77 | 83 |
| | | fair | 36 | 60 | 73 | 79 |
| | | good | 25 | 53 | 70 | 77 |
| Farm steads | | – | 59 | 74 | 82 | 86 |
| Road (dirt) | | – | 72 | 82 | 87 | 87 |
| Roads (Hard surface) | | – | 74 | 84 | 90 | 92 |

A value of CN for AMC-II condition is selected from Table 8.5 depending on the land use pattern. If the area consists of patches of land used then a composite curve number (CN) for the watershed can be obtained by weighing them in proportion of the area. For example for a watershed of 100 km², if soil of 70 sq. km (70% of area) has CN of 60 and 30 sq. km (30%) has CN 80, the weighted CN for the catchment is $0.7 \times 60 + 0.3 \times 80 = 66$. The value of CN is lower for soils with high infiltration, than for soils with low infiltration. A poor condition refers to heavily grazed pastures with almost no vegetation. A good condition refers to pastures, lightly grazed with 75% or more land covered with plants.

In India, most of rain gauges are non-recording type. Rainfall data obtained from such non-recording gauges have been used to establish rainfall runoff relationships. The following relation is normally used for small watersheds:

**Table 8.6 Antecedent Rainfall Conditions and Curve Number (for $I_a = 0.2S$)**

($I_a$ = initial abstraction consisting of interception, depression storage and infiltration)

| Curve number for condition II | Multiply the curve number by the factor to arrive at | |
|:---:|:---:|:---:|
| | Condition I | Condition III |
| 5 | 0.40 | 2.60 |
| 10 | 0.40 | 2.22 |
| 15 | 0.40 | 2.00 |
| 20 | 0.45 | 1.85 |
| 25 | 0.48 | 1.72 |
| 30 | 0.50 | 1.72 |
| 35 | 0.51 | 1.57 |
| 40 | 0.55 | 1.50 |
| 45 | 0.58 | 1.50 |
| 50 | 0.62 | 1.40 |
| 55 | 0.64 | 1.35 |
| 60 | 0.67 | 1.30 |
| 65 | 0.69 | 1.26 |
| 70 | 0.73 | 1.21 |
| 75 | 0.76 | 1.17 |
| 80 | 0.79 | 1.14 |
| 85 | 0.82 | 1.10 |
| 90 | 0.87 | 1.07 |
| 95 | 0.92 | 1.03 |
| 100 | 1.00 | 1.00 |

$$\frac{\text{Actual retention of rainfall}}{\text{Potential maximum retention}} = \frac{\text{Direct runoff}}{\text{Rainfall} - \text{Initial abstraction}}$$

or

$$\frac{P - I_a - Q_d}{S} = \frac{Q_d}{P - I_a}$$

or

$$Q_d = \frac{(P - I_a)^2}{(P - I_a + S)} = \frac{(P - 0.2S)^2}{(P + 0.8S)} \tag{8.29}$$

where $Q_d$ is the runoff depth in cm over the basin, $P$ is the mean rainfall in cm over the basin, $S$ the potential maximum retention in cm, $I_a$ the initial losses consisting of interception, depression storage and infiltration. For black soil, $I_a$ is taken as 0.1 $S$ for AMC II and III type, $I_a$ is taken as 0.3$S$ for AMC-I type. For all other types, $I_a$ is 0.2 $S$. In general the value of $I_a$ is take as 0.2$S$. For all soil regions of India, except black soil region of AMC II and III, $Q_d$ is calculated as

$$Q_d = \frac{(P - 0.3S)^2}{(P + 0.7S)} \tag{8.30a}$$

For black soil of AMC II and III

$$Q_d = \frac{(P - 0.1S)^2}{(P + 0.9S)} \tag{8.30b}$$

Now, $S$ (cm) is related to curve number (CN) as follows:

$$CN = \frac{2540}{25.4 + S} \tag{8.31}$$

The peak discharge in $m^3$/sec can be calculated from the following relation

$$Q_p (m^3/sec) = \frac{0.0208 \times A \times Q_d}{t_p} \tag{8.32}$$

where for small watersheds time to peak $t_p$ in h is obtained as

$$t_p = 0.6t_c + t_c^{1/2} \tag{8.33}$$

where $A$ is the area of the watershed (Ha), $Q_d$ the depth of runoff (cm) is obtained from equation (8.29) or (8.30) depending on the type of soil and $Q_P$ is the peak discharge ($m^3$/sec). Graphical solution of the runoff depth for small watershed from the knowledge of the curve number and rainfall is given in Fig. 8.2.

$Q = (P - 0.1S)^2/(P + 0.9S)$
$P = 1.0$ to $200.0$ mm
$Q = 0.0$ to $200.0$ mm

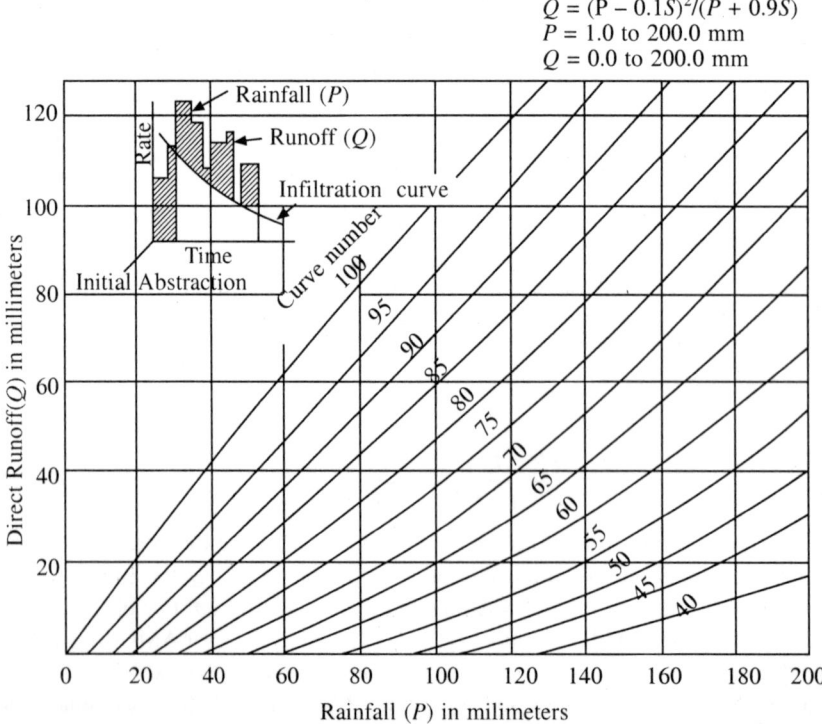

**Fig. 8.2   Direct Rainfall Depth as Function of Rainfall and Curve Number**

**Example 8.2:** A cross drainage structure is to be constructed at a place in Orissa 65 km away from the Bay of Bengal. Calculate peak discharge for the catchment area of 15 km$^2$. Use any four empirical relations and tabulate their results. Calculate the coefficients for Ryve's and Ali Nawaz Jung Bahadur formula if they are to be used to give flood peak of same magnitude as Dicken's formula. Also calculate the flood peaks for return period of 50 years if the flood is considered as the result of 24 h rainfall.

**Solution**

(i) Dicken's formula $Q_P = CA^{3/4}$

For Orissa, $C = 28$ (maximum for Orissa region)

$$Q_P = 28 \times (15)^{0.75} = 213.4 \text{ m}^3/\text{sec}$$

(ii) Ryve's formula $= CA^{2/3}$

For use of Ryve's formula in Orissa, the value of $C$ should be properly selected. To equate the flood peak it with Dicken's formula we have to choose $C$ as 35

$$Q_P = 35 \times (15)^{2/3} = 213.4 \text{ m}^3/\text{sec}$$

(iii) Inglis formula,

$$Q_p = \frac{124\,A}{\sqrt{(A + 10.4)}} = \frac{124 \times 15}{\sqrt{(15 + 10.4)}} = 369.1 \text{m}^3/\text{sec}$$

(iv) Ali-Nawaz Jung Bahadur formula.

$$Q_P = C\,(0.39006\,A)^{0.925\,-\,1/14\,\log\,(0.3906A)}$$

$$= 49 \times (5.859)^{0.925-0.0548} = 49 \times (5.859)^{0.8701} = 49 \times 4.657 = 228.2 \text{ m}^3/\text{sec}.$$

To equate it with Dicken's formula, we have to choose $C$ as 45.9.

**Formula with return periods**

(i) Fuller's formula $Q_p = C\,A^{0.8}\,[1+ 0.3474 \ln T]$

$$=1.8(15)^{0.8}\,[1+ 0.3474 \ln 50] = 37 \text{ m}^3/\text{sec}$$

(ii) Horton's Formula $Q_p= A71.2\,T^{1/4}\,/\,A^{0.5} = 71.2 \times 50^{1/4} \times 15^{0.5} = 733.3 \text{ m}^3/\text{sec}$

As can be seen from Table 8.7, Inglis formula gives higher peak than the other methods. Horton's formula taking a 50 year return period flood peak is much higher than the Fuller's 50 year return period flood peak. Fuller's method gives low estimate of $Q_p$.

**Table 8.7   Comparison of Peak Discharge for Example 8.2**

| Sl.No. | Formula Name | $Q_P$ (m$^3$/sec) | Value of $C$ | Remark |
|--------|-------------|-------------------|--------------|--------|
| (1) | (2) | (3) | (4) | (5) |
| 1. | Dicken's formula | 213.4 | 28 | $C_1$-range is 22–28 |
| 2. | Ryve's formula | 213.4 | 35 | $C_R$ range is upto 40 |
| 3. | Inglis formula | 369.0 | – | No coefficient. This formula holds good for the study region |
| 4. | Ali Nawaz Jung Bahadur formula | 228.2 | 49 | $C_A$ is 49 for South India |
| 5. | Horton's formula | 733.3 | – | Return period of the flood $T = 50$ years. |
| 6. | Fuller's | 37.0 | – | Return period of the flood $T = 50$ years |

**Example 8.3:** Estimate the peak runoff for a 6-h, 10-year recurrence interval flood taking antecedent moisture condition-III for a watershed consisting of 400 hectare of straight row good crop and the remaining 200 ha. of good woodland. The soil is hydrologic soil group C. Take $t_c$ for the watershed as 50 min. The 6 h 10-year recurrence interval rainfall recorded from IMD map is found to be 20.0 cm.

**Solution**
From Table 8.5, the curve number for straight row good crop is 85 and for the good woodland, the curve number is 70. Therefore the composite curve number for the watershed is found as

$$CN = \frac{83 \times 400 + 70 \times 200}{400 + 200} = 80$$

The CN from II is converted to AMC III obtained by multiplying a factor 1.14 as given in Table 8.6.

∴                    CN for AMC type III = $80 \times 1.14 = 91.2$

Now $S$ can be computed from equation (8.31) as

$$CN = \frac{2540}{25.4 + S} \text{ or } 91.2 = \frac{2540}{25.4 + S}$$

or        $S = 2.45$ cm. Rainfall $P = 20.0$ cm

Runoff    $Q_p = \frac{(P - I_a)^2}{(P - I_a) + S} = \frac{(20 - 0.2S)^2}{20 - 0.2S + S} + \frac{(20 - 0.2 \times 2.45)^2}{20 + 0.8 \times 2.45} = 17.3.$

Time to peak for such small catchment is given by $t_p = 0.6\, t_c + \sqrt{t_c}$

or                    $t_p = 0.6 \times 0.833 + \sqrt{0.833} = 1.41$ h.

Peak discharge, $Q_p = \frac{(0.0208 \times A \times Q_d)}{t_p} = \frac{(0.0208 \times 600 \times 17.3)}{1.41} = 153.1 \text{ m}^3/\text{sec}.$

**Example 8.4:** Obtain the runoff volume from a watershed receiving 12 cm of rainfall. Assume AMC-II condition. Land use pattern are given as follows:

        Residential area 30%, soil group C
        Meadow, good condition 40% area, soil group D
and    Open space good condition area 30% soil group D

**Solution**
From Table 8.5, the curve number $CN$ for the groups of soils are 80, 78 and 80. The weighted or composite $CN = 0.3 \times 80 + 0.4 \times 78 + 0.3 \times 80 = 79.2$

Value of $S$ is found out as $CN = \frac{2540}{(25.4 + S)}$ or $79.9 = \frac{2540}{(25.4 + S)}$ or $S = 6.67$ cm

$$Q_d = \frac{(12 - 0.2 \times 6.67)^2}{12 + 0.8 \times 6.67} = \frac{(10.67)^2}{17.34} = 6.57 \text{ cm}$$

Runoff depth from the storm is 6.57 cm over the catchment.

### 8.3.3   Envelope Curves
Attempts were made to correlate the peak discharge with the catchment area of a region under the assumption of fairly homogeneity in the hydrometeorological

and basin characteristics. Investigators of various countries tried to develop and put forth curves correlating catchment area with peak discharges for the regions under their study. Such curves are known as *envelope curves* in which a log -log paper is used to plot various drainage area (km$^2$) as abscissa against the corresponding discharges (m$^3$/sec) as ordinate. A smooth curve enveloping all data points on the log-log plot should necessarily be a straight line. The following envelope curves are normally referred.

### 8.3.3.1  Kanwar Sain and Karpov Curves

Kanwar Sain and Karpov (1967) prepared two such enveloping curves, one for northern and central parts and the other for the southern part of India (Fig. 8.3). The curves were drawn covering drainage areas from 768 to 5,12,000 sq.km. (300 to 2,00,000 sq. miles). The curve was used to arrive at the maximum flood discharge for Nagarjun Sagar project in Andhra Pradesh.

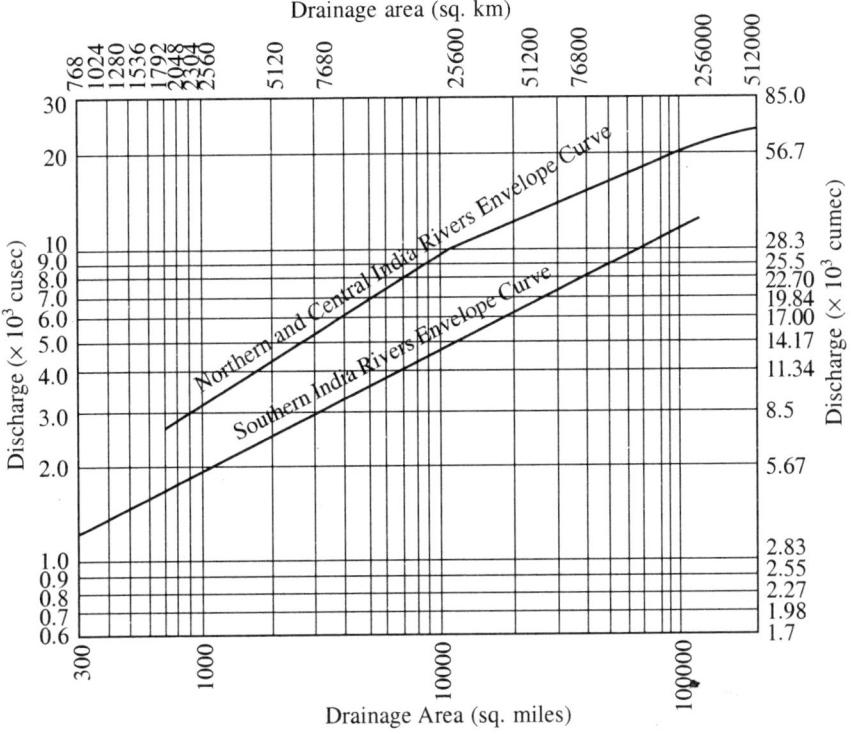

**Fig. 8.3  Envelope Curves for Indian Rivers to Determine the Peak Flow Rates (After Kanwar Sain and Karpov 1967)**

### 8.3.3.2  Baird's Envelope Curve

The following enveloping regression equation proposed by Baird (1951) for computation of maximum flood discharge for any part of the world is still in use.

$$Q_{mf} = 3025 \, A \, (278 + A)^{-0.78} \tag{8.34}$$

where $Q_{mf}$ is the maximum flood discharge (m³/sec) and $A$ the catchment area (km²).

### 8.3.3.3    Creager's Curve

Creager (Table 8.8) proposed an envelope curve which gives higher value for small catchments and lower value for large catchments.

**Table 8.8    Creager's Enveloping Curve Points**

| Catchment area (sq. km) | Flood peak/sq. km (m³/sec) |
|:---:|:---:|
| 10 | 20.0 |
| 50 | 10.0 |
| 100 | 7.50 |
| 200 | 5.10 |
| 400 | 3.80 |
| 600 | 3.00 |
| 1000 | 1.30 |
| 2000 | 1.50 |
| 3000 | 1.00 |
| 5000 | 0.76 |
| 8000 | 0.50 |
| 10000 | 0.42 |
| 20000 | 0.225 |

### 8.3.3.4    Meyer's Modified Envelope Curve

Meyer proposed the following modified envelope curve

$$Q_P = 5682.7 \ PA^{1/2} \tag{8.35}$$

where $Q_P$ is the peak discharge (m³/sec), $A$ the area of catchment (km²) and $P$ the numerical percentage depending on the basin characteristics.

## 8.4    FLOOD ESTIMATION FOR GAUGED CATCHMENT

Computing the peak rate of runoff from a watershed introduces an element of probability. This is due to the scope of predicting the storm patterns and the factors affecting runoff from catchment rainfall. The antecedent moisture condition of the catchment, initial losses, catchment characteristics and rainfall characteristics affect the runoff process each of which depend on number of other characteristics. The system is complex and a better way to solve the problem is when the basin under question is gauged for its runoff data. Basing on the availability of data the two approaches for predicting the peak flow from a given gauged watershed are:

(i) Approach-I    :  When the peak river discharges are available.
(ii) Approach-II  :  When complete runoff hydrograph along with the corresponding rainfall data and the maximum probable precipitation over the watershed are available.

Approach (i) is solved by the use of flood-frequency analysis and (ii) by unit hydrograph approach.

### 8.4.1 Flood Frequency Analysis

Flood frequency analysis considers the annual peak flows at a site for all the years. The method of analysis and predicting flood from the data of runoff peaks is called flood frequency analysis. It gives only the magnitude of flood peak of desired recurrence interval or return period, but doesn't provide information about the complete hydrograph or the flood volume. Predication of flood peaks from the flood data of recorded maximum series are reliable when analysis is carried out for return periods of less than the data length. However, when data is to be extrapolated, for example when flood peaks of 1000 or 10000 years are required to be predicted from an annual maximum services of say 30–40 years, then the prediction should be carried out with caution as the sample data may not be true representative of the population. There may be long term trend or a cycle associated with the system. For such predictions, confidence bends or limits are to be estimated at 95% or other acceptable percentages, depending on the precision requirement.

The method of predicting an event of return period of $T$ years is discussed in Chapter 2. The application of the method to predict the rainfall for various recurrence intervals are discussed in Chapter 3. Some important methods used widely for flood frequency analysis are discussed in Example 8.5. It may be mentioned here that the missing data if any, may not be filled in before fitting any distribution as the series need not be a continuous one to fit a theoretical distribution.

---

**Example 8.5:** The annual flood peaks at a gauging site from 1970–90 are given below.
(i) Determine flood magnitudes for 100 and 1000 years by various distributions.
(ii) What is the recurrence interval of flood magnitude of 1000 m³/sec ?

| Year 1970 | 71 | 72 | 73 | 74 | 75 | 76 | 77 | 78 | 79 | 80 | 81 | 82 |
|---|---|---|---|---|---|---|---|---|---|---|---|---|
| | 83 | 84 | 85 | 86 | 87 | 88 | 89 | 90 | | | | |

| Flood (m³/sec) | 1065 | 645 | 1005 | 1350 | 860 | 150 | 2260 | 650 | 2840 | 990 | 870 | 910 |
|---|---|---|---|---|---|---|---|---|---|---|---|---|
| | 750 | 930 | 750 | 1070 | 830 | 1095 | 384 | 2230 | 3210 | | | |

**Solution**

Method of frequency factors are used to determine flood peaks of return periods of 100 and 1000 year. The statistical parameters are calculated in Table 8.9.

From Table 8.9 we get

Standard deviation $\qquad \sigma_{n-1}^2 = \dfrac{\Sigma(x - x_{av})^2}{N-1} = \dfrac{12361221}{(21-1)} = 786.2$ or $\sigma_{n-1} = 786.2$

Coefficient of variation $\quad C_v = \dfrac{\sigma_{n-1}}{1183} = \dfrac{786.2}{1183} = 0.665$

Coefficient of skew $\qquad C_s = \dfrac{\Sigma(X - X_{av})^3 \, N}{\sigma^3(N-1)(N-2)} = \dfrac{1.3 \times 10^{10} \times 21}{(786.2)^3 \times 20 \times 19} = 1.478$

Chow's general equation for flood frequency analysis is given as

$$X_T = X_{av} + K\sigma$$

**Table 8.9  Statistical Parameters for the Data of Example 8.5**

| Year | Annual Runoff $x_t$ | $x - x_{av} = \Delta x$ | $\Delta x^2$ | $\Delta x^3$ | $\log x_i = y_i$ | col. (6) $-2.9876$ | col. (7)* col(7) | col. (7)³ |
|------|------|------|------|------|------|------|------|------|
| (1) | (2) | (3) | (4) | (5) | (6) | (7) | (8) | (9) |
| 1970 | 1065 | −118 | 13924 | −1643032 | 3.02735 | 0.03972 | 0.00158 | 0.00006 |
| 1971 | 645 | −5382 | 289444 | −1.6E08 | 2.809559 | −0.17806 | 0.03171 | −0.00564 |
| 1972 | 1005 | −178 | 31684 | −5639752 | 3.002166 | 0.01454 | 0.00021 | 0.000003 |
| 1973 | 1350 | 167 | 27889 | 4657463 | 3.130333 | 0.142707 | 0.02036 | 0.002906 |
| 1974 | 860 | −323 | 104329 | −3.4E07 | 2.934498 | −0.05312 | 0.00282 | −0.00014 |
| 1975 | 150 | −1033 | 1067089 | −1.1E09 | 2.17609 | −0.81153 | 0.65858 | −0.53446 |
| 1976 | 2260 | 1077 | 1159929 | 1.2E09 | 3.354108 | 0.36648 | 0.13431 | 0.049222 |
| 1977 | 650 | −533 | 284089 | −1.5E08 | 2.81213 | −0.17471 | 0.03052 | −0.00533 |
| 1978 | 2840 | 1657 | 2745649 | 4.5E09 | 3.453318 | 0.45692 | 0.21687 | 0.10099 |
| 1979 | 990 | −193 | 37249 | −7189057 | 2.995635 | 0.00801 | 0.00006 | 0.0000 |
| 1980 | 870 | −313 | 97969 | −3.1E07 | 2.93951 | −0.04810 | 0.00231 | −0.00011 |
| 1981 | 910 | −273 | 74529 | −2.0E07 | 2.95904 | −0.02858 | 0.00081 | −0.00002 |
| 1982 | 750 | −433 | 187489 | −8.1E07 | 2.87506 | −0.11256 | 0.02167 | −0.00142 |
| 1983 | 930 | −253 | 64009 | −1.6E07 | 2.96848 | −0.01914 | 0.00037 | −0.0000 |
| 1984 | 750 | −433 | 187489 | −8.1E07 | 2.87506 | −0.11256 | 0.01267 | −0.00142 |
| 1985 | 1070 | −113 | 12769 | −1442897 | 3.029383 | 0.041757 | 0.00174 | 0.00007 |
| 1986 | 830 | −353 | 124609 | −4.4E07 | 2.919078 | −0.06854 | 0.00469 | −0.0003 |
| 1987 | 1095 | −88 | 7744 | −681472 | 3.039414 | 0.051788 | 0.00268 | 0.00014 |
| 1988 | 384 | −799 | 638401 | −5.1E08 | 2.58433 | −0.40329 | 0.16264 | −0.0656 |
| 1989 | 2230 | 1047 | 1096209 | 1.1E09 | 3.348304 | 0.360678 | 0.13009 | 0.04692 |
| 1990 | 3210 | 2027 | 4108729 | 8.3E09 | 3.506505 | 0.518879 | 0.26923 | 0.13970 |
| Sum = | 24844 | | 12361221 | 1.3E10 | 62.74015 | 0.000011 | 1.696976 | −0.27448 |
| Mean = | 1183 | | | | 2.9876 | | | |

where $K$ (same as $k$ in section 2.4) is the frequency factor, $T$ the recurrence interval in years, $X_T$ the flood magnitude of return period $T$ years, $X_{av}$ the mean of the maximum series, $\sigma$ is the standard deviation. Using the derived statistical parameters, the values of flood peaks for 100 and 1000 year return periods for various methods are discussed below.

**(i) Gumbel's Method**

(a) From the frequency factor table for Gumbel distribution (Table 2.7), the frequency factors for return periods of 100 and 1000 years for the sample length of 21 are read as

$$K_{100} = 3.815 \text{ and } K_{1000} = 5.96$$

$$X_{100} = 1183 + 3.815 \times 786.2 = 4182 \text{ m}^3/\text{sec}$$

$$X_{1000} = 1183 + 5.960 \times 786.2 = 5869 \text{ m}^3/\text{sec}$$

(b) Using the relation of Gumbel modified by R.V. Powel (1943)

$$K_{100} = - \sqrt{6}/\pi[0.5772 + \ln \cdot \ln(T/T-1)]$$

$$= - \sqrt{6}/\pi \, [0.5772 + \ln \cdot \ln (100/99)]$$

$$= - \sqrt{6}/\pi \, (-)4.023 = 3.137$$

$$K_{1000} = - \sqrt{6}/\pi[0.5772 + \ln.\ln(1000/999)]$$

$$= - \sqrt{6}/\pi \, (-6.33) = 4.935$$

The values of frequency factors using Gumbel-Powel equation is less than the frequency factors given by Gumbel's table because of the data length. Powel method is used when data length is large (more than 100 years). Using Gumbel-Powel relation, 100 and 1000 year return period floods are computed below.

$$X_{100} = 1183 + 3.137 \times 786.2 = 3649 \text{ m}^3/\text{sec}$$

$$X_{1000} = 1163 + 4.935 \times 786.2 = 5063 \text{ m}^3/\text{sec}$$

*Confidence limits*
Confidence bands are drawn using equations (2.53) through (2.56).

For flood of return period of 100 years $K_{100} = 3.815$

Probable error is given by $S_e = (a \, \sigma_{n-1})/\sqrt{N}$

where $a = (1+1.3K +1.1K^2)^{1/2} = (1+ 1.3 \times 3.815 + 1.1 \times 3.815^2)^{1/2}$

$$= (1 + 4.9595 + 16.01)^{1/2} = 4.687$$

$S_e = (4.687 \times 786.2)/\sqrt{21} = 804.1 \text{ m}^3/\text{sec}$

For plotting the confidence limits, the upper and lower bound points at 95% level are obtained from the relation

$$X_{UL} = X_T \pm 1.96 \, S_e$$

From the equation, we get

upper bound of $X_{100} = 4182 + 1.96 \times 804.1 = 5758 \text{ m}^3/\text{sec}$

and    lower bound of $X_{100} = 4182 - 1.96 \times 804.1 = 2606 \text{ m}^3/\text{sec}$

We can say with 95% confidence that the estimated discharge of $X_{100}$ of 4182 m³/sec lies between the ranges 5758 and 2606 m³/sec. In other words we can say that the probability of $X_{100}$ of 4182 m³/sec lies between 5758 to 2606 m³/sec for 95% of times. Similarly for the flood of 1000 year return period, the frequency factor is $K_{1000} = 5.96$

The departure coefficient $a = (1 + 1.3 \times 5.96 + 1.1 \times 5.96^2)^{1/2} = 6.915$

Probable error $\qquad S_e = (6.915 \times 786.2)/\sqrt{21} = 1187.2$

Upper bound of $\qquad X_{100} = 5869 + 1.96 \times 1187.2 = 8196$ m³/sec

lower bound $\qquad X_{1000} = 5869 - 1.96 \times 1187.2 = 3542$ m³/sec

The 1000 year return period flood of 5869 lies between 8196 and 3542 m³/sec for 95% of times.

### (ii) Pearson Type-III
The sample has the coefficient of skewness $C_s$ as 1.478. The frequency factors for 100 and 1000 year return periods with $C_s$ as 1.478 is obtained from Table 2.8A.

$\qquad T = 100, K_{100} = 3.32 ; \qquad X_{100} = 1183 + 3.32 \times 786.2 = 3793$ m³/sec

$\qquad T = 1000, K_{1000} = 5.26; \qquad X_{1000} = 1183 + 5.26 \times 786.2 = 5318$ m³/sec

As per US practice, $C_s$ may or may not be multiplied by the factor $(1 + 8.5/N)$. In the present problem the factor is not multiplied to the values of $C_s$.

### (iii) Log-Pearson Type III
For Log-Pearson type-III distribution, the series are first transferred to logarithmic scale. The mean, standard deviation and coefficient of skewness are computed from the log transferred series given in Table 8.9.

$$\text{Mean } (y_i = \log x_i) = 2.987626$$

$$\text{Standard deviation } \left[\frac{\Sigma(y - y_{av})^2}{(N - 1)}\right]^{1/2} = \left[\frac{1.696976}{(21 - 1)}\right]^{1/2} = 0.2913$$

Coefficient of variation $C_v = 0.2913/2.9876 = 0.0975$

$$\text{Coefficient of skewness} = C_s = \frac{(y - y_{av})^3 \times N}{\sigma^3 \times (N - 1)(N - 2)} = \frac{-0.2744 \times 21}{(0.2913)^3 \times 20 \times 19}$$

$$= (-)0.613$$

Corresponding to $C_s = -0.613$, the frequency factors are read from the Table 2.8B as

$$K_{100} = 1.8704 \text{ and } K_{1000} = 2.2587$$

Flood peaks are calculated as follows

$\qquad K_{100} = 1.8704 : \qquad Y_{100} = 2.987626 + 1.8704 \times 0.2913 = 3.5324735$

$\qquad K_{1000} = 2.2587 : \qquad Y_{1000} = 2.987626 + 2.2587 \times 0.2913 = 3.6455853$

taking antilogarithms $\qquad X_{100} = 3408$ m³/sec and $X_{1000} = 4422$ m³/sec

### (iv) Lognormal Distribution
Log normal distribution is a particular case of Log Pearson type-III distribution with zero skew.

We cannot assume $C_s = 0$ here as its value is $-0.613$. For such a situation, Chow's frequency factor table for log normal distribution is used to calculate the frequency factors for 100 and 1000 year return periods. For fitting this distribution, the coefficient of skewness is calculated as

$$C_s = 3C_v + C_v^3 = 3 \times 0.665 + 0.665^3 = 2.29$$

From Table 2.9 which gives theoretical log normal frequency factors, for $C_s = 2.29$, the frequency factors are calculated

for $\qquad T = 100, P = 1/T = 0.01 = 1\%, K_{100} = 3.62,$

$$X_{100} = 1183 + 3.62 \times 786 = 4028.3 \text{ m}^3/\text{sec}$$

for $\qquad T = 1000, P = (1/1000) \times 100 = 0.1\%, K_{1000} = 6.64,$

$$X_{1000} = 1183 + 6.64 \times 786 = 6402 \text{ m}^3/\text{sec}.$$

### 8.4.1.1 Fosters Method

Foster developed two types of curves for computation of frequency factors $K$ (Table 8.10 and 8.11) for use in hydrologic analysis. To use the tables, the coefficient of skewness $C_s$ is required to be calculated first. If Foster type I curve is used, then $C_s$ is multiplied by $(1 + 6/N)$ to get $C_{s1}$ to be used for reading frequency factors $K$ from the above tables. Thus $C_{s1} = C_s (1 + 6/N)$. When type -III curve is to be used then $C_s$ is multiplied by $(1 + 8.5/N)$ to get the changed value of $C_s$ as $C_{s2} = C_s (1 + 8.5/N)$. The value of $C_{s2}$ is used to read the value of frequency factors from the table. The flood peaks computed by both the methods are given below.

**Example 8.6.** Solve example 8.5 using Fosters method
Foster Type I Distribution

$$C_s = 1.478, \ C_{s1} = C_s\left(1 + \frac{6}{N}\right) = 1.478\left(1 + \frac{6}{21}\right) = 1.900$$

For $C_{s1} = 1.90$, from the table for Foster type-I distribution, the frequency factors are read from Table 8.10.

$T = 100 : K_{100} = 3.60 \qquad : X_{100} = 1183 + 3.60 \times 786.2 = 4013 \text{ m}^3/\text{sec}$

$T = 1000 : K_{1000} = 5.20 \quad .: X_{1000} = 1183 + 5.20 \times 786.2 = 5271 \text{ m}^3/\text{sec}$

**Foster Type III Distribution**
Assuming the sample to follow Foster type III distribution, the frequency factors are obtained from Table 8.11.

$C_s = 1.478, \ C_{s2} = C_s(1 + 8.5/N) = 1.478(1 + 8.5/21) = 2.08$

for $C_{s2} = 2.08$ from Table 8.11 the frequency factors are read as

$T = 100, \quad K_{100} = 3.645 : \ X_{100} = 1183 + 3.645 \times 786.2 = 4049 \text{ m}^3/\text{sec}$

$T = 1000, \quad K_{1000} = 6.05 : \ X_{1000} = 1183 + 6.05 \times 786.2 = 5940 \text{ m}^3/\text{sec}$

Table 8.10   Skew Curve Factors for Foster Type-I Curve (Frequency in percent)

| $C_s$ | 99 | 95 | 80 | 50 | 20 | 5 | 1 | 0.1 | 0.01 | 0.001 | 0.0001 |
|---|---|---|---|---|---|---|---|---|---|---|---|
| 0.0 | -2.08 | -1.64 | -0.92 | -0.00 | 0.92 | 1.64 | 2.08 | 2.39 | 2.53 | 2.59 | 2.62 |
| 0.2 | -1.91 | -1.56 | -0.93 | -0.05 | 0.89 | 1.72 | 2.25 | 2.66 | 2.83 | 2.94 | 3.00 |
| 0.4 | -1.75 | -1.47 | -0.93 | -0.09 | 0.87 | 1.79 | 2.42 | 2.95 | 3.18 | 3.35 | 1.44 |
| 0.6 | -1.59 | -1.38 | -0.92 | -0.13 | 0.85 | 1.85 | 2.58 | 3.24 | 3.59 | 3.80 | 3.92 |
| 0.8 | -1.44 | -1.30 | -0.91 | -0.17 | 0.83 | 1.90 | 2.75 | 3.55 | 4.00 | 4.27 | 4.43 |
| 1.0 | -1.30 | -1.21 | -0.89 | -0.21 | 0.80 | 1.95 | 2.92 | 3.85 | 4.42 | 4.75 | 4.95 |
| 1.2 | -1.17 | -1.12 | -0.86 | -0.25 | 0.77 | 1.99 | 3.09 | 4.15 | 4.83 | 5.25 | 5.50 |
| 1.4 | -1.06 | -1.03 | -0.83 | -0.29 | 0.73 | 2.03 | 3.25 | 4.45 | 5.25 | 5.75 | 6.05 |
| 1.6 | -0.96 | -0.95 | -0.80 | -0.32 | 0.69 | 2.07 | 3.40 | 4.75 | 5.67 | 6.25 | 6.65 |
| 1.8 | -0.87 | -0.87 | -0.76 | -0.35 | 0.64 | 2.10 | 3.54 | 5.05 | 6.08 | 6.75 | 7.2 |
| 2.0 | -0.80 | -0.79 | -0.71 | -0.37 | -0.58 | 1.13 | 3.67 | 5.35 | 6.50 | 7.25 | 7.8 |

**Table 8.11  Skew Curve Factors for Foster Type-III Curve (Frequeeny in Percent)**

| $C_s$ | 99 | 95 | 80 | 50 | 20 | 5 | 1 | 0.1 | 0.001 | 0.01 | 0.0001 |
|------|-------|-------|-------|-------|------|------|------|------|-------|------|--------|
| 0.0  | −2.33 | −1.64 | −0.84 | −0.00 | 0.84 | 1.64 | 2.33 | 3.09 | 4.27  | 3.73 | 4.26   |
| 0.2  | −2.18 | −1.58 | −0.85 | −0.03 | 0.83 | 1.69 | 2.48 | 3.38 | 4.89  | 4.16 | 4.48   |
| 0.4  | −2.03 | −1.51 | −0.85 | −0.06 | 0.82 | 1.74 | 2.62 | 3.67 | 5.42  | 4.60 | 6.24   |
| 0.6  | −1.88 | −1.45 | −0.86 | −0.09 | 0.80 | 1.79 | 2.77 | 3.96 | 6.01  | 5.04 | 7.02   |
| 0.8  | −1.74 | −1.38 | −0.86 | −0.13 | 0.78 | 1.83 | 2.90 | 4.25 | 6.61  | 5.48 | 7.82   |
| 1.0  | −1.59 | −1.31 | −0.85 | −0.16 | 0.76 | 1.87 | 3.03 | 4.54 | 7.22  | 5.92 | 8.63   |
| 1.2  | −1.45 | −1.25 | −0.84 | −0.19 | 0.74 | 1.70 | 3.15 | 4.82 | 7.85  | 6.37 | 9.45   |
| 1.4  | −1.32 | −1.18 | −0.82 | −0.22 | 0.71 | 1.93 | 3.28 | 5.11 | 8.50  | 6.82 | 10.28  |
| 1.6  | −1.19 | −1.11 | −0.80 | −0.25 | 0.68 | 1.96 | 3.40 | 5.39 | 9.17  | 7.28 | 11.12  |
| 1.8  | −1.08 | −1.03 | −0.78 | −0.28 | 0.64 | 1.98 | 3.50 | 5.68 | 9.84  | 7.75 | 11.96  |
| 2.0  | −0.99 | −0.95 | −0.75 | −0.31 | 0.61 | 2.00 | 3.60 | 5.91 | 10.51 | 8.21 | 12.81  |
| 2.0  | −0.90 | −0.89 | −0.71 | −0.33 | 0.58 | 2.01 | 3.70 | 6.20 |       |      |        |
| 2.4  | −0.83 | −0.82 | −0.68 | −0.35 | 0.54 | 2.01 | 3.78 | 6.47 |       |      |        |
| 2.6  | −0.77 | −0.76 | −0.65 | −0.37 | 0.51 | 2.01 | 3.87 | 6.73 |       |      |        |
| 2.8  | −0.71 | −0.71 | −0.62 | −0.38 | 0.47 | 2.02 | 3.95 | 6.99 |       |      |        |
| 3.0  | −0.67 | −0.66 |       | −0.40 | 0.42 | 2.02 | 4.02 | 7.25 |       |      |        |

### 8.4.1.2　Chow's Regression Method

Chow proposed a linear equation of the form

$$Q_t = a + bX_t \tag{8.36}$$

where $a$ and $b$ are the regression constants, $Q_t$ the flood of return period $T$ years and $X_t$ is given as

$$X_t = \log \log \left\{ \frac{T}{(T-1)} \right\} \tag{8.37}$$

By the method of least square, the constants $a$ and $b$ can be computed as

$$\Sigma \, Q_t = aN + b \, \Sigma \, X_t$$

and

$$\Sigma \, Q_t X_t = a \, \Sigma \, X_t + b \, \Sigma \, X^2_t$$

When the data are arranged in decreasing order , the return period $T$ is given as $T = (N + 1)/m$. Substituting $T$ in equation (8.37) the value of $X_t$ is obtained as

$$X_t = \log \left[ \log \left( \frac{(N+1)}{(N+1-m)} \right) \right] \tag{8.38}$$

The above equation is solved to get $Q_t$ of the desired return period.

---

**Example 8.7:** Solve example 8.5 by Chow's approach.

**Solution**

Calculation are shown in Table 8.12 from which

$$\Sigma \, Q_t = 24843 \qquad\qquad \Sigma \, X_t = -12.396$$

$$\Sigma \, Q_t X_t = -21717.5 \qquad\qquad \Sigma \, X^2_t = 11.847$$

Therefore　　$24843 = 21a + b(-12.396)$　or　$a = \dfrac{(24843 + 12.396 \, b)}{21}$

Again　　　　$-21717.5 = -a \times 12.396 + b \times 11.847$

Substituting　　$-21717.5 = -12.396 \, \dfrac{(24843 + 12.396b)}{21} + 11.847b$

or　　　　　　$-21717.5 = -14664.5 - 7.3172b + 11.847b$

or　　　　　　$-7053 = 4.53b$　or　$b = -1556.95$

$$a = \frac{24843 - 12.396 \times 1556.95}{21} = 263.95$$

The equation becomes

$$Q_t = 263.95 - 1556.95 \, X_t = 263.95 - 1556.95 \times \log\{\log \{T/(T-1)\}$$

Now for T = 100 and 1000 years, the flood peaks are obtained as

$$Q_{100} = 263.95 - 1556.95 \log (\log 100/99) = 3938 \text{ m}^3 \text{ /sec}$$

$$Q_{1000} = 263.95 - 1556.95 \log (\log 1000/99) = 5498 \text{ m}^3/\text{sec}$$

**Table 8.12 Chow's Method of Flood Frequency Analysis for Example 8.7**

| Year | Runoff (m³/sec) $Q_t$ | Runoff in decreasing order | Rank (m) | log $[(N+1)/(N+1-m)]$ | (log (col. 5)) | col. (6) col. (3) $X_t Q_t$ | col. (6)× col. (6) $X_t^2$ |
|------|------|------|------|------|------|------|------|
| (1) | (2) | (3) | (4) | (5) | (6) | (7) | (8) |
| 1970 | 1065 | 3210 | 1 | 0.0202 | −1.6946 | −5439.7 | 2.871 |
| 1971 | 645 | 2840 | 2 | 0.0414 | −1.3831 | −3928.0 | 1.912 |
| 1972 | 1005 | 2260 | 3 | 0.0637 | −1.1961 | −2703.2 | 1.431 |
| 1973 | 1350 | 2230 | 4 | 0.1120 | −1.0597 | −2363.1 | 1.123 |
| 1974 | 860 | 1350 | 5 | 0.1383 | −0.9509 | −1283.2 | 0.904 |
| 1975 | 150 | 1095 | 6 | 0.1663 | −0.8591 | −940.7 | 0.738 |
| 1976 | 2260 | 1070 | 7 | 0.1963 | −0.7790 | −833.5 | 0.607 |
| 1977 | 650 | 1065 | 8 | 0.2285 | −0.7071 | −753.1 | 0.500 |
| 1978 | 2840 | 1005 | 9 | 0.2632 | −0.6412 | −644.4 | 0.411 |
| 1979 | 990 | 990 | 10 | 0.3010 | −0.5796 | −573.8 | 0.336 |
| 1980 | 870 | 930 | 11 | 0.3424 | −0.5214 | −484.9 | 0.272 |
| 1981 | 910 | 910 | 12 | 0.3882 | −0.4654 | −423.5 | 0.217 |
| 1982 | 750 | 870 | 13 | 0.4393 | −0.4110 | −357.6 | 0.169 |
| 1983 | 930 | 860 | 14 | 0.4973 | −0.3572 | −307.2 | 0.128 |
| 1984 | 750 | 830 | 15 | 0.5643 | −0.3033 | −251.7 | 0.092 |
| 1985 | 1070 | 750 | 16 | 0.6435 | −0.2485 | −186.4 | 0.062 |
| 1986 | 830 | 750 | 17 | 0.7404 | −0.1915 | −143.6 | 0.037 |
| 1987 | 1095 | 650 | 18 | 0.8653 | −0.1306 | − 84.9 | 0.017 |
| 1988 | 384 | 645 | 19 | 0.0414 | −0.0628 | −40.5 | 0.004 |
| 1989 | 2230 | 384 | 20 | 1.0414 | 0.0176 | + 6.8 | 0.000 |
| 1990 | 3210 | 150 | 21 | 1.3424 | 0.1279 | + 19.2 | 0.016 |
| Total = | 24844 | 24844 | | | −12.396 | −21717.5 | 11.85 |

### 8.4.1.3 *Stochastic Method*

A well known equation for computation of annual flood peak using probability law is written the form

$$Q_t = Q_{min} + 2.303(Q_{av} - Q_{min})\log (nT/N) \qquad (8.39)$$

The equation uses the theory of sums of random numbers where $T = N/m$ which is the California plotting position formula, $n$ the number of recorded floods which assigns only one value to the same event and $N$ total number of data. For example if a flood peak of 300 occurs 3 times in a series of 10 values then $n$ is taken as 8 and $N$ as 10, $Q_{min}$ is the minimum value of flood in the series, $Q_{av}$ the average of the series and $Q_t$ is the flood of return period $t$ years.

**Example 8.8:** Solve example 8.5 by stochastic approach and tabulate the results of the flood frequencies of all the methods used under the examples (8.5), (8.6), (8.7) and (8.8).

**Solution**

Flood peaks of 100 years and 1000 years are computed below. Here $n = 20$ and $N = 21$ (as there is one time repetition of the flood magnitude of 750 m³/sec).

$$Q_{av} = 1183 \text{ m}^3/\text{sec and } Q_{min} = 150 \text{ m}^3/\text{sec}$$

Floods of 100 and 1000 years are calculated from equation (8.39). For return periods of $T = 100$ years

$$Q_{100} = 150 + 2.303(1183 - 150) \log \left[ \frac{(20 \times 100)}{21} \right]$$

$$= 150 + 2379 \log (95.23) = 4857 \text{ m}^3/\text{sec}$$

$$Q_{1000} = 150 + 2.303(1183 - 150) \log \left[ \frac{(20 \times 1000)}{21} \right]$$

$$= 150 + 2379 \log (952.3) = 7237 \text{ m}^3/\text{sec}.$$

**Table 8.13   Comparison of the Results of Examples 8.5, 8.6, 8.7 and 8.8**

| Sl.No | Method | 100 year peak | 1000 year peak |
|-------|--------|---------------|----------------|
| (1) | (2) | (3) | (4) |
| 1. | Gumbel's ( K-T table) | 4182 | 5869 |
| 2. | Gumbel - Powel Relation | 3649 | 5063 |
| 3. | Pearson type - III | 3793 | 5318 |
| 4. | Log Pearson type - III | 3408 | 4422 |
| 5. | Log Normal distribution | 4028 | 6402 |
| 6. | Fosters method type I | 4013 | 5271 |
| 7. | Fosters type III | 4049 | 5940 |
| 8. | Chow's regression method | 3938 | 5498 |
| 9. | Stochastic method | 4857 | 7237 |

It is to be emphasised here that *no flood-frequency studies should be carried out for a basin when records available are less than 10 years*. Note that there is a wide variation in the 100 year flood peak from 3408 to 4857 m$^3$/sec and the 1000 year flood varies from 4422 to 7237 m$^3$/sec. Graphical solution to the above type of problem has been discussed in Chapter-3.

### 8.4.2   Unit Hydrograph Approach

When the safety of large dams or barrages are involved, the interest is to know the peak flood as well as the probable hydrograph resulting from the probable storm. For undertaking such a study the catchment at the project should drain more than 250 sq. km or storage of the structure should be higher than 60 million cubic meters as outlined in Table 8.1. All the foregoing approaches can give only the flood peak but cannot give the shape of complete hydrograph resulting from a probable storm. Use of unit hydrograph of Chapter-7 and the concept of probable maximum precipitation (PMP) or standard project storm (SPS) of Chapter 3 is used to obtain a complete flood hydrograph. The procedure to obtain probable maximum flood (PMF) from probable maximum precipitation are given below.

#### (1) Computation of PMP

For computation of probable maximum precipitation, the storm experience of a basin should be the maximum. To achieve this, all the severemost storms experienced by the basin and the adjoining areas which are hydrometeorologically homogeneous are considered. When the adjoining basins experience greater storm then this storm is applied to the basin in question. This exercise increases the storm experience of the basin and is achieved by storm transposition. Storm transposition should be carried out only from meteorologically homogeneous regions. The main factors affecting meteorological homogeneity between two basins are (a) distance from sea, (b) mean annual temperature, (c) direction of prevailing wind and (d) topography.

This means that two basins must be affected by the same moisture source, experience the same type of storms, having the similar topographic features and orientation to the seasonal winds. Storm transposition therefore considers not only the storms which have occurred over or near the basin in the past but also those storms which have resulted in heavy rainfall on areas that are meteorologically homogeneous. Storm transposition is carried out for the basins having inadequate rainfall data or where no severe storm has occurred over the region. Steps in storm transposition are

(i) Identify when and where the heaviest storms occurred in meteorologically homogeneous regions of the project catchment.

(ii) Identify the synoptic situations associated with the storm.

(iii) Isohyetal patterns of the selected storms are prepared on a map.

(iv) The scale of the isohyetal maps and the problem basin should preferably be the same.

(v) The isohyetal patterns of the storms are transferred to a transparent sheet. They are superimposed on the problem basin in the most critical manner so as to yield maximum depths of rainfall over the basin.

(vi) Depth–Area-Duration analysis are carried out for the storms.

(vii) The storms yielding maximum depths of precipitation over the basin are considered for further maximisation.

(viii) Precautions to be taken in maximisations are:

(a) Both areas should be meteorologically homogeneous.

(b) Rainstorms from a hilly orographic region should not be transposed to a place far away from it.

(c) Large latitudinal shift may involve considerable change in air mass characteristics of the storm.

(d) Axis of the storms to be transposed should not be rotated exceeding 20°.

*(2) Storm Maximisation*
Storm maximisation is carried out to ascertain how much the rainfall from a particular storm would have increased by physically possible increase in meteorological factors which produce the storm. The maximisations carried are:

(a) Wind maximisation

(b) Maximisations due to transposition, barrier and topography

(c) Moisture maximisations.

Chapter 3 gives details of storm transposition and maximisations. Usually moisture maximisation is carried out as the other two are considered as small and therefore neglected.

*(3) Time Distribution*
For estimation of PMF or SPF using unit hydrograph, increments of the design storm at short intervals are required. For watersheds where self-recording rain gauges (SRRG) are not installed, the procedure is to determine time distribution of the design storm on the basis of rainfall data observed at other SRRG stations during the severe storm in meteorological homogeneous area. If there is SRRG station in the catchment in operation during storm, the time distribution of the catchment average storm is carried out on percentile basis with respect to SRRG data. To do this the average catchment precipitation representing PMP is multiplied by the time percentages of SSRG-data. Time distribution is carried out as per the following steps.

  (i)   Select the SRRG station and let 24h or 48h severe most storm precipitation at the station be $P$ cm.
  (ii)  For say 48 h storm, obtain the rainfall totals of SRRG rainfall at the end of 1, 2, 3, 6, 9, 12, 15, 18, 24, 27, ... , 48 h. Let they be $P_1, P_2, P_3, ...$ , $P_{48} (\Sigma P_i = P)$.
  (iii) Get the cumulative percentages as

$$Z_1 = \frac{100 P_1}{P}, Z_2 = \frac{100 P_2}{P}, ... , Z_{48} = \frac{100 P_{48}}{P}.$$

  (iv)  Distribute the given 48h PMP as per the percentage of $Z_1, Z_2, ... , Z_{48}$.
  (v)   Calculate the PMP increments for the hours of 1, (2 – 1), (3 – 2), (6 – 3), ... hours by subtraction.
  (vi)  When more than one storm is transposed then the above steps should be repeated for other storms.
  (vii) Percentages of different durations are plotted and an enveloping curve through maximum percentages of each durations from amongst the different storms are drawn.
        Time distribution of PMP is obtained at the end of step (v) or (vii).

*(4) Design Unit Hydrograph*
Chapter 7 discusses the derivation of unit hydrograph from different approaches. Each of them exhibit different peaks and time base. This is due to the violation of conditions and assumptions inherited in deriving the unit hydrograph. The usual practice to obtain a design UH from the derived UH is to increase its peak by 25 or 50% and adjust the area of UH such that the volume of the unit hydrograph is 1 cm for the catchment.

*(5) Critical Sequencing of PMP or SPS*
In computing maximum probable flood, the sequence of rainfall increments to be adopted is usually the one which gives the most critical condition of runoff. Such a sequence is called critical sequence and can be determined in the following way

1. Time distribution of design storm depths of step (3) are arranged opposite to the ordinates of design UH in such a way that the largest increment is opposite to the largest UH ordinates, the second largest is opposite to the second largest UH ordinate and so on. The procedure is repeated to cover the entire rainfall intensities.
2. The arrangement of the rainfall increments entered in the column is reversed to obtain the critical sequence.

The details are shown in example 8.9.

### (6) Losses from Storm Precipitation
Infiltration and other losses are usually known or they can be derived from the available hydrologic records of a basin. The initial loss is subtracted from the first rainfall increment and thereafter a uniform loss rate equal to infiltration rate is applied. Losses like interception, depression storage and evapotranspiration during storm are called initial losses. In absence of any data, the initial loss of 10 mm and infiltration loss rate of 1 mm/h is quite acceptable for Indian condition. These losses are subtracted from the time distribution of PMP to get the rainfall excess in the same critical sequence.

### (7) Computation of Design Flood
The critical sequence of effective rainfall (rainfall excess) blocks of step (6) are applied to the design UH to obtain the design direct runoff hydrograph (DDRH). To do this, the first effective rainfall increment is successively multiplied by each of the UH ordinates and the resulting quantities represent the ordinates of direct surface runoff hydrograph produced by the first increment of rainfall excess. Similarly direct surface runoff hydrograph resulting from 2nd, 3rd, 4th and other incremental blocks of effective rainfall are also computed. The total direct runoff from the design storm is obtained by adding the direct surface runoff resulting from each increment of rainfall blocks lagged successively by a time interval equal to unit duration of UH (for details see Example 8.9).

### (8) Addition of Base Flow to Direct Runoff Hydrograph
To obtain the design flood hydrograph, a base flow must be added to the DRH. The base flow is estimated on the basis of analysis of previous hydrographs. In absence of any data, a base flow of 0.05 $m^3$/sq. km of catchment is assumed. This rate of base flow is adopted by Central Water Commission for Indian catchments.

---

**Example 8.9:** A 24-h probable maximum precipitation obtained for a project area of 1100 sq. km (after transposition and maximisations) is found to be 372.6 mm. Calculate the design flood at the project site using a 3-h UH. The ordinates of 3-h UH obtained from Clark's model are given below. Maximum 24-h rainfall data for a self recording rain gauge (SRRG) located close to the basin is also given.

| Time (h) | 0 | 3 | 6 | 9 | 12 | 15 | 18 | 21 | 24 | 27 | 30 | 33 |
|---|---|---|---|---|---|---|---|---|---|---|---|---|
| UH ordinates ($m^3$/sec) | 0 | 3.3 | 11.4 | 27.5 | 20 | 14 | 10 | 6.5 | 4.3 | 2.8 | 2.0 | 0 |
| 24-h SRRG records (mm) | 0 | 31.8 | 81.6 | 91.2 | 133.6 | 196.1 | 224.2 | 253.3 | 265 | | | |

**Solution**

*(1) Time distribution*
Probable maximum precipitation (Design storm) for the project catchment is 372.6 mm.
There is no SRRG within the project catchment. The recording type rain gauge located
close to the basin gives the time percentage of storm rainfall on the storm day on the basis
of which time distribution of PMP is carried out in Table 8.14.

Table 8.14    Time Distribution of PMP for Example 8.9

| Time (h) | Cumulative 24th SRRG rainfall (mm) | Cumulative % rainfall $100 \times (2)/265$ | Cumulative catchment rainfall = $372.6 \times (3)$ mm | 3h rainfall increment (mm) |
|---|---|---|---|---|
| (1) | (2) | (3) | (4) | (5) |
| 0 | 0 | 0 | 0 | 0 |
| 3 | 31.8 | 12 | 44.7 | 44.7 |
| 6 | 81.6 | 30.8 | 114.8 | 70.1 |
| 9 | 91.2 | 34.4 | 128.2 | 13.4 |
| 12 | 133.6 | 50.4 | 187.8 | 59.6 |
| 15 | 196.1 | 74.0 | 275.7 | 87.9 |
| 18 | 224.2 | 84.6 | 315.2 | 39.5 |
| 21 | 253.3 | 95.6 | 356.2 | 41.0 |
| 24 | 265.0 | 100 | 372.6 | 16.4 |

*(2) Design Unit Hydrograph*
For the example, the UH has been obtained using Clark's model. As per Central
Water Commission practice, the peak of UH may be increased by 50% and the
area of UH is rearranged on trial and error basis to yield 1 cm of rainfall depth
from the catchment. The design UH is shown in Fig. 8.4. The ordinates of design
UH from the figure are given in Table 8.15.

*(3) Critical Sequencing and incorporation of loses from the storm*
Critical sequencing of PMP is carried out in Table 8.16. Highest storm increment
of 87.9 mm is placed against the highest UH of 41.25 m³/sec and the second
highest of 70.1 against the second highest of 20 of UH ordinates and so on. The
col.(3) is reversed in col.(4) to get critical sequencing of the rainfall. Initial loss
of 10 mm and subsequent loss rate of 1 mm/h i.e. 3 mm in 3 hours is subtracted
from the design storm to get the excess rainfall of the critical storm, as given in
col.(6). This storm critical PMP is to be convoluted with the design UH ordinates
given in Table 8.15 to get PMF.

*(4) Computation of design flood using design UH and rainfall excess
hyetograph of PMP*
The UH ordinates (Table 8.17; col. (2)) are multiplied by the rainfall excess
hyetograph blocks and each column is lagged by time unit equal to the unit

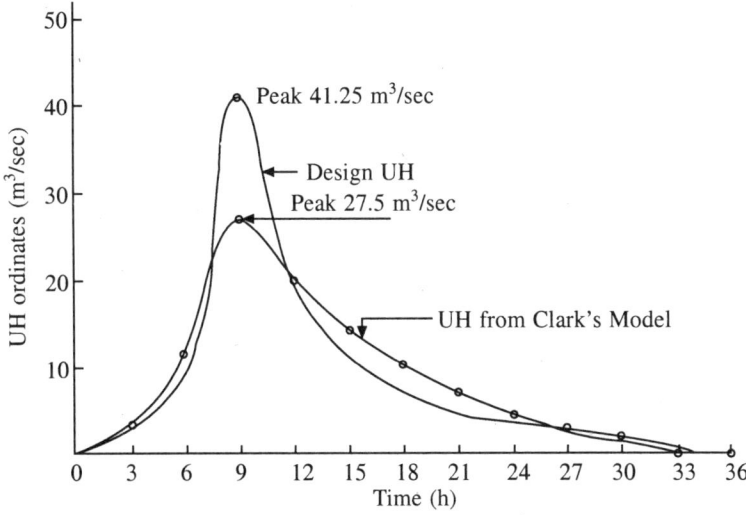

Fig. 8.4   Design Unit Hydrograph for Example 8.9

Table 8.15   Design Unit Hydrograph Ordinates (m³/sec)

| Time (h) | 0 | 3 | 6 | 9 | 12 | 15 | 18 | 21 | 24 | 27 | 30 | 33 |
|---|---|---|---|---|---|---|---|---|---|---|---|---|
| UH ordinates (m³/sec) | 0.0 | 3.3 | 11.4 | 27.5 | 20.0 | 14.0 | 10.0 | 6.5 | 4.3 | 2.8 | 2.2 | 0.0 |
| Design UH | 0.00 | 3.30 | 10.00 | 41.25 | 20.00 | 10.00 | 6.00 | 4.00 | 3.00 | 2.55 | 2.00 | 0.00 |

Table 8.16   Critical Sequencing of Design Storm

| Time (h) | Design UH Ordinates (m³/sec) (from Fig 8.4) | First arrangement of time distribution of PMP (mm) | Design/ Critical sequence of rainfall (mm) | Loss (mm) | Rainfall excess of PMP. (mm) |
|---|---|---|---|---|---|
| (1) | (2) | (3) | (4) | (5) | (6) |
| 0 | 0.00 | 0.0 | 13.4 | 10 | 3.4 |
| 3 | 3.30 | 16.4 | 39.5 | 3 | 36.5 |
| 6 | 10.00 | 44.7 | 41.0 | 3 | 38.0 |
| 9 | 41.25 | 87.9 | 59.6 | 3 | 56.6 |
| 12 | 20.00 | 70.1 | 70.1 | 3 | 67.1 |
| 15 | 10.00 | 59.6 | 87.9 | 3 | 84.9 |
| 18 | 6.00 | 41.0 | 44.7 | 3 | 41.7 |
| 21 | 4.00 | 39.5 | 16.4 | 3 | 13.4 |
| 24 | 3.00 | 13.4 | | | |
| 27 | 2.55 | | | | |
| 30 | 2.00 | | | | |
| 33 | 0.00 | | | | |

**Table 8.17  Design Flood Computation Using PMP for Example 8.9**

| Time (h) | Design UH Ordinates (m³/sec) | DRH from 0.34 | DRH from 3.65 | DRH from 3.8 | DRH from 5.66 | DRH from 6.71 | DRH from 8.49 | DRH from 4.17 | DRH from 1.34 | Sum of DRH (3) to (10) | Base flow 0.05 × 1100 | PMF (m³/sec) |
|---|---|---|---|---|---|---|---|---|---|---|---|---|
| (1) | (2) | (3) | (4) | (5) | (6) | (7) | (8) | (9) | (10) | (11) | (12) | (13) |
| 0 | 0.00 | 0 | | | | | | | | 0.0 | 55 | 55.0 |
| 3 | 3.30 | 1.12 | 0 | | | | | | | 1.1 | 55 | 56.1 |
| 6 | 10.00 | 3.4 | 12.0 | 0 | | | | | | 15.4 | 55 | 70.4 |
| 9 | 41.25 | 14.0 | 36.5 | 12.5 | 0 | | | | | 63.1 | 55 | 118.1 |
| 12 | 20.00 | 6.8 | 150.0 | 38.0 | 18.7 | 0 | | | | 214.0 | 55 | 269.0 |
| 15 | 10.00 | 3.4 | 73.0 | 156.8 | 56.6 | 22.1 | 0 | | | 311.9 | 55 | 366.9 |
| 18 | 6.00 | 2.04 | 36.5 | 76.0 | 233.5 | 67.1 | 28.0 | 0 | | 443.1 | 55 | 498.1 |
| 21 | 4.00 | 1.36 | 21.9 | 38.0 | 113.2 | 276.8 | 84.9 | 13.8 | 0 | 550.0 | 55 | 605.0 |
| 24 | 3.00 | 1.02 | 14.6 | 22.8 | 56.6 | 134.2 | 350.2 | 41.7 | 4.4 | 625.6 | 55 | 680.6 |
| 27 | 2.55 | 0.87 | 10.95 | 15.2 | 34.0 | 67.1 | 169.8 | 172.0 | 13.4 | 483.3 | 55 | 538.3 |
| 30 | 2.00 | 0.68 | 9.3 | 11.4 | 22.6 | 40.3 | 84.9 | 83.4 | 55.3 | 307.9 | 55 | 362.9 |
| 33 | 0.00 | 0.00 | 7.3 | 9.7 | 17.0 | 26.8 | 51.0 | 41.7 | 26.8 | 180.3 | 55 | 235.3 |
| 36 | | | 0.0 | 7.6 | 14.4 | 20.1 | 34.0 | 25.0 | 13.4 | 114.5 | 55 | 169.5 |
| 39 | | | | 0.0 | 11.3 | 17.1 | 25.5 | 16.7 | 8.0 | 78.6 | 55 | 133.6 |
| 42 | | | | | 0.0 | 13.4 | 21.7 | 12.5 | 5.4 | 53.0 | 55 | 108.0 |
| 45 | | | | | | 0.0 | 17.0 | 10.6 | 4.0 | 31.6 | 55 | 86.6 |
| 48 | | | | | | | 0.0 | 8.3 | 3.4 | 11.7 | 55 | 66.7 |
| 51 | | | | | | | | 0.0 | 2.7 | 2.7 | 55 | 57.7 |
| 54 | | | | | | | | | 0.0 | 0.0 | 55 | 55.0 |

duration of the UH, i.e., 3 h here as given in the problem. A base flow of 55 cumecs at the rate of 0.05 cumecs per sq. km is added to the DRH of col. (11) resulting from the PMP. This is done as per the ongoing practice accepted by CWC. In col. (13), probable maximum flood resulting from PMP is obtained. The procedure becomes easier if the time interval of col. (1) is read at the unit duration of UH.

## 8.5 REGIONAL FLOOD FREQUENCY ANALYSIS

For any hydrologic (i.e. flood frequency) analysis the available data should be more than 10 years. To overcome this deficiency, the available stream records of hydrologic homogenous regions are utilised. The first step in regional flood frequency study is to carry out the *homogeneity test* for all the neighbouring stations. This helps to define the region as a whole. There are many methods available to carry out regional homogeneity test, but the method proposed by Darymple (1960) is widely used. After homogeneity test the next step is to carry out the regional flood frequency analysis. The method developed by Darymple and Benson (1960) for U.S. Geological Survey is adopted here. Thus in the second step, the development of a dimensionless frequency curve is attempted. The curve is plotted between the recurrence interval in abscissa versus the ratio of peak flood to mean flow as ordinate. The third step is the development of relation between the catchment area and the mean flood ($Q_{2.33}$). This helps to predict the mean annual flood of an ungauged catchment of the region. Stepwise procedure of the homogeneity test and regional flood frequency analysis are outlined below.

1. Data for all the gauge and discharge stations neighbouring the river gauge in question with available data of 10 years or more are collected.
2. For all the selected stations, flood of 10 year and 2.33 year return periods are computed by Gumbel's method. Flood of 2.33 year return period is the average flood as discussed earlier.
3. A table is prepared in which the first column should contain the station names. In the second column, enter the length of available data corresponding to col.(1). In col. (3), enter $Q_{2.33}$ for all the stations of first column.
4. In the fourth column, the corresponding 10 year return period floods are calculated and entered.
5. The ratio $q_r = Q_{10} / Q_{2.33}$ for each station is determined and entered in the fifth column. Compute $q_{rav} = \Sigma q_r/N$ which represents the average ratio. N is the number of stations used in the analysis.
6. In the sixth column values of ($Q_{2.33} \times q_{r\,av}$) are entered.
7. Return periods of magnitudes under step (6) for all the stations of step (1) are computed and entered in the seventh column.
8. For plotting 95% confidence bands, the upper and lower limits of 10-year flood at 95% level are computed from the relation, $Q_U = Q_{10} + S_e f$ and $Q_L = Q_{10} - S_e f$, where $S_e = \sqrt{1 + 1.3\,K + 1.1\,K^2} \times (\sigma_{n-1}/\sqrt{N})$, $f$ is 1.96 at 95% confidence level. Values $S_e$, $Q_U$ and $Q_L$ are centered in cols. (8), (9) and (10), respectively.

9. $Q_U$ and $Q_L$ values, frequency factors $K$ are found out as the relation between $X_{av}$ and $\sigma$ are already known. For the sample length and Gumbel's $K$-$T$ table, $T$ is found out. For $Q_U$ the values of frequency factors $K_u$ and return period $T$ is given in col. (11) and the corresponding values of $Q_L$ in col. (12).

10. A graph is plotted taking the year of record of col. (2) in abscissa in ordinary scale and return period of col. (7) as ordinate in log scale. In the graph all the stations are marked as points in the scatter diagram.

11. All the upper bound return periods obtained under step (9) and entered in col. (11) for different stations are plotted in the graph of step (10) and joined together giving a line called *upper limit* of 95% confidence curve. Similarly all the return periods of lower bound $Q_L$ floods of col. (12) are joined to give a line for *lower limit* of 95% confidence curve. Such a curve is shown in Fig. 8.5.

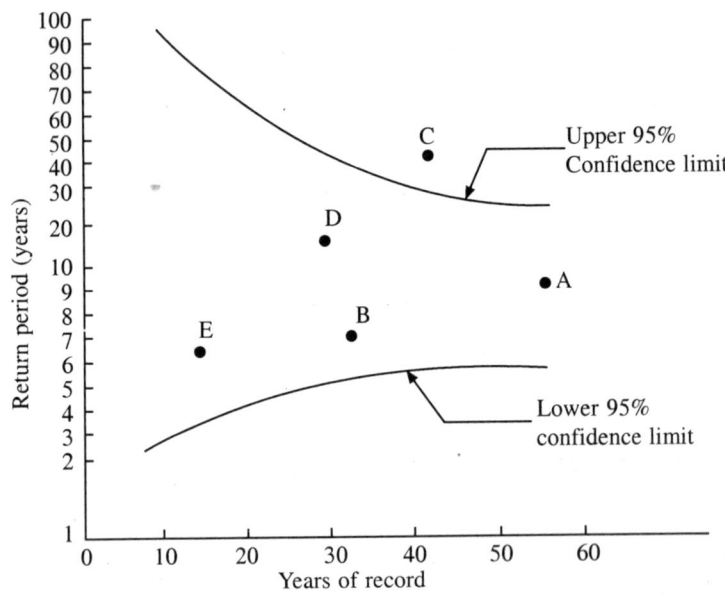

Fig. 8.5 **Homogenity Test for Regional Flood Frequency Analysis**

12. The stations falling outside the 95% upper and lower limit curves are considered *non-homogeneous* and therefore discarded for further analysis.

Since all stations should have at least 10 years of flood record to accept it for flood-frequency analysis, the homogeneity test is considered for $Q_{10}$ and $Q_{2.33}$ floods. After conducting homogeneity test, the next step is

(i) Calculate flood for different return periods say $Q_{10}$, $Q_{50}$, $Q_{100}$, $Q_{1000}$ and $Q_{10,000}$ for all stations selected after homogeneity test from their respective flood records.

(ii) Compute the ratios $q_{r\,av10} = \frac{1}{N}\Sigma Q_{10}/Q_{2.33}$, $q_{r\,av50} = \frac{1}{N}\Sigma Q_{50}/Q_{2.33}$,

$q_{rav100} = \dfrac{1}{N} \Sigma \, Q_{100}/Q_{2.33,}$ and so on for all the stations. These ratios can
be entered in col. (13) onwards in the same table. Average of all $q_{r10}, q_{r50},$
$q_{r100}, q_{r1000}$ and $q_{r10000}$ are calculated.

(iii) Plot a curve between recurrence interval $T$ in abscissa vs. the average
ratios ($q_{r\,av}$) of step (ii) as ordinate. A straight line is fitted through them.
The abscissa is in log scale. Such a curve obtained from example 8.9 is
shown in Fig. 8.6.

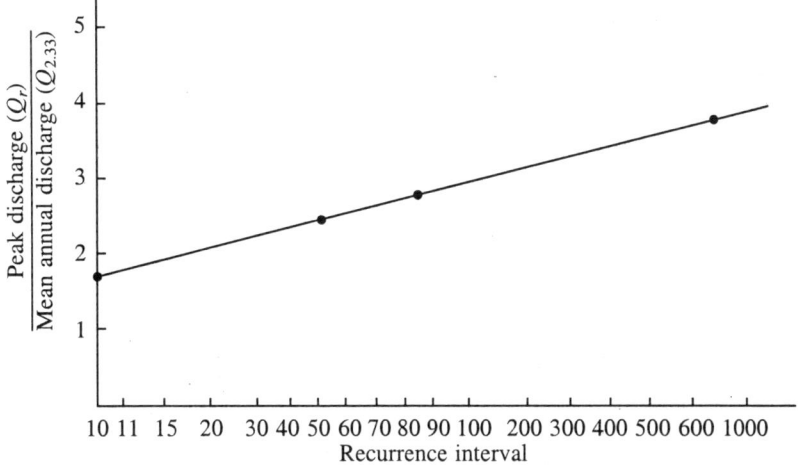

**Fig. 8.6 Regional Flood Frequency Curve**

(iv) The upper bound 95% and lower bound 95% confidence limit curves
may be fitted to the curve obtained in step (iii) in Section 2.6. The plot
is known as *regional curve*.

(v) Now from the data of the drainage area of various ($N$) stations and their
mean annual floods $Q_{2.33}$, a regression relation can be established for the
region between $A_i$ and $Q_{2.33}$ using $N$ stations data.

(vi) From the regression relation of step (v) knowing the area of the ungauged
basin, $Q_{2.33}$ can be computed. Knowing $Q_{2.33}$ and the desired return
period $T$, the value of $Q_T / Q_{2.33}$ is obtained from the graph of step (iii)
from which $Q_T$ is estimated.

The following example illustrates the above procedure.

---

**Example 8.10:** Table 8.18 gives data of 5 stations in 4 tributaries of river Mahanadi.
Carry out regional flood frequency analysis of the basin after carrying out the homogeneity
test. Plot the regional frequency curve. Gumbel's method can be used to compute flood
peak of different periods.

**Solution**
Calculation as outlined above are carried out in Table 8.19. which is self-explanatory.
Computations are carried out in line with the stepwise procedure given. Homogeneity test
is shown in Fig. 8.5 and the required regional frequency curve in Fig. 8.6.

**Table 8.18 Data for Regional Flood Frequency for Example 8.10**

| Station name | Data length(years) | Mean $Q_{2.33}$ | $Q_{10}$ | $Q_{50}$ | $Q_{100}$ | $Q_{1000}$ | Standard deviation |
|---|---|---|---|---|---|---|---|
| (1) | (2) | (3) | (4) | (5) | (6) | (7) | (8) |
| A | 55 | 388 | 630 | 870 | 970 | 1300 | 166.2 |
| B | 32 | 350 | 680 | 1000 | 1135 | 1590 | 215.4 |
| C | 41 | 610 | 840 | 1070 | 1165 | 1480 | 154.1 |
| D | 29 | 455 | 680 | 900 | 1010 | 1310 | 158.6 |
| E | 13 | 153 | 300 | 440 | 505 | 706 | 86.5 |

From the homogeneity test (Fig. 8.5), it can be seen that the station C is beyond the limit of 95% confidence band. Therefore, the station is omitted from the regional group of stations for further analysis. Columns (13) through (16) gives the regional values of $q_{rav10}$, $q_{rav\,50}$, $q_{rav100}$, and $q_{rav1000}$ as 1.766, 2.488, 2.816 and 3.847, respectively. The return period versus the $q_{rav}$ values are plotted to yield the regional curve as shown in Fig. 8.6.

## 8.6 ANALYSIS OF PARTIAL DURATION SERIES

As defined earlier, a partial duration series is obtained when all events above a selected minimum base value ($P_t$) are considered for analysis regardless of their sequence of occurrence. For such a series, it may so happen that for some years there may be number of events (floods) above a selected base value while some other years may go fully unrepresented. For such a series, correct statistical analysis cannot be carried out. However the advantage of such a series is that all larger floods are incorporated. Many a times the second, third and fourth highest floods of a year are higher than the highest flood of other years. The characteristics of annual maximum and partial duration series are

1. Both series (annual maximum series and partial duration series) give almost the same result for large floods, i.e., floods of higher return periods.
2. Partial duration series gives higher floods for short return periods.
3. For design of spillways annual maximum series are to be considered.
4. During construction of large dams, partial duration series is the best suited as the construction period is limited to 5–10 years only. Partial duration series are invariably used for design of coffer dams during construction period of large dams.
5. Partial duration series should be used for independent events like rainfall analysis and should be discouraged for computation of flood.

**Table 8.19  Homogeneity Test for Regional Flood Frequency Analysis**

| (1) Station name | (2) Data Length | (3) Mean $Q_{2.33}$ | (4) $Q_{10}$ | (5) $q_r =$ (4)/(3) | (6) $q_{rav}$ $\times Q_{2.33}$ | (7) $X_T$ of floods of (5) | (8) $S_e$ | (9) $Q_U=Q_{10}+$ $1.96\times(8)$ | (10) $Q_L=Q_{10}-$ $1.96\times(8)$ | (11) $K_U/T$ | (12) $K_L/T$ | (13) $\dfrac{Q_{10}}{Q_{2.33}}$ | (14) $\dfrac{Q_{50}}{Q_{2.33}}$ | (15) $\dfrac{Q_{100}}{Q_{2.33}}$ | (16) $\dfrac{Q_{1000}}{Q_{2.33}}$ |
|---|---|---|---|---|---|---|---|---|---|---|---|---|---|---|---|
| A | 55 | 388 | 630 | 1.623 | 655* | 12** | 5.8*** | 739 | 520 | 2.11/25 | 0.79/5.5 | 1.62 | 2.24 | 2.50 | 3.35 |
| B | 32 | 350 | 680 | 1.943 | 591 | 07 | 89.9 | 756 | 504 | 1.88/20.5 | 0.71/4.9 | 1.94 | 2.86 | 3.24 | 4.54 |
| C | 41 | 610 | 840 | 1.377 | 1030 | 41 | 55.9 | 950 | 730 | 2.21/31 | 0.78/5.4 | discarded....... | | | |
| D | 29 | 455 | 680 | 1.538 | 768 | 17 | 69.9 | 837 | 563 | 2.41/40 | 0.68/4.8 | 1.54 | 1.98 | 2.22 | 2.88 |
| E | 13 | 153 | 300 | 1.961 | 258 | 06 | 58.4 | 415 | 185 | 3.02/87 | 0.37/3.4 | 1.96 | 2.88 | 3.30 | 4.61 |
| Sum= | | | | 8.443 | | | | | | | | 7.066 | 9.95 | 12.27 | 15.39 |

Mean = $q_{rav10}$        1.6886

$q_{r\,av10} = 1.77$  $q_{r\,av50} = 2.49$  $q_{r\,av100} = 2.816$  $q_{r\,av1000} = 3.847$

\* $655 = 1.6886 \times 388$, $K = \dfrac{(665 - 388)}{166.2} = 1.666$ (Standard Deviation is given as 166.2)

\*\* For $K = 1.666$ the return period $T$ is 12 years (from Gumbel's $K$-$T$ table for $n = 55$)

\*\*\* $S_e = \sqrt{(1 + 1.3K + 1.1K^2)} \times \left(\dfrac{\sigma_{n-1}}{\sqrt{N}}\right) = \sqrt{(1 + 1.3 \times 1.666 + 1.1 \times 1.666^2)} \times \left(\dfrac{166.2}{\sqrt{55}}\right) = 5.8$

## PROBLEMS

8.1   (a)   What do you mean by design flood? Why its computation is so important for water resource projects?
      (b)   Enumerate various methods for estimating design flood.
      (c)   Give five formulae for estimating flood peak in India.

8.2   A flood of 1000 cumec exceeded 60 times during a period of 30 years. A flood of 3500 cumecs exceeded twice. Determine the annual probability and average recurrence interval for both floods.

8.3   What do understand by Probable Maximum Flood? Write stepwise procedure to determine it.

8.4   A 15 km$^2$ watershed is composed of 30% wood in fair hydrologic condition of B type, 30% of native pasture in good condition of D type and 40% contoured row crop in poor condition of A type. A storm of 20 cm falls over the watershed. Compute the runoff.

8.5   A minor irrigation project has no gauge-discharge station at its upstream. How would you proceed to develop an inflow hydrograph to the reservoir?

8.6   What is flood frequency analysis? How will you carry out flood frequency analysis at a project site? What are the data requirements and the limitations?

8.7   Determine the coefficients $C$ and $n$ of an empirical formula $Q = CA^n$, where $Q$ is the peak discharge in m$^3$/sec, $A$ is the catchment area in km$^2$ from the following data.

| Area (km$^2$) | 20 | 35 | 75 | 110 | 156 | 220 | 280 | 315 |
|---|---|---|---|---|---|---|---|---|
| Discharge (m$^3$/sec) | 310 | 450 | 530 | 600 | 700 | 760 | 805 | 850 |

8.8   Analysis of annual flood peaks for 50 years at a river site in Orissa shows the following relations
      10 year flood - 20200 m$^3$/sec ; 100 year flood-38500 m$^3$/sec
      Calculate (i) the magnitude of 200 year flood and (ii) probability of having a 50 year flood in the next 10 years.

8.9   Explain the terms: (a) Annual series, (b) Partial duration series, (c) Recurrence interval, (d) Probable maximum precipitation, (d) Rational formula for flood estimation.

8.10  Mean and standard deviation from annual peak of a river covering 80 years of data are 4100 and 1600 m$^3$/sec respectively. Using Gumbel's method, calculate the return period of the flood of 9100 m$^3$/sec.

8.11  A series of 40 years of flood peaks was arranged in decreasing order and the fifth largest flood was found to be 360 m$^3$/sec. Determine (i) Chance percent of this flood to occur on any one year, (ii) Chance percent that it may not occur in the next 10 years, (iii) Chance percent that its occurrence in next 10 years and (iv) Recurrence interval of this flood.

8.12  The following information are available at a gauging site

| River | Data length | Mean of the flood | Standard deviation |
|---|---|---|---|
| A | 80 years | 6200 | 2850 |
| B | 50 years | 5400 | 3210 |

      (i) Estimate 200 year and 500 year floods for the two rivers using Gumbel's method,
      (ii) Construct 99% confidence bands on the predicted 200 and 500 year floods.

8.13  Discuss critically the method of computing peak discharge for small catchments using rational formula.

8.14 A semi-urban catchment area of 1 km$^2$ has 60% of its area occupied by ordinary houses, 10% by roads, 8% by multistorey buildings, 10% parks and the rest are light industrial area. Time of concentration of the area is 40 min. The area is flat. There is rainfall of 67 mm occurring for a period of 60 min. Estimate the peak discharge.

8.15 The following information are available from the analysis of 40 years of data from a river gauging site taken on logarithm scale.
Mean of the series = 2.37; Standard deviation = 0.21; Coefficient of skewness = 0.64
Estimate the peak discharge for 50, 100 and 200 years using (i) Log-Pearson type III distribution (ii) Log normal distribution.

8.16 Maximum annual floods at a river site in m$^3$/sec are given below. Calculate 20, 100. 200 and 500 years flood using Foster's method, Chow's method and Gumbel's method. Obtain the return period of 650 m$^3$/sec.

| Year | 1972 | 73 | 74 | 75 | 76 | 77 | 78 | 79 | 80 | 81 | 82 |
|------|------|----|----|----|----|----|----|----|----|----|----|
|      | 1983 | 84 | 85 | 86 | 87 | 88 | 89 | 90 | 91 | 92 | 93 |
|      | 1994 | 95 |    |    |    |    |    |    |    |    |    |
| Flood | 730 | 380 | 452 | 810 | 460 | 800 | 320 | 445 | 390 | 440 | 470 |
|       | 872 | 350 | 415 | 960 | 435 | 270 | 300 | 390 | 1075 | 440 | 500 |
|       | 560 | 790 |    |    |    |    |    |    |    |    |    |

8.17 Dew point temperature at 1800 m above mean sea level is 8°C. Reduce this temperature to mean sea level.

8.18 Calculate a 10-year, 30 min design rainfall intensity for the following stations:
(a) Jharsuguda, (b) Dehradun, (c) Mangalore, (d) Mahabaleswar and (e) Indore.

8.19 Storm rainfall over a catchment along with the design UH are given below. Assuming standard loss rate from the rainfall, compute the design flood. Take base flow as 40 m$^3$/sec. Unit duration of the unit hydrograph is 6 h.

| Time (h) | 12 | 18 | 24 | 6 | 12 | 18 | 24 | 6 | 12 |
|----------|----|----|----|---|----|----|----|---|----|
|          | 18 | 24 | 6 |   |    |    |    |   |    |
| Storm rainfall (mm) | 0 | 110 | 250 | 320 | 380 |   |    |   |    |
| UH ordinates (m$^3$/sec) | 0 | 35 | 90 | 145 | 130 | 100 | 75 | 50 | 35 |
|          | 20 | 8 | 0 |   |    |    |    |   |    |

8.20 Mean and standard deviation of annual flood series from a record of 50 years are 5232 and 2410 m$^3$/sec respectively. Using Gumbel's frequency factors calculate flood peak for 200 year return period. Estimate 99%, 95% and 80% confidence limits for the 200 year flood.

8.21 Obtain the peak discharge for a drainage area of 60 sq. km located in (a) Orissa, (b) Uttar Pradesh, (c) Western Ghat of Maharashtra, (d) Kerala and (e) coastal Andhra pradesh. Use appropriate equations.

8.22 A barrage has the discharge capacity of 220 m$^3$/sec. Analysis of the available 24 years data show that mean and standard deviation are 72 and 32 m$^3$/sec respectively. Calculate return period for the design flood for the barrage.

# Chapter Nine

# Flood Routing

## 9.1   INTRODUCTION

Flood routing is the process of determining the flood hydrograph at a location downstream of a reservoir or a channel section from the knowledge of the upstream inflow hydrograph. The net effect of the reservoir (or the channel) is to modify the flood hydrograph. When the interest is to know the flood hydrograph from an existing or proposed reservoir then, it is called reservoir routing. On the other hand, if the interest is to know the probable effect of the channel at a downstream location to an inflow hydrograph while it propagates, the process is called channel routing. Hence, routing of flood is broadly classified into two groups; viz. (1) channel routing and (2) reservoir routing.

This chapter discussess various techniques used in routing floods either through a channel or a reservoir. The resultant hydrograph at the downstream location is called *outflow hydrograph*. Routing helps to fix the capacity of the spillway of reservoirs, water control structures, protection works and forecasting of floods for evacuation and other operations. In both the routings, the two distinct modifications that take place to the inflow hydrograph are:

(a)   The peak of outflow hydrograph is less than the peak of inflow hydrograph. We call this phenomena as *attenuation of the peak*. The time base of the hydrograph is increased. It is due to the combined effect of storage and channel friction.

(b)   The peak of outflow hydrograph occurs sometime later than the inflow hydrograph. This is called *translation* or *lag* of the peak and is due to the travel time of the flood waves in the channel or reservoir.

The flood is therefore said to be *moderated* while passing through a reservoir or a channel. Outflow from a reservoir is directly related to the head over spillway, more is the head, higher is the discharge over spillway. When flood volume accumulates over a spillway, its head increases and so is the outflow from it. The rate of outflow is always less than the inflow values during the initial period. A plot of the shapes of the inflow and outflow hydrographs are shown in Fig. 9.1. Initially there is accumulation of inflow flood volumes to build the storage in the section. As storage increases from A to C through B (Fig. 9.1) the outflow also increases slowly, but with lesser rate than the increase of inflow. A stage comes (at point C), when the two curves cross each other. At this point the outflow is maximum as the inflow hydrograph has already reached its peak. After point C, the accumulated storage represented by the area ABCA is gradually discharged from the system.

The shape of Fig. 9.1 holds equally good for channel and reservoir routing. The shape of outflow hydrograph in channels depends on the geometry, bed slope, the shape of inflow hydrograph, length of the reach and the volume of water already contained by the channel before the inflow hydrograph arrives at the section. For peaked inflow hydrographs, the attenuation is higher than moderate floods. Channels having wider section at the outlet or for longer routing length of the channels, the reduction in the peak is higher. For steep channels or channels with good storage of water initially, the reduction of peak will be less.

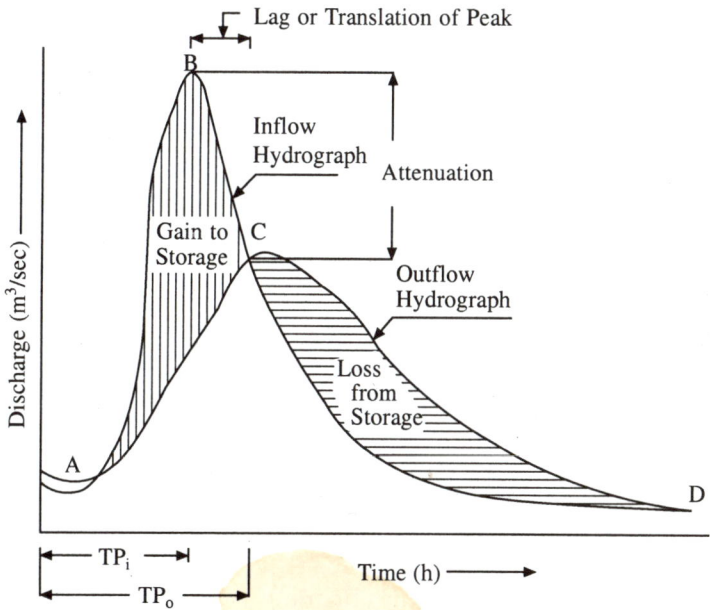

**Fig. 9.1   Hydrograph Shapes Before and After Routing**

Since the basic idea of routing is to determine the flood hydrograph at a location downstream of a reservoir or a channel from the knowledge of the complete inflow hydrograph upstream of it, some important methods of routings are discussed in this chapter.

## 9.2   ROUTING METHODS

All the methods available for routing floods in channels and reservoirs are broadly classified into three groups: (1) Hydrologic routing, (2) Hydraulic routing and (3) Routing machines.

Figure 9.2 illustrates various routing methods employed under each group. Natural channels are so complex that a true picture of flow hydraulics is difficult to obtain through mathematics. Flow through a river reach always involves a varied flow problem and the flow is unsteady. Hydraulic routing methods use two conservation equations. The model using the conservation of mass and momentum equations to the flow problems always give better results. However

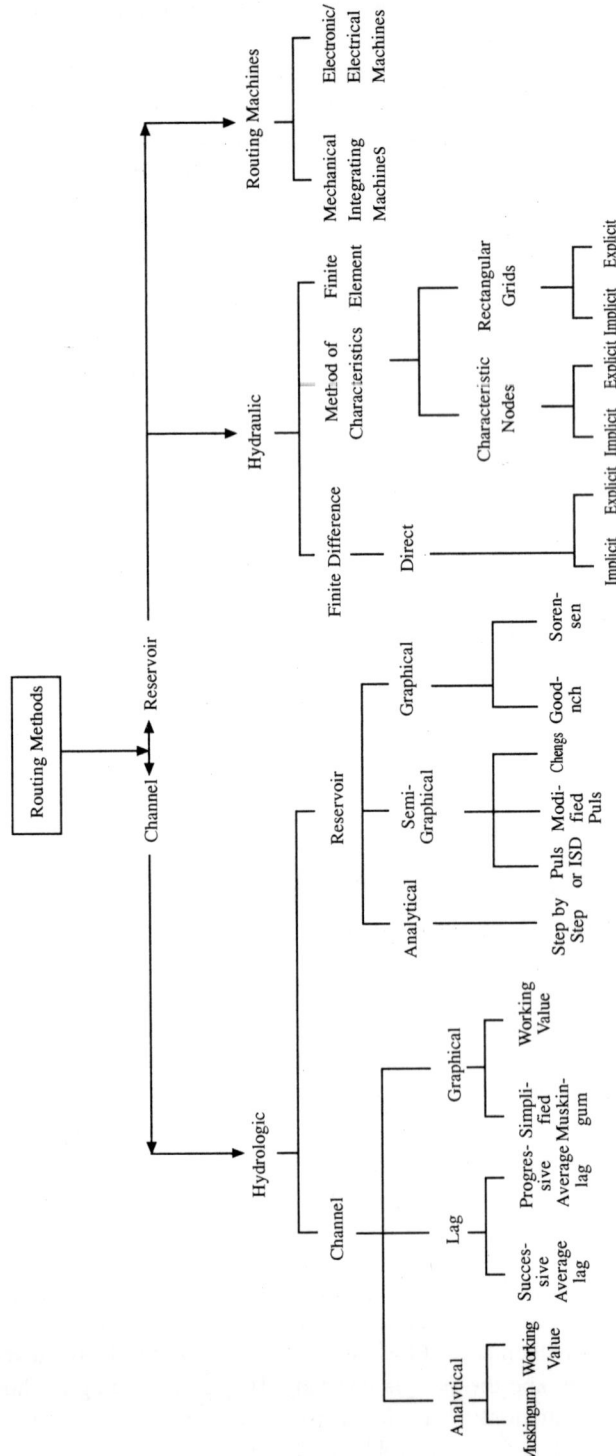

**Fig. 9.2   Routing Methods**

this method of routing lacks the interest of hydrologist due to the use of difficult mathematics, high quality of input data, need for good computing technology and involvement of number of informations while routing. On the other hand, the hydrologic routing oversimplifies the procedure such that the method of routing becomes very simple and the results are acceptable to a hydrologist for all practical purposes. Routing machines takes care of the problem by correlating the channel or reservoir system with mechanical gears or fluctuation of voltage in an instrument. The hydrologic routing is discussed in this chapter with a brief introduction to the hydraulic routing and routing machines.

For unsteady flow, the continuity equation in differential form is

$$\frac{dq}{dx} + \frac{dA}{dt} = 0 \tag{9.1a}$$

or in conservation form the equation can be written as

$$v\frac{dy}{dx} + y\frac{dv}{dx} + \frac{dy}{dt} = 0 \tag{9.1b}$$

The momentum equation per unit width of flow, neglecting eddy losses, wind shear affect and lateral inflow can take the following form

$$\frac{dv}{dt} + v\frac{dv}{dx} + g\left(\frac{dy}{dx} - S_0 + S_f\right) = 0 \tag{9.2}$$

where $dq/dx$ is the rate of change of channel flow with distance $x$, $A$ the average cross sectional area, $t$ the time, $v$ the average flow velocity, $g$ the acceleration due to gravity, $y$ the depth of flow in the channel, $S_0$ the bed slope and $S_f$ the friction slope. Equations (9.1) together with (9.2) are called *St. Venant equation*. Hydraulic flood routing use these equations for calculating propagation of flood wave with time. Solution to the problem is carried out by numerical methods using a computing machine. For hydrologic routing, equations (9.1) can be written as

$$I - O = dS/dt \tag{9.3}$$

where $S$ is the storage of water, $I$ the inflow rate, $O$ the outflow rate and $dS/dt$ is the change in storage. Momentum is not considered for hydrologic problems but a relation between storage-outflow and storage-inflow is used for routing. Using these equations, some important hydrologic routing methods are discussed as follows.

## 9.3   HYDROLOGIC CHANNEL ROUTING

In hydrologic routing, the basic continuity equation (9.3) along with a relationship between the storage-inflow or storage-outflow is used. The approach is highly simplified to suit field situations, where the flow is essentially nonsteady, nonuniform and the properties of channel and flood plain vary from stage to stage and from section to section. The lateral inflow at places make the situation very complicated. The problem is simplified by assuming no lateral inflow and the changes occurring gradually with time. During the rising stage, a flood wave propagates in the channel as shown in Fig. 9.3. In the rising stage the inflow is always higher than the outflow. This gives rise to additional storage between section 3 – 3′ and 4 – 4′ as shown in the form of a wedge. Thus, two storages are considered between channel sections 3 – 3′ and 4 – 4′.

**Prism Storage:** The storage of water bound between any two sections of a channel by an imaginary plane passing parallel to the channel bottom as shown in Fig. 9.3 is called prism storage. It is shown as *AB34′4′* in Fig. (9.3 b).

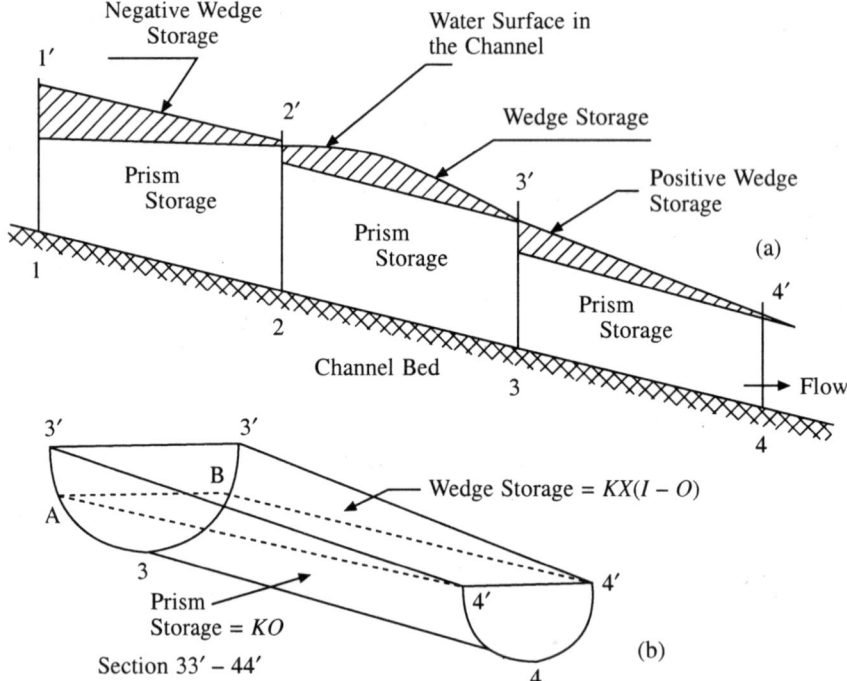

**Fig. 9.3　Possible Profile of Flood Wave in a Channel**

**Wedge Storage:** It is the storage of water between the top of the prism storage and the actual water surface during the passage of flood wave in a channel. It is represented as 3′3′4′4′ AB in Fig. 9.3 b.

Consider the sections 1, 2, 3 and 4 of the channel in Fig. (9.3a). When the flood wave propagates, there is negative wedge at the recession and a positive wedge at the advance flood. Another positive wedge is always present at the wave peak. The prism storage is dependent on the outflow $O$ alone and the wedge storage is dependant on the inflow and outflow, i.e. ($I - O$). Different methods of routing differ in their approach to account for their relation between storage and outflow.

G.T. McCarthy, while studying the Muskingum conservancy district, Ohio flood control project of U.S. Army Corps of Engineers (1934) proposed the idea of storage as a function of both inflow and outflow of a channel between two sections. Thus total storage can be expressed as

$$S = KX\,I^n + KO^n - KXO^n$$

$$= K[X\,(I^n - O^n) + O^n]$$

or
$$S = K[XI^n + (1 - X)O^n] \tag{9.4}$$

where $K$ and $X$ are coefficients and $n$ is an exponent whose value is taken as unit for natural channels. For a defined rectangular channel, its value can be as low as 0.60.

### 9.3.1 Muskingum Equation

For natural channels ($n = 1$) equation (9.4) is written as

$$S = K[X\,I + (1 - X)O] \tag{9.5}$$

where $K$ is called the *storage time constant* having the dimension of time. It can be approximated to travel time of flood wave between two given sections of the channel. The parameter $X$ is called *weighing factor*, whose value varies between 0.1 and 0.5. For most of the natural channel its value lies between 0.1 and 0.3. The above equation is called Muskingum equation due to its origin. For $X = 0$, equation (9.5) gives $S = KO$. This means there is no weighing between inflow and outflow, the contribution of inflow to the storage being zero, the storage is a function of outflow only. We known that the equation $S = KO$ is the definition of a linear reservoir.

Consider a time interval $\Delta t$. Let the storage, inflow and outflow at the beginning of $\Delta t$ are $S_1$, $I_1$, $O_1$ and at the end of $\Delta t$ are $S_2$, $I_2$ and $O_2$, respectively. Applying equation (9.5) for the period $\Delta t$

$$S_1 = K\,[I_1 X + (1 - X)O_1]$$

$$S_2 = K\,[I_2 X + (1 - X)O_2]$$

and subtracting one from the other, we get

$$S_2 - S_1 = K[X(I_2 - I_1) + (1 - X)(O_2 - O_1)] \tag{9.6}$$

Continuity equation defined in (9.3) is rewritten as

$$\text{Inflow }(I) - \text{Outflow }(O) = \text{Change in storage} \tag{9.7}$$

$$\frac{I_1 + I_2}{2} - \frac{O_1 + O_2}{2} = \frac{S_2 - S_1}{\Delta t} \tag{9.8}$$

or
$$S_2 - S_1 = \frac{I_1 + I_2}{2}\,\Delta t - \frac{O_1 + O_2}{2}\,\Delta t \tag{9.9}$$

Combining equations (9.6) and (9.9) and solving for $O_2$

$$O_2 = C_0 I_2 + C_1 I_1 + C_2 O_1 \tag{9.10}$$

where

$$C_0 = \frac{-KX + 0.5\Delta t}{K - KX + 0.5\Delta t}, \quad C_1 = \frac{KX + 0.5\Delta t}{K - KX + 0.5\Delta t}, \quad C_2 = \frac{K - KX - 0.5\Delta t}{K - KX + 0.5\Delta t}$$

$$(9.11)$$

Therefore
$$C_0 + C_1 + C_2 = 1.0 \tag{9.12}$$

Equation (9.10) is popularly called *Muskingum routing equation*. Though the routing interval $\Delta t$ is chosen arbitrarily, smaller is its value, more accurate is the routing results. On the other hand, if $\Delta t$ is taken large enough (say $\Delta t = K$), then for $X = 0.5$ equation (9.10) gives $O_2 = I_1$. This means the peak is only translated with a lag time of $\Delta t = K$. The limiting value of $\Delta t$ should be such that

1. $\Delta t \geq 2 KX$
2. The weighing factor $X$ should always be less than 0.5 as for values greater than 0.5, $O_2$ becomes negative.

### 9.3.1.1 Determination of K and X
The two approaches to determine the coefficients, $K$ and $X$ for a channel are:

*Approach 1*

1. Between the given two sections of the channel note two observed hydrographs. Let $I$ is the inflow hydrograph at the upstream section 1–1, then $O$ is the corresponding outflow hydrograph at the downstream section 2–2.
2. Chose a trial value of $X$.
3. Plot a graph between storage $S$ against $[XI + (1-X)O]$ covering the entire hydrograph using an ordinary graph paper. If the plot forms a loop then change the value of $X$ and repeat the plot till the loop closes to a straight line between the rising and falling curves. This is because $S$ and $[XI + (1-X)O]$ are assumed to be linearly related. To begin with, $X$ may be taken as 0.05 and increased gradually by a suitable step of say 0.05. Graphs for each increment may be plotted separately. A typical plot of the loop for $X = 0.20, 0.25$ and $0.30$ are shown in Figs. 9.4(a), (b) and (c).

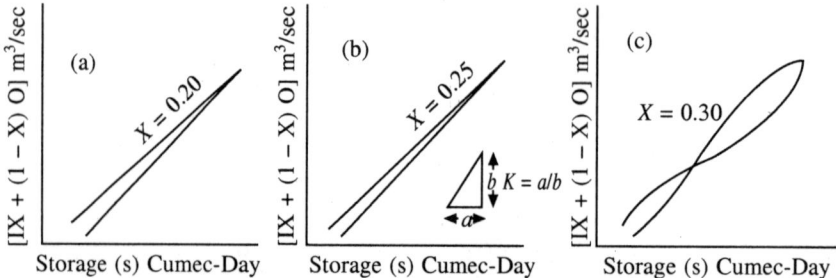

**Fig. 9.4   Determination of Muskingum Storage Constants K and X.**

4. Storage time constant $K$ is the reciprocal of slope of the line formed for the selected value of $X$. Thus, if $S$ is plotted in abscissa and $\{XI + (1-X)O\}$ as

ordinate than the reciprocal of slope of the line $S/[IX(1-X)O]$ is the value of $K$ for which $X$ is already selected from the straight line plot of the curve. In Fig. 9.4(b), $K$ is shown as the ratio $a/b$ for $X = 0.25$.

---

**Example 9.1:** The inflow and outflow hydrographs for a river reach are given below. Determine Muskingum's coefficients $K$ and $X$ for the reach.

| Time (h) | 0 | 12 | 24 | 36 | 48 | 60 | 72 | 84 | 96 | 108 | 120 |
|---|---|---|---|---|---|---|---|---|---|---|---|
| Inflow (cumecs) | 15 | 195 | 255 | 170 | 115 | 80 | 65 | 50 | 35 | 30 | 20 |
| Outflow (cumecs) | 10 | 28 | 115 | 175 | 165 | 140 | 120 | 90 | 70 | 50 | 30 |

**Solution**
All pertinent calculation for the determination of Muskingum coefficients $K$ and $X$ are carried in Table 9.1.

**Table 9.1  Determination of Muskingum Coefficients $K$ and $X$**

| Time (h) | Inflow (cumecs) | Outflow (cumecs) | $I - O$ | Mean storage cumecs-day | Σ col. (5) | For $X = 0.2$ $0.8O + 0.2I$ | For $X = 0.25$ $0.75O + 0.25I$ | For $X = 0.3$ $0.7O + 0.3I$ |
|---|---|---|---|---|---|---|---|---|
| (1) | (2) | (3) | (4) | (5) | (6) | (7) | (8) | (9) |
| 0 | 15 | 10 | 5 | 1.25* | 1.25 | 11 | 11.25 | 11.5 |
| 12 | 195 | 28 | 167 | 43.0** | 44.5 | 61.4 | 69.75 | 78.1 |
| 24 | 255 | 115 | 140 | 76.75 | 121.3 | 143.0 | 150.0 | 157.0 |
| 36 | 170 | 175 | −5 | 33.75 | 155.0 | 174.0 | 173.7 | 173.0 |
| 48 | 115 | 165 | −50 | −13.75 | 141.3 | 155.0 | 152.5 | 150.0 |
| 60 | 80 | 140 | −60 | −27.50 | 113.8 | 128.0 | 125.0 | 122.0 |
| 72 | 65 | 120 | −55 | −28.75 | 85.0 | 109.0 | 106.3 | 103.5 |
| 84 | 50 | 90 | −40 | −23.75 | 61.3 | 82.0 | 80.0 | 78.0 |
| 96 | 35 | 70 | −35 | −18.75 | 42.3 | 63.0 | 61.3 | 59.5 |
| 108 | 30 | 50 | −20 | −13.75 | 28.8 | 46.0 | 45.0 | 44.0 |
| 120 | 20 | 30 | −10 | −7.5 | 21.3 | 28.0 | 27.5 | 27.0 |

$$*\left\{\frac{(5+0)}{2}\right\}\frac{12}{24} = 1.25; \qquad **\left\{\frac{167+5}{2}\right\}\frac{12}{24} = 43.0$$

Table 9.1 is self-explanatory. Routing interval $dt$ is taken as 12 h. Using equation (9.3) mean storage is calculated by multiplying the storage $(I - O)$ by 12/24. The values of $X$ is taken initially as 0.20 and increased to 0.25 and 0.30 subsequently. Values of $[IX + (1-X)O]$ are calculated and given in col. (7) through (9). Plot of cumulative storages vs. $[IX + (1-X)O]$ for all the three trial values of $X$ are shown in Fig. 9.5. As can be seen from the figure $X = 0.20$ gives a closed loop. Therefore $X$ is taken as 0.20 for the problem.

*Approach 2*
Combining equations (9.6) and (9.8) and solving for the value of $K$ we get

$$K = \frac{0.5\Delta t\,[I_2 + I_1) - (O_2 + O_1)]}{X(I_2 - I_1) + (1 - X)(O_2 - O_1)} \qquad (9.13)$$

In this approach $X$ is assumed and equation (9.13) is evaluated for $K$. Since $O_i$ and $I_i$ are known for the problem, the values of the numerator and denominator

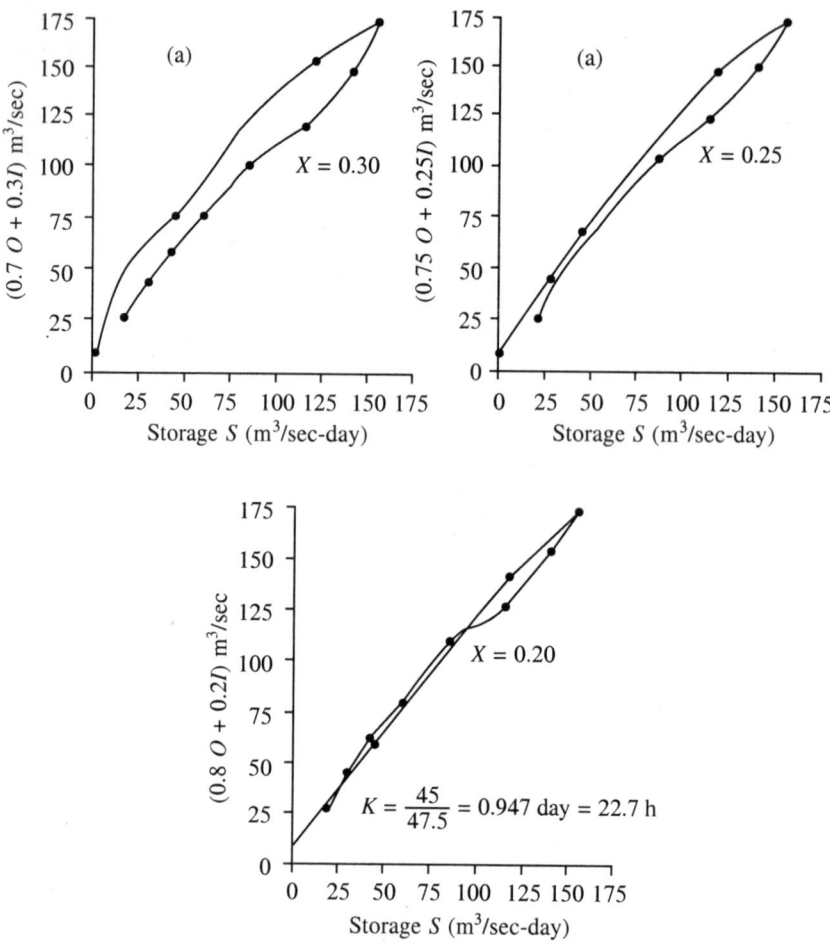

**Fig. 9.5    Calculation of K and X for Example 9.1**

are plotted in a graph. Cumulative values of numerator are taken in abscissa and the cumulative values of denominator are taken as ordinate. For a particular value of $X$, one loop will be plotted. The procedure is repeated with changed value of $X$ such that the loop closes to form a straight line. Select the value of $X$, when the loop appears to be close to a straight line. Reciprocal of the slope of the straight line gives the value of $K$ for the problem. Numerator of equation (9.13) represents the storage and the denominator, the weighted flow increments in the channel. Plot of the accumulated numerator and accumulated denominator is shown in Fig. 9.6.

---

**Example 9.2:** Solve Example 9.1 by Approach 2.

**Solution**
Computation are carried out in Table 9.2.

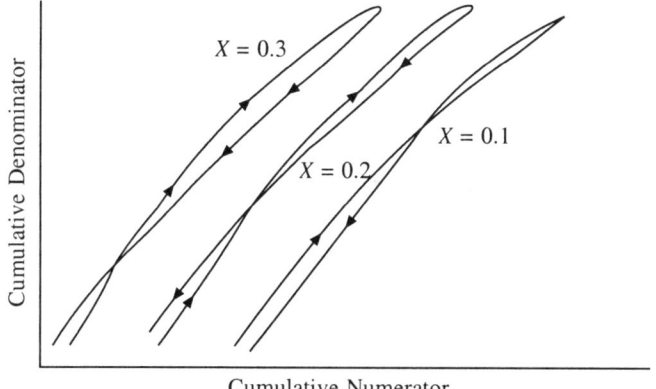

**Fig. 9.6   Estimation of Muskingum's K and X using Approach 2**

Successive values of numerator of equation (9.13) are computed by taking $\Delta t$ as 12 h. The cumulative numerator values are given in col. (5). Similarly for values of $X = 0.15$, 0.20 and 0.25, the denominator of equation (9.13) are calculated in the cols. (6), (8) and (10) respectively with their cumulative values in the cols. (7), (9) and (11). For the three cases of $X$, the values of numerator vs. denominator are plotted as shown in Fig. 9.7a, b and c respectively. It can be seen from the plots that for $X = 0.20$, the loop closes near to a straight line. Therefore $X$ is taken as 0.20 for the problem. The storage time constant $K$ is obtained from Fig. 9.7b as 23.5 h.

### 9.3.1.2   Routing Procedure
While using Muskingum method for channel routing the following steps are followed:

1. Choose two sites, $A$ and $B$ along the channel such that there is no loss or gain of water between the two reaches due to tributaries, effluent or influent conditions of the stream. The site $A$ is in the upstream where the inflow hydrograph is available and the site $B$ is where the routed outflow is to be determined.
2. Collect data of a complete flood (inflow) hydrograph at site $A$. This data is represented as inflow $I$. For site $B$, the hydrograph $O$ is to be computed when the flood from $A$ reaches $B$.
3. Compute $K$ and $X$ for channel between sections $A$ and $B$ as outlined in Approaches 1 or 2 from the data of previously observed hydrographs at $A$ and $B$.
4. Select the time interval of routing $\Delta t$. The relation used to determine $\Delta t$ is $K > \Delta t > 2\,KX$.
5. Calculate $C_0$, $C_1$ and $C_2$ and check the values through equation (9.12).
6. For the initial time, i.e., $t = 0$, $I_1$, $I_2$, and $O_1$ are known. Use equation (9.10) to calculate $O_2$.
7. Now $O_2$ of step (6) becomes $O_1$ in the next step of routing. Repeat the procedure to cover the entire inflow hydrograph.

Calculation is usually carried out in a tabular form in which computation is

Table 9.2  Determination of Muskingum Coefficients K and X by Approach 2

| Time (h) | Inflow m³/sec | Outflow m³/sec | Numerator | Σ col.(4) | Denominator | | | | | |
|---|---|---|---|---|---|---|---|---|---|---|
| | | | | | $X = 0.15$ | Σ col. (6) | $X = 0.20$ | Σ col. (8) | $X = 0.25$ | Σ col. (10) |
| (1) | (2) | (3) | (4) | (5) | (6) | (7) | (8) | (9) | (10) | (11) |
| 0 | 15 | 10 | 1032 | 1032 | 42.3 | 42.3 | 50.4 | 50.4 | 58.5 | 58.8 |
| 12 | 195 | 28 | 1842 | 2874 | 83.0 | 125.3 | 81.6 | 132.0 | 80.3 | 138.8 |
| 24 | 255 | 115 | 810 | 3684 | 38.3 | 163.6 | 31.0 | 163.0 | 23.7 | 162.5 |
| 36 | 170 | 175 | −330 | 3354 | −12.5 | 151.1 | −15.0 | 148.0 | −17.6 | 144.9 |
| 48 | 115 | 165 | −660 | 2694 | −26.5 | 124.6 | −27.0 | 121.0 | −27.6 | 117.3 |
| 60 | 80 | 140 | −690 | 2094 | −19.3 | 105.3 | −19.0 | 102.0 | −18.8 | 98.5 |
| 72 | 65 | 120 | −570 | 1524 | −27.3 | 78.0 | −27.0 | 75.0 | −26.3 | 72.2 |
| 84 | 50 | 90 | −450 | 1074 | −19.3 | 58.7 | −19.0 | 56.0 | −1.8 | 53.4 |
| 96 | 35 | 70 | −330 | 744 | −17.8 | 40.9 | −17.0 | 39.0 | −16.3 | 37.1 |
| 108 | 30 | 50 | −180 | 564 | −18.5 | 22.4 | −18.0 | 21.0 | −17.5 | 19.6 |
| 120 | 20 | 30 | | | | | | | | |

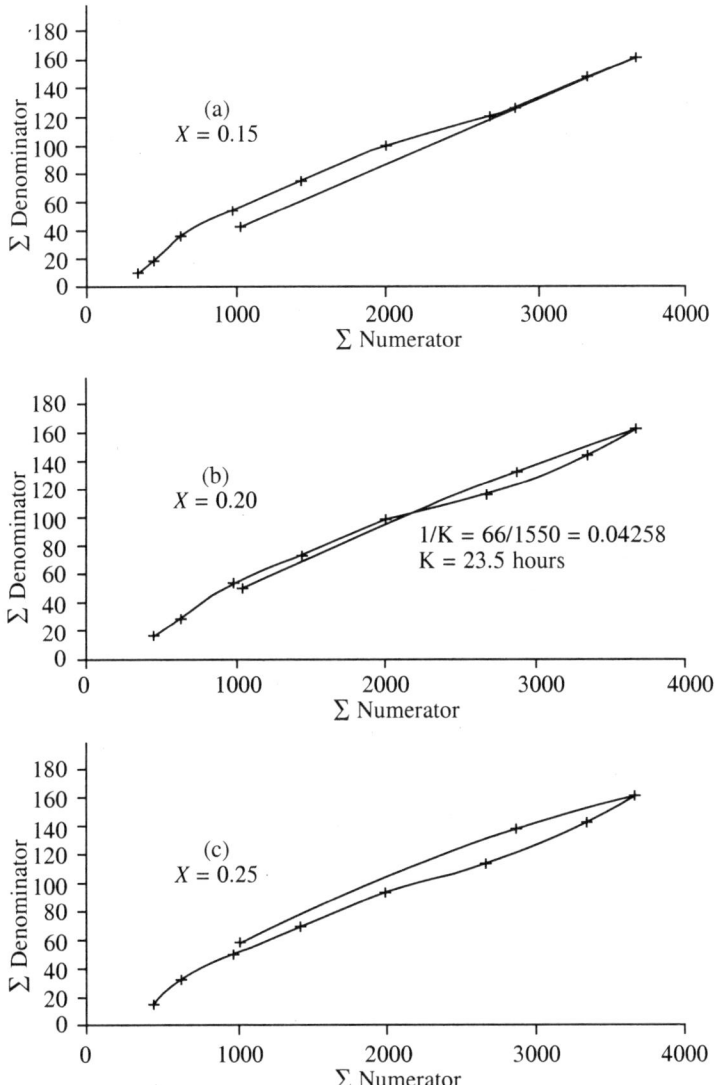

**Fig. 9.7** **Computation of Muskingum's Coefficients *K* and *X* by Approach 2 for Example 9.2**

done row-wise. Use of LOTUS 1–2–3 reduces the routing work to minimum. Standard computer packages are also available for Muskingum routing.

---

**Example 9.3:** Between two reaches A and B of a river, the values of Muskingum coefficients determined are $K = 24$ h and $X = 0.20$. Take outflow at the begining of routing step equal to inflow. Find the outflow hydrograph at *B*.

| Time (h) | 12 | 24 | 36 | 48 | 60 | 72 | 84 | 96 | 108 | 120 | 132 | 144 | 156 | 168 |
|---|---|---|---|---|---|---|---|---|---|---|---|---|---|---|
| Inflow ($m^3$/sec) | 14 | 22 | 36 | 93 | 141 | 102 | 86 | 73 | 61 | 50 | 38 | 26 | 20 | 16 |

**Solution**

Given that $K = 24$ and $X = 0.20$

$$2 K X = 2 \times 24 \times 0.2 = 9.6$$

Therefore $\Delta t$ should be between 9.60 to 24.0 h.

Take $\Delta t = 12.0$ h to suite the time ordinates given in the problem. From equation (9.11)

$$C_0 = \frac{-KX + 0.5\Delta t}{K - KX + 0.5\Delta t} = \frac{-24 \times 0.2 + 0.5 \times 12}{24 - 24 \times 0.2 + 0.5 \times 12} = 0.0476$$

$$C_1 = \frac{KX + 0.5\Delta t}{K - KX + 0.5\Delta t} = \frac{24 \times 0.2 + 0.5 \times 12}{24 - 24 \times 0.2 + 0.5 \times 12} = 0.4286$$

$$C_2 = \frac{K - KX - 0.5\Delta t}{K - KX + 0.5 \Delta t} = \frac{24 - 24 \times 0.2 - 0.5 \times 12}{24 - 24 \times 0.2 + 0.5 \times 12} = 0.5238$$

Routing equation is given as $\qquad O_2 = C_0 I_2 + C_1 I_1 + C_2 O_1$

where $C_0 + C_1 + C_2 = 0.0467 + 0.4286 + 0.5238 = 1.0$

Calculation is carried out in Table 9.3.

Table 9.3   Channel Routing by Muskingum Method

| Time (h) | Inflow $I_i$ (cumecs) | $0.0467 I_2$ (cumecs) | $0.4286 I_1$ (cumecs) | $0.5238 O_1$ (cumecs) | $O_2 = C_0 I_2 + C_1 I_1 + C_2 O_1$ (cumecs) |
|---|---|---|---|---|---|
| (1) | (2) | (3) | (4) | (5) | (6) |
| 12 | 14 | – | 6.0 | 7.33 | 14.00* |
| 24 | 22 | 1.03 | 9.43 | 7.52 | 14.36** |
| 36 | 36 | 1.68 | 15.43 | 9.76 | 18.63 |
| 48 | 93 | 4.34 | 39.86 | 15.47 | 29.53 |
| 60 | 41 | 6.58 | 60.43 | 32.43 | 61.91 |
| 72 | 102 | 4.76 | 43.72 | 51.13 | 97.62 |
| 84 | 86 | 4.02 | 36.86 | 51.79 | 98.87 |
| 96 | 73 | 3.41 | 31.29 | 48.22 | 92.06 |
| 108 | 61 | 2.85 | 26.41 | 43.14 | 82.36 |
| 120 | 50 | 2.34 | 21.43 | 7.51 | 71.62 |
| 132 | 38 | 1.77 | 16.29 | 31.80 | 60.71 |
| 144 | 26 | 1.21 | 11.14 | 25.82 | 49.3 |
| 156 | 20 | 0.93 | 8.57 | 19.85 | 37.90 |
| 168 | 16 | 0.75 | 6.86 | 15.28 | 29.17 |

*14 = Assumed to match $I_1 = 14.0$ initially

**14.36 = {0.0467 × 22 + 0.4286 × 14 + 0.5238 × 14}

$O_2$ of 14.36 is taken as $O_1$ for the second step of routing to get $O_3$ as 18.63.

For the first iteration $I_1 = 14$ and $I_2 = 22$

For the second iteration $I_1 = 22$ and $I_2 = 36$ and so on till the entire inflow hydrograph is covered.

The routing steps are repeated to get the routed outflow hydrograph as given in the col.(6) of Table 9.3.

### 9.3.2 Working Value Method

Equation (9.5) can be written as

$$S = KXI + KO - KXO = [KX(I - O) + KO] \qquad (9.14)$$

where $KX(I - O)$ is the *wedge storage* and $KO$ the *channel storage*. If $KX(I - O)$ is replaced by $K(D - O)$, where $D$ is the *working discharge* as shown in Fig. 9.8, then storage equation (9.14) can be rewritten in terms of working value $D$ as

$$D = IX + (1 - X)O \qquad (9.15)$$

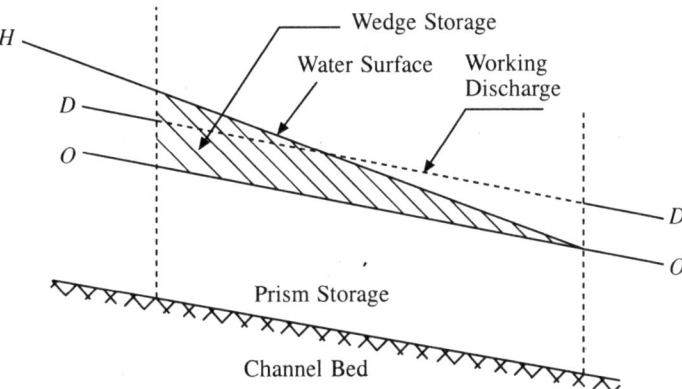

**Fig. 9.8   Schematic Diagram Showing Working Discharge *D***

For outflow at time step $\Delta t$

$$O_1 = D_1 - \frac{X}{1 - X}(I_1 - D_1) \qquad (9.16\ a)$$

and

$$O_2 = D_2 - \frac{X}{1 - X}(I_2 - D_2) \qquad (9.16\ b)$$

Combining equations (9.16 a), (9.16 b) with (9.9), we get

$$S_2 - S_1 = \frac{(I_2 + I_1)\Delta t}{2(1 - X)} - \frac{(D_1 + D_2)\Delta t}{2(1 - X)} \qquad (9.17)$$

Let

$$R_1 = S_1(1 - X) + 0.5\ \Delta t\ D_1 \qquad (9.18)$$

$$R_2 = S_2(1 - X) + 0.5\ \Delta t\ D_2 \qquad (9.19)$$

$$= D_2[K(1 - X) + 0.5\ \Delta t] \qquad \{\text{as } S_2 = KD_2\}$$

Then from equation (9.14), the value of $R_2$ can be obtained as

$$R_2 = R_1 + 0.5(I_1 + I_2)\ \Delta t - \Delta t\ D_1 \qquad (9.20)$$

$R_2$ and $R_1$ are called *working values*. By combining equations (9.14) and (9.15) and solving for $S$, we get $S = KD$. Therefore, the storages $S$ and $D$ are also related as

$$S_1 = KD_1 \text{ and } S_2 = KD_2 \qquad\qquad (9.21)$$

The steps followed in the working value method of routing are:

1. Compute $K$ and $X$ as before for the channel reach.
2. Compute $D_1$ from known $I_1$, $O_1$ and $X$ from equation (9.16a).
3. Using equation (9.21) compute $S_1$.
4. Compute $R_1$ from equation (9.18).
5. Compute $R_2$ from equation (9.20).
6. Use equation (9.19) and $S_2 = KD_2$ and solve for $D_2$.
7. Compute $O_2$ from equation (9.16b).
8. Repeat the steps to cover the entire inflow hydrograph.

---

**Example 9.4:** Solve Example (9.3) by working value method.

**Solution**

Take $X = 0.2$, The value of $O_1$, i.e., the initial outflow is taken equal to the inflow of 14 $m^3$. $D_1$ is calculated as 14 $m^3$. Flood routing by working value method is carried out in Table 9.4.

**Table 9. 4   Routing by Working Value Method**

| Time (h) | Inflow $I_i$ (cumecs) | $D_1 = IX$ $+(1-X)O$ (cumecs) | $S_1 = KD_1$ (cumecs) | $R_1 = S_10.8$ $+0.5\times12\times D$ (cumecs) | $R_2$ (cumec h) | $D_2$ (cumecs) | $O_2$ (cumecs) |
|---|---|---|---|---|---|---|---|
| (1) | (2) | (3) | (4) | (5) | (6) | (7) | (8) |
| 0 | 14 | – | – | – | – | – | 14.00 |
| 12 | 22 | 14.00 | 336.00 | 352.80 | 400.80[*] | 15.90[**] | 14.38[***] |
| 36 | 36 | 15.90 | 381.71 | 400.80 | 557.94 | 22.14 | 18.68 |
| 48 | 93 | 22.14 | 531.37 | 557.94 | 1066.26 | 42.31 | 29.64 |
| 60 | 141 | 42.31 | 1015.48 | 1066.26 | 1962.51 | 77.88 | 62.10 |
| 72 | 102 | 77.88 | 1869.06 | 1962.51 | 2485.98 | 98.65 | 97.81 |
| 84 | 86 | 98.65 | 2367.60 | 2485.98 | 2430.18 | 96.44 | 99.04 |
| 96 | 73 | 96.44 | 2314.46 | 2430.18 | 2226.95 | 88.37 | 92.21 |
| 108 | 61 | 88.37 | 2120.91 | 2226.95 | 1970.50 | 78.19 | 82.49 |
| 120 | 50 | 78.19 | 1876.67 | 1970.50 | 1698.17 | 67.39 | 71.73 |
| 132 | 38 | 67.39 | 1617.30 | 1698.17 | 1417.52 | 56.25 | 60.81 |
| 144 | 26 | 56.25 | 1350.01 | 1417.52 | 1126.51 | 44.70 | 49.38 |
| 156 | 20 | 44.70 | 072.86 | 1126.51 | 866.08 | 34.37 | 37.96 |
| 168 | 16 | 34.37 | 824.83 | 866.08 | 669.66 | 26.57 | 29.22 |
| 180 | 12 | 26.57 | 637.77 | 669.66 | 446.77 | 17.73 | 22.16 |

[*]$R_2 = R_1 + 0.5(I_1 + I_2) \Delta t - \Delta t \Delta_1 = 352.8 + (14 + 22) \times 0.5 \times 12 - 12 \times 14 = 400.80$

[**] $R_2 = S_2 (1 - X) + 0.5 \Delta t D_2 = KD_2 (1 - X) + 0.5 \Delta t D_2$

or $400.80 = 24 D_2 (1 - 0.20) + 0.5 \times 12 \times D_2 = 25.2 D_2$ or $D_2 = 15.90$

[***] $O_2 = D_2 - \dfrac{X}{1 - X}(I_2 - D_2) = 15.9 - \dfrac{0.20}{(1 - 0.2)}(22 - 15.9) = 14.38$ cumecs

The routing process is continued to cover the entire inflow hydrograph. The routed outflow is shown in col. (8) of Table 9.4.

## 9.4 HYDRAULIC CHANNEL ROUTING

In this routing the flow is considered as unsteady. The two partial deferential equations for routing unsteady flow are obtained from equations (9.1) and (9.2) by substituting $S_0 = S_f$, $u = v$, $h = y$, the equations are

1. Momentum equation

$$\frac{\partial u}{\partial t} + u \frac{\partial u}{\partial x} + g \frac{\partial h}{\partial x} = 0 \qquad (9.22)$$

2. Continuity equation

$$\frac{\partial h}{\partial t} + h \frac{\partial u}{\partial x} + u \frac{\partial h}{\partial x} = 0 \qquad (9.23)$$

where $u$ is the velocity, $h$ the head of water, $x$ the direction of movement of water, $t$ the time and $g$ acceleration due to gravity. The two equations represent gradually varied unsteady flow equations proposed by St. Venant (1871), where the momentum diffusing terms $g(S_0 - S_f)$ are equated to zero. Solution of equations (9.22) and (9.23) are possible by using various numerical techniques, coupled with a high speeded digital computer. Muskingum channel routing is an approximate hydrologic solution of the continuity equation (9.23) with a drastically reduced equation of motion in the form of *storage-inflow-outflow*. The numerical methods available to solve St. Venant equation are detailed in Fig. 9.2. Method of Characteristics (MOC) are popularly used to convert the equations (9.22) and (9.23) to ordinary differential form (characteristic form) and a finite difference method is applied to solve it.

Let $c^2 = gh$, differentiating it, we get

$$g \frac{\partial h}{\partial x} = 2c \frac{\partial c}{\partial x} \qquad (9.24)$$

Multiplying equation (9.23) by $g$ and substituting $c^2 = gh$, the equation reduces to

$$\frac{\partial c^2}{\partial t} + c^2 \frac{\partial u}{\partial x} + u \frac{\partial c^2}{\partial x} = 0 \qquad (9.25)$$

or

$$\frac{\partial(2c)}{\partial t} + c \frac{\partial u}{\partial x} + u \frac{\partial(2c)}{\partial x} = 0 \qquad (9.26)$$

Similarly from equation (9.22) we get

$$\frac{\partial u}{\partial t} + u \frac{\partial u}{\partial x} + c \frac{\partial(2c)}{\partial x} = 0 \qquad (9.27)$$

Adding equations (9.26) and (9.27)

$$\frac{\partial(u + 2c)}{\partial t} + (u + c) \frac{\partial u}{\partial x} + (u + c) \frac{\partial(2c)}{\partial x} = 0$$

or

$$\frac{\partial(u + 2c)}{\partial t} + (u + c) \frac{\partial(u + 2c)}{\partial x} = 0 \qquad (9.28)$$

Similarly subtracting (9.26) from (9.27), we get

$$\frac{\partial (u - 2c)}{\partial t} + (u - c)\frac{\partial (u - 2c)}{\partial x} = 0 \tag{9.29}$$

Equations (9.28) and (9.29) are the flow equations in characteristic form, where the quantity $(u \pm 2c)$ is being transported with quantity $(u \pm c)$. From the definition of total differential of the function $f(x, t)$

$$df = \left(\frac{\partial f}{\partial t}\right)dt + \left(\frac{\partial f}{\partial x}\right)dx$$

or

$$\frac{\partial f}{\partial t} = \frac{\partial f}{\partial t} + \left(\frac{dx}{dt}\right)\frac{\partial f}{\partial x} \tag{9.30}$$

If $f$ is constant then $df/dt = 0$ and so

$$\frac{\partial f}{\partial t} + \left(\frac{dx}{dt}\right)\frac{\partial f}{\partial x} = 0$$

Comparing equation (9.30) with (9.28) and (9.29) along the lines we get

$$\left(\frac{dx}{dt}\right)_{\pm} = (u \pm c) = u \pm (gh)^{1/2} = C_{\pm} \tag{9.31}$$

This is the characteristic form with slope in $x$-$t$ plane with $dx/dt = C_{\pm}$. It is to be emphasised that $C_{\pm}$ is not the same as $c$. The quantity

$$(u \pm 2c) = u \pm (2gh)^{1/2} = J_{\pm} \tag{9.32}$$

$J_{\pm}$ is called the *Reimann Invarient*.

Equation (9.31) provides an expression for slopes of the characteristic lines and equation (9.32) provides a means to find the constants carried along by each characteristics. Assuming horizontal bed $S_0 = 0$ and no friction $S_f = 0$, equations (9.31) and (9.32) can be used to solve (9.22) and (9.23). A general but pure computational introduction to MOC for solving flow problems is given by Abbott (1966). Application of the above method of characteristics to flow problems can be referred from that book.

### 9.4.1   Finite Difference Method

The unsteady flow equations can be solved numerically using the method of finite difference. Grid points covering the domain of interest (shown in Fig. 9.9(a)) are constructed. To illustrate the process a differential equation of the following form may be considered

$$\frac{\partial z}{\partial t} + \frac{\partial z}{\partial x} = 0 \tag{9.33}$$

A 'backward difference' approximation to the above equation at $P$ (Fig. 9.9(b)) in $z$-$x$ plane takes the slope AP in the curve

$$\left(\frac{\partial z}{\partial x}\right)^n_j = \frac{\{z^n_j - z^n_{j-1}\}}{\Delta x} \tag{9.34}$$

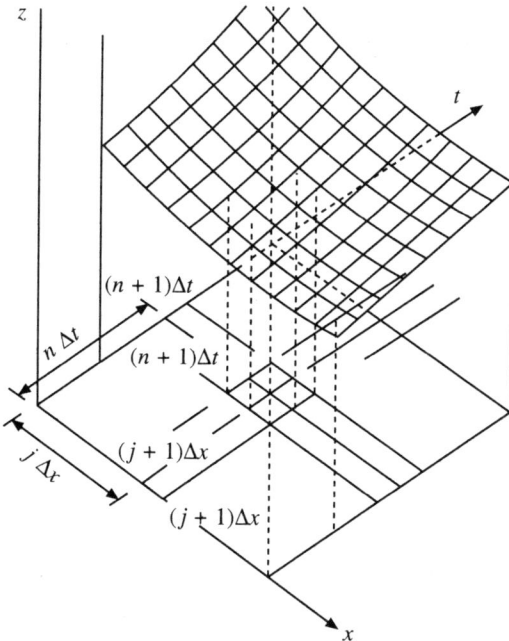

**Fig. 9.9(a). Three Dimensional visualization of the discretization of the 'hydraulic surface' in $(z, x, t)$ space**

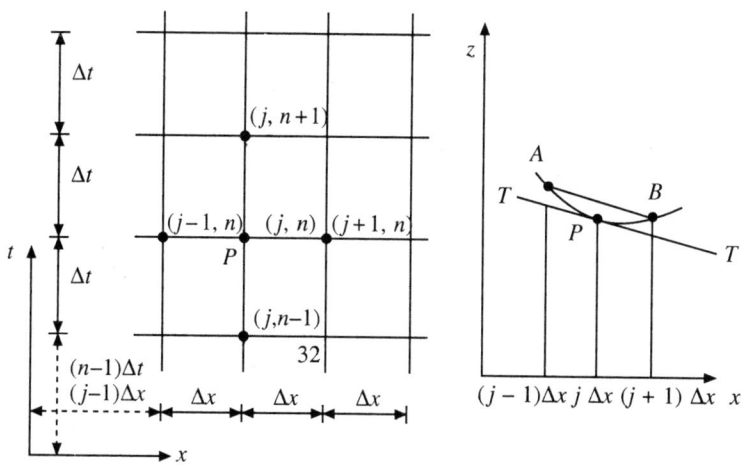

**Fig. 9.9(b). A plane view showing the grid notations and the corresponding view in $z$–$x$ plane**

A 'forward difference' approximation using the slope PB takes the form

$$\left(\frac{\partial z}{\partial x}\right)_j^n = \frac{\{z_{j+1}^n - z_j^n\}}{\Delta x} \tag{9.35}$$

Taking the entire curve AB, the 'central difference' approximation can be written as

$$\left(\frac{\partial z}{\partial x}\right)^n_j = \frac{\{z^n_{j+1} - z^n_{j-1}\}}{(2\Delta x)} \tag{9.36}$$

On the same basis using $z$-$t$ plane $\left(\dfrac{\partial z}{\partial t}\right)^n_j$ can be approximated (Fig. 9.9a) to the above three types of possibilities.

Using a backward difference in $x$-direction and a forward difference in $t$-direction equation (9.33) can be written as

$$\frac{\{z^{n+1}_j - z^n_j\}}{(\Delta t)} + \frac{\{z^n_j - z^n_{j-1}\}}{(\Delta x)} = 0$$

or $\qquad z^{n+1}_j = z^n_j - \left(\dfrac{\Delta t}{\Delta x}\right)\{z^n_j - z^n_{j-1}\} = \left(1 - \dfrac{\Delta t}{\Delta x}\right)z^n_j + \left(\dfrac{\Delta t}{\Delta x}\right)z^n_{j-1}$ (9.37)

In the above difference scheme, the value of dependent variable at the time step $(n + 1)$ is expressed as an explicit function of the values of the dependent function of earlier schemes of $z^n_j$ and $z^n_{j-1}$.. Thus $z^{n+1}_j$ level is calculated from $z^n_j$ and $z^n_{j-1}$. An implicit scheme relates more values of $(n + 1)$ level to known values at $n$th level.

As an example of implicit scheme, a hyperbolic equation of the type

$$\frac{\partial z}{\partial t} = a^2\left(\frac{\partial^2 z}{\partial x^2}\right) - b^2\left(\frac{\partial^2 z}{\partial t}\right)$$

can be written as

$$\frac{\{z^{n+1}_j - z^n_j\}}{(\Delta t)} = a^2 \frac{\{z^{n+1}_{j+1} - 2z^{n+1}_j + z^{n+1}_{j-1}\}}{(\Delta x^2)} + \frac{\{z^n_{j+1} - 2z^n_j + z^n_{j-1}\}}{(\Delta t^2)} \tag{9.38}$$

There are many difference schemes available. The solution to the difference schemes are carried out by assuming that approximations of the lower rows in the grid system are known. To solve the problem, initial and boundary conditions should be known. Many numerical techniques are available to solve the difference schemes, but it is important that the technique employed should be stable. Using the finite difference schemes, solution of the momentum and continuity equation given in (9.22) and (9.23) can be obtained.

### 9.4.2  Numerical Methods in Routing

All the methods of routing discussed earlier used the continuity equation which is a differential equation of the first order and is expressed as

$$I_t - O_h = \frac{dS}{dt} \tag{9.39}$$

where the inflow $I_t$ is the function of time $t$, the outflow $O_h$ is a function of head *h* over the spillway crest. Reservoir storage is the product of reservoir area $A_h$ and reservoir depth *h* over the spillway crest. The area is a function of head *h*. Therefore the change in storage *dS* is equal to $A_h \, dh$ . The continuity equation can be expressed in a differential form as

$$A_h \frac{dh}{dt} = I_t - O_h$$

or
$$\frac{dh}{dt} = \frac{(I_t - O_h)}{A_h} \tag{9.40}$$

The slope *dh/dt* is approximated to $\Delta H / \Delta t$. Standard methods of solving such a differential equation is obtained by taking forward a small increment of the independent variables of time *t* and depth *h*. A third or fourth order Runge-Kutta method can be successfully employed to evaluate the $(j + 1)$th step of elevation *h*. For a third order scheme there are three such increments in each interval $\Delta t$ and three such approximations are made for change in head $\Delta h$. Routing begins with known initial conditions of time, inflow, reservoir storage and elevation. At the end of time step $\Delta t$, the water elevation at $(j + 1)$th step is given as

$$hj_{+1} = h_j + \Delta h \tag{9.41}$$

where $\Delta h$ is evaluated from the order of the Runge-Kutta scheme. Computer programs for solving various order Runge-Kutta schemes are available. Use of such methods have become popular in hydrologic routing.

## 9.5 HYDROLOGIC RESERVOIR ROUTING

Whenever a flood wave enters a reservoir, the water level rises. Spillway and undersluices are the outlets from a reservoir through which the excess water from the reservoir may be released. In a reservoir, the head over the spillway (elevation or stage) is related to its storage and discharge. Therefore, for any stage of the reservoir, the storage *S* and the outflow *O* can be read from the plot of the elevation-storage and elevation-discharge relations. This makes the routing process simpler. Typical elevation-storage, elevation-discharge and storage-discharge curves are given in Fig. 9.10. The basic equation governing the routing process in a reservoir is the continuity equation which states that "the difference between inflow and outflow for any time step $\Delta t$ equals to the change in the storage". To solve the continuity equation for the time step $\Delta t$, during routing the following information are collected at the reservoir site:

1. A complete inflow hydrograph $I_i$ entering into the reservoir.
2. Reservoir elevation-storage curve or their relation above the spillway crest.
3. Reservoir elevation-outflow or storage-outflow curve above crest.
4. Values of $O_1, I_1$ and $S_1$ at the beginning of routing.

Various methods of hydrologic routing (Fig. 9.2) use the continuity equation in some form or the other. Some important methods of routing are discussed as follows.

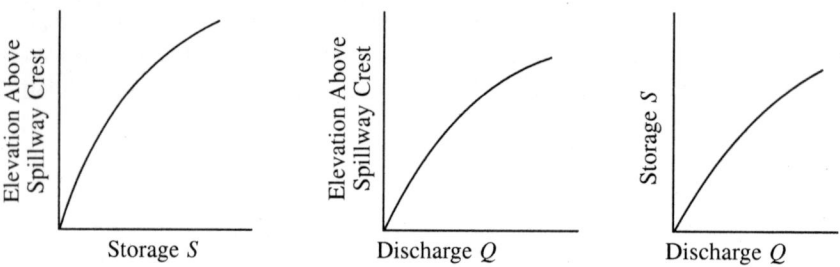

**Fig. 9.10    Typical Elevation-Storage, Elevation-Discharge and Storage-Discharge Curves for a Reservoir**

### 9.5.1    Analytical (Trial and Error) Method

The continuity equation is written as

$$\frac{(I_1 + I_2)}{2}\Delta t - \frac{(O_1 + O_2)}{2}\Delta t = S_2 - S_1. \tag{9.42}$$

At the beginning of routing values of $I_1$, $I_2$, $\Delta t$, $O_1$ and $S_1$ are known. The two unknowns are the outflow $O_2$ and storage $S_2$ at the end of time step $\Delta t$ of routing. Care should be taken to properly choose the time step of routing $\Delta t$ such that the peak of the hydrograph is not missed. On the other side, if the time step $\Delta t$ is chosen less than 1 h then it will involve too much of calculations. Depending on the time base of hydrograph, the value of $\Delta t$ between 1 and 6 h may be selected. For very large catchments, higher $\Delta t$ upto 24 h may be accepted. Stepwise procedure of this method is as follows:

1. Decide the time step of routing $\Delta t$ in h.
2. Plot the hydrograph and read the ordinates at $\Delta t$ h interval. This gives the ordinates $I_i$ of the inflow hydrograph. Care should be taken so that the peak of the inflow hydrograph is not missed due to selection of $\Delta t$ h.
3. Compute from the reservoir elevation $E$ the corresponding storage $S$ for all ranges of $E$ above the crest elevation from the level of routing.
4. Collect the stage-discharge curve developed for the spillway of the reservoir. If such a relation is not available then develop such a curve for the problem by the following relation for discharge over an uncontrolled spillway

$$O = \left(\frac{2}{3}\right)C_d(2g)^{1/2} L_e H^{3/2} \tag{9.43a}$$

where        $L_e = (L - 0.1 \times n \times 2 \times H)$ \tag{9.43b}

$L_e$ the effective length of the spillway in m, $H$ the head over the spillway crest, $L$ the net length of the spillway (= gross length − all pier widths), $C_d$ the coefficient of discharge of the spillway which can be taken as 2.2, $g$ the gravitational acceleration and $n$ is the numerator of piers in the spillway.

5. Assume trial reservoir level and read the outflow $O_2$ from the elevation-discharge curve corresponding to the assumed level. $O_1$ is already known from the data.

6. Since all other data are known, compute $S_2$ from equation (9.42).
7. Corresponding to $S_2$, find the reservoir level which should match the assumed reservoir level of step (5). If the two levels do not match then try another reservoir level as in step (5) and repeat steps (6) and (7) unless they are equal.
8. For the second step of routing $I_3$, $I_2$, $S_2$ and $O_2$ are known. Steps (5)-(7) are repeated and the process is continued till reservoir level is back to the original level after routing the entire inflow flood hydrograph. The routing steps are iterative and is easy for programming in the computer. The method is popularly applied to most of the field problems.

---

**Example 9.5:** Elevation-storage-discharge data at a proposed reservoir are given below which includes the discharge from an undersluice located at RL 260.0 m and the spillway crest at RL 329.0 m. Flood impinges when the reservoir level is at RL 260.30 m. Route the flood hydrograph through the reservoir and calculate the maximum water level (MWL) by analytical method.

| Elevation (m) | 260.0 | 264.5 | 270.5 | 276.5 | 282.5 | 288.5 | 294.5 | 300.5 | 306.5 | 312.5 | 318.5 | 324.5 |
|---|---|---|---|---|---|---|---|---|---|---|---|---|
| | 329.0 | 329.9 | 330.5 | | | | | | | | | |

| Storage (M.m³) | 0.009 | 0.15 | 0.80 | 2.20 | 4.40 | 8.20 | 13.6 | 21.0 | 30.0 | 40.0 | 52.0 | 66.0 |
|---|---|---|---|---|---|---|---|---|---|---|---|---|
| | 77.0 | 79.0 | 80.5 | | | | | | | | | |

| Discharge (m³/sec) | 4 | 30 | 46 | 58 | 67 | 75 | 83 | 90 | 96 | 103 | 108 | 114 |
|---|---|---|---|---|---|---|---|---|---|---|---|---|
| | 117 | 130 | 290 | | | | | | | | | |

The following inflow hydrograph is expected at the reservoir.

| Time (h) | 12 | 18 | 24 | 30 | 36 | 42 | 48 | 54 | 60 | 66 | 72 | 78 |
|---|---|---|---|---|---|---|---|---|---|---|---|---|
| Inflow (m³/sec) | 12 | 60 | 160 | 250 | 230 | 200 | 140 | 90 | 60 | 30 | 20 | 15 |

**Solution**

Calculations are carried out in Table 9.5. The reservoir elevation at the end of first step is assumed as 266.5 m. Corresponding outflow $O_2$ is read as 35.23 m³/sec. The mean inflow volume $(I_1 + I_2) \times 0.5 \times \Delta t$ is 0.78 M.m³ and mean outflow volume $(O_1 + O_2) \times 0.5 \times \Delta t$ is 0.42 M.m³. Equation (9.42) gives $(S_2 - S_1) = 0.36$ M.m³. As $S_1 = 0.009$ M.m³ we get $S_2 = 0.369$ M.m³. From storage - elevation curve, for storage of 0.369 the elevation is found as 266.5 m. This completes the first step of routing. The procedure is repeated to cover the entire inflow hydrograph and till the initial reservoir level is reached back.

---

## 9.5.2 Modified Puls Method

The continuity equation (9.39) is arranged as

$$(I_1 + I_2)\, 0.5\, \Delta t + \{S_1 - (0.5 O_1 \Delta t)\} = \{S_2 + 0.5 O_2\, \Delta t)\} \qquad (9.44)$$

At the beginning of routing $I_1$, $I_2$, $\Delta t$, $S_1$ and $O_1$ are known. Therefore, the terms of the left hand side of the equation (9.44) are known. The procedure to compute $O_2$ at the end of time step $\Delta t$ from knowledge of elevation-storage-discharge relation are given as follows:

1. Decide the time step $\Delta t$ of routing. This should be chosen such that the

Table 9.5   Reservoir Routing by Analytical Method

| Reservoir Elevation (m) | Outflow from Reservoir (m³/sec) | Storage in Reservoir (M. m³) | Time (h) | Inflow (m³/sec) | $(I_1 + I_2)$ × 0.5Δt (M. m³) | Initial Elevation (m) | Reservoir Outflow (m³/sec) $(O_1)$ | Final Elevation (m) | Reservoir Outflow (m³/sec) $(O_2)$ | $(O_1 + O_2)$ × 0.5Δt (M. m³) | Final Storage (M. m³) |
|---|---|---|---|---|---|---|---|---|---|---|---|
| (1) | (2) | (3) | (4) | (5) | (6) | (7) | (8) | (9) | (10) | (11) | (12) |
| 260.0 | 4.0 | 0.009 | 12 | 12 | 0.78 | 260.0 | 4.0 | 266.5 | 35.2 | 0.42 | 0.37 |
| 264.5 | 30.0 | 0.150 | 18 | 60 | 2.38 | 266.5 | 35.2 | 274.4 | 53.8 | 0.96 | 1.79 |
| 270.5 | 46.0 | 0.800 | 24 | 160 | 4.43 | 274.4 | 53.8 | 282.9 | 67.6 | 1.31 | 4.81 |
| 276.5 | 58.0 | 2.200 | 30 | 250 | 5.18 | 282.9 | 67.6 | 288.6 | 75.6 | 1.54 | 8.44 |
| 282.5 | 67.0 | 4.400 | 36 | 230 | 4.64 | 288.6 | 75.6 | 292.2 | 79.8 | 1.68 | 11.53 |
| 288.5 | 75.0 | 8.200 | 42 | 200 | 3.67 | 292.2 | 79.8 | 294.6 | 83.2 | 1.76 | 13.65 |
| 294.5 | 83.0 | 13.600 | 48 | 140 | 2.48 | 294.6 | 83.2 | 295.0 | 83.6 | 1.80 | 14.03 |
| 300.5 | 90.0 | 21.000 | 54 | 90 | 1.62 | 295.0 | 83.6 | 294.8 | 83.3 | 1.81 | 13.02 |
| 306.5 | 96.0 | 30.00 | 60 | 60 | 0.97 | 294.8 | 83.3 | 293.7 | 82.3 | 1.78 | 13.04 |
| 312.5 | 103.0 | 40.000 | 66 | 30 | 0.54 | 293.7 | 82.3 | 292.5 | 80.8 | 1.76 | 11.77 |
| 318.5 | 108.0 | 52.000 | 72 | 20 | 0.38 | 292.5 | 80.8 | 291.1 | 78.4 | 1.71 | 10.43 |
| 324.5 | 114.0 | 66.000 | 78 | 15 | 0.16 | 291.1 | 78.4 | 289.2 | 75.9 | 1.66 | 8.92 |
| 329.0 | 117.0 | 77.000 | 84 | | | 289.2 | 75.9 | 287.0 | 73.3 | 1.61 | 7.31 |
| 329.9 | 130.0 | 79.000 | 90 | | | 287.0 | 73.3 | 284.9 | 70.2 | 1.54 | 5.75 |
| 330.5 | 290.0 | 80.500 | 96 | | | 284.9 | 70.2 | 281.9 | 66.3 | 1.47 | 4.28 |
| | | | 102 | | | 281.9 | 66.3 | 277.9 | 60.2 | 1.37 | 2.91 |
| | | | 108 | | | 277.9 | 60.2 | 274.1 | 53.2 | 1.22 | 1.69 |
| | | | 114 | | | 274.1 | 53.2 | 273.2 | 51.7 | 1.13 | 0.56 |
| | | | | | | 273.3 | 51.7 | 267.6 | 38.4 | 0.97 | 0 |
| | | | | | | 267.6 | 38.4 | 260.0 | 0 | – | 0 |

peak of inflow hydrograph is not missed. If the time to peak is say 15 h, then $\Delta t$ may be selected as 1, 3 or 5 h depending on the time base $T_b$ of the hydrograph.

2. Plot a curve between elevation as ordinate and storage $(S + 0.5\Delta t \, O)$ in abscissa. On the same paper plot a curve between elevation vs. discharge. The abscissa thus have two scales one for storage $(S + 0.5\Delta t \, O)$ and other for discharge $O$.

3. Plot the inflow hydrograph and read ordinates of inflow hydrograph at $\Delta t$ intervals. Let the ordinates be represented as $I_i$.

4. Routing starts with the computation of left hand side of equation (9.44). All the terms of the equation $I_1$, $I_2$, $\Delta t$, $S_1$ and $O_1$ are known from which $(S_2 + 0.5\Delta t \, O_2)$ is obtained.

5. From the plot of step (2) reservoir elevation $E$ corresponding to $(S_2 + 0.5\Delta t \, O_2)$ is read.

6. From the plot of the same graph of step (2) between reservoir elevation $(E)$ and discharge $(O)$, the value of outflow $(O)$ corresponding to the elevation of step (5) is read. This outflow is the value of $O_2$ at the end of time step $\Delta t$.

7. Subtracting $O_2 \Delta t$ from $(S_2 + 0.5\Delta t \, O_2)$ of step (4) gives $(S_2 - 0.5\Delta t \, O_2)$ which is the second bracket term of the left hand side of equation (9.44). This is taken for the next step of routing.

8. For the second step, i.e., for the 2nd $\Delta t$ interval, the subscripts of all $I$, $S$, and $O$ in equation (9.44) are increased by 1. This brings the routing to step (4). The procedure is repeated till the entire inflow hydrograph is covered and till the reservoir is back to the original elevation.

---

**Example 9.6:** A reservoir has the following storage-elevation-discharge data

| Elevation (m) | 480 | 490 | 500 | 501 | 502 | 503 | 504 | 505 | 506 | 507 | 508 | 509 | 510 |
|---|---|---|---|---|---|---|---|---|---|---|---|---|---|
| Storage (H.m) | 100 | 115 | 160 | 170 | 180 | 195 | 210 | 235 | 260 | 290 | 330 | 380 | 450 |
| Discharge (m³/sec) | 0 | 0 | 0 | 10 | 30 | 60 | 80 | 110 | 140 | 170 | 200 | 250 | 300 |

Route the inflow hydrograph given below. Find the maximum elevation of water in the reservoir by Modified Puls method. This is the first flood entering the reservoir due to monsoon, when the water level is at the lowest at 480.0 m.

| Time (h) | 6 | 7 | 8 | 9 | 10 | 11 | 12 | 13 | 14 | 15 | 16 | 17 |
|---|---|---|---|---|---|---|---|---|---|---|---|---|
| | 18 | 19 | 20 | 21 | 22 | 23 | 24 | | | | | |
| Inflow (m³/sec) | 30 | 30 | 60 | 120 | 240 | 420 | 450 | 430 | 400 | 300 | 240 | 180 |
| | 120 | 90 | 60 | 40 | 30 | 20 | 10 | | | | | |

**Solution**

Computations are carried out in Table 9.6. To solve the problem by Modified Puls method, a graph between $(S+0.5\Delta t \, O)$ vs. elevation and elevation vs. discharge is plotted on the same paper as shown in Fig. 9.11. Time interval of routing $\Delta t$ is taken as 1 h. The inflow hydrograph arrives at the reservoir when water level is at the lowest level of 480.0 m. The outflow from the reservoir begins at RL 500.0 m. Therefore, the initial inflows are used to buildup storage upto RL 500.0 m.

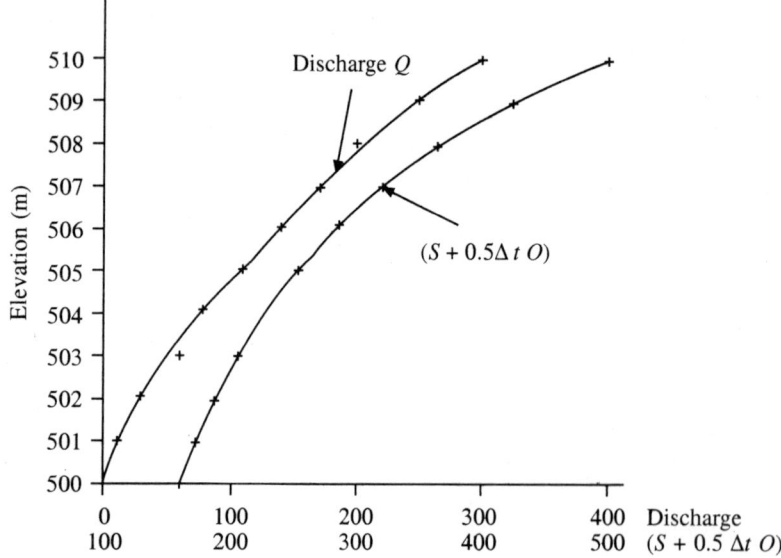

**Fig. 9.11   Plot between Elevation-Discharge and Elevation-(S+0.5 Δt O) for Example 9.6**

At the beginning of routing at 6th h, the outflow $O_1$ is zero. For the first step $(I_1 + I_2)$ 0.5 Δ$t$ = 10.8 Ha.m. and $(S - 0.5$ Δ$t$ $O) = 100.0$. Therefore $(S + 0.5$ Δ$t$ $O)$ from equation (9.44) = 10.8 + 100 = 110.8 Ha.m. The level of water corresponding to RL 110.8 is 487.2 m. At this level the reservoir outflow is still zero. Therefore $(S_2 + 0.5$ Δ$t$ $O) - ($Δ$t$ $O)$ is again 110.8 Ha.m, which equals to the left side of the quantity $(S_2 - 0.5$ Δ$t$ $O)$. The procedure is repeated for $(I_1 + I_2)$ 0.5 Δ$t$ = 16.2 and $(S_2 - 0.5$ Δ$t$ $O) = 110.8$ giving rise to right hand side of equation (9.44) as (16.2 + 110.8) = 127.0 Ha.m. Routing is continued till a level of reservoir storage of 501.0 m is reached back. The inflow and outflow hydrographs are plotted in Fig. 9.12 for comparison.

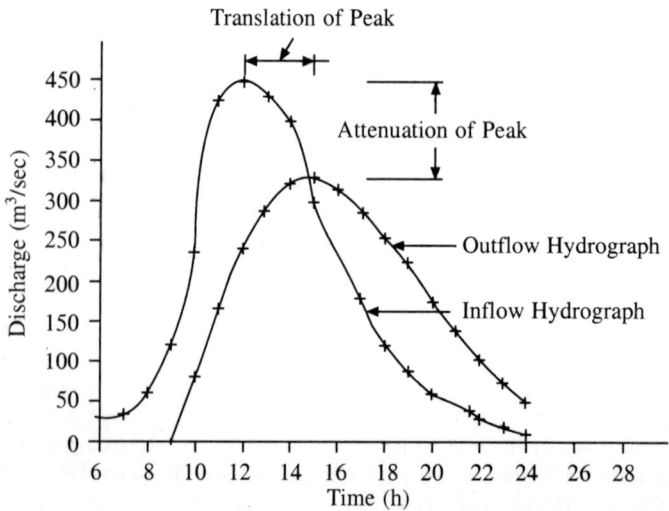

**Fig. 9.12   Inflow and Outflow Hydrograph of Reservoir Routing for Example 9.6**

**Table 9.6 Reservoir Routing by Modified Puls Method**

| Curve Points from data | | | | Routing Points | | | | | | |
|---|---|---|---|---|---|---|---|---|---|---|
| Elevation (m) | Outflow (m³/sec) | Storage (Ha. m) | $(S + 0.5 \Delta tO)$ (Ha. m) | Time (h) | Inflow (m³/sec) | Average Inflow volume | $(S - 0.5\Delta tO)$ (Ha. m) | $(S + 0.5 \Delta tO)$ (Ha. m) | Elevation (m) | Outflow (m3/sec) |
| (1) | (2) | (3) | (4) | (5) | (6) | (7) | (8) | (9) | (10) | (11) |
| 480 | 0 | 100 | 100.0 | 6 | 30 | | | 110.8 | 480 | 0 |
| 490 | 0 | 115 | 115.0 | 7 | 30 | 10.8 | 100.0 | 127.0 | 487.2 | 0 |
| 500 | 0 | 160 | 160.0 | 8 | 60 | 16.2 | 110.8 | 159.4 | 492.7 | 0 |
| 501 | 10 | 170 | 171.8 | 9 | 120 | 32.4 | 127.0 | 224.2 | 449.9 | 0 |
| 502 | 30 | 180 | 185.4 | 10 | 240 | 64.8 | 159.4 | 314.2 | 504.0 | 80 |
| 503 | 60 | 195 | 205.8 | 11 | 420 | 118.8 | 195.4 | 411.5 | 506.82 | 1 64.6 |
| 504 | 80 | 210 | 224.4 | 12 | 450 | 156.6 | 254.9 | 484.0 | 508.77 | 238.5 |
| 505 | 110 | 235 | 254.8 | 13 | 430 | 158.4 | 325.6 | 530.0 | 509.75 | 287.3 |
| 506 | 140 | 260 | 285.2 | 14 | 400 | 149.4 | 380.6 | 540.2 | 510.27 | 321.7 |
| 507 | 170 | 290 | 320.6 | 15 | 300 | 126.0 | 414.2 | 518.6 | 510.38 | 330.0 |
| 508 | 200 | 330 | 366.0 | 16 | 240 | 97.2 | 421.4 | | 510.15 | 312.0 |

*(Contd)*

| (1) | (2) | (3) | (4) | (5) | (6) | (7) | (8) | (9) | (10) | (11) |
|---|---|---|---|---|---|---|---|---|---|---|
| 509 | 250 | 380 | 425.0 | 17 | 180 | 75.6 | 406.2 | 481.8 | 509.70 | 286.0 |
| 510 | 300 | 450 | 504.0 | 18 | 120 | 54.0 | 378.8 | 432.8 | 509.10 | 255.0 |
| 511 | 380 | 540 | 600.0 | 19 | 90 | 38.8 | 341.0 | 378.8 | 508.2 | 210.0 |
|  |  |  |  | 20 | 60 | 27.0 | 303.2 | 330.2 | 507.2 | 176.0 |
|  |  |  |  | 21 | 40 | 18.0 | 266.8 | 284.8 | 506.0 | 140.0 |
|  |  |  |  | 22 | 30 | 12.6 | 234.4 | 247.0 | 504.85 | 105.0 |
|  |  |  |  | 23 | 20 | 9.0 | 209.2 | 218.2 | 503.7 | 74.0 |
|  |  |  |  | 24 | 10 | 5.4 | 191.6 | 197.0 | 502.7 | 48.0 |

### 9.5.3  Inflow-Storage-Discharge (ISD) Method

This method, developed by I.G. Puls of US Army Corps of Engineers, assumes that the outflow from a reservoir is a function of the elevation or head of water over the spillway. This is true for ungated spillway but for gated (controlled) spillway, the gates are so operated that the condition of constant function between elevation and discharge is satisfied. The procedure for this method is outlined as follows:

The continuity equation is reproduced as

$$(I_1 + I_2)\, 0.5\Delta t + (S_1 - 0.5\Delta t\, O_1) = (S_2 + 0.5\Delta t\, O_2) \tag{9.45}$$

To solve the equation by ISD method, plot the following curves in a single graph paper and follow the procedure:

1. Plot a curve between storage $S$ in abscissa vs. reservoir elevation $E$ as ordinate on a graph paper. This is given as data for any reservoir routing problem.
2. On the same paper plot $(S + 0.5\Delta t\, O)$ as storage vs. elevation. This graph can easily be plotted as follows: For a known reservoir elevation, the outflow $O$ can be computed from the discharge over spillway formula and by adding storage $(0.5\Delta t\, O)$ to $S$, we can always compute $(S + 0.5\Delta t\, O)$.
3. Also plot elevation vs. $(S - 0.5\Delta t\, O)$ as outlined in step (2) on the same paper.
4. At the beginning $I_1$, $I_2$, $\Delta t$, $S_1$ and $O_1$ are known.
5. From the continuity equation (9.45), calculate the left hand side as all the terms are known at the beginning of routing. Therefore $(S_2 + 0.5\Delta t\, O_2)$ representing the right side of equation (9.45) is known.
6. From the plots of steps (1), (2) and (3) knowing $(S_2 + 0.5\Delta t\, O_2)$ the reservoir elevation and therefore $(S_2 - 0.5\Delta t\, O_2)$ and $O$ are read. $O$ is the outflow $O_2$ here.
7. Adding $(S_2 - 0.5\Delta t\, O_2)$ of step (6) and $(I_2 + I_3)\, 0.5\, \Delta t$, the right side of equation (9.45) for the second step of routing is obtained, after the second time interval $\Delta t$. Thus $(S_3 + O_3\, 0.5\Delta t)$ is obtained. The process from steps (5) to (7) is repeated till the entire inflow hydrograph is covered.

---

**Example 9.7** Route the following inflow hydrograph for the reservoir given below

| Time in (h) | 0 | 3 | 6 | 9 | 12 | 15 | 18 | 21 | 24 | 27 | 30 |
|---|---|---|---|---|---|---|---|---|---|---|---|
| | 33 | 36 | 39 | 42 | 45 | 48 | 51 | 54 | 57 | | |
| Inflow (m³/sec) | 14 | 30 | 55 | 92 | 112 | 148 | 131 | 118 | 104 | 92 | 80 |
| | 69 | 60 | 52 | 44 | 38 | 32 | 26 | 21 | 16 | | |

Elevation-Storage-Discharge data of the reservoir are

| Elevation (m) | 361 | 362 | 363 | 364 | 365 | 366 | 367 | 368 | 369 | 370 |
|---|---|---|---|---|---|---|---|---|---|---|
| Storage (Mm³) | 52.5 | 149 | 273 | 420 | 588 | 766 | 962 | 1080 | 1430 | 1680 |
| Outflow (m³/sec) | 2.3 | 8.8 | 14.1 | 22.0 | 41.2 | 67.8 | 91.5 | 126.0 | 162.0 | 211.0 |

## Solution

For the solution of the above problem using ISD method, the first step is to plot curves between elevation-storage, elevation-$(S + 0.5\Delta t\, O)$ and elevation-$(S-0.5\Delta t\, O)$ on the same paper. The plotted curve is shown in Fig. 9.13. Another curve between elevation-discharge is plotted in Fig. 9.14. Routing is carried out in Table 9.7.

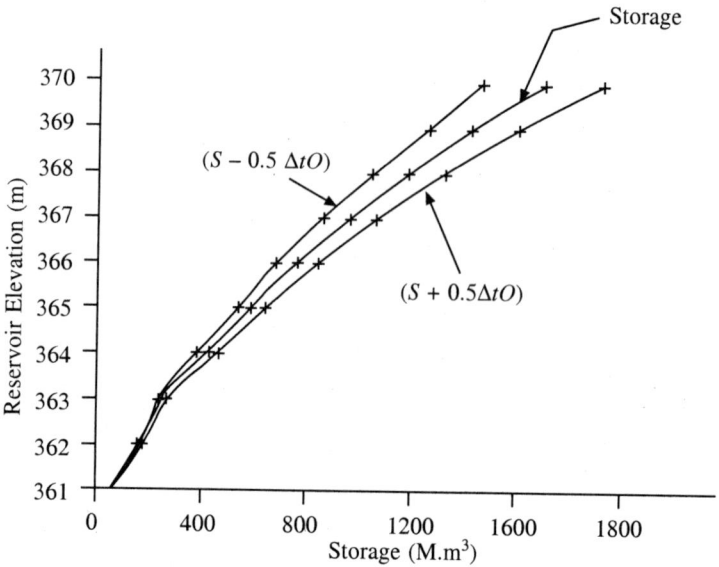

**Fig. 9.13** **Plot between Elevation-Storage, Elevation-$(S + 0.5 \Delta t O)$ and Elevation $(S - 0.5\Delta t O)$ for Example 9.7**

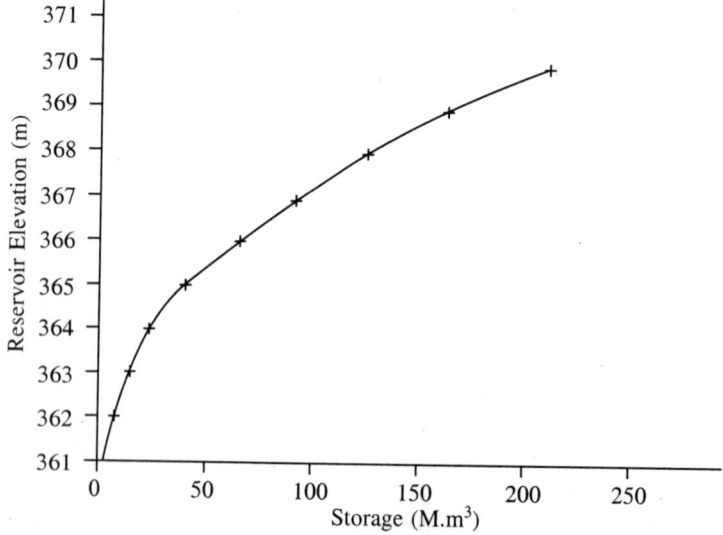

**Fig. 9.14** **Plot between Elevation-Discharge for Example 9.7**

Routing continuous till the reservoir elevation is reached back to RL. 361.0 m. Outflow from the reservoir at the beginning of routing is taken as 2.3 m³/sec and the elevation of

Table 9.7  Reservoir Flood Routing by ISD Method

| Time (h) | Inflow $I$ (m³/sec) | $(I_1+I_2)\,0.5\Delta t$ (Ha.m) | $(S-0.5\Delta tO)$ (Ha.m) | $(S+0.5\Delta tO)$ (Ha.m) | Elevation from graph (Fig. 9.13) (m) | Outflow from (Fig. 9.14) (m³/sec) |
|---|---|---|---|---|---|---|
| (1) | (2) | (3) | (4) | (5) | (6) | (7) |
| 0 | 14 | | | | 361.0 | 2.3 |
| | | 23.8 | 50 | 73.8 | | |
| 3 | 30 | | | | 361.2 | 3.3 |
| | | 45.9 | 70.0 | 115.9 | | |
| 6 | 55 | | | | 361.65 | 5.5 |
| | | 79.4 | 109.9 | 189.3 | | |
| 9 | 92 | | | | 362.3 | 8.0 |
| | | 110.2 | 180.0 | 290.2 | | |
| 12 | 112 | | | | 363.0 | 13.0 |
| | | 140.4 | 260.0 | 400.4 | | |
| 15 | 148 | | | | 363.65 | 20.0 |
| | | 150.7 | 360.0 | 510.7 | | |
| 18 | 131 | | | | 364.3 | 28.0 |
| | | 134.5 | 460.0 | 594.5 | | |
| 21 | 118 | | | | 364.8 | 38.0 |
| | | 120.0 | 520.0 | 640.0 | | |
| 24 | 104 | | | | 365.0 | 40.0 |
| | | 105.8 | 550.0 | 655.8 | | |
| 27 | 92 | | | | 365.1 | 42.0 |
| | | 92.9 | 560.0 | 652.9 | | |
| 30 | 80 | | | | 365.05 | 41.5 |
| | | 80.5 | 550.0 | 630.5 | | |
| 33 | 69 | | | | 364.95 | 38.5 |
| | | 69.7 | 530.0 | 599.7 | | |
| 36 | 60 | | | | 364.75 | 35.0 |
| | | 60.5 | 510.0 | 570.5 | | |
| 39 | 52 | | | | 364.6 | 32.5 |
| | | 51.8 | 490.0 | 541.8 | | |
| 42 | 44 | | | | 364.45 | 30.5 |
| | | 44.3 | 466.0 | 510.3 | | |
| 45 | 38 | | | | 364.25 | 26.5 |
| | | 37.8 | 435.0 | 472.8 | | |
| 48 | 32 | | | | 364.1 | 25.0 |
| | | 31.3 | 410.0 | 441.3 | | |
| 51 | 26 | | | | 363.9 | 22.0 |
| | | 25.4 | 380.0 | 405.4 | | |
| 54 | 21 | | | | 363.65 | 20.0 |
| | | 20.0 | 340.0 | 360.0 | | |
| 57 | 16 | | | | 363.4 | 17.5 |
| | | | 300.0 | 300.0 | | |
| 60 | | | | | 363.1 | 15.0 |
| | | | 260.0 | 260.0 | | |
| 63 | | | | | 362.8 | 13.0 |
| | | | 230.0 | 230.0 | | |

water level in the reservoir as 361.0 m. Corresponding to 361.0 m, $(S-0.5\Delta t\,O)$ is read from the graph as 50.0

$$(I_1 + I_2)0.5\Delta t + (S-0.5\Delta t\,O) = 23.8 + 50 = 73.8.$$

This should be equal to the value of $(S + 0.5\Delta t\,O)$ according to equation (9.45). For $(S + 0.5\Delta t\,O) = 73.8$ the new elevation, outflow and $(S - 0.5\Delta t\,O)$ is measured from the plotted curves between elevation-storage, elevation-$(S + 0.5\,\Delta t\,O)$ and elevation-$(S - 0.5\,\Delta t\,O)$. The process is repeated till the reservoir level routed back to RL 361.0 m.

**Example 9.8:** For the reservoir of example 9.7, route the following inflow hydrograph.

| Time (h) | 12 | 15 | 18 | 21 | 24 | 27 | 30 | 33 | 36 | 39 | 42 | 45 |
|---|---|---|---|---|---|---|---|---|---|---|---|---|
|  | 48 | 51 | 54 | 57 |  |  |  |  |  |  |  |  |
| Inflow | 14 | 22 | 48 | 77 | 97 | 124 | 107 | 86 | 62 | 50 | 39 | 30 |
| (m³/sec) | 22 | 19 | 17 | 15 |  |  |  |  |  |  |  |  |

Determine the maximum water level. What is the time lag between the peak of inflow and outflow hydrograph.

**Solution**
Routing interval is taken as $\Delta t = 3\,\text{h} = 1.08 \times 10^4$ sec.
The Elevation-Storage-Discharge and $(S + 0.5\Delta t\,O)$ relations are as follows:

**Table 9.8   Elevation-Storage-Discharge Relation of Example 9.8**

| Elevation $(E)$ (m) | Discharge $(O)$ (m³/sec) | $(S + 0.5\Delta t\,O)$ (Ha.m) | Storage $(S)$ (Ha.m) |
|---|---|---|---|
| (1) | (2) | (3) | (4) |
| 361 | 2.3 | 55.0 | 52.5 |
| 362 | 8.8 | 158.5 | 149.0 |
| 363 | 14.1 | 288.2 | 273.0 |
| 364 | 22.0 | 443.8 | 420.0 |
| 365 | 41.2 | 632.5 | 588.0 |
| 366 | 67.8 | 839.2 | 766.0 |
| 367 | 91.5 | 1060.8 | 962.0 |
| 368 | 126.0 | 1316.0 | 1180.0 |
| 369 | 162.0 | 1605.0 | 1430.0 |
| 370 | 211.0 | 1908.0 | 1680.0 |

Routing is carried out in Table 9.9. At the beginning of routing reservoir elevation is 361.0 m and the corresponding discharge $O_1$ is 2.3 m³/sec. The quantity $(S - 0.5\Delta t\,O) = 50.0$. For the first step of routing $(I_1 + I_2)\Delta t + (S - 0.5\Delta t\,O) = 19.4 + 50.0 = 69.4$ M.m³. This equals to $(S + 0.5\Delta t\,O)$. From the relation between elevation and $(S + 0.5\Delta t\,O)$, the elevation is read as 361.15 m and $O_2$ as 3.2 m³/sec. Routing is continued till the level 361.0 m in the reservoir is reached back.

Table 9.9 Reservoir Routing for Example 9.8

| Time (h) | Inflow $I$ (m³/sec) | $(I_1+I_2)0.5\Delta t$ (Ha.m) | $(S-0.5\Delta tO)$ (Ha.m) | $(S+0.5\Delta tO)$ (Ha.m) | Elevation from graph (m) | Outflow from equation (m³/sec) |
|---|---|---|---|---|---|---|
| (1) | (2) | (3) | (4) | (5) | (6) | (7) |
| 12 | 14 | | | | 361.0 | 2.3 |
| | | 19.4 | 50 | 69.4 | | |
| 15 | 22 | | | | 361.15 | 3.2 |
| | | 37.8 | 65.9 | 103.7 | | |
| 18 | 48 | | | | 361.50 | 5.0 |
| | | 67.5 | 98.3 | 165.8 | | |
| 21 | 77 | | | | 362.07 | 9.5 |
| | | 94.0 | 155.5 | 249.5 | | |
| 24 | 97 | | | | 362.7 | 12.0 |
| | | 119.3 | 236.5 | 355.8 | | |
| 27 | 124 | | | | 363.45 | 17.0 |
| | | 124.7 | 337.4 | 462.1 | | |
| 30 | 107 | | | | 364.5 | 23.5 |
| | | 104.2 | 436.7 | 540.9 | | |
| 33 | 86 | | | | 364.5 | 30.0 |
| | | 80.0 | 508.5 | 588.5 | | |
| 36 | 62 | | | | 364.8 | 36.0 |
| | | 60.5 | 549.6 | 610.1 | | |
| 39 | 50 | | | | 364.9 | 38.0 |
| | | 48.0 | 569.1 | 617.1 | | |
| 42 | 39 | | | | 364.95 | 40.0 |
| | | 37.3 | 573.9 | 611.2 | | |
| 45 | 30 | | | | 364.9 | 38.0 |
| | | 28.1 | 570.2 | 598.3 | | |
| 48 | 22 | | | | 364.85 | 37.0 |
| | | 22.1 | 558.3 | 580.4 | | |
| 51 | 19 | | | | 364.75 | 33.5 |
| | | 19.4 | 544.2 | 563.6 | | |
| 54 | 17 | | | | 364.6 | 30.5 |
| | | 17.3 | 530.7 | 548.0 | | |
| 57 | 15 | | | | 364.5 | 30.0 |
| 60 | | | 515.6 | 515.6 | 364.4 | 28.0 |
| 63 | | | 485.4 | 485.4 | 364.2 | 25.0 |
| 66 | | | 458.2 | 458.2 | 364.05 | 23.0 |
| 69 | | | 433.6 | 433.6 | 363.95 | 22.0 |
| 72 | | | 409.8 | 409.8 | 363.75 | 18.5 |
| 75 | | | 389.8 | 389.8 | 363.6 | 17.5 |
| 78 | | | 370.9 | 370.9 | 363.5 | 17.0 |
| 81 | | | 352.5 | 352.5 | 363.4 | 16.0 |

### 9.5.4   Goodrich Method

In this method, the continuity equation used is of the form.

$$(I_1 + I_2) + \left(\frac{2S_1}{\Delta t} - O_1\right) = \left(\frac{2S_2}{\Delta t} + O_2\right) \tag{9.46}$$

The routing steps are:

1. At the beginning of routing all terms of left hand side of equation (9.46) are known from which the right side of the equation is obtained at the end of time step $\Delta t$.

2. From the known elevation-storage-discharge relation, plot two curves between elevation-discharge and elevation-$\left(\frac{2S}{\Delta t} - O\right)$ on the same paper or plot an outflow-$\left(\frac{2S}{\Delta t} + O\right)$ curve.

3. From $\left(\frac{2S}{\Delta t} + O\right)$-elevation and elevation-discharge relation or outflow-$\left(\frac{2S}{\Delta t} + O\right)$ curve compute $O_2$, corresponding to the value of $\left(\frac{2S_2}{\Delta t} + O_2\right)$ of step (1).

4. From $\left(\frac{2S_2}{\Delta t} + O_2\right)$ subtract ($2O_2$) of step (3) to obtain $\left(\frac{2S}{\Delta t} - O\right)$.

5. The new $\left(\frac{2S}{\Delta t} - O\right)$ is used as second bracket term of equation (9.46) to go for the next step of routing for the second time step $\Delta t$.

6  The process is repeated to complete the entire inflow hydrograph.

---

**Exercise 9.9** Inflow hydrograph into a reservoir are given as follows:

| Time (min) | 0 | 10 | 20 | 30 | 40 | 50 | 60 | 70 | 80 | 90 |
|---|---|---|---|---|---|---|---|---|---|---|
| | 100 | 110 | 120 | 130 | 140 | 150 | | | | |

| Inflow ($m^3$/sec) | 0 | 1.7 | 3.4 | 5.1 | 6.8 | 8.5 | 10.2 | 9.06 | 7.93 | 6.8 |
|---|---|---|---|---|---|---|---|---|---|---|
| | 5.66 | 4.5 | 3.4 | 2.26 | 1.13 | 0 | | | | |

Elevation-Discharge-Storage relation of the reservoir above the crest of spillway are given in Table 9.10.

Use a level pool routing method to calculate the reservoir outflows from the given inflow hydrograph.

**Solution**

The inflow hydrograph is at 10 min interval. The first step in this routing is to calculate elevation and ($2S/\Delta t + O$) relation for all the elevations and plot a curve between outflow and ($2S/\Delta t + O$) from the given storage-Elevation-Outflow values. For the curve, the following Elevation-Storage-($2S/\Delta t + O$) data are utilised.

**Table 9.10   Elevation-Storage-Discharge Data for Example 9.9**

| Elevation (m) | Discharge (m³/sec) | Storage (m³) |
|---|---|---|
| (1) | (2) | (3) |
| 0.000 | 0.000 | 0.0 |
| 0.305 | 0.227 | 1233.6 |
| 0.610 | 0.850 | 2467.3 |
| 0.914 | 1.700 | 3700.9 |
| 1.219 | 2.747 | 4934.6 |
| 1.524 | 3.880 | 6168.2 |
| 1.829 | 4.900 | 7401.9 |
| 2.134 | 5.806 | 8635.5 |
| 2.438 | 6.542 | 9869.2 |
| 2.743 | 7.165 | 11102.8 |
| 2.896 | 7.477 | 11719.6 |
| 3.050 | 7.788 | 12336.4 |

| Elevation (m) | 0 | 0.305 | 0.61 | 0.914 | 1.219 | 1.524 | 1.829 | 2.134 | 2.438 | 2.743 | 2.896 | 3.05 |
|---|---|---|---|---|---|---|---|---|---|---|---|---|
| Discharge (m³/sec) | 0 | 0.227 | 0.85 | 1.700 | 2.747 | 3.880 | 4.900 | 5.806 | 6.542 | 7.165 | 7.477 | 7.788 |
| $(2S/\Delta t + O)$ (m³/sec) | 0 | 4.33* | 9.06 | 14.05 | 19.20 | 24.11 | 29.57 | 34.58 | 39.45 | 44.18 | 46.53 | 48.91 |

$$* \ 4.33 = \frac{(1233.6 \times 2)}{(10 \times 60)} + 0.227.$$

The plot of $\left( \dfrac{2S}{\Delta t} + O \right)$ vs. discharge curve is shown in Fig. 9.15. Routing of the flow through the reservoir is carried out in Table 9.11.

Equation (9.46) is used for routing calculations. Storage and outflow is taken as zero

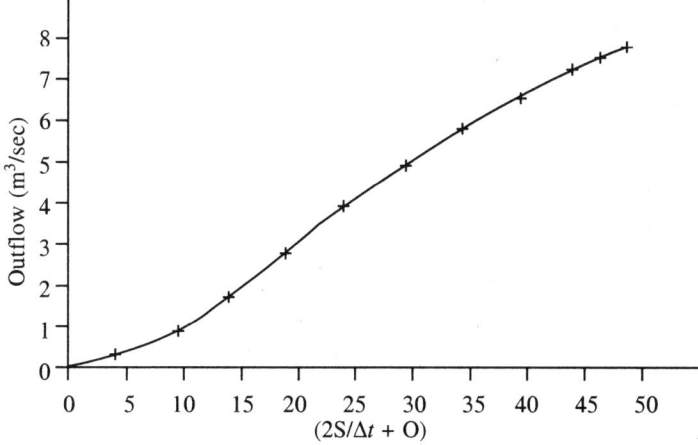

**Fig. 9.15   Plot Between $(2S/\Delta t + O)$ and Discharge for Example 9.9**

**Table 9.11　Reservoir Routing by Level Pool Method for Example 9.9**

| Time (min) | Inflow ($m^3$/sec) | $I_j + I_{j+1}$ ($m^3$/sec) | $(2S_j/\Delta t - Q_j)$ ($m^3$/sec) | $(2S_{j+1}/\Delta t + Q_{j+1})$ ($m^3$/sec) | Outflow ($m^3$/sec) |
|---|---|---|---|---|---|
| (1) | (2) | (3) | (4) | (5) | (6) |
| 0 | 0.0 | – | 0.0 | – | 0 |
| 10 | 1.7 | 1.7 | 1.56 | 1.70 | 0.068 |
| 20 | 3.4 | 5.1 | 5.70 | 6.66 | 0.484 |
| 30 | 5.1 | 8.5 | 10.73 | 14.19 | 1.730 |
| 40 | 6.8 | 11.9 | 15.65 | 22.62 | 3.489 |
| 50 | 8.5 | 15.3 | 20.62 | 30.94 | 5.160 |
| 60 | 10.2 | 18.7 | 26.27 | 39.31 | 6.522 |
| 70 | 9.06 | 19.26 | 30.84 | 45.53 | 7.343 |
| 80 | 7.93 | 16.99 | 32.54 | 47.83 | 7.646 |
| 90 | 6.8 | 14.73 | 32.12 | 47.27 | 7.573 |
| 100 | 5.66 | 12.46 | 30.14 | 44.59 | 7.219 |
| 110 | 4.5 | 10.16 | 27.02 | 40.34 | 6.661 |
| 120 | 3.4 | 7.9 | 23.22 | 34.95 | 5.880 |
| 130 | 2.26 | 5.66 | 19.35 | 28.89 | 4.772 |
| 140 | 1.13 | 3.39 | 15.72 | 22.75 | 3.515 |
| 150 | 0 | 1.13 | 12.33 | 16.85 | 2.260 |
| 160 | | | 9.58 | 12.33 | 1.376 |
| 170 | | | 7.73 | 9.58 | 0.926 |
| 180 | | | 6.44 | 7.73 | 0.646 |
| 190 | | | 5.52 | 6.44 | 0.459 |
| 200 | | | 4.81 | 5.52 | 0.357 |

for the first step. The value of $\left( \dfrac{2S_1}{\Delta t} - O_1 \right)$ is taken as zero. The outflow $(I_1 + I_2)$ is 1.70 for the first step. At the end of $\Delta t = 10$ min, the storage outflow function is given as

$$(I_1 + I_2) + \left( \frac{2S_1}{\Delta t} - O_1 \right) = \left( \frac{2S_2}{\Delta t} + O_2 \right) \quad \text{or} \quad 1.70 + 0.0 = 1.70$$

Outflow corresponding to $\left( \dfrac{2S_2}{\Delta t} + O_2 \right) = 1.70$ is calculated as 0.068 by linear interpolation. For the next step of routing $\left( \dfrac{2S_2}{\Delta t} - O_2 \right)$ is calculated as

$$\left( \frac{2S_2}{\Delta t} - O_2 \right) = \left( \frac{2S_2}{\Delta t} + O_2 \right) - 2O_2 = 1.70 - 2 \times 0.068 = 1.56$$

For the second step $(2S_3/\Delta t + O_3)$ is calculated as

$$\left( \frac{2S_3}{\Delta t} + O_3 \right) = (I_3 + I_2) + \left( \frac{2S_2}{\Delta t} - O_2 \right) = (1.7 + 3.4) + 1.56 = 6.66$$

For $\left( \dfrac{2S_3}{\Delta t} + O_3 \right) = 6.66$, the outflow $O_3 = 0.484$. Again $2O_3$ is subtracted from $\left( \dfrac{2S_3}{\Delta t} + O_3 \right)$

to give $\left(\dfrac{2S_3}{\Delta t} - O_3\right) = 5.70$. The complete routed hydrograph is shown in the col. (6) of Table 9.11.

## 9.6 FLOOD ROUTING MACHINES

### 9.6.1 Mechanical Flood Routers

There are a number of machines developed for routing floods through reservoirs or channels. The integrating flood router developed by J.F. Tarpley (1939) consists of five drum-mounted graph charts driven continuously by motor arrangement. In three drums, input data like inflow hydrograph, elevation-discharge and elevation-storage curves are fed as input graph chart, whereas the rest two drums plot time-elevation and time-outflow graphs continuously while routing is in operation. To read input data from three drums, pointers are kept in position touching the graphs or charts. To plot the output graphs pen pointers are attached to the last two drums say (4) and (5). Movement of the pointers, drums and pens are synchronised such that as the drums rotate all the five pointers move, the two output graphs are automatically plotted. The machine needs to be calibrated to fit to the given system.

The rolling type mechanical flood router developed by F.B. Harkness (1951) of U.S. Army Corps of Engineers is used for channel routing where Muskingum method is used with time step of $\Delta t = 2KX$. In this instrument, an undercarriage with wheel, pen and pointer arrangement moves over the plotted inflow hydrograph. The undercarriage is attached to a $T$-shaped frame at such distance that as the undercarriage moves over the plotted inflow hydrograph, the outflow hydrograph is automatically plotted by a pen arrangement attached to it at fixed distance to the inflow hydrograph pointer reader. As the system is a mechanical device, the instrument must be set properly to fit the river system. Thus knowing $K$ and $X$ the $T$-frame and undercarriage is set in position. Substantial improvements have been made to this instrument since Harkeness proposed the device. The present day motor-driven device is an easy handling, fast tracing and more accurate instrument with good sensitivity. A sketch of the instrument is shown in Fig. 9.16.

### 9.6.2 Electric Analog Routing Machine

Such a routing machine (Fig. 9.17) was developed by U.S. Weather Bureau (1955) in which the inflow, outflow and storage of water is made analogous with that of electric current. A condenser capable of storing current is made analogous to storage $S$. The inflow $I$ and outflow $O$ from a reservoir or channel is made similar to the inflow and outflow of current from the condenser. The routing equation in channel is analogous to the electric circuit. By such a machine, variation of storage with time constant $K$ value can be incorporated. With the addition of an oscilloscope and other electronic devices, the instrument can be made more handy, accurate and fast. At several flood forecasting sites, routing is carried out by such machines.

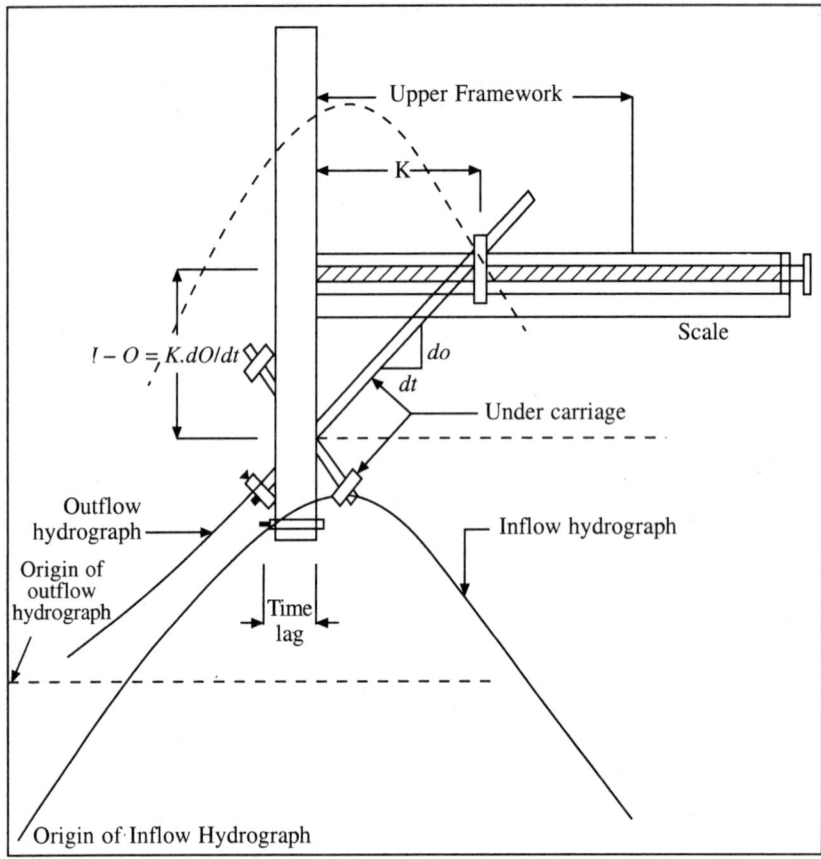

Fig. 9.16 Mechanical Flow Router (After Harkness)

### 9.6.3 Digital Computers

With the advent of digital computers, the problem of routing has become much simplified. A unique computer program can be written for the system of channel or reservoir for routing various inflow hydrographs. The program can be verified from the knowledge of previous inflow-outflow hydrographs. Once the program is validated, it takes a few seconds to compute the outflow hydrograph from the information of inflow hydrograph. Large number of watershed simulation models have been developed which takes into account the rainfall, meteorological parameters and basin characteristics as input data, computes an unit hydrograph, convolutes it with rainfall excess hyetograph to obtain flood hydrograph and routes the flood hydrograph through the channel or reservoir giving directly the outflow for the system. Use of telemetry gauges in a basin and a validated computer program at the desired site of the river makes the process quick and accurate.

### 9.7 FLOOD FORECASTING

The objective of various channel and reservoir routing is to know the shape of

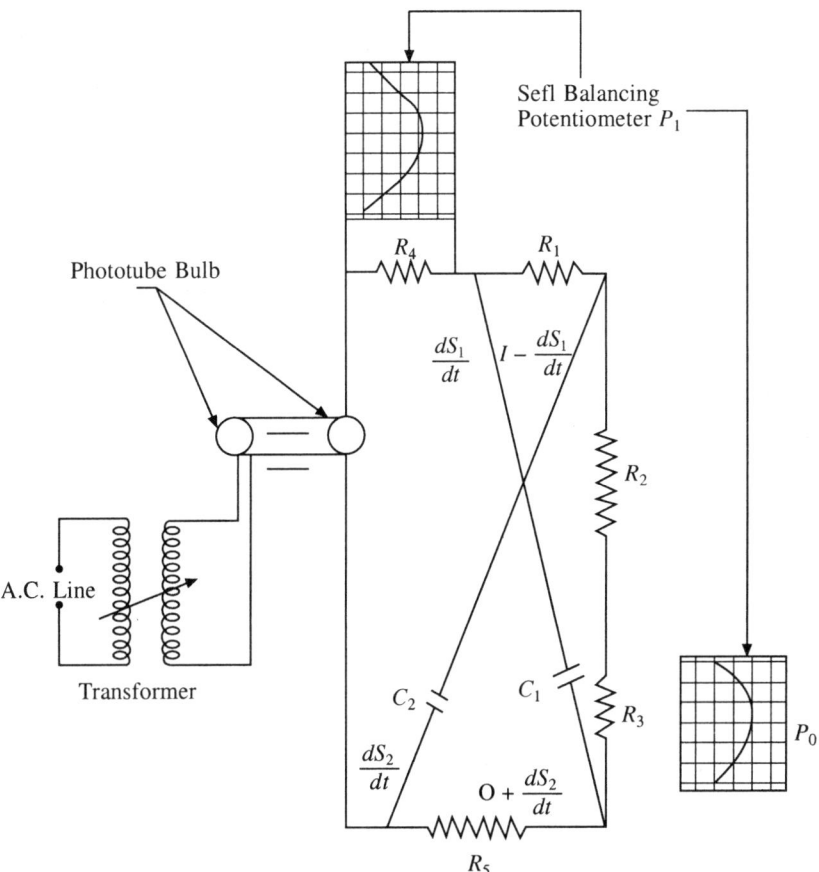

**Fig. 9.17 Electric Analog Routing Machine**

the outflow hydrograph at a downstream location from the knowledge of the flood inflows at one or more locations upstream of it. Floods in the flood plains of down valley regions are difficult to be contained completely due to the following reasons:

1. Developing countries cannot afford to construct cascade of reservoirs for containing all the floods expected in a basin.
2. Reservoirs or other water resource structures are designed to contain floods up to certain magnitude.
3. Flood plains are encroached increasingly by habitations, agriculture and industries due to rapid growth of population. Due to this even small floods cause pronounced effects at its delta head.
4. Due to gradual deforestation of the catchment, infiltration losses have reduced. This has increased the flood producing capacity of the same catchment from the same intensity of rainfall.

Hence, only alternative left is to forecast the flood much in advance so that people from the probable affected areas can be evacuated. This saves enormous life and property.

Various terms used in forecasting are (a) *River forecasting* which means the determination of flood stages at a selected downstream location of definite time in advance from the knowledge of the information of the upstream locations. Such forecast may be either short-term or long-term. The term *Flood forecasting* means the forecast of floods for a particular event on short-term basis. *Hydrologic forecasting* includes both the long-and short- term forecasting. This term also includes river and flood forecasting. Therefore such forecast techniques use all types of information like rainfall, infiltration losses, antecedent precipitation index (API), river stages and other data as input and runoff hydrograph or the peak flow as output. A hydrologic forecast also includes daily stream flows in the rivers. A short-term forecast ranges for a few hours only whereas a medium forecast lasts for a few days.

It is to be emphasised here that the forecast system should be accurate and meaningful. An erroneous warning leads to loss of faith of the people with the system. Therefore, factors like (i) reliability, (ii) advance prediction, (iii) discrimination of the forecast information and (iv) evacuation of life and movable property from the affected area are the important factors associated with an effective forecasting system. This can be achieved by the following forecasting methods.

*Method I*

1. Computation of flood hydrograph at an upstream location is made by either of the following:

   (i)   Using antecedent precipitation index (API) and a coaxial graph method, compute the storm runoff depth. From the knowledge of unit hydrograph for the basin, the runoff can be distributed over the time span.

   (ii)  The average storm rainfall over the basin can be used to construct a rainfall histogram. A loss rate when separated on the basis of the past experience of the floods gives an effective rainfall hyetograph (ERH). The ERH is convoluted with the unit hydrograph for the basin to give the necessary direct runoff hydrograph. Addition of base 'flow gives the computed storm runoff from the basin.

   (iii) From various rainfall runoff models, the flood for the basin at the desired outlet can be obtained.

2. Route the flood hydrograph to the desired location downstream of the river by Muskingum or Working Value method to get the complete flood hydrograph at the place of interest.

All except the API and coaxial graphical method of computing storm runoff are discussed in the previous chapters. The API method is discussed below.

### Ancedent Precipitation Index and Coaxial Graph to Compute Runoff

Antecedent precipitation index (API) is represented by the equation

$$\text{API} = a_1 p_1 + a_2 p_2 + \dots + a_t p_t \tag{9.47}$$

where $a_1, a_2, \dots, a_t$ are the coefficients with values less than unit and $p_1, p_2, \dots, p_t$ are the precipitation at 1-day, 2-days, $\dots$, $t$-days before the present day. It is assumed that $a_t$ decrease with time. It can be represented as $a_t = K^t$, where $t$ is time in days and $K$ a constant varying between 0.30 and 0.95. Therefore, for any given day, the influence of ancedent moisture of past several days of precipitation is represented as

$$\text{API} = K^1 p_1 + K^2 p_2 + \ldots + K^t p_t \qquad (9.48)$$

Computation of antecedent precipitation index is best illustrated in the following example.

---

**Example 9.10:** Compute API from the given rainfall records on 20/09/1998 and 25/09/1999 assuming $K$ as 0.80. Take the following rainfall data for the month of September 1999.

| Date of 9/99 | 11 | 12 | 13 | 14 | 15 | 16 | 17 | 18 | 19 | 20 | 21 |
|---|---|---|---|---|---|---|---|---|---|---|---|
| | 22 | 23 | 24 | 25 | | | | | | | |
| Rainfall (mm) | 10 | 0 | 0 | 5 | 0 | 0 | 20 | 0 | 0 | 30 | 0 |
| | 0 | 0 | 60 | 10 | | | | | | | |

**Solution**
API on 12/9/99 due to rainfall of 10 mm on 11/9/1999 is

$$\text{API } (12/9/99) = 0.80^1 \times 10 = 8.0 \text{ mm}$$

and its value on 13/9/99 is

$$\text{API } (13/9/99) = 0.80^2 \times 10 = 6.4 \text{ mm}$$

Likewise rainfall of 11, 14, 17 and 20/09/99 converted to 20/09/99 as

$$\text{API } (20/9/99) = 30 + 0.8 \times 0 + 0.8^2 \times 0 + 0.8^3 \times 20 + 0.8^4 \times 0 + 0.8^5 \times 0 + 0.8^6 \times 5$$
$$+ \ 0.8^7 \times 0 + 0.8^8 \times 0 + 0.8^9 \times 10$$

$$= 30 + 0.8^3 \times 20 + 0.8^6 \times 5 + 0.8^9 \times 10$$

$$= 30 + 14.24 + 1.31 + 1.34 = 42.90 \text{ mm}$$

API for 25/09/1999 is given by

$$\text{API } (25/9/99) = 10 + 0.8^1 \times 60 + 0.8^5 \times 42.90 = 10 + 48 + 14.0 = 72 \text{ mm}.$$

To compute runoff from a known API value, a typical coaxial graph shown in Fig. 9.18 can be used. Block-1 of the coaxial graph gives a relation between the week of the year vs. API. Knowing API and the week of the year we locate a point in block-1. Moving from the block-1 to block-2 in a line vertically down from the point located in the block-1 and knowing the storm durations in h, we locate a point in block-2. From block-2, we proceed horizontally to block-3, where the plot of storm precipitation depth is given. This horizontal line from known storm duration (of block-2) touches the storm precipitation depth line of block-3. From block-3, moving up to the block-4, we get storm runoff in cm. Data needed for this procedure is (a) the API, (b) storm duration, (c) storm precipitation in cm, provided that the coaxial chart is available for the basin. Such chart can also be prepared easily from the data of past records.

By knowing the runoff depths from the storm, the next problem is to distribute it over time. This can be achieved from the knowledge of the data of time distribution of storm rainfall from a self recording rain gauge. Time distribution of storm rainfall is convoluted over the unit hydrograph of the basin to get the flood hydrograph. The method is discussed

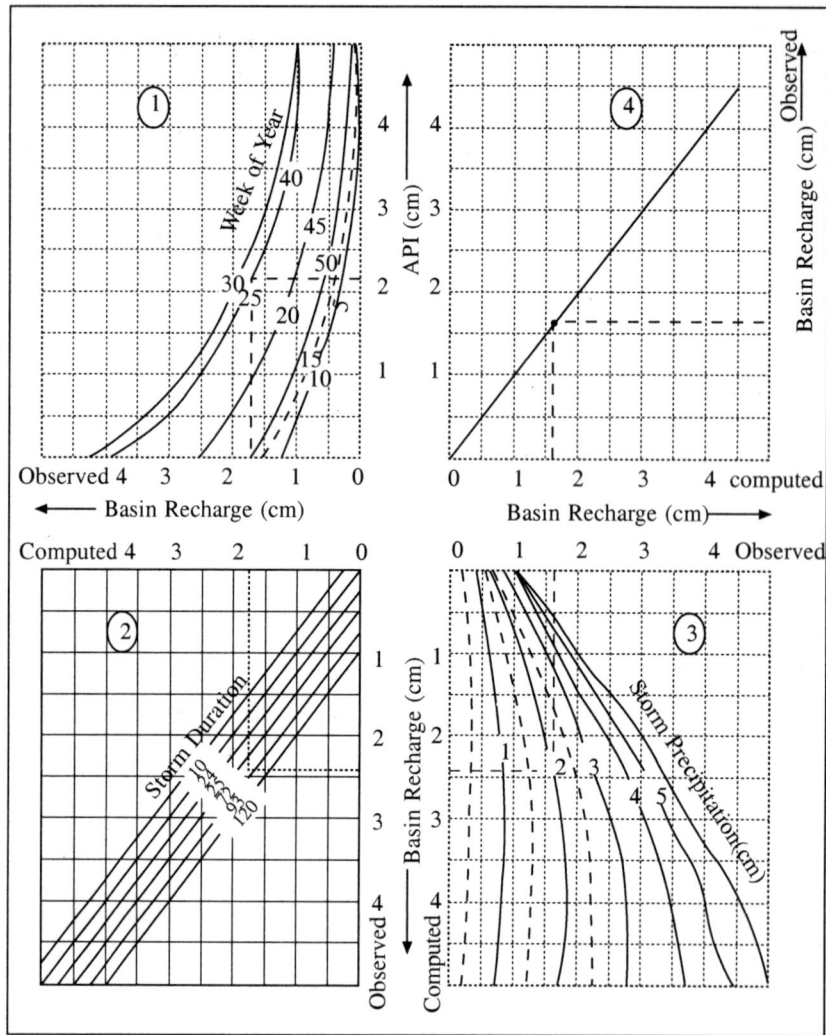

**Fig. 9.18  Multivariate Coaxial Graphical Method of Basin Recharge Relationship**

in detail in Chapter 8. Further, the runoff hydrograph is routed using Muskingum equation to obtain the peak and the flood hydrograph at the other desired downstream location in the river.

*Method II*

Multiple correlation between the stages of upstream main river-tributary system with the downstream forecasting river station can be established. Using such a correlation between the stages, flood at the forecasting station can be predicted at any particular time from the knowledge of upstream gauges. Multiple correlation between stages are discussed here.

*Multiple Stage Correlation*

A simple gauge to gauge correlation between the upstream base station and the downstream forecasting station is the best way to predict flood, when such a relation can be established for all stages. In India, most of the flood forecasting are carried out on the basis of such developed relations. Other parameters like rainfall and API for the catchment between the base station and forecasting station can be incorporated in the system to give a more realistic forecast of the flood. Various types of such multiple correlation are described as follows:

(1) A simple relation between the $N$th hour gauge of the base station and $(N + T)$th hour gauge of the forecasting station is best suited where the river section is uniform and there is no addition or loss of water between the two sections, $T$ is the actual travel time between the stations on the basis of which the flood forecasting is carried out. Such a simple relation between Panposh near Rourkela and Jenapur near Bhadrak for Brahmani river in Orissa is in use to predict flood at Jenapur. The coaxial correlation is shown in Fig. 9.19. To predict the flood at Jenapur, the stage at Panposh is marked. Let it be the point A as shown in the figure. Moving from point $A$ to $D$ through $B$ and $C$ is all the information required for forecasting. The correlation graph gives the stages at Talcher and Jenapur after a lag of 24 and 44 h, respectively.

(2) A correlation between the change in gauge of the upstream base station and change in gauge of the forecasting station in $T$ h is plotted. For a river with permanent control, such a relation gives good results. When a change in the stage at the base station takes place, the corresponding change at the forecasting station after the lag time can easily be predicted from such relations. Fig. 9.20 shows such a correlation.

(3) An improved way to correlate the gauges is to plot a relation between the $N$th and $(N + T)$th hour gauges of the forecasting station with change in the gauge of the upstream base station during $T$-h as shown in Fig. 9.21.

(4) Large fluctuations in the gauge of base station can also be incorporated by correlating the $N$th and $(N + T)$th hour stage of the forecasting station with change in stage in the past $T$-h as variable of the forecasting station. In the adjoining quadrant, the average gauge of base station is considered as variable. Such a graph is quite useful when, wide fluctuations in the upstream reaches are reported with reduced fluctuations in the lower forecasting reach due to large scale storage between the two reaches. The rising and falling stages can also be incorporated along with the above as shown in Fig. 9.22.

(5) The principle of multiple correlation can be extended to predict the flood of a river when one or more major tributaries join upstream of the forecasting station. To solve such a problem, a graph paper is divided into quadrants and in each quadrant, change in the gauge levels of different tributaries are plotted forming a *multiple correlation curve*. To incorporate time of travel of water in different tributaries which vary according to the distance of the base station from the forecasting station and the magnitude of the flood in the tributaries the parameter is considered as third variable in each quadrant. Thus, when three tributaries join a main river, the plot covers three quadrants of the graph paper. Such a system of flood forecasting is quite common in India. Central Water

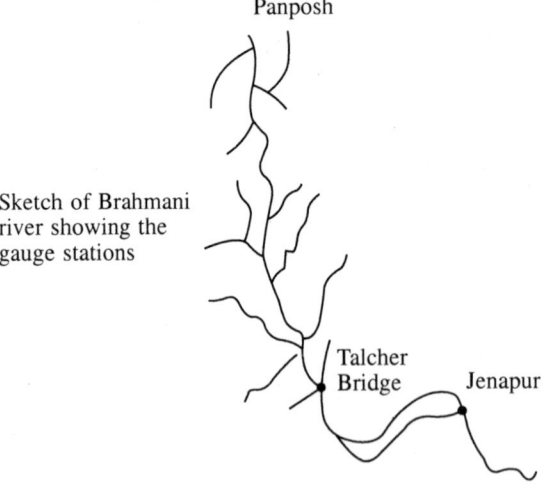

Panposh

Sketch of Brahmani
river showing the
gauge stations

Talcher
Bridge

Jenapur

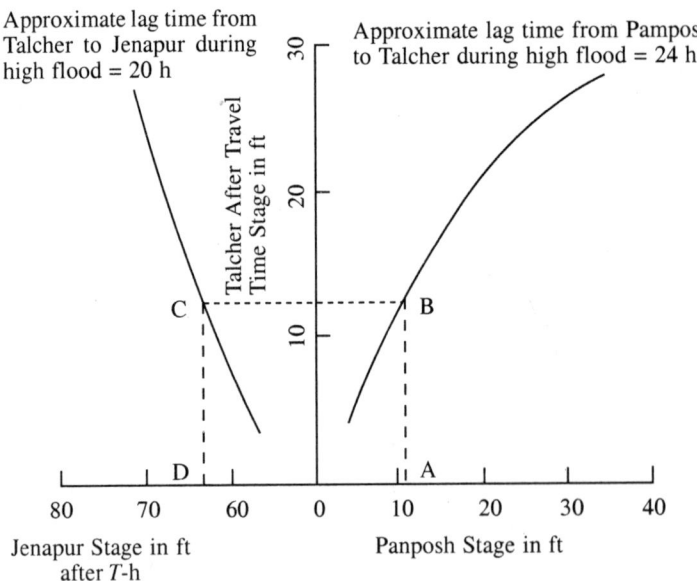

Approximate lag time from
Talcher to Jenapur during
high flood = 20 h

Approximate lag time from Pamposh
to Talcher during high flood = 24 h

Talcher After Travel
Time Stage in ft

C

B

D

A

80　　70　　60　　0　　10　　20　　30　　40

Jenapur Stage in ft
after $T$-h

Panposh Stage in ft

**Fig. 9.19　Coaxial Correlation Graph Between Panposh, Talcher and
Jenapur for River Brahmani, Orissa**

Commission uses this technique for flood forecasting at Gandhighat near Patna
from the knowledge of the base stations of: (i) river Sone at Chopan, (ii) river
Gondak at Revaghat, (iii) river Ghagra at Darauli and (iv) river Ganga at Buxer.
The correlation graph is produced in Fig. 9.23.

*Multiple Correlation with Hydrological Data*
In a river system, the catchment characteristics between the upstream base station
and the downstream forecasting station are always different. Therefore, the

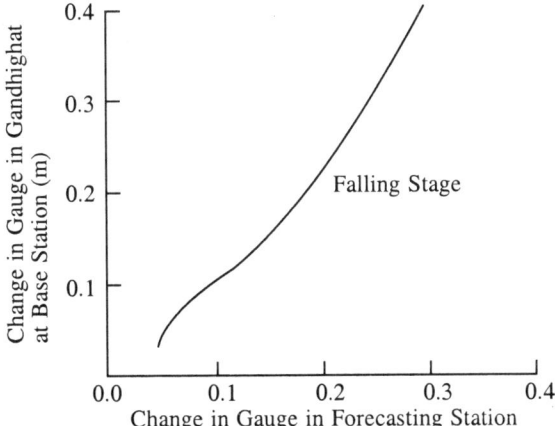

**Fig. 9.20   Correlation Graph for Change in Guge of U/S and D/S Stations**

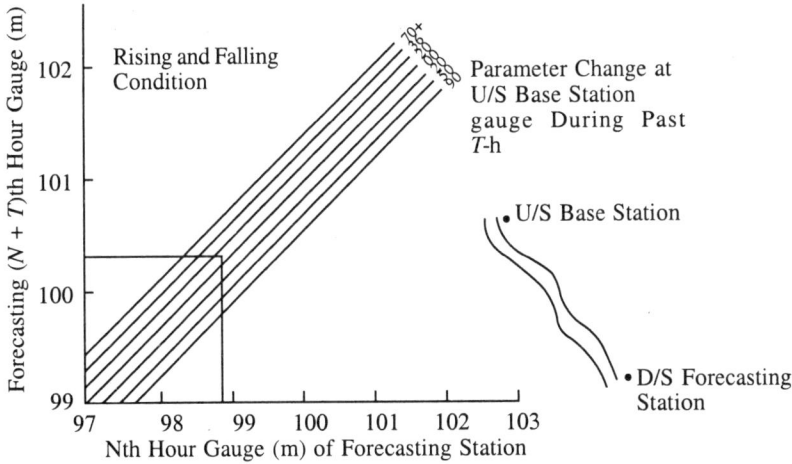

**Fig. 9.21   Correlation Graph Between *N*th Hour Stage and (*N* + *T*)th Hour Stage at Forecasting Station**

additional catchment between the two must be properly incorporated for the accuracy of the flood forecasting model. To do this, the procedure discussed above along with a relation between API and rainfall to predict flood from the balance catchment should be incorporated in a quadrant of the plot. Thus knowing the stages of upstream stations and the magnitude of flood from the interim catchment, the forecasting can be carried out accurately. Such a system is used for many forecasting stations in India. Currently flood forecasting at Delhi for river Yamuna is done on the basis of such a developed relation (Fig. 9.24).

---

**Example 9.11:** A major tributary joins a main river upstrem of a flood forecasting station. Flood stages of previous storms for the upstream main river, its tributaries and the downstream forecasting station are given. Develop a simple forecasting model for the system. Find the correlation coefficient between the three gauge stations.

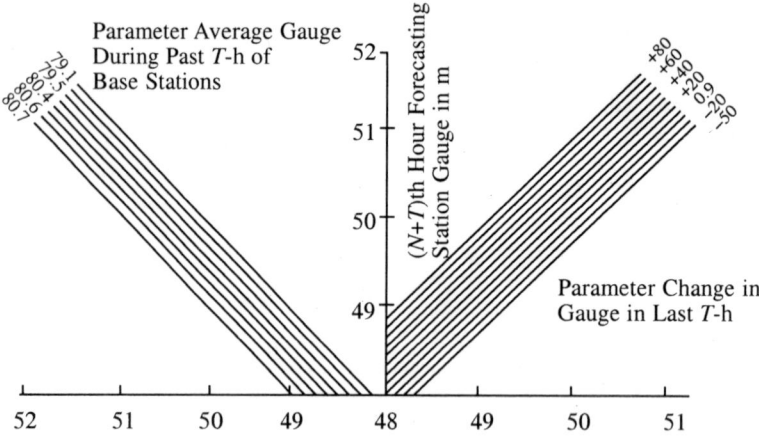

(N+T)th hour forecasting station gauge (m)   Nth hour forecasting station gauge (m)

**Fig. 9.22   Correlation Graph for Large Fluctuations in Gauge of Base Station**

| Stage U/S main river (m) | 165.3 | 165.8 | 166.6 | 167.2 | 168.0 | 168.4 | 168.8 | 169.3 | 170.2 |
|---|---|---|---|---|---|---|---|---|---|
| Stage U/S tributary (m) | 112.2 | 112.2 | 112.7 | 113.3 | 114.0 | 114.1 | 114.3 | 114.3 | 114.6 |
| Stage Forecasting station (m) | 80.3 | 80.5 | 82.2 | 84.3 | 85.6 | 85.7 | 86.0 | 86.3 | 87.1 |

**Solution**

The equation for the problem can be written as

$$X_1 = a + bX_2 + cX_3$$

where $X_1$ is the stage of U/S main river, $X_2$ is stage of U/S tributary and $X_3$ the stage of the forecasting station. Coefficients $a$, $b$ and $c$ can be estimated by the method of least square.

The above equation can be written in the following forms

$$\Sigma X_1 = aN + b \Sigma X_2 + c \Sigma X_3$$

$$\Sigma X_1 X_2 = a \Sigma X_2 + b \Sigma X_2^2 + c \Sigma X_2 X_3$$

$$\Sigma X_1 X_3 = a \Sigma X_3 + b \Sigma X_2 X_3 + c X_3^2$$

The values are calculated in Table 9.12.

Substituting the values from the above table we get

$$1509.6 = 9a + 1021.7b + 758c$$

$$171385 = 1021.7a + 115993b + 86069c$$

$$127174 = 758.0a + 86069b + 63894c$$

Solving the above three equations we get

$$a = 0.719, b = 1.393, c = 0.105$$

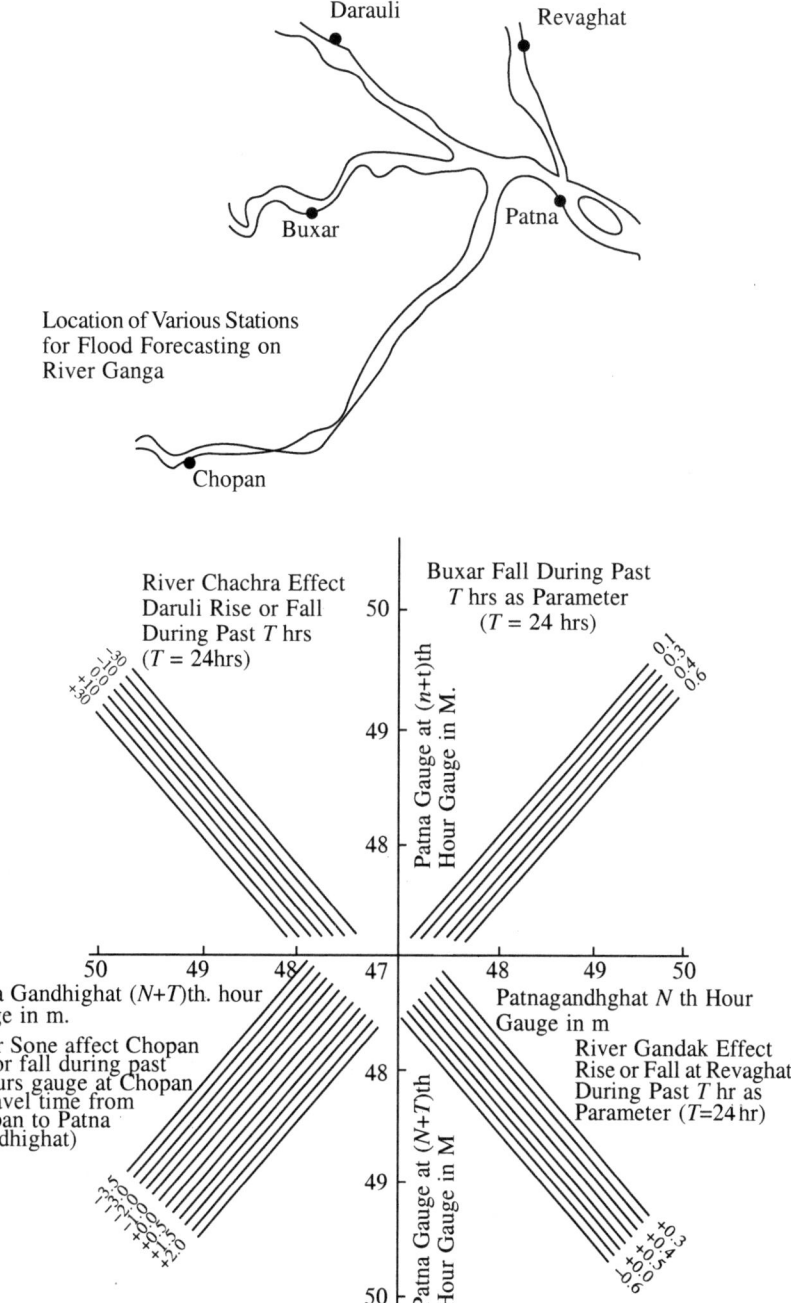

**Fig. 9.23** **Correlation Graph for River Ganga at Patna (Gandhighat Site)**

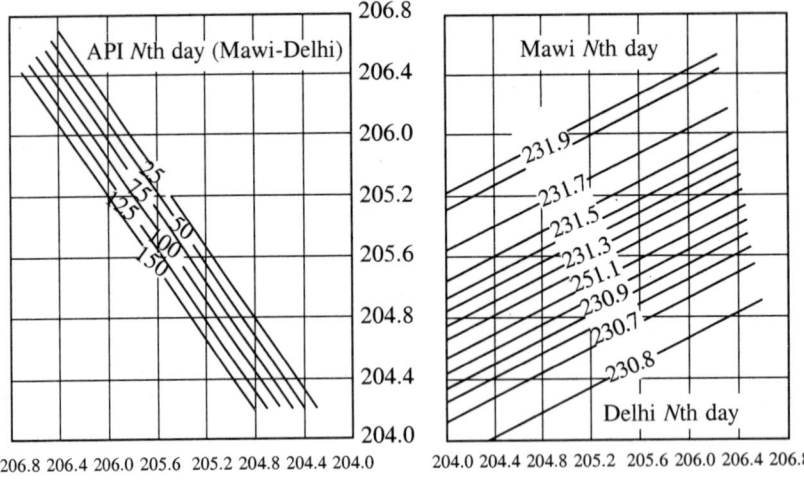

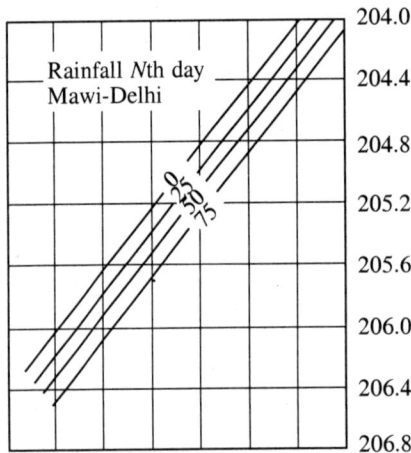

**Fig. 9.24   Correlation Graph for Yamuna River Between Mawi and Delhi**

Thus the equation becomes

$$X_1 = 0.719 + 1.393X_2 + 0.105X_3$$

Standard deviation is given by

$$\sigma_1 = \left\{ \frac{\Sigma (x_1 - x_{1av})^2}{(n-1)} \right\}^{1/2} = \left\{ \frac{21.42}{8} \right\}^{1/2} = 1.63$$

$$\sigma_2 = \left\{ \frac{\Sigma (x_2 - x_{2av})^2}{(n-1)} \right\}^{1/2} = \left\{ \frac{7.16}{8} \right\}^{1/2} = 0.95$$

$$\sigma_3 = \left\{ \frac{\Sigma (x_3 - x_{3av})^2}{(n-1)} \right\}^{1/2} = \left\{ \frac{53.18}{8} \right\}^{1/2} = 2.58$$

Table 9.12 Multiple Correlation Between Forecasting Station and Base Station

| $x_1$ | $x_2$ | $x_3$ | $x_1x_2$ | $x_2x_2$ | $x_2x_3$ | $x_1x_3$ | $x_3x_3$ | $(x_1 - 167.7)^2$ | $(x_2 - 113.5)^2$ | $(x_3 - 84.2)^2$ |
|---|---|---|---|---|---|---|---|---|---|---|
| (1) | (2) | (3) | (4) | (5) | (6) | (7) | (8) | (9) | (10) | (11) |
| 165.3 | 112.2 | 80.3 | 18547 | 12589 | 9009 | 13274 | 6448 | 5.92 | 1.75 | 15.38 |
| 165.8 | 112.2 | 80.5 | 18602 | 12589 | 9032 | 13347 | 6480 | 3.74 | 1.75 | 13.85 |
| 166.6 | 112.7 | 82.2 | 18776 | 12701 | 9264 | 13694 | 6757 | 1.28 | 0.67 | 4.09 |
| 167.2 | 113.3 | 84.3 | 18944 | 12837 | 9551 | 14095 | 7106 | 0.28 | 0.05 | 0.006 |
| 168.0 | 114.0 | 85.6 | 19152 | 12996 | 9758 | 14381 | 7327 | 0.07 | 0.23 | 1.89 |
| 168.4 | 114.1 | 85.7 | 19214 | 13019 | 9778 | 14432 | 7344 | 0.44 | 0.33 | 2.18 |
| 168.8 | 114.3 | 86.0 | 19294 | 13064 | 9830 | 14517 | 7396 | 1.14 | 0.60 | 3.16 |
| 169.3 | 114.3 | 86.3 | 19351 | 13064 | 9864 | 14610 | 7448 | 2.45 | 0.60 | 4.31 |
| 170.2 | 114.6 | 87.1 | 19505 | 13133 | 9982 | 14824 | 7586 | 6.08 | 1.16 | 8.28 |
| Sum 1509.6 | 1021.7 | 758 | 171385 | 115993 | 86069 | 127174 | 63894 | 21.42 | 7.156 | 53.175 |
| Mean 167.7 | 113.5 | 84.2 | | | | | | | | |

$$r_{12} = \frac{(\Sigma x_1 x_2 - n x_{1av} x_{2av})}{\{(n-1)\sigma_1\sigma_2\}} = \frac{(171385 - 9 \times 167.7 \times 113.5)}{(8 \times 1.63 \times 0.95)} = 0.972$$

$$r_{13} = \frac{(\Sigma x_1 x_3 - n x_{1av} x_{3av})}{\{(..-1)\sigma_1\sigma_3\}} = \frac{(127174 - 9 \times 167.7 \times 84.2)}{(8 \times 1.63 \times 2.58)} = 0.966$$

$$r_{23} = \frac{(\Sigma x_2 x_{23} - n x_{2av} x_{3av})}{\{(n-1)\,\sigma_2\sigma_3\}} = \frac{(86069 - 9 \times 113.5 \times 84.2)}{(8 \times 0.95 \times 2.58)} = 0.993$$

Other types of regression relation like $X_3 = d + eX_2 + fX_1$ or $X_2 = g + hX_1 + kX_3$ can be developed between the stations to predict the behaviour of one station from the data of other two.

## 9.8   FLOOD CONTROL MEASURES

One of the main function of the State Water Resources Department is to take up various measures to control the probable floods likely to arrive at important strategic locations in the river course. A river basin is unique of its kind and so is its flood problems. Therefore, the measures to control the flood should be decided specific to the location. There are various methods available to control the flood and the suitability of any particular method to the location and flood magnitude depends on the consideration of large number of factors. Therefore, selecting a particular method out of the following depend on the experience, intelligence, design consideration and suitability of the structure to fit the locality.

**(1) Construction of Storage Reservoirs:** Reservoirs are constructed when an obstruction is created by means of a dam across a river so that it stores water upstream of it. It is very expensive to construct a reservoir for the sole purpose of flood control. In India such a single purpose reservoir is yet to be built.

Depending on the storage provided, a reservoir can be classified as small (tank), medium or major. Construction of large number of small reservoirs have been done in India in the past for the purpose of irrigation. These tanks are effective in reducing flood magnitude to a certain extent but their effectiveness in controlling flood in a basin is now in question as there cannot be a definite regulation pattern fixed to correlate thousands of such tanks to contain flood. Most of such tanks gets filled up by the time flood impinges the tanks. These tanks usually do not have an operation schedule.

A large reservoir constructed across the main river with multi-purpose objectives like flood control, irrigation, hydro-power generation and other uses, also controls flood to a great extent during rainy season. India has such reservoirs in large numbers. These reservoirs increase the economy of the region due to indirect benefits.

**(2) Construction of Dykes:** Dykes or Levees are simple earth dam sections, constructed parallel to a river flow in its flood plain at suitable distance away from the main channel. The structure controls flood very effectively in the reach, where it is constructed. Depending on the area to be protected, dykes may be constructed to one or both sides of a river. When a masonary wall of reinforced cement concrete is constructed instead of an earthen section, the structure is known as *flood wall*. The structure must be properly designed to withstand water pressure and sliding. At places a second dyke is constructed on the same side of the river parallel to the first one to guard very important places and/or a city or a town in case of the failure of the first one. Dykes are designed like earth dam sections whereas a flood wall is designed as retaining wall. There must be seepage holes in the body of the masonary dykes. Gated outlets must also be provided to drain out water collected from the city to the river. Dykes are advantages as they are less costly, simple to construct and the construction materials are easily available at the site. However, they are likely to breach, the maintenance cost is high and the failure chance due to wave action, flood impact and seepage are high. These structures create problem of drainage to its protected area. A simple dyke section is shown in Fig. 9.25.

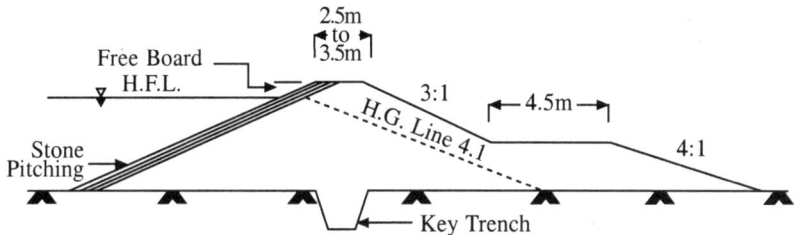

**Fig. 9.25   Typical Section of a Levee**

**(3) Improvement of Channel Capacity:** By improving the channel capacity, the velocity of water increases in a given reach and can be achieved by (i) lining the river for the cross section, (ii) providing cutoff to a meander channel and (iii) straightening the channel. Such works lower the river stage at the locality and to some distance upstream of it. There is every possibility of back water affect at the cross section just downstream of this section. Such improvements are undertaken in the areas where levees or dykes cannot be constructed due to non-availability of space and storage reservoirs cannot be constructed upstream of such places. A cutoff to a meander channel is shown in Fig. 9.26.

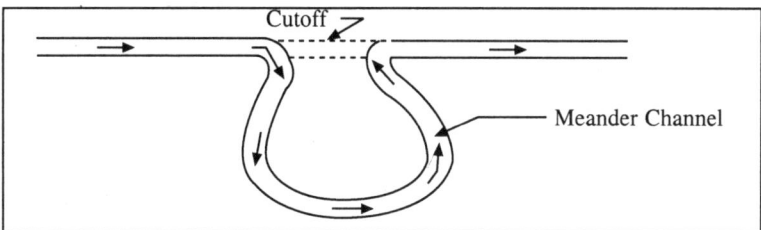

**Fig. 9.26   A Cutoff to a Meander Channel**

**(4) Construction of Diversion Channels:** Sometimes a branch channel is taken out from the upstream of a main river to reduce the flood peak of the area located just downstream of it. Thus a part of flood water is diverted to the diversion channel. Such practice is followed at number of places to protect important cities. In India, a diversion channel taken off at 5 km upstream of Srinagar city from river Jhelum protects the city. This diversion channel outfalls in the famous Wullar lake of Srinagar. Normally a regulator is provided at the offtake location.

**(5) Construction of Spurs or Groynes:** Spurs or groynes are sometimes constructed projecting from the bank into the river at the side to be protected from flood. There are many types of groynes constructed. When a solid groyne is constructed by earth material, the section is designed like an earth dam section. One of the main problem with groynes is to protect its nose which projects into the river. Large stones are normally dumped at the nose of the groynes to protect it from high water velocities and the section is also pitched by large stones. There are three types of groynes:

   (i) *Repelling groynes* projects upstream into the river from the point of its origin from the bank (Fig. 9.27a).

  (ii) *Attracting groynes* projects downstream into the river from the point of its origin from the bank (Fig. 9.27b).

 (iii) *Perpendicular groynes* projects normal to the river from the point of its origin from the bank (Fig. 9.27c).

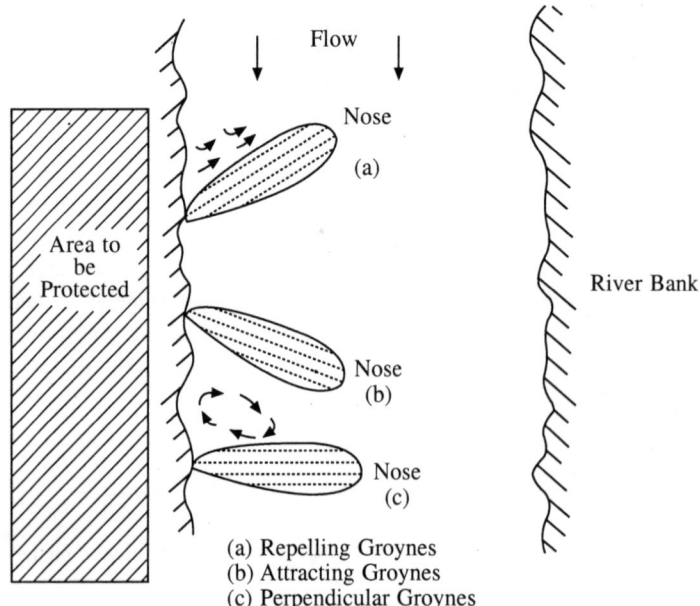

(a) Repelling Groynes
(b) Attracting Groynes
(c) Perpendicular Groynes

**Fig. 9.27    A Plan View of Location of Groynes for Flood Protection Works in a River reach.**

Layout of the above groynes at a river reach along with the area to be protected is shown in Fig. 9.27. Groynes are also classified into various types, depending on the material used for the construction and the type of construction. Some are constructed permeable initially, thus allowing water to flow through them. In course of time, flow debris and silt accumulates, resulting in the formation a solid wall.

**(6) Construction of Terraces:** Terraces at the slopping hills can store good quantity of water for short time, thus forcing the overland flowing water to infiltrate into the ground. Such terraces are in the form of cascades of 1.5 to 2.0 m width with small banks of 0.20 to 0.25 m high at the downfall edge. Such structures detain water in series and can be very affective for small basins in controlling flood.

**(7) Upstream Soil Conservation Measures:** Another affective way to reduce flood in a river is to go for aggressive soil conservation measures at the upstream catchment. It is stated that a thick grass cover reduces the surface runoff above 90%, whereas 67% reduction can be achieved by row crops and 33% reduction can be achieved in close growing crops, taking zero percent reduction as the standard for a barren impervious land. Such plantation measures not only save the important soil from wastage but help to increase infiltration rate thereby charging to ground water potential.

Another important impact of such a measure is that silt load of the flood water is reduced. This silt, which otherwise gets deposited in the reservoirs or river beds at the delta head is avoided. Deposit of silt reduces the life of the reservoir and flood carrying capacity in a river section.

Flood control measures vary from place to place. Even for the same river it varies from one location to another depending on the economy and other factors. A particular method found suitable at a location may not be suitable at other and hence use of a particular method to suit other places may be considered after a detailed study.

## PROBLEMS

9.1 What do you understand by routing of a flood? Write the basic equations in hydrologic reservoir routing. How is hydrologic routing different from hydraulic routing?

9.2 Describe in steps the Muskingum method of flood routing if the values of $K$ and $X$ are known for a given channel reach. What should be the limiting value of routing step $\Delta t$?

9.3 Derive the Muskingum channel routing equation

$$O_2 = C_0 I_2 + C_1 I_1 + C_2 O_1$$

where $I_i$ are the inflows, $C_i$ the constants and $O_i$ the outflows.

9.4 What do you understand by Antecedent Precipitation Index? How is API used to determine runoff from a given catchment?

9.5 What are the various methods of reservoir routing? Write down the steps involved in ISD method.

9.6 Explain in brief the various methods of multiple correlations used for river stage forecasting for different rivers in India.

9.7 What are the types of flood control measures that can be undertaken for a river system?

9.8 A reservoir has the following Elevation-Storage-Discharge relations.

$$S = 0.015h^2$$
$$Q = 50\, h^{0.5}$$

where $h$ is the depth of water above the pool level of 202.0 m, $Q$ the discharge in m³/sec and $S$ the storage in Ha.m. If the following inflow hydrograph arrives at the minimum pool level of 202.0 m, calculate the maximum outflow rate from the reservoir.

| Time (h) | 0 | 4 | 8 | 12 | 16 | 20 | 24 | 28 | 32 |
|---|---|---|---|---|---|---|---|---|---|
| Inflow (m³/sec) | 30 | 80 | 120 | 200 | 160 | 129 | 100 | 60 | 30 |

9.9 Observed values of inflows and outflows at two sections of a river reach are given below. Determine $K$ and $X$ between the reach to be used for Muskingum method of routing.

| Time (h) | 0 | 6 | 12 | 18 | 24 | 30 | 36 | 42 | 48 | 54 |
|---|---|---|---|---|---|---|---|---|---|---|
| | 60 | 66 | 72 | 78 | | | | | | |
| Inflow (m³/sec) | 15 | 70 | 200 | 250 | 220 | 180 | 140 | 90 | 70 | 50 |
| | 35 | 25 | 16 | 13 | | | | | | |
| Outflow (m³/sec) | 15 | 17 | 45 | 140 | 190 | 205 | 180 | 150 | 125 | 100 |
| | 60 | 40 | 25 | 16 | | | | | | |

9.10 For a river reach $K$ is 28 h and $X$ is 0.25. Route the following inflow hydrograph. Take $O_1 = I_1$ for the beginning step. Determine the values of attenuation and translation of the peak.

| Time (h) | 0 | 6 | 12 | 18 | 24 | 30 | 36 | 42 | 48 | 54 | 60 |
|---|---|---|---|---|---|---|---|---|---|---|---|
| Inflow (m³/sec) | 30 | 62 | 242 | 170 | 114 | 78 | 56 | 44 | 38 | 34 | 30 |

9.11 The elevation-storage-discharge data obtained from a reservoir are:

| Elevation (m) | 250.5 | 257 | 262 | 265 | 268 | 270 | 272 | 273.5 | 275 | 276 | 277 |
|---|---|---|---|---|---|---|---|---|---|---|---|
| Storage (Ha.m) | 20 | 80 | 136 | 208 | 310 | 400 | 520 | 650 | 700 | 760 | 800 |
| Discharge (m³/sec) | 4 | 12 | 20 | 32 | 45 | 65 | 92 | 115 | 113 | 160 | 200 |

Determine the maximum pool elevation of the reservoir when the following flood impinges at minimum reservoir level.

| Time (h) | 0 | 6 | 12 | 18 | 24 | 30 | 36 | 42 | 48 | 54 | 60 | 66 |
|---|---|---|---|---|---|---|---|---|---|---|---|---|
| Inflow (m³/sec) | 12 | 50 | 160 | 270 | 240 | 205 | 150 | 95 | 68 | 36 | 23 | 14 |

9.12 A reservoir has the following storage characteristics:

| $(Q + 2S/\Delta t)$(m³) | 85 | 114 | 142 | 172 | 198 | 227 | 254 | 283 |
|---|---|---|---|---|---|---|---|---|
| Outflow (m³/sec) | 26 | 31 | 35 | 38 | 41 | 43 | 45 | 46.5 |

How much is the flood wave modified in the reservoir ?

9.13 A reservoir has the following elevation-storage-discharge relation:

| Elevation (m) | 105 | 108 | 110 | 111 | 112 | 113 | 114 |
|---|---|---|---|---|---|---|---|
| Storage (Ha.m) | 24 | 66 | 99 | 127 | 155 | 185 | 200 |

Discharge (m³/sec) is given by the relation $Q = 16 \times H^{1.5}$, where $H$ is the head over the crest level of 110.0 m. The following inflow hydrograph is arrived when the reservoir level was 107.0 m. Compute the highest elevation and maximum discharge from the reservoir due to the inflow.

| Time (h) | 0 | 6 | 12 | 18 | 24 | 30 | 36 | 42 | 48 | 54 |
|---|---|---|---|---|---|---|---|---|---|---|
| Inflow (m³/sec) | 7 | 24 | 60 | 75 | 62 | 36 | 26 | 18 | 12 | 7 |

9.14 Storage-elevation-outflow data from a reservoir are as follows:

| Elevation (m) | 20.0 | 20.8 | 21.4 | 22.0 | 22.6 | 23.2 | 23.8 |
|---|---|---|---|---|---|---|---|
| Storage (Mm³) | 0.80 | 1.50 | 2.00 | 2.70 | 3.30 | 4.0 | 4.9 |
| Outflow (m³/sec) | 0 | 12 | 35 | 70 | 115 | 165 | 215 |

Spillway crest elevation is at 20.0 m. The design flood is given as follows:

| Time (h) | 0 | 2 | 4 | 6 | 8 | 10 | 12 | 14 | 16 | 18 | 20 | 22 | 24 |
|---|---|---|---|---|---|---|---|---|---|---|---|---|---|
| Inflow (m³/sec) | 6 | 15 | 30 | 52 | 60 | 70 | 54 | 46 | 38 | 32 | 22 | 14 | 8 |

Reservoir is at 20.8 m when the flood impinges. Plot the outflow hydrograph.

9.15 A reservoir has the following Elevation-Area curve

| Elevation (m) | 127 | 128 | 129 | 130 | 131 | 132 | 133 | 134 | 135 | 136 | 137 |
|---|---|---|---|---|---|---|---|---|---|---|---|
| Area (Ha) | 10 | 20 | 60 | 90 | 110 | 125 | 170 | 200 | 230 | 280 | 340 |

Plot the storage-elevation curve for the reservoir.

9.16 Storage-Outflow data from a reservoir are as follows :

| Storage (Mm³) | 12.0 | 18.0 | 24.5 | 37.0 | 47.5 |
|---|---|---|---|---|---|
| Outflow (m³/sec) | 12 | 174 | 470 | 1280 | 2220 |

Route the following inflow hydrograph. Take spillway crest level from the storage of 12.0 Mm³. What is the peak outflow from the reservoir?

| Time (h) | 8 | 10 | 12 | 14 | 16 | 18 | 20 | 22 | 24 | 2 | 4 |
|---|---|---|---|---|---|---|---|---|---|---|---|
|  | 6 | 8 | 10 | | | | | | | | |
| Inflow (m³/sec) | 40 | 80 | 210 | 280 | 500 | 1280 | 1900 | 1450 | 900 | 600 | 400 |
|  | 250 | 150 | 100 | | | | | | | | |

# Chapter Ten

# Reservoir and Sedimentation

## 10.1 INTRODUCTION

Availability of water in tropical countries like India is highly uneven both in space and time. Rainfall period in such regions is limited to 4–5 months in a year and rest of the months are almost dry. The demand for water is increasing continuously to meet the needs of the municipality, industry, agriculture, power generation and other mandatory releases at various locations in a river course. To meet the demand of water at these locations arising out of the variability of the resource, storage reservoirs provide the only viable alternative. That was the reason, tanks were built during ancient civilizations at various parts of the world even before the concept of water cycle was known to them.

A dam is built across a river to reserve sufficient water behind it so that water can supplement the various demands during lean seasons. With the present day technology, an engineer needs to be fully equipped to select the best alternative, keeping in mind the factors like (1) dependable yield at the proposed dam site, (2) targeted pattern of demands for various uses like irrigation, drinking water, industrial, hydropower and other uses at various periods of the year, (3) suitability of the site for constructing such a reservoir, (4) rehabilitation problems, (5) foundation suitability, (6) availability of economy, (7) sedimentation problems and (8) operation of the system. Here certain aspects of reservoir and its life due to sedimentation are discussed. Cross section of a dam is shown in Fig. 10.1.

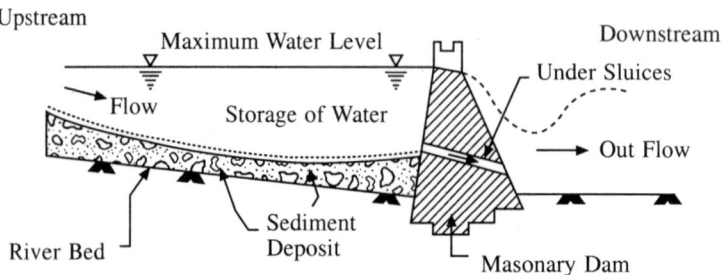

**Fig. 10.1   Cross–Section of a Masonary Dam**

## 10.2   FIXATION OF RESERVOIR CAPACITY

Storage capacity of a reservoir is the maximum difference between the cumulative supply and demand during the period of the driest year of the available records. Several methods are available to calculate the storage requirement for a reservoir. Two popular methods are:

### 10.2.1 Ripples Mass Curve

Chapter 6 discussed different approaches for computing runoff from a river at any perspective location in its course. The runoff series at a proposed reservoir site gives a dependable yield on the basis of which the storage requirement of the project to meet the targeted demands are ascertained. This is achieved by constructing a flow mass curve representing the cumulative inflow into the reservoir vs. time. The time scale is usually taken in months. This mass curve is called *Ripples Mass Curve*. A flow mass curve at a potential project site is shown in Fig. 10.2. The curve represents one of the driest period of stream flow at a potential reservoir site in Orissa. Such a curve at a reservoir site gives the following information.

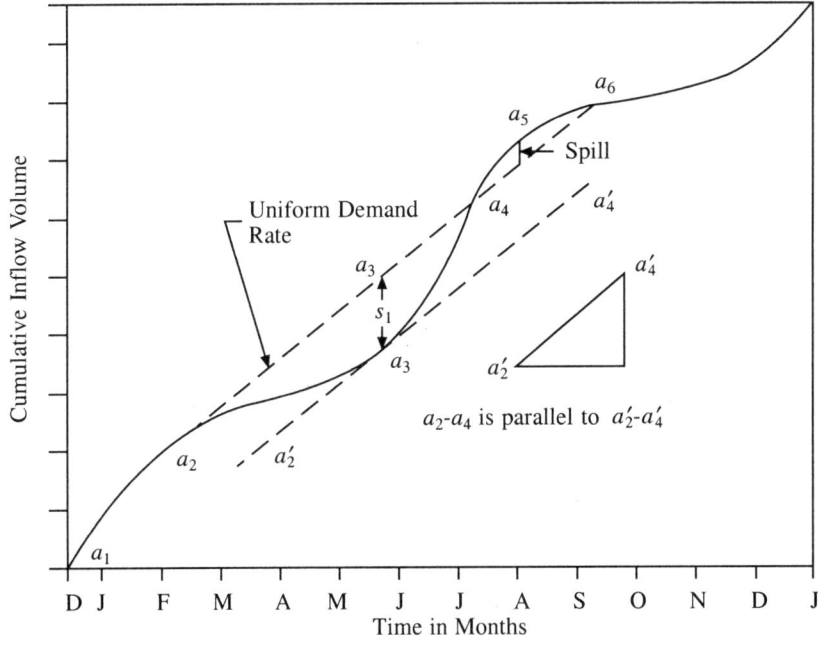

**Fig. 10.2  A Flow Mass Curve**

1. Slope of the mass curve at any point is a measure of inflow rate at that time.
2. Required rates of drawl of water from the reservoir is computed by drawing tangents with slopes equal to the demand rate at the convex point $a_2$ or $a_5$ of the mass curve. Slope of the demand line is shown as $a_2 - a_4$ in the figure.
3. Maximum departure between the demand line and the mass curve ($S_1$) gives the required storage capacity for the reservoir to meet the targeted demands.
4. The demand line must cut the mass curve when it is extended forward. If it does not cut, then the reservoir is not going to refill again during the period. For the curve shown the demand line cuts the mass inflow curve at $a_4$.

5. Vertical distance between the tangent line (demand line) and the next peak gives the peak rate of water spilled out. Volume of water spilled out is shown as $a_4a_5a_6$ (Fig. 10.2).
6. The mass curve should include the driest period to meet the most reliable demand pattern from the reservoir.

Reservoir filling starts at $a_3$ and at $a_4$ the process is complete. The storage capacity required to meet the uniform demand rate $a_2a_4$ is obtained by drawing a parallel line $a_2'$-$a_4'$ to the line $a_2$-$a_4$ at $a_3$ and measuring the distance $S_1$ between $a_2$-$a_4$ and $a_2'$-$a_4'$. If more such dry periods are available then the maximum of such departures is to be considered as storage. The reservoir is just full at $a_2$ and $a_4$. When the demand patterns are known, the mass curves for inflow and demand are plotted from the same coordinate system for the driest period. The plotting is made till the two curves again meet forming a loop. The maximum departure for such mass curves forming the loop is shown in Fig. 10.3, which is the storage capacity required for the reservoir.

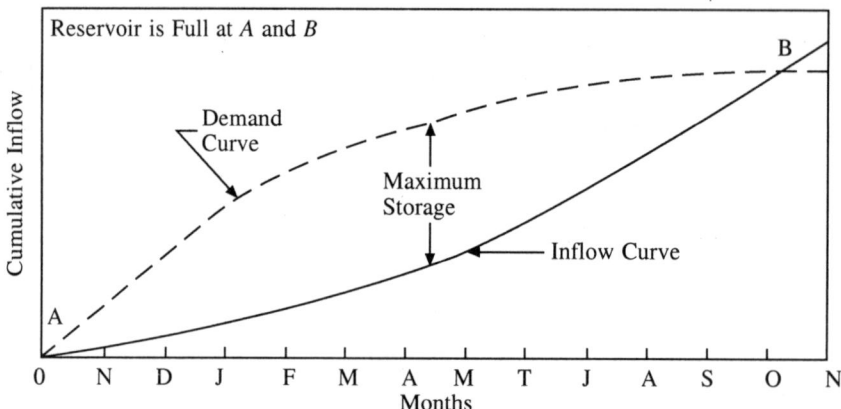

**Fig. 10.3   Mass Curve Between Monthly Inflows and Demand**

Care must be taken to correct the observed inflows for the monthly losses of evaporation from the reservoir, seepage through the dam, other mandatory releases for the downstream and the volume of sediment inflow into the reservoir. Consequently, the losses may be accounted for by considering them as the demands from the reservoir for the respective months.

### 10.2.2   Sequent Peak Algorithm

Another method to determine the storage requirement from a reservoir is the use of sequent peak algorithm (Fig. 10.4), which can easily be programmed in a digital computer. Normal and maximum storage through sequent peak algorithm is calculated as follows:

1. Convert the monthly inflows into the volume units for the period of available data.
2. Estimate the monthly volumes of all the outflows from the reservoir. This should include the losses from evaporation, seepage and other losses.

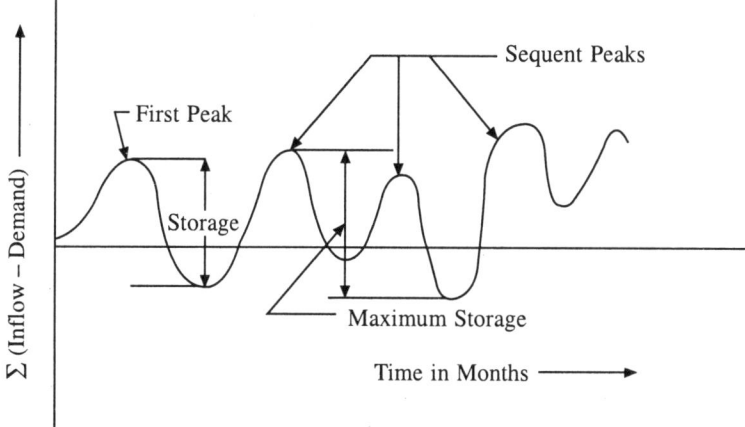

**Fig. 10.4   Sequent Peak Algorithm**

3. Compute the cumulative values of $\Sigma$ (inflow – outflow) from the reservoir.
4. Plot a graph by taking months as abscissa and $\Sigma$ $(I - O)$ of step (3) as ordinate on a ordinary graph paper.
5. The data will plot with peaks and troughs. The second and subsequent peaks are called *sequent peaks*. The maximum difference between any sequent peak and the just following trough is the maximum storage required for the reservoir. The difference between the first peak and the trough following it, is the storage required under normal inflows.

---

**Example 10.1:** River flows at a proposed reservoir site for a drought period of 15 months are given. The targeted demands for all releases are found out from a working table. Compute the storage required by sequent peak analysis. All units are in $Mm^3$.

| Months | Jun | Jul | Aug | Sep | Oct | Nov | Dec | Jan | Feb | Mar | Apr | May | Jun | Jul | Aug |
|---|---|---|---|---|---|---|---|---|---|---|---|---|---|---|---|
| River flows | 500 | 700 | 800 | 400 | 300 | 300 | 200 | 100 | 300 | 600 | 800 | 900 | 300 | 400 | 900 |
| Demands | 300 | 300 | 400 | 500 | 700 | 800 | 500 | 400 | 300 | 300 | 200 | 500 | 700 | 600 | 200 |

**Solution**
Calculation are carried out in Table 10.1. The difference between inflow and outflow are entered in col. (6) and their cumulative volumes in col. (7). Curves between cumulative volumes of inflow and outflow are plotted against time in months (Fig. 10.5). From the plot, the storage required is measured from the difference between the first peak and the following trough.

The storage is found as 1600 $Mm^3$. From the tabular calculation, the value of 1600 represents the cumulative of all negative values of inflow and outflow in col. (6). It is the sum of all ordinates of col. (6) from September to January.

**Example 10.2:** How much storage is required to maintain a minimum demand of 30 cumecs from the monthly inflows into the reservoir given below ?

| Months | Jan | Feb | Mar | Apr | May | Jun | Jul | Aug | Sep | Oct | Nov | Dec |
|---|---|---|---|---|---|---|---|---|---|---|---|---|
| River flows $(m^3/sec)$ | 48 | 42 | 23 | 11 | 14 | 22 | 45 | 55 | 43 | 35 | 27 | 21 |

**Table 10.1    Reservoir Storage by Sequent Peak Method**

| Time | Inflow $I$ (M.m³) | Demand $D$ (M.m³) | $\Sigma$ Inflow (M.m³) | $\Sigma$Demand (M.m³) | $(I - D)$ (M.m³) | $\Sigma(I - D)$ (M.m³) |
|------|------|------|------|------|------|------|
| (1) | (2) | (3) | (4) | (5) | (6) | (7) |
| Jun | 500 | 300 | 500 | 300 | +200 | +200 |
| Jul | 700 | 300 | 1200 | 600 | +400 | +600 |
| Aug | 800 | 400 | 2000 | 1000 | +400 | +1000 |
| Sep | 400 | 500 | 2400 | 1500 | −100 | +900 |
| Oct | 300 | 700 | 2700 | 2200 | −400 | +500 |
| Nov | 300 | 800 | 3000 | 3000 | −500 | 0 |
| Dec | 200 | 500 | 3200 | 3500 | −300 | −300 |
| Jan | 100 | 400 | 3300 | 3900 | −300 | −600 |
| Feb | 300 | 300 | 3600 | 4200 | 000 | −600 |
| Mar | 600 | 300 | 4200 | 4500 | +300 | −300 |
| Apr | 800 | 200 | 5000 | 4700 | +600 | +300 |
| May | 900 | 500 | 5900 | 5200 | +400 | +700 |
| Jun | 300 | 700 | 6200 | 5900 | −400 | +300 |
| Jul | 400 | 600 | 6600 | 6500 | −200 | +100 |
| Aug | 900 | 200 | 7500 | 6700 | +700 | +800 |

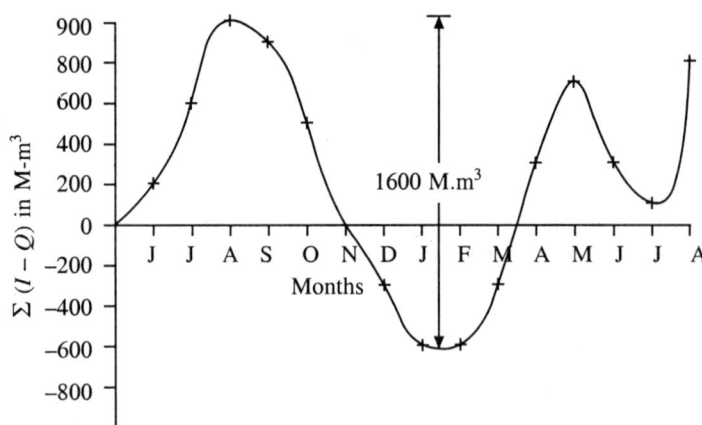

**Fig. 10.5.    Cumulative Values of Inflow-Outflow for the Computation of Storage for Example 10.1**

**Solution**

The mean monthly flows are converted to monthly volumes in M.m³. This is done by multiplying the number of seconds of the month to the mean monthly inflows. Calculation are carried out in Table 10.2.

The inflow volume of col. (4) are cumulated in col. (5). The cumulative flow volumes are plotted in a graph against month as abscissa (Fig.10. 6). For simplification each month is taken on same scale even though they have different number of days making them. Demand is 30 cumec or $(30 \times 30 \times 24 \times 60 \times 60)/10^6 = 77.76$ M.m³ in a month. A demand line $xy$ is drawn at 30 m³. Line $x'y'$ is parallel to $xy$ at the bottom of the valley $P$ on the curve. $S_1$ is the vertical distance between the two parallel lines. Storage is the length $S_1$ measured in the vertical scale. This is found as 140.0 M.m³.

**Table 10.2 Reservoir Storage due to Constant Demand Rate**

| Month | No. of days in the month | Inflow $I$ (mean) (m³/sec) | Inflow volume (M.m³) | Cumulative volume (M.m³) |
|-------|------|------|------|------|
| (1) | (2) | (3) | (4) | (5) |
| Jan | 31 | 48 | 128.6* | 128.6 |
| Feb | 28 | 42 | 101.6 | 230.2 |
| Mar | 31 | 23 | 61.6 | 291.8 |
| Apr | 30 | 11 | 28.5 | 320.3 |
| May | 31 | 14 | 37.5 | 357.8 |
| Jun | 30 | 22 | 57.0 | 414.8 |
| Jul | 31 | 45 | 120.5 | 535.3 |
| Aug | 31 | 55 | 147.3 | 682.6 |
| Sep | 30 | 43 | 111.5 | 794.1 |
| Oct | 31 | 35 | 93.7 | 887.8 |
| Nov | 30 | 27 | 70.0 | 957.8 |
| Dec | 31 | 21 | 56.2 | 1014.0 |

$$*\,128.6 = \frac{(48 \times 31 \times 24 \times 60 \times 60)}{10^6}$$

**Example 10.3:** Use the mass curve of Example 10.2 and obtain the maximum demand that can be maintained by the reservoir if the possible storage at the dam site is 100 M.m³ due to submergence problems.

**Solution**
Solution to the problem is carried out by locating point $E$ in Fig. 10.6 such that the vertical distance $PE$ is 100 M.m³. Points $x$ representing the first hump of the mass curve and $E$ are joined by a straight line. The line $xE$ is extended to meet the mass curve at $Z$. A parallel line $X'Z'$ is drawn such that it passes through $P$. The slope of $XZ$ is the demand that can be met from the reservoir. The slope is found as 28 cumecs.

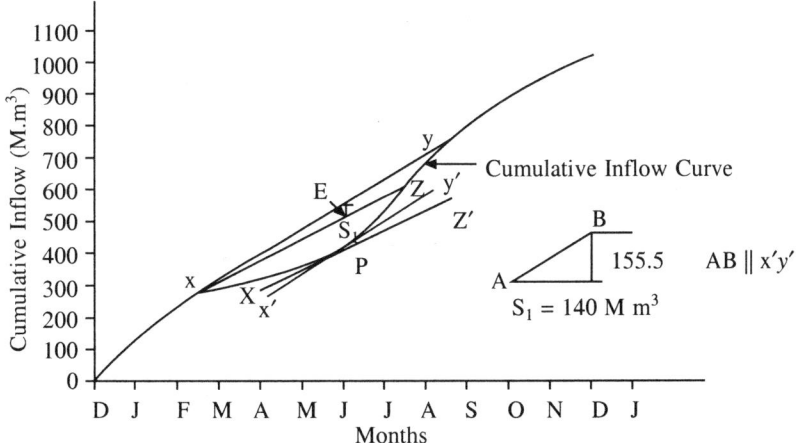

**Fig. 10.6 Mass Curve for Examples 10.2 and 10.3**

**Example 10.4:** The following inflow and demand volumes (M.m$^3$) information are available at a reservoir site located downstream of another reservoir.

**Table 10.3   Inflow and Demand Patterns at a Reservoir Site (unit in M.m$^3$)**

| Month | Outflow from u/s reservoir | Flow from intermediate catchment | Irrigation demand | Other demands | Evaporation losses |
|-------|------|------|------|------|------|
| (1) | (2) | (3) | (4) | (5) | (6) |
| Jan | 3.2 | 0.6 | 1.3 | 0.1 | 0.3 |
| Feb | 2.4 | 0.4 | 1.4 | 0.1 | 0.4 |
| Mar | 2.0 | 0.4 | 1.5 | 0.2 | 0.5 |
| Apr | 1.4 | 0.2 | 1.8 | 0.2 | 0.6 |
| May | 0.5 | 0.1 | 1.3 | 0.2 | 0.8 |
| Jun | 0.9 | 0.1 | 1.6 | 0.2 | 0.8 |
| Jul | 15.8 | 0.9 | 2.5 | 0.1 | 0.8 |
| Aug | 18.0 | 1.1 | 2.1 | 0.1 | 0.7 |
| Sep | 13.1 | 0.8 | 1.7 | 0.1 | 0.6 |
| Oct | 5.2 | 0.8 | 1.0 | 0.1 | 0.5 |
| Nov | 4.1 | 0.7 | 0.6 | 0.1 | 0.4 |
| Dec | 3.3 | 0.5 | 0.8 | 0.1 | 0.3 |

Calculate the storage needed to meet the demands.

**Solution**

Calculations are carried out in Table 10.4.

Table 10.4 shows that maximum cumulative demand is −4.30 M.m$^3$. Therefore, a storage of 4.3 M.m$^3$ is required to meet the targeted demands of irrigation, evaporation and other demands.

## 10.3 DETERMINATION OF SPILLWAY SIZE

One of the important objective of any storage reservoir is to moderate the inflow flood hydrograph to the extent the people living at the downstream are protected from the hazards. To achieve this, the peak outflow and reservoir storage should be properly designed to take care of the design flood likely to occur during the project life. The peak rate of outflow from reservoir should be less than the maximum channel carrying capacity of the downstream reach with a factor of safety for free board and local inflows.

For a reservoir the worst situation is expected when the largest flood arrives at the full reservoir level (FRL). During reservoir routing it is seen that the level of water rises above FRL during routing and comes back to original level at the end of routing. This new height is called Maximum Water Level (MWL). The terms FRL and MWL will be defined later. If the project is large and important for which probable maximum flood (PMF) is considered in its design, then the PMF is routed through the reservoir to determine the maximum water level (MWL) to which the reservoir is to be built with allowance for free board. All reservoirs store more water per unit depth at higher elevations because the

**Table 10.4  Estimation of Storage for Example 10.4**

| Month | Inflow = outflow from u/s reservoir (M·m³) | Inflow from other C.A (M·m³) | Total inflow (M·m³) | Demands | | | | Col.(4) – (8) (M·m³) | ΣCol.(9) | |
|-------|------|------|------|------|------|------|------|------|------|------|
| | | | | irrigation (M·m³) | evaporation (M·m³) | others (M·m³) | Total (M·m³) | | −ve (M·m³) | +ve (M·m³) |
| (1) | (2) | (3) | (4) | (5) | (6) | (7) | (8) | (9) | (10) | (11) |
| Jan | 3.2 | 0.6 | 3.8 | 1.3 | 0.3 | 0.1 | 1.7 | 2.1 | – | 2.1 |
| Feb | 2.4 | 0.4 | 2.8 | 1.4 | 0.4 | 0.1 | 1.9 | 0.9 | – | 3.0 |
| Mar | 2.0 | 0.4 | 2.4 | 1.5 | 0.5 | 0.2 | 2.2 | 0.2 | – | 3.2 |
| Apr | 1.4 | 0.2 | 1.6 | 1.8 | 0.6 | 0.2 | 2.6 | −1.0 | −1.0 | – |
| May | 0.5 | 0.1 | 0.6 | 1.3 | 0.8 | 0.2 | 2.3 | −1.7 | −2.7 | – |
| Jun | 0.9 | 0.1 | 1.0 | 1.6 | 0.8 | 0.2 | 2.6 | −1.6 | −4.3 | – |
| Jul | 15.8 | 0.9 | 16.7 | 2.5 | 0.8 | 0.1 | 3.4 | 13.3 | – | 13.3 |
| Aug | 18.0 | 1.1 | 19.1 | 2.1 | 0.7 | 0.1 | 2.9 | 16.2 | – | 29.5 |
| Sep | 13.1 | 0.8 | 13.9 | 1.7 | 0.6 | 0.1 | 2.4 | 11.5 | – | 41.0 |
| Oct | 5.2 | 0.8 | 6.0 | 1.0 | 0.5 | 0.1 | 1.6 | 4.4 | – | 45.4 |
| Nov | 4.1 | 0.7 | 4.8 | 0.6 | 0.4 | 0.1 | 1.1 | 3.7 | – | 49.1 |
| Dec | 3.3 | 0.5 | 3.8 | 0.8 | 0.3 | 0.1 | 1.2 | 2.6 | – | 51.7 |

C.A = Catchment area.

reservoirs are usually *V*-shaped in their vertical section. Since the inflow (PMF or SPF or flood of design return period) and outflow are defined with respect to the location of the reservoir in the basin, a safe routed outflow for the down stream may require an increase in the crest of spillway and therefore, the top of the dam. The size or length of spillway, the height of crest and the top level of dam with respect to the river bed level are three interrelated parameters which should be computed and decided judiciously and optimally from cost benefit, social necessity, riparian rights, law, justice and political necessity of the area. For an ungated spillway a typical equation of flood discharge *Q* over the spillway crest in m³/sec is

$$Q = 0.67 \, C_d \sqrt{2g} \, L_e \cdot H_a^{3/2} \qquad (10.1)$$

where $C_d$ is the coefficient of discharge, $L_e$ the effective length of spillway in meters taking care of piers and abutment end contractions, $H_a$ the height of water in meters over spillway crest and $g$ the acceleration due to gravity. Equation (10.1) shows a relation between the spillway length, the height over the spillway and the outflow from the spillway. There can be a large number of alternatives between $L_e$ and $H$ for the same $Q$, but the best one to suit the site conditions is decided judiciously and economically.

## 10.4 ALLOCATION OF STORAGE SPACE FOR VARIOUS USES

Knowledge about the different levels of a reservoir is necessary before we discuss the allocation of storage space for various uses like flood control, conservation purposes and other needs. The different water levels are discussed as follows.

**Full Reservoir Level (FRL):** It is the level of spillway crest (for ungated spillway) or the top of the spillway gate (for gated spillway) to which the reservoir is usually filled. This level represents the top of conservation storage of a reservoir to which filling is scheduled immediately after the rainy season.

**Maximum Water Level (MWL):** During flood period, a reservoir while routing the inflow hydrograph, increases its water level to a position above the FRL. This new elevation to which water in the reservoir rises when design flood impinges at FRL is called maximum water level. After the routing of the flood, water level is lowered back to FRL. Therefore, the head between FRL and MWL is the routed head necessary for the flood to outflow from the spillway. The storage between FRL and MWL is called *surcharge storage.*

**Dead Storage Level (DSL):** The dead storage level or the low water level (LWL) is the minimum reservoir level below which water is not allowed to be drawn for conservation purposes. This also represents a level below which the silt carried by the river is expected to be deposited. Usually about 10% of the total reservoir storage is taken as dead storage volume.

**Live Storage (LS):** It is also known as the *useful* or *conservation* storage of a reservoir and is the difference between the storages at FRL and DSL. This is released for purposes like irrigation, domestic and industrial water supply, power generation, recreational uses and other mandatory demands by the reservoir regulating authorities round the year.

**Bank Storage (BS):** The storage of water in the permeable reservoir banks is called bank storage. The water gets released to the reservoir gradually with lowering of water levels in the reservoir.

### 10.4.1 Reservoir Elevation-Area-Capacity Curves

To plot area-elevation and elevation-capacity curves for a reservoir, the area is surveyed and a contour map at 1 or 0.50 m interval is drawn upto the expected top level of the proposed reservoir. Starting from the minimum contour corresponding the deepest water level in the reservoir, area between successive contours are planimetered and recorded as $A_1, A_2, A_3, \ldots, A_n$. The plot between the elevations and area $A_i$ results in a parabolic curve called elevation-area curve. From the knowledge of the areas at different heights, the volumes between successive contours having areas $A_1, A_2, A_3, \ldots A_n$ for contour interval of $d$ m is calculated by using either of the following:

(1) Prismoidal formula $\qquad V = \dfrac{d}{6}\{A_1 + A_2 + 4A_m\}$ (10.2)

where $\qquad\qquad\qquad A_m = 0.5(A_1 + A_2)$

(2) Simpson formula

$$V = \frac{2d}{3}\{A_2 + A_4 + \ldots + A_m + 2(A_1 + A_3 + \ldots + A_{m-1})\}$$ (10.3)

(3) Trapezoidal formula:

for two successive contours $\qquad V = \dfrac{d}{3}\{(A_1 + A_2 + \sqrt{(A_1 A_2)})\}$ (10.4a)

for three successive contours $V = \dfrac{d}{3}(A_1 + 4A_2 + A_3)$ (10.4b)

For better accuracy, the contour interval should possibly be small. Among the three equations, the prismoidal formula gives better results. A typical Elevation-Area-Capacity curve is shown in Fig. 10.7.

### 10.4.2 Reservoir Operation

Allocation of storage space for various uses is done by reservoir operation study. Depending on the expected availability of water during various periods of the year, the operation policy changes its priority of water releases for various sectors. Many a times a group of reservoirs of a basin are operated simultaneously for flood control and other beneficial releases so as to yield optimum result. Reservoir operation policy is, therefore, a complex problem and the situation becomes complicated when a group of multipurpose reservoirs are to be operated simultaneously. Optimisation-cum-Simulation model can be effectively used to

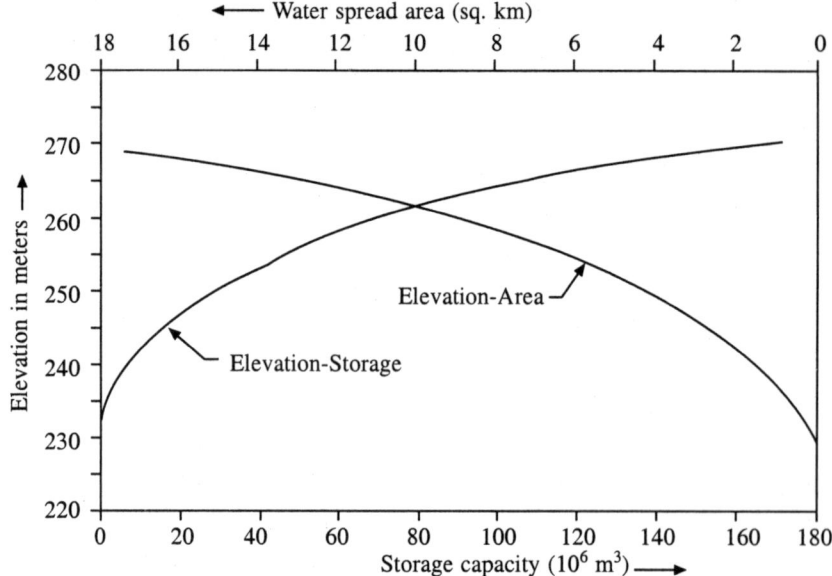

**Fig. 10.7   An Elevation-Area and Elevation-Capacity Curve of a Reservoir**

decide the operation schedule for a single or a group of multipurpose reservoirs. A *rule curve* defining various limiting values of storages for each month of the year in the form of a chart or graph is generally drawn for each reservoir which guides the engineers controlling the reservoir operation. Depending on various dependable flows, various rule curves can be drawn. For a year when 75% dependable flow in a river is expected then the rule curve drawn for such a flow condition may be followed. A rule curve for a multipurpose reservoir is given in Fig. 10.8.

### 10.4.3   Reservoir Working Table

For any proposed reservoir, the percentages of its success or failure to meet the targeted demands are worked out from the output of reservoir working table. Data required to carry out a working table are:

1. Monthly volumes of inflows.
2. Monthly demands for various purposes like irrigation, hydropower generation, municipality requirements, evaporation losses, seepage losses and other mandatory requirements.
3. Storage informations like dead storage level, top of conservation storage for various months, full reservoir level and limits on reservoir from flood operation considerations.

Knowing the information on inflows, demand patterns, storage constraints and reservoir losses, a trial working table is prepared by balancing the water budgets for all the 12 months of a year and for all the years subsequently. The percentage of success should be (a) 100% for municipality (drinking water supply) demand, (b) 90% for hydropower generation demand, (c) 75% for irrigation

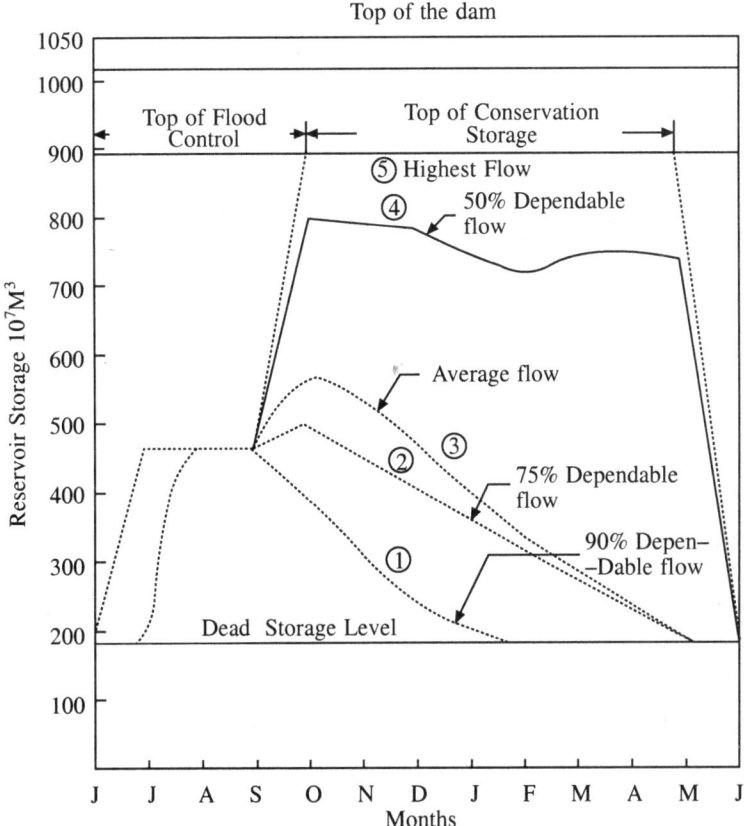

**Fig. 10.8** **A Rule Curve for a Multipurpose Reservoir in Orissa. Depending on the Inflows, Release Patterns for the Reservoir Follow Curves 1-5**

needs and (d) 100% for other mandatory downstream requirements. If the available yield series do not meet such demands then the demand pattern needs to be changed to meet the targeted success. As an example, an approved reservoir working table for a medium irrigation project located in Orissa for the year 1985–86 is given in Table 10.5.

**Example 10.5:** Carry out a reservoir working table for the year 1985–1986 for Manjore irrigation project. Monthwise inflows into the reservoir, its capacity and demand patterns are given in the cols. (1) through (6) of Table 10.5.

**Solution**
The following steps are followed while carrying out the reservoir working table.

1. Reservoir operation starts from the first month of a water-year. This is usually taken as June for India.
2. Initial reservoir level of 112.66 m at the beginning of June in col.(2) of the table is the closure of the reservoir level at the end of May of the previous water year (1984–1985).

Table 10.5    Reservoir Working Table for Manjore Irrigation Project for 1985–86

| Month of the year | Initial reservoir level (m) | Initial capacity for the month beginning (Ha. m) | Inflow (Ha. m) | Evaporation losses (Ha. m) | Irrigation drawl (Ha. m) | Total drawl (Ha. m) | Final capacity (Ha. m) | Final level (m) | Short fall (Ha. m) | Spill over loss (Ha. m) |
|---|---|---|---|---|---|---|---|---|---|---|
| (1) | (2) | (3) | (4) | (5) | (6) | (7) | (8) | (9) | (10) | (11) |
| June | 112.66 | 2914 | 184 | 107 | 243 | 350 | 2748 | 112.40 | 0 | 0 |
| July | 112.40 | 2748 | 1670 | 95 | 1077 | 1172 | 3246 | 113.11 | 0 | 0 |
| Aug | 113.11 | 3246 | 3048 | 124 | 121 | 245 | 6048 | 116.52 | 0 | 0 |
| Sept. | 116.52 | 6048 | 6369 | 184 | 784 | 968 | 10077 | 120.00 | 0 | 1372 |
| Oct. | 120.00 | 10077 | 1325 | 185 | 768 | 954 | 10077 | 120.00 | 0 | 371 |
| Nov. | 120.00 | 10077 | 0 | 144 | 486 | 630 | 9447 | 119.49 | 0 | 0 |
| Dec. | 119.49 | 9447 | 0 | 134 | 636 | 770 | 8678 | 118.87 | 0 | 0 |
| Jan | 118.87 | 8678 | 0 | 116 | 1700 | 1817 | 6861 | 117.38 | 0 | 0 |
| Feb | 117.38 | 6861 | 0 | 99 | 1168 | 1267 | 4326 | 114.60 | 0 | 0 |
| March | 116.04 | 5594 | 0 | 156 | 1112 | 1268 | 4326 | 114.60 | 0 | 0 |
| April | 114.60 | 4326 | 0 | 174 | 615 | 789 | 3537 | 113.51 | 0 | 0 |
| May | 113.51 | 3537 | 0 | 172 | 174 | 347 | 3190 | 113.04 | 0 | 0 |
| Total | | | 12596 | 1690 | 8887 | | | | 0 | 1743 |

[N.B: For June of next year extra storage = 3190 – 2914 = 276 Ha. m is carried over. therefore, 12596 – 276 – 10577 = 1743 Ha. m, is the total spill over quantity shown in col.(11)].

3. The initial reservoir capacity of 2914 Ha.m of col. (3) corresponds to the reservoir level of 112.66 m.
4. Inflow of 184 Ha.m is taken as the river inflow volume for the entire month of June of the year 1985–86.
5. Evaporation loss of 107.0 Ha.m, entered in col. (5) is taken from the average reservoir water spread area for the month. This takes into account the pan evaporation data multiplied with the pan coefficient to get lake evaporation. At places where the daily lake evaporation data are available, the evaporation is multiplied by the water spread area and is summed up for the whole month to get the monthly evaporation volume.
6. For dry months, the irrigation drawl capacity from the reservoir is decided on the basis of inflow, evaporation loss and minimum reservoir level fixed by intake level for irrigation and the DSL. If the irrigation demand is more than the drawl volume available in the reservoir, then there will be shortfall for the month. Demands for irrigation from cropping pattern of the cultivable command area governs the irrigation release for the month. Therefore, shortfall is decided with respect to the fixed demands.
7. The percentage of success is calculated on the basis of run for all years for which inflow data are available. The percentage success should be atleast 75% for irrigation demands, for an exclusive irrigation project like this. This is calculated on the basis of the number of years $m$, the project can supply water to its full targeted demands out of $n$ years of working tables carried out in the analysis, i.e., $(100 \, m/n) > 75$. All other terms in the table are self explanatory.

## 10.5 RESERVOIR SEDIMENTATION

Sediment is defined as the fragmental earth materials eroded, transported and deposited elsewhere naturally by agents like air and water. Sediment transport is a natural process and therefore, it cannot be stopped completely. There have been instances of a reservoir (like reservoir Soloman, Kansas, USA) that got completely filled in by sediment within one year of its completion, while the reservoir Yasuka lost 85% of its capacity on the thirteenth year of its construction. The concern is the time lapse between the construction of a reservoir and its filling to the extent its useful storage is completely filled in by sediment. For any country it poses a grave problem. Millions of rupees worth useful storages of reservoirs in India gets filled up by sediments. Table 10.6 gives the annual rates of sediment inflows for various important reservoirs in India.

With such large sediment inflows as shown in col.(4) of Table 10.6, it is likely that some of the reservoirs will get completely filled in before its design life. The problem of sedimentation, which involves (1) erosion at the place of origin, (2) transportation through the river water and (3) deposition in the reservoir are discussed here. We define below some of the terms connected with sediment to understand the subject.

**Sediment Yield:** It is the total flow of sediment from a watershed measured at a location in a river at a specified time.

**Erosion:** It is the process of detachment and transportation of sediment by the erosive agents.

**Table 10.6    Annual Sedimentation Rate for Various Reservoirs in India**

| Sl. No. | Reservoir name and location | Catchment area (km$^2$) | Annual sediment rate (Ha.m/ 100 km$^2$) | Capacity of reservoir ($\times 10^6$ m$^3$) | Surface area at MRL $\times 10^6$ m$^2$ |
|---|---|---|---|---|---|
| (1) | (2) | (3) | (4) | (5) | (6) |
| 1. | Mayura Kashi (WB) | 1860 | 20.09 | 616 | – |
| 2. | Ramaganga | 3076 | 17.30 | – | – |
| 3. | Shivaji Sagar | 819 | 15.20 | 2987 | – |
| 4. | Maithon (Bihar) | 6300 | 13.02 | 1360 | – |
| 5. | Ukair (Gujarat) | – | 10.90 | – | – |
| 6. | Gandhi Sagar (MP) | 22600 | 10.05 | 8450 | 660 |
| 7. | Panchet | 11,000 | 9.92 | 1497 | 153 |
| 8. | Tawa | – | 8.10 | – | – |
| 9. | Nizam sagar (AP) | 21,694 | 6.57 | 715 | 130 |
| 10. | Dantiwada | – | 6.32 | – | – |
| 11. | Tungabhadra (Karnataka) | 28,200 | 6.00 | 4040 | 378 |
| 12. | Bhakra(Punjab) | 56,800 | 6.00 | 9868 | 169 |
| 13. | Lower Bhawani (TN) | 6,150 | 4.10 | 930 | – |
| 14. | Hirakund | 83,400 | 3.89 | 8141 | 725 |
| 15. | Matutala (UP) | 20,750 | 3.50 | 1135 | – |
| 16. | Machkund | – | 2.33 | – | – |

**Sediment Delivery Ratio:** It is the ratio of sediment delivered at a gauging site in a river to the total erosion from the entire area upstream of it.

**Bed Load:** The coarse sediment materials moving close to the river bed by rolling or sliding is called bed load. Sometimes, materials moving within 10 –15 cm from the bed is taken as bed load.

**Suspended Load:** Relatively finer sediment particles which mix and move with river water in suspension and are found throughout the channel, are called suspended load.

**Wash Load:** Fine, very fine and electrochemically charged soil particles carried by river water are called wash load and do not ordinarily settle at the bottom of the container even after keeping it undisturbed for hours.

## 10.6    DETERMINATION OF SEDIMENT YIELD AT A RIVER SITE

### 10.6.1    Sheet Erosion

Flowing water is the most active agent for erosion of soil. Other helping agents are the action of wind, gravity, ice and human activities. The impact of rain drops loosen the soil particles or break the soil lumps. Action of flowing sheet of water on the land surface helps to erode away  top soil from the surface and transport it down to the channels. Hence, the problem of water erosion is: (a) *Sheet-erosion* that includes the detachment of the geological material from the land surface by rain drop impact and its subsequent removal by overland flow and (b) *channel erosion* which includes river bank erosion and transportation of materials by concentrated flow.

Geologic erosion occurs today as it did in the past but the disturbance of the soil cover by human activities like overgrazing of grass land, cutting forests, forest fires, ploughing of land and various mining and other excavations have magnified the problem to many fold. Factors affecting the erodibility of soil are:

**Particle Size of Soil**: Larger the size, lesser is its scope for sheet erosion.

**Land Slope**: Higher is the land slope more is the action of erosive agents, the optimum being at 40° slope.

**Vegetation**: Thicker is the soil covered by vegetation, lesser is the scope of soil erosion from the area.

**Presence of Salt and other Colloidal Materials**: Binding minerals like kaolinite, montmorillonite, biotite etc. help to increase the force of cohesion between soil particles thereby reducing the erodibility of soil.

**Moisture Content of Soil**: More is the moisture content in the soil, less is the scope of its erosion.

**Soil Compaction**: Higher is the compaction of soil less is the chance of its erosion.

**Soil Properties and Occurrences**: Soil texture, structure, stratification, permeability and composition affect the erosion to the extent that they affect the soil binding and neutralize the force of weathering agents.

**Human Activities**: Agriculture operations, land use, construction of projects and mining are examples of some of the human activities that help soil erosion.

**Rainfall Characteristics**: Intensity, duration, quantity and distribution of rainfall over space and time are the important factors affecting sediment yield.

The first step in estimating quantity of sediment is the computation of sheet erosion by various formulae proposed by various investigators and some of them are:

*(1) Musgrave Equation*
Annual gross sheet erosion rates proposed by Musgrave (1947) on the basis of data from 19 widely scattered research stations in USA by considering more than 40,000 storms is written as

$$E = FR\left(\frac{S}{10}\right)^{1.35} \left(\frac{L}{22.1}\right)^{0.35} \left(\frac{P_{30}}{3.5}\right)^{1.75} \tag{10.5}$$

where $E$ is the soil loss by sheet erosion in tons/year/acre, $F$ the soil erosion rate which varies from 65 to 79 tons/acre/year depending on the soil type (which depend on texture and permeability of the soil), $R$ the cover factor which varies from 0.1 for poorly covered land to 0.95 for row crops, $S$ the land slope in percentage, the default being 10%, $L$ the length of land slope in feet and $P_{30}$ is the maximum 30 min., $-2$ year frequency rainfalls in inches.

*(2) Universal Equation*

Agricultural Research Service of U.S. Department of Agriculture developed an universal equation (1961) for predicting erosion from small catchments. The equation is

$$E_a = R_f\, S_e\, S_{ls}\, C_m\, S_{cp} \tag{10.6}$$

where $E_a$ is the average soil loss in tons per unit area, $R_f$ the rainfall erosion factor which depends on the kinetic energy of raindrops, $S_e$ the soil erodibility factor, $S_{ls}$ the land slope, length and stiffness factor, the value of which are obtained mostly from test plots, $C_m$ is the cropping and management factor with a maximum value of 1.0 and $S_{cp}$ is the supporting conservation practice factor to be decided from land covering, terracing and cropping practices. The value of the factor $S_{cp}$ ranges from 0.25 to 1.00, less for small slopping land. $S_{ls}$ is obtained by dividing the existing length and stiffness to a 9% slope, 22.25 m length plot. $R_f$ is calculated by multiplying storm energy of rain drops at terminal velocity $(0.005 \times \text{mass} \times V_t^2)$ of a maximum 30 min intensity rainfall and summing them for all intensities. $V_t$ is the terminal velocity of rain drops. The soil erodibility factor may be taken on the basis of test plot results. It is difficult to get proper data for use of equation (10.6) for a developing country like India. Therefore, the universal equation is not popular amongst developing countries.

### 10.6.2   Sediment Measurement by Sample Recorder

The sediment resulting from sheet erosion from a catchment may not always reach at the point of measurement. Some part of the sediment may be deposited enroute. The ratio between the yield of sediment at the measuring site and the gross erosion in the watershed is called *sediment delivery ratio*. Thus, the sediment yield is the gross yield minus the quantity deposited en route. This sediment gets deposited in the reservoirs affecting its useful life. It is essential to find the sediment load that are likely to reach a reservoir. Methods of calculating sediment yield of a watershed are:

**(a) Measuring Suspended Sediment Load:** Sediment measurement carried out by sediment sampler at a site gives the most reliable results of the sediment yield. The bed load must be calculated and added to it to give the total sediment yield at the river gauging site. A typical sampler used in India is shown in Fig. 10.9. A depth integrating hand sampler used for small streams is shown in Fig. 10.10.

The sampler is taken to a depth of 0.6*d* below the surface or at two depths at 0.2*d* and 0.8*d*. Sample collected in the sampler bottle is analysed in the laboratory either by *gravimetric* or *hydrometric* method.

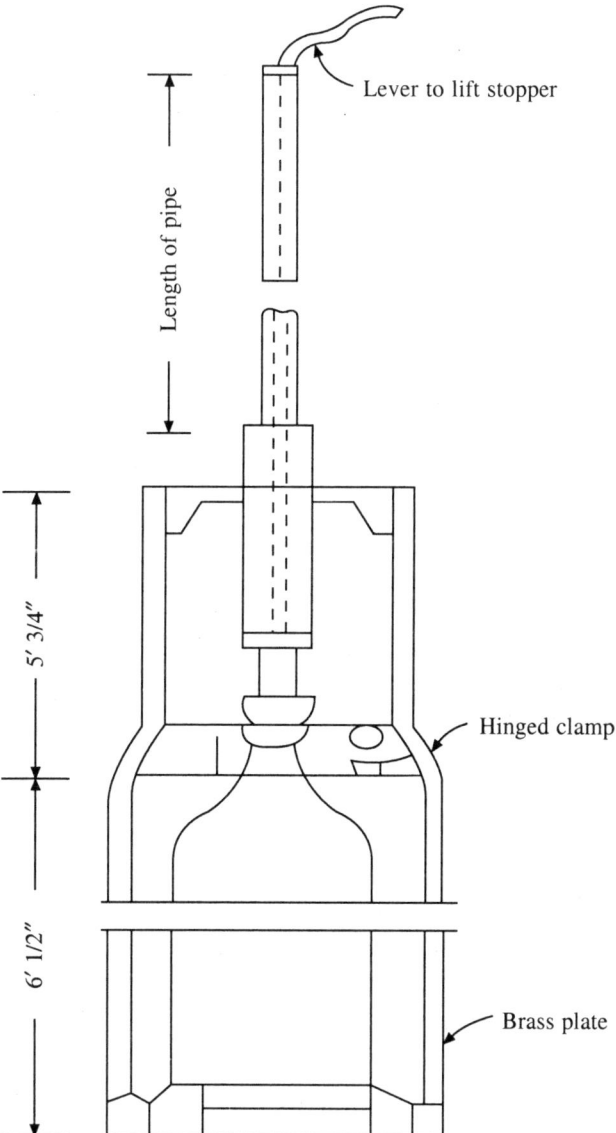

**Fig. 10.9   Line Diagram of a Typical Sediment Sampler (Punjab Bottle Sampler)**

The sample collected by the sampler is first passed through a BSS-100 sieve and the coarse particles retained are taken out and oven dried. Thus, the quantity of sediment higher than 0.2 mm diameter in size representing the coarse sediment are obtained. Sediment water is allowed to stand still for 20 min. so that the finer particles settle down. This mass is removed by the process of decantation (pouring out water from the settled tank). The settled residue is dried and weighed. This gives the sediment load of medium size soil grain between 0.075 and 0.2 mm. Poured water contains sediment of finer particles. To remove this, the sample is filtered through a filter paper and the quantity retained therein is dried and weighed. This gives the fine sediment load below 0.075 mm.

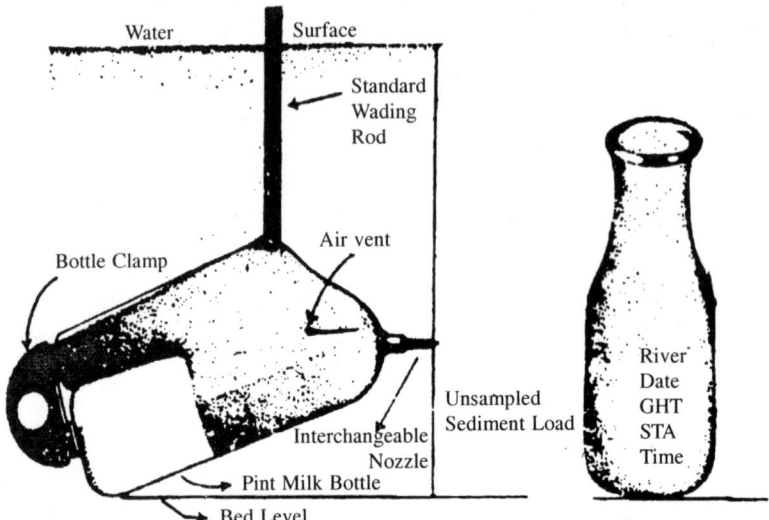

**Fig. 10.10   AUS DH-48 depth-integrating hand sampler for small streams**

Sediment load is expressed in parts per million (PPM), which is obtained by dividing the sediment weight by the total weight of original sample (water + sediment) and multiplying by $10^6$. This figure is converted to tons/day as

$$\text{Sediment load in PPM} = \frac{\text{Dry weight of sediment} \times 10^6}{\text{Total weight of original sample}} \qquad (10.7)$$

When a large number of such sample records are available at a site, then a curve between sediment load in tons/day in abscissa and daily discharge in m³/sec is plotted which is popularly known as *sediment rating curve*. Such a plotting is usually carried out using a log-log paper.

A relation between stream flow $Q$ (m³/sec) and suspended sediment load $q_s$ in tons/day is of the type

$$q_s = K Q^n \qquad (10.8)$$

Taking logarithmic of equation (10.8), it can be reduced to the following straight line form.

$$\log q_s = \log K + n \log Q \qquad (10.9)$$

The equation is similar to the form $Y = a + bX$ which can be solved by the method of least square. When such a relation is established at a site, the rate of sediment yield can be computed simply by knowing the discharge rate only.

However, care must be taken to see that for different seasons of the year different curves are prepared as sediment yield of a basin for the month of say February is different to that of July. A typical curve on day to day plotting may look like the one given in Fig. 10.11. The figure is plotted from actual field data collected from a CWC gauging site in Orissa and is valid for the monsoon season. It plots a straight line for higher ranges between B and C but for lower ranges, i.e., from A to B another curve may be developed.

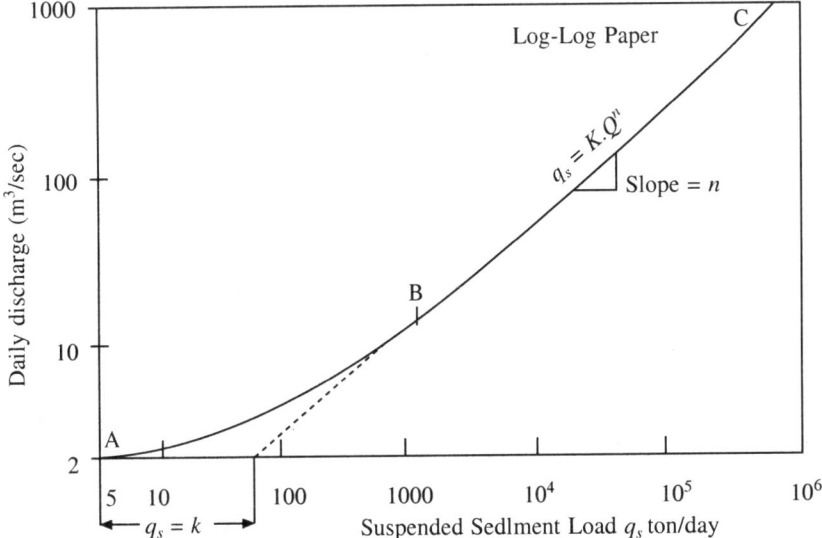

**Fig. 10.11 A Rating Curve for River Mahanadi (Monsoon)**

### 10.6.3   Bed Load Estimation

The bed load usually vary between 2.5 and 25% of the total suspended load which should be added to the suspended sediment load to get the total sediment yield of the basin. A *Russian type bed-load sampler* gives better results for alluvial rivers but the use of the following empirical relations are quite acceptable for calculating bed load.

*Schoklitsch Formula*

On the basis of experimental data the bed load equation is represented as

$$G_i = \frac{63050\,B}{\sqrt{d_m}}\left(10.76q - \frac{0.00021\,d_m}{S_e^{1.333}}\right)S_e^{1.5} \qquad (10.10)$$

where $G_i$ is the bed load in tons/day.

*Mayer-Peter Formula*

$$G_i = (39.25\,q^{2/3}\,S_0 - 9.95d_m)^{2/3} \qquad (10.11)$$

where $G_i$ is the bed load in pound/sec/ft of channel width, $q$ water discharge in cusecs/ft of channel and $d_m$ the effective grain diameter in feet.

*Haywood Formula*

$$G_i = \left( \frac{q^{2/3} S_0 - 1.2 d_m^{4/3}}{0.117 d^{1/3}} \right)^{3/2}$$  (10.12)

where $G_i$ is in pound/sec/ft of channel width, $q$ water discharge in cusecs/ft of channel and $d$ the effective grain diameter in feet.

*Mayer-Peter and Muller Formula*

$$G_i = B \left[ 32.53 \left( \frac{Q_s}{Q_0} \right) \left( \frac{d^{1/6}}{n_s} \right)^{3/2} DS_e - 1.881\, d_m \right]^{3/2}$$  (10.13)

where $G_i$ is the bed load in ton/day, $q$ the observed water discharge per unit width in m³/sec/m, $Q_s$ the water discharge rate at the bed load depth in m³/sec, $Q_0$ the water flowing in the river section in m³/sec, $d$ the grain diameter of size 90% bed material finer than this, $D$ the mean channel depth in m, $S_e$ the hydraulic gradient, $B$ the width of the channel in m, $S_0$ the bed slope, $d_m$ the mean diameter of silt grain in mm and $n_s$ is the Manning's rugosity coefficient. For Orissa catchments bed load of 10% of suspended load is usually taken for reservoir planning.

---

**Example 10.6:** Observed sediment load by a silt sampler at Sundargarh site of Mahanadi river for 6 years are available. Calculate the average sediment load/year/100 sq. km. of the catchment at the site. Calculate total load for 50 years which is to be distributed in the reservoir. Develop a regression type relation between the silt load and discharge at the site and predict the silt for the previous 4 years for which there was no silt observation. Drainage area at the sampling site is 5870 sq. km.

| Year | 1978–79 | 79–80 | 80-81 | 81-82 | 82–83 | 83-84 | 84–85 | 85-86 | 86–87 | 87–88 |
|---|---|---|---|---|---|---|---|---|---|---|
| Discharge (M.m³) | 4250 | 8324 | 3750 | 12920 | 2184 | 5708 | 13060 | 4145 | 7800 | 2846 |
| Sediment (M Tonnes) | – | – | – | – | 3.826 | 10.309 | 14.182 | 6.895 | 11.214 | 4.92 |

**Solution**

$$\text{Average sediment load} = \frac{(3.862 + 10.309 + 14.182 + 6.892 + 11.214 + 4.92)}{6}$$

$$= 51.382/6 = 8.563 \text{ M. tons/year}$$

Since the above is suspended load measured by silt sampler, assume 10% of suspended load as bed load.

$$\therefore \text{ Total load} = \text{Suspended load} + \text{Bed load} = 8.5637 + \left( \frac{10}{100} \right) \times 8.5637$$

$$= 9.420 \text{ M. tons/year}$$

Sediment deposited in a reservoir consolidates gradually due to the increasing silt load on it every year and the weight of water above it. Assuming the average sediment unit weight (sediment in a reservoir consists of sand, silt and clay in water) as 1.2 T/m³.

Total silt load/year = 9.420 M. Tons = 9.42 Ton/1.2 T/m³ = 7.85 M. m³ = 785 Ha.m.

Catchment area at the site = 5870 sq. km.

∴          Sediment load/100 sq. km/year = $\dfrac{785}{58.70}$ = 13.373 Ha. m

∴    In 50 years life of the reservoir, total sediment load = 13.373 × 50 × $\dfrac{5870}{100}$

         = 39250 Ha.m. = 392.5 M.m³, which is to be distributed in the reservoir.

Estimation of sediment load for four years (1978–79 to 1981–82) can be done by developing a regression type non-linear equation between sediment load and discharge. The calculations for the development of the relation are carried out in Table 10.7.

**Table 10.7    Development of Regression Relation Between Sediment Yield and Discharge**

| Year | Discharge Q (M.m³) | $q_s$ (M.Ton) | $q_s$ (Ha.m) | log (Q) = x | $x^2$ | log ($q_s$) = y | xy (5) × (7) |
|------|------|------|------|------|------|------|------|
| (1) | (2) | (3) | (4) | (5) | (6) | (7) | (8) |
| 82–83 | 2184 | 3.862 | 354* | 3.33925 | 11.15059 | 2.549 | 8.512 |
| 83–84 | 5708 | 10.309 | 945 | 3.75648 | 14.11117 | 2.975 | 11.177 |
| 84–85 | 13060 | 14.182 | 1300 | 4.11594 | 16.94099 | 3.114 | 12.817 |
| 85–86 | 4145 | 6.895 | 632 | 3.61752 | 13.08648 | 2.801 | 10.136 |
| 86–87 | 7800 | 11.214 | 1028 | 3.89209 | 15.14840 | 3.012 | 11.729 |
| 87–88 | 2846 | 4.920 | 451 | 3.45423 | 11.93174 | 2.654 | 9.168 |
| Total | | | 4710 | 22.17551 | 82.36937 | 17.087 | 63.533 |

$* \; 354 = \dfrac{3.862 \times 1.1 \times 10^6}{1.2 \times 10^4}$

$$\log a = \frac{17.087 \times 82.369 - 2.176 \times 63.533}{6 \times 82.369 - 22.176^2} = \frac{1407.439 - 1408.908}{494.214 - 491.775} = -0.60226.$$

Therefore $a$ = 0.2499

$$b = \frac{(6 \times 63.533 - 22.1755 \times 17.087)}{(6 \times 82.369 - 22.176^2)} = 0.93695$$

Thus the equation becomes     $q_s = 0.2499 \, Q^{0.93695}$

Using the equation, the sediment for the remaining four years from 1978–79 to 1981–82 are calculated in Table 10.8.

## 10.6.4   Empirical Relations for Total Sediment Load

Many investigators have proposed empirical relations to compute the total sediment load of a river by correlating various parameters. The following two equations are presented for computation of total silt load for a basin.

**Table 10.8   Estimation of Sediment for Example 10.6**

| Year | Discharge $Q$ (M. m³) | Sediment (M. Ton) | Yield $q_s$ (Ha. m) |
|------|------------------------|---------------------|---------------------|
| (1) | (2) | (3) | (4) |
| 78–79 | 4250 | 627.17 | 6.842 |
| 79–80 | 8324 | 1177.40 | 12.844 |
| 80–81 | 3750 | 557.77 | 6.085 |
| 81–82 | 12920 | 1777.52 | 19.39 |

*Swamy's Regression Relation*

Swamy and Garde (1977) proposed a relation correlating the volume of sediment deposited in a reservoir with the cumulative volume of water inflow and initial bed slope of the channel as

$$V_s = CBV_{ci}^{0.94} S_b^{0.84} \tag{10.14}$$

where $V_s$ is the volume of sediment deposited in a reservoir in M.m³, $B$ the width of the reservoir at full reservoir level in m, $V_{ci}$ the cumulative volume of inflows per unit width $B$ of the reservoir, $S_b$ the bed slope of the river and $C$ the regression constant with safe value of the order of 1.16. However, a value less than 1.16 may be adopted depending on the reservoir.

*Jogelkar's Relation*

A relation proposed by Jogelkar (1960) between the annual silting rate $Q_s$ in M.m³ per 100 sq.km of the catchment area and the catchment area $A$ in sq.km is

$$Q_s^{\cdot} = 0.59 \, A^{-0.24} \tag{10.15}$$

*Study of Similar Catchment*

Sediment yield of an unmeasured watershed $Q_{um}$ can be computed from sediment yield of measured watershed $Q_m$ of similar topography, land cover and land use on area proportion basis.

$$Q_{um} = \left( \frac{A_{um}}{A_m} \right) Q_m \tag{10.16}$$

where $A_{um}$ and $A_m$ are the areas of unmeasured and measured catchments, respectively.

*Varshney's Equations*

Varshney and Raichur proposed the following enveloping equations for calculating sediment yield for any ungauged basin:

(a) Upto 130 sq. km for mountainous rivers

$$V = 0.395 \, A^{-0.311} \tag{10.17}$$

(b) Rivers draining plain area up to 130 sq. km

$$V = 0.392\ A^{-0.302} \qquad (10.18)$$

(c) For areas greater than 130 sq. km, for North India catchment

$$V = 1.534\ A^{-0.264} \qquad (10.19)$$

(d) For South India rivers upto 130 sq. km

$$V = 0.46\ A^{-0.468} \qquad (10.20)$$

(e) For South India catchment greater than 130 sq. km area

$$V = 0.277\ A^{-0.194} \qquad (10.21)$$

where $V$ is the sediment yield rate in M. m$^3$ per 100 sq. km of catchment, $A$ the catchment area in km$^2$. For calibration of a region, where sediment yield is known, the proposed equations should not yield more than double or less than half of the measured watershed sediment yield.

*Khosla's Method*
Khosla (1953) proposed an equation for the estimation of annual silting rate ($Q_s$) in M. m$^3$/100 sq. km/year by correlating with catchment area (A) in sq. km as

$$Q_s = 0.323\ A^{-0.28}$$

The equation always gives lower estimate of sediment yield at a site.

*From Gross Erosion and Sediment Delivery Ratio*

$$\text{Sediment yield} = (\text{Gross erosion}) \times \text{delivery ratio} \qquad (10.22)$$

---

**Example 10.7:** Estimate sediment load of Ib irrigation project at Sundargarh site in Orissa from the following data

Catchment area = 5870 sq. km
River slope at the site = 0.0052
Width of reservoir at FRL = 4500 m
Average inflows for 7 years are given as

| Year | 1988 | 89 | 90 | 91 | 92 | 93 | 94 |
|---|---|---|---|---|---|---|---|
| Yield (M.m$^3$) | 3827 | 898 | 4299 | 2937 | 2094 | 2927 | 3970 |

Use various methods and propose the most acceptable sediment yield for the project.

**Solution**
Most of the drainage are of the project lies in Bihar. Therefore, the equation for sediment yield proposed by Varshney for North Indian catchment may be used.
(a) Varshney's equation for North India catchments

$$Q_s = 1.534\ (A)^{-0.264} = 1.534\ (5870)^{-0.264}$$

$$= 0.1552\ \text{M.m}^3\ \text{per}\ 100\ \text{km}^2$$

$$= 15.52\ \text{Ha.m/100 km}^2/\text{year}$$

(b) Jogelkar's equation

$$Q_s = 0.59 \ (A)^{-0.24} = 0.59 \ (5870)^{-0.24}$$

$$= 0.0735 \ \text{M. m}^3/100 \ \text{sq. km/year}$$

$$= 7.35 \ \text{Ha. m}/100 \ \text{sq. km/year}$$

(c) From study of similar catchments

River Ib drains water to Hirakund reservoir. Therefore, rate of sedimentation for this project can be taken the same as Hirakund dam project. From Table 10.6, it can be taken as 3.89 Ha. m/100 sq. km/year.

(d) Swamy's equation

$$V_s = CBV_{ci}^{0.94} \ S_b^{0.84} = 1.16 \times 4500 \times (0.0052)^{0.84} \ (V_{ci})^{0.94} = 63.9(V_{ci})^{0.94}$$

Using the equation the sediment rate is calculated as in Table 10.9.

**Table 10.9   Swamy's Method of Sediment Yield**

| Sl. No. | Year | Discharge $(\text{M.m}^3)$ | Cumulative discharge$(\text{Mm}^3)$ | $V_{ci} = (4)/4500$ | $Q_s = 63.9(V_{ci})^{0.94}$ |
|------|------|------|------|------|------|
| (1) | (2) | (3) | (4) | (5) | (6) |
| 1 | 1988 | 3827 | 3827 | 0.850 | 54.90 |
| 2 | 1989 | 898 | 4725 | 1.050 | 66.90 |
| 3 | 1990 | 4299 | 9024 | 2.005 | 122.90 |
| 4 | 1991 | 2937 | 11961 | 2.658 | 160.2 |
| 5 | 1992 | 2094 | 14055 | 3.123 | 186.4 |
| 6 | 1993 | 2927 | 16982 | 3.774 | 222.7 |
| 7 | 1994 | 3970 | 20952 | 4.656 | 271.3 |

Sedimentation rate/100 sq. km. $\dfrac{271.3 \times 100}{7 \times 5870} = 0.66 \ \text{Mm}^3 = 66 \ \text{Ha.m./year}$

(e) From Example 10.6, sediment yield data at a nearby project site is available from 1982–83 to 1987–88. Average sediment yield in Ha.m calculated in Example is 9.42 $\text{M.m}^3 = 942 \ \text{Ha.m}$.

Therefore

sediment yield per 100 sq. km of catchment per year $= \dfrac{942 \times 100}{5870} = 16.05 \ \text{Ha.m}$

The results of the above five methods are tabulated as follows:

(a) Varshney's equation gives        $Q_s = 15.52 \ \text{Ha.m}/100 \ \text{sq.km/year}$
(b) Joglekar's equation              $Q_s = 7.35 \ \text{Ha. m}/100 \ \text{sq. km/year}$
(c) From similar catchment study     $Q_s = 3.89 \ \text{Ha.m}/100 \ \text{sq.km/year}$
(d) Swamy's equation                 $Q_s = 66.0 \ \text{Ha. m}/100 \ \text{sq. km./year}$
(e) Observed data from average
    of 1982–83 to 1987–88            $Q_s = 16.05 \ \text{Ha.m}/100 \ \text{sq.km/year}$

For the project Varshney's equation gives comparable results to the observed values. Sediment yield of 16 Ha.m/100 sq. km/year can be accepted for the design of the project, though the observed data is limited to a short period of six years. Other methods give either too high or low results.

## 10.7  RESERVOIR SEDIMENTATION

Transportation of sediment by flowing water and their deposition in the reservoirs depend on flow conditions, sediment composition and their interaction with each other. When river water enters a reservoir, the velocity decreases because of the increased cross sectional area through which it passes. Reservoir water is usually less turbid than inflowing stream water. The two fluids have different densities. The river water is usually cooler than reservoir water. Due to these factors, the heavy river water flows along the bottom of the reservoir towards the dam under influence of gravity. Thus a stratified flow condition is developed as shown in Fig. 10.12. This water reaches the dam and then rises vertically along the wall of the dam. In the process, mixing also takes place between the reservoir and river water and part of turbid water spills over the dam. With time, the sediment usually settles at the bottom of the reservoir.

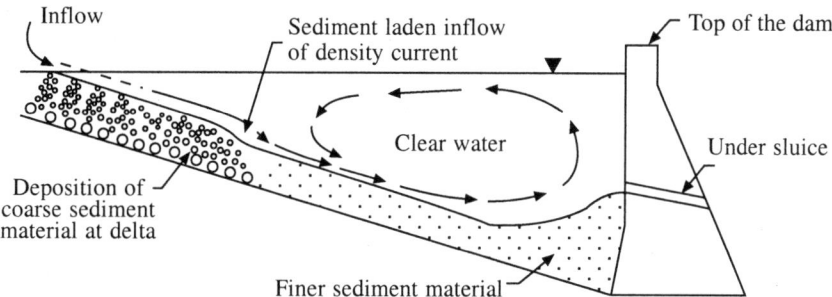

**Fig. 10.12   A Typical Section of Reservoir Showing a Pattern of Sediment Deposit**

It is estimated that by 2000 A.D. about 20% of our reservoir capacities have been filled up by silt reducing irrigation scope for 4 million hectares. Results of sedimentation surveys for Indian reservoirs show that the rate of sedimentation vary from 4.75 to 14 Ha. m/100 sq. km/year. All the reservoirs loose 0.5 to 1.0% of their storage capacity every year. IS: 6518–1972 recommends a provision of 10–20 Ha. m/100 sq. km/year of sedimentation for the entire economic life of the reservoir. Table 10.10 gives the estimated and observed sediment rates of various reservoirs of India.

The complexity of various problems in the process of sedimentation of reservoirs due to its inflowing are:

1. Uncertainty regarding the space occupied by sediment in the reservoir.
2. Lack of knowledge regarding the contribution of sediment by water from the catchment.
3. Variability of sediment inflow from year to year and from season to season.
4. Inability to evaluate accurately the inflow of sediment and the bed load.
5. Reservoir operation schedule.
6. Variability in reservoir capacity to inflow ratio from year to year.

**Table 10.10   Annual Rates of Siltation of some Indian Reservoirs in Ha.m/100 km²/year**

| Name of reservoir | Annual rate assumed while designing | Observed rate of sediment | % Change | Year of observation |
|---|---|---|---|---|
| (1) | (2) | (3) | (4) | (5) |
| Bhakra | 4.29 | 6.00 | 140 | 1975 |
| Nizam Sagar | 0.20 | 6.57 | 3300 | 1967 |
| Panchet | 2.47 | 9.02 | 365 | 1974 |
| Maithan | 1.62 | 13.02 | 804 | 1971 |
| Beas | 4.29 | 15.10 | 352 | 1975 |
| Ghod | 3.61 | 15.51 | 430 | 1970 |
| Ramganga | 4.29 | 17.30 | 403 | 1973 |
| Mayurakshi | 3.61 | 20.09 | 557 | 1975 |
| Hirakund | 4.91 | 3.58 | 73 | 1977 |
| Average | 3.25 | 11.80 | 713.80 | |

The density of the sediment laden river water is always higher than the clearer reservoir water. At a reservoir, the inflow water therefore descends below the reservoir water and moves close to the river bed towards the dam from the delta as shown in Fig. 10.12. The coarse materials are deposited at the delta and finer materials are carried away towards the reservoir and deposited in a pattern shown in the same figure. The specific gravity of soil materials forming sediment vary between 2.4 and 2.7. This allows larger particles to settle immediately at the junction of the river with the reservoir forming a delta. The densities of the sediment observed for several reservoirs show a figure between 500 and 1750 kg/m³. Increase in the clay content of the sediment decreases the density of the sediment in the reservoir. Density of the sediment deposited at delta of the reservoirs are always higher than at the dam section. The density is also greatly affected by the reservoir operation. For Indian reservoirs a density of 1200 kg/m³ for water laden sediment is quite acceptable. From this, the sediment load observed by weight can be converted to volume of the reservoir space.

### 10.7.1   Reservoir Classification

The sediment deposit is not confined to a particular elevation in a reservoir. It may also be deposited above the dead storage level. There is no exact solution to the problem of predicting the pattern of sediment distribution in a reservoir. The sediment deposit pattern survey of the existing reservoirs help to formulate distribution of sediment at various elevations of a proposed project. To help this, Borland and Miller (1960) classified reservoirs into 4-standard types on the basis of slope of plot on a log-log paper between the reservoir depth $H$ as ordinate and reservoir capacity $Q$ in abscissa. Thus a factor $m$ defining the type of reservoir is the inverse of the slope of linear plot between $H$ and $Q$ above as shown in Fig. 10.13. Depending on the type of the reservoir, the sediment is distributed in a reservoir close to a fixed pattern. Therefore, a reservoir needs to be classified first to its type. From this plot between the progressive capacity and depth on a log-log paper, the value of the slope $m$ can be obtained, on the basis of which the sediment distribution can be carried out. Table 10.11 gives the classification of the reservoirs into types.

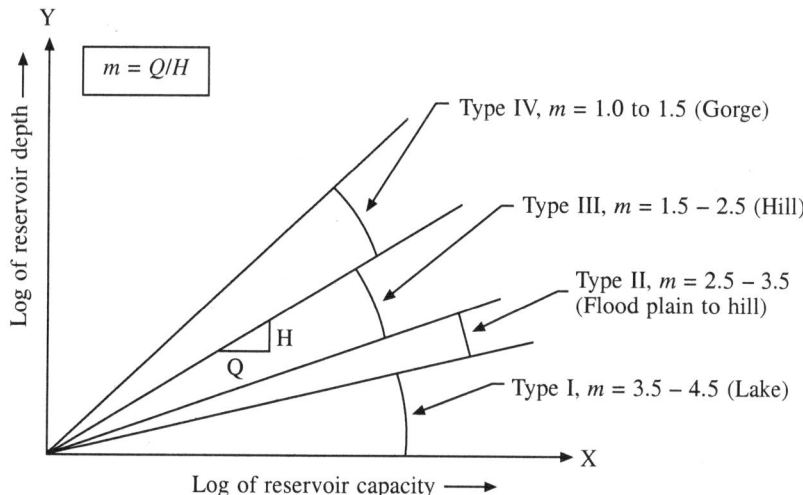

**Fig. 10.13   Reservoir Type Curve (After Borland and Miller)**

**Table 10.11   Classification of Reservoirs into Types**

| Sl. No. | Reservoir type | Category | Value of $m$ | Scope of silt deposit at lower elevation | Equation for empirical area reduction method |
|---------|----------------|----------|--------------|------------------------------------------|---------------------------------------------|
| (1) | (2) | (3) | (4) | (5) | (6) |
| 1. | Lake | I | 3.5-4.5 | very low | $A_p = 3.417\ P^{1.5}(1 - P)^{0.2}$ (10.23) |
| 2. | Flood plain foot hills | II | 2.5-3.5 | low | $A_p = 2.324\ P^{0.5}(1 - P)^{0.4}$ (10.24) |
| 3. | Hill | III | 1.5-2.5 | moderate | $A_p = 15.882\ P^{1.1}(1 - P)^{2.3}$ (10.25) |
| 4. | Gorge | IV | 1.0-1.5 | good | $A_p = 4.232\ P^{0.1}(1 - P)^{2.5}$ (10.26) |

## 10.7.2   Distribution of Sediment in Reservoirs

Following Borland and Miller, the two methods in predicting the sediment distribution of a reservoir are: (i) Empirical area reduction and (ii) Area increment methods.

### (A) Empirical Area Reduction Method

Based on the observation of sediment distribution of several reservoirs, a mathematical procedure to find the sediment distribution at various elevations were proposed by Borland and Miller. Depending on the reservoir type suitable equation for the type of reservoir can be selected from Table 10.11. The sediment distribution curve proposed by Borland and Miller is shown in Fig. 10.14. Stepwise procedure to compute sediment distribution at various elevations of a reservoir is as follows:

1. From the data of proposed reservoir, obtain information on full reservoir level, stream bed level at just upstream of the dam, total volume of sediment to be distributed during the life of the project and the river bed slope at the site.
2. Plot the elevation capacity curve of the reservoir on a log-log paper to find *m*. This defines the type of the reservoir.
3. Prepare a table and enter the elevation of the reservoir in the first column starting with the full reservoir level (FRL) in the first row to the lowest level of the reservoir corresponding to bed level of the river in the last row of the column.

Type I : $A_p = 3.417\ P^{1.5}\ (1 - P)^{0.2}$
Type II : $A_p = 2.324\ P^{0.5}\ (1 - P)^{0.4}$
Type III : $A_p = 15.882\ P^{1.1}\ (1 - P)^{2.3}$
Type IV : $A_p = 4.232\ P^{0.1}(1 - P)^{2.5}$

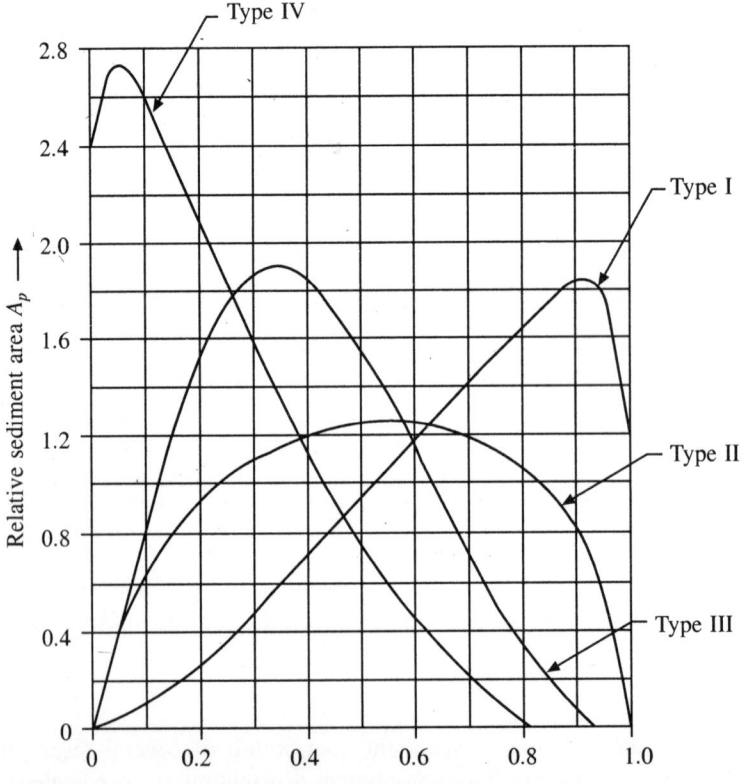

**Fig. 10.14   Sediment Distribution Curve (after Borland and Miller)**

4. Enter the area and capacities of the corresponding elevations of step (3) in the next two columns.

5. Enter the depth of reservoir corresponding to the various elevations of step (3) in the next col. (4). It is obtained by subtracting the river bed level corresponding to the lowest reservoir level from the elevations of step (3).

6. Compute the relative depth *P* by dividing the reservoir depths of step (5) by the maximum depth of the reservoir. Thus, in this column, the relative depth *P* should vary from 1.00 corresponding to FRL to 0.00 at river bed level.

7. Choose the equation for the reservoir type from Table 10.11. Compute $A_P$ for various *P* of step (6) and enter them in the next col. (6).

8. Assume a trial elevation ET to which the reservoir is expected to fill up by sediments in its design life. All the area below this level will be completely filled up by sediments. Now compute a constant *K* for the sediment load distribution as follows

$$K = \frac{\text{Area corresponding to the trial elevation (ET)}}{A_P \text{ corresponding to the trial elevation (ET)}} \qquad (10.27)$$

9. Compute the sediment area $A_i$ as $K \times A_P$ and enter them in the next col. (7). Since all the area below the trial elevation ET are assumed to be completely filled up by sediment, the area below this is entered from the data of the elevation-area curve as of step (4).

10. Sediment volumes are computed as:

$$S_v = \frac{(A_1 + A_2) \times H}{2} \qquad (10.28)$$

where $A_1$, $A_2$ are the area in Ha. of two adjacent elevations of difference H, $A_1$ and $A_2$ are obtained as in step (9) above for various elevations. The values of $S_v$ are calculated in Ha.m and entered in the next col. (8).

11. In the next column, the accumulated sediment volumes are calculated. It is calculated from the elevation starting with river bed level to FRL which are the cumulative figures of step (10).

12. If the total accumulated sediment volume of col.(9) is different to the sediment volume expected in the life of the reservoir, then the whole process is repeated with the assumption of another higher new zero elevation (ET) of step (8). If the accumulated sediment of step (11) is higher than the total sediment to be distributed, then a lower zero elevation (ET) is assumed or vice versa.

13. In the next column the revised area for the reservoir is entered. This is obtained by subtracting the area of step (9) from step (4).

14. The modified and revised reservoir capacity is obtained by subtracting the sediment distribution volumes of step (11) from the original volumes of step (4).

15. The revised elevation area capacity curve after the designed life of the project is plotted.

**Example 10.8 :** Accepting the silt load as 9.0 Ha.m/100 sq. km/year, distribute the pattern of sediment in the reservoir after 50 years by empirical-area reduction method if the net catchment area at the dam site is 5574 sq. km. Assume type-II reservoir. Elevation-Area-Capacity curve for the reservoir are given in the col. 1 through col. 3 of Table 10.12.

**Solution**

Silt load accepted by Central Water Commission and Government of Orissa for the project is $Q_s = 9.00$ Ha. m/100 sq. km/year. Accepting this rate as the sediment inflow, the distribution in the reservoir is carried out in Table 10.12.

$$\text{Total silt in 50 years} = \frac{9.00 \times 50 \times 5574}{100} = 25084 \text{ Ha.m} = 250.84 \text{ M.m}^3.$$

For calculation of $K$ assume new elevation at level 240.0 m.
Area corresponding to 240.0 m is 687 Ha.

$$\text{Relative depth} \quad p = \frac{(240 - 230)}{42.5} = 0.235$$

$$A_p = 2.324 \times (0.235)^{0.5} \times (1 - 0.235)^{0.40} = 1.012$$

Therefore

$$K = \frac{687}{1.012} = 678.8$$

The revised elevation-area and elevation-capacity curve along with the original elevation-area-capacity curves are plotted in Fig. 10.15. The columns in Table 10.12 are made as per the procedure outlined.

---

*(B) Area-Increment Method*

In this method it is assumed that the reservoir area curve before and after sedimentation is parallel to each other. Equation for this procedure is

$$V_s = \frac{A_{TC}(E - E_T)}{100} + V_T \tag{10.29}$$

where $V_s$ is the total sediment volume to be distributed in the designed life of the project in $M.m^3$ and $V_T$ the volume of sediment in $M.m^3$ below the trial zero elevation with the corresponding elevation $E_T$ in m. Reservoir is assumed to be completely filled upto the height $E_T$ in its life; $E$ being the total depth of reservoir from full reservoir level to the river bed level. $V_s$ is distributed as (i) volume $V_T$ completely gets filled up to the depth $E_T$ and (ii) distributing the elevations above $E_T$ thereof by a constant area rate $A_{TC}$ in Ha. The method holds good for all those reservoirs where the sediment volume is not substantial. Stepwise procedure is:

1. Prepare a table in which the first three columns should represent the elevation, original area and capacity of the proposed reservoir. They are to be entered starting with the full reservoir (FRL) in the first row down to the deepest bed level of the reservoir in the last row.
2. Compute the reservoir depths for various elevations by subtracting elevations from the deepest reservoir level (usually the river bed level) and enter it in col. (4).
3. Assume a zero elevation $E_T$ on trial and enter the area $A_{TC}$ corresponding to $E_T$ in the next column for all reservoir elevations above it, while for

**Table 10.12   Sediment Distribution in the Reservoir by Empirical Area Reduction Method for Example 10.8**

| Elevation (m) | Original Area (Ha.) | Original Capacity (M.m³) | Depth of Reservoir col. (1) −230 | Relative Depth(P) .(4)/42.5 | $A_P = 2.324P^{0.5}(1-P)^{0.4}$ | Sediment Area = $KA_P$ = 678.8$A_p$ | Sediment Volume $0.5(A_1+A_2)$ H/100 | Sediment Accumulated | Revised Area (Ha.) | Revised Volume (M. m³) |
|---|---|---|---|---|---|---|---|---|---|---|
| (1) | (2) | (3) | (4) | (5) | (6) | (7) | (8) | (9) | (10) | (11) |
| 272.5 | 13562 | 1875.00 | 42.5 | 1.00 | 0.0 | 0.0 | 6.17 | 250.84 | 13562 | 1624.2 |
| 270.0 | 11512 | 1561.44 | 40.0 | 0.941 | 0.727 | 493.5 | 23.57 | 244.62 | 11018 | 1316.8 |
| 266.0 | 9579 | 1134.95 | 36.0 | 0.847 | 1.009 | 684.9 | 29.35 | 221.10 | 8894 | 913.9 |
| 262.0 | 7780 | 736.00 | 32.0 | 0.753 | 1.153 | 782.6 | 32.311 | 91.75 | 6997 | 544.3 |
| 258.0 | 5822 | 532.42 | 28.0 | 0.659 | 1.227 | 832.9 | 33.66 | 159.44 | 4989 | 373.0 |
| 254.0 | 4246 | 332.81 | 24.0 | 0.565 | 1.252 | 849.9 | 33.78 | 125.78 | 3396 | 207.0 |
| 250.0 | 2957 | 188.98 | 20.0 | 0.471 | 1.236 | 839.0 | 32.80 | 92.00 | 2118 | 96.98 |
| 246.0 | 1904 | 95.11 | 16.0 | 0.376 | 1.180 | 801.0 | 30.71 | 59.20 | 1103 | 35.91 |
| 242.0 | 1085 | 38.00 | 12.0 | 0.282 | 1.082 | 734.5 | 20.81 | 28.49 | 350 | 9.51 |
| 238.0 | 306 | 6.81 | 8.0 | 0.188 | 0.927 | 306.0 | 6.90 | 7.68* | 0 | 0.0 |
| 234.0 | 39 | 0.57 | 4.0 | 0.094 | 0.685 | 39.0 | 0.78 | 0.78* | 0 | 0.0 |
| 230.0 | 0 | 0.00 | 0.0 | 0.000 | 0.000 | 0.0 | 0.00 | 0.00 | 0 | 0.0 |

Depth of reservoir = 272.5 − 230.0 = 42.5 m.

* Sediment Capacity cannot exceed the original capacity of col.(3).

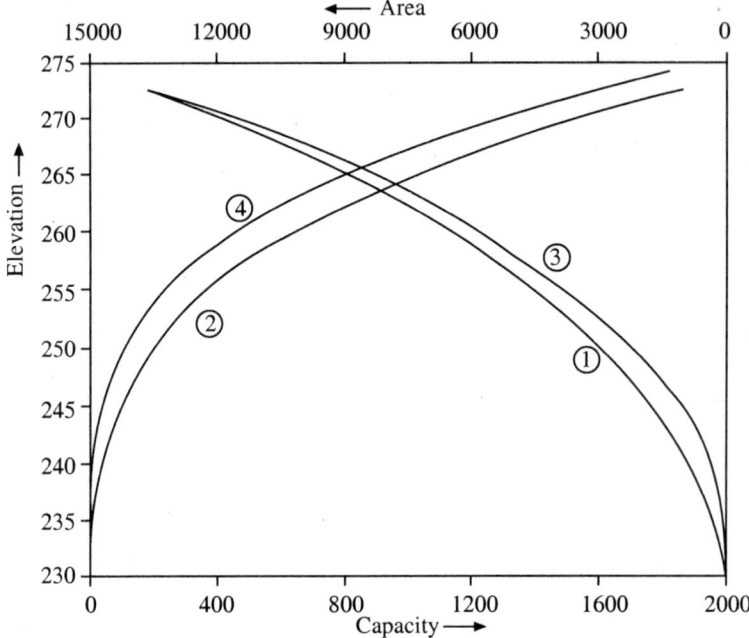

(1) Elevation-Area Curve
(2) Elevation-Capacity curve
(3) Revised area by Empirical Area-Reduction method
(4) Revised capacity by Empirical Area-Reduction method

**Fig. 10.15  Original and Revised Elevation-Area-Capacity Curves for Example 10.8**

    elevations below this reservoir level $(E_T)$ enter the original areas of step (1) as they are less than the area $A_{TC}$.

4. Compute the sediment volume from equation (10.29). For this $E$, values are taken as in step (2). Values of $V_T$ and $A_{TC}$ are found with respect to $E_T$.

5. The sediment volume of step (4) for the highest elevation should agree with the total sediment to be distributed within the life of the reservoir. If it does not, then a new zero elevation is to be assumed and the steps (3) and (4) are to be repeated till they agree.

6. The revised area is found by subtracting the sediment area of step (3) from that of the area of step (1).

7. The revised volume is found by subtracting the sediment volume of step (4) from the original volumes of step (1).

8. Plot the revised elevation area capacity curve which is expected after the end of the designed life of the reservoir.

---

**Example 10.9:** Solve example 10.8 by area increment method.

**Solution**

Computations are carried out in Table 10.13.

Take sediment yield as 9.0 Ha.m/100 sq. km/year.

**Table 10.13   Sediment Distribution in Reservoir by Area Increment Method**

| Elevation (m) | Original Area (Ha.) | Original Capacity (M.m³) | Depth of Reservoir col.(1) −230 | $A_{TC}$ (Ha.) | Sediment Volume 24+ 6.87 × {col.(4) − 10} | Revised Area (Ha.) | Revised Volume (M.m³) |
|---|---|---|---|---|---|---|---|
| (1) | (2) | (3) | (4) | (5) | (6) | (7) | (8) |
| 272.5 | 13562 | 1875.00 | 42.5 | 687.0 | 247.30 | 12875 | 1627.7 |
| 270.0 | 11512 | 1561.4 | 40.0 | 687.0 | 230.10 | 10825 | 1331.3 |
| 266.0 | 9579 | 1135.0 | 36.0 | 687.0 | 202.60 | 8892 | 932.4 |
| 262.0 | 7780 | 736.0 | 32.0 | 687.0 | 175.10 | 7093 | 560.9 |
| 258.0 | 5822 | 532.4 | 28.0 | 687.0 | 147.70 | 5135 | 384.7 |
| 254.0 | 4246 | 332.8 | 24.0 | 687.0 | 120.20 | 3559 | 212.6 |
| 250.0 | 2957 | 189.0 | 20.0 | 687.0 | 92.70 | 2270 | 96.3 |
| 246.0 | 1904 | 95.1 | 16.0 | 687.0 | 65.20 | 1217 | 29.9 |
| 242.0 | 1085 | 38.0 | 12.0 | 687.0 | 37.70 | 398 | 0.3 |
| 238.0 | 306 | 6.8 | 8.0 | 306.0 | 6.80 | 0 | 0.0 |
| 234.0 | 39 | 0.57 | 4.0 | 39.0 | 0.57 | 0 | 0.0 |
| 230.0 | 0 | 0.00 | 0.0 | 0.0 | 0.00 | 0 | 0.0 |

$$\text{Total Sediment for 50 years} = 9.0 \times 50 \times \frac{5574}{100} = 250.84 \text{ M.m}^3$$

$$\text{Depth of reservoir} = 272.50 - 230.0 = 42.50 \text{ m}$$

By trial assume the new zero elevation as 240.0 m
Height of zero elevation $E_T$ is 240.0 − 230.0 = 10.0 m
Area $A_{TC}$ corresponding to the RL 240.0 m = 687.0 Ha
Volume $V_T$ corresponding to the RL 240.0 m = 24.0 M.m³
Volume of sediment to be distributed by using equation (10.29) is

$$V_s = \frac{A_{TC}(E - E_T)}{100} + V_T = 24 + 687 \frac{(42.5 - 10.0)}{100} = 247.3 \text{ Mm}^3$$

This volume is close to the total quantity of sediment to be distributed for the reservoir. Therefore the zero elevation of 240.0 m is acceptable. Taking $A_{TC}$ as 687.0 Ha, distribution of sediment is carried out in Table 10.13. The equation for sediment distribution is taken as 24 + 687{col.(4)−10.0}/100 M.m³. Using the equation, volume of sediment of col. (6) is calculated. Revised area and capacity of the reservoir are calculated in col.(7) and col.(8), respectively. The calculations carried out in the table are self explanatory.

## 10.8   RESERVOIR SEDIMENT CONTROL

Survey of some of the important reservoirs in India (Tables 10.6 and 10.10) show that the reservoirs lose their capacity between 0.5 and 1.0% every year due to sedimentation. With this rate, the reservoirs are likely to be completely filled up some day with the incoming sediment, i.e., the usefulness of the reservoirs gradually decreases with the passage of time from the date of compounding water in the reservoir. Sedimentation is a major parameter in deciding the life of a reservoir. All the reservoirs are designed for storing of the incoming sediment upto the dead storage level. The space between the dead storage level and full reservoir level is the live storage capacity. There are many instances when the live storage of the reservoirs are encroached by the silts much before the designed life of the project. This is due to disturbances in the catchment area by activities like mining, urbanisation, industrialisation, deforestation, earthquakes, forest fires, human activities and other reasons. Sedimentation is a natural process that cannot be stopped completely. Therefore, the targets from the reservoirs are to be redefined with the passage of time. For an irrigation or hydropower project the reduced capacity can still be utilised for the intended purposes with reducing targets but for the projects constructed for the purpose of flood control or municipality water supply, the reduction in its capacity leads to other problems. Therefore, certain terms connected with the life of reservoirs are defined. This helps us to understand the reservoir engineering more correctly.

**Useful Life:** It is the period of a reservoir in years from its construction till the time it can serve to its design commitments without any assistance from a new reservoir. The useful life of a reservoir depends on the rate of silting. When 20% of the designed capacity of a reservoir is reduced, we can assume that the useful life of the reservoir is over.

**Economic Life:** When the economic returns from a reservoir from its various purposes like flood control, irrigation, power generation, municipal and other water supplies are no more beneficial with respect to its service requirements, then we say that the economic life of the reservoir is over.

**Design Life:** It is assumed to last for either 50 or 100 years depending on the expected useful life to which it can serve.

**Full Life:** The period in years from the construction of a reservoir till when it gets completely filled up by sediments to its full reservoir level is called the full life. When this happens, the reservoir is to be blasted off and a new dam is constructed to replace it.

**Capacity Inflow Ratio:** It is defined as the ratio between the reservoir capacity to the total inflow into the reservoir calculated annually, monthly, daily or for a storm. It represents the period of retention of runoff in the reservoir.

**Trap Efficiency of a Reservoir:** It is defined as the ratio of sediment retained in a reservoir to the total sediment inflow into the reservoir. For design of reservoirs, the trap efficiency considered varies between 90 to 95% as finer sediment particles always remain in suspension and come out of the reservoir through the under sluice or spillway. The trap efficiency increases with increase in storage capacity-catchment area ratio and decreases with the age of the reservoir. It also depends on the sediment characteristics, detention of inflow in the reservoir, the shape of the reservoir, operation of the reservoir, position of sluice gate and spillway crest and the relative characteristics of reservoir and inflow water. Under any measure, sediment entry into a reservoir cannot be fully prevented but it can be controlled. The following are some of the measures to control sediment yield into a reservoir.

1. A site with equal scope of the reservoir facilities but less charge of sediment load with the river inflow is to be studied and preferred.
2. Location of under sluice below the probable height of deposition of silt helps to remove some quantity of sediment from a reservoir.
3. By constructing minor check dams in the upstream catchment where soil erosion potential is maximum, the sediment inflow into a downstream main reservoir can be checked.
4. By adopting soil conservation measures like plantations, control grazing, terracing, benching, cover cropping like grassing and contour bunding, the problem of sheet erosion is reduced.

   Table 10.14 gives the effectiveness of soil-conservation measures for two reservoirs under Damodar Valley Corporation (DVC) from 1962 to 1979.

**Table 10.14 Effectiveness of Soil Conservation Measures in Reducing Sediment Inflow into Reservoir**

| Reservoir | Year | | | | | | | Design life (years) | Life fixed after 1st survey (years) | Life of reservoir refixed (years) |
| --- | --- | --- | --- | --- | --- | --- | --- | --- | --- | --- |
| | 1962 | 1964 | 1965 | 1966 | 1971 | 1974 | 1979 | | | |
| Maithan | 768* | – | 738* | – | 677* | – | 640* | 246 | 74 | 110 |
| Panchet | 1275* | 1157* | – | 1012* | – | 713* | – | 75 | 27 | 96 |

*Annual rate of siltation in Ha.m.

There is a declining trend in the rate of siltation in both the reservoirs due to effective soil conservation measures, thereby increasing the life expectancy of reservoirs by 36 and 69 years respectively for Maithan and Panchet reservoirs. The best way to use a dam for more years is to go for aggressive soil conservation measures, which gives a better cost-benefit ratio than going for a new project.

5. Removal of silt by excavation or dredging can be carried out for small reservoirs or tanks. However, the economics of such work vis-a-vis the construction of a new reservoir are to be studied before implementing such a scheme.

*Procedure of Reservoir Sediment Analysis*

Central Water Commission (CWC) accepts the life of medium or major reservoirs as 100 years. The total sediment load for this period is calculated. The distribution of this load is made for both 50 and 100 years. The off taking levels for irrigation canals, power house intakes and others are fixed above the zero elevation of the reservoir for 100 year sedimentation. For calculation of reservoir working tables, the revised elevation-area-capacity curve on the basis of 50 year siltation is considered.

### 10.9   RESERVOIR ECONOMICS

Any water resource project involves substantial monetary involvement which must be justified in terms of its objectives. A project besides serving to its social needs, must also be economically viable and productive. There can always be alternative sites to choose from and therefore the investment envisaged should be the minimum for the patterns of targets at the chosen site. The steps involved in an economic study are:

    (i)   Identify all possible alternative sites.

   (ii)   The expected life of the project and the purposes the project is contemplated to serve are to be ascertained.

  (iii)   Estimate the cost involved for each of the alternative project sites for the envisaged purposes.

  (iv)   Total cost involved for each project and the returns to be received back from each of them should be evaluated.

   (v)   Cost and benefit ratio for each project is to be evaluated and compared.

  (vi)   The project site which gives the maximum cost-benefit ratio should be selected and recommended for construction.

 (vii)   Sometimes political influence, fund availability, social needs, law and justice overrule the engineering decisions. These constraints are to be considered right from the early planning stage.

(viii)   All the projects should be tested for an optimisation-cum-simulation study to fix the optimum economic output from the project and to see the performance of the system to the sequence of flows expected in future.

### 10.9.1   Cost-Benefit Ratio

For a viable project, ratio of the financial benefit to its cost should be higher than unit, i.e. the benefit should be more than the cost on an annual basis. This is called Benefit-Cost (or BC) ratio and is the ratio between annual benefits and annual costs.

For clearance of water resource projects from planing commission or finance department, the BC ratio should be higher than 1.5, unless the project is meant for flood control purposes for which BC ratio can be as low as 1.1. However, for economically and socially backward areas, clearance for a project with benefit-cost ratio just exceeding unit are accepted by the planning commission. This is done to remove the regional imbalances to meet the targeted demands of drinking water, food and power demands of the people which otherwise needs to be supplied from neighboring regions. The economy of transportation of these materials needs to be incorporated. A project needs huge initial investment for its construction which is called the *capital cost*. For the amount invested, the country has to pay certain annual interest for the money it has borrowed. Once the project is complete, it requires certain annual expenses to meet its operation and maintenance. The capital cost is recovered over a certain period of years. The period is fixed on the basis of life of the project and the depreciation percentages applicable to the type of civil structures. Therefore the annual cost is the sum of : (i) annual capital cost, (ii) annual interest and (iii) annual operation, maintenance and replacement (OMR) cost. The principle of compound rate is usually applied to all water resource projects on annual basis. Thus the annual cost is computed from the equation

$$\text{Annual cost} = \frac{\{i(1 + i)^n\}(R - S)}{\{(1 + i)^n - 1\}} \tag{10.30}$$

where $i$ is the rate of interest which is taken between 7 and 17%. Consideration of a lower rate of interest needs justification for the importance of the project in the regional and national level as this gives a better cost-benefit ratio. The interest rate is sometimes taken as sum of (i) annual interest of say 10% (at present) and (ii) depreciation rate of 2%. The annual operation, maintenance and replacement cost between 2 and 5% may be calculated separately and added to the annual cost. In the equation (10.30), $n$ is the estimated period during which the cost is to be recovered or paid back. $R$ is the total initial investment or cost of the project and $S$ is the scrap value or the salvage value of the project after the completion of $n$ years. If the salvage value is zero, then $S$ is taken as zero. The annual cost should be recovered on each year. To simplify the procedure of annual recovery, a simple annual recovery rate of 16–18% is considered for cost benefit analysis.

All the benefits from various purposes are calculated annually on the basis of the prevailing rates. For hydropower generation, the cost per unit and the total units of power generated from a project gives the annual return from power. Similarly the benefits from irrigation, municipality water supply and other releases are calculated on the basis of units of water supplied for the purposes and their unit rates. For a single purpose project, the calculation is easy but for multipurpose project it needs careful considerations. Annual benefits for flood control project can be calculated by the following two ways.

In the first method, the damages due to all previous floods are recorded and the average damages are calculated. The damages that are likely to occur even after the construction of the flood control reservoir due to flood discharge to the downstream are estimated and its average computed. The difference between the two gives the annual benefit due to the project. Records of revenue department during the last 35–40 years are to be searched and all these values are to be brought to the present day worth due to cost escalation. This method is generally followed by all states for their project clearance from Central Water Commission.

In the second method, damages caused in the last 35–40 years of floods are calculated and the frequency of all these floods are computed. A graph is plotted taking frequency of the floods in abscissa and the corresponding damages as ordinate and a smooth curve is passed through these points. Area under the curve from 0 to 50 years frequency flood is calculated and averaged for 50 years (=area under the curve/50). This is taken as the average benefit from a flood control reservoir expecting the life of the project to be 50 years.

---

**Example 10.10:** A project is estimated to cost Rs. 12 crores. The life of the project is expected as 50 years. The interest on capital is 8% and the operation and maintenance cost is 3 %. Salvage from the project will be nil after 50 years. Find the annual cost of the project. Find the benefit-cost ratio for the project. The project is targeted to serve the irrigation demand only. The annual net benefit for the project considering the pre and post irrigation facilities as per the procedure outlined in Example 10.11 is found to be Rs. 224.3 lakhs.

**Solution**
The annual cost is worked out by calculating separately the interest rate and the operation and maintenance cost. For the problem the interest rate $i = 8\% = 0.08$.
Take life of the project equal to the recovery period of $n$ years.
Therefore $n = 50$.

Capital cost $R = $ Rs. $12 \times 10^7$

Salvage of the reservoir   $S = 0.0$

Substituting the values in equation (10.30), the annual cost is obtained as

$$\text{Annual cost} = \frac{\{i(1+i)^n\}(R-S)}{\{(1+i)^n - 1\}} = \frac{\{0.08(1+0.08)^{50}\ (12 \times 10^7 - 0)\}}{\{(1+0.08)^{50} - 1\}}$$

$$= .9809143\ = \text{Rs. } 98.09 \text{ lakhs}$$

$$\text{Annual maintenance cost} = 3\% \text{ of Rs. } 12 \text{ crores} = \text{Rs. } 36.0 \text{ lakhs}$$

$$\text{Total annual cost} = 98.09 + 36 = \text{say Rs. } 134.1 \text{ lakhs}$$

The net annual benefit from irrigation = Rs. 224.30 lakhs.

$$\text{Benefit-Cost ratio} = \frac{\text{Annual benefit}}{\text{Annual cost}} = \frac{224.3}{134.1} = 1.673$$

The project with benefit-cost ratio of 1.673 is economically viable.

**Example 10.11:** Calculate the cost benefit of a medium size irrigation project of the region of Orissa.

**Solution**
As a practical example, the calculation sheet of benefit-cost of a medium irrigation project with catchment area of 452.0 sq. km in Orissa is given. The general abstract of cost of the project for different items of expenses are

| Expenses (Rs. in lakhs) | | Recoveries (Rs. in lakhs) | |
|---|---|---|---|
| A. Preliminary | 28.00 | Recovery on buildings @ 15% | = – Rs.  5.61 |
| B. Land | 316.12 | (as salvage of certain | |
| C. Work Expenses | 1168.93 | items of the building) | |
| D. Regulators | 45.78 | | |
| E. Falls | 17.79 | Recoveries on sales of tools | |
| F. C.D. Works | 83.50 | and plants | = – Rs. 31.40 |
| G. Bridges | 17.5 | Recoveries from vehicles | = – Rs. 01.61 |
| H. Escape | 3.0 | | |
| K. Buildings | 72.17 | Total recoveries | = – Rs. 38.62 |
| L. Earth Work | 190.5 | | |
| L1. Service & | | | |
| boundary road | 15.0 | | |
| L2. Canal lining | 5.0 | | |

O. Miscellaneous
   Expenses                               84.0
O2. Canal Outlets                         2.29
P. Maintenance                           18.47
Q. Special T and P                       41.97
R. Special                               47.50
   Communication
V. Drainage and
   Protective works                      11.46
W. Water courses                         34.39
X. Ecology and                            7.00
   Environment
Y. Loss on stocks                        14.59
   0.25% of all                        ————————
Total = Rs.                            2224.95
Establishment                           166.57
charges (@7.5%
except land cost)
Ordinary tools and                       22.25
plants (@ 1% of works)

                                       ————————
TOTAL =                           Rs. 2413.77
Recoveries =                      Rs. −38.62

Capital value of land for offices and other places after the project is over

$$= + \text{Rs } 27.17 \text{ lakhs}$$

Total Cost of the project = 2413.77 − 38.62 + 27.17 = Rs. 2402.23 lakhs.

**Benefit from Irrigation**

|                                                                          | Before Irrigation (Rs. in lakhs) | After Irrigation (Rs. in lakhs) |
|--------------------------------------------------------------------------|:---:|:---:|
| Gross value of farm produce as per present growth of crops and market rate | 159.26 | 1390.15 |
| Add dung receipt (3%of fodder expenses after irrigation and 4.5% before irrigation) | 7.17 | 41.71 |
|                                                                          | + 166.43 | + 1431.86 |
| Expenses                                                                 |  |  |
| (a) On seeds, manure and fertilizer                                      | 13.10 | 107.65 |
| (b) Labour charges (10%of produce)                                       | 15.93 | 139.02 |
| (c) Fodder (15% of farm produce)                                         | 23.89 | 139.02 |
| (d) Depriciation (2.7% of farm produce)                                  | 4.30 | 37.53 |
| (e) Share and cash rent (5 % of farm produce)                            | 7.96 | 41.71 |
| (f) Land revenue (2 % of farm produce)                                   | 3.19 | 27.8 |
|                                                                          | − 68.37 | − 465.2 |
|                                                       Net Receipt =      | 98.06* | 966.66** |

Net receipt less from 1015 Ha. of land to be submersed after irrigation which is available now = 98.06 × 1015/5134 = 19.38 lakhs

*Rs. 98.06 lakhs = 166.43 − 68.37;    ** 966.66 = 1431.86 − 465.2

**Cost Benefit Ratio**

| Annual benefit | (Rs. in lakhs) |
|---|---|
| (a) Before Canal Irrigation | 98.06 |
| (b) After Canal Irrigation | 966.66 |
| (c) Net Profit due to Irrigation (b −.a) | 868.60 |
| (d) Deduct less due to submergence | − 19.38 |
| (e) Net Profit | 849.22 |

| Annual Cost | |
|---|---|
| (a) Capital Cost of the Project | 2402.23 |
| (b) Interest (10 % of capital outlay) | 240.22 |
| (c) Depriciation charges(2% of capital cost) | 48.04 |
| (d) Administrative expenses @ Rs.75/Ha. of annual irrigation of 8252 Ha. | 6.19 |

$$\text{Total Cost (b + c + d)} = \text{Rs. 294.45 lakhs}$$

$$\text{Benefit Cost Ratio (BC ratio)} = 849.22 / 294.45 = 2.88$$

The project is viable from the consideration of economic returns.

### 10.9.2 Optimisation of Benefits

Whenever a multipurpose water resource project is planned, its objectives and location are made variable factors due to wider national interest and large economy involved in its construction. The best site with the most profitable purposes from the project is to be chosen. Again, any possible increase in storage capacity of a project gives a multidimensional benefit from flood control, hydropower generation, irrigation, domestic water supply and other releases. This is because the increase in reservoir space is used for better flood control during the flood periods and at the end of this period, the space is filled up for conservation purposes. The additional storage is released through hydroelectric turbines generating more power and the same water can be channelised for irrigating more area. Therefore, the benefit is increased in proportion to the capacity increase in the reservoir. The limits on storages due to submergence and the yield of the basin, restrict the storages at the upper end, while the cost and benefits put the lower limit on storages to be build at a site. Usually the cost and benefits attain a maximum at certain capacity and then decline gradually. The problem to choose the most economically viable site for a reservoir with the limits for various purposes of irrigation, flood control, power generation, domestic water supply and other demands to yield the system optimally is a complex process which should be carried out under the supervision of an experienced engineer .

To evaluate the best alternative site a linear-programming-optimisation-model is normally applied to water resources projects. A simple example of linear programming is discussed here.

---

**Example 10.12:** A reservoir has 6 units of water to be supplied in 28 days. Two groups of crops are to be grown in the command area. For the first group of crop two units of water is required in 7 days time while the second group requires one unit of water in 8 days. The price of irrigation revenue is Rs 120/- for the first crop requiring two units of water and Rs 80 /- for the second crop. If the revenue collection is to be the maximum, then how many units of water for each crop should be supplied? When water is supplied to one unit, the canal system cannot supply water to the second unit.

**Solution**

The solution to the problem can be obtained by formulating the following linear equations.

Maximize : $Z = 120 X_1 + 80 X_2$ (1)

Subjected to the following constraints

Water supply constraint : $2X_1 + X_2 \leq 6$ (2)

Time constraint : $7X_1 + 8X_2 \leq 28$ (3)

Hidden constraint : $X_1 \geq 0$ and $X_2 \geq 0$

From equation (2) we get $X_1 = 3$ when $X_2 = 0$ and $X_1 = 0$ when $X_2 = 6$.
From equation (3) we get $X_1 = 4$ when $X_2 = 0$ and $X_1 = 0$ when $X_2 = 3.5$.

Taking these points, a graph can be plotted between $X_1$ and $X_2$ as shown in Fig. 10.16. The physical domain is bounded by the area ABCO. Let us discuss the maximisation at the points A, B, C and O.

At point O : $X_1 = X_2 = 0$ which gives Maximum of $Z = 0$
At point A : $X_2 = 3.5$ and $X_1 = 0$ which gives Maximum of $Z = 280$

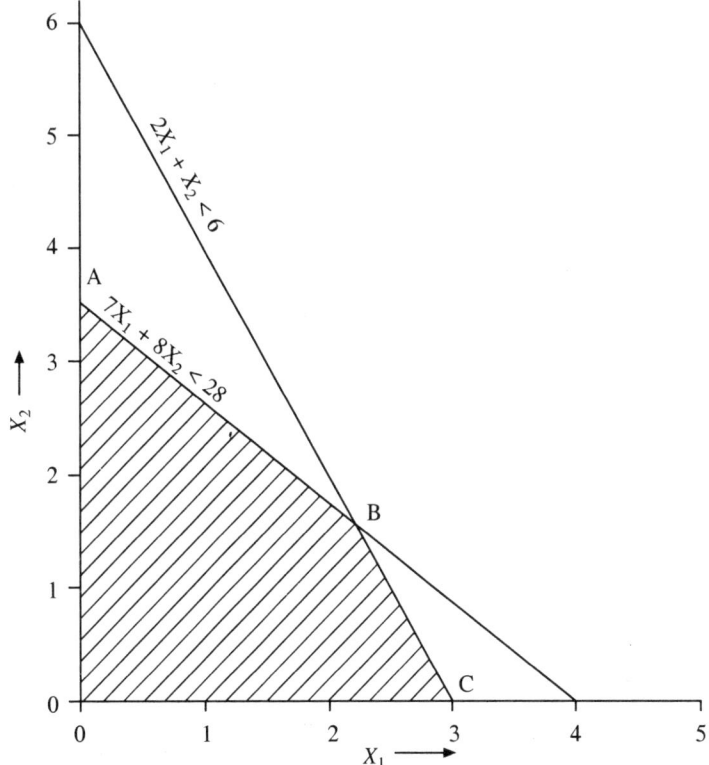

**Fig. 10.16   Graphical Solution to Linear Programming for Example 10.12**

At point C : $X_1 = 3.0$ and $X_2 = 0$ which gives Maximum of $Z = 360$
At point B : $X_1 = 2.23$ and $X_2 = 1.54$ which gives Maximum of $Z = 390.8$.

Other points in the physical domain will not give optimum results as it will be less than 390.8. Therefore maximum profit is obtained when 2.23 units of water is supplied to crop-I and 1.54 units of water is supplied to crop-II.

There are number of methods available to solve such linear optimisation problems and can be programed in digital computers and solved. The general approach to solve is either by graphical or simplex method. For computer programing, simplex method is usually adapted as it is easy to program. Such programs are available commercially.

### 10.9.3   Linear Programming in Multipurpose Water Resource Projects

Use of linear programming model to a multipurpose water resources project optimises the available water at the dam site for uses like irrigation, flood control, hydropower generation, domestic and industrial water supply and downstream water requirements to get maximum benefit. When the constraints are viable and clear, it gives an optimal solution to the objective function. In system engineering (dealing with the complex water resource projects), the physical system is mathematically represented as objective function with constraints that takes care of various aspects of the problem, identifies the functional relationship between all the system components and establishes its effectiveness and usefulness. Due to simplicity, the linear programming model (LP model ) has been widely used in water resources management as a screening model subject to linear constraints. Other models like dynamic programming and nonlinear programming are in restricted use due to the involvement of complicated mathematics, and their incapability to handle stochastic nature of inflows in the system. The LP model for a multipurpose water resource project can be written in the following form.

Maximise :   {Gross Annual Benefit – Annual Capital Cost – Annual Operation, Maintenance and Replacement Cost}

or Maximise : $\{(\beta_1 I_r + \beta_2 E + \Sigma a_4 Y_{ft}) - (C_1 I_r + C_2 Y + C_3 H) - (O_1 I_r + O_2 Y + O_3 H)$

$$(10.31)$$

The constraints are:

(a) Water target equation:

$Q_t + I_t'' = \Sigma A_j W_{j,t} + Q'_t$ (for all $t$ and for $j = 1$ to $n$)   $\qquad$ (10.32)

(b) Land area availability constraint

$\Sigma XLAM_{j \cdot t} A_j \le CCA$ (for $j = 1$ to $n$ )   $\qquad$ (10.33)

(c) Continuity equation

$S_{t+1} = S_t + I_t + P_t + I_{avt} - El_t - Q_t - Q'_t$   $\qquad$ (10.34)

(d) Upper bound reservoir content : $S_t \le Y$   $\qquad$ (10.35)

(e) Storage capacity for flood conservation purposes: $S_t \le YMAX_t$   (10.36)

(f) Upper bound capacity: $YMAX_t \le Y$   $\qquad$ (10.37)

(g) Lower bound reservoir content : $S_t \geq DS$            (10.38)

(h) Storage constraints (minimum) : $S_t \geq YMIN_t$        (10.39)

(i) Water availability constraint : $Q_t \leq S_t + F_t - WS_t$     (10.40)

(j) Turbine releases : $\eta_t E \leq Ce\ Ha_t\ Q_t$              (10.41)

(k) Load factor : $\eta_t E = \alpha_t h_t H$                 (10.42)

(l) Constraints for intensity of irrigation : $\sum A_{j \cdot t} = CINTA$   (10.43)

where *CINTA* is irrigation intensity, $A_{j+t}$ the crop area of the *j*th crop at time *t*, $\eta_t E$ the monthly energy requirement, $\alpha_t$ the load factor, $h_t$ the number of hours, $H$ the hydro power plant capacity, $S_t$ the storage at any time, $YMIN_t$ the minimum storage capacity, $DS$ is the dead storage, $YMAX_t$ the maximum capacity of the reservoir, $Y$ the gross capacity of reservoir, $XLAM_{j \cdot t}$ is the land area coefficient of *j*th crop during the month, $A_j$ the area occupied by the *j*th crop, $W_{j,t}$ the water requirement of *j*th crop, $Q_t$ the water released from reservoir, $F_t = I_t$ is the water that joins the main stream just above irrigation diversion canal from the reservoir, $C_1$, $C_2$ and $C_3$ are the annual capital cost for irrigation, reservoir storage and power, $O_1$, $O_2$ and $O_3$ are the operation, maintenance and replacement cost, $Y_{ft}$ is the flood control space $= (Y - YMAX_t)$, $\beta_1$ and $\beta_2$ are the gross annual benefit functions for irrigation and power, $a_4$ the gross annual benefit function for flood control for 4 months say from June to September, *CCA* is the culturable commendable area, $I_t$ the inflow into the reservoir, $El_t$ the reservoir evaporation, $P_t$ the precipitation directly on the reservoir, $I_{avt}$ the local inflow into the reservoir, $Q_t'$ the secondary water released from the reservoir in time *t*, *e* the turbine efficiency, $C$ is a conversion factor for hydropower, $WS_t$ the municipality and industrial demands and $S_{t+1}$ the storage at next time step.

Taking the minimum or average flow as the input, a computer program can be developed to get the maximum targeted output in terms of the economic return. While doing so, it is necessary that 100% demand for drinking water, 90% for power generation and 75% demand for irrigation are satisfied. In other words, the targets for the power, irrigation and drinking water should be so fixed that the flow sequence gives the necessary percentage of success. For example, out of say 40 years of flow sequence, the irrigation demand should be satisfied for at least 75% of time, i.e., 30 years. A Linear Programming model can successfully be used to fix the targets on the consideration of economic optimisation. Computer program developed by using the above equations have been used at the planning stage for a number of multipurpose projects to select the best alternative site. The model can also be used to test the performance of a single or a cascade of multipurpose projects in a basin. Operation schedule for a group of existing multipurpose projects can be drawn to suit various flow conditions to yield the maximum return from a water resources system.

## PROBLEMS

10.1   How you are going to use flow mass curve to determine the storage needed for a constant demand rate?

10.2   Observed runoff data (M.m$^3$) from 1970 to 1995 are available at a potential project site. Determine the 50%, 75% and 90% dependable yields at the site.

280.5, 580.4, 240.5, 360.7, 190.2, 110.3, 105.3, 400.6, 360.3, 280.4, 240.3, 460.4, 552.8, 120.8, 325.8, 40.3, 51.4, 130.7 ,326.4, 176.3, 240.8, 230.3, 187.0, 260.0, 240.1, 555.5

10.3   The following flows are recorded in the driest year at a project site. What storage will be required to draw water from the project at a constant rate of 60 m$^3$/sec.

| Month | J | F | M | A | M | J | J | A | S | O | N | D | J |
|-------|----|----|----|----|----|----|----|-----|-----|-----|----|----|----|
| Av.flow (m$^3$/sec) | 82 | 67 | 56 | 44 | 38 | 43 | 71 | 103 | 128 | 114 | 99 | 94 | 76 |

10.4   Consider the following inflow and demands at a reservoir site

| Month | J | F | M | A | M | J | J | A | S | O | N | D |
|-------|----|----|----|----|----|----|----|----|----|----|----|----|
| Inflow (M.m$^3$) | 14 | 12 | 8 | 5 | 3 | 4 | 53 | 58 | 45 | 22 | 16 | 15 |
| Demand (M.m$^3$) | 10 | 9 | 11 | 14 | 15 | 14 | 18 | 16 | 17 | 18 | 19 | 21 |
| Evaporaton (mm/day) | 3.4 | 3.4 | 6.0 | 7.6 | 8.5 | 6.0 | 5.0 | 5.0 | 5.0 | 4.2 | 3.4 | 3.4 |

The reservoir water spread area varies from 6 sq. km for the month of May to 26 sq. km in November and reaches back to 6 sq. km in May next. Calculate the storage required to meet the targeted demands. As per agreement the reservoir must release an additional 2 m$^3$/sec of water always to the downstream barrage.

10.5   Estimate the sediment load at a project site using various approaches from the following data:
Catchment area = 1830 sq. km
Width of reservoir at FRL = 560.0 m
River slope at the dam site = 0.006
Average inflows at the site are as follows

| Year | 82 | 83 | 84 | 85 | 86 | 87 | 88 | 89 | 90 |
|------|------|------|------|------|------|------|------|------|------|
| Inflow (M.m$^3$) | 2210 | 1290 | 1640 | 1780 | 2150 | 1980 | 2540 | 1285 | 1620 |

10.6   Discuss various methods of estimation of (a) suspended load, (b) bed load, (c) total load and (d) sheet erosion.

10.7   What do you understand by sediment rating curve? How it is going to be useful to estimate sediment at a place? Plot a typical sediment rating curve and discuss.

10.8   A reservoir has the following sediment and discharge data.

| Year | 81 | 82 | 83 | 84 | 85 | 86 | 87 | 88 |
|------|------|------|------|------|------|------|------|------|
| Discharge(M.m$^3$) | 1430 | 3850 | 2050 | 6510 | 2880 | 1120 | 6050 | 2220 |
| Sediment yield(M.ton) | 2.65 | 5.82 | 3.6 | 7.15 | 5.22 | 1.95 | 6.88 | 3.94 |

Calculate the average sediment load/year/100 sq. km of the catchment at the site.

Develop a regression relation and predict the sediment yield for the inflow of 3450 M.m$^3$ for the year 1978. Take the catchment area at the site as 3050 sq. km. What is the total sediment yield for 100 years?

10.9 What is meant by the following terms:
(a) Useful life, (b) Economic life, (c) Design life, (d) Trap efficiency, (e) Various levels connected with storage of a reservoir?

10.10 What are the steps to consider to control sediment inflow into a reservoir? Comment on the alternative.

10.11 Discuss the empirical- area reduction method of sediment distribution in a reservoir.

10.12 At a proposed site, the silt load is 6.5 Ha.m/year/100 sq. km. Distribute 50 year sediment by empirical area reduction method. Assume type III reservoir. Elevation-area-capacity curve for the reservoir are as follows:

| Elevation (m) | 112 | 116 | 120 | 124 | 128 | 132 | 136 | 140 | 144 | 148 | 150.5 |
|---|---|---|---|---|---|---|---|---|---|---|---|
| Original area (Ha.) | 0 | 40 | 250 | 930 | 1810 | 2780 | 4040 | 5760 | 7500 | 9200 | 10100 |
| Original Capacity (M. m$^3$) | 0 | 0.45 | 5.90 | 36.5 | 92.05 | 181.6 | 320.6 | 510.1 | 705.4 | 1090.8 | 1480.2 |

Take drainage area at the project site as 4585 sq. km.

10.13 Solve Problem 10.12 by Area increment method.

10.14 What do you understand by cost-benefit ratio? How optimisation methods are going to help in reservoir planning?

10.15 A project is estimated for Rs. 36.6 crores. Taking the life of the project as 50 years, obtain the benefit-cost ratio. Take annual interest rate as 7% and operation-maintenance cost as 2% of the project cost. Net annual benefit for the project is Rs. 910.20 lakhs. Salvage from the project at the end of 50 years life is 5% of the initial investment.

10.16 Write notes on the following: (a) Reservoir classification, (b) Khosla and Joglekar's equation of sediment yield, (c) Factors attributing silt in a natural stream, (d) Suspended load and bed load and (e) Life of a reservoir.

01. The science which deals with occurrence, distribution and circulation of water is known as
    (a) Meteorology                          (b) Metrology
    (c) Hydrology                            (d) Hydrometry
02. Relative humidity of air is the ratio of
    (a) actual vapour pressure to the saturation vapour pressure at $0°C$.
    (b) weight of water in unit volume of air to the weight of air of the same volume
    (c) actual vapour pressure to the saturation vapour pressure at the same temperature.
    (d) actual vapour pressure to the saturation vapour pressure at $4°C$.
03. When average annual rainfall and evaporation over land masses and oceans of the earth are considered,
    (a) the annual evaporation from land area is found to be the same as the annual precipitation.
    (b) more water evaporates from the oceans than what falls back on them as precipitation.
    (c) more rain falls on the oceans than what is evaporated.
    (d) more precipitation occurs over land area than what is evaporated back from the same area.
04. In the hydrological cycle the average time of residence of water in the atmosphere is
    (a) 15 days        (b) 1 day        (c) 8.9 days        (d) less than 1 day
05. When expressed in m depth of water column, the average atmospheric pressure is equal to
    (a) 1.013          (b) 10.13        (c) 101.3        (d) 1013.0
06. Which of the following is different form the rest
    (a) rain           (b) drizzle      (c) hail         (d) fog.
07. During anticyclones in Northern Hemisphere, wind blows
    (a) clockwise inward                     (b) anticlockwise inward
    (c) clockwise outward                    (d) anticlockwise outward
08. Isotherm is a line which joins the points of equal
    (a) rainfall depth                       (b) temperature
    (c) humidity                             (d) atmospheric pressure
09. The convective precipitation is caused due to
    (a) a mountainous slope
    (b) heating and subsequent vertical instability of moist air
    (c) a disturbance on the air front develops into a cyclone
    (d) the colder air mass forms a wedge and lifts the warm air mass
10. Orographic precipitation occurs due to air masses being lifted to higher altitudes by
    (a) the density difference of air masses
    (b) unequal heating of air masses
    (c) the presence of mountain barriers
    (d) large pressure difference in air masses

11. The following rain gauge which do not produce a mass curve of precipitation as record
    (a) Symon's rain gauge  (b) Tipping-bucket type gauge
    (c) Weighing-bucket type gauge  (d) Syphon gauge.

12. If the maximum depth of a 25 years-10 h rainfall depth at Bhubaneswar is 250 mm, then a 50 years-8 h-maximum rainfall depth at the same place is
    (a) < 250 mm  (b) > 250 mm
    (c) = 250 mm  (d) none of the above

13. By DAD analysis, the maximum average depth over an area due to one day storm is found to be 60 cm. For the same area, the maximum average depth for a 2-day storm can be expected to be
    (a) < 60 cm  (b) > 60 cm  (c) 60 cm  (d) 120 cm

14. For Indian conditions, IS recommends self recording raingauge of
    (a) weighing bucket type  (b) Syphon type
    (c) tipping bucket type  (d) telemetry type

15. Where $n$ is the number of data and $m$ the rank, return period using Weibul's formula is given by
    (a) $T_r = (m + 1)/n$  (b) $T_r = m/(n + 1)$
    (c) $T_r = (n - 1)/m$  (d) $T_r = n/(m - 1)$

16. If $x_1, x_2$ and $x_3$ have the return periods of 50, 75 and 100 years respectively, which of the following statements is true.
    (a) $x_1 < x_2 < x_3$  (b) $x_1 > x_2 > x_3$
    (c) $x_1 < x_2$ and $x_2 > x_3$  (d) $x_1 > x_2$ and $x_2 < x_3$

17. An unbiased estimate of standard deviation is given by
    (a) $\sqrt{\dfrac{\Sigma (x_i - \bar{x})^2}{(n - 1)}}$  (b) $\sqrt{\dfrac{\Sigma (x_i - \bar{x})^2}{n}}$

    (c) $\sqrt{\dfrac{\Sigma (x_i - x)^2}{n(n - 1)}}$  (d) $\sqrt{\dfrac{\Sigma (x_i)^2}{(n - 1)}}$

18. An appropriate distribution to describe maximum annual hydrologic series is
    (a) Gumbel distribution  (b) Chi-square distribution
    (c) Pearson type III distribution  (d) Normal distribution.

19. The minimum distance a object should maintain from a rain gauge should be
    (a) one times its height  (b) two times its height
    (c) three times its height  (d) four times its height.

20. Double mass curve technique is applied
    (a) to check the distribution of rainfall data in the catchment
    (b) to check the variability of rainfall data
    (c) to check the consistency of record at a suspected rain gauge station
    (d) to prepare isohyetal maps

21. Rainfall hyetograph is a graph drawn between
    (a) cumulative rainfall and time
    (b) rainfall intensity and time
    (c) rainfall depth and area
    (d) rainfall intensity and area of the catchment

22. From a recording type raingauge the graph chart is a
    (a) rainfall hyetograph  (b) map showing isohyetal pattern
    (c) rainfall mass curve  (d) intensity-duration curve

23. An isohyet is a line joining points of
    (a) equal rainfall intensity  (b) equal storm duration
    (c) equal rainfall depths  (d) equal excess rainfall depths

24. The average depth of rainfall resulting from a storm
    (a) decreases with increase in area
    (b) increases with increase in area
    (c) has no relation with area
    (d) remains constant with increase in area
25. A lysimeter is used to measure
    (a) evaporation                    (b) transpiration
    (c) evapotranspiration             (d) losses from rainfall
26. Wind velocity measured at a height of 2 m above the ground is 20 km/h. What would be the velocity at a height of 6 m above the ground ?
    (a) 16.6 km/h      (b) 23.4 km/h      (c) 60 km/h      (d) 40 km/h
27. When $E$ is the evaporation from a lake and $Ep$ the evaporation from a pan, then the pan coefficient is defined as
    (a) $E/E_p$          (b) $E_p/E$          (c) $E_p - E$          (d) $E_p \cdot E$
28. Salinity in water
    (a) reduces evaporation            (b) does not affect evaporation
    (c) increases evaporation
    (d) may increase or decrease evaporation depending on salt
29. Dalton's evaporation law in terms of the saturation vapour pressure $e_s$ at water surface temperature and the actual vapour pressure in air $e_a$ may be written as
    (a) $E = C(e_s + e_a)$                (b) $E = C(e_s - e_a)$
    (c) $E = C(e_s/e_a)$                  (d) $E = C(e_s \times e_a)$
30. Pan coefficient for the standard class A pan is taken in the range of
    (a) 0.9–1.0        (b) 0.6–0.80        (c) 1.0–1.2        (d) 0.8–0.90
31. The chemical generally used to reduce evaporation from ponds/lakes is
    (a) ethyl alcohol                  (b) methyl alcohol
    (c) cetyl alcohol                  (d) butyl alcohol
32. Isobar is a line joining points of equal
    (a) temperature    (b) pressure    (c) humidity    (d) sunshine hours
33. Potometer is used to measure
    (a) transpiration                  (b) evaporation
    (c) evapotranspiration             (d) vapour pressure
34. Evapotranspiration is a phenomena of
    (a) daylight hours                 (b) night-time only
    (c) land surfaces only             (d) a continuous process
35. A 4 h storm had 4 cm of rainfall and the resulting runoff was 2 cm. If the $\phi$-index is maintained at the same rate, the runoff due to 10 cm of rainfall in 6 h in the catchment is
    (a) 5.0 cm         (b) 7 cm         (c) 6 cm         (d) 3 cm
36. If N is the speed of the current meter in rpm, the velocity measured by it is proportional to
    (a) $N^{1/2}$                (b) N                (c) $N^{3/2}$                (d) $N^2$
37. While recording velocity across a vertical in a stream using a current meter, it is sufficient to measure velocities above the stream bed at locations
    (a) 0.25 and 0.75 depths           (b) 0.20 and 0.80 depths
    (c) 0.40 and 0.60 depths           (d) 0.10 and 0.90 depths
38. In single point method of finding mean velocity across a vertical, the velocity is measured above the stream bed at a location
    (a) 0.4 depth      (b) 0.6 depth    (c) 0.7 depth    (d) 0.8 depth
39. The rating curve of a stream gauging station gives the
    (a) variation of discharge in the stream with area of flow

(b) variation of discharge in the stream with stage

(c) variation of discharge in the stream with depth of flow

(d) ` variation of discharge in the stream with velocity of flow

40. The largest unit of volume is

(a) hectare-metre    (b) km$^2$-cm    (c) million-m$^3$    (d) cumec-day

41. The runoff from a drainage basin of area 8640 sq. km is estimated as 10000 cumec-day, the depth of runoff is

(a) 20 cm    (b) 40 cm    (c) 10 cm    (d) 84.4 cm

42. The volume of 1 cumec-day equals to

(a) 86400 m$^3$    (b) 8.64 ha-m    (c) 24 cumec-h    (d) all the above

43. The area of lowest infiltration capacity among the following is the

(a) forest land    (b) grazed pasture

(c) concrete pavement    (d) hard soil

44. Depression storage when expressed as % of storm rainfall

(a) increases with increase in storm rainfall

(b) decreases with increase in storm rainfall

(c) decreases with decrease in storm rainfall

(d) in dependent of storm rainfall

45. Infiltration capacity of the soil is the

(a) the depth of water absorbed by the soil during the storm

(b) rainfall intensity above which the rainfall volume equals the observed runoff volume

(c) the maximum rate at which the soil absorbs water

(d) the permeability of the soil

46. Whereas $f, f_c, f_0, e, k$ and $t$ has their usual meaning, Horton's infiltration equation is given by

(a) $f = f_c + (f_0 - f_c)e^{kt}$    (b) $f = f_0 + (f_c - f_0)e^{-kt}$

(c) $f = f_c + (f_0 + f_c)e^{-kt}$    (d) $f = f_c + (f_0 - f_c)e^{-kt}$

47. $\phi$-index is defined as

(a) difference between the total rainfall and the total runoff divided by the duration of storm

(b) rainfall intensity above which the rainfall volume equals to the observed runoff volume

(c) minimum infiltration rate during the storm

(d) maximum infiltration rate during the storm

48. A 4 h storm with a uniform intensity of 1.5 cm/h produced a runoff depth of 40 mm. The average infiltration rate during this storm is

(a) 4 mm/h    (b) 5 mm/h    (c) 6 mm/h    (d) 15 mm/h

49. The antecedent precipitation index on a certain day is $I_0$. If the recession factor is $k$, and if there is no rainfall for the next $n$ days, the API on $n$th day is given by

(a) $I_n = I_0/k^n$    (b) $I_n = I_0 k^n$    (c) $I_n = k^n/I_0$    (d) $I_n = 1/I_0 k^n$

50. The ground water flow in a stream is also known as

(a) interflow    (b) seepage    (c) base flow    (d) runoff

51. Mass curve of flow is a graph drawn between

(a) flow rate and time

(b) cumulative volume of flow and time

(c) cumulative volume of flow and cumulative time

(d) cumulative discharge and time

52. Base flow is determined from

(a) total runoff and the infiltration rate

(b) total runoff and the delayed subsurface runoff

    (c)  prompt subsurface runoff and the delayed subsurface runoff

    (d)  total runoff and the prompt subsurface runoff

53. At a section in a wide rectangular channel the depth of flow is found to increase by 40% and the water-surface slope reduces by 33% of its original value. This marks an approximate change in the discharge of

    (a)  33%        (b)  43%        (c)  40%        (d)  – 43%

54. Slope-area method is used for

    (a)  accurate estimation of discharge

    (b)  estimation of flood discharge based on high-water marks

    (c)  development of rating curve

    (d)  estimation of discharge where backwater effect is predominant

55. Dilution method of stream gauging is used for measuring discharges in

    (a)  large alluvial rivers

    (b)  flood flow in a mountain stream

    (c)  steady flow in a small turbulent stream

    (d)  in a river stretch having heavy industrial pollution loads

56. A relation between the stage $G$ and discharge $Q$ at a gauging site is usually expressed as

    (a)  $Q = \alpha\, G^{\beta}$                   (b)  $Q = \alpha\,(G - G_0) \times \beta$

    (c)  $Q = \alpha\,(G - G_0)^{\beta}$           (d)  $G = \alpha\,(Q - G_0)^{\beta}$

57. At a gauging site, if $Q_r$ is the discharge at a stage when the water surface is rising and $Q_f$ the discharge at the same stage when the water surface is falling, then

    (a)  $Q_r > Q_f$                   (b)  $Q_f/Q_r$ is constant

    (c)  $Q_r < Q_f$                   (d)  $Q_f = Q_r$

58. A section is said to have permanent control when

    (a)  stage-discharge relation change with time

    (b)  stage-discharge relation do not change with time

    (c)  there is variation of flow parameters with depth of flow

    (d)  there is variation of channel parameters with depth of flow

59. The surface velocity at any vertical section of a stream is

    (a)  a measure of mean velocity at the vertical

    (b)  smaller than the mean velocity in that vertical

    (c)  larger than the mean velocity in that vertical

    (d)  equal to the velocity in that vertical at 0.8 times the depth

60. The total rainfall in a catchment of area 300 km$^2$ during a 6 h storm is 16 cm while the surface runoff due to the storm is $0.3 \times 10^8$ m$^3$. The $\phi$ index is

    (a)  0.1 cm/h      (b)  1.0 cm/h      (c)  0.2 cm/h      (d)  10.0 cm/h

61. In a flow-mass curve study the demand line drawn from a ridge in the curve did not interest the mass curve again. This represents that

    (a)  the reservoir was not full at the beginning

    (b)  the storage was not adequate

    (c)  the demand cannot be met by the inflow as the reservoir will not refill

    (d)  the reservoir is wasting water by spill

62. The flow-duration curve is

    (a)  also known as discharge-frequency curve

    (b)  a plot between discharge and percent of time

    (c)  represents a curve of cumulative frequency distribution of flow

    (d)  all the above types of curves

63. Water year in India starts from the first day of

    (a)  January      (b)  April      (c)  June      (d)  March

64. Direct runoff is made up of
    (a) Surface runoff + Prompt interflow + Channel precipitation
    (b) Surface runoff − Prompt interflow − Channel precipitation
    (c) Overland flow only
    (d) Total rainfall volume

65. 100 cm of effective rainfall from a catchment of area 100 km² represents an runoff of
    (a) 100 M.m³    (b) 10 M.m³    (c) 1 M.m³      (d) 1000 M.m³

66. A geological formation which has good porosity but very less permeability is known as
    (a) aquifer      (b) aquifuge      (c) acquitard      (d) aquiclude

67. Which of the following formations is good for ground water exploration?
    (a) aquifer      (b) aquifuge      (c) aquitard      (d) aquiclude

68. The permeability of an aquifer
    (a) increases with increase in temperature
    (b) increases with decreases in temperature
    (c) is independent of temperature
    (d) decreases with decrease in temperature

69. An isolated unconfined aquifer is also known as
    (a) an artesian aquifer      (b) a leaky aquifer
    (c) a perched aquifer      (d) a water table aquifer

70. Specific yield of an aquifer is defined as the ratio of the
    (a) volume of pore space to the volume of soil
    (b) volume of water freely drained from a saturated soil to the volume of soil
    (c) volume of water retained, when a saturated soil is freely drained to the volume of soil.
    (d) volume of pore space to the volume of soil

71. The ratio of volume of water retained per unit volume of formation under free drainage is known as
    (a) specific yield      (b) specific retention
    (c) specific storage      (d) specific gravity

72. The surface obtained by joining the water levels in several observation well's penetrating an unconfined aquifer represents
    (a) piezometric surface      (b) water table surface
    (c) capillary fringes      (d) cone of depression

73. An influent stream
    (a) contributes runoff to groundwater
    (b) derives runoff from groundwater
    (c) neither contributes nor derives runoff from groundwater
    (d) flows only below the ground

74. The horizontal distance between the centre of the pumping well and the point of zero drawdown is
    (a) radius of influence      (b) effective length
    (c) length of aquifer      (d) diameter of influence

75. When a sample has a porosity of 28%, the specific yield of the aquifer will be
    (a) = 0.28      (b) > 0.28      (c) < 0.28      (d) difficult to predict

76. Darcy's law for ground water movement states that the velocity is proportional to
    (a) hydraulic gradient
    (b) square of the hydraulic gradient
    (c) logarithm of the hydraulic gradient
    (d) reciprocal of the hydraulic gradient

77. For Darcy's law to be valid for ground water flow, the upper limit of Reynold's number is
    (a) 0.01      (b) 0.10      (c) 1.00      (d) 10.00
78. Discharge per unit drawdown is called
    (a) specific yield            (b) specific storage
    (c) storage coefficient      (d) specific capacity
79. While considering flow between two water bodies, the piezometric head is taken as
    (a) a straight line           (b) a parabola
    (c) an arc of a circle      (d) a part of an ellipse
80. Flowing artesian wells are expected in areas where the
    (a) water table is very close to the soil surface
    (b) aquifer is confined and is in hyperpiestic condition
    (c) elevation of the piezometric head line is above the ground surface
    (d) area with abundant springs
81. The unit of intrinsic permeability is
    (a) cm/day     (b) $cm^2$/day     (c) darcy/day     (d) $cm^2$
82. Specific capacity is
    (a) constant for a well
    (b) varies with season
    (c) varies with aquifer characteristics
    (d) increases with discharge rate
83. The dimension of storage coefficient $S$ is
    (a) $L^3$                (b) $LT^{-1}$
    (c) $L^3/T$           (d) a dimensionless parameter
84. Peak discharge using Snyder's synthetic unit hydrograph is computed from the equation
    (a) $Q_p = \dfrac{2.78 C_p A}{t_p}$          (b) $Q_p = \dfrac{2.78 C_p}{t_p A}$

    (c) $Q_p = \dfrac{2.78 C_p t_p}{A}$          (d) $Q_p = \dfrac{2.78 t_p}{C_p A}$

85. The lag time of the basin is the time interval between the
    (a) centroid of the rainfall histogram and the peak of the hydrograph
    (b) beginning and end of direct runoff
    (c) beginning and end of effective rainfall
    (d) centroid of the rainfall histogram and the centroid of the hydrograph
86. The $S$-hydrograph is used to
    (a) derive IUH for the catchment
    (b) develop synthetic unit hydrograph
    (c) convert unit hydrograph of given duration into a unit hydrograph of other desired duration
    (d) to derive the unit hydrograph from complex storms
87. Using a 6 h unit hydrograph, an S-hydrograph is derived for a catchment of 254 $km^2$. The equilibrium discharge of the S-curve is
    (a) 117.8 $m^3$/s          (b) 254 $m^3$/s
    (c) $254 \times 10^6$ $m^3$/s      (d) $254 \times 6$ $m^3$/s
88. A 6 h storm occurring over a basin with uniform intensity produced an effective rainfall of 10 cm and a peak flow of 800 $m^3$/s. What will be the peak flow over the same basin, if another storm of same duration but with an effective rainfall of 15 cm occurs?
    (a) 800 $m^3$/s     (b) 1200 $m^3$/s     (c) 400 $m^3$/s     (d) 1600 $m^3$/s

89. The peak discharges in 3 and 6 h unit hydrographs of a basin occur at $t_1$ and $t_2$ h respectively. Then
    (a) $t_1 = t_2$                                 (b) $t_1 > t_2$
    (c) $t_1 < t_2$                                 (d) insufficient data to predict

90. A DRH is observed to have peak ordinate at 15 h after the commencement of effective rainfall. What is the duration of the storm if the base period is 80 h?
    (a) less than 15 h                          (b) between 15 and 30 h
    (c) more than 80 h                          (d) between 0 and 80 h

91. The basic principles of unit hydrograph theory are
    (a) non-linear response and time invariance
    (b) linear response and time invariance
    (c) linear response and time variance
    (d) nonlinear response and time variance

92. Point of inflection on the recession side of a hydrograph indicates
    (a) the beginning of baseflow              (b) the end of runoff
    (c) the end of overland flow               (d) end of rainfall

93. A return period of 100 years means
    (a) the event is to come on every 100th year
    (b) record of the maximum observed flood in the past 100 years
    (c) occurrence of atleast once in every 100 years on an average
    (d) the event is expected only after 100 years

94. A structure is designed for a $T$ year flood with an estimated useful life period of $n$ years. The probability that the event will not occur during its life period is given by
    (a) $1 - (1 - p)^n$                          (b) $(1 - 1/T)^n$
    (c) $(1 - 1/n)^T$                            (d) $1 - (1/T)^n$

95. Risk using the theory of probability is estimated from the relation
    (a) $1 - (1 - 1/T)^n$                        (b) $(1 - 1/T)^n$
    (c) $(1 - 1/n)^T$                            (d) $1 - (1/T)^n$

96. Base-flow separation is performed on
    (a) an unit hydrograph to get the direct-runoff hydrograph
    (b) a flood hydrograph to obtain the magnitude of rainfall
    (c) a flood hydrographs to obtain the rainfall hyetograph
    (d) a hydrograph to obtain the direct surface runoff

97. For a basin
    (a) higher is the drainage density larger is the flood peak
    (b) lower is the drainage density larger is the flood peak
    (c) higher is the drainage density lower is the flood peak
    (d) none of the above

98. A unit hydrograph has
    (a) 1 $m^3$/sec of peak discharge
    (b) one hour of rainfall duration
    (c) one cm of rainfall excess over the basin
    (d) one unit of base flow

99. A D-h UH for a catchment is obtained when a storm of D-h duration produces a single peak DRH,
    (a) the ordinates of DRH are divided by total runoff volume (cm)
    (b) the ordinates of DRH are divided by direct runoff volume (cm)
    (c) the ordinates of DRH are divided by duration of DRH (h)
    (d) the ordinates of DRH are divided by total rainfall depth (cm)

100. A storm hydrograph resulting from a 2 h of effective rainfall has 5 cm of direct runoff. To obtain UH from the DRH of this storm.

    (a)   divide the ordinates of DRH by 3

    (b)   divide the ordinates of DRH by 2

    (c)   divide the ordinates of DRH by 5

    (d)   divide the ordinates of DRH by 1

101. A 5-h UH of a catchment is triangular in shape with base width of 100 h and a peak discharge of 20 m³/s. The catchment of area of this hydrograph is

    (a)   180 km²    (b)   360 km²    (c)   200 km²    (d)   100 km²

102. The ordinate of an IUH of a catchment at any time is the

    (a)   slope of an 1-hour UH at that time

    (b)   slope of the ERH at that time

    (c)   slope of the S-curve of intensity 1 cm/h

    (d)   slope of DRH at that time.

103. Instantaneous unit hydrograph (IUH) is a hydrograph of

    (a)   infinitely small rainfall excess occurring for a unit duration

    (b)   unit rainfall excess of infinitely small duration

    (c)   effective rainfall of 1-h duration

    (d)   unit hydrograph of infinite rainfall excess

104. When a S-curve is derived from an UH of D-h duration

    (a)   the equilibrium discharge is dependent on D

    (b)   the time base of UH represents the time at which the S-curve attains its maximum value

    (c)   the equilibrium discharge of S-curve is dependent on the catchment area

    (d)   all the above

105. If a 3-h UH of a catchment has a peak ordinate of 50 m³/sec, the peak ordinate of a 6 h UH for the same catchment will be

    (a)   > 50 m³/s    (b)   = 50 m³/s    (c)   < 50 m³/s    (d)   inadequate data

106. Dicken's formula for flood peak is given by

    (a)   $Q = CA^{1/3}$    (b)   $Q = CA^{2/3}$    (c)   $Q = CA^{1/4}$    (d)   $Q = CA^{3/4}$

107. A catchment is made of 70% area with runoff coefficient 0.30 and remaining 30% area with runoff coefficient 0.70. What is the weighted runoff coefficient to be used in Rational formula?

    (a)   0.21    (b)   0.24    (c)   0.42    (d)   0.70

108. A straight line is fitted to a plot between the return period on logarithmic scale and flood peak on ordinary scale. The 10 year and 100 year floods are obtained as 110 m³/s and 220 m³/s respectively. What would be the magnitude of 1000 year flood.

    (a)   330 m³/s    (b)   440 m³/s    (c)   550 m³/s    (d)   660 m³/s

109. Using Gumbel's distribution the mean annual flood is equal to (when $n$ is the length of record).

    (a)   2.33 years return period flood    (b)   $n/2$ years return period flood

    (c)   50 years return period flood    (d)   $n$ years return period flood.

110. If $x_T$ and $K_T$ are the flood and frequency factor corresponding to a return period of $T$ years and $x_{av}$ and $s_x$ are the mean and standard deviation of the floods, then

    (a)   $x_T = \dfrac{x_{av}}{K_T\,S_x}$                      (b)   $x_T = \dfrac{K_T S_x}{x_{av}}$

    (c)   $x_T = x_{av} + K_T S_x$           (d)   $x_T = x_{av} - K_T S_x$

111. Which of the following form of equation is used in Modified Pul's method of routing?

    (a)   $\left(\dfrac{I_1 + I_2}{2}\right)\Delta t + \left(\dfrac{O_1 + O_2}{2}\right)\Delta t = S_2 - S_1$

(b) $\left(\dfrac{I_1 + I_2}{2}\right)\Delta t - \left(\dfrac{O_1 + O_2}{2}\right)\Delta t = S_2 - S_1$

(c) $\left(\dfrac{I_1 + I_2}{2}\right)\Delta t + \left(S_1 - \dfrac{O_1 \Delta t}{2}\right) = \left(S_2 + \dfrac{O_2 \Delta t}{2}\right)$

(d) $\left(\dfrac{I_1 + I_2}{2}\right)\Delta t - \left(S_1 - \dfrac{O_1 \Delta t}{2}\right) = \left(S_2 + \dfrac{O_2 \Delta t}{2}\right)$

112. The peak of the outflow hydrograph intersects the recession limb of inflow hydrograph in case of
    (a) reservoir routing with controlled spillway
    (b) reservoir routing with uncontrolled spillway
    (c) channel routing using Muskingum method
    (d) any flood routing method

113. Hydraulic routing methods make use of
    (a) a form of energy equation      (b) a form of continuity equation
    (c) a form of momentum equation
    (d) continuity and momentum equations

114. Which of the following is the Muskingum storage equation?
    (a) $S = K[xQ + (1 - x)I]$      (b) $S = K[xQ - (1 - x)I]$
    (c) $S = K[xI + (1 - x)Q]$      (d) $S = K[x\,I - (1 - x)Q]$

115. Muskingum method of routing satisfies the equation
    (a) $C_0 + C_1 + C_2 = 0$      (b) $C_0 + C_1 + C_2 = 1$
    (c) $C_0 C_1 C_2 = 1$      (d) $C_0/C_1 = C_1/C_2$

116. For natural channels the value of Muskingum's parameter $x$ is generally taken
    (a) between 0.1 and 0.3      (b) between 0.3 and 0.6
    (c) more than 0.5      (d) between 0.1 and 1.0

117. Pick up the wrong statement from the following:
    (a) Transport of sediment particles by bouncing along the bed is known as saltation
    (b) Trap efficiency of the reservoir decreases with increase in the capacity inflow ratio
    (c) A bed load at one section may move as suspended load at another section
    (d) Deposition of coarse sediment takes place at the entrance to the reservoir.

118. If $d$ and $v$ are the diameter and velocity of a raindrop respectively, then its erosive power is proportional to
    (a) $d^2 v^3$      (b) $d^3 v^2$      (c) $d^2 v^2$      (d) $d^3 v^3$

119. Rational formula is used to arrive at the designed peak flow of $Q_p$. If a storm intensity is doubled and the duration is increased three times, the resulting peak discharge is
    (a) $Q_p$      (b) $2\,Q_p$      (c) $3Q_p$      (d) $Q_p/2$

120. A culvert is designed for a 100 year flood. The probability that exactly one flood of the design capacity will occur in the 50 year life of the structure is
    (a) 0.200      (b) 0.400      (c) 0.395      (d) 0.780

121. A probable maximum flood is
    (a) the standard project flood of an extremely large size
    (b) a flood adopted in the design of spillways
    (c) a flood adopted in the design of any type of structures
    (d) an extremely large but physically possible flood in the region.

122. Use of UH for estimating floods is limited to catchments of size less than
    (a) 500 km$^2$      (b) 50 km$^2$      (c) 5 km$^2$      (d) 5000 km$^2$

123. To use Log-Pearson type III distribution, essential data required are
    (a) mean, standard deviation and coefficient of skew of discharge data
    (b) mean and standard deviation of the log of discharges and the number of years record
    (c) mean, standard deviation and coefficient of skew of the log of discharges
    (d) mean and standard deviation of the log of discharges.

124. Probability that a hundred year flood may not occur at all during the 25 years life of the project is
    (a)  0.95 (b)  0.990 (c)  0.778 (d)  0.999

125. For an annual flood series arranged in decreasing order of magnitude, the return period $T$ for a magnitude listed at position $m$ in a total of $N$ records is
    (a)  $m/N$ (b)  $m/(N + 1)$ (c)  $(N + 1)/m$ (d)  $N/(m + 1)$

126. Routing equation using Muskingum method is written as $Q_2 = C_0 I_2 + C_1 I_1 + C_2 Q_1$. If $K = 10$ h, $x = 0.20$ and the time step for routing $\Delta t = 4$ h, the coefficient $C_0$ is
    (a)  0.0 (b)  0.01 (c)  0.2 (d)  0.03

127. In Muskingum method of channel routing if $x = 0.5$, it represents an outflow hydrograph that has
    (a)  reduced peak
    (b)  amplified peak
    (c)  exactly the same as the inflow hydrograph
    (d)  a peak of any magnitude

128. Nash model for IUH given by $U(t) = \dfrac{1}{K\Gamma(n)} (t/K)^{n-1} (e)^{-t/K}$, the usual units of $U(t)$, $n$, $K$ and $t$ are, respectively,
    (a)  cm/h, h, h and h
    (b)  $h^{-1}$, h, h and h
    (c)  $h^{-1}$, dimensionless number, h and h
    (d)  cm/h, dimensionless number, h and h

129. In Nash model for IUH, if $M_{ERHI}$ is the first moment of ERH about the time origin divided by the total effective rainfall and $M_{DRH1}$ the first moment of DRH about the time origin divided by the total direct runoff, then
    (a)  $M_{DRH1} - M_{ERHI} = nK$ (b)  $M_{DRH1} - M_{ERHI} = nK^2$
    (c)  $M_{DRH1} - M_{ERHI} = n (n + 1) K$ (d)  $M_{DRH1} - M_{ERHI} = 2nK$

130. An isochrone is a line on a basin map
    (a)  joining raingauge stations with equal rainfall duration
    (b)  joining points having equal rainfall intensity
    (c)  joining points having equal time of travel of the surface runoff with respect to the outlet
    (d)  joining points of equal rainfall depth in a given time interval

131. In a linear reservoir the storage varies linearly with
    (a)  elevation
    (b)  the outflow rate
    (c)  time
    (d)  the inflow rate

# ANSWERS TO OBJECTIVE QUESTIONS

| | | | | |
|---|---|---|---|---|
| 1. (c) | 28. (a) | 55. (b) | 82. (c) | 109. (a) |
| 2. (c) | 29. (b) | 56. (c) | 83. (d) | 110. (c) |
| 3. (d) | 30. (b) | 57. (a) | 84. (a) | 111. (c) |
| 4. (c) | 31. (c) | 58. (b) | 85. (d) | 112. (d) |
| 5. (b) | 32. (b) | 59. (c) | 86. (c) | 113. (d) |
| 6. (d) | 33. (a) | 60. (b) | 87. (a) | 114. (c) |
| 7. (c) | 34. (d) | 61· (c) | 88. (b) | 115. (b) |
| 8. (b) | 35. (b) | 62. (d) | 89. (c) | 116. (a) |
| 9. (b) | 36. (b) | 63. (c) | 90. (a) | 117. (a) |
| 10. (c) | 37. (b) | 64. (a) | 91. (b) | 118. (c) |
| 11. (a) | 38. (a) | 65. (a) | 92. (c) | 119. (b) |
| 12. (b) | 39. (b) | 66. (d) | 93. (c) | 120. (c) |
| 13. (b) | 40. (c) | 67. (a) | 94. (b) | 121. (d) |
| 14. (b) | 41. (c) | 68. (a) | 95. (a) | 122. (d) |
| 15. (b) | 42. (d) | 69. (c) | 96. (d) | 123. (c) |
| 16. (a) | 43. (c) | 70. (b) | 97. (a) | 124. (c) |
| 17. (a) | 44. (b) | 71. (b) | 98. (c) | 125. (c) |
| 18. (a) | 45. (c) | 72. (b) | 99. (b) | 126. (a) |
| 19. (b) | 46. (d) | 73. (a) | 100. (c) | 127. (c) |
| 20. (c) | 47. (b) | 74. (a) | 101. (b) | 128. (c) |
| 21. (b) | 48. (b) | 75. (c) | 102. (c) | 129. (a) |
| 22. (c) | 49. (b) | 76. (a) | 103. (b) | 130. (c) |
| 23. (c) | 50. (c) | 77. (d) | 104. (d) | 131. (b) |
| 24. (a) | 51. (d) | 78. (c) | 105. (c) | |
| 25. (c) | 52. (b) | 79. (a) | 106. (d) | |
| 26. (b) | 53. (b) | 80. (b) | 107. (c) | |
| 27. (a) | 54. (b) | 81. (d) | 108. (a) | |

# References

Bear, J., (1979), "*Hydraulics of Groundwater*", McGraw-Hill Book Company, New York.

Biswas, A.K., (1972), "*History of Hydrology*", New York: American Elsevier.

Bower, H. (1978), "*Ground Water Hydrology*", McGraw-Hill Book Company, Tokyo.

Butler, S.C., (1957), "*Engineering Hydrology*", Prentice-Hall Inc., USA.

Byers, R.H., (1959), "*General Meteorology*", McGraw-Hill Book Company.

Central Water Commission, India, (1973), "*Estimation of Design Flood Peak*" Flood Estimation Directorate, Report No. 1/73, New Delhi.

Central Water Commission, India, (1988), "*Water Resources of India*", Publication No. 30/88, CWC, New Delhi, India.

Chow, V.T., Maidment, D.R. and L.W. Mays, (1988), "*Applied Hydrology*" McGraw-Hill Book company, New York. NY.

Chow, V.T. (1964), "*Hand Book of Applied Hydrology*" McGraw-Hill Book Company, New York, NY.

Chow, V.T., (1973) "*Open Channel Hydraulics*", McGraw-Hill Book Company.

Cunge, J.A., et. al., (1980), "*Practical Aspects of Computational River Hydraulics*", Pitman Publishing Limited, London.

Davis, S.N. and R.J.M. Dewiest, (1966) "*Hydrology*", John Wiley & Sons, NY.

Davis, E.V., and K.E. Sorensen, "*Handbook of Applied Hydraulics*", McGraw-Hill Book Company.

Deju, R.A., (1971), "*Regional Hydrology Fundamentals*", Gordon and Breach Science Publications, Inc., New York, NY.

Desmukh, M.M. (1978), "*Water Power Engineering*", Dhandpat Rai and Sons., Nai Sarak, Delhi.

Dhar, O.N. and B.K. Bhattacharaya (1975), "*A study of the Depth-Area-Duration Statistics of the Severe most Storms over Different Meteorological Divisions of North India*", Proc., National symposium on Hydrology, Roorkee, India. pp. G-4–11.

Dhar, O.N. and Bhattacharya, B.K., (1977), "*Relationship Between Central Rainfall and its Aerial Extent*", Jr. of Irrigation & Power 34(2), pp. 224–250.

French, R.H., (1986), "*Open Channel Hydraulics*", McGraw-Hill Book Company.

Garg, S.K., (1977), "*Water Resources and Hydrology*", Khanna Publishers, New Delhi.

Garg, S.K., (1978), "*Irrigation Engineering and Hydraulic Structures*", Khanna Publishers, New Delhi.

Garg, S.K., (1993), "*Hydrology and Flood Control Engineering*", Khanna Publishers, New Delhi.

Graf, W.H., (1971), "*Hydraulics of Sediment Transport*", McGraw-Hill Book Company.

Gray, D.M., (1973) "*Principles of Hydrology*", McGraw Hill Book Company.

Indian Bureau Standards, "*Guidelines for Fixing Spillway Capacity*" I.S: 11223, 1985.

Jain, M.K., et. al., (1993), "*Numerical Methods for Scientific and Engineering Computation*", Wiley Eastern Limited, New Delhi.

Joglekar, D.V., (1971), "*Manual on River Behavior Control and Training*", Central Board of Irrigation & Power, New Delhi.

Joglekar, D.V. (1961), "*Density currents in Reservoir and their Simulation Models*", Jr. of Institution of Engineers (India), Vol. 42.

Karanth, K.R., (1987), *"Ground Water Assessment Development and Management"*, Tata McGraw-Hill, New Delhi.

Khosla, A.N., *"Silting of Reservoirs"*, Central Board of Irrigation and Power, New Delhi.

Khushalani, K.B. and M. Khushalani (1971), *"Irrigation Practice and Design, Vol. I"*, Oxford & IBH, New Delhi.

Kulandaiswamy, V.C., (1960), *"A note on the Muskingum Method of Flood Routing"*, Jr. of Hydrology, 4, pp 273–276.

Linsely, R.K., Kohler, M.A. and J.L.H. Paulhus (1975), *"Applied Hydrology"*, Tata McGraw-Hill, New Delhi.

Linsely, R.K., Kohler, M.A. and J.L.H. Paulhus (1988), *"Hydrology for Engineers"*, 2nd Edition, McGraw Hill Book Company, Singapore.

Linsley, R.K. and J.B. Franzni, *"Water Resources Engineering"*, McGraw-Hill Book Company.

Mahamood, K. and V. Yevjevich, (1975), *"Unsteady Flow in Open Channels"*, Vol. 1 and 2, Water Rcs. Pub., Fort Collins, Colarado, USA.

Mathews, J.H., (1994), *"Numerial Methods for Mathematics, Science and Engineering"*, Prentice Hall of India Private Ltd., New Delhi.

Michael, A.M., (1978), *"Irrigation: Theory and Practice"*, Vikas Publishing, House, New Delhi.

Murthy, V.V.N., (1985), *"Land and Water Management Engineering"*, Kalyani Publishers, New Delhi.

Murty, B.N., (1977), *"Life of Reservoirs"*, Tech. Report No. 19, CBIP, New Delhi.

Mutreja, K.N., (1986), *"Applied Hydrology"*, Tata McGraw-Hill, New Delhi.

Nemec, J., (1973), *"Engineering Hydrology"*, Tata McGraw-Hill, New Delhi.

Ponce, V.M.,(1989), *"Hydrology-Principles and Practices"*, Prentice-Hall Eaglewood Cliffs, New Jersey.

Rao, K.L., *"Indias' Water Wealth"*, Orient Longman, New Delhi. India.

Raghunath, H.M.,(1985), *"Hydrology"*, Wiley Eastern Limited, New Delhi.

Raghunath, H.M., (1987), *"Groundwater"*, Wiley Eastern Ltd., New Delhi.

Reddi, P.J.R., (1987), *"Stochastic Hydrology"*, Laksmi Publications, New Delhi,

Reddi, P. J.R., (1992), *"A Text Book of Hydrology"*, Laksmi Publications, New Delhi.

Sahu, D (1990), *"Landforms, Hydrology and Sedimentation"*, Naya Prakash, Calcutta.

Sharma, R.K., (1993), *"A Text Book of Hydrology and Water Resources"*, Dhanpat Rai and Sons, Delhi.

Schulz, E.F.,(1973), *"Problems in Applied Hydrology"*, Water Resources Publications, Fort Collins, Colorado.

Singh, Bharat, (1975), *"Irrigation Engineering"* Nem Chand and Sons, Roorkee.

Singh, V.P., (1994), *"Elementary Hydrology"*, Prentice Hall of India, Pvt. Ltd., New Delhi.

Subramanya, K. (1991), *"Flow in Open Channels"*, Tata McGraw-Hill, New Delhi.

Subramanya, K., (1984) *"Engineering Hydrology"*, Tata McGraw-Hill, New Delhi.

Todd, D.K., (1980), *"Ground Water Hydrology"* (2nd Edition) John Wiley & Sons, New York, NY.

U.S Bureau of Reclamation, *"Water Management Manual"*.

Varshney, R.S., (1979), *"Engineering Hydrology"*, Nem Chand and Bros., Roorkee.

Ward, R.C. (1975), *"Principles of Hydrology, 2nd edition"*, McGraw-Hill.

Weisner, C.J., (1983), *"Hydrometeorology"*, Chapman & Hall Ltd. London, U.K.

Wisler, C.O. and E.F. Brater (1959) *"Hydrology"*, Joh Wiley, New York, USA.

Wilson, E.M. (1983),*"Engineering Hydrology"*, Macmillan, ELBS., London.

# Conversion Factors
## (FPS to MKS Units and Vice Versa)

**Length**

| | | |
|---|---|---|
| 1 in | = | 2.54 cm |
| 1 ft | = | 0.305 m |
| 1 mile | = | 1.609 km |
| 1 m | = | 3.281 ft |
| 1 yd | = | 0.9144 m |
| 1 km | = | 0.6214 mile |
| | = | 0.54 nautical mile |

**Area**

| | | |
|---|---|---|
| 1 in$^2$ | = | 6.452 cm$^2$ |
| 1 ft$^2$ | = | 0.0929 m$^2$ |
| 1 cm$^2$ | = | 0.155 in$^2$ |
| 1 m$^2$ | = | 10.76 ft$^2$ |
| | = | 1.094 sq. yd |
| 1 acre | = | 0.4047 ha |
| | = | 4047 m$^2$ |
| | = | 43560 sft |
| 1 ha | = | 10$^4$ m$^2$ |
| | = | 2.471 acres |
| 1 mile$^2$ | = | 2.59 km$^2$ |
| | = | 640 acres |
| 1 km$^2$ | = | 100 ha |
| | = | 0.3861 mile$^2$ |
| | = | 247 acres |

**Volume**

| | | |
|---|---|---|
| 1 cft | = | 28.32 litres |
| | = | 0.02832 m$^2$ |
| | = | 6.24 imp.gal |
| | = | 7.48 US gal |
| 1 litre | = | 1000 cc |
| | = | 0.22 imp.gal |
| 1 barrel | = | 42 US gal |
| 1 US gal | = | 3.79 litres |
| | = | 0.833 imp. gal |
| 1 m$^3$ | = | 35.31 cft |
| | = | 220 imp. gal |
| | = | 264 US gal |
| | = | 1,000 lit |
| 1 cc (ml) | = | 0.061 in$^3$ |

| | | |
|---|---|---|
| 1 acre-ft | = | 43,560 cft |
| | = | 0.1234 ha-m |
| | = | 1234 m$^3$ |
| 1 km$^3$ | = | 0.811 million aft |
| | | (M. aft) |
| 1 ha-cm | = | 100 m$^3$ |
| 1 M. m$^3$ | = | 810.7 acre-ft |
| 1 ha-m | = | 10$^4$ m$^3$ |
| | = | 8.14 acre-ft |

**Velocity**

| | | |
|---|---|---|
| 1 ft/sec | = | 30.48 cm/sec |
| 1 m/sec | = | 3.281 ft/sec |
| 1 mph | = | 1.467 ft/see (fps) |
| | = | 1.609 km/hr |
| | | (kmph) |
| | = | 0.8684 knot |
| 1 knot | = | 1.69 fps |
| | = | 0.515 m/sec |
| 1 km/hr | = | 0.2778 m/sec |
| | = | 0.9113 fps |
| | = | 0.6214 mph |
| 1 m/day | = | 22.9 gpd/ft$^2$ |

**Acceleration Due to Gravity (g)**

| | | |
|---|---|---|
| g | = | 32.2 ft/sec$^2$ |
| | = | 9.81 m/sec$^2$ |

**Flowrate (Discharge)**

| | | |
|---|---|---|
| 1 cfs (cusec) | = | 0.0283 m$^3$/sec |
| | | (cumec) |
| | = | 28.3 lps |
| | = | 374.03 imp. gpm |
| | = | 449 US gpm |
| | = | 1.983 acre-ft/day |
| | = | 724 acre-ft/year |
| 1 imp. gpm | = | 0.0757 lps |
| | = | 0.0757 × 10$^{-3}$ m$^3$/sec |
| | = | 1.2 US gpm |
| 1 US gpm | = | 0.063 lps |
| | = | 0.063 × 10$^{-3}$ m$^3$/sec |

| 1 m³/sec | = 35.31 cfs |
| | = $19.01 \times 10^6$ gpd (imp) |
| | = 13200 imp. gpm |
| | = 15800 US gpm |
| | = 70 acre-ft/day |
| | = $3.05 \times 10^6$ ft/day |
| 1 lps | = 0.03531 cfs |
| 1 m³/day | = 2190 imp. gpd |
| 1 acre-ft/day | = 271542 imp. gpd |
| | = 1233.5 m³/day |

**Pressure**

| 1 kg/cm² | = 14.23 psi |
| 1 atm | = 14.7 psia |
| | = 34 ft of water |
| | = 30 in of mercury |
| | = 76 cm of mercury |
| | = 101.32 kN/m² |
| | = 1013.2 mb (millibars) |
| 1 bar | = $10^5$ N/m² |
| 1 millibar | = 0.0143 psi |
| 1 psi | = 6.895 kN/m² |
| | = 0.7031 m of water |
| 1 psf | = 47.88 N/m² |
| 1 m-atm | = 14.223 psia |
| | = 0.9678 atm |
| | = 10 m of water |
| | = 0.967 atm |
| | = 73.5 cm of mercury |
| | = 32.8 ft of water |
| | = 982 mb |

**Mass**

| 1 slug | = 32.2 lb |
| | = 14.6 kg |
| 1 lb | = 453.6 gm |
| 1 kg | = 2.205 lb |
| | = 0.06852 slug |
| 1 ton | = 1000 kg |

**Force (Weight)**

| 1 lb | = 4.448 N |
| | = 16 oz |
| | = 7,000 grains |
| 1 gm | = 1543 grains |
| 1 kN | = 224.8 lb |
| 1 kgf | = 9.81 N |
| | = 2.205 lb |

**Mass Density**

| 1 gm/cc | = 1.94 slugs/cft |
| | = 1000 kg/m³ |
| 1 slug/ft³ | = 515.4 kg/m³ |

**Specific Weight**

| 1 gm/cc | = 62.4 pcf |
| 1 pcf | = 157.1 N/m³ |

**Dynamic Viscosity**

| 1 lb-sec/ft² | = 47.88 N. s/m² |
| | = 478.8 poise |
| 1 centipoise | = 0.01 poise |
| 1 N s/m² | = 10 P |
| 1 Kgf-sec/m² | = 9.81 N-s/m² |
| | = 98.1 poise |

**Kinematic Viscosity**

| 1 ft²/sec | = 0.093 m²/sec |
| | = 929 stokes |
| 1 m²/s | = $10^4$ Stoke |
| 1 centi Stoke | = $10^{-6}$ m²/s |

**Permeability**

| 1 darcy | = $0.966 \times 10^{-3}$ cm/sec |
| | = $0.987 \times 10^{-8}$ cm² |
| 1 m/day | = $1.16 \times 10^{-3}$ cm/sec |
| | = 1,000 lpd/m² |
| | = 20.44 imp. gpd/ft² |
| | = 24.54 US gpd/ft² |

**Power**

| 1 metric hp | = 75 m-kg/sec |
| | = 736 watts |
| | = 542.8 ft-lb |
| | = 0.986 hp |

**Temperature**

| (°F − 32) (5/9) | = °C |
| 273 + °C | = °K (Metric system) |

**Transmissibility**

| 1 m²/day | = 67.05 imp. gpd/ft |
| | = 80.52 US gpd/ft |
| | = 0.056 US gpm/ft |

**Other Factors**

| $\pi$ | = 3.1416 ... |
| $e$ | = 2.7183 ... |
| $\log_{10}e$ | = 0.4343 |

| | | | |
|---|---|---|---|
| $\log_e 10$ | = 2.303 | 1 bar | = $10^5$ Pa |
| $\log_e x$ | = 2.303 $\log_{10} x$ | 1 Joule | = $9.48 \times 10^{-4}$ BTU |
| milli | = $10^{-3}$ | | (Energy—FPS) |
| micro | = $10^{-6}$ | 1 kilo-cal | = 3.97 BTU |
| kilo | = $10^3$ | | (Heat energy—FPS) |
| mega | = $10^6$ | 1 Watt | = 3.413 BRU/hr |
| 1 micron | = $10^{-3}$ mm | | (power—FPS) |
| | = $10^{-6}$ m | 1 BTU | = 1,054 joule |
| 1 million | = $10^6$ | 1 cal | = 4.18 J |
| 1 lakh | = $10^5$ | Universal Gas | |
| 1 N | = $10^5$ dynes | Constant (R) | = 287 N-m/kg °K |
| 1 kg | = 980.665 dynes | (or j/kg °K) | |
| 1 Pa | = 1 N/m$^2$ | $\rho_{\text{fresh water}}$ | = 1000 kg/m$^3$, |
| 1 bar | = $10^5$ N/m$^2$ | $\gamma_{\text{fresh water}}$ | = 9.81 kN/m$^3$ |
| | = 100 kN/m$^2$ | $\rho_{\text{sea water}}$ | = 1025 kg/m$^3$ |
| | = 1 atm | Geothermal | |
| | = 76 cm of Hg | gradient | = 0.3 to 1°C per 100 m |

# Index